$49.95

Basic Electronics

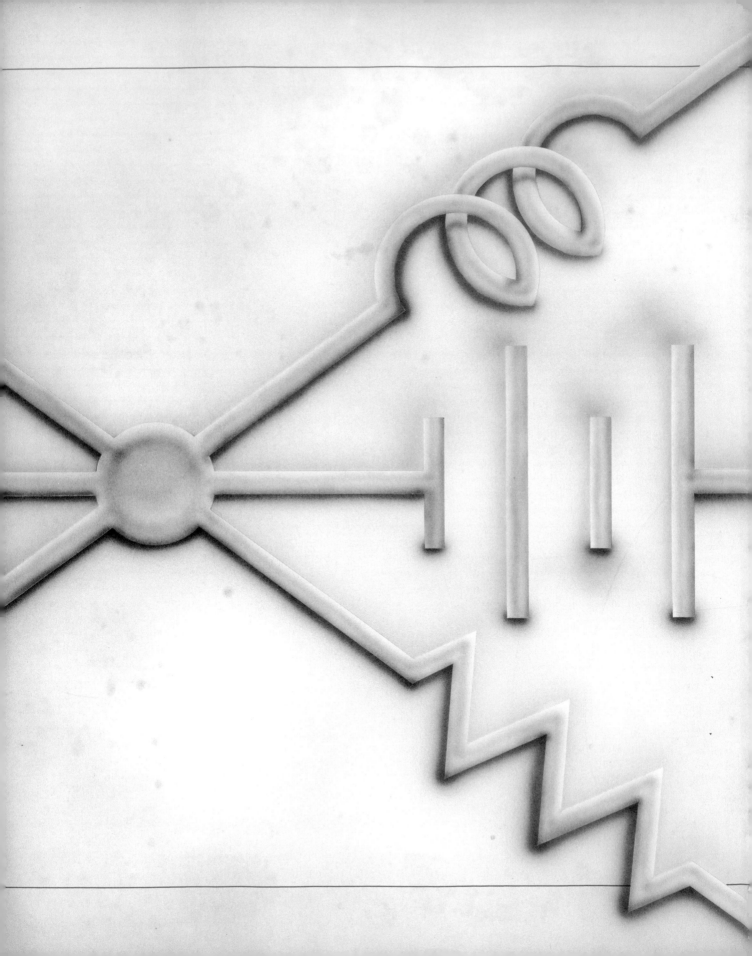

Basic Electronics

Fifth Edition

Bernard Grob

Instructor, Technical Career Institutes, Inc.

McGraw-Hill Book Company

New York Atlanta Dallas St. Louis San Francisco Auckland Bogotá
Guatemala Hamburg Lisbon London Madrid Mexico Montreal
New Delhi Panama Paris San Juan São Paulo Singapore
Sydney Tokyo Toronto

Sponsoring Editor: Gordon Rockmaker
Editing Supervisor: Larry Goldberg
Design and Art Supervisor: Caryl Valerie Spinka
Production Supervisor: S. Steven Canaris

Interior Design: Sharkey Design
Cover Photography: Robert Wolfson
Chapter Opener Art: Jon Weiman
Technical Art: Fine Line, Inc.

Library of Congress Cataloging in Publication Data

Grob, Bernard.
  Basic electronics.

  Bibliography: p.
  Includes index.
  1. Electronics.   I. Title.
TK7816.G75   1984   621.381   84-920
ISBN 0-07-024928-8

9 0 DOWDOW 8 9 1 0 9 8

ISBN 0-07-024928-8

Contents

Contents

In memory of
my father and mother

Author of
ELECTRONIC CIRCUITS AND APPLICATIONS
BASIC TELEVISION AND VIDEO SYSTEMS
MATHEMATICS: OUTLINE AND REVIEW
PROBLEMS FOR BASIC ELECTRONICS

Coauthor of
APPLICATIONS OF ELECTRONICS

Preface

This basic text is for beginning students without any experience in electricity and electronics. The first chapter is on elementary electricity; the last chapters cover transistors, integrated circuits, and digital electronics. Between these two points, the topics progress through Ohm's law, series and parallel dc circuits, networks, meters, magnetism, ac circuits with inductance and capacitance, and the subject of resonance. These fundamentals form the basis for the study of advanced applications, including communications electronics in general, and radio and television in particular. The chapter on digital electronics serves as an introduction to microprocessors and computers.

For each subject, the basic principles are explained first, followed by typical applications. Where appropriate, troubleshooting techniques are covered. This practical approach has proved effective in helping students to learn electronics in a way that is interesting and useful. While the technology has advanced dramatically over the years, the method of presentation used so successfully in the classroom has remained intact with only fine tuning taking place from edition to edition.

Mathematics is held to a minimum. Some numerical problems require powers of 10 because of the metric units. Trigonometric functions are used for ac circuits, where the phase angle is important. Also, Chap. 25 explains the details of how to apply complex numbers to ac circuits.

The order of topics follows a typical one-year course in electricity and electronics fundamentals. Typically such courses include dc and ac circuits and circuit theorems. However, the material on Kirchhoff's laws and other circuit theorems may be used for a separate course on network analysis, possibly combined with the details of RC and L/R time constants.

In the fifth edition, network analysis has been expanded to include applications of Kirchhoff's laws and network theorems to ac circuits. Chapter 12 on batteries includes state-of-the-art coverage of alkaline, silver-oxide, and lithium batteries. The chapter on resistors and color coding appears early in the book to relate more closely to laboratory work at the start of the course.

Finally, the last four chapters in the fifth edition emphasize transistors, integrated circuits, operational amplifiers, and digital electronics for computers.

The SI standard of V and v for voltage is used throughout the book, eliminating the use of E and e. Also, the SI unit of the siemens is used for conductance. For magnetic units, the SI system is emphasized, with conversion factors for cgs units. However, certain non-SI magnetic units are still widely used in industry, and these are referred to, as appropriate, in the text.

Organization The text is divided into 32 chapters for step-by-step development of the topics. Each chapter is designed to develop a key concept and build upon that concept. For example, individual chapters on Ohm's law, series circuits, and parallel circuits build up to more advanced chapters on series-parallel circuits, voltage dividers, current dividers, and networks.

The chapters on magnetism and electromagnetic induction lead into the development of sine-wave alternating voltage and current. There are separate chapters on inductance and its ac reactance before these fundamentals are combined for inductive circuits. The same sequential development is used for capacitive circuits. Then, these principles of L and C are combined with resistance for ac circuits. The chapters on resonance and filters describe these applications for sine-wave ac circuits.

The important details for RC and L/R time constants are reserved for Chap. 23. Here, the effects on dc transients and nonsinusoidal waveforms in capacitive and inductive circuits can be compared.

The four chapters on electronic devices, transistor amplifiers, digital electronics, and integrated circuits provide an introduction to electronic devices and circuits, including rectifiers, oscillators, operational amplifiers, and digital logic.

Practical approach Each chapter concludes with discussions of common troubles in components and the circuits they effect. For instance, the effects of an open circuit and a short circuit are explained in the first few chapters on dc circuits. Typical troubles in resistors, coils, capacitors, and transistors are explained in their respective chapters. Methods of testing for an open or short circuit are included.

The choke coil is explained as an application of inductive reactance, while the action of coupling and bypass capacitors is described in detail as an application of capacitive reactance.

Programmed questions An important feature is the inclusion of practice problems at the end of each main section. The purpose of these questions is to have students check their understanding of each section after reading the material. The purpose is not to see how many answers are right or wrong but rather to provide immediate reinforcement of the section just completed. Answers are given at the end of the chapter.

Learning aids The entire book is written with the student in mind. Numerous subheads, brief but substantive paragraphs, and readable sentences encourage students to study the text and figures. Illustrative examples with step-by-step solutions and a highlighted format clarify the calculations and the theory behind them. The two-color design is inviting but not merely cosmetic; it aids student comprehension.

The practice problems and answers for each main section serve the purpose of applying principles of programmed learning. This self-testing is in short units and can be reinforced immediately with correct answers.

Each chapter starts with an introduction that states the objective, followed by a list of important terms and a list of sections. This format enables the student to obtain an overall view of the material in each chapter.

At the end of each chapter a short summary lists the main points covered in the chapter. The short answer questions for self-examination are based on the summary. Also, summaries for groups of chapters are given as a review, with additional self-examination questions. This organized structure of ideas in the listing of sections, important terms, summaries, and self-testing with review motivates the student to study and learn.

There are many tables in the text for comparisons and summaries. The tables provide a concise listing of important points and help to compare similar or opposite characteristics. Examples of such tables are the comparisons between series and parallel circuits, inductive and capacitive reactance, and dc and ac circuits.

Each review summary has a short list of reference books for the topics in those chapters. A more complete bibliography on pages 714 to 716 lists books on electricity and electronics, transistors and integrated circuits, digital electronics and computers, and mathematics. The listing also includes manuals for transistors and integrated circuits, trade publications, and sources of other training materials.

Answers to all self-examination questions and to odd-numbered problems are given at the end of the book.

Glossaries A listing of technical abbreviations is given at the end of Chaps. 31 and 32. The expanding fields of digital electronics and integrated circuits have their own terminology, especially alphabetical abbreviations such as MOSFET, LSI, and TTL. It is helpful to have these summarized for convenient reference. Furthermore, a comprehensive glossary of technical terms in alphabetical order is included on pages 706 to 713.

Credits The photographs of components and equipment have been provided by many manufacturers, as noted in the legend accompanying the photograph. The fifth edition continues the original material developed in previous editions with the help of my colleagues Harry G. Rice and Philip Stein.

Finally, it is a pleasure to thank my wife, Ruth, for her help in preparation of the manuscript.

Bernard Grob

Basic Electronics

Survey of Electronics

Electronics, radio, and television are practical applications of the general principles of electricity. The same electric current produced by a battery for a flashlight can also be used in many different ways. Some examples are running a motor and producing heat and light for electric power applications, while electronic calculators and computers illustrate more advanced applications in digital electronics. In addition, radio and television are considered as communications electronics. They all are based on the fundamental laws of electricity and magnetism. Magnetic effects are always associated with an electric current.

The name *electronics* comes from the electron, which is a very small, invisible quantity of electricity present in all materials. In terms of its many uses, electronics can be defined to include all applications involving the control of electricity in a vacuum, such as vacuum tubes, in gas or vapor, and in the solid semiconductor materials used for transistors and integrated circuits. Typical semiconductor devices are shown in Fig. S-1. The main factor in their operation is control of individual

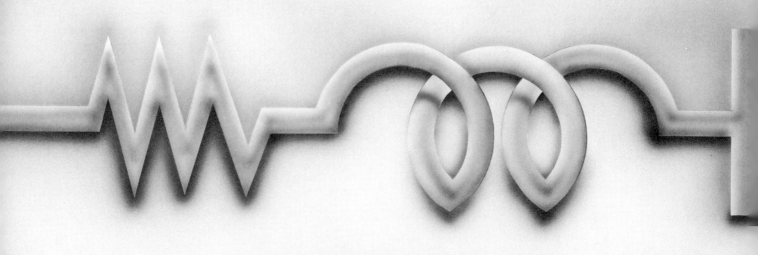

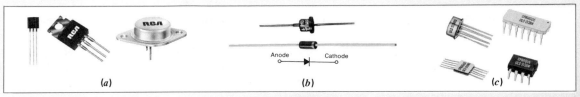

Fig. S-1 Semiconductor devices. (*a*) Transistor amplifiers. (*b*) Diode rectifiers, with schematic symbol. (*c*) Integrated circuit (IC) packages.

electrons for the desired effect. In a transistor, as an example, the input can control a larger output current, to provide amplification. The integrated circuit (IC) contains many transistors on a small semiconductor chip. Integrated-circuit chips are used for the small personal computer in Fig. S-2.

Fig. S-2 Personal computer. (*Texas Instruments*)

More details are explained in the following sections:

S-1 Development of Electronics
S-2 Wireless Broadcasting
S-3 Radio Broadcast Services
S-4 Applications of Electronics
S-5 Electronic Components
S-6 Electronic Circuits

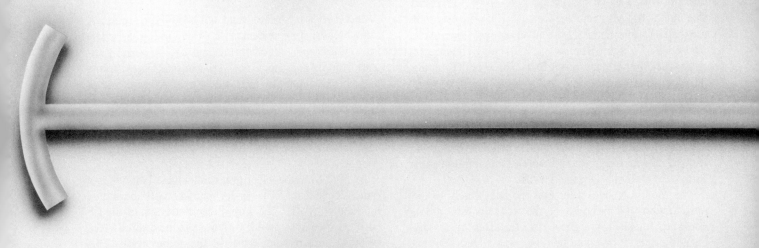

S-1 Development of Electronics

History shows that electronics started in the pioneer days of radio communications. Television developed from radio. Furthermore, radio itself is based on earlier experiments in electricity and magnetism. The start of wireless transmission for radio communications can be taken from the work of Heinrich Hertz, a German physicist. In 1887, he was the first to demonstrate the effect of electromagnetic radiation through space. The distance of transmission was only a few feet. However, the experiment demonstrated that radio waves could travel from one place to another without the need for any connecting wires between the transmitting and receiving equipment.

Hertz proved that radio waves, although invisible, travel with the same velocity as light waves. In fact, radio waves and light waves are two examples of electromagnetic radiation. This form of energy combines the effects of electricity and magnetism. An electromagnetic wave transmits electric energy through space.

The work of Hertz followed earlier experiments on electricity and magnetism. In 1820, a Danish physicist, H. C. Oersted, showed that an electric current produces magnetic effects. Then, in 1831, a British physicist, Michael Faraday, discovered that a magnet in motion can generate electricity. The motion provides the requirement of a change in the magnetic field. In 1864, the British physicist James Clark Maxwell, on the basis of earlier work in electricity and magnetism, predicted the electromagnetic waves demonstrated later by Hertz.

The importance of this work can be judged by the fact that basic units are named after these scientists. The maxwell (Mx) and the oersted (Oe) are units of magnetism. The hertz (Hz) unit is equal to one cycle per second, which is the measure for the frequency of any alternating voltage or current. The farad (F) is the unit of capacitance, indicating how much electric charge can be stored in a capacitor.

In 1895, Guglielmo Marconi used a long wire as an antenna to develop a practical radio system for long distances. The antenna is needed for efficient radiation. He succeeded in producing wireless communication across the Atlantic Ocean in 1901.

The rapid advances after that are due largely to the introduction and progress of the vacuum tube as an amplifier for electric signals. Dr. Lee DeForest, with his audion tube, invented in 1906, was a leader in this field. As the design of vacuum tubes advanced, radio broadcasting progressed rapidly. Regularly scheduled programs were broadcast in 1920 by station KDKA in the standard amplitude modulation (AM) radio band. The commercial frequency modulation (FM) radio broadcast service was started in 1939. Stereo broadcasting in this band began in 1961.

Commercial television broadcasting was started officially in 1941, but its popular use did not begin until 1945. Our present color television system was adopted in 1953.

Since the invention of transistors in 1948 at Bell Telephone Laboratories, solid-state devices have replaced tubes for most uses in electronics, radio, and television. The transistor is an application of controlled electron flow in solid semiconductor materials such as silicon (Si) and germanium (Ge). Transistors and tubes have similar uses for the control of electron flow and amplification of signals. The transistor is much smaller, however, and more efficient, as it does not have the heater used in tubes.

Solid-state electronic devices include transistors, diodes, and integrated circuits. A diode is not an amplifier but is used as a one-way conductor to convert alternating current to direct current. Solid-state devices have made new applications practicable because of their small size and the economy of IC packages. One example is the rapid growth of digital electronics for electronic calculators, personal computers, and many other uses.

S-2 Wireless Broadcasting

Broadcasting means to send out in all directions. A radio broadcasting system is illustrated in Fig. S-3. The transmitter sends out electromagnetic radio waves radiated from its antenna. Receivers can pick up the transmitted radio signal by means of a receiving antenna or aerial. The receiver reproduces the desired signal transmitted by the broadcast station. There are many radio signals in space from different transmitters, but the receiver can be tuned to the frequency of the station we want.

In Fig. S-3, the electromagnetic wave shown is a radio-frequency (RF) carrier signal with amplitude modulation (AM). The amplitude or strength of the RF carrier varies in step with variations in the desired voice or music information, which is the audio signal. This technique of modulating a carrier wave is necessary

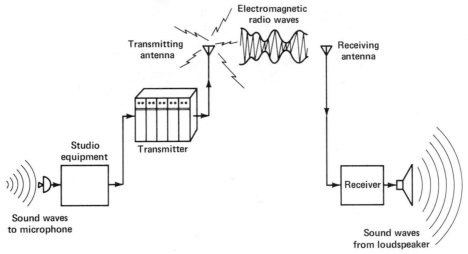

Fig. S-3 Radio broadcasting between transmitter and receiver. An amplitude-modulated (AM) carrier wave is shown here.

because the audio signal itself cannot be used for wireless transmission. The variations are too slow for effective radiation from an antenna. A higher-frequency carrier wave is chosen for the best radio transmission. Its modulation provides the desired signal information.

In the method of frequency modulation (FM), the modulating signal varies the frequency of the RF carrier wave. Either AM or FM can be used for any type of modulating signal.

Frequency is an important characteristic of any varying voltage or current, to specify how fast the amplitudes change. A complete set of changes is one cycle. The number of cycles repeated in a second is the frequency. The unit for frequency is the hertz (Hz), equal to one cycle per second (cps). As an example, the 60-cycle ac power line has a frequency of 60 Hz.

Radio frequencies are generally considered to be about 30,000 Hz and above. Radio-frequency carrier frequencies for wireless transmission are usually much higher. They are usually specified in kilohertz (kHz), equal to 1000 Hz, and megahertz (MHz), where 1 MHz = 1,000,000 Hz.

S-3 Radio Broadcast Services

Radio is an abbreviated form of radiotelegraph and radiotelephone. The word *radio* means radiation for wireless transmission. At first, communication was by radiotelegraph, using short dots and longer dashes in the Morse code. Now radiotelephone is used more for voice communications and for broadcasting voice and music programs for entertainment. This application can be considered as communications electronics, which includes television. Practically all radio and television receivers are now solid-state, with transistors and integrated circuits instead of vacuum-tube amplifiers. See Fig. S-4 for a comparison of the old and the new in radio.

The transmission distance for wireless communication can be less than a mile or as much as 5000 miles, depending on the type of service. There are many different uses, such as radio broadcasting of voice and music, television broadcasting, amateur radio, and citizen's band (CB) radio. In addition, radio communication is used for specific services, such as police radio. Finally, radio navigation for ships and planes is another important application.

All radio services in the United States are regulated by the Federal Communications Commission (FCC). The FCC assigns the RF carrier frequencies for transmission and monitors use of the airwaves. A few of the most important radio applications are described here briefly, but a more complete list, including FCC frequency allocations from 30 kHz to 300,000 MHz, is given in App. B.

Standard AM Radio Broadcast Band This service is the original system of broadcasting for what we generally call radio. Amplitude modulation is used

(a)   *(b)*

Fig. S-4 (*a*) Old radio. (*b*) New radio. Width is 16 in. (*Tandy*)

in transmission of the assigned RF carrier wave. Stations are assigned every 10 kHz in the band of 540 to 1600 kHz. The last digit is usually omitted from these numbers on the tuning dial for the carrier frequencies of different stations.

FM Radio Broadcast Band   This band is 88 to 108 MHz, with stations assigned every 200 kHz or 0.2 MHz. The system of frequency modulation reduces static and interference. For this reason, the FM band is used for broadcasting high-fidelity audio signals.

Television Broadcasting   Television is just another application of wireless radio communications, but with picture information in addition to the sound signal. Two separate carrier waves are transmitted by the station in its assigned channel. One carrier is an AM picture signal, modulated by video signal with the picture information. The other carrier is an FM sound signal modulated by the audio.

A television channel is 6 MHz wide to include both the picture and sound signals for each broadcast station. Channel 2, for instance, is 54 to 60 MHz.

Amateur Radio   This field is one of the largest noncommercial radio services. Amateur radio operators, or "hams," usually build and operate their own transmitters and receivers to call one another in one of

the assigned bands. A popular band is 7 to 7.3 MHz. Their main organization is the American Radio Relay League (ARRL), Newington, Connecticut.

Citizen's Band (CB) Radio   Forty 10-kHz channels from 26.965 to 27.405 MHz are available for public use of two-way radio. These CB channels are for class D service, with maximum power output of 4 watts (W). The CB transceiver includes a transmitter and a receiver. No operator's license is required for these personal radio services, often used in cars and boats.

S-4   Applications of Electronics

In addition to its use in radio and television, electronics is used in almost all industries for control functions, automation, and computing. There are so many applications that the broad field of electronics must be considered in smaller areas. Three logical groupings of electronics applications are defined here. Also included is a brief description of some important divisions with some typical job titles for working in the electronics business.

1. *Communications electronics.* This field includes AM radio, FM radio with stereo, and television with color. The equipment is divided between transmitters and receivers. Also, transmitters can be divided between radio-frequency equipment to produce the carrier wave radiated from the antenna and

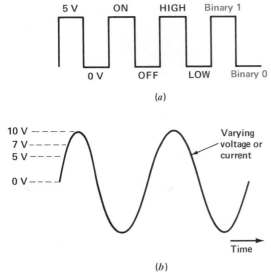

(a)

(b)

Fig. S-5 (a) Pulse waveform that is either HIGH or LOW for digital circuits. (b) Analog waveform with continuous variations.

the audio and video equipment in the studio that supplies the modulating signal with the desired information.

High-fidelity audio equipment can be considered with radio receivers. The receiver itself has audio amplifiers to drive the loudspeakers that reproduce the sound.

Satellite communications is also a transmit-receive system using electromagnetic radio waves. The satellite just happens to be orbiting around the earth at a height of above 22,000 miles in order to

have a tremendous field of view. Actually, the satellite is a relay station for transmitter and receiver earth stations.

2. *Electric power.* These applications are in the generation and distribution of 60-Hz ac power, as the source of energy for electrical equipment. Included are lighting, heating, motors, and generators.

3. *Digital electronics.* We see the digits 0 to 9 on an electronic calculator or digital watch, but digital electronics has a much broader meaning. The circuits for digital applications operate with pulses of voltage or current, as shown in Fig. S-5a. A pulse waveform is either completely ON or OFF because of the sudden changes in amplitude. In-between values have no function. Note that the ON and OFF states can also be labeled HIGH and LOW, or 1 and 0 in *binary notation,* which uses only two digits. Effectively, the digital pulses correspond to the action of switching circuits that are either on or off.

Voltage or current variations with a continuous set of values form an *analog waveform,* as shown in Fig. S-5b. The 60-Hz power line and audio and video signals are common examples. In Fig. S-5b, note that values between 0 and 10 V are marked to indicate that all the in-between values are an essential part of the waveform.

Actually, all the possible variations in types of electronic circuits can be divided into just two types—digital circuits that recognize pulses when they are HIGH or LOW, and analog circuits that use all values in the waveform. The applications of digital electronics, including calculators (Fig. S-6), computers, data proc-

(a)

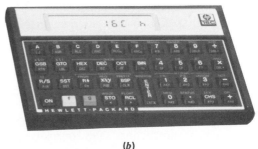

(b)

Fig. S-6 Two examples of programmable electronic calculators. Width is about 6 in. (a) Scientific type with exponents, logarithms, and trigonometric functions for sine (SIN), cosine (COS), and tangent (TAN) of angles. (b) Type for digital electronics and computer science. Functions included are for decimal (DEC), hexadecimal (HEX), octal (OCT), or binary (BIN) numbers. (*Hewlett-Packard*)

essing, and data communications, possibly form the largest branch of electronics. Two typical calculators are shown in Fig. S-6. In addition, many other applications, including radio and television, use both analog and digital circuits. Analog-to-digital (A/D) converters can change the signal from one form to the other.

In addition to all the general applications in communications, digital equipment, and electric power services, several fields that could be of specific interest include automotive electronics, industrial electronics, and medical electronics. Both digital and analog techniques are used.

In *automotive electronics,* more and more electronic equipment is used in cars for charging the battery, power-assist functions, measuring gages, and monitoring and control of engine performance. Perhaps the most important application is electronic ignition. This method provides better timing of the ignition spark, especially at high speeds.

Industrial electronics includes control of welding and heating processes; the use of elevator control; operation of copying machines; metal detectors and smoke detectors; moisture control; and computer-controlled machinery. In addition, there are many types of remote-control functions, such as automatic garage door openers and burglar alarms. Closed-circuit television is often used for surveillance.

Medical electronics combines electronics with biology. Medical research, diagnosis, and treatment all use electronic equipment. Examples are the electron microscope and the electrocardiograph machine. In hospitals, oscilloscopes are commonly used as the display to monitor the heartbeat of patients in intensive care.

Job Titles Different specialties in electronics are indicated by the following titles for engineers: antenna, audio, computer, digital, illumination, information theory, magnetics, microwave, motors and generators, packaging, power distribution, radio, semiconductor, television, and test equipment. Many of these fields combine physics and chemistry, especially for semiconductors.

The types of jobs in these fields include engineer for research, development, production, sales, or management; teacher; technician; technical writer; computer programmer; drafter; service worker; tester; and inspector. Technicians and service workers are needed for testing, maintenance, and repair of all the different types of electronic equipment.

Fig. S-7 A vacuum-tube amplifier.

S-5 Electronic Components

Considering the many applications of electronics, we can be a little surprised that only a few basic types of components are used for all the different kinds of equipment. Each type has variations for specific uses, but still the following is a short list.

1. *Transistors.* This is probably the most important use of solid-state semiconductor devices. They include diode rectifiers and transistor amplifiers, either as separate, discrete components or as part of an IC chip. In addition, the silicon controlled rectifier (SCR) and triac are used for power-control circuits.
2. *Electron tubes.* These include vacuum-tube diodes, amplifiers, and the cathode-ray tube (CRT) for a visual display device. A glass envelope encloses the

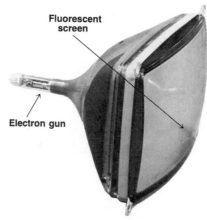

Fig. S-8 Television picture tube as example of a CRT. (*Zenith Radio Corporation*)

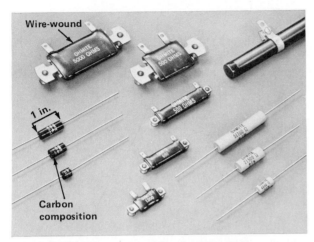

Fig. S-9 Typical resistors. (*Ohmite Mfg. Co.*)

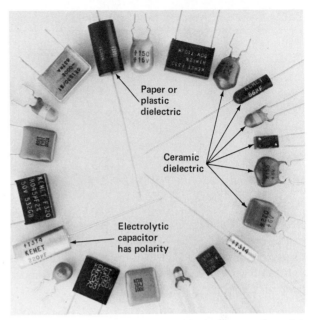

Fig. S-10 Typical small capacitors or condensers. Length is ¼ to 1 in. (*Union Carbide Corp.*)

required electrodes in a vacuum. An amplifier tube is shown in Fig. S-7.

3. *Visual display devices.* In addition to the CRT as a vacuum tube, solid-state types of displays include the light-emitting diode (LED) and liquid-crystal display (LCD). As examples, electronic calculators and digital watches use an LED or LCD for the digital display. Figure S-8 shows a television picture tube as an example of a CRT.
4. *Resistors* (Fig. S-9).
5. *Capacitors or condensers* (Fig. S-10).
6. *Inductors or coils* (Fig. S-11).

Note that transistors, tubes, and display devices (items 1, 2, and 3 above) are considered *active components*. This classification means that the device can be used for amplification, rectification, or to change energy from one form to another. For instance, the display device converts electrical input to light output.

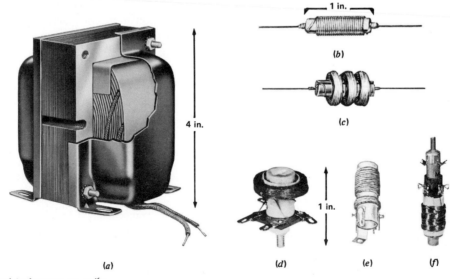

Fig. S-11 Typical inductors or coils.

Table S-1. Schematic Symbols for Resistors

Type	Symbol	Notes
Fixed	R ──\/\/\/──	Limits current and completes circuit
Variable	R ──\/\/\/── or R ──\/\/\/──	Varies current or voltage; used for volume, contrast, and tone controls

Resistors, capacitors, and inductors (items 4, 5, and 6) can be considered *passive components*. Their electrical characteristics are needed to form a complete circuit with voltage and current sources and active components in electronic equipment.

Resistors These can be the carbon-composition type or wound with special resistance wire. Their function is to limit the amount of current or divide the voltage in a circuit.

The unit of resistance R is the ohm (Ω). As an example, $R = 100\ \Omega$ is a common value for a resistor. Schematic symbols for R are shown in Table S-1. More details are given in Chap. 6, "Resistors."

Capacitors A capacitor is constructed of two conductor plates separated by an insulator (called a dielectric). Its basic function is to concentrate the electric field of voltage applied across the dielectric. As a result, the capacitor can accumulate and store electric charge from the voltage source. Furthermore, the dielectric can discharge the stored energy when the charging source is replaced by a conducting path.

When ac voltage is applied, the capacitor charges and discharges as the voltage varies. The practical application of this effect is the use of capacitors to pass an ac signal but to block a steady dc voltage. The higher the frequency, the easier it is for the capacitor to couple ac signal from one amplifier circuit to the next. Smaller values can be used, therefore, for a coupling capacitor at higher frequencies.

The unit of capacitance C is the farad (F). However, this is a very large unit that requires metric fractions, often millionths of a farad. This millionths unit is abbreviated with the Greek letter mu, μ, for micro. As an example, $C = 5\ \mu F$ or 5 microfarads is a common value of an audio coupling capacitor. Schematic symbols for C are shown in Table S-2. More details of capacitance are explained in Chaps. 21 to 23.

Inductors An inductor is just a coil of wire. Its basic function is to concentrate the magnetic field of electric current in the coil. Most important, an induced voltage is generated when the current with its associated magnetic field changes in value or direction.

In the practical application of a *choke,* the inductor can pass a steady direct current better than alternating

Table S-2. Schematic Symbols for Capacitors

Type	Symbol	Uses
Fixed	C ──┤├──	Stores charge in dielectric; passes ac voltage but blocks dc voltage
Electrolytic	C ──┤├+──	Fixed value with large C but has polarity
Variable	C ──┤├──	Variable capacitor; used for tuning

Table S-3. Schematic Symbols for Inductors or Coils

Type	Symbol	Uses
Air core	L ⟿	Concentrates magnetic field; for radio frequencies
Iron core	L ⟿	For 60-Hz and audio frequencies
Variable powdered-iron core	L --→ ⟿	Variable inductance; used for tuning
Iron-core transformer	L_P ⟿ / ⟿ L_S	Increases or decreases ac voltage from primary L_P to secondary L_S

current. The reason is that a steady current cannot produce induced voltage. Note that the effect of a choke, passing a steady current, is the opposite of that of a coupling capacitor, which blocks dc voltage.

A *transformer* consists of two or more coil windings in the same magnetic field. Induced voltage is produced when the current changes in any winding. The purpose of a transformer is to increase or decrease the amount of ac voltage coupled between the windings. However, the transformer operates only with alternating current.

The symbol for inductance is L, for linkages of the lines of force in the magnetic field. The unit is the henry (H). As an example, $L = 250 \ \mu H$ or 250 microhenrys is a typical value of a choke for radio frequencies. Schematic symbols for L are shown in Table S-3. An iron core is used at audio frequencies for greater concentration of the magnetic field of the coil.

S-6 Electronic Circuits

Applications of electronics are basically electric circuits with semiconductor devices and their associated R, C, and L components. The semiconductor component is used for control purposes, amplification, or rectification. Where necessary, a visual display shows the results.

The R, C, and L are needed with the semiconductor device in order to apply its internal effect to an external component. Resistors and capacitors are very common. They are small physically and relatively inexpensive, for most of the values required. Furthermore, R and C can easily be integrated with transistors on the same chip.

A dc circuit uses a dc voltage source, such as a battery, which has fixed polarity. In an ac circuit, the voltage periodically reverses in polarity.

Most electronic circuits combine dc and ac values. The reason is that the desired signal with the required information is usually an ac voltage or current, but a transistor amplifier, as an example, needs dc voltage applied in order to conduct any current at all. For instance, the 9-V battery in a small transistor radio is the dc power supply for the transistors amplifying the ac signal.

In conclusion, all types of electronic circuits are practical applications of the basic principles of electricity and magnetism.

Chapter 1
Electricity

Electricity is an invisible force that can produce heat, light, motion, and many other physical effects. The force is an attraction or repulsion between electric charges. More specifically, electricity can be explained in terms of electric charge, current, voltage, and resistance. The corresponding electrical units are the coulomb for measuring charge, the ampere for current, the volt for potential difference, and the ohm for resistance. A basic element of electricity is the electric circuit. A circuit is a closed path that allows for the movement of charges. *Current* is the name given to the movement of charges. The study of electricity involves the behavior of charges, current, and voltage with the components that make up the circuit.

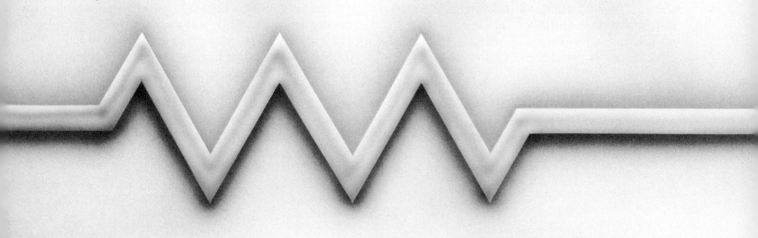

Important terms in this chapter are:

alternating current   electron flow
ampere of current   hole charge
closed circuit   ion charge
conventional current   insulator
coulomb of charge   ohm of resistance
conductance   polarity
conductor   proton
difference of potential   static electricity
direct current

More details are explained in the following sections:

1-1   **Negative and Positive Polarities**
1-2   **Electrons and Protons in the Atom**
1-3   **Structure of the Atom**
1-4   **The Coulomb Unit of Charge**
1-5   **The Volt Unit of Potential Difference**
1-6   **Charge in Motion Is Current**
1-7   **Resistance Is Opposition to Current**
1-8   **The Closed Circuit**
1-9   **Direct Current (DC) and Alternating Current (AC)**
1-10 **Sources of Electricity**

1-1   Negative and Positive Polarities

We see the effects of electricity in a battery, static charge, lightning, radio, television, and many other applications. What do they all have in common that is electrical in nature? The answer is basic particles of electric charge with opposite polarities. All the materials we know, including solids, liquids, and gases, contain two basic particles of electric charge: the *electron* and the *proton*. An electron is the smallest amount of electric charge having the characteristic called *negative polarity*. The proton is a basic particle with *positive polarity*.

Actually, the negative and positive polarities indicate two opposite characteristics that seem to be fundamental in all physical applications. Just as magnets have north and south poles, electric charges have the opposite polarities labeled negative and positive. The opposing characteristics provide a method of balancing one against the other to explain different physical effects.

It is the arrangement of electrons and protons as basic particles of electricity that determines the electrical characteristics of all substances. As an example, this paper has electrons and protons in it. There is no evidence of electricity, though, because the number of electrons equals the number of protons. In that case the opposite electrical forces cancel, making the paper electrically neutral. The neutral condition means that opposing forces are exactly balanced, without any net effect either way.

When we want to use the electrical forces associated with the negative and positive charges in all matter, work must be done to separate the electrons and protons. Changing the balance of forces produces evidence of electricity. A battery, for instance, can do electrical work because its chemical energy separates electric charges to produce an excess of electrons at its negative terminal and an excess of protons at its positive terminal. With separate and opposite charges at the two terminals, electric energy can be supplied to a circuit connected to the battery. Figure 1-1 shows a battery with its negative ($-$) and positive ($+$) terminals marked to emphasize the two opposite polarities.

Practice Problems 1-1
Answers at End of Chapter

a. Is the charge of an electron positive or negative?
b. Is the charge of a proton positive or negative?
c. Is it true or false that the neutral condition means equal positive and negative charges?

1-2   Electrons and Protons in the Atom

Although there are any number of possible methods by which electrons and protons might be grouped, they assemble in specific combinations that result in a stable arrangement. Each stable combination of electrons and protons makes one particular type of atom. For example, Fig. 1-2 illustrates the electron and proton structure of one atom of the gas hydrogen. This atom consists of a central mass called the *nucleus* and 1 electron outside. The proton in the nucleus makes it the massive

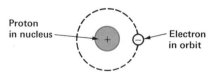

Proton in nucleus — Electron in orbit

Fig. 1-1 Negative and positive polarities on 1.5-V battery.

Fig. 1-2 Electron and proton in hydrogen atom.

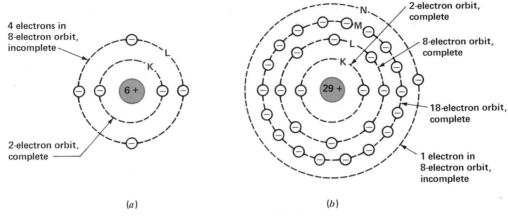

4 electrons in
8-electron orbit,
incomplete

2-electron orbit,
complete

2-electron orbit,
complete

8-electron orbit,
complete

18-electron orbit,
complete

1 electron in
8-electron orbit,
incomplete

(a)

(b)

Fig. 1-3 Atomic structure showing nucleus and orbital rings. (a) Carbon atom with 6 protons in nucleus and 6 orbital electrons. (b) Copper atom with 29 protons in nucleus and 29 orbital electrons.

and stable part of the atom because a proton is 1840 times heavier than an electron.

In Fig. 1-2, the 1 electron in the hydrogen atom is shown in an orbital ring around the nucleus. In order to account for the electrical stability of the atom, we can consider the electron as spinning around the nucleus, as planets revolve around the sun. Then the electrical force attracting the electrons in toward the nucleus is balanced by the mechanical force outward on the rotating electron. As a result, the electron stays in its orbit around the nucleus.

In an atom that has more electrons and protons than hydrogen, all the protons are in the nucleus, while all the electrons are in one or more outside rings. For example, the carbon atom illustrated in Fig. 1-3a has 6 protons in the nucleus and 6 electrons in two outside rings. The total number of electrons in the outside rings must equal the number of protons in the nucleus in a neutral atom.

The distribution of electrons in the orbital rings determines the atom's electrical stability. Especially important is the number of electrons in the ring farthest from the nucleus. This outermost ring requires 8 electrons for stability, except when there is only one ring, which has a maximum of 2 electrons.

In the carbon atom in Fig. 1-3a, with 6 electrons, there are just 2 electrons in the first ring because 2 is its maximum number. The remaining 4 electrons are in the second ring, which can have a maximum of 8 electrons.

As another example, the copper atom in Fig. 1-3b has only 1 electron in the last ring, which can include 8 electrons. Therefore, the outside ring of the copper atom is less stable than the outside ring of the carbon atom.

When there are many atoms close together in a copper wire, the outermost orbital electrons are not sure which atoms they belong to. They can migrate easily from one atom to another at random. Such electrons that can move freely from one atom to the next are often called *free electrons*. This freedom accounts for the ability of copper to conduct electricity very easily. It is the movement of free electrons that provides electric current in a metal conductor.

The net effect in the wire itself without any applied voltage, however, is zero because of the random motion of the free electrons. When voltage is applied, it forces all the free electrons to move in the same direction to produce electron flow, which is an electric current.

Conductors, Insulators, and Semiconductors When electrons can move easily from atom to atom in a material, it is a *conductor*. In general, all the metals are good conductors, with silver the best and copper second. Their atomic structure allows free movement of the outermost orbital electrons. Copper wire is generally used for practical conductors because it costs much less than silver. The purpose of using conductors is to allow electric current to flow with minimum opposition.

The wire conductor is used only as a means of delivering current produced by the voltage source to a device that needs the current in order to function. As an example, a bulb lights only when current is made to flow through the filament.

A material with atoms in which the electrons tend to stay in their own orbits is an *insulator* because it cannot conduct electricity very easily. However, the insulators are able to hold or store electricity better than the conductors. An insulating material, such as glass, plastic, rubber, paper, air, or mica, is also called a *dielectric*, meaning it can store electric charge.

Insulators can be useful when it is necessary to prevent current flow. In addition, for applications requiring the storage of electric charge, as in capacitors, a dielectric material must be used because a good conductor cannot store any charge.

Carbon can be considered a semiconductor, conducting less than the metal conductors but more than the insulators. In the same group are germanium and silicon, which are commonly used for transistors and other semiconductor components.

Elements   The combinations of electrons and protons forming stable atomic structures result in different kinds of elementary substances having specific characteristics. A few familiar examples are the elements hydrogen, oxygen, carbon, copper, and iron. An *element* is defined as a substance that cannot be decomposed any further by chemical action. The atom is the smallest particle of an element that still has the same characteristics as the element. *Atom* itself is a Greek word meaning a particle too small to be subdivided. As an example of the fact that atoms are too small to be visible, a particle of carbon the size of a pinpoint contains many billions of atoms. The electrons and protons within the atom are even smaller.

Table 1-1 lists some more examples of elements. These are just a few out of a total of 106.[1] Notice how the elements are grouped. The metals listed across the top row are all good conductors of electricity. Each has an atomic structure with an unstable outside ring that allows many free electrons.

The semiconductors have 4 electrons in the outermost ring. This means they neither gain nor lose electrons but share them with similar atoms. The reason is that 4 is exactly halfway to the stable condition of 8 electrons in the outside ring.

The inert gases have a complete outside ring of 8 electrons, which makes them chemically inactive. Remember that 8 electrons in the outside ring is a stable structure. An example is neon.

---

[1] A more complete listing of the elements, in alphabetical order, is given in App. C.

Table 1-1. Examples of the Chemical Elements

Group	Element	Symbol	Atomic number	Electron valence
Metal conductors, in order of conductance	Silver	Ag	47	+1
	Copper	Cu	29	+1*
	Gold	Au	79	+1*
	Aluminum	Al	13	+3
	Iron	Fe	26	+2*
Semiconductors	Carbon	C	6	±4
	Silicon	Si	14	±4
	Germanium	Ge	32	±4
Active gases	Hydrogen	H	1	±1
	Oxygen	O	8	−2
Inert gases	Helium	He	2	0
	Neon	Ne	10	0

*Some metals have more than one valence number in forming chemical compounds. Examples are cuprous or cupric copper, ferrous or ferric iron, and aurous or auric gold.

Molecules and Compounds A group of two or more atoms forms a molecule. For instance, two atoms of hydrogen (H) form a hydrogen molecule (H_2). When hydrogen unites chemically with oxygen, the result is water (H_2O), which is a compound. A compound, then, consists of two or more elements. The molecule is the smallest unit of a compound with the same chemical characteristics. We can have molecules for either elements or compounds. However, atoms exist only for the elements.

Practice Problems 1-2
Answers at End of Chapter

a. Which have more free electrons: metals or insulators?
b. Which is the best conductor: silver, carbon, or iron?
c. Which is a semiconductor: copper, silicon, or neon?

1-3   Structure of the Atom

Although nobody has ever seen an atom, its hypothetical structure fits experimental evidence that has been measured very exactly. The size and electric charge of the invisible particles in the atom are indicated by how much they are deflected by known forces. Our present planetary model of the atom was proposed by Niels Bohr in 1913. His contribution was joining the new ideas of a nuclear atom developed by Lord Rutherford (1871–1937) with the quantum theory of radiation developed by Max Planck (1858–1947) and Albert Einstein (1879–1955).

As illustrated in Figs. 1-2 and 1-3, the nucleus contains protons for all the positive charge in the atom. The number of protons in the nucleus is equal to the number of planetary electrons. Thus, the positive and negative charges are balanced, as the proton and electron have equal and opposite charges. The orbits for the planetary electrons are also called *shells* or *energy levels*.

Atomic Number This gives the number of protons or electrons required in the atom for each element. For the hydrogen atom in Fig. 1-2, the atomic number is 1, which means the nucleus has 1 proton balanced by 1 orbital electron. Similarly, the carbon atom in Fig. 1-3 with atomic number 6 has 6 protons in the nucleus and 6 orbital electrons. Also, the copper atom has 29 protons and 29 electrons because its atomic number is 29. The atomic number is listed for each of the elements in Table 1-1 to indicate the atomic structure.

Table 1-2. Shells of Orbital Electrons in the Atom

Shell	Maximum electrons	Inert gas
K	2	Helium
L	8	Neon
M	8 (up to calcium) or 18	Argon
N	8, 18, or 32	Krypton
O	8 or 18	Xenon
P	8 or 18	Radon
Q	8	—

Orbital Rings The planetary electrons are in successive shells called K, L, M, N, O, P, and Q at increasing distances outward from the nucleus. Each shell has a maximum number of electrons for stability. As indicated in Table 1-2, these stable shells correspond to the inert gases, like helium and neon.

The K shell, closest to the nucleus, is stable with 2 electrons, corresponding to the atomic structure for the inert gas helium. Once the stable number of electrons has filled a shell, it cannot take any more electrons. The atomic structure with all its shells filled to the maximum number for stability corresponds to an inert gas.

Elements with a higher atomic number have more planetary electrons. These are in successive shells, tending to form the structure of the next inert gas in the periodic table.[1] After the K shell has been filled with 2 electrons, the L shell can take up to 8 electrons. Ten electrons filling the K and L shells is the atomic structure for the inert gas neon.

The maximum number of electrons in the remaining shells can be 8, 18, or 32 for different elements, depending on their place in the periodic table. The maximum for an outermost shell, though, is always 8.

---

[1]For more details of the periodic table of the elements, developed in 1869 by Dmitri Mendelyeev, refer to a textbook on chemistry or physics, or see "Periodic Chart of the Atoms," Sargent Welch Scientific Co., Skokie, Ill. 60076.

To illustrate these rules, we can use the copper atom in Fig. 1-3b as an example. There are 29 protons in the nucleus balanced by 29 planetary electrons. This number of electrons fills the K shell with 2 electrons, corresponding to the helium atom, and the L shell with 8 electrons. The 10 electrons in these two shells correspond to the neon atom, which has an atomic number of 10. The remaining 19 electrons for the copper atom then fill the M shell with 18 electrons and 1 electron in the outermost N shell. These values can be summarized as follows:

K shell = 2 electrons

L shell = 8 electrons

M shell = 18 electrons

N shell = 1 electron

Total = 29 electrons

For most elements, we can use the rule that the maximum number of electrons in a filled inner shell equals $2n^2$, where n is the shell number in sequential order outward from the nucleus. Then the maximum number of electrons in the first shell is $2 \times 1 = 2$; for the second shell $2 \times 2^2 = 8$, for the third shell $2 \times 3^2 = 18$, and for the fourth shell $2 \times 4^2 = 32$. These values apply only to an inner shell that is filled with its maximum number of electrons.

Electron Valence   This value is the number of electrons in an incomplete outermost shell. Copper, for instance, has a valence of 1 because there is 1 electron in the last shell, after the inner shells have been completed with their stable number. Similarly, hydrogen has a valence of 1, and carbon has a valence of 4. The number of outer electrons is considered positive valence, as these electrons are in addition to the stable shells.

Except for H and He, the goal of valence is 8 for all the atoms, as each tends to form the stable structure of 8 electrons in the outside ring. For this reason, valence can also be considered as the number of electrons in the outside ring needed to make 8. This value is the negative valence. As examples, the valence of copper can be considered +1 or −7; carbon has the valence of ±4. The inert gases have a valence of 0, as they all have a complete stable outer shell of 8 electrons.

The valence indicates how easily the atom can gain or lose electrons. For instance, atoms with a valence of +1 can lose this 1 outside electron, especially to atoms with a valence of +7 or −1, which need 1 electron to complete the outside shell with 8 electrons.

Subshells   Although not shown in the illustrations, all the shells except K are divided into subshells. This subdivision accounts for different types of orbits in the same shell. For instance, electrons in one subshell may have elliptical orbits, while other electrons in the same main shell have circular orbits. The subshells indicate magnetic properties of the atom.

Particles in the Nucleus   A stable nucleus, which is not radioactive, contains protons and neutrons. A neutron is electrically neutral without any net charge. Its mass is almost the same as a proton.

A proton has the positive charge of a hydrogen nucleus. The charge is the same amount as that of an orbital electron but of opposite polarity. There are no electrons in the nucleus. Table 1-3 lists the charge and mass for these three basic particles in all atoms.

Practice Problems 1-3
Answers at End of Chapter

a. An element with 14 protons and 14 electrons has what atomic number?

b. What is the electron valence of an element with atomic number 3?

Table 1-3. Stable Particles in the Atom

Particle	Charge	Mass
Electron, in orbital shells	0.16×10^{-18} C, negative	9.108×10^{-28} g
Proton, in nucleus	0.16×10^{-18} C, positive	1.672×10^{-24} g
Neutron, in nucleus	None	1.675×10^{-24} g

1-4   The Coulomb Unit of Charge

If you rub a hard rubber pen or comb on a sheet of paper, the rubber will attract a corner of the paper if it is free to move easily. The paper and rubber then give evidence of a static electric charge. The work of rubbing resulted in separating electrons and protons to produce a charge of excess electrons on the surface of the rubber and a charge of excess protons on the paper.

Because paper and rubber are dielectric materials, they hold their extra electrons or protons. As a result, the paper and rubber are no longer neutral, but each has an electric charge. The resultant electric charges provide the force of attraction between the rubber and the paper. This mechanical force of attraction or repulsion between charges is the fundamental method by which electricity makes itself evident.

Any charge is an example of *static electricity* because the electrons or protons are not in motion. There are many examples. When you walk across a wool rug, your body becomes charged with an excess of electrons. Similarly, silk, fur, and glass can be rubbed to produce a static charge. This effect is more evident in dry weather, because a moist dielectric does not hold its charge so well. Also, plastic materials can be charged easily, which is why thin, lightweight plastics seem to stick to everything.

The charge of many billions of electrons or protons is necessary for common applications of electricity. Therefore, it is convenient to define a practical unit called the *coulomb* (C) as equal to the charge of 6.25×10^{18} electrons or protons stored in a dielectric (see Fig. 1-4). The analysis of static charges and their forces is called *electrostatics*.

The symbol for electric charge is Q or q, standing for quantity. For instance, a charge of 6.25×10^{18} electrons[1] is stated as $Q = 1$ C. This unit is named after Charles A. Coulomb (1736–1806), a French physicist, who measured the force between charges.

Negative and Positive Polarities

Historically, the negative polarity has been assigned to the static charge produced on rubber, amber, and resinous materials in general. Positive polarity refers to the

---

[1]For an explanation of how to use powers of 10, see B. Grob, *Mathematics: Outline and Review Problems for Basic Electronics*, McGraw-Hill Book Company, New York.

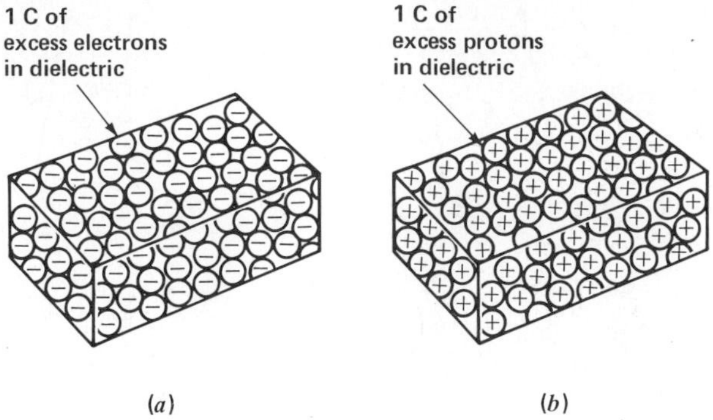

Fig. 1-4 The coulomb (C) unit of charge. (*a*) Quantity of 6.25×10^{18} excess electrons for a negative charge. (*b*) Same amount of protons for a positive charge, caused by removing electrons.

static charge produced on glass and other vitreous materials. On this basis, the electrons in all atoms are basic particles of negative charge because their polarity is the same as the charge on rubber. Protons have positive charge because the polarity is the same as the charge on glass.

Charges of Opposite Polarity Attract

If two small charged bodies of light weight are mounted so that they are free to move easily and are placed close to each other, one can be attracted to the other when the two charges have opposite polarity (Fig. 1-5*a*). In terms of electrons and protons, they tend to be attracted to each other by the force of attraction between opposite charges. Furthermore, the weight of an electron is only about $\frac{1}{1840}$ the weight of a proton. As a result, the force of attraction tends to make electrons move to protons.

Charges of the Same Polarity Repel

In Fig. 1-5*b* and *c*, it is shown that when the two bodies have an equal amount of charge with the same polarity, they repel each other. The two negative charges repel in Fig. 1-5*b*, while two positive charges of the same value repel each other in Fig. 1-5*c*.

Polarity of a Charge

An electric charge must have either negative or positive polarity, labeled $-Q$ or $+Q$, with an excess of either electrons or protons. A

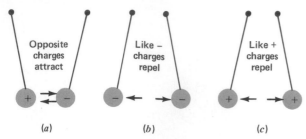

Fig. 1-5 Force between electric charges. (*a*) Opposite charges attract. (*b*) Two negative charges repel each other. (*c*) Two positive charges repel.

neutral condition is considered zero charge. On this basis, consider the following examples, remembering that the electron is the basic particle of charge and the proton has exactly the same amount, although of opposite polarity.

Example 1   A neutral dielectric has added to it 12.5×10^{18} electrons. What is its charge in coulombs?

Answer   This number of electrons is double the charge of 1 C. Therefore, $-Q = 2$ C.

Example 2   A dielectric has a positive charge of 12.5×10^{18} protons. What is its charge in coulombs?

Answer   This is the same amount of charge as in Example 1 but positive. Therefore, $+Q = 2$ C.

Example 3   A dielectric with $+Q$ of 2 C has 12.5×10^{18} electrons added. What is its charge then?

Answer   The 2 C of negative charge added by the electrons cancels the 2 C of positive charge, making the dielectric neutral, for $Q = 0$.

Example 4   A neutral dielectric has 12.5×10^{18} electrons removed. What is its charge?

Answer   The 2 C of electron charge removed allows an excess of 12.5×10^{18} protons. Since the proton and electron have exactly the same amount of charge, now the dielectric has a positive charge of $+Q = 2$ C.

Note that we generally consider the electrons moving, rather than the heavier protons. However, a loss of a given number of electrons is equivalent to a gain of the same number of protons.

Charge of an Electron   Fundamentally, the quantity of any charge is measured by its force of attraction or repulsion. The extremely small force of an electron or proton was measured by Millikan[1] in experiments done from 1908 to 1917. Very briefly, the method consisted of measuring the charge on vaporized droplets of oil, by balancing the gravitational force against an electrical force that could be measured very precisely.

A small drop of oil sprayed from an atomizer becomes charged by friction. Furthermore, the charges can be increased or decreased slightly by radiation. These very small changes in the amount of charge were measured. The three smallest values were 0.16×10^{-18} C, 0.32×10^{-18} C, and 0.48×10^{-18} C. These values are multiples of 0.16. In fact, all the charges measured were multiples of 0.16×10^{-18} C. Therefore, we conclude that 0.16×10^{-18} C is the basic charge from which all other values are derived. This ultimate charge of 0.16×10^{-18} C is the charge of 1 electron or 1 proton. Then

1 electron or $Q_e = 0.16 \times 10^{-18}$ C

The reciprocal of 0.16×10^{-18} gives the number of electrons or protons in 1 C. Then

1 C $= 6.25 \times 10^{18}$ electrons

Note that the factor 6.25 equals exactly $\frac{1}{0.16}$ and the 10^{18} is the reciprocal of 10^{-18}.

---

[1]Robert A. Millikan (1868–1953), an American physicist. Millikan received the Nobel prize in physics for this oil-drop experiment.

The Electric Field of a Static Charge The ability of an electric charge to attract or repel another charge is actually a physical force. To help visualize this effect, lines of force are used, as shown in Fig. 1-6. All the lines form the electric field. The lines and the field are imaginary, since they cannot be seen. Just as the field of the force of gravity is not visible, however, the resulting physical effects prove the field is there.

Each line of force in Fig. 1-6 is directed outward to indicate repulsion of another charge in the field with the same polarity as Q, either positive or negative. The lines are shorter further away from Q to indicate that the force decreases inversely as the square of the distance. The larger the charge, the greater is the force. These relations describe Coulomb's law of electrostatics.

The electric field in the dielectric between two plates with opposite charges is the basis for the ability of a capacitor to store electric charge. More details are explained in Chap. 20, "Capacitance." In general, any charged insulator has capacitance. Practical capacitors, though, are constructed in a form that concentrates the electric field.

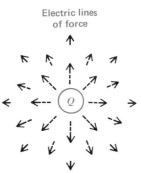

Electric lines of force

Fig. 1-6 Electric field around a stationary charge Q.

produce a neutral condition. We could consider that the accumulated electrons are drawn tight and are straining themselves to be attracted toward protons in order to return to the neutral condition. Similarly, the work of producing the charge causes a condition of stress in the protons, which are trying to attract electrons and return to the neutral condition. Because of these forces, the charge of electrons or protons has potential, as it is ready to give back the work put into producing the charge. The force between charges is in the electric field.

Practice Problems 1-4
Answers at End of Chapter

a. How many electron charges are there in the practical unit of one coulomb?

b. How much is the charge in coulombs for a surplus of 25×10^{18} electrons?

1-5   The Volt Unit of Potential Difference

Potential refers to the possibility of doing work. Any charge has the potential to do the work of moving another charge, by either attraction or repulsion. When we consider two unlike charges, they have a difference of potential.

A charge is the result of work done in separating electrons and protons. Because of the separation, there is stress and strain associated with opposite charges, since normally they would be balancing each other to

Potential between Different Charges When one charge is different from the other, there must be a difference of potential between them. For instance, consider a positive charge of 3 C, shown at the right in Fig. 1-7a. The charge has a certain amount of potential, corresponding to the amount of work this charge can do. The work to be done is moving some electrons, as illustrated.

Assume a charge of 1 C can move 3 electrons. Then the charge of +3 C can attract 9 electrons toward the right. However, the charge of +1 C at the opposite side can attract 3 electrons toward the left. The net result, then, is that 6 electrons can be moved toward the right to the more positive charge.

In Fig. 1-7b, one charge is 2 C, while the other charge is neutral with 0 C. For the difference of 2 C, again 2×3 or 6 electrons can be attracted to the positive side.

In Fig. 1-7c, the difference between the charges is still 2 C. The +1 C attracts 3 electrons to the side. Also, the −1 C repels 3 electrons. This effect is really the same as attracting 6 electrons.

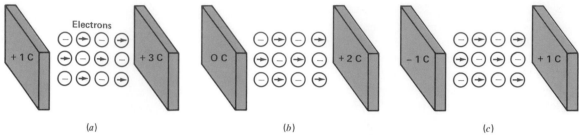

Fig. 1-7 The work required to move electrons between two charges depends on their difference of potential (PD). This difference is equivalent to 2 C of charge for the examples in (a), (b), and (c).

Therefore, the net number of electrons moved in the direction of the more positive charge depends on the difference of potential between the two charges. This difference corresponds to 2 C for all three cases in Fig. 1-7. Potential difference is often abbreviated *PD*.

The only case without any potential difference between charges is where they both have the same polarity and are equal in amount. Then the repelling and attracting forces cancel, and no work can be done in moving electrons between the two identical charges.

The Volt Unit of Potential Difference This unit is named after Alessandro Volta (1754–1827). Fundamentally, the volt is a measure of the work needed to move an electric charge. When 0.7376 foot-pound (ft·lb) of work is required to move 6.25×10^{18} electrons between two points, each with its own charge, the potential difference is 1 V.

Note that 6.25×10^{18} electrons make up one coulomb. Therefore the definition of a volt is for a coulomb of charge.

Also, 0.7376 ft·lb of work is the same as 1 joule (J), which is the practical metric unit of work[1] or energy. Therefore, we can say briefly that *one volt equals one joule of work per coulomb of charge.*

The symbol for potential difference is *V* for voltage. In fact, the volt unit is used so often that potential difference is often called voltage. Remember, though, that voltage is the potential difference between two points. Two terminals are necessary to measure a potential difference.

Consider the 2.2-V lead-acid cell in Fig. 1-8. Its output of 2.2 V means that this is the amount of potential

difference between the two terminals. The cell then is a voltage source, or a source of electromotive force (*emf*).

Sometimes the symbol *E* is used for emf, but the standard symbol is *V* for any potential difference. This applies either to the voltage generated by a source or to the voltage drop across a passive component, such as a resistor.

In a practical circuit, the voltage determines how much current can be produced.

Practice Problems 1-5
Answers at End of Chapter

a. How much potential difference is there between two identical charges?
b. Which applies a greater PD, a 1.5-V battery or a 12-V battery?

1-6   Charge in Motion Is Current

When the potential difference between two charges forces a third charge to move, the charge in motion is an *electric current*. To produce current, therefore, charge must be moved by a potential difference.

In solid materials, such as copper wire, the free electrons are charges that can be forced to move with relative ease by a potential difference, since they require relatively little work to be moved. As illustrated in Fig. 1-8, if a potential difference is connected across two ends of a copper wire, the applied voltage forces the free electrons to move. This current is a drift of electrons, from the point of negative charge at one end,

---

[1]See App. D, "Physics Units."

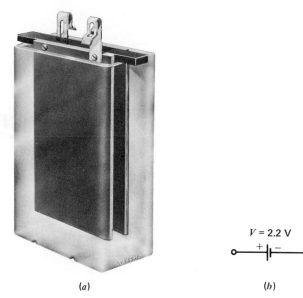

(a)   *(b)*

Fig. 1-8 Chemical cell as a voltage source. (a) Voltage output is potential difference between the two terminals. (b) Schematic symbol of any dc voltage source. Longer line indicates positive side.

moving through the wire, and returning to the positive charge at the other end.

To illustrate the drift of free electrons through the wire shown in Fig. 1-9, each electron in the middle row is numbered, corresponding to a copper atom to which the free electron belongs. The electron at the left is labeled S to indicate that it comes from the negative charge of the source of potential difference. This one electron S is repelled from the negative charge $-Q$ at the left and is attracted by the positive charge $+Q$ at the right. Therefore, the potential difference of the voltage source can make electron S move toward atom 1. Now

atom 1 has an extra electron. As a result, the free electron of atom 1 can then move to atom 2. In this way, there is a drift of free electrons from atom to atom. The final result is that the one free electron labeled 8 at the extreme right in Fig. 1-9 moves out from the wire to return to the positive charge of the voltage source.

Considering this case of just one electron moving, note that the electron returning to the positive side of the voltage source is not the electron labeled S that left the negative side. All electrons are the same, however, and have the same charge. Therefore, the drift of free electrons resulted in the charge of one electron moving through the wire. This charge in motion is the current. With more electrons drifting through the wire, the charge of many electrons moves, resulting in more current.

The current is a continuous flow of electrons. Only the electrons move, not the potential difference. For ordinary applications, where the wires are not long lines, the potential difference produces current instantaneously through the entire length of wire.

Furthermore, the current must be the same at all points of the wire at any time. Although a point nearer to the negative terminal of the voltage source has a greater repelling force on the free electrons, at this point the free electrons are farther from the positive terminal and have less attracting force. At the middle of the wire, the free electrons have equal forces of attraction and repulsion. Near the positive terminal, there is a greater attracting force on the free electrons but less repelling force from the negative terminal of the voltage source. In all cases, the total force causing motion of the free electrons is the same at any point of the wire, therefore resulting in the same current through all parts of the wire.

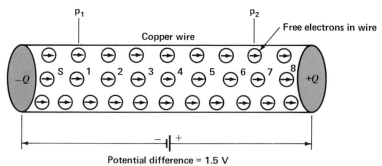

Fig. 1-9 Potential difference across two ends of wire conductor causes drift of free electrons through the wire, producing electric current.

Potential Difference Is Necessary to Produce Current   The number of free electrons that can be forced to drift through the wire to produce the moving charge depends upon the amount of potential difference across the wire. With more applied voltage, the forces of attraction and repulsion can make more free electrons drift, producing more charge in motion. A larger amount of charge moving with the same speed means a higher value of current. Less applied voltage across the same wire results in a smaller amount of charge in motion, which is a smaller value of current. With zero potential difference across the wire, there is no current.

Two cases of zero potential difference and no current can be considered in order to emphasize the important fact that potential difference is needed to produce current. Assume the copper wire to be by itself, not connected to any voltage source, so that there is no potential difference across the wire. The free electrons in the wire can move from atom to atom, but this motion is random, without any organized drift through the wire. If the wire is considered as a whole, from one end to the other, the current is zero.

As another example, suppose that the two ends of the wire have the same potential. Then free electrons cannot move to either end, because both ends have the same force, and there is no current through the wire. A practical example of this case of zero potential difference would be to connect both ends of the wire to just one terminal of a battery. Each end of the wire would have the same potential, and there would be no current. The conclusion, therefore, is that two connections to two points at different potentials are needed in order to produce the current.

The Ampere of Current   Since current is the movement of charge, the unit for stating the amount of current is defined in rate of flow of charge. When the charge moves at the rate of 6.25×10^{18} electrons flowing past a given point per second, the value of the current is one *ampere* (A). This is the same as one coulomb of charge per second. The ampere unit of current is named after André M. Ampère (1775–1836).

Referring back to Fig. 1-9, note that if 6.25×10^{18} free electrons move past p_1 in 1 s, the current is 1 A. Similarly, the current is 1 A at p_2 because the electron drift is the same throughout the wire. If twice as many electrons moved past either point in 1 s, the current would be 2 A.

The symbol for current is I or i for intensity, since the current is a measure of how intense or concentrated the electron flow is. Two amperes of current in a copper wire is a higher intensity than 1 A; a greater concentration of moving electrons results because of more electrons in motion, although all the electrons move with the same speed. Sometimes current is called *amperage*. However, the current in electronic circuits is usually in smaller units, milliamperes and microamperes.

How Current Differs from Charge   Charge is a quantity of electricity accumulated in a dielectric. The charge is static electricity, at rest, without any motion. When the charge moves, usually in a conductor, the current I indicates the intensity of the electricity in motion. This characteristic is a fundamental definition of current:

$$I = \frac{Q}{T} \tag{1-1}$$

where I is the current in amperes, Q is in coulombs, and the time T is in seconds (s). It does not matter whether the moving charge is positive or negative. The only question is how much charge moves and what its rate of motion is.

Example 5   The charge of 12 C moves past a given point every second. How much is the intensity of charge flow?

Answer $I = \dfrac{Q}{T} = \dfrac{12\text{ C}}{1\text{ s}}$

$I = 12$ A

Example 6   The charge of 5 C moves past a given point in 1 s. How much is the current?

Answer $I = \dfrac{Q}{T} = \dfrac{5\text{ C}}{1\text{ s}}$

$I = 5$ A

Table 1-4. Types of Electric Charges for Current

Type of charge	Amount of charge	Polarity	Type of current	Applications
Electron	$Q_e = 0.16 \times 10^{-18}$ C	Negative	Electron flow	In wire conductors, vacuum tubes, and N-type semiconductors
Ion	Q_e or multiples of Q_e	Positive or negative	Ion current	In liquids and gases
Hole	$Q_e = 0.16 \times 10^{-18}$ C	Positive	Hole current	In P-type semiconductors

This fundamental definition of current can also be used to consider the charge as equal to the product of the current multiplied by the time. Or

$$Q = I \times T \qquad (1\text{-}2)$$

For instance, we can have a dielectric connected to conductors with a current of 0.4 A. If the current can deposit electrons for the time of 0.2 s, the accumulated charge in the dielectric will be

$$Q = I \times T$$
$$= 0.4 \times 0.2$$
$$Q = 0.08 \text{ C}$$

The formulas $Q = IT$ for charge and $I = Q/T$ for current illustrate the fundamental nature of Q as an accumulation of static charge in an insulator, while I measures the intensity of moving charges in a conductor.

The General Nature of Current The moving charges that provide current in metal conductors like a copper wire are the free electrons of the copper atoms.

In this case, the moving charges have negative polarity. The direction of motion between two terminals for this *electron current,* therefore, is toward the more positive end. It is important to note, however, that there are examples of positive charges in motion. Common applications include current in liquids, gases, and semiconductors. For the case of current resulting from the motion of positive charges, its direction is opposite from the direction of electron flow. Whether negative or positive charges move, though, the current is still defined fundamentally as Q/T.

Types of Electric Charges for Current See Table 1-4. The most common charge is the electron. In metal conductors and solid materials in general, free electrons in the atoms can be forced to move by setting up a potential difference. Therefore current is produced. The direction of electron flow is from the negative terminal of the voltage source, through the external circuit, and returning to the positive source terminal (Fig. 1-10a). Also, electrons are released by thermionic emission from the heated cathode in a vacuum

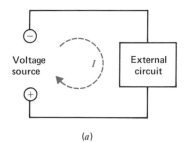
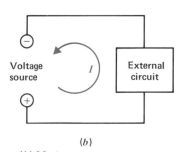

(a)   (b)

Fig. 1-10 Direction of current. (a) Dotted lines show electron flow. (b) Motion of positive charges in the opposite direction for hole current or ion current.

tube. Finally, N-type semiconductors such as silicon and germanium have unbound electrons as a result of doping with impurity elements that can provide valence electrons from the added atoms.

For P-type semiconductors, the silicon and germanium are doped with impurity elements that cause a deficiency of electrons in the bonds between atoms. Each vacant space where an electron is missing is called a *hole charge*. The polarity is positive, opposite from the electron, but the amount of charge is exactly the same. In short, a hole charge is a deficiency of one valence electron in semiconductors. When hole charges move in a P-type semiconductor, they provide hole current. The direction of flow for the positive charges is from the positive terminal of the voltage source, through the external circuit, and returning to the negative source terminal (Fig. 1-10b).

An ion is an atom that has either lost or gained one or more valence electrons to become electrically charged. Therefore, the ion charge may be either positive or negative. The amount may be the charge of 1 electron Q_e, $2Q_e$, $3Q_e$, etc. Ions can be produced by applying voltage to liquids and gases to produce ionization of the atoms. The ions are much less mobile than electrons or hole charges because an ion includes a complex atom with its nucleus.

Note that protons are not included as charge carriers for current in Table 1-4. The reason is that the protons are bound in the nucleus. They cannot be released except by nuclear forces. Therefore, a current of positive charges is a flow of either hole charges or positive ions. The hole charge has the same amount of charge as the proton, which is the same amount as an electron. However, the positive hole charge is in the valence structure of the atoms, not in the nucleus.

Magnetic Field Around an Electric Current

When any current flows, it has an associated magnetic field. Figure 1-11 shows how iron filings line up in a circular field pattern corresponding to the magnetic lines of force. The magnetic field is in a plane perpendicular to the current. It should be noted that the iron filings are just a method of making the imaginary lines of force visible. The filings become magnetized by the magnetic field. Both magnetic and electric fields can do the physical work of attraction or repulsion.

The magnetic field of any current is the basis for many electromagnetic applications, including magnets,

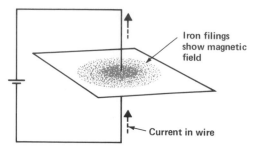

Fig. 1-11 Magnetic field around any electric current.

relays, loudspeakers, transformers, and coils in general. Winding the conductor in the form of a coil concentrates the magnetic field. More details of electromagnetism and magnetic fields are explained in Chap. 15, "Electromagnetic Induction."

Practice Problems 1-6
Answers at End of Chapter

a. The flow of 2 C/s of electron charges is how many amperes of current?

b. The flow of 2 C/s of hole charges is how many amperes of current?

c. How much is the current with zero potential difference?

1-7 Resistance Is Opposition to Current

The fact that a wire conducting current can become hot is evidence of the fact that the work done by the applied voltage in producing current must be accomplished against some form of opposition. This opposition, which limits the amount of current that can be produced by the applied voltage, is called *resistance*. Conductors have very little resistance; insulators have a large amount of resistance.

The atoms of a copper wire have a large number of free electrons, which can be moved easily by a potential difference. Therefore, the copper wire has little opposition to the flow of free electrons when voltage is applied, corresponding to a low value of resistance.

Carbon, however, has fewer free electrons than copper. When the same amount of voltage is applied to the

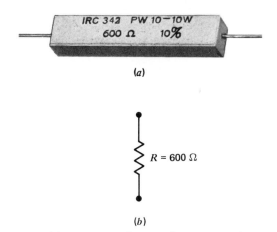

Fig. 1-12 (a) Wire-wound type of resistor, with cement coating for insulation. (b) Schematic symbol for any type of resistance.

carbon as to the copper, fewer electrons will flow. It should be noted that just as much current can be produced in the carbon by applying more voltage. For the same current, though, the higher applied voltage means that more work is necessary, causing more heat. Carbon opposes the current more than copper, therefore, and has a higher value of resistance.

The Ohm The practical unit of resistance is the *ohm* (Ω). A resistance that develops 0.24 calorie[1] of heat when one ampere of current flows through it for one second has one ohm of opposition. As an example of a low resistance, a good conductor like copper wire can have a resistance of 0.01 Ω for a 1-ft length. The resistance-wire heating element in a 600-W toaster has a resistance of 24 Ω, and the tungsten filament in a 100-W light bulb has a resistance of 144 Ω, with a 120-V source.

Figure 1-12 shows a wire-wound resistor. This type of resistance can be manufactured with a value from a few ohms to millions of ohms. The abbreviation for resistance is R or r. The symbol used for the ohm is the Greek letter *omega*, written as Ω. In diagrams, resistance is indicated by a zigzag line, as shown by R in Fig. 1-12.

---

[1]One calorie is the quantity of heat that will raise the temperature of one gram of water by one degree Celsius. See App. D, "Physics Units."

Conductance The opposite of resistance is conductance. The less the resistance, the higher the conductance. Its symbol is G, and the unit is the *siemens* (S), named after Ernst von Siemens, a European inventor. (The old, equivalent unit name for siemens is *mho*, which is *ohm* spelled backward.) Specifically, G is the reciprocal of R, or $G = 1/R$. For example, 5 Ω of resistance is equal to ⅕ S of conductance.

Whether to use R or G for components is usually a matter of convenience. In general, R is easier to use in series circuits, because the series voltages are proportional to the resistances; G can be more convenient in parallel circuits, because the parallel currents are proportional to the conductances. (Series and parallel circuits are explained in Chaps. 3 and 4.)

Practice Problems 1-7
Answers at End of Chapter

a. Which has more resistance, carbon or copper?
b. With the same voltage applied, which resistance will allow more current, 4.7 Ω or 5000 Ω?
c. What is the conductance value in siemens units for a 10-Ω R?

1-8 The Closed Circuit

In electrical applications requiring the use of current, the components are arranged in the form of a circuit, as shown in Fig. 1-13. A circuit can be defined as a path for current flow. The purpose of this circuit is to light the incandescent bulb. The bulb lights when the tungsten-filament wire inside is white hot, producing an incandescent glow.

By itself the tungsten filament cannot produce current. A source of potential difference is necessary. Since the battery produces a potential difference of 1.5 V across its two output terminals, this voltage is connected across the filament of the bulb by means of the two wires so that the applied voltage can produce current through the filament.

In Fig. 1-13c the schematic diagram of the circuit is shown. Here the components are represented by shorthand symbols. Note the symbols for the battery and resistance. The connecting wires are shown simply as straight lines because their resistance is small enough to

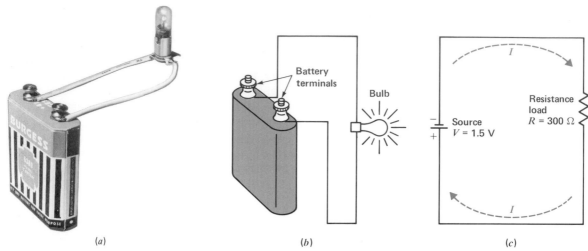

Fig. 1-13 An electric circuit with voltage source connected to a resistance load. (*a*) Photo of circuit. (*b*) Wiring diagram. (*c*) Schematic diagram.

be neglected. A resistance of less than 0.01 Ω for the wire is practically zero compared with the 300-Ω resistance of the bulb. If the resistance of the wire must be considered, the schematic diagram includes it as additional resistance in the same current path.

It should be noted that the schematic diagram does not look like the physical layout of the circuit. The schematic shows only the symbols for the components and their electrical connections.

Any electric circuit has three important characteristics:

1. There must be a source of potential difference. Without the applied voltage, current cannot flow.
2. There must be a complete path for current flow, from one side of the applied voltage source, through the external circuit, and returning to the other side of the voltage source.
3. The current path normally has resistance. The resistance is in the circuit for the purpose of either generating heat or limiting the amount of current.

How the Voltage Is Different from the Current
It is the current that moves through the circuit. The potential difference does not move.

In Fig. 1-13 the voltage across the filament resistance makes electrons flow from one side to the other. While the current is flowing around the circuit, however, the potential difference remains across the fila-

ment to do the work of moving electrons through the resistance of the filament.

The circuit is redrawn in Fig. 1-14 to emphasize the comparison between *V* and *I*. The voltage is the potential difference across the two ends of the resistance. If you want to measure the PD, just connect the two leads of the voltmeter across the resistor. However, the current is the intensity of the electron flow past any one point in the circuit. Measuring the current is not as easy. You would have to break open the path, at any point, and then insert the current meter to complete the circuit.

To illustrate the difference between *V* and *I* another way, suppose the circuit in Fig. 1-13 is opened by disconnecting the bulb. Now no current can flow because there is no closed path. Still, the battery has its potential difference. If you measure across the two terminals, the voltmeter will read 1.5 V even though the current is zero.

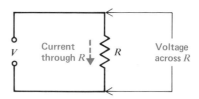

Fig. 1-14 Comparison of voltage and current.

The Voltage Source Maintains the Current

As current flows in the circuit, electrons leave the negative terminal of the cell or battery in Fig. 1-13, and the same number of free electrons in the conductor are returned to the positive terminal. As electrons are lost from the negative charge and gained by the positive charge, the two charges would tend to neutralize each other. The chemical action inside the battery, however, continuously separates electrons and protons to maintain the negative and positive charges on the outside terminals that provide the potential difference. Otherwise, the current would neutralize the charges, resulting in no potential difference, and the current would stop. Therefore, the battery keeps the current flowing by maintaining the potential difference across the circuit. For this reason the battery is the voltage source, or generator, for the circuit.

The Circuit Is a Load on the Voltage Source

We can consider the circuit as a means whereby the energy of the voltage source is carried by means of the current through the filament of the bulb, where the electric energy is used in producing heat energy. On this basis, the battery is the *source* in the circuit, since its voltage output represents the potential energy to be used. The part of the circuit connected to the voltage source is the *load resistance,* since it determines how much work the source will supply. In this case, the bulb's filament is the load resistance for the battery.

The resistance of the filament determines how much current the 1.5-V source will produce. Specifically, the current here is 0.005 A, equal to 1.5 V divided by 300 Ω. With more opposition, the same voltage will produce less current. For the opposite case, less opposition allows more current.

The current that flows through the load resistance is the *load current*. Note that a lower value of ohms for the load resistance corresponds to a higher load current. Unless noted otherwise, the term *load* by itself can be assumed generally to mean the load current. Therefore, a heavy or big load electrically means a high value of load current, corresponding to a large amount of work supplied by the source.

Direction of the Electron Flow

As shown in Fig. 1-13c, the direction of the electron drift is from the negative side of the battery, through the load resistance *R,* and back to the positive terminal of the voltage source. Note that this is the direction in the external circuit connected across the output terminals of the voltage source.

Inside the battery, the electrons move to the negative terminal because this is how the voltage source produces its potential difference. The battery is doing the work of separating charges, accumulating electrons at the negative terminal and protons at the positive terminal. Then the potential difference across the two output terminals can do the work of moving electrons around the external circuit. In the circuit outside the voltage source, however, the direction of the electron flow is from a point of negative potential to a point of positive potential.

Conventional Current

The direction of moving positive charges, opposite from electron flow, is considered the conventional direction of current. In electrical engineering, circuits are usually analyzed with conventional current. The reason is based on the fact that, by the positive definitions of force and work, a positive potential is considered above a negative potential. So conventional current is a motion of positive charges "falling downhill" from a positive to a negative potential. The direction of conventional current, therefore, is the direction of positive charges in motion.

An example of positive charges moving in the direction of conventional current is hole current in P-type semiconductors. Also, a current of positive ions in liquids or gases moves in the opposite direction from electron flow. For instance, the current through the electrolyte inside a battery is ionization current.

Actually, either a positive or negative potential of the same value can do the same amount of work in moving charge. Any circuit can be analyzed either with electron flow or by conventional current in the opposite direction. In this book the current is considered as electron flow in the applications in which electrons are the moving charges.

In summary, we can say that the closed circuit, normal circuit, or just a circuit is a closed path that has V to produce I with R to limit the amount of current. The circuit provides a means of using the energy of the battery as a voltage source. The battery has its potential difference V with or without the circuit. However, the battery alone is not doing any work in producing load current. The bulb alone has its resistance, but without current the bulb does not light. With the circuit, the voltage source is used for the purpose of producing current to light the bulb.

Open Circuit   When any part of the path is open or broken, the circuit is open because there is no continuity in the conducting path. The open circuit can be in the connecting wires or in the bulb's filament as the load resistance. The resistance of an open circuit is infinitely high. The result is no current in an open circuit.

Short Circuit   In this case, the voltage source has a closed path across its terminals, but the resistance is practically zero. The result is too much current in a short circuit. Usually, the short circuit is a bypass across the load resistance. For instance, a short across the conducting wires for a bulb produces too much current in the wires but no current through the bulb. Then the bulb is shorted out. The bulb is not damaged, but the wires can become hot enough to burn unless the line has a fuse as a safety precaution against too much current.

Practice Problems 1-8
Answers at End of Chapter

Answer True or False for the circuit in Fig. 1-13.
- **a.** The bulb has a PD of 1.5 V across its filament only when connected to the voltage source.
- **b.** The battery has a PD of 1.5 V across its terminals only when connected to the bulb.

1-9   Direct Current (DC) and Alternating Current (AC)

The electron flow illustrated in the circuit of Fig. 1-13c is direct current because it has just one direction. The reason for the unidirectional current is that the battery maintains the same polarity of output voltage.

It is the flow of charges in just one direction and the fixed polarity of applied voltage that are the characteristics of a dc circuit. Actually, the current can be a

motion of positive charges, rather than electrons, but the conventional direction of current does not change the fact that direct current has just one direction. Furthermore, the dc voltage source can change the amount of its output voltage, but if the same polarity is maintained, direct current will flow in just one direction, meeting the requirements of a dc circuit. A battery is a steady dc voltage source because it has fixed polarity and its output voltage is a steady value.

An alternating voltage source periodically reverses or alternates in polarity. The resulting alternating current, therefore, periodically reverses in direction. In terms of electron flow, the current always flows from the negative terminal of the voltage source, through the circuit, and back to the positive terminal, but when the generator alternates in polarity, the current must reverse its direction. The 60-cycle ac power line used in most homes is a common example. This frequency means that the voltage polarity and current direction go through 60 cycles of reversal per second.

The unit for 1 cycle per second is 1 hertz (Hz). Therefore 60 cycles per second is a frequency of 60 Hz.

The details of ac circuits are explained in Chap. 16 and the remainder of the book. Direct-current circuits are analyzed first because they usually are simpler. However, the principles of dc circuits also apply to ac circuits. Both types are important, as most electronic circuits include ac voltages and dc voltages. The waveforms for these two types of voltages are illustrated in

Table 1-5. Comparison of DC Voltage and AC Voltage

DC voltage	AC voltage
Fixed polarity	Reverses in polarity
Can be steady or vary in magnitude	Varies between reversals in polarity
Steady value cannot be stepped up or down by a transformer	Can be stepped up or down for electric power distribution
Electrode voltages for transistor amplifiers	Signal input and output for amplifiers
Easier to measure	Easier to amplify
Heating effect the same for direct or alternating current	

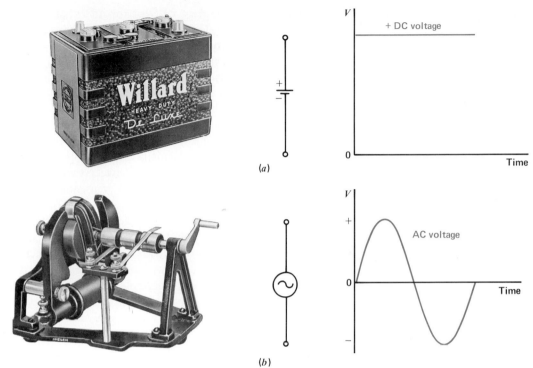

Fig. 1-15 Comparison of dc and ac voltages. Source shown at left, symbol at center, and graph of waveform at the right. (a) Steady dc voltage of one polarity from a battery. (b) Sine-wave ac voltage with alternating polarity from a small laboratory-type rotary ac generator. (*Sargent Welch Scientific Co.*)

Fig. 1-15. Their uses are compared in Table 1-5. Note that transistors require dc electrode voltages in order to amplify an ac signal voltage.

Practice Problems 1-9
Answers at End of Chapter

Answer True or False.
a. When the polarity of the applied voltage reverses, the direction of current flow also reverses.
b. A battery is a dc voltage source because it cannot reverse the polarity across its output terminals.

1-10 Sources of Electricity

There are electrons and protons in the atoms of all materials, but to do useful work the charges must be separated to produce a potential difference that can make current flow. Some of the more common methods of providing electrical effects are listed here.

Static Electricity by Friction In this method, electrons in an insulator can be separated by the work of rubbing to produce opposite charges that remain in the dielectric.

Conversion of Chemical Energy Wet or dry cells and batteries are the applications. Here a chemical reaction produces opposite charges on two dissimilar metals, which serve as the negative and positive terminals.

Ions Atoms can be made to gain or lose orbital electrons, by either a chemical reaction or the electric field of an applied voltage. Such atoms with a net charge are ions. The ion charges can be either negative or positive. Ions are generally the charge carriers that provide ionization current in liquids and gases.

Electromagnetism

Electricity and magnetism are closely related. Any moving charge has an associated magnetic field; also, any changing magnetic field can produce current. A motor is an example of how current can react with a magnetic field to produce motion; a generator produces voltage by means of a conductor rotating in a magnetic field.

Photoelectricity

Some materials are photoelectric, meaning they can emit electrons when light strikes the surface. The element cesium is often used as a source of *photoelectrons*. Also, photovoltaic cells or solar cells use silicon to generate output voltage from the light input. In another effect, the resistance of the element selenium changes with light. When this is combined with a fixed voltage source, wide variations between *dark current* and *light current* can be produced. Such characteristics are the basis of many photoelectric devices, including television camera tubes, photoelectric cells, and phototransistors.

Thermal Emission

Some materials when heated can "boil off" electrons from the surface. Then these emitted electrons can be controlled to provide useful applications of electric current. The emitting electrode is called a *cathode,* while an *anode* is used to collect the emitted electrons.

Doping of Semiconductors

Because they have a valence of ±4, combined with a crystalline molecular structure, semiconductors such as germanium and silicon can be changed from the inert condition by adding small amounts of an impurity element. This technique of adding charges to a semiconductor is called *doping*. When electrons are added, the result is negative or N-type material; a deficiency of electrons makes the material positive or P-type. A deficiency of 1 electron in the semiconductor is called a *hole charge*. With one type of semiconductor between two opposite types, the result is a PNP or NPN transistor. The details of transistors are explained in Chaps. 29 and 30.

Although there are so many different applications, remember that all electrons are the same, with identical charge and mass. Whether the electron flow results from a battery, rotary generator, or photoelectric device, and is controlled by a vacuum tube or transistor, the analysis of voltage, current, and resistance in the different types of circuits must follow the basic principles described here.

Practice Problems 1-10
Answers at End of Chapter

a. The excess charges at the negative terminal of a battery are _____ .
b. The charges emitted from a heated cathode are _____ .
c. The charges in a P-type semiconductor are _____ .
d. In a liquid that is conducting current, the moving charges are _____ .

Summary

1. Electricity is present in all matter in the form of electrons and protons.
2. The electron is the basic quantity of negative electricity, the proton of positive electricity. Both have the same amount of charge but opposite polarities. The charge of 6.25×10^{18} electrons or protons equals the practical unit of one coulomb.
3. Charges of the same polarity tend to repel each other; charges of opposite polarities attract. There must be a difference of charges for any force of attraction or repulsion.
4. Electrons tend to move toward protons because an electron has $1/1840$ the weight of a proton. Electrons in motion provide an electron current.

Table 1-6. Electrical Characteristics

Characteristic	Symbol	Unit	Description
Charge	Q or q^1	Coulomb (C)	Quantity of electrons or protons; $Q = I \times T$
Current	I or i^1	Ampere (A)	Charge in motion; $I = Q/T$
Voltage	V or $v^{1,2}$	Volt (V)	Potential difference between two unlike charges; makes charge move to produce I
Resistance	R or r^3	Ohm (Ω)	Opposition that reduces amount of current
Conductance	G or g^3	Siemens (S)	Reciprocal of R, or $G = 1/R$

[1] Small letter q, i, or v is used for an instantaneous value of a varying charge, current, or voltage.
[2] E or e is sometimes used for a generated emf, but the standard symbol for any potential difference is V or v in the international system of units (SI).
[3] Small letter r or g is used for internal resistance or conductance of transistors and tubes.

5. The atomic number of an element gives the number of protons in the nucleus of its atom, balanced by an equal number of orbital electrons.
6. The number of electrons in the outermost orbit is the valence of the element.
7. Table 1-6 summarizes the main features of electric circuits. In the symbols, the small letters q, v, and i are used when the characteristic varies with respect to time. Also, the small letters r and g indicate internal characteristics of a source.
8. Types of negative charges include electrons and negative ions. Types of positive charges include protons, positive ions, and hole charges.
9. An electric circuit is a closed path for electron flow. Potential difference must be connected across the circuit to produce current. In the external circuit outside the voltage source, electrons flow from the negative terminal toward the positive terminal.
10. Direct current has just one direction, as the dc voltage source has a fixed polarity. Alternating current periodically reverses in direction as the ac voltage source reverses its polarity.

Self-Examination
Answers at Back of Book

Answer True or False.

1. All matter has electricity in the form of electrons and protons in the atom.
2. The electron is the basic unit of negative charge.
3. A proton has the same amount of charge as the electron but opposite polarity.
4. Electrons are repelled from other electrons but are attracted to protons.
5. The force of attraction or repulsion between charges is in their electric field.
6. The nucleus is the massive stable part of an atom, with positive charge.
7. Neutrons add to the weight of the atom's nucleus but not to its electric charge.
8. An element with atomic number 12 has 12 orbital electrons.
9. This element has an electron valence of +2.
10. To produce current in a circuit, potential difference is connected across a closed path.
11. A dc voltage has fixed polarity, while ac voltage periodically reverses its polarity.
12. The coulomb is a measure of the quantity of stored charge.

13. If a dielectric has 2 C of excess electrons, removing 3 C of electrons will leave the dielectric with the positive charge of 1 C.
14. A charge of 5 C flowing past a point each second is a current of 5 A.
15. A 7-A current charging a dielectric will accumulate charge of 14 C after 2 s.
16. A voltage source has two terminals with different charges.
17. An ion is a charged atom.
18. The resistance of a few feet of copper wire is practically zero.
19. The resistance of the rubber or plastic insulation on the wire is also practically zero.
20. A resistance of 600 Ω has a conductance of 6 S.

Essay Questions

1. Briefly define each of the following, giving its unit and symbol: charge, potential difference, current, resistance, and conductance.
2. Name two good conductors, two good insulators, and two semiconductors.
3. Explain briefly why there is no current in a light bulb unless it is connected across a source of applied voltage.
4. Give three differences between voltage and current.
5. In any circuit: (a) state two requirements for producing current; (b) give the direction of electron flow.
6. Show the atomic structure of the element sodium (Na), with atomic number 11. What is its electron valence?
7. Make up your own name for direct current and dc voltage to indicate how it differs from alternating voltage and current.
8. State the formulas for each of the following two statements: (a) Current is the time rate of change of charge. (b) Charge is current accumulated over a period of time.
9. Refer to Table 1-4. (a) Name two types of moving charges that provide current in the direction of electron flow. (b) Name two types that provide conventional current.
10. Why is it that protons are not considered a source of moving charges for current flow?
11. Give one difference and one similarity in comparing electric and magnetic fields.
12. Give three methods of providing electric charges, and give their practical applications.
13. Give one way in which electricity and electronics are similar and one way in which they are different.
14. What kind of a meter would you use to measure the potential difference of a battery?
15. Give at least two voltage sources you have used and classify them as either dc or ac sources.
16. Compare the functions of a good conductor like copper wire with an insulator like paper, air, and vacuum.
17. How would you define *electromagnetism*?
18. Why is the characteristic of frequency used with alternating current and voltage?

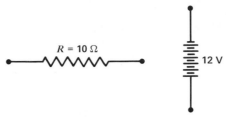

Fig. 1-16 For Prob. 8.

Problems
Answers to Odd-numbered Problems at Back of Book

1. The charge of 8 C flows past a given point every 2 s. How much is the current in amperes?
2. The current of 4 A charges an insulator for 2 s. How much charge is accumulated?
3. Convert the following to siemens or mhos of conductance: (**a**) 1000 Ω; (**b**) 500 Ω; (**c**) 10 Ω; (**d**) 0.1 Ω.
4. Convert the following to ohms of resistance: (**a**) 0.001 S; (**b**) 0.002 S; (**c**) 0.1 S; (**d**) 10 S.
5. A battery can supply 11 J of energy to move 5 C of charge. How much is the voltage of the battery? (Hint: One volt equals one joule per coulomb.)
6. A material with a deficiency of 25×10^{18} electrons gains 31.25×10^{18} electrons. The excess electrons are then made to flow past a given point in 1 s. How much current is produced by the resultant electron flow?
7. Convert 5 S of conductance to ohms of resistance.
8. Connect the components in Fig. 1-16 to form an electric circuit. Label the source voltage, with polarity, and the load resistance. Show the direction of electron flow. Determine the amount of current I by Ohm's law.

Answers to Practice Problems

1-1 **a.** negative
    b. positive
    c. true

1-2 **a.** metals
    b. silver
    c. silicon

1-3 **a.** 14
    b. 1

1-4 **a.** 6.25×10^{18}
    b. 4 C

1-5 **a.** zero
    b. 12 V

1-6 **a.** 2 A
    b. 2 A
    c. zero

1-7 **a.** carbon
    b. 4.7 Ω
    c. $\frac{1}{10}$ S

1-8 **a.** T
    b. F

1-9 **a.** T
    b. T

1-10 **a.** electrons
    b. electrons
    c. holes
    d. ions

Chapter 2
Ohm's Law

This unit explains how the amount of current I in a circuit depends on its resistance R and the applied voltage. Specifically, $I = V/R$, determined in 1828 by the experiments of George Simon Ohm. If you know any two of the factors V, I, and R, you can calculate the third. As an example, with 6 V applied across an R of 2 Ω, the I is $\frac{6}{2} = 3$ A. Ohm's law also determines the amount of electric power in the circuit. The amount of power consumed in a circuit is equal to the voltage times the current. The power used by the R above is therefore equal to $6 \times 3 = 18$ W. These relations apply to both dc and ac circuits.

Important terms in this chapter are:

volt	micro
ampere	kilo
ohm	mega
watt	linear graph
joule	power
milli	voltampere characteristic

More details are explained in the following sections:

2-1 The Current $I = V/R$

If we keep the same resistance in a circuit but vary the voltage, the current will vary. The circuit in Fig. 2-1 demonstrates this idea. The applied voltage V can be varied from 0 to 12 V, as an example. The bulb has a 12-V filament, which requires this much voltage for its normal current to light with normal intensity. The meter I indicates the amount of current in the circuit for the bulb.

With 12 V applied, the bulb lights, indicating normal current. When V is reduced to 10 V, there is less light because of less I. As V decreases, the bulb becomes dimmer. For zero volts applied there is no current and the bulb cannot light. In summary, the changing brilliance of the bulb shows that the current is varying with the changes in applied voltage.

For the general case of any V and R, Ohm's law is

$$I = \frac{V}{R} \qquad \qquad (2\text{-}1)$$

where I is the amount of current through the resistance R connected across the source of potential difference V. With volts as the practical unit for V and ohms for R, the amount of current I is in amperes. Therefore,

$$\text{Amperes} = \frac{\text{volts}}{\text{ohms}}$$

This formula says to simply divide the voltage across R by the ohms of resistance between the two points of potential difference to calculate the amperes of current through R. In Fig. 2-2, for instance, with 6 V applied across a 3-Ω resistance, by Ohm's law the amount of current I equals $6/3$ or 2 A.

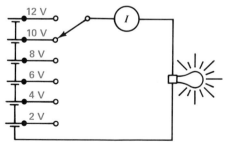

Fig. 2-1 Increasing the applied voltage produces more current I to light the bulb with more intensity.

High Voltage but Low Current It is important to realize that with high voltage, the current can have a low value when there is a very large amount of resistance in the circuit. For example, 1000 V applied across 1,000,000 Ω results in a current of only $1/1000$ A. By Ohm's law,

$$I = \frac{V}{R}$$
$$= \frac{1000 \text{ V}}{1,000,000 \text{ } \Omega}$$
$$= \frac{1}{1000}$$
$$I = 0.001 \text{ A}$$

The practical fact is that high-voltage circuits usually do have a small value of current in electronic equipment. Otherwise, tremendous amounts of power would be necessary.

Low Voltage but High Current At the opposite extreme, a low value of voltage in a very low resistance circuit can produce a very large amount of current. A 6-V battery connected across a resistance of 0.01 Ω produces 600 A of current:

$$I = \frac{V}{R}$$
$$= \frac{6 \text{ V}}{0.01 \text{ } \Omega}$$
$$I = 600 \text{ A}$$

Less I with More R Note the values of I in the following two examples also:

Example 1 A heater with a resistance of 8 Ω is connected across the 120-V power line. How much is the current I?

Answer $I = \dfrac{V}{R} = \dfrac{120 \text{ V}}{8 \text{ } \Omega}$

$$I = 15 \text{ A}$$

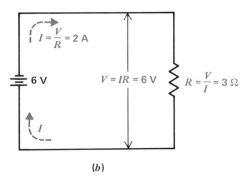

(a)   (b)

Fig. 2-2 Example of using Ohm's law. (a) Voltage source applied across resistance R. (b) Schematic diagram with values calculated by Ohm's law.

Example 2   A small light bulb with a resistance of 2400 Ω is connected across the same 120-V power line. How much is the current I?

Answer $I = \dfrac{V}{R} = \dfrac{120 \text{ V}}{2400 \text{ }\Omega}$

$I = 0.05$ A

Although both cases have the same 120 V applied, the current is much less in Example 2 because of the higher resistance.

Typical V and I   Transistors and integrated circuits generally operate with a dc supply of 3, 5, 6, 9, 12, 20, or 28 V. The current is usually in millionths or thousandths of one ampere up to about 5 A.

Practice Problems 2-1
Answers at End of Chapter

a. Calculate I for 24 V applied across 8 Ω.
b. Calculate I for 12 V applied across 8 Ω.
c. Calculate I for 24 V applied across 12 Ω.
d. Calculate I for 6 V applied across 1 Ω.

2-2   The Voltage $V = IR$

Referring to Fig. 2-2, the amount of voltage across R must be the same as V because the resistance is connected directly across the battery. The numerical value of this V is equal to the product $I \times R$.[1] For instance, the IR voltage in Fig. 2-2 is 2 A × 3 Ω, which equals the 6 V of the applied voltage. The formula is

$$V = IR \qquad\qquad (2\text{-}2)$$

With I in ampere units and R in ohms, their product V is in volts. Actually, this must be so because the I value equal to V/R is the amount that allows the IR product to be the same as the voltage across R.

Besides the numerical calculations possible with the IR formula, it is useful to consider that the IR product means voltage. Whenever there is current through a resistance, it must have a potential difference across its two ends equal to the IR product. If there were no potential difference, no electrons could flow to produce the current.

---

[1]For an explanation of how to invert factors from one side of an equation to the other side, see B. Grob, *Mathematics Outline and Review Problems for Basic Electronics*, McGraw-Hill Book Company, New York.

Practice Problems 2-2
Answers at End of Chapter

a. Calculate V for 0.002 A through 1000 Ω.
b. Calculate V for 0.004 A through 1000 Ω.
c. Calculate V for 0.002 A through 2000 Ω.

2-3 The Resistance $R = V/I$

As the third and final version of Ohm's law, the three factors V, I, and R are related by the formula

$$R = \frac{V}{I} \qquad (2\text{-}3)$$

In Fig. 2-2, R is 3 Ω because 6 V applied across the resistance produces 2 A through it. Whenever V and I are known, the resistance can be calculated as the voltage across R divided by the current through it.

Physically, a resistance can be considered as some material with elements having an atomic structure that allows free electrons to drift through it with more or less force applied. Electrically, though, a more practical way of considering resistance is simply as a V/I ratio. Anything that allows 1 A of current with 10 V applied has a resistance of 10 Ω. This V/I ratio of 10 Ω is its characteristic. If the voltage is doubled to 20 V, the current will also double to 2 A, providing the same V/I ratio of a 10-Ω resistance.

Furthermore, we do not need to know the physical construction of a resistance to analyze its effect in a circuit, so long as we know its V/I ratio. This idea is illustrated in Fig. 2-3. Here, a box with some unknown material in it is connected into a circuit where we can measure the 12 V applied across the box and the 3 A of current through it. The resistance is 12 V/3 A, or 4 Ω.

There may be liquid, gas, metal, powder, or any other material in the box, but electrically it is just a 4-Ω resistance because its V/I ratio is 4.

Practice Problems 2-3
Answers at End of Chapter

a. Calculate R for 12 V with 0.003 A.
b. Calculate R for 12 V with 0.006 A.
c. Calculate R for 12 V with 0.001 A.

2-4 Practical Units

The three forms of Ohm's law can be used to define the practical units of current, potential difference, and resistance as follows:

$$1 \text{ ampere} = \frac{1 \text{ volt}}{1 \text{ ohm}}$$

$$1 \text{ volt} = 1 \text{ ampere} \times 1 \text{ ohm}$$

$$1 \text{ ohm} = \frac{1 \text{ volt}}{1 \text{ ampere}}$$

One ampere is the amount of current through a one-ohm resistance that has one volt of potential difference applied across it.

One volt is the potential difference across a one-ohm resistance that has one ampere of current through it.

One ohm is the amount of opposition in a resistance that has a V/I ratio of 1, allowing one ampere of current with one volt applied.

In summary, the circle diagram in Fig. 2-4 for $V = IR$ can be helpful in using Ohm's law. Note that V is always at the top for $V = IR$, $V/R = I$, or $V/I = R$.

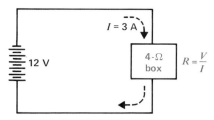

Fig. 2-3 The resistance R of any component is its V/I ratio.

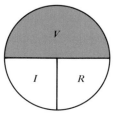

Fig. 2-4 A circle diagram that can be used to memorize $V = IR$, $V/R = I$, and $V/I = R$. The V is always at the top.

Practice Problems 2-4
Answers at End of Chapter

a. Calculate V for 0.007 A through 5000 Ω.
b. Calculate the amount of I for 12,000 V across 6,000,000 Ω.
c. Calculate R for 8 V with 0.004 A.

2-5 Multiple and Submultiple Units

The basic units—ampere, volt, and ohm—are practical values in most electric power circuits, but in many electronics applications these units are either too small or too big. As examples, resistances can be a few million ohms, the output of a high-voltage supply in a television receiver is about 20,000 V, and current through tubes and transistors is generally thousandths or millionths of an ampere.

In such cases, it is helpful to use multiples and submultiples of the basic units. As shown in Table 2-1, these units are based on the decimal system of tens, hundreds, thousands, etc. The common conversions for V, I, and R are given here, but a complete listing of all the prefixes is in App. F. Note that capital M is used for 10^6 to distinguish it from small m for 10^{-3}.

Example 3 The I of 8 mA flows through a 5-kΩ R. How much is the IR voltage?

Answer $V = IR = 8 \times 10^{-3} \times 5 \times 10^3$
$= 8 \times 5$
$V = 40$ V

In general, milliamperes multiplied by kilohms results in volts for the answer, as 10^{-3} and 10^3 cancel.

Example 4 How much current is produced by 60 V across 12 kΩ?

Answer $I = \dfrac{V}{R} = \dfrac{60}{12 \times 10^3} = 5 \times 10^{-3}$
$I = 5$ mA

Note that volts across kilohms produces milliamperes of current. Similarly, volts across megohms produces microamperes.

In summary, common combinations to calculate the current I are

$$\frac{V}{k\Omega} = mA \qquad \text{and} \qquad \frac{V}{M\Omega} = \mu A$$

Also, common combinations to calculate IR voltage are

$$mA \times k\Omega = V$$
$$\mu A \times M\Omega = V$$

These relations occur often in electronic circuits because the current is generally in units of milliamperes or microamperes.

Practice Problems 2-5
Answers at End of Chapter

a. Change the following to basic units with powers of 10: 6 mA, 5 kΩ, and 3 μA.
b. Change the following to units with metric prefixes: 6×10^{-3} A, 5×10^3 Ω, and 3×10^{-6} A.

Table 2-1. Conversion Factors

Prefix	Symbol	Relation to basic unit	Examples
mega	M	1,000,000 or 1×10^6	5 MΩ (megohms) = 5,000,000 ohms = 5×10^6 ohms
kilo	k	1000 or 1×10^3	18 kV (kilovolts) = 18,000 volts = 18×10^3 volts
milli	m	0.001 or 1×10^{-3}	48 mA (milliamperes) = 48×10^{-3} ampere = 0.048 ampere
micro	μ	0.000 001 or 1×10^{-6}	15 μV (microvolts) = 15×10^{-6} volt = 0.000 015 volt

2-6 The Linear Proportion between *V* and *I*

The Ohm's law formula $I = V/R$ states that V and I are directly proportional for any one value of R. This relation between V and I can be analyzed by using a fixed resistance of 2 Ω for R_L, as in Fig. 2-5. Then when V is varied, the meter shows I values directly proportional to V. For instance, with 12 V, I equals 6 A; for 10 V, the current is 5 A; an 8-V potential difference produces 4 A.

All the values of V and I are listed in the table in Fig. 2-5*b* and plotted in the graph in Fig. 2-5*c*. The I values are one-half the V values because R is 2 Ω. However, I is zero with zero volts applied.

Plotting the Graph The voltage values for V are
marked on the horizontal axis, called the *x axis* or *abscissa*. The current values I are on the vertical axis, called the *y axis* or *ordinate*.

Because the values for V and I depend on each other, they are variable factors. The independent variable here is V because we assign values of voltage and note the resulting current. Generally, the independent variable is plotted on the *x* axis, which is why the V values are shown here horizontally while the I values are on the ordinate.

The two scales need not be the same. The only requirement is that equal distances on each scale represent equal changes in magnitude. On the *x* axis here 2-V steps are chosen, while the *y* axis has 1-A scale divisions. The zero point at the origin is the reference.

The plotted points in the graph show the values in the table. For instance, the lowest point is 2 V horizontally from the origin, and 1 A up. Similarly, the next point is at the intersection of the 4-V mark and the 2-A mark.

A line joining these plotted points includes all values of I, for any value of V, with R constant at 2 Ω. This also applies to values not listed in the table. For instance, if we take the value of 7 V, up to the straight line and over to the I axis, the graph shows 3.5 A for I.

Voltampere Characteristic The graph in Fig.
2-5*c* is called the voltampere characteristic of R. It shows how much current the resistor allows for different voltages. Multiple and submultiple units of V and I can be used, though. For transistors and tubes the units of I are often milliamperes or microamperes.

Linear Resistance The straight-line graph in Fig.
2-5 shows that R is a linear resistor. A linear resistance has a constant value of ohms. Its R does not change with the applied voltage. Then V and I are directly pro-

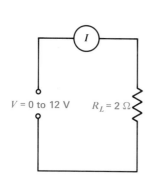

Volts	Ohms	Amperes
0	2	0
2	2	1
4	2	2
6	2	3
8	2	4
10	2	5
12	2	6

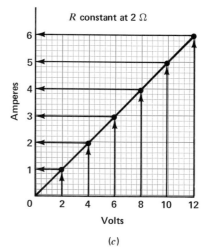

(*a*)   (*b*)   (*c*)

Fig. 2-5 Experiment to show that I increases in direct proportion to V. (*a*) Circuit with variable V but constant R. (*b*) Table of increasing I for higher V. (*c*) Graph of V and I values. This linear voltampere characteristic shows a direct proportion between V and I.

portional. Doubling the value of V from 4 to 8 V results in twice the current, from 2 to 4 A. Similarly, three or four times the value of V will produce three or four times I, for a proportional increase in current.

Nonlinear Resistance

This type has a nonlinear voltampere characteristic. As an example, the resistance of the tungsten filament in a light bulb is nonlinear. The reason is that R increases with more current as the filament becomes hotter. Increasing the applied voltage does produce more current, but I does not increase in the same proportion as the increase in V.

Inverse Relation between I and R

Whether R is linear or not, the current I is less for more R, with the applied voltage constant. This is an inverse relation, meaning that I goes down as R goes up. Remember that in the formula $I = V/R$, the resistance is in the denominator. A higher value of R actually lowers the value of the complete fraction.

As an example, let V be constant at 1 V. Then I is equal to the fraction $1/R$. As R increases, the values of I decrease. For R of 2 Ω, I is $\frac{1}{2}$ or 0.5 A. For a higher R of 10 Ω, I will be smaller at $\frac{1}{10}$ or 0.1 A.

Practice Problems 2-6
Answers at End of Chapter

Refer to the graph in Fig. 2-5c.
a. Are the values of I on the y or x axis?
b. Is this R linear or nonlinear?

2-7   Electric Power

The unit of electric power is the *watt* (W), named after James Watt (1736–1819). One watt of power equals the work done in one second by one volt of potential difference in moving one coulomb of charge.

Remember that one coulomb per second is an ampere. Therefore power in watts equals the product of amperes times volts.

Power in watts = volts × amperes
$$P = V \times I \qquad (2-4)$$

When a 6-V battery produces 2 A in a circuit, for example, the battery is generating 12 W of power.

The power formula can be used in three ways:

$$P = V \times I$$

$$I = P \div V \text{ or } \frac{P}{V}$$

$$V = P \div I \text{ or } \frac{P}{I}$$

Which formula to use depends on whether you want to calculate *P, I,* or *V*. Note the following examples:

Example 5   A toaster takes 10 A from the 120-V power line. How much power is used?

Answer $P = V \times I = 120 \text{ V} \times 10 \text{ A}$
$\qquad\qquad P = 1200 \text{ W}$

Example 6   How much current flows in the filament of a 300-W bulb connected to the 120-V power line?

Answer $P = V \times I$ or $I = P/V$. Then
$$I = \frac{300 \text{ W}}{120 \text{ V}}$$
$$I = 2.5 \text{ A}$$

Example 7   How much current flows in a 60-W bulb connected to the 120-V power line?

Answer $P = V \times I$ or $I = P/V$. Then
$$I = \frac{60 \text{ W}}{120 \text{ V}}$$
$$I = 0.5 \text{ A}$$

Note that the lower-wattage bulb uses less current.

Work and Power

Work and energy are essentially the same with identical units. Power is different, however, because it is the time rate of doing work.

As an example of work, if you move 100 lb a distance of 10 ft, the work is 100 lb × 10 ft or 1000 ft·lb, regardless of how fast or how slowly the work is done. Note that the unit of work is foot-pounds, without any reference to time.

However, power equals the work divided by the time it takes to do the work. If it takes 1 s, the power in this example is 1000 ft·lb/s; if the work takes 2 s, the power is 1000 ft·lb in 2 s, or 500 ft·lb/s.

Similarly, electric power is the time rate at which charge is forced to move by voltage. This is why the power in watts is the product of volts and amperes. The voltage states the amount of work per unit of charge; the current value includes the time rate at which the charge is moved.

Watts and Horsepower Units

A further example of how electric power corresponds to mechanical power is the fact that

746 W = 1 hp = 550 ft·lb/s

This relation can be remembered more easily as 1 hp equals approximately ¾ kilowatt (kW). One kilowatt = 1000 W.

Practical Units of Power and Work

Starting with the watt, we can develop several other important units. The fundamental principle to remember is that power is the time rate of doing work, while work is power used during a period of time. The formulas are

$$\text{Power} = \frac{\text{work}}{\text{time}} \qquad (2\text{-}5)$$

and

$$\text{Work} = \text{power} \times \text{time} \qquad (2\text{-}6)$$

With the watt unit for power, one watt used during one second equals the work of one joule. Or one watt is one joule per second. Therefore, 1 W = 1 J/s. The joule is a basic practical unit of work or energy.[1]

To summarize these practical definitions,

1 joule = 1 watt·second
1 watt = 1 joule/second

---

[1]See App. D, "Physics Units."

In terms of charge and current,

1 joule = 1 volt·coulomb
1 watt = 1 volt·ampere

Remember that the ampere unit has time in the denominator as one coulomb/second.

Electron Volt (eV)

This unit of work can be used for an individual electron, rather than the large quantity of electrons in a coulomb. An electron is charge, while the volt is potential difference. Therefore, 1 eV is the amount of work required to move an electron between two points that have a potential difference of one volt.

The number of electrons in one coulomb for the joule unit equals 6.25×10^{18}. Also, the work of one joule is a volt-coulomb. Therefore, the number of electron volts equal to one joule must be 6.25×10^{18}. As a formula,

$$1 \text{ J} = 6.25 \times 10^{18} \text{ eV}$$

Either the electron volt or joule unit of work is the product of charge times voltage, but the watt unit of power is the product of voltage times current. The division by time to convert work to power corresponds to the division by time that converts charge to current.

Kilowatthours

This is a unit commonly used for large amounts of electrical work or energy. The amount is calculated simply as the product of the power in kilowatts multiplied by the time in hours during which the power is used. As an example, if a light bulb uses 300 W or 0.3 kW for 4 hours (h), the amount of energy is 0.3 × 4, which equals 1.2 kWh.

We pay for electricity in kilowatthours of energy. The power-line voltage is constant at 120 V. However, more appliances and light bulbs require more current because they all add in the main line to increase the power.

Suppose the total load current in the main line equals 20 A. Then the power in watts from the 120-V line is

$P = 120 \text{ V} \times 20 \text{ A}$
$P = 2400 \text{ W or } 2.4 \text{ kW}$

If this power is used for 5 h, then the energy or work

supplied equals $2.4 \times 5 = 12$ kWh. The cost at 6 cents/kWh is $0.06 \times 12 = 0.72$ or 72 cents. This charge is for a 20-A load current from the 120-V line during the time of 5 h.

Practice Problems 2-7
Answers at End of Chapter

a. An electric heater takes 15 A from the 120-V power line. Calculate the power.
b. How much is the load current for a 100-W bulb connected to the 120-V power line?

2-8 Power Dissipation in Resistance

When current flows in a resistance, heat is produced because friction between the moving free electrons and the atoms obstructs the path of electron flow. The heat is evidence that power is used in producing current. This is how a fuse opens, as heat resulting from excessive current melts the metal link in the fuse.

The power is generated by the source of applied voltage and consumed in the resistance in the form of heat. As much power as the resistance dissipates in heat must be supplied by the voltage source; otherwise, it cannot maintain the potential difference required to produce the current.

The correspondence between electric power and heat is indicated by the fact that 1 W used during the time of 1 s is equivalent to 0.24 calorie of heat energy. The electric energy converted to heat is considered to be dissipated or used up because the calories of heat cannot be returned to the circuit as electric energy.

Since power is dissipated in the resistance of a circuit, it is convenient to express the power in terms of the resistance R. The $V \times I$ formula can be rearranged as follows:

Substituting IR for V,

$$P = V \times I = IR \times I$$
$$P = I^2 R \qquad (2\text{-}7)$$

This is a common form for the power formula because of the heat produced by current in a resistance.

For another form, substitute V/R for I. Then

$$P = V \times I = V \times \frac{V}{R}$$

$$P = \frac{V^2}{R} \qquad (2\text{-}8)$$

In all the formulas, V is the voltage across R in ohms, producing the current I in amperes, for power in watts.

Any one of the three formulas can be used to calculate the power dissipated in a resistance. The one to be used is just a matter of convenience, depending on which factors are known.

In Fig. 2-6, for example, the power dissipated with 2 A through the resistance and 12 V across it is $2 \times 12 = 24$ W.

Or, calculating in terms of just the current and resistance, the power is the product of 2 squared, or 4, times 6, which equals 24 W.

Using the voltage and resistance, the power can be calculated as 12 squared, or 144, divided by 6, which also equals 24 W.

No matter which formula is used, 24 W of power is dissipated, in the form of heat. This amount of power must be generated continuously by the battery in order to maintain the potential difference of 12 V that produces the 2-A current against the opposition of 6 Ω.

In some applications, the electric power dissipation is desirable because the component must produce heat in order to do its job. For instance, a 600-W toaster must dissipate this amount of power to produce the necessary amount of heat. Similarly, a 300-W light bulb must dissipate this power to make the filament white-hot so that it will have the incandescent glow that furnishes the light. In other applications, however, the heat may be just an undesirable byproduct of the need to provide current through the resistance in a circuit. In any case, though, whenever there is current in a resistance, it dissipates power equal to I^2R.

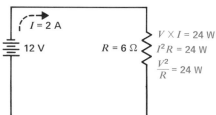

Fig. 2-6 Calculating the electric power in a circuit as $P = V \times I$, $P = I^2R$, or $P = V^2/R$.

Example 8 Calculate the power in a circuit where the source of 100 V produces 2 A in a 50-Ω R.

Answer $P = I^2R = 4 \times 50$

$\qquad P = 200$ W

This means the source generates 200 W of power while the resistance dissipates 200 W in the form of heat.

Example 9 Calculate the power in a circuit where the same source of 100 V produces 4 A in a 25-Ω R.

Answer $P = I^2R = 16 \times 25$

$\qquad P = 400$ W

Note the higher power in Example 9 because of more I, even though R is less than in Example 8.

Components that utilize the power dissipated in their resistance, such as light bulbs and toasters, are generally rated in terms of power. The power rating is at normal applied voltage, which is usually the 120 V of the power line. For instance, a 600-W 120-V toaster has this rating because it dissipates 600 W in the resistance of the heating element when connected across 120 V.

Note this interesting point about the power relations. The lower the source voltage, the higher the current required for the same power. The reason is that $P = V \times I$. For instance, an electric heater rated at 240 W from the 120-V power line takes 240 W/120 V = 2 A of current from the source. However, the same 240 W from a 12-V source, as in a car or boat, requires 240 W/12 V = 20 A. More current must be supplied by a source with lower voltage, to provide a specified amount of power.

Practice Problems 2-8
Answers at End of Chapter

a. Current I is 2 A in a 5-Ω R. Calculate P.
b. Voltage V is 10 V across a 5-Ω R. Calculate P.
c. Resistance R has 10 V with 2 A. Calculate the values for P and R.

2-9 Power Formulas

In order to calculate I or R for components rated in terms of power at a specified voltage, it may be convenient to use the power formulas in different forms. There are three basic power formulas, but each can be in three forms for nine combinations, as follows:

$$P = VI \qquad P = I^2R \qquad P = \frac{V^2}{R}$$

$$\text{or} \quad I = \frac{P}{V} \quad \text{or} \quad R = \frac{P}{I^2} \quad \text{or} \quad R = \frac{V^2}{P}$$

$$\text{or} \quad V = \frac{P}{I} \quad \text{or} \quad I = \sqrt{\frac{P}{R}} \quad \text{or} \quad V = \sqrt{PR}$$

Example 10 How much current is needed for a 600-W 120-V toaster?

Answer $I = \dfrac{P}{V} = \dfrac{600}{120}$

$\qquad I = 5$ A

Example 11 How much is the resistance of a 600-W 120-V toaster?

Answer $R = \dfrac{V^2}{P} = \dfrac{14,400}{600}$

$\qquad R = 24\ \Omega$

Example 12 How much current is needed for a 24-Ω R that dissipates 600 W?

Answer $I = \sqrt{\dfrac{P}{R}} = \sqrt{\dfrac{600}{24}} = \sqrt{25}$

$\qquad I = 5$ A

Note that all these formulas are based on Ohm's law $V = IR$ and the power formula $P = V \times I$. The following example with a 300-W bulb also illustrates this idea. The bulb is connected across the 120-V line. Its 300-W filament requires current of 2.5 A, equal to P/V. These calculations are

$$I = \frac{P}{V} = \frac{300 \text{ W}}{120 \text{ V}} = 2.5 \text{ A}$$

The proof is that the VI product then is 120×2.5, which equals 300 W.

Furthermore, the resistance of the filament, equal to V/I, is 48 Ω. These calculations are

$$R = \frac{V}{I} = \frac{120 \text{ V}}{2.5 \text{ A}} = 48 \text{ Ω}$$

If we use the power formula $R = V^2/P$, the answer is the same 48 Ω. These calculations are

$$R = \frac{V^2}{P} = \frac{(120^2)}{300}$$

$$R = \frac{14,400}{300} = 48 \text{ Ω}$$

In any case, when this bulb is connected across 120 V so that it can dissipate its rated power, the bulb draws 2.5 A from the power line and the resistance of the white-hot filament is 48 Ω.

Furthermore, it is important to note that the Ohm's-law calculations can be used for just about all types of circuits. As an example, Fig. 2-7 shows three resistors in a series-parallel circuit. However, if we consider just R_3, the I, V, and P values can be calculated as shown. Actually, these values are the same as for the one resistor in Fig. 2-6. As far as this resistor is concerned, it could be in either circuit and not know any difference.

Practice Problems 2-9
Answers at End of Chapter

a. How much is the R of a 100-W 120-V light bulb?
b. How much power is dissipated by a 2-Ω R with 10 V across it?

2-10 Electric Shock

While you are working on electric circuits, there is often the possibility of receiving an electric shock by touching the ''live'' conductors when the power is on. The shock is a sudden involuntary contraction of the

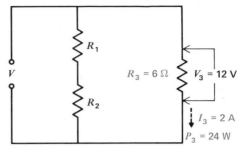

Fig. 2-7 Ohm's law can be applied to the entire circuit or any one component such as R_3. Complete solution for this circuit is in Fig. 5-3 in Chap. 5.

muscles, with a feeling of pain, caused by current through the body. If severe enough, the shock can be fatal. Safety first, therefore, should always be the rule.

The greatest shock hazard is from high-voltage circuits that can supply appreciable amounts of power. The resistance of the human body is also an important factor. If you hold a conducting wire in each hand, the resistance of the body across the conductors is about 10,000 to 50,000 Ω. Holding the conductors tighter lowers the resistance. If you hold only one conductor, your resistance is much higher. It follows that the higher the body resistance, the smaller the current that can flow through you.

A safety rule, therefore, is to work with only one hand if the power is on. Also, keep yourself insulated from earth ground when working on power-line circuits, since one side of the line is usually connected to earth. In addition, the metal chassis of radio and television receivers is often connected to the power-line ground. The final and best safety rule is to work on the circuits with the power disconnected if at all possible and make resistance tests.

Note that it is current through the body, not through the circuit, which causes the electric shock. This is why high-voltage circuits are most important, since sufficient potential difference can produce a dangerous amount of current through the relatively high resistance of the body. For instance, 500 V across a body resistance of 25,000 Ω produces 0.02 A, or 20 mA, which can be fatal. As little as 10 μA through the body can cause an electric shock. In an experiment[1] on electric shock to determine the current at which a person could

---

[1]C. F. Dalziel and W. R. Lee, "Lethal Electric Currents," *IEEE Spectrum*, February 1969.

release the live conductor, this value of "let-go" current was about 9 mA for men and 6 mA for women.

In addition to high voltage, the other important consideration in how dangerous the shock can be is the amount of power the source can supply. The current of 0.02 A through 25,000 Ω means the body resistance dissipates 10 W. If the source cannot supply 10 W, its output voltage drops with the excessive current load. Then the current is reduced to the amount corresponding to how much power the source can produce.

In summary, then, the greatest danger is from a source having an output of more than about 30 V with enough power to maintain the load current through the body when it is connected across the applied voltage. In general, components that can supply high power are physically big because of the need for dissipating heat.

Practice Problems 2-10
Answers at End of Chapter

Answer True or False.
a. One hundred twenty volts is more dangerous than 12 V for electric shock.
b. Resistance tests with an ohmmeter can be made with power off in the circuit.

Summary

1. The three forms of Ohm's law are $I = V/R$, $V = IR$, and $R = V/I$.
2. One ampere is the amount of current produced by one volt of potential difference across one ohm of resistance. This current of 1 A is the same as 1 C/s.
3. With constant R, the amount of I increases in direct proportion as V increases. This linear relation between V and I is shown by the graph in Fig. 2-5.
4. With constant V, the current I decreases as R increases. This is an inverse relation.
5. Power is the time rate of doing work or using energy. The unit is the watt. One watt equals $1 \text{ V} \times 1 \text{ A}$. Also, watts = joules per second.
6. The unit of work or energy is the joule. One joule equals $1 \text{ W} \times 1 \text{ s}$.
7. The most common multiples and submultiples of the practical units are listed in Table 2-1.
8. Voltage applied across your body can produce a dangerous electric shock. Whenever possible, shut off the power and make resistance tests. If the power must be on, use only one hand. Do not let the other hand rest on a conductor.
9. Table 2-2 summarizes the practical units used with Ohm's law.

Table 2-2. Practical Units of Electricity

Coulomb	Ampere	Volt	Watt	Ohm	Siemens
6.25×10^{18} electrons	$\dfrac{\text{Coulomb}}{\text{second}}$	$\dfrac{\text{Joule}}{\text{coulomb}}$	$\dfrac{\text{Joule}}{\text{second}}$	$\dfrac{\text{Volt}}{\text{ampere}}$	$\dfrac{\text{Ampere}}{\text{volt}}$

Self-Examination
Answers at Back of Book

Fill in the missing answers.

1. With 10 V across 5 Ω R, the current I is _____ A.
2. When 10 V produces 2.5 A, R is _____ Ω.

3. With 8 A through a 2-Ω R, the IR voltage is __16__ V.
4. The resistance of 500,000 Ω is __0.5__ MΩ.
5. With 10 V across 5000 Ω R, the current I is __2__ mA.
6. The power of 50 W = 2 A × __25__ V.
7. The energy of 50 J = 2 C × __25__ V.
8. The current drawn from the 120-V power line by a 1200-W toaster = __10__ A.
9. The current of 400 μA = __0.4__ mA.
10. With 12 V across a 2-Ω R, its power dissipation = __72__ W.
11. A circuit has a 4-A I. If V is doubled and R is the same, I = __8__ A.
12. A circuit has a 4-A I. If R is doubled and V is the same, I = __2__ A.
13. A television receiver using 240 W from the 120-V power line draws current I = __2__ A.
14. The rated current for a 100-W 120-V bulb = __0.83__ A.
15. The resistance of the bulb in question 14 is __144__ Ω.
16. The energy of 12.5 × 10^{18} eV = __2__ J.
17. The current of 1200 mA = __1.2__ A.
18. In an amplifier, the load resistor R_L of 5 kΩ has 15 V across it. Through R_L, then, the current = __3__ mA.
19. In a transistor circuit, a 1-kΩ resistor R_1 has 200 μA through it. Across R_1, then, its voltage = __0.2__ V.
20. In a transistor circuit, a 50-kΩ resistor R_2 has 6 V across it. Through R_2, then, its current = _____ mA.

Essay Questions

1. State the three forms of Ohm's law relating V, I, and R.
2. (a) Why does higher applied voltage with the same resistance result in more current? (b) Why does more resistance with the same applied voltage result in less current?
3. Calculate the resistance of a 300-W bulb connected across the 120-V power line, using two different methods to arrive at the same answer.
4. State which unit in each of the following pairs is larger: (a) volt or kilovolt; (b) ampere or milliampere; (c) ohm or megohm; (d) volt or microvolt; (e) siemens or microsiemens; (f) electron volt or joule; (g) watt or kilowatt; (h) kilowatthour or joule; (i) volt or millivolt; (j) megohm or kilohm.
5. State two safety precautions to follow when working on electric circuits.
6. Referring back to the resistor shown in Fig. 1-12 in Chap. 1 on page 27, suppose that it is not marked. How could you determine its resistance by Ohm's law? Show your calculations that result in the V/I ratio of 600 Ω. However, do not exceed the power rating of 10 W.
7. Give three formulas for electric power.
8. What is the difference between work and power? Give two units for each.
9. Prove that 1 kWh is equal to 3.6 × 10^6 J.
10. Give the metric prefixes for 10^{-6}, 10^{-3}, 10^3, and 10^6.
11. Which two units in Table 2-2 are reciprocals of each other?
12. A circuit has a constant R of 5000 Ω, while V is varied from 0 to 50 V in 10-V steps. Make a table listing the values of I for each value of V. Then draw a

graph plotting these values of milliamperes vs. volts. (This graph should be similar to Fig. 2-5c.)

13. Give the voltage and power rating for at least two types of electrical equipment.

14. Which uses more current from the 120-V power line, a 600-W toaster or a 200-W television receiver?

15. Refer to the two resistors in series with each other in Fig. 3-1 on page 54. How much would you guess is the current through R_2?

16. Consider the two resistors R_1 and R_2 shown before in Fig. 2-7. Would you say they are in one series path or separate parallel paths?

Problems

Answers to Odd-Numbered Problems at Back of Book

1. A 60-V source is connected across a 30-kΩ resistance. (a) Draw the schematic diagram. (b) How much current flows through the resistance? (c) How much current flows through the voltage source? (d) If the resistance is doubled, how much is the current in the circuit?

2. A 12-V battery is connected across a 2-Ω resistance. (a) Draw the schematic diagram. (b) Calculate the power dissipated in the resistance. (c) How much power is supplied by the battery? (d) If the resistance is doubled, how much is the power?

3. A picture-tube heater has 800 mA of current with 6.3 V applied. (a) Draw the schematic diagram, showing the heater as a resistance. (b) How much is the resistance of the heater?

4. Convert the following units, using powers of 10 where necessary: (a) 12 mA to amperes; (b) 5000 V to kilovolts; (c) ½ MΩ to ohms; (d) 100,000 Ω to megohms; (e) ½ A to milliamperes; (f) 9000 μS to siemens; (g) 1000 μA to milliamperes; (h) 5 kΩ to ohms; (i) 8 nanoseconds (ns) to seconds.

5. A current of 2 A flows through a 6-Ω resistance connected across a battery. (a) How much is the applied voltage of the battery? (b) Calculate the power dissipated in the resistance. (c) How much power is supplied by the battery?

6. (a) How much resistance allows 30 A current with 6 V applied? (b) How much resistance allows 1 mA current with 10 kV applied? Why is it possible to have less current in (b) with the higher applied voltage?

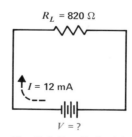

$R_L = 820\ \Omega$

$I = 12$ mA

$V = ?$

Fig. 2-8 For Prob. 11.

7. A source of applied voltage produces 1 mA through a 10-MΩ resistance. How much is the applied voltage?

8. Calculate the current *I*, in ampere units, for the following examples: (**a**) 45 V applied across 68 kΩ; (**b**) 250 V across 12 MΩ; (**c**) 1200 W dissipated in 600 Ω.

9. Calculate the *IR* voltage for the following examples: (**a**) 68 μA through 22 MΩ; (**b**) 2.3 mA through 47 kΩ; (**c**) 237 A through 0.012 Ω.

10. Calculate the resistance *R*, in ohms, for the following examples: (**a**) 134 mA produced by 220 V; (**b**) 800 W dissipated with 120 V applied; (**c**) a conductance of 9000 μS.

11. Find the value of *V* in Fig. 2-8.

Answers to Practice Problems

2-1 a. 3 A
  b. 1.5 A
  c. 2 A
  d. 6 A

2-2 a. 2 V
  b. 4 V
  c. 4 V

2-3 a. 4000 Ω
  b. 2000 Ω
  c. 12,000 Ω

2-4 a. 35 V
  b. 0.002 A
  c. 2000 Ω

2-5 a. See Prob. **b**
  b. See Prob. **a**

2-6 a. *y* axis
  b. linear

2-7 a. 1.8 kW
  b. 0.83 A

2-8 a. 20 W
  b. 20 W
  c. 20 W and 5 Ω

2-9 a. 144 Ω
  b. 50 W

2-10 a. T
  b. T

Chapter 3
Series Circuits

When the components in a circuit are connected in successive order with an end of each joined to an end of the next, they form a series circuit. The resistors R_1 and R_2 are in series with each other and the battery. The result is only one path for electron flow. Therefore, the current I is the same in all the series components.

Important terms in this chapter are:

aiding voltages   potential difference
applied voltage   proportional parts
chassis ground   series components
IR drop   series string
negative potential   total power
open circuit   total resistance
opposing voltages   voltage drop
positive potential   voltage polarity

More details are described in the following sections:

3-1 Why *I* Is the Same in All Parts of a Series Circuit

An electric current is a movement of charges between two points, produced by the applied voltage. When components are connected in successive order as in Fig. 3-1, they form a series circuit. The resistors R_1 and R_2 are in series with each other and the battery. In Fig. 3-2, the battery supplies the potential difference that forces electrons to drift from the negative terminal at A, toward B, through the connecting wires and resistances R_1, R_2, and R_3, back to the positive battery terminal at J.

At the negative battery terminal, its negative charge repels electrons. Therefore, free electrons in the atoms of the wire at this terminal are repelled from A toward B. Similarly, free electrons at point B can then repel adjacent electrons, producing an electron drift toward C in the direction away from the negative battery terminal.

At the same time, the positive charge of the positive battery terminal attracts free electrons, causing electrons to drift toward I and J. As a result, the free electrons in the resistances R_1, R_2, and R_3 are forced to drift toward the positive terminal.

The positive terminal of the battery attracts electrons just as much as the negative side of the battery repels electrons. Therefore, the motion of free electrons in the circuit starts at the same time at the same speed in all parts of the circuit.

The electrons returning to the positive battery termi-nal are not the same electrons as those leaving the negative terminal. Free electrons in the wire are forced to move to the positive terminal because of the potential difference of the battery.

The free electrons moving away from one point are continuously replaced by free electrons flowing from an adjacent point in the series circuit. All electrons have the same speed as those leaving the battery. In all parts of the circuit, therefore, the electron drift is the same, with an equal number of electrons moving at one time with the same speed. That is why the current is the same in all parts of the series circuit.

In Fig. 3-2b, when the current is 2 A, for example, this is the value of the current through R_1, R_2, R_3, and the battery. Not only is the amount of current the same throughout, but in all parts of a series circuit the current cannot differ in any way because there is just one current path for the entire circuit.

The order in which components are connected in series does not affect the current. In Fig. 3-3b, resistances R_1 and R_2 are connected in reverse order compared with Fig. 3-3a, but in both cases they are in series. The current through each is the same because there is only one path for the electron flow. Similarly, R_3, R_4, and R_5 are in series and have the same current for the connections shown in Fig. 3-3c, d, and e. Furthermore, the resistances need not be equal.

The question of whether a component is first, second, or last in a series circuit has no meaning in terms of current. The reason is that I is the same amount at the same time in all the series components.

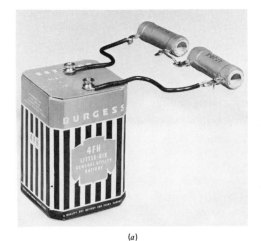

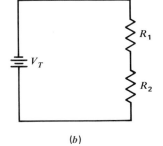

(a) (b)

Fig. 3-1 A series circuit. (a) Photo of wiring. (b) Schematic diagram.

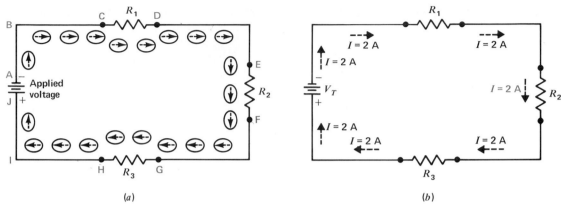

Fig. 3-2 (*a*) Electron drift is the same in all parts of a series circuit. (*b*) The current *I* is the same at all points in a series circuit.

In fact, series components can be defined as those in the same current path. The path is from one side of the voltage source, through the series components, and back to the other side of the applied voltage. However, the series path must not have any point where the current can branch off to another path in parallel. This feature of series circuits applies not only to direct current, but also to alternating current of any frequency and for any waveshape.

Practice Problems 3-1
Answers at End of Chapter

a. In Fig. 3-2, name five parts that have the *I* of 2 A.
b. In Fig. 3-3*e*, when *I* in R_5 is 5 A, then *I* in R_3 is _____ A.

3-2   Total *R* Equals the Sum of All Series Resistances

When a series circuit is connected across a voltage source, as shown in Fig. 3-3, the free electrons forming the current must drift through all the series resistances. This path is the only way the electrons can return to the battery. With two or more resistances in the same current path, therefore, the total resistance across the voltage source is the opposition of all the resistances.

Specifically, the total resistance R_T of a series string is equal to the sum of the individual resistances. This rule is illustrated in Fig. 3-4. In Fig. 3-4*b*, 2 Ω is added in series with the 3 Ω of Fig. 3-4*a*, producing the total resistance of 5 Ω. The total opposition of R_1 and R_2 limiting the amount of current is the same as though a 5-Ω resistance were used, as shown in the equivalent circuit in Fig. 3-4*c*.

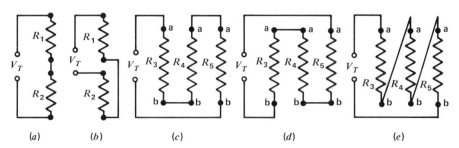

Fig. 3-3 Examples of series connections. Resistances R_1 and R_2 are in series in both (*a*) and (*b*). Also, R_3, R_4, and R_5 are in series in (*c*), (*d*), and (*e*).

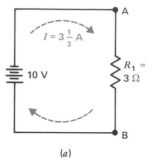
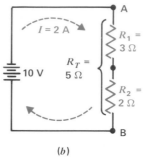
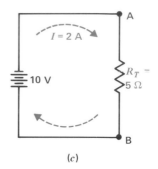

(a)   (b)   (c)

Fig. 3-4 Series resistances are added for the total R_T. (a) Resistance R_1 alone is 3 Ω. (b) The R_1 and R_2 in series total 5 Ω. (c) The R_T of 5 Ω is the same as one resistance of 5 Ω between points A and B.

Series String A combination of series resistances is often called a *string*. The string resistance equals the sum of the individual resistances. For instance, R_1 and R_2 in Fig. 3-4 form a series string having the R_T of 5 Ω. A string can have two or more resistors.

By Ohm's law, the amount of current between two points in a circuit equals the potential difference divided by the resistance between these points. As the entire string is connected across the voltage source, the current equals the voltage applied across the entire string divided by the total series resistance of the string. Between points A and B in Fig. 3-4, for example, 10 V is applied across 5 Ω in Fig. 3-4b and c to produce 2 A. This current flows through R_1 and R_2 in one series path.

Series Resistance Formula In summary, the total resistance of a series string equals the sum of the individual resistances. The formula is

$$R_T = R_1 + R_2 + R_3 + \cdots + \text{etc.} \qquad \text{(3-1)}$$

where R_T is the total resistance and R_1, R_2, and R_3 are individual series resistances. This formula applies to any number of resistances, whether equal or not, as long as they are in the same series string.

Note that R_T is the resistance to use in calculating the current in a series string. Then Ohm's law is

$$I = \frac{V_T}{R_T} \qquad \text{(3-2)}$$

where R_T is the sum of all the resistances, V_T is the voltage applied across the total resistance, and I is the current in all parts of the string.

Example 1 Two resistances R_1 and R_2 of 5 Ω each and R_3 of 10 Ω are in series. How much is R_T?

Answer $R_T = R_1 + R_2 + R_3 = 5 + 5 + 10$
$$R_T = 20 \ \Omega$$

Example 2 With 80 V applied across the series string of Example 1, how much is the current in R_3?

Answer $I = \dfrac{V_T}{R_T} = \dfrac{80 \text{ V}}{20 \ \Omega}$
$$I = 4 \text{ A}$$

This 4-A current is the same in R_3, R_2, R_1, or any part of the series circuit.

Note that adding series resistance reduces the current. In Fig. 3-4a the 3-Ω R_1 allows 10 V to produce 3⅓ A. However, I is reduced to 2 A when the 2-Ω R_2 is added for a total series resistance of 5 Ω opposing the 10-V source.

Practice Problems 3-2
Answers at End of Chapter

a. V is 10 V and R_1 is 5 kΩ, calculate I.
b. A 2-kΩ R_2 and 3-kΩ R_3 are added in series with R_1. Calculate R_T.
c. Calculate I in R_1, R_2, and R_3.

3-3   Series *IR* Voltage Drops

With current I through a resistance, by Ohm's law the voltage across R is equal to $I \times R$. This rule is illustrated in Fig. 3-5 for a string of two resistors. In this circuit, I is 1 A because the applied V_T of 10 V is across the total R_T of 10 Ω, equal to the 4-Ω R_1 plus the 6-Ω R_2. Then I is 10V/10 Ω = 1 A.

For each IR voltage in Fig. 3-5, multiply each R by the 1 A of current in the series circuit. Then

$$V_1 = IR_1 = 1 \text{ A} \times 4 \text{ } \Omega = 4 \text{ V}$$
$$V_2 = IR_2 = 1 \text{ A} \times 6 \text{ } \Omega = 6 \text{ V}$$

The V_1 of 4 V is across the 4 Ω of R_1. Also, the V_2 of 6 V is across the 6 Ω of R_2.

The IR voltage across each resistance is called an *IR drop*, or a *voltage drop*, because it reduces the potential difference available for the remaining resistance in the series circuit. Note that the symbols V_1 and V_2 are used for the voltage drops across each resistor to distinguish them from the source V_T applied across both resistors.

In Fig. 3-5, the V_T of 10 V is applied across the total series resistance of R_1 and R_2. However, because of the

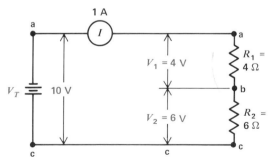

Fig. 3-5 An example of IR voltage drops V_1 and V_2 in a series circuit.

IR voltage drop of 4 V across R_1, the potential difference across R_2 is only 6 V. The negative potential drops from 10 V at point a, with respect to the common reference point at c, down to 6 V at point b. The potential difference of 6 V between b and the reference at c is the voltage across R_2.

Similarly, there is an IR voltage drop of 6 V across R_2. The negative potential drops from 6 V at point b with respect to point c, down to 0 V at point c with respect to itself. The potential difference between any two points on the return line to the battery must be zero because the wire has practically zero resistance and therefore no IR drop.

It should be noted that voltage must be applied by a source of potential difference such as the battery in order to produce current and have an IR voltage drop across the resistance. With no current through a resistor, it has resistance only, but there is no potential difference across the two ends.

The IR drop of 4 V across R_1 in Fig. 3-5 represents that part of the applied voltage used to produce the current of 1 A through the 4-Ω resistance. Also, across R_2 the IR drop is 6 V because this much voltage allows 1 A in the 6-Ω resistance. The IR drop is more in R_2 because more potential difference is necessary to produce the same amount of current in the higher resistance. For series circuits, in general, the highest R has the largest IR voltage drop across it.

Practice Problems 3-3
Answers at End of Chapter

Refer to Fig. 3-5.
a. How much is the sum of V_1 and V_2?
b. Calculate I as V_T/R_T.
c. How much is I through R_1?
d. How much is I through R_2?

3-4   The Sum of Series *IR* Drops Equals the Applied V_T

The whole applied voltage is equal to the sum of its parts. For example, in Fig. 3-5, the individual voltage drops of 4 V and 6 V total the same 10 V produced by the battery. This relation for series circuits can be stated

$$V_T = V_1 + V_2 + V_3 + \cdots + \text{etc.} \qquad (3\text{-}3)$$

where V_T is the applied voltage equal to the total of the individual IR drops.

Example 3   A voltage source produces an IR drop of 40 V across a 20-Ω R_1, 60 V across a 30-Ω R_2, and 180 V across a 90-Ω R_3, all in series. How much is the applied voltage?

Answer $V_T = 40 + 60 + 180$

$V_T = 280$ V

Note that the IR drop across each R results from the same current of 2 A, produced by 280 V across the total R_T of 140 Ω.

Example 4   An applied V_T of 120 V produces IR drops across two series resistors R_1 and R_2. If the drop across R_1 is 40 V, how much is the voltage across R_2?

Answer   Since V_1 and V_2 must total 120 V, and one is 40 V, the other must be the difference between 120 and 40 V. Or $V_2 = V_T - V_1$, which equals $120 - 40$. Then $V_2 = 80$ V.

It really is logical that V_T is the sum of the series IR drops. The current I is the same in all the series components. For this reason, the total of all the series voltages V_T is needed to produce the same I in the total of all the series resistances R_T as the I that each resistor voltage produces in its R.

A practical application of voltages in a series circuit is illustrated in Fig. 3-6. In this circuit, two 120-V light bulbs are operated from a 240-V line. If one bulb were connected to 240 V, the filament would burn out. With the two bulbs in series, however, each has 120 V for proper operation. The two 120-V drops across the bulbs in series add to equal the applied voltage of 240 V.

Practice Problems 3-4
Answers at End of Chapter

a. A series circuit has IR drops of 10 V, 20 V, and 30 V. How much is the applied voltage V_T of the source?

b. One hundred volts is applied to R_1 and R_2 in series. If V_1 is 25 V, how much is V_2?

3-5   Polarity of IR Voltage Drops

When an IR voltage drop exists across a resistance, one end must be either more positive or more negative than the other end. Otherwise, without a potential difference no current could flow through the resistance to produce the IR drop. The polarity of this IR voltage can be associated with the direction of I through R. In brief, electrons flow into the negative side of the IR voltage and out the positive side (Fig. 3-7a).

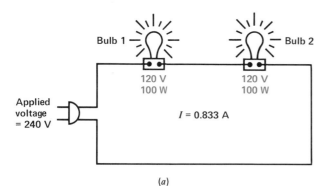

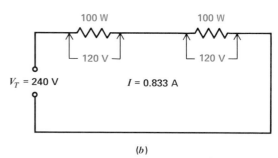

Fig. 3-6 Series string of two 120-V light bulbs operating from 240-V line. (a) Wiring diagram. (b) Schematic diagram.

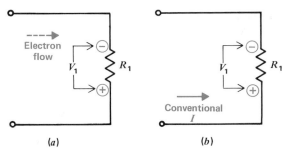

Fig. 3-7 Polarity of *IR* voltage drops. (*a*) Electron flow is into negative side of V_1. (*b*) Same polarity of V_1 with positive charges into positive side.

If we want to consider conventional current, with positive charges moving in the opposite direction from electron flow, the rule is reversed for the positive charges. See Fig. 3-7*b*. Here the positive charges for *I* are moving into the positive side of the *IR* voltage.

However, for either electron flow or conventional current the actual polarity of the *IR* drop is the same. In both *a* and *b* of Fig. 3-7, the top end of *R* in the diagrams is negative since this is the negative terminal of the source producing the current. After all, the resistor does not know which direction of current we are thinking of.

A series circuit with two *IR* voltage drops is shown in Fig. 3-8. We can analyze these polarities in terms of electron flow. The electrons move from the negative terminal of the source V_T through R_1 from point c to d. Electrons move into c and out from d. Therefore c is the negative side of the voltage drop across R_1. Similarly, for the *IR* voltage drop across R_2, point e is the negative side, compared with point f.

A more fundamental way to consider the polarity of *IR* voltage drops in a circuit is the fact that between any two points the one nearer to the positive terminal of the voltage source is more positive; also, the point nearer

to the negative terminal of the applied voltage is more negative. A point nearer to the terminal means there is less resistance in its path.

In Fig. 3-8 point c is nearer to the negative battery terminal than point d. The reason is that c has no resistance to a, while the path from d to a includes the resistance of R_1. Similarly, point f is nearer to the positive battery terminal than point e, which makes f more positive than e.

Notice that points d and e in Fig. 3-8 are marked with both plus and minus polarities. The plus polarity at d indicates that it is more positive than c. This polarity, however, is shown just for the voltage across R_1. Point d cannot be more positive than points f and b. The positive terminal of the applied voltage must be the most positive point because the battery is generating the positive potential for the entire circuit.

Similarly, points a and c must have the most negative potential in the entire string, since point a is the negative terminal of the applied voltage. Actually, the plus polarity marked at d only means that this end of R_1 is less negative than c by the amount of voltage drop across R_1.

Consider the potential difference between e and d in Fig. 3-8, which is only a piece of wire. This voltage is zero because there is no resistance between these two points. Without any resistance here, the current cannot produce the *IR* drop necessary for a difference in potential. Points e and d are, therefore, the same electrically since they have the same potential.

When we go around the external circuit from the negative terminal of V_T, with electron flow, the voltage drops are drops in negative potential. For the opposite direction, starting from the positive terminal of V_T, the voltage drops are drops in positive potential. Either way, the voltage drop of each series *R* is its proportional part of the V_T needed for the one value of current in all the resistances.

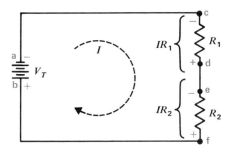

Fig. 3-8 Example of two *IR* voltage drops in series. Electron flow is shown for direction of *I*.

Practice Problems 3-5
Answers at End of Chapter

Refer to Fig. 3-8.
a. Which point in the circuit is the most negative?
b. Which point in the circuit is the most positive?
c. Which is more negative, point d or f?

3-6 Polarities to Chassis Ground

In practical circuits, one side of the voltage source V_T is usually connected to chassis ground. The purpose is to simplify the wiring. On a plastic board with printed wiring, a rim of solder around the edge serves as the chassis ground return, as illustrated in Fig. 3-9a. Then only one terminal of the source voltage V_T is used for the *high side* of the wiring. The circuit components have return connections to the opposite side of V_T through the chassis ground conductor.

Either the negative or the positive terminal of V_T can be connected to the chassis ground return line. With the negative side grounded, V_T supplies positive voltage for the high side of the circuit (Fig. 3-9b). For the opposite case in Fig. 3-9c, the high side of the circuit has negative voltage with respect to chassis ground.

In Fig. 3-9, the two equal resistances divide the applied voltage equally. Then R_1 and R_2 each have a voltage drop of 10 V, equal to one-half the 20 V of V_T. The sum of the *IR* drops is $10 + 10 = 20$ V, equal to the total applied voltage.

Positive Voltages to Ground In Fig. 3-9b, point S is at $+20$ V. However, point J at the junction of R_1 and R_2 is at $+10$ V. The potential of $+10$ V at point J is 10 V less than at S because of the 10-V drop across R_1. All these voltages are positive to chassis ground because the negative side of V_T and the ground return are really the same.

Negative Voltages to Ground In Fig. 3-9c everything is the same as in Fig. 3-9b but with negative instead of positive voltages. Point S is at -20 V, since the positive side of the source voltage is grounded.

Practice Problems 3-6
Answers at End of Chapter

a. In Fig. 3-9b, give each voltage to ground at points S, J, and G.

b. In Fig. 3-9c, give each voltage to ground at points S, J, and G.

3-7 Total Power in a Series Circuit

The power needed to produce current in each series resistor is used up in the form of heat. Therefore, the total power used is the sum of the individual values of power dissipated in each part of the circuit. As a formula,

$$P_T = P_1 + P_2 + P_3 + \cdots + \text{etc.} \tag{3-4}$$

As an example, in Fig. 3-10, R_1 dissipates 40 W for P_1, equal to 20 V $\times$ 2 A for the *VI* product. Or, the P_1 calculated as I^2R is $4 \times 10 = 40$ W. Also, the P_1 is V^2/R, or $400/10 = 40$ W.

Similarly, P_2 for R_2 is 80 W. This value is 40×2 for *VI*, 4×20 for I^2R, or $^{1600}/_{20}$ for V^2/R.

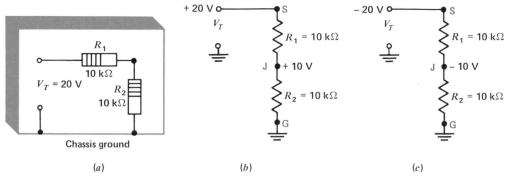

Fig. 3-9 Polarity of *IR* voltage drops to chassis ground. (*a*) Wiring diagram with ground wire around printed-circuit board. (*b*) Schematic diagram with $+20$ V for V_T with respect to chassis ground. (*c*) Here V_T is -20 V, as the positive side is grounded.

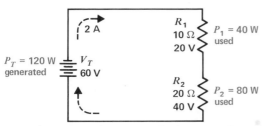

Fig. 3-10 The sum of the individual powers P_1 and P_2 used in each resistance equals the total power P_T produced by the source.

The total power dissipated by R_1 and R_2, then, is $40 + 80 = 120$ W. This power is generated by the source of applied voltage.

The total power can also be calculated as $V_T \times I$. The reason is that V_T is the sum of all the series voltages and I is the same in all the series components. In this case, then, P_T is $60 \times 2 = 120$ W, calculated as $V_T \times I$.

The total power here is 120 W, calculated either from the total voltage or from the sum of P_1 and P_2. This is the amount of power produced by the battery. The voltage source produces this power, equal to the amount used by the load.

Practice Problems 3-7
Answers at End of Chapter

a. Each of three equal resistances dissipates 2 W. How much is P_T supplied by the source?
b. A 1-kΩ R_1 and 40-kΩ R_2 are in series with a 50-V source. Which R dissipates more power?

3-8   Series-Aiding and Series-Opposing Voltages

Series-aiding voltages are connected with polarities that allow current in the same direction. In Fig. 3-11a, the 6 V of V_1 alone could produce 3 A electron flow from the negative terminal, with the 2-Ω R. Also, the 8 V of V_2 could produce 4 A in the same direction. The total I then is 7 A.

Instead of adding the currents, however, the voltages V_1 and V_2 can be added, for a V_T of $6 + 8 = 14$ V. This 14 V produces 7 A in all parts of the series circuit with a resistance of 2 Ω. Then I is $^{14}/_2 = 7$ A.

Voltages are connected series-aiding when the plus terminal of one is connected to the negative terminal of the next. They can be added for a total equivalent voltage. This idea applies in the same way to voltage sources, such as batteries, and to voltage drops across resistances. Any number of voltages can be added, as long as they are connected with series-aiding polarities.

Series-opposing voltages are subtracted, as shown in Fig. 3-11b. Notice here that the positive terminals of V_1 and V_2 are connected. Subtract the smaller from the larger value, and give the net V the polarity of the larger voltage. In this example, V_T is $8 - 6 = 2$ V. The polarity of V_T is the same as V_2 because it is larger than V_1.

If two series-opposing voltages are equal, the net voltage will be zero. In effect, one voltage balances out the other. The current I also is zero, without any net potential difference.

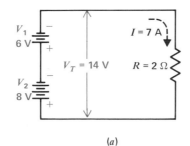

(a)

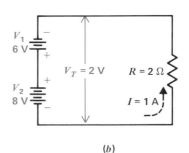

(b)

Fig. 3-11 Example of voltages V_1 and V_2 in series. (a) Note the connections for series-aiding polarities. Here $8 \text{ V} + 6 \text{ V} = 14 \text{ V}$ for V_T. (b) Connections for series-opposing polarities. Now $8 \text{ V} - 6 \text{ V} = 2 \text{ V}$ for V_T.

Practice Problems 3-8
Answers at End of Chapter

a. Voltage V_1 of 40 V is series-aiding with V_2 of 60 V. How much is V_T?

b. The same V_1 and V_2 are connected series-opposing. How much is V_T?

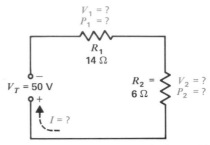

Fig. 3-12 Analyzing a series circuit to find I, V_1, V_2, P_1, and P_2. See text for solution.

3-9 Analyzing Series Circuits

Refer to Fig. 3-12. Suppose that the source V_T of 50 V is known, with the 14-Ω R_1 and 6-Ω R_2. The problem is to find R_T, I, the individual voltage drops V_1 and V_2 across each resistor, and the power dissipated.

We must know the total resistance R_T to calculate I because the total applied voltage V_T is given. This V_T is applied across the total resistance R_T. In this example, R_T is 14 + 6 = 20 Ω.

Now I can be calculated as V/R_T, or $^{50}/_{20}$, which equals 2.5 A. This 2.5-A I flows through R_1 and R_2.

The individual voltage drops are

$$V_1 = IR_1 = 2.5 \times 14 = 35 \text{ V}$$
$$V_2 = IR_2 = 2.5 \times 6 = 15 \text{ V}$$

Note that V_1 and V_2 total 50 V, equal to the applied V_T.

To find the power dissipated in each resistor,

$$P_1 = V_1 \times I = 35 \times 2.5 = 87.5 \text{ W}$$
$$P_2 = V_2 \times I = 15 \times 2.5 = 37.5 \text{ W}$$

These two values of dissipated power total 125 W. The power generated by the source equals $V_T \times I$ or 50 × 2.5, which is also 125 W.

General Methods for Series Circuits For other types of problems with series circuits it is useful to remember the following:

1. When you know the I for one component, use this for I in all the components, as the current is the same in all parts of a series circuit.
2. To calculate I, the total V_T can be divided by the total R_T, or an individual IR drop can be divided by its R. For instance, the current in Fig. 3-12 could be

calculated as V_2/R_2 or $^{15}/_6$, which equals the same 2.5 A for I. However, do not mix a total value for the entire circuit with an individual value for only part of the circuit.

3. When you know the individual voltage drops around the circuit, these can be added to equal the applied V_T. This also means that a known voltage drop can be subtracted from the total V_T to find the remaining voltage drop.

These principles are illustrated by the problem in Fig. 3-13. In this circuit R_1 and R_2 are known but not R_3. However, the current through R_3 is given as 3 mA.

With just this information, all values in this circuit can be calculated. The I of 3 mA is the same in all three series resistances. Therefore,

$$V_1 = 3 \text{ mA} \times 10 \text{ k}\Omega = 30 \text{ V}$$
$$V_2 = 3 \text{ mA} \times 30 \text{ k}\Omega = 90 \text{ V}$$

The sum of V_1 and V_2 is 30 + 90 = 120 V. This 120 V plus V_3 must total 180 V. Therefore, V_3 is 180 − 120 = 60 V.

With 60 V for V_3, equal to IR_3, then R_3 must be $^{60}/_{0.003}$, equal to 20,000 Ω or 20 kΩ. The total circuit resistance is 60 kΩ, which results in the current of 3 mA with 180 V applied, as specified in the circuit.

Another way of doing this problem is to find R_T first. The equation $I = V_T/R_T$ can be inverted to calculate the R_T as V_T/I. With a 3-mA I and 180 V for V_T, the value of R_T must be 180 V/3 mA = 60 kΩ. Then R_3 is 60 kΩ − 40 kΩ = 20 kΩ.

The power dissipated in each resistance is 90 mW in R_1, 270 mW in R_2, and 180 mW in R_3. The total power is 540 mW.

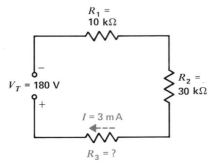

Fig. 3-13 Find the resistance of R_3. See text for analysis of this series circuit.

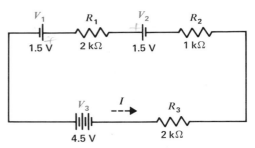

Fig. 3-15 Find I for this series circuit with three voltage sources. See text for solution.

Series Voltage-Dropping Resistors A common application of series circuits is to use a resistance to drop the voltage from the source V_T to a lower value, as in Fig. 3-14. The load R_L here represents a transistor radio that operates normally with a 9-V battery. When the radio is on, the dc load current with 9 V applied is 18 mA. Therefore, the requirements are 9 V, at 18 mA as the load.

To operate this radio from 12.6 V, the voltage-dropping resistor R_S is inserted in series to provide a voltage drop V_S that will make V_L equal to 9 V. The required voltage drop across V_S is the difference between V_L and the higher V_T. As a formula,

$$V_S = V_T - V_L$$
$$V_S = 12.6 - 9 = 3.6 \text{ V}$$

Furthermore, this voltage drop of 3.6 V must be provided with a current of 18 mA, as the current is the same through R_S and R_L. To calculate R_S, then, it is 3.6 V/18 mA, which equals 0.2 kΩ or 200 Ω.

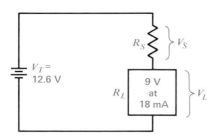

Fig. 3-14 An example of a series voltage-dropping resistor R_S to drop V_T of 12.6 V to 9 V for R_L. See text for calculations.

Circuit with Voltage Sources in Series See Fig. 3-15. Note that V_1 and V_2 are series-opposing, with + to + through R_1. Their net effect, then, is 0 V. Therefore, V_T consists only of V_3, equal to 4.5 V. The total R is $2 + 1 + 2 = 5$ kΩ for R_T. Finally, I is V_T/R_T or 4.5 V/5 kΩ, which is equal to 0.9 mA.

Practice Problems 3-9
Answers at End of Chapter

Refer to Fig. 3-13.
a. Calculate V_1 across R_1.
b. Calculate V_2 across R_2.
c. How much is V_3?

3-10   Effect of an Open Circuit in a Series Path

An open circuit is a break in the current path. The resistance of the open is very high because an insulator like air takes the place of a conducting part of the circuit. Remember that the current is the same in all parts of a series circuit. Therefore, an open in any part results in no current for the entire circuit. As illustrated in Fig. 3-16, the circuit is normal in Fig. 3-16a, but in Fig. 3-16b there is no current in R_1, R_2, or R_3 because of the open in the series path.

The open between P_1 and P_2, or at any other point in the circuit, has practically infinite resistance because its opposition to electron flow is so great compared with the resistance of R_1, R_2, and R_3. Therefore, the value of current is practically zero, even though the battery produces its normal applied voltage of 40 V.

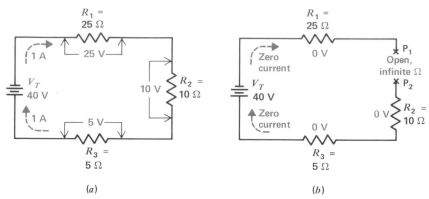

Fig. 3-16 Effect of an open in a series circuit. (*a*) Normal closed circuit with current of 1 A. (*b*) Open in any part of the path results in no current in the entire circuit.

To take an example, suppose that the open circuit between P_1 and P_2 has a resistance of 40×10^9 Ω (40 gigaohms). The resistance of the entire circuit is essentially 40×10^9 Ω, since the resistance of R_1, R_2, and R_3 can then be neglected compared with the resistance of the open. Such a high resistance is practically infinite ohms.

By Ohm's law, the current that results from 40 V applied across 40×10^9 Ω is 1×10^{-9} A, which is practically zero. This is the value of current in all parts of the series circuit. With practically no current, the *IR* voltage drop is practically zero across the 25 Ω of R_1, the 10 Ω of R_2, and the 5 Ω of R_3.

In summary, with an open in any part of a series circuit, the current is zero in the entire circuit. There is no *IR* voltage drop across any of the series resistances, although the generator still maintains its output voltage.

The Case of Zero *IR* Drop

In Fig. 3-16*b*, each of the resistors in the open circuit has an *IR* drop of zero. The reason is that current of practically zero is the value in all the series components. Each *R* still has its resistance. However, with zero current the *IR* voltage is zero.

The Source Voltage V_T Is Still Present with Zero *I*

The open circuit in Fig. 3-16*b* illustrates another example of how *V* and *I* are different forms of electricity. There is no current with the open circuit because there is no complete path outside the battery between its two terminals. However, the battery is generating a potential difference across the positive and negative terminals. This source voltage is present with or without current in the external circuit. If you measure V_T, the meter will read 40 V with the circuit closed or open.

The same idea applies to the 120-V ac voltage from the power line in the home. The 120-V potential difference is across the two terminals of the wall outlet. If you connect a lamp or appliance, current will flow in the circuit. When nothing is connected, though, the 120-V potential difference is still there at the outlet. If you should touch it, you would get an electric shock. The generator at the power station is maintaining the 120 V at the outlets as a source to produce current in any circuits that are plugged in.

The Applied Voltage Is Across the Open Terminals

It is useful to note that the entire applied voltage is present across the open circuit. Between P_1 and P_2 in Fig. 3-16*b*, there is 40 V. The reason is that essentially all the resistance of the series circuit is between P_1 and P_2. Therefore, the resistance of the open circuit develops all the *IR* voltage drop.

The extremely small current of one-billionth of an ampere is not enough to develop any appreciable *IR* drop across R_1, R_2, and R_3. Across the open[1], however,

---

[1]The voltage across an open circuit equals the applied voltage, even without any current, after the capacitance between the open terminals becomes charged by *V*, as described in Chap. 20, "Capacitance."

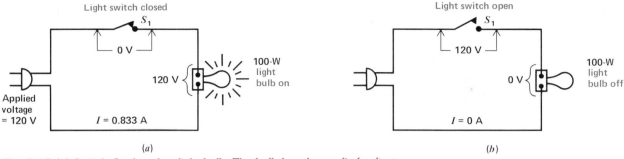

Fig. 3-17 (a) Switch S_1 closed to light bulb. The bulb has the applied voltage and normal current. (b) Switch S_1 open. Current is zero, and applied voltage is across the open switch.

the resistance is 40 billion Ω. Therefore, the IR voltage across the open here is one-billionth of an ampere multiplied by 40 billion Ω, which equals 40 V.

We could also consider the open circuit as a proportional voltage divider. Since practically all the series resistance is between P_1 and P_2, all the applied voltage is across the open terminals.

The fact that the open terminals have the entire applied voltage indicates a good way to find an open component in a series string. If you measure the voltage across each good component, zero voltage will be normal. However, the component that has the full source voltage is the one that is open.

A practical example is a fuse in a circuit. A good fuse has the normal current, practically zero resistance, and zero voltage. If the fuse blows, the result is no current in the circuit but the applied voltage is across the open fuse.

Another example is illustrated in Fig. 3-17 to empha-

size again the difference between voltage and current in a circuit. In Fig. 3-17a, the circuit is normal, with the switch closed to light the bulb. The 120 V of the source is across the bulb to produce the normal current of 0.833 A for the 100-W bulb. Across the switch, the voltage is zero because it has practically no resistance in the closed position. When the switch is opened in Fig. 3-17b, however, the results are: (1) There is no current to light the bulb; (2) there is no voltage across the bulb because the IR drop is zero without any current; and (3) the open switch has the applied voltage of 120 V.

Practice Problems 3-10
Answers at End of Chapter

a. Which component has 120 V in Fig. 3-17a?
b. Which component has 120 V in Fig. 3-17b?

Summary

1. There is only one current I in a series circuit. $I = V_T/R_T$, where V_T is the voltage applied across the total series resistance R_T. This I is the same in all the series components.
2. The total resistance R_T of a series string is the sum of the individual resistances.
3. The applied voltage V_T equals the sum of the series IR voltage drops.
4. The negative side of an IR voltage drop is where electrons flow in, attracted to the positive side at the opposite end.
5. The sum of the individual values of power used in the individual resistances equals the total power supplied by the source.
6. Series-aiding voltages are added; series-opposing voltages are subtracted.
7. An open results in no current in all parts of the series circuit.

8. In an open circuit, the voltage across the two open terminals is equal to the applied voltage.

Self-Examination
Answers at Back of Book

Choose (a), (b), (c), or (d).

1. When two resistances are connected in series, (a) they must both have the same resistance value; (b) the voltage across each must be the same; (c) they must have different resistance values; (d) there is only one path for current through both resistances.
2. In Fig. 3-3c, if the current through R_5 is 1 A, then the current through R_3 must be (a) ⅓ A; (b) ½ A; (c) 1 A; (d) 3 A.
3. With a 10-kΩ resistance in series with a 2-kΩ resistance, the total R_T equals (a) 2 kΩ; (b) 8 kΩ; (c) 10 kΩ; (d) 12 kΩ.
4. With two 45-kΩ resistances in series across a 90-V battery, the voltage across each resistance equals (a) 30 V; (b) 45 V; (c) 90 V; (d) 180 V.
5. The sum of series IR voltage drops (a) is less than the smallest voltage drop; (b) equals the average value of all the voltage drops; (c) equals the applied voltage; (d) is usually more than the applied voltage.
6. Resistances R_1 and R_2 are in series with 90 V applied. If V_1 is 30 V, then V_2 must be (a) 30 V; (b) 90 V; (c) 45 V; (d) 60 V.
7. With a 4-Ω resistance and a 2-Ω resistance in series across a 6-V battery, the current (a) in the larger resistance is 1⅓ A; (b) in the smaller resistance is 3 A; (c) in both resistances is 1 A; (d) in both resistances is 2 A.
8. When one resistance in a series string is open, (a) the current is maximum in the normal resistances; (b) the current is zero in all the resistances; (c) the voltage is zero across the open resistance; (d) the current increases in the voltage source.
9. The resistance of an open series string is (a) zero; (b) infinite; (c) equal to the normal resistance of the string; (d) about double the normal resistance of the string.
10. A source of 100 V is applied across a 20-Ω R_1 and 30-Ω R_2 in series, and V_1 is 40 V. The current in R_2 is (a) 5 A; (b) 3⅓ A; (c) 1⅓ A; (d) 2 A.

Essay Questions

1. Show how to connect two resistances in series with each other across a voltage source.
2. State three rules for the current, voltage, and resistance in a series circuit.
3. For a given amount of current, why does more resistance have a bigger voltage drop across it?
4. Two 300-W 120-V light bulbs are connected in series across a 240-V line. If the filament of one bulb burns open, will the other bulb light? Why? With the open circuit, how much is the voltage across the source and across each bulb?
5. Prove that if $V_T = V_1 + V_2 + V_3$, then $R_T = R_1 + R_2 + R_3$.

6. State briefly a rule for determining polarity of the voltage drop across each resistor in a series circuit.
7. Redraw the circuit in Fig. 3-13, marking the polarity of V_1, V_2, and V_3.
8. State briefly a rule to determine when voltages are series-aiding.
9. Derive the formula $P_T = P_1 + P_2 + P_3$ from the fact that $V_T = V_1 + V_2 + V_3$.
10. In a series string, why does the largest R dissipate the most power?
11. Give one application of series circuits.
12. Why do the rules for series components apply to either dc or ac circuits?

Problems
Answers to Odd-Numbered Problems at Back of Book

1. A circuit has 10 V applied across a 10-Ω resistance R_1. How much is the current in the circuit? How much resistance R_2 must be added in series with R_1 to reduce the current one-half? Show the schematic diagram of the circuit with R_1 and R_2.
2. Draw the schematic diagram of 20-, 30-, and 40-Ω resistances in series. (a) How much is the total resistance of the entire series string? (b) How much current flows in each resistance, with a voltage of 18 V applied across the series string? (c) Find the voltage drop across each resistance. (d) Find the power dissipated in each resistance.
3. An R_1 of 90 kΩ and an R_2 of 10 kΩ are in series across a 6-V source. (a) Draw the schematic diagram. (b) How much is V_2?
4. Draw a schematic diagram showing two resistances R_1 and R_2 in series across a 100-V source. (a) If the IR voltage drop across R_1 is 60 V, how much is the IR voltage drop across R_2? (b) Label the polarity of the voltage drops across R_1 and R_2. (c) If the current is 1 A through R_1, how much is the current through R_2? (d) How much is the resistance of R_1 and R_2? How much is the total resistance across the voltage source? (e) If the voltage source is disconnected, how much is the voltage across R_1 and across R_2?
5. Three 20-Ω resistances are in series across a voltage source. Show the schematic diagram. If the voltage across each resistor is 10 V, how much is the applied voltage? How much is the current in each resistance?
6. How much resistance R_1 must be added in series with a 100-Ω R_2 to limit the current to 0.3 A with 120 V applied? Show the schematic diagram. How much power is dissipated in each resistance?
7. Find the total R_T of the following resistances in series: 2 MΩ, 0.5 MΩ, 47 kΩ, 5 kΩ, and 470 Ω.
8. Referring to Fig. 3-14, calculate the power dissipated in the voltage-dropping resistor R_S.
9. Draw the circuit with values for three equal series resistances across a 90-V source, where each R has one-third the applied voltage and the current in the circuit is 6 mA.
10. A 100-W bulb normally takes 0.833 A, and a 200-W bulb takes 1.666 A from the 120-V power line. If these two bulbs were connected in series across a 240-V power line, prove that the current would be 1.111 A in both bulbs, assuming the resistances remain constant.

11. Calculate I in R_1 and R_2 for the diagrams in Fig. 3-9a, b, and c.

12. In Fig. 3-18, calculate I, V_1, V_2, P_1, P_2, and P_T. (Note: R_1 and R_2 are in series with V_T even though the source is shown at the right instead of at the left.)

13. If R_1 is increased to 8 kΩ in Fig. 3-18, what will be the new I?

14. In Fig. 3-19, find R_1. Why is I in the direction shown?

15. In Fig. 3-20, find R_2.

16. Figure 3-21 shows the circuit for keeping a 12.6-V car battery charged from a 15-V dc generator. Calculate I and show the direction of electron flow between points A and B.

17. In Fig. 3-22, find V_2. Show polarity for V_1, V_2, and V_3.

18. In Fig. 3-23, find V_T. Show polarity for V_T, V_1, V_2, and V_3.

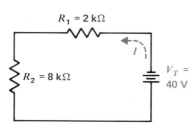

Fig. 3-18 For Probs. 12 and 13.

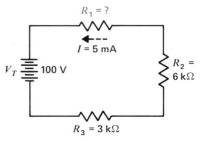

Fig. 3-19 For Prob. 14.

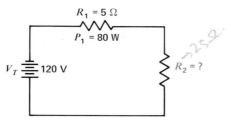

Fig. 3-20 For Prob. 15.

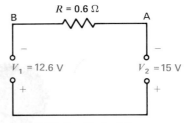

Fig. 3-21 For Prob. 16.

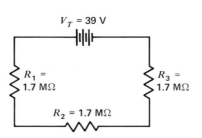

Fig. 3-22 For Prob. 17.

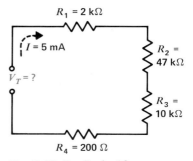

Fig. 3-23 For Prob. 18.

Answers to Practice Problems

3-1 a. R_1, R_2, R_3, V_T
    and the wires
  b. 5 A

3-2 a. 2 mA
  b. 10 kΩ
  c. 1 mA

3-3 a. 10 V
  b. 1 A
  c. 1 A
  d. 1 A

3-4 a. 60 V
  b. 75 V

3-5 a. point a or c
  b. point b or f
  c. point d

3-6 a. S is +20 V
    J is +10 V
    G is 0 V
  b. S is −20 V
    J is −10 V
    G is 0 V

3-7 a. 6 W
  b. 40-kΩ R_2

3-8 a. 100 V
  b. 20 V

3-9 a. $V_1 = 30$ V
  b. $V_2 = 90$ V
  c. $V_3 = 60$ V

3-10 a. Bulb
  b. Switch

Chapter 4
Parallel Circuits

When two or more components are connected across one voltage source, they form a parallel circuit. Each parallel path is then a branch, with its own individual current. Parallel circuits, therefore, have one common voltage across all the branches but individual branch currents that can be different. These characteristics are opposite from series circuits that have one common current but individual voltage drops that can be different.

Important terms in this chapter are:

inverse relation   parallel conductances
main-line current   positive terminal
negative terminal   reciprocal resistance formula
open branch   short circuit
parallel bank of resistors   total current method
parallel branch currents   total power

More details are explained in the following sections:

4-1   The Applied Voltage V_A Is the Same Across Parallel Branches

A parallel circuit is formed when two or more components are connected across a voltage source, as shown in Fig. 4-1. In this figure, R_1 and R_2 are in parallel with each other and with the battery. In Fig. 4-1b, the points a, b, c, and e are really equivalent to a direct connection at the negative terminal of the battery because the connecting wires have practically no resistance. Similarly, points h, g, d, and f are the same as a direct connection at the positive battery terminal. Since R_1 and R_2 are directly connected across the two terminals of the battery, both resistances must have the same potential difference as the battery. It follows that the voltage is the same across components connected in parallel. The parallel circuit arrangement is used, therefore, to connect components that require the same voltage.

A common application of parallel circuits is typical house wiring to the power line, with many lights and appliances connected across the 120-V source (Fig. 4-2). The wall receptacle has the potential difference of 120 V across each pair of terminals. Therefore, any resistance connected to an outlet has the applied voltage of 120 V. The light bulb is connected to one outlet and the toaster to another outlet, but both have the same applied voltage of 120 V. Therefore, each operates independently of any other appliance, with all the individual branch circuits connected across the 120-V line.

Practice Problems 4-1
Answers at End of Chapter

a. In Fig. 4-1, how much is the common voltage across R_1 and R_2?

b. In Fig. 4-2, how much is the common voltage across the bulb and the toaster?

4-2   Each Branch I Equals V_A/R

In applying Ohm's law, it is important to note that the current equals the voltage applied across the circuit divided by the resistance between the two points where that voltage is applied. In Fig. 4-3, 10 V is applied across the 5 Ω of R_2, resulting in the current of 2 A between points e and f through R_2. The battery voltage is also applied across the parallel resistance of R_1, applying 10 V across 10 Ω. Through R_1, therefore, the current is 1 A between points c and d. The current has a different value through R_1, with the same applied voltage, because the resistance is different. These values are calculated as follows:

$$I_1 = \frac{V_A}{R_1} = \frac{10}{10} = 1 \text{ A}$$

$$I_2 = \frac{V_A}{R_2} = \frac{10}{5} = 2 \text{ A}$$

Just as in a circuit with just one resistance, any branch that has less R allows more I. If R_1 and R_2 were equal, however, the two branch currents would have the same value. For instance, in Fig. 4-1b each branch has its own current equal to 1.5 V/5 Ω = 0.3 A.

The I can be different in parallel circuits that have different R because V is the same across all the branches. Any voltage source generates a potential dif-

(a)

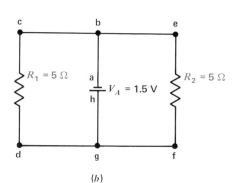

(b)

Fig. 4-1 Example of a parallel circuit with two resistors. (a) Photograph of wiring. (b) Schematic diagram.

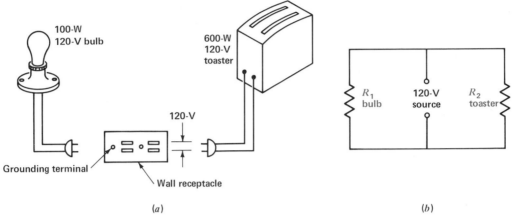

Fig. 4-2 Light bulb and toaster connected in parallel to the 120-V line. (*a*) Wiring diagram. (*b*) Schematic diagram.

ference across its two terminals. This voltage does not move. Only I flows around the circuit. The source voltage is available to make electrons move around any closed path connected to the generator terminals. How much I is in each separate path depends on the amount of R in each branch.

Practice Problems 4-2
Answers at End of Chapter

Refer to Fig. 4-3.
a. How much is the voltage across R_1?
b. How much is I_1 through R_1?
c. How much is the voltage across R_2?
d. How much is I_2 through R_2?

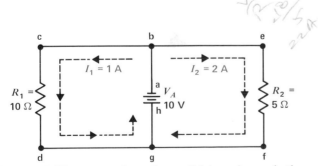

Fig. 4-3 The current in each parallel branch equals the applied voltage V_A divided by each branch R.

4-3 The Main-Line I_T Equals the Sum of the Branch Currents

Components to be connected in parallel are usually wired directly across each other, with the entire parallel combination connected to the voltage source, as illustrated in Fig. 4-4. This circuit is equivalent to wiring each parallel branch directly to the voltage source, as shown in Fig. 4-1, when the connecting wires have essentially zero resistance.

The advantage of having only one pair of connecting leads to the source for all the parallel branches is that usually less wire is necessary. The pair of leads connecting all the branches to the terminals of the voltage source is the *main line*. In Fig. 4-4, the wires from g to a on the negative side and from b to f in the return path form the main line.

In Fig. 4-4*b*, with 20 Ω of resistance for R_1 connected across the 20-V battery, the current through R_1 must be 20 V/20 Ω = 1 A. This current is electron flow from the negative terminal of the source, through R_1, and back to the positive battery terminal. Similarly, the R_2 branch of 10 Ω across the battery has its own branch current of 20 V/10 Ω = 2 A. This current flows from the negative terminal of the source, through R_2, and back to the positive terminal, since it is a separate path for electron flow.

All the current in the circuit, however, must come from one side of the voltage source and return to the opposite side for a complete path. In the main line,

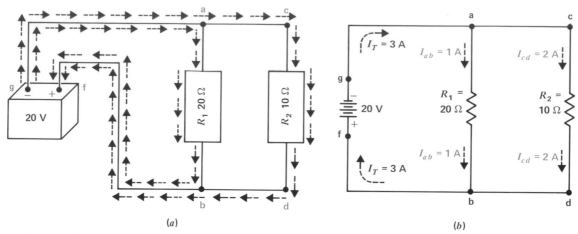

Fig. 4-4 The main line current equals the sum of all the branch currents. From g to a is the negative side of the main line, and from b to f is the positive side. (*a*) Wiring diagram. Arrows inside the lines indicate current for R_1 in the main line; arrows outside indicate current for R_2. (*b*) Schematic diagram. Current I_T is total line current for both R_1 and R_2.

therefore, the amount of current is equal to the total of the branch currents.

For example, in Fig. 4-4*b*, the total current in the line from point g to point a is 3 A. The total current at branch point a subdivides into its component branch currents for each of the branch resistances. Through the path of R_1 from a to b the current is 1 A. The other branch path acdb through R_2 has a current of 2 A. At the branch point b, the electron flow from both parallel branches combines, so that the current in the main-line return path from b to f has the same value of 3 A as in the other side of the main line.

The formula for the total current I_T in the main line is

$$I_T = I_1 + I_2 + I_3 + \cdots + \text{etc.} \qquad (4\text{-}1)$$

This rule applies for any number of parallel branches, whether the resistances are equal or unequal.

Example 1   An R_1 of 20 Ω, an R_2 of 40 Ω, and an R_3 of 60 Ω are connected in parallel across the 120-V power line. How much is the total line current I_T?

Answer   Current I_1 for the R_1 branch is $^{120}\!/_{20}$ or 6 A. Similarly, I_2 is $^{120}\!/_{40}$ or 3 A, and I_3 is $^{120}\!/_{60}$ or 2 A. The total current in the main line is

$$I_T = I_1 + I_2 + I_3 = 6 + 3 + 2$$
$$I_T = 11 \text{ A}$$

Example 2   Two branches R_1 and R_2 across the 120-V power line draw a total line current I_T of 15 A. The R_1 branch takes 10 A. How much is the current I_2 in the R_2 branch?

Answer $I_2 = I_T - I_1 = 15 - 10$
$$I_2 = 5 \text{ A}$$

With two branch currents, one must equal the difference between I_T and the other branch current.

Example 3   Three parallel branch currents are 0.1 A, 500 mA, and 800 μA. Calculate I_T.

Answer   All values must be in the same units to be added. Converted to milliamperes, therefore, 0.1 A = 100 mA and 800 μA = 0.8 mA. Then

$$I_T = 100 + 500 + 0.8$$
$$I_T = 600.8 \text{ mA}$$

You can add the currents in A, mA, or μA units, as long as the same unit is used for all the currents.

Practice Problems 4-3
Answers at End of Chapter

a. Parallel branch currents are 1 A for I_1, 2 A for I_2, and 3 A for I_3. Calculate I_T.
b. Assume $I_T = 6$ A for three branch currents, I_1 is 1 A, and I_2 is 2 A. Calculate I_3.

4-4   Resistances in Parallel

The total resistance across the main line in a parallel circuit can be found by Ohm's law: *Divide the common voltage across the parallel resistances by the total current of all the branches*. Referring to Fig. 4-5a, note that the parallel resistance of R_1 with R_2, indicated by the combined resistance R_T, is the opposition to the total current in the main line. In this example, V_A/I_T is 60 V/3 A = 20 Ω for R_T.

The total load connected to the source voltage is the same as though one equivalent resistance of 20 Ω were connected across the main line. This is illustrated by the equivalent circuit in Fig. 4-5b. For any number of parallel resistances of any value, therefore,

$$R_T = \frac{V_A}{I_T} \qquad (4\text{-}2)$$

where I_T is the sum of all the branch currents and R_T is the equivalent resistance of all the parallel branches across the voltage source V_A.

Example 4   Two branches, each with a 5-A current, are connected across a 90-V source. How much is the equivalent total resistance R_T?

Answer   The total line current I_T is $5 + 5 = 10$ A. Then,

$$R_T = \frac{V_A}{I_T} = \frac{90}{10}$$

$$R_T = 9\ \Omega$$

Parallel Bank   A combination of parallel branches is often called a *bank*. In Fig. 4-5, the bank consists of the 60-Ω R_1 and 30-Ω R_2 in parallel. Their combined parallel resistance R_T is the bank resistance, equal to 20 Ω in this example. A bank can have two or more parallel resistors.

When a circuit has more current with the same applied voltage, this greater value of I corresponds to less R because of their inverse relation. Therefore, the combination of parallel resistances R_T for the bank is always less than the smallest individual branch resistance. The reason is that I_T must be more than any one branch current.

Why R_T Is Less Than Any Branch R   It may seem unusual at first that putting more resistance into a circuit lowers the equivalent resistance. This feature of parallel circuits is illustrated in Fig. 4-6. Note that equal resistances of 30 Ω each are added across the source voltage, one branch at a time. The circuit in Fig. 4-6a has just R_1, which allows 2 A with 60 V applied.

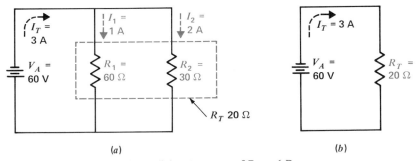

(a)   (b)

Fig. 4-5 Resistances in parallel. (a) Combined parallel resistances of R_1 and R_2 is the total R_T in the main line. (b) Equivalent circuit showing combined R_T drawing the same 3-A I_T as the parallel combination of R_1 and R_2.

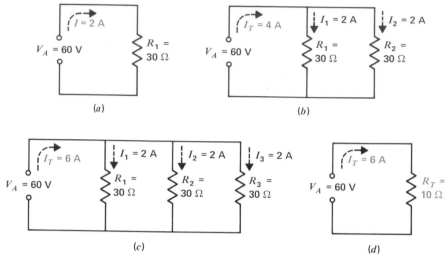

Fig. 4-6 How adding parallel branches of resistors increases I_T but reduces R_T. (a) One resistor. (b) Two branches. (c) Three branches. (d) Equivalent circuit of the three branches in (c).

In Fig. 4-6b the R_2 branch is added across the same V_A. This branch also has 2 A. Now the parallel circuit has a 4-A total line current because of $I_1 + I_2$. Then the third branch, which also takes 2 A for I_3, is added in Fig. 4-6c. The combined circuit with three branches therefore requires a total load current of 6 A, which is supplied by the voltage source.

The combined resistance across the source, then, is V_A/I_T, which is 6% or 10 Ω. This equivalent resistance R_T, representing the entire load on the voltage source, is shown in Fig. 4-6d. More resistance branches reduce the combined resistance of the parallel circuit because more current is required from the same voltage source.

Reciprocal Resistance Formula   We can derive this formula from the fact that I_T is the sum of all the branch currents, or,

$$I_T = I_1 + I_2 + I_3 + \cdots + \text{etc.}$$

However, I_T is V/R_T. Also, each I is V/R. Substituting V/R_T for I_T on the left side of the equation and V/R for each branch I on the right side, the result is

$$\frac{V}{R_T} = \frac{V}{R_1} + \frac{V}{R_2} + \frac{V}{R_3} + \cdots + \text{etc.}$$

Dividing by V because it is the same across all the resistances,

$$\frac{1}{R_T} = \frac{1}{R_1} + \frac{1}{R_2} + \frac{1}{R_3} + \cdots + \text{etc.} \tag{4-3}$$

This reciprocal formula applies to any number of parallel resistances of any value. Using the values in Fig. 4-7a as an example,

$$\frac{1}{R_T} = \frac{1}{20} + \frac{1}{10} + \frac{1}{10}$$

$$\frac{1}{R_T} = \frac{1}{20} + \frac{2}{20} + \frac{2}{20} = \frac{5}{20}$$

$$R_T = \frac{20}{5}$$

$$R_T = 4 \ \Omega$$

Notice that the value for $1/R_T$ must be inverted to obtain R_T when using Formula (4-3) because the formula gives the reciprocal of R_T.

Total-Current Method   Figure 4-7b shows how this same problem can be calculated in terms of total current instead of by the reciprocal formula, if it is easier to work without fractions. Although the applied voltage is not always known, any convenient value can be assumed because it cancels in the calculations. It is

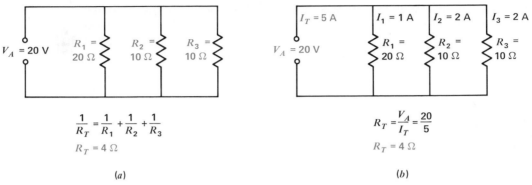

$$\frac{1}{R_T} = \frac{1}{R_1} + \frac{1}{R_2} + \frac{1}{R_3}$$

$$R_T = 4\ \Omega$$

$$R_T = \frac{V_A}{I_T} = \frac{20}{5}$$

$$R_T = 4\ \Omega$$

(a)   (b)

Fig. 4-7 Combining parallel resistances to find R_T by two different methods. (a) Using the reciprocal resistance formula to calculate R_T as 4 Ω. (b) Using the total line current method with an assumed line voltage of 20 V gives the same R_T of 4 Ω.

usually simplest to assume an applied voltage of the same numerical value as the highest resistance. Then one assumed branch current will automatically be 1 A and the other branch currents will be more, eliminating fractions less than 1 in the calculations.

For the example in Fig. 4-7b, the highest branch R is 20 Ω. Therefore, assume 20 V for the applied voltage. Then the branch currents are 1 A in R_1, 2 A in R_2, and 2 A in R_3. Their sum is $1 + 2 + 2 = 5$ A for I_T. The combined resistance R_T across the main line is V_A/I_T, or 20 V/5 A = 4 Ω. This is the same value calculated with the reciprocal resistance formula.

Special Case of Equal R in All Branches If R is equal in all branches, the combined R_T equals the value of one branch resistance divided by the number of branches. This rule is illustrated in Fig. 4-8, where three 60-kΩ resistances in parallel equal 20 kΩ.

The rule applies to any number of parallel resistances, but they must all be equal. As another example, five 60-Ω resistances in parallel have the combined resistance of 60/5, or 12 Ω. A common application is two equal resistors wired in a parallel bank for R_T equal to one-half R.

Special Case of Only Two Branches When there are two parallel resistances and they are not equal, it is usually quicker to calculate the combined resistance by the method shown in Fig. 4-9. This rule says that the combination of two parallel resistances is their product divided by their sum.

$$R_T = \frac{R_1 \times R_2}{R_1 + R_2} \qquad\qquad (4\text{-}4)$$

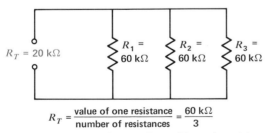

$$R_T = \frac{\text{value of one resistance}}{\text{number of resistances}} = \frac{60\ \text{k}\Omega}{3}$$

Fig. 4-8 For the special case of equal branch resistances, divide R by the number of branches to find R_T. Here, R_T is 60 kΩ/3 = 20 kΩ.

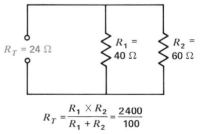

$$R_T = \frac{R_1 \times R_2}{R_1 + R_2} = \frac{2400}{100}$$

Fig. 4-9 For the special case of just two branch resistances, of any values, R_T equals their product divided by the sum. Here, R_T is 2400/100 = 24 Ω.

where R_T is in the same units as all the individual resistances. For the example in Fig. 4-9,

$$R_T = \frac{R_1 \times R_2}{R_1 + R_2} = \frac{40 \times 60}{40 + 60} = \frac{2400}{100}$$

$$R_T = 24 \ \Omega$$

Each R can have any value, but there must be only two resistances. Note that this method gives R_T directly, not its reciprocal. If you use the reciprocal formula for this example, the answer will be $1/R_T = \frac{1}{24}$, which is the same value as R_T equals 24 Ω.

Short-Cut Calculations

Figure 4-10 shows how these special rules can help in reducing parallel branches to a simpler equivalent circuit. In Fig. 4-10a, the 60-Ω R_1 and R_4 are equal and in parallel. Therefore, they are equivalent to the 30-Ω R_{14} in Fig. 4-10b. Similarly, the 20-Ω R_2 and R_3 are equivalent to the 10 Ω of R_{23}. The circuit in Fig. 4-10a is equivalent to the simpler circuit in Fig. 4-10b with just the two parallel resistances of 30 and 10 Ω.

Finally, the combined resistance for these two equals their product divided by their sum, which is 300/40 or 7.5 Ω, as shown in Fig. 4-10c. This value of R_T in Fig. 4-10c is equivalent to the combination of the four branches in Fig. 4-10a. If you connect a voltage source across either circuit, the generator current in the main line will be the same for both cases.

The order of connections for parallel resistances does not matter in determining R_T. There is no question as to which is first or last because they are all across the same voltage source.

Finding an Unknown Branch Resistance

In some cases with two parallel resistors, it is useful to be able to determine what size R_x to connect in parallel with a known R in order to obtain a required value of R_T. Then the factors can be transposed as follows:

$$R_x = \frac{R \times R_T}{R - R_T} \tag{4-5}$$

This formula is derived from Formula (4-4).

Example 5 What R_x in parallel with 40 Ω will provide an R_T of 24 Ω?

Answer $R_x = \dfrac{R \times R_T}{R - R_T} = \dfrac{40 \times 24}{40 - 24} = \dfrac{960}{16}$

$\qquad\qquad R_x = 60 \ \Omega$

This problem corresponds to the circuit shown before in Fig. 4-9.

Note that Formula (4-5) for R_x has a product over a difference. The R_T is subtracted because it is the smallest R. Remember that both Formulas (4-4) and (4-5) can be used with only two parallel branches.

Example 6 What R in parallel with 50 kΩ will provide an R_T of 25 kΩ?

Answer $R = 50$ kΩ

Two equal resistances in parallel have R_T equal to one-half R.

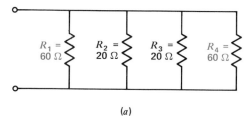
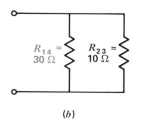
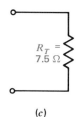

(a) (b) (c)

Fig. 4-10 An example of parallel resistance calculations with four branches. (a) Original circuit. (b) Resistors combined into two branches. (c) Equivalent circuit reduces to one R_T for all the branches.

Practice Problems 4-4
Answers at End of Chapter

a. Find R_T for three 4.7-MΩ resistances in parallel.
b. Find R_T for 3 MΩ in parallel with 2 MΩ.
c. Find R_T for two 20-Ω resistances in parallel with 10 Ω.

Practice Problems 4-5
Answers at End of Chapter

a. If G_1 is 2 S and G_2 in parallel is 4 S, calculate G_T.
b. If G_1 is 0.05 μS, G_2 is 0.2 μS, and G_3 is 0.5 μS, all in parallel, find G_T and R_T.

4-5 Conductances in Parallel

Since conductance G is equal to $1/R$, the reciprocal resistance Formula (4-3) can be stated for conductance as

$$G_T = G_1 + G_2 + G_3 + \cdots + \text{etc.} \qquad (4\text{-}6)$$

With R in ohms, G is in siemens. For the example in Fig. 4-11, G_1 is $\frac{1}{20} = 0.05$, G_2 is $\frac{1}{5} = 0.2$, and G_3 is $\frac{1}{2} = 0.5$. Then

$$G_T = 0.05 + 0.2 + 0.5 = 0.75 \text{ S}$$

Notice that adding the conductances does not require reciprocals. Actually, each value of G is the reciprocal of R.

Working with G may be more convenient than working with R in parallel circuits since it will avoid using the reciprocal formula for R_T. Each branch current is directly proportional to its conductance. This idea corresponds to the fact that in series circuits each voltage drop is directly proportional to each series resistance.

The reason why parallel conductances are added directly can be illustrated by assuming a 1-V source across all the branches. Then calculating the values of $1/R$ for the conductances is the same as calculating the branch currents. These values are added for the total I_T or G_T.

4-6 Total Power in Parallel Circuits

Since the power dissipated in the branch resistances must come from the voltage source, the total power equals the sum of the individual values of power in each branch. This rule is illustrated in Fig. 4-12. We can also use this circuit as an example of how to apply the rules of current, voltage, and resistance for a parallel circuit.

The applied 10 V is across the 10-Ω R_1 and 5-Ω R_2 in Fig. 4-12. The branch current I_1 then is V_A/R_1 or $\frac{10}{10}$, which equals 1 A. Similarly, I_2 is $\frac{10}{5}$, or 2 A. The total I_T is $1 + 2 = 3$ A. If we want to find R_T, it equals V_A/I_T or $\frac{10}{3}$, which is $3\frac{1}{3}$ Ω.

The power dissipated in each branch R is $V_A \times I$. In the R_1 branch, I_1 is $\frac{10}{10} = 1$ A. Then P_1 is $V_A \times I_1$ or $10 \times 1 = 10$ W.

For the R_2 branch, I_2 is $\frac{10}{5} = 2$ A. Then P_2 is $V_A \times I_2$ or $10 \times 2 = 20$ W.

Adding P_1 and P_2, the answer is $10 + 20 = 30$ W. This P_T is the total power dissipated in both branches.

This value of 30 W for P_T is also the total power supplied by the voltage source by means of its total line current I_T. With this method, the total power is $V_A \times I_T$ or $10 \times 3 = 30$ W for P_T. The 30 W of power supplied by the voltage source is dissipated or used up in the branch resistances.

Note that in both parallel and series circuits the sum of the individual values of power dissipated in the cir-

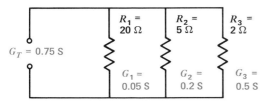

Fig. 4-11 Conductances in parallel are added for the total G_T.

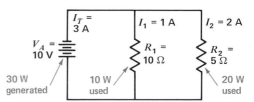

Fig. 4-12 The sum of P_1 and P_2 used in each branch equals the total power P_T produced by the source.

cuit equals the total power generated by the source. This can be stated as a formula

$$P_T = P_1 + P_2 + P_3 + \cdots + \text{etc.} \qquad \text{(4-7)}$$

The series or parallel connections can alter the distribution of voltage or current, but power is the rate at which energy is supplied. The circuit arrangement cannot change the fact that all the energy in the circuit comes from the source.

Practice Problems 4-6
Answers at End of Chapter

a. Two parallel branches each have 2 A at 120 V. Calculate P_T.
b. Three parallel branches of 10, 20, and 30 Ω have 60 V applied. Calculate P_T.

4-7   Analyzing Parallel Circuits

For many types of problems with parallel circuits it is useful to remember the following points:

1. When you know the voltage across one branch, this voltage is across all the branches. There can be only one voltage across branch points with the same potential difference.
2. If you know I_T and one of the branch currents I_1, you can find I_2 by subtracting I_1 from I_T.

The circuit in Fig. 4-13 illustrates these points. The problem is to find the applied voltage V_A and the value of R_3. Of the three branch resistances, only R_1 and R_2 are known. However, since I_2 is given as 2 A, the I_2R_2 voltage must be $2 \times 60 = 120$ V.

Although the applied voltage is not given, this must also be 120 V. The voltage across all the parallel branches is the same 120 V that is across the R_2 branch.

Now I_1 can be calculated as V_A/R_1. This is $^{120}\!/_{30} = 4$ A for I_1.

Current I_T is given as 7 A. The two branches take $2 + 4 = 6$ A. The third branch current through R_3 must be $7 - 6 = 1$ A for I_3.

Now R_3 can be calculated as V_A/I_3. This is $^{120}\!/_1 = 120 \Omega$ for R_3.

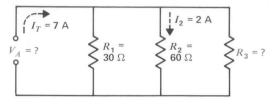

Fig. 4-13 Analyzing a parallel circuit. What are the values for V_A and R_3? See solution in text.

Practice Problems 4-7
Answers at End of Chapter

Refer to Fig. 4-13.
a. How much is V_2 across R_2?
b. Calculate I_1 through R_1.

4-8   Effect of an Open Branch in Parallel Circuits

An open in any circuit is an infinite resistance that results in no current. However, in parallel circuits there is a difference between an open circuit in the main line and an open circuit in a parallel branch. These two cases are illustrated in Fig. 4-14. In Fig. 4-14a the open circuit in the main line prevents any electron flow in the line to all the branches. The current is zero in every branch, therefore, and none of the bulbs can light.

However, in Fig. 4-14b the open is in the branch circuit for bulb 1. The open branch circuit has no current, then, and this bulb cannot light. The current in all the other parallel branches is normal, though, because each is connected to the voltage source. Therefore, the other bulbs light.

These circuits show the advantage of wiring components in parallel. An open in one component opens only one branch, while the other parallel branches have their normal voltage and current.

Practice Problems 4-8
Answers at End of Chapter

a. How much is the R of an open filament or heater?
b. In Fig. 4-14b, if only bulb 3 is open, which bulbs will light?

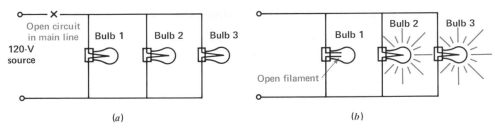

(a)   *(b)*

Fig. 4-14 Effect of an open with parallel circuits. (*a*) Open circuit in the main line—no current and no light for all bulbs. (*b*) Open circuit in one branch—bulb 1 does not light, but the two other bulbs operate normally.

4-9   Effect of a Short Circuit Across Parallel Branches

A short circuit has practically zero resistance. Its effect, therefore, is to allow excessive current. Consider the example in Fig. 4-15. In Fig. 4-15*a*, the circuit is normal, with 1 A in each branch and 2 A for the total line current. However, suppose the conducting wire at point H should accidentally contact the wire at point G, as shown in Fig. 4-15*b*. Since the wire is an excellent conductor, the short circuit results in practically zero resistance between points H and G. These two points are connected directly across the voltage source. With no opposition, the applied voltage could produce an infinitely high value of current through this current path.

The Short-Circuit Current   Practically, the amount of current is limited by the small resistance of the wire. Also, the source usually cannot maintain its output voltage while supplying much more than its rated load current. Still, the amount of current can be dangerously high. For instance, the short-circuit cur-

rent might be more than 100 A instead of the normal line current of 2 A in Fig. 4-15*a*. Because of the short circuit, excessive current flows in the voltage source, in the line to the short circuit at point H, through the short circuit, and in the line returning to the source from G. Because of the large amount of current, the wires can become hot enough to ignite and burn. There should be a fuse that would open if there is too much current in the main line because of a short circuit across any of the branches.

The Short-Circuited Components Have No Current   For the short circuit in Fig. 4-15*b*, the *I* is 0 A in the parallel resistors R_1 and R_2. The reason is that the short circuit is a parallel path with practically zero resistance. Then all the current flows in this path, bypassing the resistors R_1 and R_2. Therefore R_1 and R_2 are short-circuited or *shorted out* of the circuit. They cannot function without their normal current. If they were filament resistances of light bulbs or heaters, they would not light without any current.

The short-circuited components are not damaged, however. They do not even have any current passing

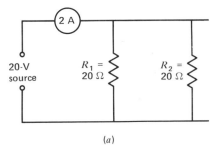   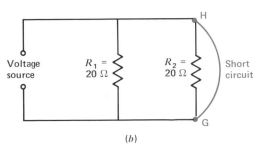

(a)   *(b)*

Fig. 4-15 Effect of a short circuit across parallel branches. (*a*) Normal circuit. (*b*) Short circuit across points H and G shorts out all the branches.

through them. Assuming that the short circuit has not damaged the voltage source and the wiring for the circuit, the components can operate again when the circuit is restored to normal by removing the short circuit.

All the Parallel Branches Are Short-Circuited   If there were only one R in Fig. 4-15 or any number of parallel components, they would all be shorted out by the short circuit across points H and G. Therefore, a short circuit across one branch in a parallel circuit shorts out all the parallel branches.

This idea also applies to a short circuit across the voltage source in any type of circuit. Then the entire circuit is shorted out.

Practice Problems 4-9
Answers at End of Chapter

Refer to Fig. 4-15.
a. How much is the R of the short circuit between H and G?
b. How much is I_1 in R_1 with the short circuit across R_2?

Summary

1. There is only one voltage V_A across all components in parallel.
2. The current in each branch I_b equals the voltage V_A across the branch divided by the branch resistance R_b. Or $I_b = V_A/R_b$.
3. The total line current equals the sum of all the branch currents. Or $I_T = I_1 + I_2 + I_3 + \cdots +$ etc.
4. The equivalent resistance R_T of parallel branches is less than the smallest branch resistance, since all the branches must take more current from the source than any one branch.
5. For only *two* parallel resistances of any value, $R_T = R_1R_2/(R_1 + R_2)$.
6. For any number of *equal* parallel resistances, R_T is the value of one resistance divided by the number of resistances.
7. For the general case of any number of branches, calculate R_T as V_A/I_T or use the reciprocal resistance formula: $1/R_T = 1/R_1 + 1/R_2 + 1/R_3 + \cdots +$ etc.
8. For any number of conductances in parallel, their values are added for G_T, in the same way as parallel branch currents are added.
9. The sum of the individual values of power dissipated in parallel resistances equals the total power produced by the source.
10. An open circuit in one branch results in no current through that branch, but the other branches can have their normal current. However, an open circuit in the main line results in no current for any of the branches.
11. A short circuit has zero resistance, resulting in excessive current. When one branch is short-circuited, all the parallel paths are also short-circuited. The entire current is in the short circuit and bypasses the short-circuited branches.

Self-Examination
Answers at Back of Book

Choose (a), (b), (c), or (d).

1. With two resistances connected in parallel: (a) the current through each must be the same; (b) the voltage across each must be the same; (c) their combined

resistance equals the sum of the individual values; (d) each must have the same resistance value.

2. With 100 V applied across ten 50-Ω resistances in parallel, the current through each resistance equals (a) 2 A; (b) 10 A; (c) 50 A; (d) 100 A.

3. With three 1-kΩ resistances connected in parallel, the combined equivalent resistance equals (a) ⅓ kΩ; (b) 1 kΩ; (c) 2 kΩ; (d) 3 kΩ.

4. A 1-Ω resistance in parallel with a 2-Ω resistance provides a combined equivalent resistance of (a) 3 Ω; (b) 1 Ω; (c) 2 Ω; (d) ⅔ Ω.

5. With resistances of 100, 200, 300, 400, and 500 Ω in parallel, R_T is (a) less than 100 Ω; (b) more than 1 MΩ; (c) about 500 Ω; (d) about 1 kΩ.

6. With two resistances connected in parallel, if each dissipates 10 W, the total power supplied by the voltage source equals (a) 5 W; (b) 10 W; (c) 20 W; (d) 100 W.

7. With eight 10-MΩ resistances connected in parallel across a 10-V source, the main-line current equals (a) 0.1 μA; (b) ⅛ μA; (c) 8 μA; (d) 10 μA.

8. A parallel circuit with 20 V applied across two branches has a total line current of 5 A. One branch resistance equals 5 Ω. The other branch resistance equals (a) 5 Ω; (b) 20 Ω; (c) 25 Ω; (d) 100 Ω.

9. Three 100-W light bulbs are connected in parallel across the 120-V power line. If one bulb opens, how many bulbs can light? (a) None; (b) one; (c) two; (d) all.

10. If a parallel circuit is open in the main line, the current (a) increases in each branch; (b) is zero in all the branches; (c) is zero only in the branch that has highest resistance; (d) increases in the branch that has lowest resistance.

Essay Questions

1. Draw a wiring diagram showing three resistances connected in parallel across a battery. Indicate each branch and the main line.

2. State two rules for the voltage and current values in a parallel circuit.

3. Explain briefly why the current is the same in both sides of the main line that connects the voltage source to the parallel branches.

4. (a) Show how to connect three equal resistances for a combined equivalent resistance one-third the value of one resistance. (b) Show how to connect three equal resistances for a combined equivalent resistance three times the value of one resistance.

5. Why can the current in parallel branches be different when they all have the same applied voltage?

6. Why does the current increase in the voltage source as more parallel branches are added to the circuit?

7. Show the algebra for deriving the formula $R_T = R_1 R_2/(R_1 + R_2)$ from the reciprocal formula for two resistances.

8. Redraw Fig. 4-15 with five parallel resistors R_1 to R_5 and explain why they all would be shorted out with a short circuit across R_3.

9. State briefly why the total power equals the sum of the individual values of power, whether a series circuit or a parallel circuit is used.

10. Explain why an open in the main line disables all the branches, but an open in one branch affects only that branch current.

11. Give two differences between an open circuit and a short circuit.
12. List as many differences as you can in comparing series circuits with parallel circuits.
13. Why are household appliances connected to the 120-V power line in parallel instead of being in series?
14. Give one advantage and one disadvantage of parallel connections.

Problems

Answers to Odd-Numbered Problems at Back of Book

1. A 15-Ω R_1 and a 45-Ω R_2 are connected in parallel across a 45-V battery. (a) Draw the schematic diagram. (b) How much is the voltage across R_1 and R_2? (c) How much is the current in R_1 and R_2? (d) How much is the main-line current? (e) Calculate R_T.
2. For the circuit in question 1, how much is the total power supplied by the battery?
3. A parallel circuit has three branch resistances of 20, 10, and 5 Ω for R_1, R_2, and R_3. The current through the 20-Ω branch is 1 A. (a) Draw the schematic diagram. (b) How much is the voltage applied across all the branches? (c) Find the current through the 10-Ω branch and the 5-Ω branch.
4. (a) Draw the schematic diagram of a parallel circuit with three branch resistances, each having 30 V applied and a 2-A branch current. (b) How much is I_T? (c) How much is R_T?
5. Referring to Fig. 4-12, assume that R_2 opens. (a) How much is the current in the R_2 branch? (b) How much is the current in the R_1 branch? (c) How much is the line current? (d) How much is the total resistance of the circuit? (e) How much power is generated by the battery?
6. Two resistances R_1 and R_2 are in parallel across a 60-V source. The total line current is 10 A. The current I_1 through R_1 is 4 A. Draw a schematic diagram of the circuit, giving the values of currents I_1 and I_2 and resistances R_1 and R_2 in both branches. How much is the combined equivalent resistance of both branches across the voltage source?
7. Find the R_T for the following groups of branch resistances: (a) 10 Ω and 25 Ω; (b) five 10-kΩ resistances; (c) two 500-Ω resistances; (d) 100 Ω, 200 Ω, and 300 Ω; (e) two 5-kΩ and two 2-kΩ resistances; (f) four 40-kΩ and two 20-kΩ resistances.
8. Find R_3 in Fig. 4-16.

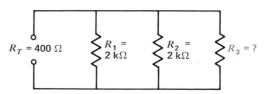

Fig. 4-16 For Prob. 8.

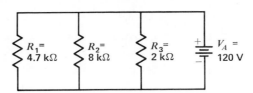

Fig. 4-17 For Prob. 12.

9. Find the total conductance in siemens for the following branches: $G_1 =$ 9000 μS; $G_2 = 7000$ μS; $G_3 = 22,000$ μS.
10. Referring to Fig. 4-11, calculate R_T by combining resistances. Show that this R_T equals $1/G_T$, where G_T is 0.75 S.
11. How much parallel R_x must be connected across a 100-kΩ resistance to reduce R_T to (**a**) 50 kΩ; (**b**) 25 kΩ; (**c**) 10 kΩ?
12. In Fig. 4-17, (**a**) find each branch current and show the direction of electron flow; (**b**) calculate I_T; (**c**) calculate R_T; (**d**) calculate P_1, P_2, P_3, and P_T.

Answers to Practice Problems

4-1 a. 1.5 V
    b. 120 V

4-2 a. 10 V
    b. 1 A
    c. 10 V
    d. 2 A

4-3 a. $I_T = 6$ A
    b. $I_3 = 3$ A

4-4 a. $R_T = 1.57$ MΩ
    b. $R_T = 1.2$ MΩ
    c. $R_T = 5$ Ω

4-5 a. $G_T = 6$ S
    b. $G_T = 0.75$ μS
       $R_T = 1.33$ MΩ

4-6 a. 480 W
    b. 660 W

4-7 a. 120 V
    b. $I_1 = 4$ A

4-8 a. Infinite ohms
    b. Bulbs 1 and 2

4-9 a. 0 Ω
    b. $I_1 = 0$ A

Chapter 5
Series-Parallel Circuits

In many circuits, some components are connected in series to have the same current, while others are in parallel for the same voltage. Such a circuit is used where it is necessary to provide different amounts of current and voltage with one source of applied voltage. In analyzing these circuits, the techniques of series circuits and parallel circuits are applied individually to produce a simplified total circuit.

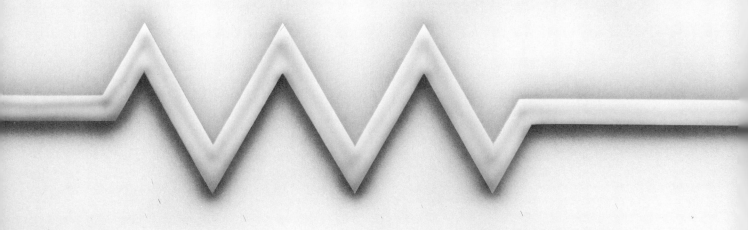

Important terms in this chapter are:

balanced bridge   strings in parallel
banks in series   voltage to chassis ground
chassis-ground connections   voltage divider
grounded tap   Wheatstone bridge

More details are explained in the following sections:

5-1 Finding R_T for Series-Parallel Resistances

In Fig. 5-1, R_1 is in series with R_2. Also, R_3 is in parallel with R_4. However, R_2 is *not* in series with R_3 and R_4. The reason is the branch point A where the current through R_2 divides for R_3 and R_4. As a result, the current through R_3 must be less than the current through R_2. Therefore, R_2 and R_3 cannot be in series because they do not have the same current. For the same reason, R_4 also cannot be in series with R_2.

The wiring is shown in Fig. 5-1a with the schematic diagram in Fig. 5-1b. To find R_T, we add the series resistances and combine the parallel resistances.

In Fig. 5-1c, the 0.5-kΩ R_1 and 0.5-kΩ R_2 in series total 1 kΩ for $R_{1\text{-}2}$. The calculations are

$$0.5 \text{ k}\Omega + 0.5 \text{ k}\Omega = 1 \text{ k}\Omega$$

Also, the 1-kΩ R_3 in parallel with the 1-kΩ R_4 can be combined, for an equivalent resistance of 0.5 kΩ for $R_{3\text{-}4}$, as shown in Fig. 5-1d. The calculations are

$$\frac{1 \text{ k}\Omega}{2} = 0.5 \text{ k}\Omega$$

This parallel $R_{3\text{-}4}$ combination of 0.5 kΩ is then added to the series $R_{1\text{-}2}$ combination for the final R_T value of 1.5 kΩ. The calculations are

$$0.5 \text{ k}\Omega + 1 \text{ k}\Omega = 1.5 \text{ k}\Omega$$

The 1.5 kΩ is the R_T of the entire circuit connected across the source V_T of 1.5 V.

With R_T known to be 1.5 kΩ, we can find I_T in the main line produced by 1.5 V. Then

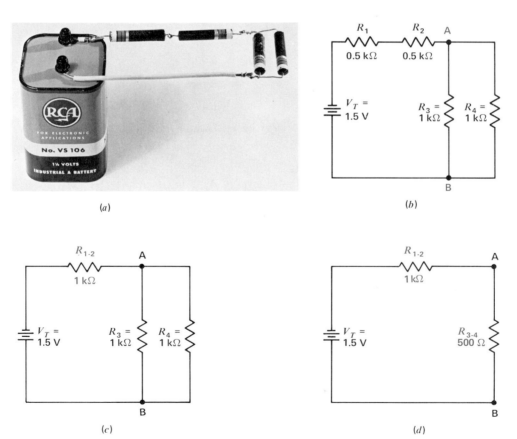

(a)

(b)

(c)

(d)

Fig. 5-1 Series-parallel circuit. (a) Photograph of wiring. (b) Schematic diagram. (c) Resistances R_1 and R_2 in series added for $R_{1\text{-}2}$. (d) Resistances R_3 and R_4 in parallel combined as $R_{3\text{-}4}$.

$$I_T = \frac{V_T}{R_T} = \frac{1.5 \text{ V}}{1.5 \text{ k}\Omega} = 1 \text{ mA}$$

This 1-mA I_T is the current through R_1 and R_2 in Fig. 5-1a and b or $R_{1\text{-}2}$ in Fig. 5-1c.

At branch point A, at the top of the diagram, the 1 mA of electron flow for I_T divides into two branch currents of 0.5 mA each for R_3 and R_4. Since these two branch resistances are equal, I_T divides into two equal parts. At branch point B at the bottom of the diagram, the two 0.5-mA branch currents combine to equal the 1-mA I_T in the main line, returning to the source V_T.

Practice Problems 5-1
Answers at End of Chapter

Refer to Fig. 5-1.
a. Calculate the series R of R_1 and R_2.
b. Calculate the parallel R of R_3 and R_4.
c. Calculate R_T across the source V_T.

5-2 Resistance Strings in Parallel

More details about the voltages and currents in a series-parallel circuit are illustrated by the example in Fig. 5-2. Suppose there are four 120-V 100-W light bulbs to be wired, with a voltage source that produces 240 V. Each bulb needs 120 V for normal brilliance. If the bulbs were connected across the source, each would have the applied voltage of 240 V. This would cause excessive current in all the bulbs that could result in burned-out filaments.

If the four bulbs were connected in series, each would have a potential difference of 60 V, or one-fourth the applied voltage. With too low a voltage, there would be insufficient current for normal operation and the bulbs would not operate at normal brilliance.

However, two bulbs in series across the 240-V line provide 120 V for each filament, which is the normal operating voltage. Therefore, the four bulbs are wired in strings of two in series, with the two strings in parallel across the 240-V source. Both strings have 240 V applied. In each string, two series bulbs divide the applied voltage equally to provide the required 120 V for the filaments.

Another example is illustrated in Fig. 5-3. This circuit has just two parallel branches where one branch includes R_1 in series with R_2. The other branch has just the one resistance R_3. Ohm's law can be applied to each branch.

Branch Currents I_1 and I_2 Each branch current equals the voltage applied across the branch divided by the total resistance in the branch. In branch 1, R_1 and R_2 total $8 + 4 = 12\ \Omega$. With 12 V applied, this branch current I_1 is $^{12}\!/_{12} = 1$ A. Branch 2 has only the 6-Ω R_3. Then I_2 in this branch is $^{12}\!/_{6} = 2$ A.

Series Voltage Drops in a Branch For any one resistance in a string, the current in the string multiplied by the resistance equals the IR voltage drop across that particular resistance. Also, the sum of the series IR drops in the string equals the voltage across the entire string.

Branch 1 is a string with R_1 and R_2 in series. The I_1R_1 drop equals 8 V, while the I_1R_2 drop is 4 V. These drops of 8 and 4 V add to equal the 12 V applied. The voltage across the R_3 branch is also the same 12 V.

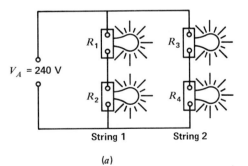
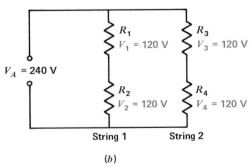

Fig. 5-2 Two identical series strings in parallel. All bulbs have a 120-V 100-W rating. (a) Wiring diagram. (b) Schematic diagram.

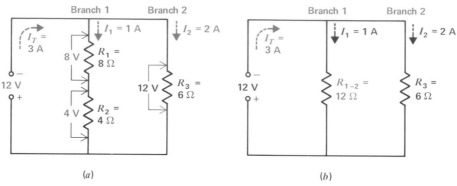

Fig. 5-3 Series string in parallel with another branch. (*a*) Schematic diagram. (*b*) Equivalent circuit.

Calculating I_T   The total line current equals the sum of the branch currents for all the parallel strings. Here I_T is 3 A, equal to the sum of 1 A in branch 1 and 2 A in branch 2.

Calculating R_T   The resistance of the total series-parallel circuit across the voltage source equals the applied voltage divided by the total line current. In Fig. 5-3, R_T equals 12 V/3 A, or 4 Ω. This resistance can also be calculated as 12 Ω in parallel with 6 Ω. For the product divided by the sum, $72/18 = 4$ Ω for the equivalent combined R_T.

Applying Ohm's Law   There can be any number of parallel strings and more than two series resistances in a string. Still, Ohm's law can be used in the same way for the series and parallel parts of the circuit. The series parts have the same current. The parallel parts have the same voltage. Remember that for V/R the R must include all the resistance across the two terminals of V.

Practice Problems 5-2
Answers at End of Chapter

Refer to Fig. 5-3*a*.
a. If I in R_2 were 6 A, what would I in R_1 be?
b. If the source voltage were 18 V, what would V_3 be across R_3?

5-3   Resistance Banks in Series

In Fig. 5-4*a*, the group of parallel resistances R_2 and R_3 is a bank. This is in series with R_1 because the total current of the bank must go through R_1.

The circuit here has R_2 and R_3 in parallel in one bank so that these two resistances will have the same potential difference of 20 V across them. The source applies 24 V, but there is a 4-V drop across R_1.

The two series voltage drops of 4 V across R_1 and 20 V across the bank add to equal the applied voltage of 24 V. The purpose of a circuit like this is to provide the same voltage for two or more resistances in a bank, where the bank voltage must be less than the applied voltage by the amount of IR drop across any series resistance.

To find the resistance of the entire circuit, combine the parallel resistances in each bank and add the series resistance. As shown in Fig. 5-4*b*, the two 10-Ω resistances R_2 and R_3 in parallel are equivalent to 5 Ω. Since the bank resistance of 5 Ω is in series with 1 Ω for R_1, the total resistance is 6 Ω across the 24-V source. Therefore, the main-line current is 24 V/6 Ω, which equals 4 A.

The total line current of 4 A divides into two parts of 2 A each in the parallel resistances R_2 and R_3. Note that each branch current equals the bank voltage divided by the branch resistance. For this bank, $20/10 = 2$ A for each branch.

The branch currents are combined in the line to provide the total 4 A in R_1. This is the same total current flowing in the main line, in the source, into the bank, and out of the bank.

There can be more than two parallel resistances in a bank and any number of banks in series. Still, Ohm's

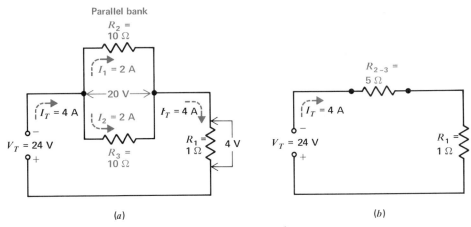

Fig. 5-4 Parallel bank of R_2 and R_3 in series with R_1. (*a*) Schematic diagram. (*b*) Equivalent circuit.

law can be applied the same way to the series and parallel parts of the circuit. The general procedure for circuits of this type is to find the equivalent resistance of each bank and then add all the series resistances.

Practice Problems 5-3
Answers at End of Chapter

Refer to Fig. 5-4*a*.
a. If V_2 across R_2 were 40 V, what would V_3 across R_3 be?
b. If I in R_2 were 4 A, with 4 A in R_3, what would I in R_1 be?

5-4 Resistance Banks and Strings in Series-Parallel

In the solution of such circuits, the most important fact to know is which components are in series with each other and what parts of the circuit are parallel branches. The series components must be in one current path without any branch points. A branch point such as point A or B in Fig. 5-5 is common to two or more current paths. For instance, R_1 and R_6 are *not* in series with each other. They do not have the same current, because the current in R_1 divides at point A into its two component branch currents. Similarly, R_5 is not in series with R_2, because of the branch point B.

To find the currents and voltages in Fig. 5-5, first find R_T in order to calculate the main-line current I_T as

V_T/R_T. In calculating R_T, start reducing the branch farthest from the source and work toward the applied voltage. The reason for following this order is that you cannot tell how much resistance is in series with R_1 and R_2 until the parallel branches are reduced to their equivalent resistance. If no source voltage is shown, R_T can still be calculated from the outside in toward the open terminals where a source would be connected.

To calculate R_T in Fig. 5-5, the steps are as follows:

1. The bank of the 12-Ω R_3 and 12-Ω R_4 in parallel in Fig. 5-5*a* is equal to the 6-Ω R_7 in Fig. 5-5*b*.
2. The 6-Ω R_7 and 4-Ω R_6 in series in the same current path total 10 Ω for R_{13} in Fig. 5-5*c*.
3. The 10-Ω R_{13} is in parallel with the 10-Ω R_5, across the branch points A and B. Their equivalent resistance, then, is the 5-Ω R_{18} in Fig. 5-5*d*.
4. Now the circuit in Fig. 5-5*d* has just the 15-Ω R_1, 5-Ω R_{18}, and 30-Ω R_2 in series. These resistances total 50 Ω for R_T, as shown in Fig. 5-5*e*.
5. With a 50-Ω R_T across the 100-V source, the line current I_T is $^{100}\!/_{50} = 2$ A.

To see the individual currents and voltages, we can use the I_T of 2 A for the equivalent circuit in Fig. 5-5*d*. Now we work from the source V out toward the branches. The reason is that I_T can be used to find the voltage drops in the main line. The IR voltage drops here are:

$$V_1 = I_T R_1 = 2 \times 15 = 30 \text{ V}$$
$$V_{18} = I_T R_{18} = 2 \times 5 = 10 \text{ V}$$
$$V_2 = I_T R_2 = 2 \times 30 = 60 \text{ V}$$

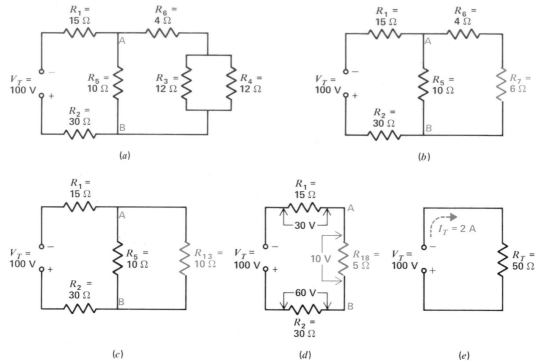

(a)

(b)

(c)

(d)

(e)

Fig. 5-5 Reducing a series-parallel circuit to an equivalent series circuit to find the R_T. (a) Actual circuit. (b) Resistances R_3 and R_4 in parallel equal R_7. (c) Resistances R_7 and R_6 in series equal R_{13}. (d) Resistances R_{13} and R_5 in parallel equal R_{18}. (e) Resistances R_{18}, R_1, and R_2 in series are added for the total circuit resistance of 50 Ω.

The 10-V drop across R_{18} is actually the potential difference between branch points A and B. This means 10 V across R_5 and R_{13} in Fig. 5-5c. The 10 V produces 1 A in the 10-Ω R_5 branch. The same 10 V is also across the R_{13} branch.

Remember that the R_{13} branch is actually the string of R_6 in series with the R_3R_4 bank. Since this branch resistance is 10 Ω, with 10 V across it, the branch current here is 1 A. The 1 A through the 4 Ω of R_6 produces a voltage drop of 4 V. The remaining 6-V IR drop is across the R_3R_4 bank. With 6 V across the 12-Ω R_3, its current is ½ A; the current is also ½ A in R_4.

Tracing all the current paths from the source, the main-line current through R_1 is 2 A. At the branch point A, this current divides into 1 A for R_5 and 1 A for the string with R_6. There is a 1-A branch current in R_6, but it subdivides in the bank, with ½ A in R_3 and ½ A in R_4. At the branch point B, the total bank current of 1 A combines with the 1 A through the R_5 branch, resulting in a 2-A total line current through R_2, the same as through R_1 in the opposite side of the line.

Practice Problems 5-4
Answers at End of Chapter

Refer to Fig. 5-5a.
a. Which R is in series with R_2?
b. Which R is in parallel with R_3?
c. Which R is in series with the R_3R_4 bank?

5-5   Analyzing Series-Parallel Circuits

The circuits in Figs. 5-6 to 5-9 will be solved now. The following principles are illustrated:

1. With parallel strings across the main line, the branch currents and I_T can be found without R_T (see Figs. 5-6 and 5-7).
2. When parallel strings have series resistance in the main line, R_T must be calculated to find I_T, assuming no branch currents are known (see Fig. 5-9).

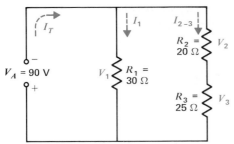

Fig. 5-6 Find all the currents and voltages. See text for solution by calculating the branch currents first.

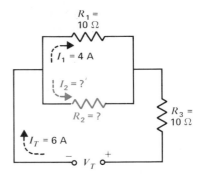

Fig. 5-8 Find R_2 and I_2. See text for solution.

3. The source voltage is applied across the R_T of the entire circuit, producing an I_T that flows only in the main line.
4. Any individual series R has its own IR drop that must be less than the total V_T. In addition, any individual branch current must be less than I_T.

Solution for Fig. 5-6 The problem here is to calculate the branch currents I_1 and I_{2-3}, total line current I_T, and the voltage drops V_1, V_2, and V_3. This order will be used for the calculations, because we can find the branch currents from the 90 V across the known branch resistances.

In the 30-Ω branch of R_1, the branch current is $^{90}/_{30} = 3$ A for I_1. The other branch resistance, with a 20-Ω R_2 and a 25-Ω R_3, totals 45 Ω. This branch current then is $^{90}/_{45} = 2$ A for I_{2-3}. In the main line, I_T is $3 + 2$, which equals 5 A.

For the branch voltages, V_1 must be the same as V_A, equal to 90 V. Or $V_1 = I_1R_1$, which is $3 \times 30 = 90$ V.

In the other branch, the 2-A I_{2-3} flows through the 20-Ω R_2 and the 25-Ω R_3. Therefore, V_2 is $2 \times 20 = 40$ V. Also, V_3 is $2 \times 25 = 50$ V. Note that these 40-V

and 50-V series IR drops in one branch add to equal the 90-V source.

If we want to know R_T, it can be calculated as V_A/I_T. Then 90 V/5 A equals 18 Ω. Or R_T can be calculated by combining the branch resistances of 30 Ω in parallel with 45 Ω. Then R_T is $(30 \times 45)/(30 + 45)$. This answer is $^{1350}/_{75}$, which equals the same value of 18 Ω for R_T.

Solution for Fig. 5-7 To find the applied voltage first, the I_1 branch current is given. This 3-A current through the 10-Ω R_1 produces a 30-V drop V_1 across R_1. The same 3-A current through the 20-Ω R_2 produces 60 V for V_2 across R_2. The 30-V and 60-V drops are in series with each other across the applied voltage. Therefore, V_A equals the sum of $30 + 60$, or 90 V. This 90 V is also across the other branch combining R_3 and R_4 in series.

The other branch current I_2 in Fig. 5-7 must be 4 A, equal to the 7-A I_T minus the 3-A I_1. With 4 A for I_2, the voltage drop across the 12-Ω R_3 equals 48 V for V_3. Then the voltage across R_4 is $90 - 48$, or 42 V for V_4, as the sum of V_3 and V_4 must equal the applied 90 V.

Finally, with 42 V across R_4 and 4 A through it, this resistance equals $^{42}/_4$, or 10.5 Ω. Note that 10.5 Ω for R_4 added to the 12 Ω of R_3 equals 22.5 Ω, which allows $^{90}/_{22.5}$ or a 4-A branch current for I_2.

Solution for Fig. 5-8 The division of branch currents also applies to Fig. 5-8, but the main principle here is that the voltage must be the same across R_1 and R_2 in parallel. For the branch currents, I_2 is 2 A, equal to the 6-A I_T minus the 4-A I_1. The voltage across the 10-Ω R_1 is 4×10, or 40 V. This same voltage is also

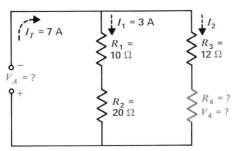

Fig. 5-7 Find the applied voltage V_A, V_4, and R_4. See text for solution by calculating I_2 and the branch voltage.

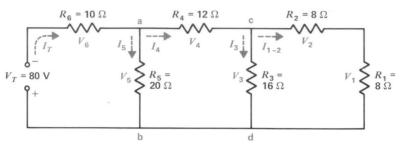

Fig. 5-9 Find all currents and voltages. See text for solution in which R_T and then I_T are calculated to find V_6 first.

across R_2. With 40 V across R_2 and 2 A through it, R_2 equals $^{40}/_2$ or 20 Ω.

If we want to find V_T in Fig. 5-8, it can be calculated as 100 V. The 6-A I_T through R_3 produces a voltage drop of 60 V for V_3. Also, the voltage across the parallel bank with R_1 and R_2 has been calculated as 40 V. This 40 V across the bank in series with 60 V across R_3 totals 100 V for the applied voltage.

Solution for Fig. 5-9

In order to find all the currents and voltage drops, we need R_T to calculate I_T through R_6 in the main line. Combining resistances for R_T, we start with R_1 and R_2 and work in toward the source. Add the 8-Ω R_1 and 8-Ω R_2 in series with each other for 16 Ω. This 16 Ω combined with the 16-Ω R_3 in parallel equals 8 Ω between points c and d. Add this 8 Ω to the series 12-Ω R_4 for 20 Ω. This 20 Ω with the parallel 20-Ω R_5 equals 10 Ω between points a and b. Add this 10 Ω in series with the 10-Ω R_6, to make R_T of 20 Ω for the entire series-parallel circuit.

Current I_T in the main line is V_T/R_T, or $^{80}/_{20}$, which equals 4 A. This 4-A I_T flows through the 10-Ω R_6, producing a 40-V IR drop for V_6.

Now that we know I_T and V_6 in the main line, we use these values to calculate all the other voltages and currents. Start from the main line, where we know the current, and work outward from the source. To find V_5, the IR drop of 40 V for V_6 in the main line is subtracted from the source voltage. The reason is that V_5 and V_6 must add to equal the 80 V of V_T. Then V_5 is 80 − 40 = 40 V.

Voltages V_5 and V_6 happen to be equal at 40 V each. They split the 80 V in half because the 10-Ω R_6 equals the combined resistance of 10 Ω between branch points a and b.

With V_5 known to be 40 V, then I_5 through the 20-Ω R_5 is $^{40}/_{20}$ = 2 A. Since I_5 is 2 A and I_T is 4 A, I_4 must

be 2 A also, equal to the difference between I_T and I_5. At the branch point a, the 4-A I_T divides into 2 A through R_5 and 2 A through R_4.

The 2-A I_4 through the 12-Ω R_4 produces an IR drop equal to 2 × 12 = 24 V for V_4. It should be noted now that V_4 and V_3 must add to equal V_5. The reason is that both V_5 and the path with V_4 and V_3 are across the same two points ab or ad. Since the potential difference across any two points is the same regardless of the path, $V_5 = V_4 + V_3$. To find V_3 now, we can subtract the 24 V of V_4 from the 40 V of V_5. Then 40 − 24 = 16 V for V_3.

With 16 V for V_3 across the 16-Ω R_3, its current I_3 is 1 A. Also I_{1-2} in the branch with R_1 and R_2 is equal to 1 A. The 2-A I_4 into branch point c divides into the two equal branch currents of 1 A each because of the equal branch resistances.

Finally, with 1 A through the 8-Ω R_2 and 8-Ω R_1, their voltage drops are $V_2 = 8$ V and $V_1 = 8$ V. Note that the 8 V of V_1 in series with the 8 V of V_2 add to equal the 16-V potential difference V_3 between points c and d.

All the answers for the solution of Fig. 5-9 are summarized below:

$R_T = 20$ Ω	$I_T = 4$ A	$V_6 = 40$ V
$V_5 = 40$ V	$I_5 = 2$ A	$I_4 = 2$ A
$V_4 = 24$ V	$V_3 = 16$ V	$I_3 = 1$ A
$I_{1-2} = 1$ A	$V_2 = 8$ V	$V_1 = 8$ V

Practice Problems 5-5
Answers at End of Chapter

a. In Fig. 5-6, which R is in series with R_2?
b. In Fig. 5-6, which R is across V_A?
c. In Fig. 5-7, how much is I_2?
d. In Fig. 5-8, how much is V_3?

5-6 Wheatstone[1] Bridge

A bridge circuit has four terminals, two for input voltage and two for output. The purpose is to have a circuit where the voltage drops can be balanced to provide zero voltage across the output terminals, with voltage applied across the input. In Fig. 5-10 the input terminals are C and D, with output terminals A and B.

The bridge circuit has many uses for comparison measurements. In the Wheatstone bridge, an unknown resistance R_X is balanced against a standard accurate resistor R_S for precise measurement of resistance.

In Fig. 5-10, S_1 applies battery voltage to the four resistors in the bridge. To balance the bridge, the value of R_S is varied. Balance is indicated by zero current in the galvanometer G. Finally, S_2 is a spring switch that is closed just to check the meter reading.

The reason for zero current in the meter can be seen by analysis of the voltage drops across the resistors. R_S in series with R_X forms a voltage divider across V_T; the parallel string of R_1 in series with R_2 is also a voltage divider across the same source. When the voltage division is in the same ratio for both strings, the voltage drop across R_S equals the voltage across R_2. Also, the voltage across R_X then equals the voltage across R_1. In this case, points A and B must be at the same potential. The difference of potential across the meter then must be zero, and there is no deflection.

At balance, the equal voltage ratios in the two branches of the Wheatstone bridge can be stated as

---

[1]Sir Charles Wheatstone (1802–1875), English physicist and inventor.

$$\frac{I_A R_X}{I_A R_S} = \frac{I_B R_1}{I_B R_2} \qquad or \qquad \frac{R_X}{R_S} = \frac{R_1}{R_2}$$

Note that I_A and I_B cancel. Now, inverting R_S to the right side of the equation,

$$R_X = R_S \times \frac{R_1}{R_2} \tag{5-1}$$

Usually, the total resistance of R_1 and R_2 is fixed, but any desired ratio can be chosen by moving point B on the ratio arm. The bridge is balanced by varying R_S for zero current in the meter. At balance, then, the value of R_X can be determined by multiplying R_S by the ratio of R_1/R_2. As an example, if the ratio is $1/100$ and R_S is 248 Ω, the value of R_X equals 248×0.01, or 2.48 Ω.

The balanced bridge circuit can be analyzed as simply two series resistance strings in parallel when the current is zero through the meter. Without any current between A and B, this path is effectively open. When current flows through the meter path, however, the bridge circuit must be analyzed by Kirchhoff's laws or network theorems, as described in Chaps. 9 and 10.

Practice Problems 5-6
Answers at End of Chapter

a. A bridge circuit has how many *pairs* of terminals?
b. In Fig. 5-10, how much is V_{AB} at balance?

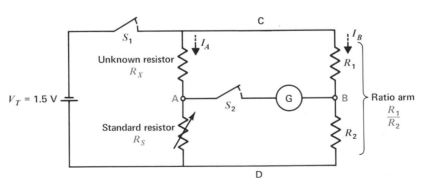

Fig. 5-10 Wheatstone-bridge circuit.

5-7   Chassis-Ground Connections

In the wiring of practical circuits, one side of the voltage source is usually grounded. For the 120-V ac power line in residential wiring, the ground is actually earth ground, usually by connection to a metal cold-water pipe. For electronic equipment, the ground just indicates a metal chassis, which is used as a common return for connections to the source. With printed wiring on a plastic board instead of a metal chassis, a conducting path around the entire board is used as a common return for chassis ground. The chassis ground may or may not be connected to earth ground. In either case the grounded side is called the "cold side" or "low side" of the applied voltage, while the ungrounded side is the "hot side" or "high side."

Grounding One Side of the Source Voltage

Three examples are shown in Fig. 5-11. In Fig. 5-11a, one side of the 120-V ac power line is grounded. Note the symbol ⏚ for earth ground. This symbol also indicates a chassis ground that is connected to one side of the voltage source. In electronic equipment, black wire is generally used for chassis ground returns and red wire for the high side of the voltage source. See Table F-1 in App. F.

In Fig. 5-11b and c the 12-V battery is used as an example of a voltage source connected to chassis ground but not to earth. For instance, in an automobile one side of the battery is connected to the metal frame of the car. In Fig. 5-11b, the negative side is grounded, while in Fig. 5-11c the positive side is grounded. Some people have the idea that ground must always be negative, but this is not necessarily so.

The reason for connecting one side of the 120-V ac power line to earth ground is to reduce the possibility of electric shock. However, chassis ground in electronic equipment is mainly a common-return connection. Where the equipment operates from the power line, the metal chassis should be at ground potential, not connected to the hot side of the ac outlet. This connection reduces the possibility of electric shock from the chassis. Also, hum from the power line is reduced in audio, radio, and television equipment.

Practice Problems 5-7
Answers at End of Chapter

a. In Fig. 5-11b, give the voltage to ground with polarity.
b. Do the same for Fig. 5-11c.

5-8   Voltages Measured to Chassis Ground

When a circuit has the chassis as a common return, we generally measure the voltages with respect to chassis. Let us consider the voltage divider without any ground in Fig. 5-12a, and then analyze the effect of grounding different points on the divider. It is important to realize that this circuit operates the same way with or without the ground. The only factor that changes is the reference point for measuring the voltages.

In Fig. 5-12a, the three 10-Ω resistances R_1, R_2, and R_3 divide the 30-V source equally. Then each voltage

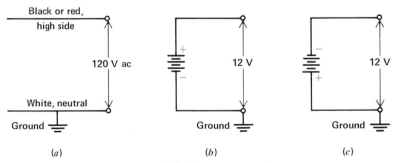

(a)   (b)   (c)

Fig. 5-11 Grounding one side of a voltage source. (a) The ac power line. (b) Negative side of battery connected to chassis ground. (c) Positive side of battery connected to chassis ground.

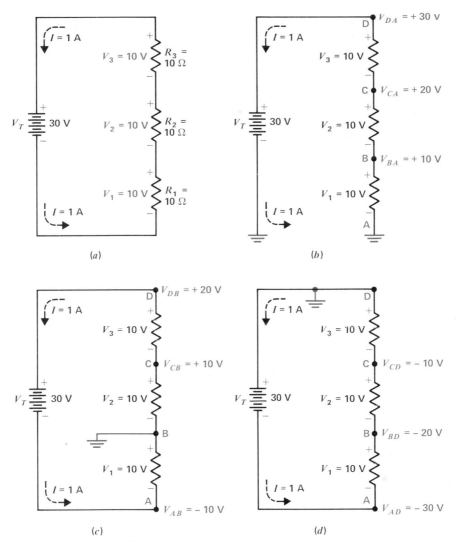

Fig. 5-12 Voltages to chassis ground. (*a*) Voltage divider without ground. (*b*) With negative side of source V_T grounded, all voltages are positive to the chassis ground. (*c*) Positive and negative voltages with respect to the ground at point B. (*d*) With positive side of source grounded, all voltages are negative to chassis ground.

drop is $^{30}/_3 = 10$ V for V_1, V_2, and V_3. The polarity is positive at the top and negative at the bottom, the same as V_T.

If we want to consider the current, I is $^{30}/_{30} = 1$ A. Each IR drop is $1 \times 10 = 10$ V for V_1, V_2, and V_3.

Positive Voltages to Negative Ground

In Fig. 5-12*b*, the negative side of V_T is grounded and the bottom end of R_1 is also grounded to complete the cir-

cuit. The ground is at point A. Note that the individual voltages V_1, V_2, and V_3 are still 10 V each. Also the current is still 1 A. The direction is also the same, from the negative side of V_T, through the metal chassis, to the bottom end of R_1. The only effect of the chassis ground here is to provide a conducting path from one side of the source to one side of the load.

With the ground in Fig. 5-12*b*, though, it is useful to consider the voltages with respect to chassis ground. In other words, the ground at point A will now be the reference for all voltages. When a voltage is indicated for only one point in a circuit, generally the other point is assumed to be chassis ground. We must have two points for a potential difference.

Let us consider the voltages at points B, C, and D. The voltage at B to ground is V_{BA}. This double subscript notation shows that we measure at B with respect to A. In general, the first letter indicates the point of measurement and the second letter is the reference point.

Then V_{BA} is +10 V. The positive sign is used here to emphasize the polarity. The value of 10 V for V_{BA} is the same as V_1 across R_1 because points B and A are across R_1. However, V_1 as the voltage across R_1 really cannot be given any polarity without a reference point.

When we consider the voltage at C, then, V_{CA} is +20 V. This voltage equals $V_1 + V_2$, connected with series-aiding polarities. Also, for point D at the top, V_{DA} is +30 V for $V_1 + V_2 + V_3$.

Positive and Negative Voltages to a Grounded Tap

In Fig. 5-12c point B in the divider is grounded. The purpose is to have the divider supply negative and positive voltages with respect to chassis ground. The negative voltage here is V_{AB}, which equals −10 V. This value is the same 10 V of V_1, but V_{AB} is the voltage at the negative end A with respect to the positive end B. The other voltages in the divider are $V_{CB} = +10$ V and $V_{DB} = +20$ V.

We can consider the ground at B as a dividing point for positive and negative voltages. For all points toward the positive side of V_T, any voltage is positive to ground. Going the other way, at all points toward the negative side of V_T, any voltage is negative to ground.

Negative Voltages to Positive Ground

In Fig. 5-12d, point D at the top of the divider is grounded, which is the same as grounding the positive side of the source V_T. The voltage source here is inverted, compared with Fig. 5-12b, as the opposite side is grounded. In Fig. 5-12d, all the voltages on the divider are negative to ground. Here, $V_{CD} = -10$ V, while $V_{BD} = -20$ V and $V_{AD} = -30$ V. Any point in the circuit must be more negative than the positive terminal of the source, even when this terminal is grounded.

Practice Problems 5-8
Answers at End of Chapter

Refer to Fig. 5-12c and give the voltage with polarity for
a. A to ground.
b. B to ground.
c. D to ground.
d. V_{DA} across V_T.

5-9   Opens and Shorts in Series-Parallel Circuits

A short circuit has practically zero resistance. Its effect, therefore, is to allow excessive current. An open circuit has the opposite effect because an open circuit has infinitely high resistance with practically zero current. Furthermore, in series-parallel circuits an open or short circuit in one path changes the circuit for the other resistances. For example, in Fig. 5-13, the series-parallel circuit in Fig. 5-13a becomes a series circuit with only R_1 when there is a short circuit between terminals A and B. As an example of an open circuit, the series-parallel circuit in Fig. 5-14a becomes a series circuit with just R_1 and R_2 when there is an open circuit between terminals C and D.

Effect of a Short Circuit   We can solve the series-parallel circuit in Fig. 5-13a in order to see the effect of the short circuit. For the normal circuit, with S_1 open, R_2 and R_3 are in parallel. Although R_3 is drawn horizontally, both ends are across R_2. The switch S_1 has no effect as a parallel branch here because it is open.

The combined resistance of the 80-Ω R_2 in parallel with the 80-Ω R_3 is equivalent to 40 Ω. This 40 Ω for the bank resistance is in series with the 10-Ω R_1. Then R_T is 40 + 10 = 50 Ω.

In the main line, I_T is $100/50 = 2$ A. Then V_1 across the 10-Ω R_1 in the main line is $2 \times 10 = 20$ V. The remaining 80 V is across R_2 and R_3 as a parallel bank. As a result, $V_2 = 80$ V and $V_3 = 80$ V.

Now consider the effect of closing switch S_1. A closed switch has zero resistance. Not only is R_2 short-

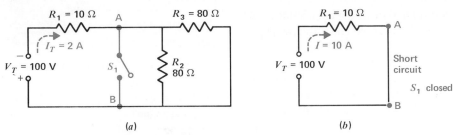

Fig. 5-13 Effect of a short circuit with series-parallel connections. (a) Normal circuit with S_1 open. (b) Circuit with short between A and B when S_1 is closed; R_2 and R_3 are short-circuited.

circuited, but R_3 in the bank with R_2 is also short-circuited. The closed switch short-circuits everything connected between terminals A and B. The result is the series circuit shown in Fig. 5-13b.

Now the 10-Ω R_1 is the only opposition to current. I equals V/R_1, which is $^{100}/_{10} = 10$ A. This 10 A flows through R_1, the closed switch, and the source. With 10 A through R_1, instead of its normal 2 A, the excessive current can cause excessive heat in R_1. There is no current through R_2 and R_3, as they are short-circuited out of the path for current.

Effect of an Open Circuit

Figure 5-14a shows the same series-parallel circuit as Fig. 5-13a, except that switch S_2 is used now to connect R_3 in parallel with R_2. With S_2 closed for normal operation, all currents and voltages have the values calculated for the series-parallel circuit. However, let us consider the effect of

opening S_2, as shown in Fig. 5-14b. An open switch has infinitely high resistance. Now there is an open circuit between terminals C and D. Furthermore, because R_3 is in the open path, its 80 Ω cannot be considered in parallel with R_2.

The circuit with S_2 open in Fig. 5-14b is really the same as having just R_1 and R_2 in series with the 100-V source. The open path with R_3 has no effect as a parallel branch. The reason is that no current flows through R_3.

We can consider R_1 and R_2 in series as a voltage divider, where each IR drop is proportional to its resistance. The total series R is $80 + 10 = 90$ Ω. The 10-Ω R_1 is $^{10}/_{90}$ or $^{1}/_{9}$ of the total R and the applied V_T. Then V_1 is $^{1}/_{9} \times 100$ V $= 11$ V and V_2 is $^{8}/_{9} \times 100$ V $= 89$ V, approximately. The 11-V drop for V_1 and 89-V drop for V_2 add to equal the 100 V of the applied voltage.

Note that V_3 is zero. Without any current through R_3, it cannot have any voltage drop.

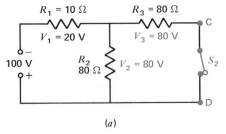

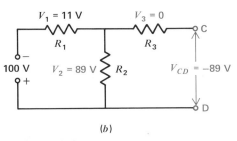

Fig. 5-14 Effect of an open in a series-parallel circuit. (a) Normal circuit with S_2 closed. (b) Series circuit with R_1 and R_2 when S_2 is open. Resistance R_3 in the open path has no current and zero IR voltage drop.

99

Furthermore, the voltage across the open terminals C and D is the same 89 V as the potential difference V_2 across R_2. Since there is no voltage drop across R_3, terminal C has the same potential as the top terminal of R_2. Terminal D is directly connected to the bottom end of resistor R_2. Therefore, the potential difference from terminal C to terminal D is the same 89 V that appears across resistor R_2.

Practice Problems 5-9
Answers at End of Chapter

a. In Fig. 5-13, the short circuit increases I_T from 2 A to what value?

b. In Fig. 5-14, the open branch reduces I_T from 2 A to what value?

Summary

1. Table 5-1 summarizes the main characteristics of series and parallel circuits. In circuits combining series and parallel connections, the components in one current path without any branch points are in series; the parts of the circuit connected across the same two branch points are in parallel.
2. To calculate R_T in a series-parallel circuit with R in the main line, combine resistances from the outside back toward the source.
3. Chassis ground is commonly used as a return connection to one side of the source voltage. Voltages measured to chassis ground can have either negative or positive polarity.

Table 5-1. Comparison of Series and Parallel Circuits

Series circuit	Parallel circuit
Current the same in all components	Voltage the same across all branches
V across each series R is $I \times R$	I in each branch R is V/R
$V_T = V_1 + V_2 + V_3 + \cdots + $ etc.	$I_T = I_1 + I_2 + I_3 + \cdots + $ etc.
$R_T = R_1 + R_2 + R_3 + \cdots + $ etc.	$G_T = G_1 + G_2 + G_3 + \cdots + $ etc.
R_T must be more than the largest individual R	R_T must be less than the smallest branch R
$P_T = P_1 + P_2 + P_3 + \cdots + $ etc.	$P_T = P_1 + P_2 + P_3 + \cdots + $ etc.
Applied voltage is divided into IR voltage drops	Main-line current is divided into branch currents
The largest IR drop is across the largest series R	The largest branch I is in the smallest parallel R
Open in one component causes entire circuit to be open	Open in one branch does not prevent I in other branches

4. When the potential is the same at the two ends of a resistance, its voltage is zero. Or if no current flows through a resistance, it cannot have any *IR* voltage drop.

Self-Examination
Answers at Back of Book

Choose (a), (b), (c), or (d).

1. In the series-parallel circuit in Fig. 5-1*b*: (a) R_1 is in series with R_3; (b) R_2 is in series with R_3; (c) R_4 is in parallel with R_3; (d) R_1 is in parallel with R_3.
2. In the series-parallel circuit in Fig. 5-2*b*: (a) R_1 is in parallel with R_3; (b) R_2 is in parallel with R_4; (c) R_1 is in series with R_2; (d) R_2 is in series with R_4.
3. In the series-parallel circuit in Fig. 5-5, the total of all the branch currents into branch point A and out of branch point B equals (a) ½ A; (b) 1 A; (c) 2 A; (d) 4 A.
4. In the circuit in Fig. 5-2 with four 120-V 100-W light bulbs, the resistance of one bulb equals (a) 72 Ω; (b) 100 Ω; (c) 144 Ω; (d) 120 Ω.
5. In the series-parallel circuit in Fig. 5-4*a*: (a) R_2 is in series with R_3; (b) R_1 is in series with R_3; (c) the equivalent resistance of the R_2R_3 bank is in parallel with R_1; (d) the equivalent resistance of the R_2R_3 bank is in series with R_1.
6. In a series circuit with unequal resistances: (a) the lowest *R* has the highest *V*; (b) the highest *R* has the highest *V*; (c) the lowest *R* has the most *I*; (d) the highest *R* has the most *I*.
7. In a parallel bank with unequal branch resistances: (a) the current is highest in the highest *R*; (b) the current is equal in all the branches; (c) the voltage is highest across the lowest *R*; (d) the current is highest in the lowest *R*.
8. In Fig. 5-14, with S_2 open, R_T equals (a) 90 Ω; (b) 100 Ω; (c) 50 Ω; (d) 10 Ω.
9. In Fig. 5-12*c*, V_{DA} equals (a) +10 V; (b) −20 V; (c) −30 V; (d) +30 V.
10. In the Wheatstone bridge of Fig. 5-10, at balance: (a) $I_A = 0$; (b) $I_B = 0$; (c) $V_2 = 0$; (d) $V_{AB} = 0$.

Essay Questions

1. In a series-parallel circuit, how can you tell which resistances are in series with each other and which are in parallel?
2. Draw a schematic diagram showing two resistances in a bank that is in series with one resistance.
3. Draw a diagram showing how to connect three resistances of equal value so that the combined resistance will be 1½ times the resistance of one unit.
4. Draw a diagram showing two strings in parallel across a voltage source, where each string has three series resistances.
5. Explain why components are connected in series-parallel, showing a circuit as an example of your explanation.
6. Give two differences between a short circuit and an open circuit.
7. Explain the difference between voltage division and current division.
8. Show an example where a voltage is negative with respect to chassis ground.

9. Draw a circuit with nine 40-V 100-W bulbs connected to a 120-V source.
10. (a) Two 10-Ω resistors are in series with a 100-V source. If a third 10-Ω R is added in series, explain why I will decrease. (b) The same two 10-Ω resistors are in parallel with the 100-V source. If a third 10-Ω R is added in parallel, explain why I_T will increase.

Problems

Answers to Odd-Numbered Problems at Back of Book

1. Refer to Fig. 5-1. (a) Calculate the total resistance of the circuit if all resistances are 10 Ω. (b) How much is the main-line current if V_T equals 100 V?
2. In Fig. 5-2, calculate the total power supplied by the source for the four 100-W bulbs.
3. Refer to the diagram in Fig. 5-15. (a) Why is R_1 in series with R_3 but not with R_2? (b) Find the total circuit resistance across the battery.

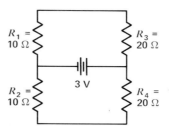

Fig. 5-15 For Prob. 3.

4. Two 60-Ω resistances R_1 and R_2 in parallel require 60 V across the bank with 1 A through each branch. Show how to connect a series resistance R_3 in the main line to drop an applied voltage of 100 V to 60 V across the bank. (a) How much is the required voltage across R_3? (b) How much is the required current through R_3? (c) How much is the required resistance of R_3? (d) If R_3 opens,

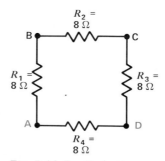

Fig. 5-16 For Prob. 5.

how much is the voltage across R_1 and R_2? **(e)** If R_1 opens, what are the voltages across R_2 and R_3?

5. Refer to the diagram in Fig. 5-16. **(a)** Calculate R across points AD. **(b)** How much is R across points AD with R_4 open?

6. Show how to connect four 100-Ω resistances in a series-parallel circuit with a combined resistance equal to 100 Ω. **(a)** If the combination is connected across a 100-V source, how much power is supplied by the source? **(b)** How much power is dissipated in each resistance?

7. The following four resistors are in series with a 32-V source: R_1 is 24 Ω, R_2 is 8 Ω, R_3 is 72 Ω, and R_4 is 240 Ω. **(a)** Find the voltage drop across each resistor. **(b)** Calculate the power dissipated in each resistor. **(c)** Which resistor has the most voltage drop? **(d)** Which resistor dissipates the most power?

8. The same four resistors are in parallel with the 32-V source. **(a)** Find the branch current in each resistor. **(b)** Calculate the power dissipated in each resistor. **(c)** Which resistor has the most branch current? **(d)** Which resistor dissipates the most power?

9. Find R_1 and R_2 for a voltage divider that takes 5 mA from a 20-V source, with 5 V across R_2.

10. Refer to Fig. 9-2 in Chap. 9 on page 171. Show the calculations for R_T, I_T, and each of the individual voltages and currents.

11. In the Wheatstone-bridge circuit of Fig. 5-17, find each voltage, label polarity, and calculate R_X. The bridge is balanced.

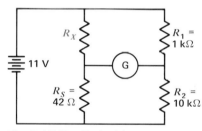

Fig. 5-17 For Prob. 11.

12. In Fig. 5-18, find each V and I for the four resistors.

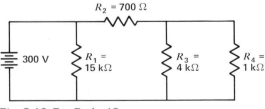

Fig. 5-18 For Prob. 12.

13. In Fig. 5-19, calculate R_T.

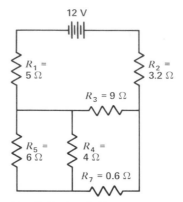

Fig. 5-19 For Prob. 13.

14. In Fig. 5-20, find V_6.

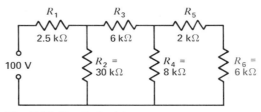

Fig. 5-20 For Prob. 14.

15. Refer to Fig. 5-21. (**a**) Calculate V_2. (**b**) Find V_2 when R_3 is open.
16. In Fig. 5-22, find I and V for the five resistors and calculate V_T.

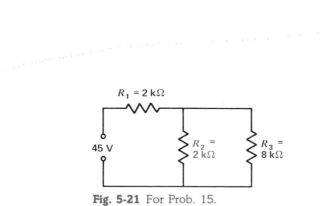

Fig. 5-21 For Prob. 15.

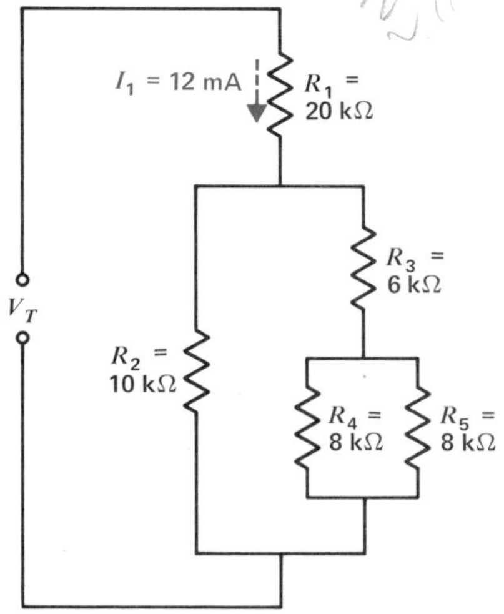

Fig. 5-22 For Prob. 16.

17. Refer to Fig. 5-23. (**a**) Find V_1, V_2, V_3, I_1, I_2, I_3, and I_T in the circuit as shown. (**b**) Now connect point G to ground. Give the voltages, with polarity, at terminals A, B, and C with respect to ground. In addition, give the values of I_1, I_2, I_3, and I_T with point G grounded.
18. In Fig. 5-24, give the voltages at points A, B, and C with polarity to ground when (**a**) point A is grounded; (**b**) point B is grounded; (**c**) point C is grounded.

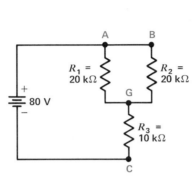

Fig. 5-23 For Prob. 17.

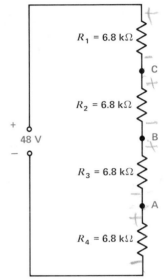

Fig. 5-24 For Prob. 18.

Answers to Practice Problems

5-1 a. $R = 1$ kΩ
   b. $R = 0.5$ kΩ
   c. $R_T = 1.5$ kΩ

5-2 a. $I = 6$ A
   b. $V_3 = 18$ V

5-3 a. $V_3 = 40$ V
   b. $I = 8$ A

5-4 a. R_1
   b. R_4
   c. R_6

5-5 a. R_3
   b. R_1

5-5 c. $I_2 = 4$ A
   d. $V_3 = 60$ V

5-6 a. Two
   b. 0 V

5-7 a. $+12$ V
   b. -12 V

5-8 a. -10 V
   b. 0 V
   c. $+20$ V
   d. $+30$ V

5-9 a. $I = 10$ A
   b. $I = 1.1$ A

Chapter 6
Resistors

Although it might seem that resistance has a disadvantage in reducing the current in a circuit, actually resistors are probably the most common components in electronic equipment. A resistor is manufactured with a specific value of ohms for R. The most common resistors are carbon-composition and wire-wound.

The purpose of using a resistor in a circuit is either to reduce I to a specific value or to provide a desired IR voltage drop. For an example, a series resistor in the output circuit of a transistor amplifier accom-

plishes both of these functions. First, the value of R affects the amount of I in the transistor. Furthermore, the IR voltage drop provides a sample of the I external to the transistor, so that the amplified voltage can be connected to the next circuit. Another feature of resistance is that the effect is the same for dc and ac circuits.

Important terms in this chapter are:

carbon-composition   potentiometer
cermet   preferred value
color code   rheostat
decade box   taper
film type   tolerance
noisy controls   wire-wound

More details are explained in the following sections:

6-1 Types of Resistors
6-2 Resistor Color Coding
6-3 Variable Resistors
6-4 Rheostats and Potentiometers
6-5 Power Rating of Resistors
6-6 Choosing the Resistor for a Circuit
6-7 Series and Parallel Combinations of Resistors
6-8 Resistor Troubles

6-1   Types of Resistors

The two main characteristics of a resistor are its resistance R in ohms and its power rating in watts, W. Resistors arc available in a very wide range of R values, from a fraction of an ohm to many megohms. The power rating may be as high as several hundred watts or as low as $\frac{1}{10}$ W.

The R is the resistance value required to provide the desired I or IR voltage drop. Also important is the wattage rating because it specifies the maximum power the resistor can dissipate without excessive heat. Dissipation means that the power is wasted as I^2R loss, since the resultant heat is not used. Too much heat can make the resistor burn. The wattage rating of the resistor is generally more than the actual power dissipation, as a safety factor.

Wire-wound resistors are used where the power dissipation is about 5 W or more. Typical wire-wound resistors are shown in Fig. 6-1. For 2 W or less, carbon resistors are preferable because they are small and cost less. Between 2 and 5 W, combinations of carbon resistors can be used. Also, small wire-wound resistors are available in a 3- or 4-W rating.

Most common in electronic equipment are carbon resistors with a power rating of 1 W or less. Typical carbon-composition resistors arc shown in Fig. 6-2. Figure 6-3 shows a group of resistors to be mounted on a printed-circuit (PC) board. The resistors can be inserted automatically by machine.

Resistors with higher R values usually have smaller wattage ratings because they have less current. As an example, a common value is 1 MΩ at $\frac{1}{4}$ W, for a resis-

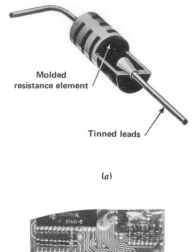

(a)

(b)

Fig. 6-2 Carbon-composition resistors. (a) Internal construction. Length in $\frac{3}{4}$ in., without leads, for 1-W power rating. Color strips give R value in ohms. Tinned leads have coating of solder. (b) Group of resistors mounted on printed-circuit (PC) board.

tor only $\frac{1}{2}$ in. long. The less the power rating, the smaller the actual physical size of the resistor. However, the resistance value is not related to physical size.

Wire-Wound Resistors   In this construction, a special type of wire called *resistance wire* is wrapped around an insulating core, as shown in Fig. 6-2. The length of wire used and its specific resistivity determine the R of the unit. Types of resistance wire include tungsten and manganin, as explained in Chap. 11, "Conductors and Insulators." The insulated core is commonly porcelain, a phenolic material like Bakelite, cement, or just plain pressed paper. Bare wire is used, but the entire unit is generally encased in an insulating material.

Since they are generally for high-current applications with low resistance and appreciable power, wire-wound resistors are available in wattage ratings from 5 W up to 100 W or more. The resistance can be less than 1 Ω up to several thousand ohms.

In addition, wire-wound resistors are used where accurate, stable resistance values are necessary. Examples are precision resistors for the function of an ammeter shunt or a precision potentiometer to adjust for an exact amount of R.

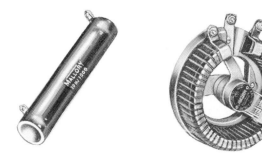

Fig. 6-1 Wire-wound resistors with 50-W power rating, (a) Fixed R, 5 in. long. (b) Variable R, 2 in. diameter. (P. R. Mallory)

Fig. 6-3 Typical carbon-type fixed resistors. Leads are cut and formed for insertion into holes with 0.5-in. spacing on PC board.

Carbon-Composition Resistors

This type of resistor is made of finely divided carbon or graphite mixed with a powdered insulating material as a binder, in the proportions needed for the desired R value. As shown in Fig. 6-1*a*, the resistor element is enclosed in a plastic case for insulation and mechanical strength. Joined to the two ends of the carbon resistance element are metal caps with leads of tinned copper wire for soldering the connections into a circuit. These are called *axial leads* because they come straight out from the ends.

Carbon resistors are commonly available in R values of 1 Ω to 20 MΩ. Examples are 10 Ω, 220 Ω, 4.7 kΩ, and 68 kΩ. The power rating is generally ¹⁄₁₀, ⅛, ¼, ½, 1, or 2 W.

Film-Type Resistors

There are two kinds. The carbon-film type has a thin coating around an insulator. Metal-film resistors have a spiral around a ceramic substrate (Fig. 6-4). Their advantage is more precise R values. The film-type resistors use metal end caps for the terminal leads, which makes the ends a little higher than the body.

Cermet Resistors

These have a carbon coating fired onto a solid ceramic substrate. The purpose is to have more precise R values and greater stability with heat. They are often made in a small square with leads to fit a printed-circuit (PC) board.

Fusible Resistors

This type is a wire-wound resistor made to burn open easily when the power rating is exceeded. It then serves the dual functions of a fuse and a resistor to limit the current.

Nonlinear Resistors

All the resistor types considered here are linear, meaning that they follow the Ohm's-law relation $I = V/R$. In some applications, special characteristics are useful. One example is a *thermistor*, which increases its R with higher temperature. Another type is the *varistor*, which is a device whose R depends on the applied voltage.

Practice Problems 6-1
Answers at End of Chapter

Answer True or False.
a. An R of 10 Ω with a 10-W rating would be a wire-wound resistor.
b. An R of 12 kΩ with a 1-W rating would be a carbon resistor.
c. Axial leads are not used for carbon resistors.

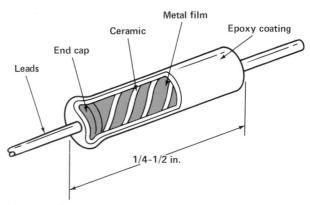

Fig. 6-4 Construction of metal-film resistor. (*Stackpole Components Co.*)

6-2   Resistor Color Coding

Because carbon resistors are small physically, they are color-coded to mark their R value in ohms. The basis of this system is the use of colors for numerical values, as listed in Table 6-1. In memorizing the colors, note that the darkest colors, black and brown, are for the lowest numbers, zero and one, through lighter colors to white for nine. The color coding is standardized by the Electronic Industries Association (EIA). These colors are also used for small capacitors, as summarized in App. G on all the color codes.

Resistance Color Stripes   The use of bands or stripes is the most common system for color-coding carbon resistors, as shown in Fig. 6-5. Color stripes are printed at one end of the insulating body, which is usually tan. Reading from left to right, the first band close to the edge gives the first digit in the numerical value of R. The next band marks the second digit. The third band is the decimal multiplier, which gives the number of zeroes after the two digits.

In Fig. 6-6a, the first stripe is red for 2 and the next stripe is green for 5. The red multiplier in the third stripe means add two zeroes to 25, or ''this multiplier is 10^2.'' The result can be illustrated as follows:

Red   Green   Red
 ↓   ↓   ↓
 2   5   ×   100   = 2500

Therefore, this R value is 2500 Ω.

The example in Fig. 6-6b illustrates that black for the third stripe just means ''do not add any zeroes to the first two digits.'' Since this resistor has red, green, and black stripes, the R value is 25 Ω.

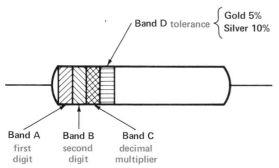

Fig. 6-5 How to read color stripes on carbon resistors.

Table 6-1. Color Code

Color	Value	Color	Value
Black	0	Green	5
Brown	1	Blue	6
Red	2	Violet	7
Orange	3	Gray	8
Yellow	4	White	9

Resistors under 10 Ω   For these values, the third stripe is either gold or silver, indicating a fractional decimal multiplier. When the third stripe is gold, multiply the first two digits by 0.1. In Fig. 6-6c, the R value is

$$25 \times 0.1 = 2.5 \ \Omega$$

Silver means a multiplier of 0.01. If the third band in Fig. 6-6c were silver, the R value would be

$$25 \times 0.01 = 0.25 \ \Omega$$

It is important to realize that the gold and silver colors are used as decimal multipliers only in the third stripe. However, gold and silver are used most often as a fourth stripe to indicate how accurate the R value is.

Resistor Tolerance   The amount by which the actual R can be different from the color-coded value is the *tolerance,* usually given in percent. For instance, a 2000-Ω resistor with ± 10 percent tolerance can have resistance 10 percent above or below the coded value. This R, therefore, is between 1800 and 2200 Ω. The calculations are as follows:

10 percent of 2000 is $0.1 \times 2000 = 200$

For $+10$ percent, the value is

$$2000 + 200 = 2200 \ \Omega$$

For -10 percent, the value is

$$1000 - 200 = 1800 \ \Omega$$

As illustrated in Fig. 6-5, silver in the fourth band indicates a tolerance of ± 10 percent; gold indicates ± 5

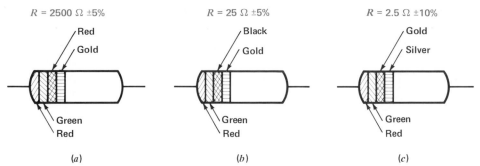

Fig. 6-6 Examples of color-coded *R* values.

percent. If there is no color band for tolerance, it is ±20 percent. The inexact value of carbon resistors is a disadvantage of their economical construction. They usually cost only about 10 to 50 cents each, or less in larger quantities. In most circuits, though, a small difference in resistance can be tolerated.

It should be noted that some resistors have five stripes, instead of four. In this case, the first three stripes give three digits, followed by the decimal multiplier in the fourth stripe and tolerance in the fifth stripe. These resistors have more precise values, with tolerances of 0.1 to 2 percent. More details are given in App. G.

Wire-Wound-Resistor Marking   Usually, wire-wound resistors are big enough physically to have the *R* value printed on the insulating case. The tolerance is generally ±5 percent, except for precision resistors, which have a tolerance of ±1 percent or less.

Some small wire-wound resistors may be color-coded with stripes, however, like carbon resistors. In this case, the first stripe is double the width of the others to indicate a wire-wound resistor. This type may have a wattage rating of 3 or 4 W.

Preferred Resistance Values   In order to minimize the problem of manufacturing different *R* values for an almost unlimited variety of circuits, specific values are made in large quantities so that they are cheaper and more easily available than unusual sizes. For resistors of ±10 percent, the preferred values are 10, 12, 15, 18, 22, 27, 33, 39, 47, 56, 68, and 82 with their decimal multiples. As examples, 47, 470, 4700, and 47,000 are preferred values. In this way, there is a preferred value available within 10 percent of any *R* value needed in a circuit.

For more accurate resistors of lower tolerance, there are additional preferred values. They are listed in Table G-3 in App. G.

Practice Problems 6-2
Answers at End of Chapter

a. Give the color for 4.
b. What tolerance does a silver stripe show?
c. Give the multiplier for red in the third stripe.
d. Give *R* and the tolerance for a resistor coded with yellow, violet, brown, and gold stripes.

6-3   Variable Resistors

Variable resistors can be wire-wound, as in Fig. 6-2*b*, or the carbon type, illustrated in Fig. 6-7. Inside the metal case of Fig. 6-7*a*, the control has a circular disc, shown in Fig. 6-7*b*, that is the carbon-composition resistance element. It can be a thin coating on pressed paper or a molded carbon disc. Joined to the two ends are the external soldering-lug terminals 1 and 3. The middle terminal is connected to the variable arm that contacts the resistor element by a metal spring wiper. As the shaft of the control is turned, the variable arm moves the wiper to make contact at different points on the resistor element. The same idea applies to the slide control in Fig. 6-8, except that the resistor element is long instead of circular.

When the contact moves closer to one end, the *R* decreases between this terminal and the variable arm.

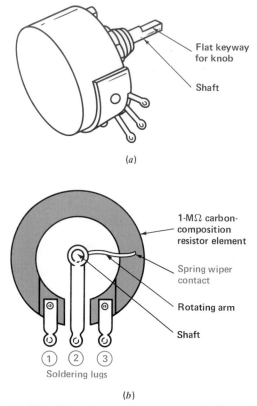

(a)

Flat keyway
for knob

Shaft

1-MΩ carbon-
composition
resistor element

Spring wiper
contact

Rotating arm

Shaft

① ② ③
Soldering lugs

(b)

Fig. 6-7 Variable carbon-resistance control. Diameter is
¾ in. (a) External view. (b) Internal view of circular resistance element.

Between the two ends, however, R is not variable but
always has the maximum resistance of the control.

Carbon controls are available with a total R from
1000 Ω to 5 MΩ, approximately. Their power rating is
usually ½ to 2 W.

A carbon control is often combined with a power
on-off switch. In Fig. 6-9, the shaft of the volume control also operates the on-off switch to turn on the receiver. However, the two units are in completely separate circuits.

Fig. 6-8 Slide control for variable R. Length is 2 in.

Tapered Controls   The way R varies with shaft
rotation is called the *taper* of the control. With a linear
taper, one-half rotation changes R by one-half the maximum value. Similarly, all values of R change in direct
proportion to rotation. For a nonlinear taper, though, R
can change more gradually at one end, with bigger
changes at the opposite end. This effect is accomplished by different densities of carbon in the resistance
element. For the example of a volume control, its audio
taper allows smaller changes in R at low settings. Then
it is easier to make changes without having the volume
too loud or too low.

Decade Resistance Box   As shown in Fig. 6-10,
the decade box is a convenient unit for providing any
one R within a wide range of values. It can be considered as test equipment for trying different R values in a
circuit. Inside the box are five series strings of resistors, with one string for each dial switch.

The first dial at the bottom connects in R of 1 to 9 Ω. It
is the *units* dial.

The second dial has units of 10 from 10 to 90 Ω. It is
the *tens* dial.

The hundreds dial has R of 100 to 900 Ω.

The thousands dial has R of 1000 to 9000 Ω.

The last dial at the top has R of 10,000 to 90,000 Ω.

On-off switch
for power line

Audio signal for
volume control

Fig. 6-9 Volume control with on-off switch mounted on
back end of shaft.

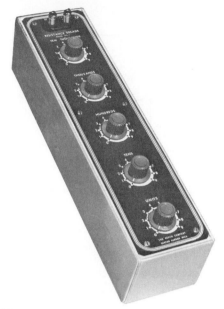

Fig. 6-10 Decade resistance box.

The five dial sections are connected in series with one another internally. Then any value from 1 to 99,999 Ω can be obtained. Note the exact values that are possible. As an example, when all five dials are on 2, the total R equals 22,222 Ω.

Practice Problems 6-3
Answers at End of Chapter

a. In Fig. 6-7, which terminal provides variable R?
b. Is an audio taper linear or nonlinear?
c. In Fig. 6-10, the top and bottom dials are on zero with the other three on 5. How much is the total R?

6-4   Rheostats and Potentiometers

These are variable resistances, either carbon or wire-wound, used to vary the amount of current or voltage for a circuit. The controls can be used in either dc or ac applications.

A rheostat is a variable R with two terminals connected in series with a load. The purpose is to vary the amount of current.

Table 6-2. Potentiometers and Rheostats

Rheostat	Potentiometer
Two terminals	Three terminals
In series with load and V source	Ends are connected across V source
Varies the I	Taps off part of V

A potentiometer, generally called a *pot* for short, has three terminals. The fixed maximum R across the two ends is connected across a voltage source. Then the variable arm is used to vary the voltage division between the center terminal and the ends. This function of a potentiometer is compared with a rheostat in Table 6-2.

Rheostat Circuit   The function of the rheostat R_2 in Fig. 6-11 is to vary the amount of current through R_1. For instance, R_1 can be a small light bulb that requires a specified I. Therefore, the two terminals of the rheostat R_2 are connected in series with R_1 and the source V in order to vary the total resistance R_T in the circuit. When R_T changes, the I changes, as read by the current meter in the series circuit.

In Fig. 6-11, R_1 is 5 Ω and the rheostat R_2 varies from 0 to 5 Ω. With R_2 at its maximum of 5 Ω, then R_T equals $5 + 5 = 10$ Ω. The I is 1.5 V/10 Ω, which equals 0.15 A or 150 mA.

When R_2 is at its minimum value of 0 Ω, the R_T equals 5 Ω. Then I is 1.5 V/5 Ω = 0.3 A or 300 mA for the maximum current. As a result, varying the rheostat changes the circuit resistance to vary the current though R_1. The I increases as R decreases.

It is important that the rheostat have a wattage rating high enough for the maximum I when R is minimum. Rheostats are often wire-wound variable resistors used to control relatively large values of current in low-resistance circuits for ac power applications.

Potentiometer Circuit   The purpose of the circuit in Fig. 6-12 is to tap off a variable part of the 100-V from the source. Consider this circuit in two parts:

1. The applied V is input to the two end terminals of the potentiometer.

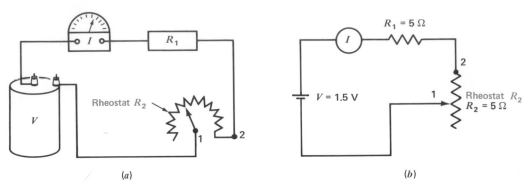

Fig. 6-11 Rheostat connected in series circuit to vary the current. (*a*) Wiring diagram with ammeter to measure *I*. (*b*) Schematic diagram.

2. The variable *V* is output between the variable arm and an end terminal.

Two pairs of connections to the three terminals are necessary, with one terminal common to the input and output. One pair connects the source *V* to the end terminals 1 and 3. The other pair of connections is between the variable arm at the center terminal and one end. This end has double connections for input and output. The other end has only an input connection.

When the variable arm is at the middle value of the 500-kΩ *R* in Fig. 6-12, the 50 V is tapped off between terminals 2 and 1 as one-half the 100-V input. The other 50 V is between terminals 2 and 3. However, this voltage is not used for output.

As the control is turned up to move the variable arm closer to terminal 3, more of the input voltage is available between 2 and 1. With the control at its maximum *R*, the voltage between 2 and 1 is the entire 100 V. Actually, terminal 2 then is the same as 3.

When the variable arm is at minimum *R*, rotated to terminal 1, the output between 2 and 1 is zero. Now all the applied voltage is across 2 and 3, with no output for the variable arm. It is important to note that the source voltage is not short-circuited. The reason is that the maximum *R* of the pot is always across the applied *V*, regardless of where the variable arm is set. Typical examples of small potentiometers used in electronic circuits are shown in Fig. 6-13.

Potentiometer Used as a Rheostat   Commercial rheostats are generally wire-wound high-wattage resistors for power applications. However, a small low-wattage rheostat is often needed in electronic circuits. One example is a continuous tone control in a receiver. The control requires the variable series resistance of a rheostat but dissipates very little power.

A method of wiring a potentiometer as a rheostat is to connect just one end of the control and the variable

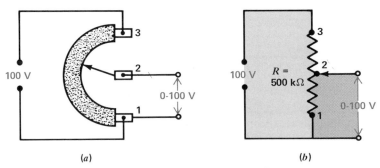

Fig. 6-12 Potentiometer connected across voltage source to function as a voltage divider. (*a*) Wiring diagram. (*b*) Schematic diagram.

Fig. 6-13 Small potentiometers used for electronic circuits. Terminal leads formed for insertion in PC board. Diameter is ½ in. for control at the left and ¾ in. at the right. (*Centralab*)

arm, using only two terminals. The third terminal is open, or floating, not connected to anything.

Another method is to wire the unused terminal to the center terminal. When the variable arm is rotated, different amounts of resistance are short-circuited. This method is preferable because there is no floating resistance.

Either end of the potentiometer can be used for the rheostat. The direction of increasing R with shaft rotation reverses, though, for connections at opposite ends. Also, the taper is reversed on a nonlinear control.

Practice Problems 6-4
Answers at End of Chapter

a. How many circuit connections to a potentiometer are needed?
b. How many circuit connections to a rheostat are needed?
c. In Fig. 6-12, with a 500-kΩ linear potentiometer, how much is the output voltage with 400 kΩ between lugs 1 and 2?

6-5 Power Rating of Resistors

In addition to having the required ohms value, a resistor should have a wattage rating high enough to dissipate the I^2R power produced by the current flowing through the resistance, without becoming too hot. Carbon resistors in normal operation are often quite warm, up to a maximum temperature of about 85°C, which is close to the 100°C boiling point of water. Carbon resis-

tors should not be so hot, however, that they ''sweat'' beads of liquid on the insulating case. Wire-wound resistors operate at very high temperatures, a typical value being 300°C for the maximum temperature. If a resistor becomes too hot because of excessive power dissipation, it can change appreciably in resistance value or burn open.

The power rating is a physical property that depends on the resistor construction, especially physical size. Note the following:

1. A larger physical size indicates a higher power rating.
2. Higher-wattage resistors can operate at higher temperatures.
3. Wire-wound resistors are physically larger with higher wattage ratings than carbon resistors.

For approximate sizes, a 2-W carbon resistor is about 1 in. long with ¼ in. diameter; a ¼-W resistor is about 0.35 in. long with diameter of 0.1 in.

For both types, a higher power rating allows a higher voltage rating. This rating gives the highest voltage that may be applied across the resistor without internal arcing. As examples for carbon resistors, the maximum voltage is 500 V for a 1-W rating, 250 V for ¼-W, and 150 V for ⅛-W. In wire-wound resistors, excessive voltage can produce an arc between turns; in carbon resistors, the arc is between carbon granules.

Shelf Life Resistors keep their characteristics almost indefinitely when not used. Without any current in a circuit to heat the resistor, it has practically no change with age. The shelf life of resistors is usually no problem, therefore. Actually, the only components that should be used fresh from manufacture are batteries and electrolytic capacitors.

Practice Problems 6-5
Answers at End of Chapter

Answer True or False.
a. A 5-Ω 50-W resistor is physically larger than a 5-MΩ 1-W resistor.
b. Resistors should not operate above a temperature of 0°C.

6-6 Choosing the Resistor for a Circuit

In deciding what size resistor to use, the first requirement is to have the amount of resistance needed. Consider the example in Fig. 6-14a. Resistor R_2 is to be inserted in series with R_1 for the purpose of limiting the current to 0.1 A with a 100-V source. The total resistance R_T needed equals 1000 Ω. This calculation is

$$R_T = \frac{V}{I} = \frac{100 \text{ V}}{0.1 \text{ A}} = 1000 \text{ Ω}$$

The value of R_1 is 900 Ω. For R_T of 1000 Ω, then, the required value of R_2 is 1000 − 900 = 100 Ω. R_2 is connected in series with R_1. The current of 0.1 A flows through both R_1 and R_2.

The I^2R power dissipated in R_2 equals 1 W. This calculation is

$$P = (0.1 \text{ A})^2 \times 100 \text{ Ω}$$
$$= 0.01 \times 100$$
$$P = 1 \text{ W}$$

However, a 2-W resistor would normally be used. This safety factor of 2 in the power rating is common practice with carbon resistors, so that they will not become too hot in normal operation.

A resistor with a higher wattage rating but the same R could also be used, if there is space on the circuit board for the larger size. The higher power rating would allow the circuit to operate normally and last longer without breaking down from excessive heat.

However, a higher wattage rating is inconvenient when the next larger size is wire-wound and physically bigger.

Wire-wound resistors can operate closer to their power rating, assuming adequate ventilation. They have a higher rating for maximum operating temperature, compared with carbon resistors.

Another example of choosing a carbon resistor for a circuit is shown in Fig. 6-14b. The 10 MΩ of R_4 is used with the 10 MΩ of R_3 to provide an IR voltage drop of 200 V, equal to one-half the source of 400 V. Since the two resistances are equal, they divide the applied voltage into two equal parts of 200 V. This IR voltage drop can be calculated from the current. I is equal to 400 V/20 MΩ = 20 μA. Then the voltage drops are

$$V_{R_3} = 20 \text{ μA} \times 10 \text{ MΩ} = 200 \text{ V}$$
$$V_{R_4} = 20 \text{ μA} \times 10 \text{ MΩ} = 200 \text{ V}$$

The I^2R power dissipated in R_4 is 4 mW. This value can be calculated as

$$P = (20 \text{ μA})^2 \times 10 \text{ MΩ}$$
$$= 400 \times 10^{-12} \times 10 \times 10^6$$
$$P = 4000 \text{ μW, 0.004 W, or 4 mW}$$

However, the wattage rating used here is ¼ W, which equals 250 mW. In this case, the wattage rating is much higher than the actual amount of power dissipated in the resistor. A ¼-W power rating is used to provide a high enough voltage rating.

Notice the small amount of power dissipated in this circuit with a relatively high applied voltage of 400 V.

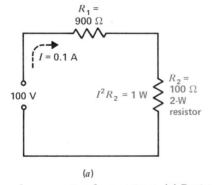

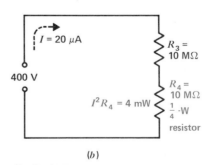

(a)   (b)

Fig. 6-14 Examples of power rating for a resistor. (a) Resistance R_2 dissipates 1 W, but a 2-W resistor is used, for a safety factor of 2. (b) Resistance R_4 dissipates 0.004 W or 4 mW, but a ¼-W resistor is used.

The reason is that the very high resistance limits the current to a low value.

In general, using a resistor with a suitable power rating provides the required voltage rating. The exception, however, is a low-current, high-voltage circuit where the applied voltage is of the order of kilovolts. An example is the special multiplier resistor used in the high-voltage probe of a meter that measures 0 to 40 kV.

Practice Problems 6-6
Answers at End of Chapter

a. In Fig. 6-14a, calculate the power dissipation in R_2 as $V_2 \times I$.
b. In Fig. 6-14b, calculate the power dissipation in R_4 as $V_4 \times I$.

6-7 Series and Parallel Combinations of Resistors

In some cases two or more resistors are combined in series or parallel to obtain a desired R value with a higher wattage rating. Several examples are shown in Fig. 6-15.

The total resistance R_T depends on the series or parallel connections. However, the combination has a power rating equal to the sum of the individual values, whether the resistors are in series or in parallel. The reason is that the total physical size increases with each added resistor. Equal resistors are generally used in order to have equal distribution of I, V, and P.

In Fig. 6-15a, the two equal resistors in series double the resistance for R_T. Also, the power rating of the combination is twice the rating for one resistor.

In Fig. 6-15b, the two equal resistors in parallel have one-half the resistance for R_T. However, the combined power rating is still twice the rating for one resistor.

In Fig. 6-15c, the series-parallel combination of four resistors makes R_T the same as each R. The total power rating, though, is four times the rating for one resistor.

Practice Problems 6-7
Answers at End of Chapter

Give R_T and the combined power rating of two 1-kΩ resistors rated at 2 W connected:
a. In series.
b. In parallel.

6-8 Resistor Troubles

The most common trouble in resistors is an open. When the open resistor is a series component, there is no current in the entire series path and it cannot operate. Defective carbon-composition resistors are often partially open, resulting in much higher resistance than the color-coded value.

Noisy Controls In applications such as volume and tone controls, carbon controls are preferred because the smoother change in resistance results in less noise when the variable arm is rotated. With use, how-

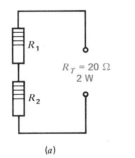
(a)

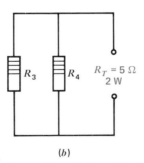
(b)

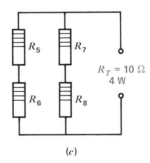
(c)

Fig. 6-15 Total R_T and wattage rating for series and parallel combinations of resistors. Each R is 10 Ω with 1-W rating. (a) For R_1 and R_2 in series, add wattage ratings and R_T is $2 \times R$. (b) For R_3 and R_4 in parallel, add wattage ratings, and R_T is ½ R. (c) For this series-parallel combination, add all wattage ratings and R_T is equal to R.

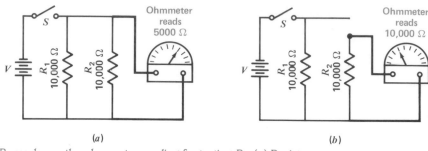

Fig. 6-16 Parallel R_1 can lower the ohmmeter reading for testing R_2. (a) Resistances R_1 and R_2 in parallel. (b) Resistance R_2 isolated by disconnecting one end of R_1.

ever, the resistance element becomes worn by the wiper contact, making the control noisy. When a volume or tone control makes a scratchy noise as the shaft is rotated, it indicates a worn-out resistance element.

Checking Resistors with an Ohmmeter

Since the ohmmeter has its own voltage source, it is always used without any external power applied to the resistance being measured. Just connect the ohmmeter leads across the resistance to be measured.

An open resistor reads infinitely high ohms. For some reason, infinite ohms is often confused with zero ohms. Remember, though, that infinite ohms means an open circuit. The current is zero, but the resistance is infinitely high. Furthermore, it is practically impossible for a resistor to become short-circuited in itself. The resistor may be short-circuited by some other part of the circuit. However, the construction of resistors is such that the trouble they develop is an open circuit, with infinitely high ohms.

The ohmmeter must have an ohms scale capable of reading the resistance value, or the resistor cannot be checked. In checking a 10-MΩ resistor, for instance, if the highest reading is 1 MΩ, on the ohmmeter it will indicate infinite resistance, even if the resistor has its normal value of 10 MΩ. An ohms scale of 100 MΩ or more should be used for checking such high resistances.

To check resistors of less than 10 Ω, a low-ohms scale of about 100 Ω or less is necessary. Center scale should be 6 Ω or less. Otherwise, the ohmmeter can read a normally low resistance value as zero ohms.

When checking resistance in a circuit, it is important to be sure there are no parallel paths across the resistor being measured. Otherwise, the measured resistance can be much lower than the actual resistor value, as illustrated in Fig. 6-16a. Here, the ohmmeter reads the resistance of R_2 in parallel with R_1. To check across R_2 alone, one end is disconnected, as in Fig. 6-16b.

For very high resistances, it is important not to touch the ohmmeter leads. There is no danger of shock, but the body resistance of about 50,000 Ω as a parallel path will lower the ohmmeter reading.

Practice Problems 6-8
Answers at End of Chapter

a. What is the ohmmeter reading for a short circuit?
b. What is the ohmmeter reading for an open resistor?

Summary

1. The two main types of resistors are carbon-composition and wire-wound. Their characteristics are compared in Table 6-3. The schematic symbols for fixed and variable resistances are summarized in Fig. 6-17.
2. A rheostat is a variable series resistance with two terminals to adjust the amount of current in a circuit.
3. A potentiometer is a variable voltage divider with three terminals.

Table 6-3. Comparison of Resistor Types

Carbon-composition resistors	Wire-wound resistors
Carbon granules in binder	Turns of resistance wire
R up to 20 MΩ	R down to a fraction of 1 Ω
Color-coded for resistance value	Resistance printed on unit
For low-current circuits; power ratings of $^1/_{10}$ to 2 W	For high-current circuits; ratings of 5 to over 100 W
Variable potentiometers and rheostats to 5 MΩ, for controls such as volume and tone in receivers	Low-resistance rheostats for varying current; potentiometers up to 50 kΩ for voltage divider in power supply

4. Carbon resistors are practically always color-coded, as in Figs. 6-5 and 6-6, to indicate the resistance value.

5. The wattage rating of carbon resistors depends mainly on their physical size, larger resistors being able to dissipate more power. The power rating is not part of the color coding but may be printed on the resistor or judged from its size.

6. With carbon resistors the wattage rating should be about double the actual I^2R power dissipation for a safety factor of 2 or more.

7. Carbon resistors can be combined for a higher wattage rating. The total power rating is the sum of the individual wattage values, whether in series or in parallel. In series, though, the total resistance increases; in parallel, the combined resistance decreases.

8. The most common trouble in resistors is an open circuit. An ohmmeter reads infinite ohms across the open resistor, assuming there is no parallel path.

Self-Examination
Answers at Back of Book

Choose (a), (b), (c), or (d).

1. Which of the following are typical resistance and power-dissipation values for a wire-wound resistor? (a) 1 MΩ $^1/_3$ W; (b) 500 Ω 1 W; (c) 50,000 Ω 1 W; (d) 10 Ω 50 W.

2. Which of the following are typical resistance and power-dissipation values for a

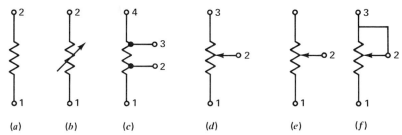

Fig. 6-17 Schematic symbols for resistors. (a) Fixed R. (b) Any type of variable R. (c) Tapped R. (d) Potentiometer. (e) Potentiometer used as rheostat. Either end terminal 1 or 3 can be open. (f) End terminal 1 or 3 connected to variable arm for rheostat.

carbon-composition resistor? (a) 6800 Ω 1 W; (b) 5 Ω 5 W; (c) 10,000 Ω 10 W; (d) 1000 Ω 100 W.

3. For a carbon-composition resistor color-coded with yellow, violet, orange, and silver stripes from left to right, the resistance and tolerance are (a) 740 Ω ±5 percent; (b) 4700 Ω ±10 percent; (c) 7400 Ω ±1 percent; (d) 47,000 Ω ±10 percent.

4. For a carbon-composition resistor color-coded with green, black, gold, and silver stripes from left to right, the resistance and tolerance are (a) 0.5 Ω ±5 percent; (b) 0.5 Ω ±10 percent; (c) 5 Ω ±10 percent; (d) 50 Ω ±10 percent.

5. A resistor with the color-coded value of 100 Ω and ±20 percent tolerance can have an actual resistance between (a) 80 and 120 Ω; (b) 90 and 110 Ω; (c) 98 and 102 Ω; (d) 100 and 120 Ω.

6. Two 1000-Ω 1-W resistors are connected in parallel. Their combined resistance value and wattage rating is (a) 500 Ω 1 W; (b) 500 Ω 2 W; (c) 1000 Ω 2 W; (d) 2000 Ω 2 W.

7. A resistor is to be connected across a 45-V battery to provide 1 mA of current. The required resistance with a suitable wattage rating is (a) 4.5 Ω 1 W; (b) 45 Ω 10 W; (c) 450 Ω 2 W; (d) 45,000 Ω ¼ W.

8. Which of the following is a preferred resistor value? (a) 47; (b) 520; (c) 43,000; (d) 54,321.

9. When checked with an ohmmeter, an open resistor reads (a) zero; (b) infinite; (c) high but within the tolerance; (d) low but not zero.

10. One precaution in checking resistors with an ohmmeter is: (a) Check high resistances on the lowest ohms range. (b) Check low resistances on the highest ohms range. (c) Disconnect all parallel resistance paths. (d) Check high resistances with your fingers touching the test leads.

Essay Questions

1. State the colors corresponding to digits 1 to 9, inclusive.
2. Give the tolerance values indicated by a gold or silver band.
3. What are the values for gold and silver as decimal multipliers in the third band?
4. Give three types of fixed resistors.
5. What are two differences between potentiometers and rheostats?
6. Show how to use a potentiometer as a rheostat.
7. Why do high-R carbon resistors, of the order of megohms, usually have a low power rating of 1 W or less?
8. Draw the diagram for connecting two 1000-Ω 1-W resistors to obtain 2000 Ω with a power rating of 2 W.
9. Give the color coding for the following R values: 1 MΩ, 33,000 Ω, 8200 Ω, 150 Ω, 68 Ω, and 2.2 Ω. (Tolerance coding not required.)
10. A 50-Ω rheostat R_1 is in series with a 25-Ω R_2 and a 50-V source. Draw a graph of I against R_1 as it is varied in 10-Ω steps.
11. Show how to connect resistors for the following examples: (a) two 20-kΩ 1-W resistors for a total R_T of 10 kΩ with a power rating of 2 W. (b) two 20-kΩ 1-W resistors for 40 kΩ and 2 W. (c) four 20-kΩ 1-W resistors for 20 kΩ and 4 W.

12. Describe briefly how you would check a 10-MΩ resistor to see if it is open. Give two precautions to make sure the test is not misleading.

13. Show the schematic diagram to tap off a variable voltage from a 45-V battery as the source.

14. Show how to connect three 10-kΩ resistors for a total R_T of 15 kΩ.

15. Give two examples of variable resistors you have seen used for controls on electronic equipment.

Problems

Answers to Odd-Numbered Problems at Back of Book

1. A current of 1 mA flows through a 1-MΩ 2-W carbon resistor. (**a**) How much power is dissipated as heat in the resistor? (**b**) How much is the maximum power that can be dissipated without excessive heat?

2. A 50-W wire-wound resistor has 28 V across it with 1.4 A of current. (**a**) Calculate the power dissipation. (**b**) How much is R?

3. Give the resistance and tolerance for R_1 to R_6 in Table 6-4.

4. Give the range of R values for a 4700-Ω resistor with tolerance of ± 5 percent.

5. An R of 12.3 Ω has a tolerance of ± 1 percent. What is the range of R values?

6. Fill in the R_T on the decade box in Fig. 6-10 for the dial settings in Table 6-5.

7. Determine the R and W ratings of a resistor to fit the following requirements: 5-V IR drop with 25-mA current and safety factor of 2 for power dissipation.

Table 6-4. Color-Coded R Values for Prob. 3

	Band A	Band B	Band C	Band D	Resistance	Tolerance
R_1	Yellow	Violet	Red	Silver		
R_2	Red	Red	Green	Silver		
R_3	Orange	Orange	Black	Gold		
R_4	White	Brown	Brown	Gold		
R_5	Red	Red	Gold	Gold		
R_6	Brown	Black	Orange	No color		

Table 6-5. Dial Settings for Decade Box in Fig. 6-10

$R \times 10^4$	$R \times 10^3$	$R \times 10^2$	$R \times 10$	R	Total R_T
9	6	7	4	2	
0	5	6	8	3	
6	7	0	5	4	
1	2	3	4	5	
5	4	3	2	1	

Answers to Practice Problems

6-1 a. T
    b. T
    c. F

6-2 a. Yellow
    b. ±10 percent
    c. 100
    d. 470 Ω ±5 percent

6-3 a. Terminal 2
    b. Nonlinear
    c. 5550 Ω

6-4 a. 4 connections to 3 lugs
    b. 2
    c. 80 V

6-5 a. T
    b. F

6-6 a. 1 W
    b. 4 mW

6-7 a. 2 kΩ 4 W
    b. 0.5 kΩ 4 W

6-8 a. 0 Ω
    b. Infinite ohms

Review Chapters 1 to 6

Summary

1. The electron is the basic quantity of negative electricity; the proton is the basic quantity of positive electricity. Both have the same charge but opposite polarities.
2. A quantity of electrons is a negative charge; a deficiency of electrons is a positive charge. Like charges repel each other; unlike charges attract.
3. Charge is measured in coulombs; 6.25×10^{18} electrons equals one coulomb. Charge in motion is current. One coulomb per second equals one ampere of current.
4. Potential difference is measured in volts. One volt produces one ampere of current against the opposition of one ohm of resistance.
5. The three forms of Ohm's law are $I = V/R$, $V = IR$, and $R = V/I$.
6. Power in watts equals VI, I^2R, or V^2/R, with V, I, and R in volts, amperes, and ohms, respectively.
7. The most common multiples and submultiples of the practical units are *mega* or M for 10^6, *micro* or μ for 10^{-6}, *kilo* or k for 10^3, and *milli* or m for 10^{-3}.
8. For series resistances: (**a**) the current is the same in all resistances; (**b**) the *IR* drops can be different with unequal resistances; (**c**) the applied voltage equals the sum of the series *IR* drops; (**d**) the total resistance equals the sum of the individual resistances; (**e**) an open circuit in one resistance results in no current through the entire series circuit.
9. For parallel resistances: (**a**) the voltage is the same across all resistances; (**b**) the branch currents can be different with unequal resistances; (**c**) the total line current equals the sum of the parallel branch currents; (**d**) the combined resistance of parallel branches is less than the smallest resistance, as determined by the reciprocal Formula (4-3); (**e**) an open circuit in one branch does not open the other branches; (**f**) a short circuit across one branch short-circuits all the branches.
10. In series-parallel circuits, the resistances in one current path without any branch points are in series; all the rules of series resistances apply. The resistances

across the same two branch points are in parallel; all the rules of parallel resistances apply.

11. The two main types of resistors are wire-wound and carbon-composition. The wire-wound type has relatively low R for power applications of 5 to 100 W or more; the carbon type has high R with power ratings of 2 W or less.

12. The color coding of carbon resistors is illustrated in Figs. 6-5 and 6-6. The stripe colors indicate R in ohms. The power rating depends on the physical size of the resistor.

13. A rheostat is a variable series resistance with two terminals to adjust the amount of current in a circuit.

14. A potentiometer is a variable voltage divider with three terminals.

15. The most common trouble in resistors is an open circuit. The ohms value then is infinitely high.

Review Self-Examination
Answers at Back of Book

Choose (a), (b), (c), or (d).

1. In which of the following circuits will the voltage source produce the most current? (a) 10 V across a 10-Ω resistance; (b) 10 V across two 10-Ω resistances in series; (c) 10 V across two 10-Ω resistances in parallel; (d) 1000 V across a 1-MΩ resistance.

2. Three 120-V 100-W bulbs are in parallel across the 120-V power line. If one bulb burns open: (a) the other two bulbs cannot light; (b) all three bulbs light; (c) the other two bulbs can light; (d) there is excessive current in the main line.

3. A circuit allows 1 mA of current to flow with 1 V applied. The conductance of the circuit equals (a) 0.002 Ω; (b) 0.005 μS; (c) 1000 μS; (d) 1 S.

4. If 2 A of current is allowed to accumulate charge for 5 s, the resultant charge equals (a) 2 C; (b) 10 C; (c) 5 A; (d) 10 A.

5. A potential difference applied across a 1-MΩ resistor produces 1 mA of current. The applied voltage equals (a) 1 μV; (b) 1 mV; (c) 1 kV; (d) 1,000,000 V.

6. A string of two 1000-Ω resistances is in series with a parallel bank of two 1000-Ω resistances. The total resistance of the series-parallel circuit equals (a) 250 Ω; (b) 2500 Ω; (c) 3000 Ω; (d) 4000 Ω.

7. In the circuit of question 6, one of the resistances in the series string opens. Then the current in the parallel bank (a) increases slightly in both branches; (b) equals zero in one branch but is maximum in the other branch; (c) is maximum in both branches; (d) equals zero in both branches.

8. With 100 V applied across a 10,000-Ω resistance, the power dissipation equals (a) 1 mW; (b) 1 W; (c) 100 W; (d) 1 kW.

9. Ten volts is applied across R_1, R_2, and R_3 in series, producing 1 A in the series circuit. R_1 equals 6 Ω and R_2 equals 2 Ω. Therefore, R_3 equals (a) 2 Ω; (b) 4 Ω; (c) 10 Ω; (d) 12 Ω.

10. A 5-V source and a 3-V source are connected with series-opposing polarities. The combined voltage across both sources equals (a) 5 V; (b) 3 V; (c) 2 V; (d) 8 V.

11. In a circuit with three parallel branches, if one branch opens, the main-line current will be (a) more; (b) less; (c) the same; (d) infinite.

12. A 10-Ω R_1 and a 20-Ω R_2 are in series with a 30-V source. If R_1 opens, the voltage drop across R_2 will be (a) zero; (b) 20 V; (c) 30 V; (d) infinite.

13. A voltage V_1 of 40 V is connected series-opposing with V_2 of 50 V. The total voltage across both components is: (a) 10 V; (b) 40 V; (c) 50 V; (d) 90 V.

14. Two series voltage drops V_1 and V_2 total 100 V for V_T. When V_1 is 60 V, then V_2 must equal: (a) 40 V; (b) 60 V; (c) 100 V; (d) 160 V.

15. Two parallel branch currents I_1 and I_2 total 100 mA for I_T. When I_1 is 60 mA, then I_2 must equal: (a) 40 mA; (b) 60 mA; (c) 100 mA; (d) 160 mA.

16. A carbon resistor is color-coded with brown, green, red, and gold stripes from left to right. Its value is (a) 1500 Ω $\pm$ 5 percent; (b) 6800 Ω $\pm$ 5 percent; (c) 10,000 Ω $\pm$ 10 percent; (d) 500,000 Ω $\pm$ 5 percent.

17. Which of the following statements is true? (a) A rheostat needs three terminals; (b) an open resistor has zero ohms of R; (c) carbon resistors can have a power rating less than 2 W; (d) wire-wound resistors cannot be used for a rheostat.

18. With 30 V applied across two equal resistors in series, 10 mA of current flows. Typical values for each resistor to be used here are (a) 10 Ω 10 W; (b) 1500 Ω ½ W; (c) 3000 Ω 10 W; (d) 30 MΩ 2 W.

References
Additional References at Back of Book

Cooke, N. M., H. F. R. Adams, and P. B. Dell: *Basic Mathematics for Electronics,* McGraw-Hill Book Company, New York.

De France, J. J.: *Electrical Fundamentals,* Prentice-Hall, Inc., Englewood Cliffs, N.J.

Grob, B.: *Mathematics Outline and Review Problems for Basic Electronics,* McGraw-Hill Book Company, New York.

Periodic Chart of the Atoms, Sargent Welch Scientific Co., Skokie, Ill. 60076.

Ridsdale, R. E.: *Electric Circuits,* McGraw-Hill Book Company, New York.

Chapter 7
Voltage Dividers and Current Dividers

Any series circuit is a voltage divider. The *IR* voltage drops are proportional parts of the applied voltage. Also, any parallel circuit is a current divider. Each branch current is part of the total line current, but in

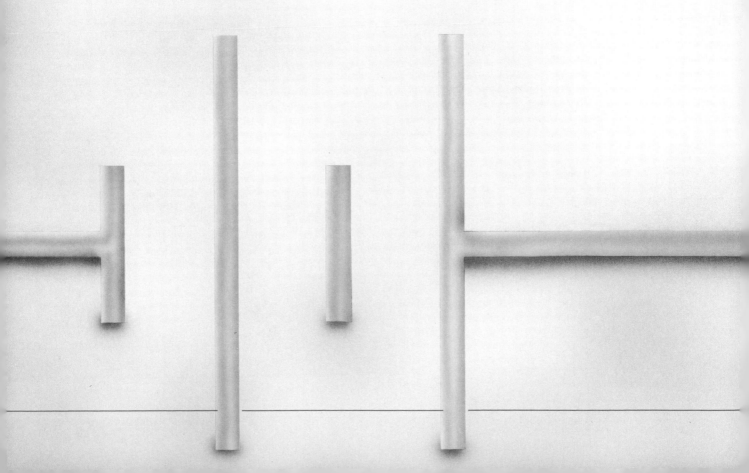

inverse proportion to the branch resistance. Special formulas can be used for the voltage and current division as short cuts in the calculations. The voltage division formula gives the series voltages even when the current is not known. Also, the current division formula gives the branch currents even when the branch voltage is not known. Finally, we consider a series voltage divider with parallel branches that have load currents. The design of such a loaded voltage divider can be applied to the important case of different voltage taps from the power supply in electronic equipment.

Important terms in this chapter are:

bleeder current   loaded voltage
current divider   voltage divider
load currents

More details are explained in the following sections:

7-1 Series Voltage Dividers
7-2 Current Divider with Two Parallel Resistances
7-3 Current Division by Parallel Conductances
7-4 Series Voltage Divider with Parallel Load Current
7-5 Design of a Loaded Voltage Divider

7-1   Series Voltage Dividers

The current is the same in all the resistances in a series circuit. Also, the voltage drops equal the product of I times R. Therefore, the IR voltages are proportional to the series resistances. A higher resistance has a greater IR voltage than a smaller resistance in the same series circuit; equal resistances have the same amount of IR drop. If R_1 is double R_2, then V_1 will be double V_2.

The series string can be considered as a *voltage divider*. Each resistance provides an IR drop V equal to its proportional part of the applied voltage. Stated as a formula,

$$V = \frac{R}{R_T} \times V_T \qquad\qquad (7\text{-}1)$$

> **Example 1**   Three 50-kΩ resistors R_1, R_2, and R_3 are in series across an applied voltage of 180 V. How much is the IR voltage drop across each resistor?
>
> **Answer**   The voltage drop is 60 V. Since R_1, R_2, and R_3 are equal, each has one-third the total resistance of the circuit and one-third the total applied voltage. Using the formula,
>
> $$V = \frac{R}{R_T} \times V_T = \frac{50\ k\Omega}{150\ k\Omega} \times 180\ V$$
>
> $$= \frac{1}{3} \times 180\ V$$
>
> $$V = 60\ V$$

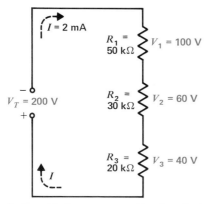

Fig. 7-1 Series string as a proportional voltage divider. Each V_R is (R/R_T) of V_T.

Note that R and R_T must be in the same units for the proportion. Then V is in the same units as V_T.

Typical Circuit   Figure 7-1 illustrates another example of a proportional voltage divider. Let the problem be to find the voltage across R_3. We can either calculate this voltage V_3 as IR_3 or determine its proportional part of the total applied voltage V_T. The answer is the same both ways. Note that R_T is 20 + 30 + 50 = 100 kΩ.

Proportional Voltage Method   Using Formula (7-1), V_3 equals $^{20}/_{100}$ of the 200 V applied for V_T because R_3 is 20 kΩ and R_T is 100 kΩ. Then V_3 is $^{20}/_{100}$ of 200 or $^1/_5$ of 200, which is equal to 40 V. The calculations are

$$V_3 = \frac{R_3}{R_T} \times V_T = \frac{20}{100} \times 200\ V = 40\ V$$

In the same way, V_2 is 60 V. The calculations are

$$V_2 = \frac{R_2}{R_T} \times V_T = \frac{30}{100} \times 200\ V = 60\ V$$

Also, V_1 is 100 V. The calculations are

$$V_1 = \frac{R_1}{R_T} \times V_T = \frac{50}{100} \times 200\ V = 100\ V$$

The sum of V_1, V_2, and V_3 in series is 100 + 60 + 40 = 200 V, which is equal to V_T.

Method of IR Drops   If we want to solve for the current in Fig. 7-1, the I is V_T/R_T or 200 V/100 kΩ = 2 mA. This I flows through R_1, R_2, and R_3 in series. The IR drops are

$$V_1 = I \times R_1 = 2\ mA \times 50\ k\Omega = 100\ V$$
$$V_2 = I \times R_2 = 2\ mA \times 30\ k\Omega = 60\ V$$
$$V_3 = I \times R_3 = 2\ mA \times 20\ k\Omega = 40\ V$$

These voltages are the same values calculated by Formula (7-1) for proportional voltage dividers.

Two Voltage Drops in Series For this case, it is not necessary to calculate both voltages. After you find one, subtract it from V_T to find the other.

As an example, assume V_T is 48 V across two series resistors R_1 and R_2. If V_1 is 18 V, then V_2 must be $48 - 18 = 30$ V.

The Largest Series R Has the Most V The fact that series voltage drops are proportional to the resistances means that a very small R in series with a much larger R has a negligible IR drop. An example is shown in Fig. 7-2. Here the 1 kΩ of R_1 is in series with the much larger 999 kΩ of R_2. The V_T is 1000 V.

The voltages across R_1 and R_2 in Fig. 7-2 can be calculated by the voltage divider formula. Note that R_T is $1 + 999 = 1000$ kΩ.

$$V_1 = \frac{R_1}{R_T} \times V_T = \frac{1}{1000} \times 1000 \text{ V} = 1 \text{ V}$$

$$V_2 = \frac{R_2}{R_T} \times V_T = \frac{999}{1000} \times 1000 \text{ V} = 999 \text{ V}$$

The 999 V across R_2 is practically the entire applied voltage. Also, the very high series resistance dissipates almost all the power.

Furthermore, the current of 1 mA through R_1 and R_2 in Fig. 7-2 is determined almost entirely by the 999 kΩ of R_2. The I for R_T is 1000 V/1000 kΩ, which equals 1 mA. However, the 999 kΩ of R_2 alone would allow 1.001 mA of current, which is very little different.

Advantage of the Voltage Divider Method
Using Formula (7-1), we can find the proportional volt-age drops from V_T and the series resistances without knowing the amount of I. For odd values of R, calculating the I may be more troublesome than finding the proportional voltages directly. Also, in many cases we can see the voltage division approximately without the need for any written calculations.

Practice Problems 7-1
Answers at End of Chapter

Refer to Fig. 7-1.
a. How much is R_T?
b. What fraction of the applied voltage is V_3?

7-2 Current Divider with Two Parallel Resistances

It is often necessary to find the individual branch currents in a bank from the resistances and I_T, but without knowing the voltage across the bank. This problem can be solved by using the fact that currents divide inversely as the branch resistances. An example is shown in Fig. 7-3. The formulas for the two branch currents are

$$I_1 = \frac{R_2}{R_1 + R_2} \times I_T \qquad (7\text{-}2)$$

and

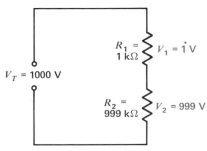

Fig. 7-2 Example of very small R_1 in series with larger R_2. Then V_1 also is very small compared with V_2.

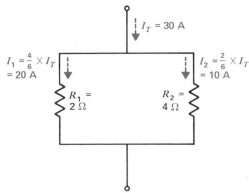

Fig. 7-3 Current division with two branch resistances. Each branch I is inversely proportional to its R. The smaller R has more I.

$$I_2 = \frac{R_1}{R_1 + R_2} \times I_T$$

Notice that the formula for each branch I has the opposite R in the numerator. The reason is that each branch current is inversely proportional to the branch resistance. The denominator is the same in both formulas, equal to the sum of the two branch resistances.

To calculate the currents in Fig. 7-3, with a 30-A I_T, a 2-Ω R_1, and a 4-Ω R_2,

$$I_1 = \frac{4}{2 + 4} \times 30$$

$$= \frac{4}{6} \times 30 = \frac{2}{3} \times 30$$

$$I_1 = 20 \text{ A}$$

For the other branch,

$$I_2 = \frac{2}{2 + 4} \times 30$$

$$= \frac{2}{6} \times 30 = \frac{1}{3} \times 30$$

$$I_2 = 10 \text{ A}$$

With all the resistances in the same units, the branch currents are in the units of I_T. For instance, kilohms of R and milliamperes of I can be used.

Actually, it is not necessary to calculate both currents. After one I is calculated, the other can be found by subtracting from I_T.

Notice that the division of branch currents in a parallel bank is opposite from the voltage division of resistance in a series string. With series resistances, a higher resistance develops a larger IR voltage proportional to its R; with parallel branches, a lower resistance takes more branch current, equal to V/R.

In Fig. 7-3, the 20-A I_1 is double the 10-A I_2 because the 2-Ω R_1 is one-half the 4-Ω R_2. This is an inverse proportion of I to R.

The inverse relation between I and R in a parallel bank means that a very large R has little effect with a much smaller R in parallel. As an example, Fig. 7-4 shows a 999-kΩ R_2 in parallel with a 1-kΩ R_1 dividing the I_T of 1000 mA. The branch currents are calculated as follows:

$$I_1 = \frac{999}{1000} \times 1000 \text{ mA} = 999 \text{ mA}$$

$$I_2 = \frac{1}{1000} \times 1000 \text{ mA} = 1 \text{ mA}$$

The 999 mA for I_1 is almost the entire line current of 1000 mA because R_1 is so small compared with R_2. Also, the smallest branch R dissipates the most power because it has the most I.

The current divider Formula (7-2) can be used only for two branch resistances. The reason is the inverse relation between each branch I and its R. In comparison, the voltage divider Formula (7-1) can be used for any number of series resistances because of the direct proportion between each voltage drop V and its R.

For more branches, it is possible to combine the branches in order to work with only two divided currents at a time. However, a better method is to use parallel conductances, because I and G are directly proportional, as explained in the next section.

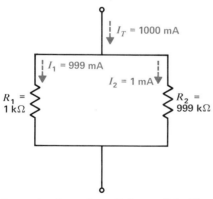

Fig. 7-4 Example of very large R_2 in parallel with small R_1. In terms of the branch currents, then, I_2 is very small compared with I_1.

Practice Problems 7-2
Answers at End of Chapter

Refer to Fig. 7-3.
a. What is the ratio of R_2 to R_1?
b. What is the ratio of I_2 to I_1?

7-3 Current Division by Parallel Conductances

Remember that the conductance G is $1/R$. Therefore, conductance and current are directly proportional. More conductance allows more current, for the same V. With any number of parallel branches, each branch current is

$$I = \frac{G}{G_T} \times I_T \qquad (7\text{-}3)$$

where G is the conductance of one branch and G_T is the sum of all the parallel conductances. The unit for G is the siemens (S).

Note that Formula (7-3), for dividing branch currents in proportion to G, has the same form as Formula (7-1), for dividing series voltages in proportion to R. The reason is that both formulas specify a direct proportion.

Two Branches As an example of using Formula (7-3), we can go back to Fig. 7-3 and find the branch currents with G instead of R. For the 2 Ω of R_1, the G_1 is ½ = 0.5. Also, the 4 Ω of R_2 has G_2 of ¼ = 0.25 S. Then G_T is 0.5 + 0.25 = 0.75 S.

The I_T is 30 A in Fig. 7-3. For the branch currents,

$$I_1 = \frac{G_1}{G_T} \times I_T$$
$$= \frac{0.50}{0.75} \times 30 \text{ A} = \frac{2}{3} \times 30 \text{ A}$$
$$I_1 = 20 \text{ A}$$

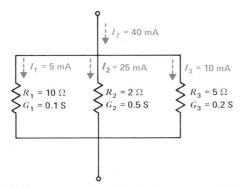

Fig. 7-5 Current divider with three branch conductances G_1, G_2, and G_3 equal to $1/R$. S is the siemens unit, formerly called the mho. Each branch I is directly proportional to the branch G.

This 20 A is the same I_1 calculated before.

For the other branch, I_2 is 30 − 20 = 10 A. Also, I_2 can be calculated as $^{0.25}\!/_{0.75}$ or ⅓ of I_T for the same 10-A value.

Three Branches A circuit with three branch currents is shown in Fig. 7-5. We can find G for the 10-Ω R_1, 2-Ω R_2, and 5-Ω R_3 as follows:

$$G_1 = \frac{1}{R_1} = \frac{1}{10 \ \Omega} = 0.1 \text{ S}$$

$$G_2 = \frac{1}{R_2} = \frac{1}{2 \ \Omega} = 0.5 \text{ S}$$

$$G_3 = \frac{1}{R_3} = \frac{1}{5 \ \Omega} = 0.2 \text{ S}$$

Remember that the siemens (S) unit is the reciprocal of the ohm (Ω) unit. The total conductance then is

$$G_T = G_1 + G_2 + G_3$$
$$= 0.1 + 0.5 + 0.2$$
$$G_T = 0.8 \text{ S}$$

The I_T is 40 mA in Fig. 7-5. To calculate the branch currents with Formula (7-3),

$$I_1 = \tfrac{1}{8} \times 40 \text{ mA} = 5 \text{ mA}$$
$$I_2 = \tfrac{5}{8} \times 40 \text{ mA} = 25 \text{ mA}$$
$$I_3 = \tfrac{2}{8} \times 40 \text{ mA} = 10 \text{ mA}$$

The sum is 5 + 25 + 10 = 40 mA for I_T.

Although three branches are shown here, Formula (7-3) can be used to find the currents for any number of parallel conductances because of the direct proportion between I and G. The method of conductances is usually easier to use than the method of resistances for three or more branches.

Practice Problems 7-3
Answers at End of Chapter

Refer to Fig. 7-5.
a. What is the ratio of G_3 to G_1?
b. What is the ratio of I_3 to I_1?

7-4   Series Voltage Divider with Parallel Load Current

The voltage dividers shown so far illustrate just a series string without any branch currents. Actually, though, a voltage divider is often used to tap off part of the applied voltage V_T for a load that needs less voltage than V_T. Then the added load is a parallel branch across part of the divider, as shown in Fig. 7-6. This example shows how the loaded voltage at the tap F is reduced below the potential it would have without the branch current for R_L.

Why the Loaded Voltage Decreases

We can start with Fig. 7-6a, which shows an R_1R_2 voltage divider alone. Resistances R_1 and R_2 in series simply form a proportional divider across the 60-V source for V_T.

For the resistances, R_1 is 40 kΩ and R_2 is 20 kΩ, making R_T equal to 60 kΩ. Also, the current I is V_T/R_T or 60 V/60 kΩ = 1 mA. For the divided voltages in Fig. 7-6a,

$$V_1 = \frac{40}{60} \times 60 \text{ V} = 40 \text{ V}$$

$$V_2 = \frac{20}{60} \times 60 \text{ V} = 20 \text{ V}$$

Note that $V_1 + V_2$ is 40 + 20 = 60 V, which is the total applied voltage.

However, in Fig. 7-6b the 20-kΩ branch of R_L changes the equivalent resistance at tap F to ground.

This change in the proportions of R changes the voltage division. Now the resistance from F to G is 10 kΩ, equal to the 20-kΩ R_1 and R_L in parallel. This equivalent bank resistance is shown as the 10-kΩ R_E in Fig. 7-6c.

Resistance R_1 is still the same 40 kΩ because it has no parallel branch. The new R_T for the divider in Fig. 7-6c is 40 kΩ + 10 kΩ = 50 kΩ. As a result, V_E from F to G in Fig. 7-6c becomes

$$V_E = \frac{R_E}{R_T} \times V_T$$

$$= \frac{10}{50} \times 60 \text{ V}$$

$$V_E = 12 \text{ V}$$

Therefore, the voltage across the parallel R_2 and R_L in Fig. 7-6b is reduced to 12 V. This voltage is at the tap F for R_L.

Note that V_1 across R_1 increases to 48 V in Fig. 7-6c. Now V_1 is $^{40}/_{50} \times 60$ V = 48 V. The V_1 increases here because there is more current through R_1.

The sum of $V_1 + V_E$ in Fig. 7-6c is 12 + 48 = 60 V. The IR drops still add to equal the applied voltage.

Path of Current for R_L

All the current in the circuit must come from the source V_T. Trace the electron flow for R_L. It starts from the negative side of V_T, through R_L, to the tap at F, and returns through R_1 in the divider to the positive side of V_T. This current I_L goes through R_1 but not R_2.

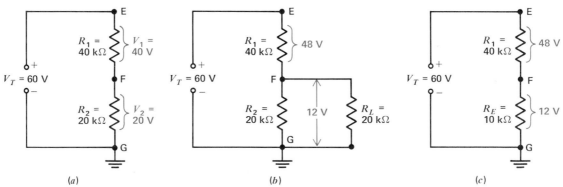

Fig. 7-6 Effect of a parallel load on part of a series voltage divider. (a) Resistances R_1 and R_2 in series without any branch current. (b) Reduced voltage across R_2 with its parallel load R_L. (c) Equivalent circuit of the loaded voltage divider.

Bleeder Current In addition, both R_1 and R_2 have their own current from the source. This current through all the resistances in the divider is bleeder current I_B. The electron flow for I_B is from the negative side of V_T, through R_2 and R_1, and back to the positive side of V_T.

The bleeder current is a steady drain on the source. However, I_B has the advantage of reducing variations in the total current in the voltage source for different values of load current.

In summary, then, for the three resistances in Fig. 7-6b, note the following currents:

1. Resistance R_L has just its load current I_L.
2. Resistance R_2 has only the bleeder current I_B.
3. Resistance R_1 has both I_L and I_B.

Practice Problems 7-4
Answers at End of Chapter

Refer to Fig. 7-6.
a. What is the proportion of R_2/R_T in Fig. 7-6a?
b. What is the proportion of R_E/R_T in Fig. 7-6c?

7-5 Design of a Loaded Voltage Divider

These principles can be applied to the design of a practical voltage divider, as shown in Fig. 7-7. This type of circuit is used for the output of a power supply in electronic equipment to supply different voltages at the taps, with different load currents. For instance, load D can represent the collector-emitter circuit for one or more power transistors that need $+100$ V for the collector supply. Also, the tap at E can be the 40-V collector supply for medium power transistors. Finally, the 18-V tap at F can be for base-emitter bias current in the power transistors and collector voltage for smaller transistors.

Note the load specifications in Fig. 7-7. Load F needs 18 V from point F to chassis ground. When the 18 V is supplied by this part of the divider, a 36-mA branch current will flow through the load. Similarly, 40 V is needed at tap E for 54 mA of I_E in load E. Also, 100 V is available at D with a load current I_D of 180 mA. The total load current here is $36 + 54 + 180 = 270$ mA.

In addition, the bleeder current I_B through the entire divider is generally specified at about 10 percent of the

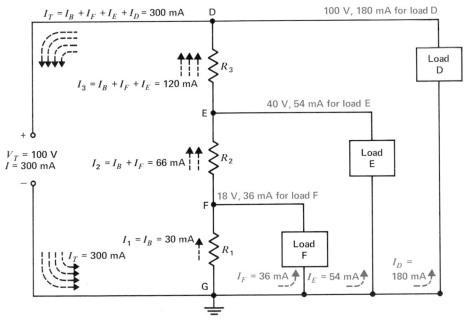

Fig. 7-7 Voltage divider for different voltages and load currents from the source V_T. See text for design calculations to find the values of R_1, R_2, and R_3.

load current. For the example here, I_B is taken as 30 mA to make a total line current I_T of $270 + 30 = 300$ mA from the source. Remember that the 30-mA I_B flows through R_1, R_2, and R_3.

The design problem in Fig. 7-7 is to find the values of R_1, R_2, and R_3 needed to provide the specified voltages. Each R is calculated as its ratio of V/I. However, the question is what are the correct values of V and I to use for each part of the divider.

Find the Current in Each R   We start with R_1 because its current is only the 30-mA bleeder current I_B. No load current flows through R_1. Therefore I_1 through R_1 equals 30 mA.

The 36-mA current I_F for load F returns to the source through R_2 and R_3. Considering just R_2 now, its current is the I_F load current and the 30-mA bleeder current I_B. Therefore, I_2 through R_2 is $36 + 30 = 66$ mA.

The 54-mA current I_E for load E returns to the source through R_3 alone. However, R_3 also has the 36-mA I_F and the 30-mA I_B. Therefore I_3 through R_3 is $54 + 36 + 30 = 120$ mA. The values for I_1, I_2, and I_3 are summarized in Table 7-1.

Note that the load current I_D for load D at the top of the diagram does not flow through R_3 or any of the resistors in the divider. However, the I_D of 180 mA is the main load current through the source of applied voltage. The 120 mA of bleeder and load currents plus the 180-mA I_D load add to equal 300 mA for I_T in the main line of the power supply.

Calculate the Voltage Across Each R   The voltages at the taps in Fig. 7-7 give the potential to chassis ground. However, we need the voltage across the two ends of each R. For R_1, the voltage V_1 is the indicated 18 V to ground because one end of R_1 is grounded. However, across R_2 the voltage is the difference between the 40-V potential at point E and the 18 V at F. Therefore V_2 is $40 - 18 = 22$ V. Similarly,

Table 7-1. Design Values for Voltage Divider in Fig. 7-7

	Current, mA	Voltage, V	Resistance, Ω
R_1	30	18	600
R_2	66	22	333
R_3	120	60	500

V_3 is calculated as 100 V at point D minus the 40 V at E, or, V_3 is $100 - 40 = 60$ V. These values for V_1, V_2, and V_3 are summarized in Table 7-1.

Calculating Each R   Now we can calculate the resistance of R_1, R_2, and R_3 as each V/I ratio. For the values listed in Table 7-1,

$$R_1 = \frac{V_1}{I_1} = \frac{18 \text{ V}}{30 \text{ mA}} = 0.6 \text{ k}\Omega = 600 \text{ }\Omega$$

$$R_2 = \frac{V_2}{I_2} = \frac{22 \text{ V}}{66 \text{ mA}} = 0.333 \text{ k}\Omega = 333 \text{ }\Omega$$

$$R_3 = \frac{V_3}{I_3} = \frac{60 \text{ V}}{120 \text{ mA}} = 0.5 \text{ k}\Omega = 500 \text{ }\Omega$$

When these values are used for R_1, R_2, and R_3 and connected in a voltage divider across the source of 100 V, as in Fig. 7-7, each load will have the specified voltage at its rated current.

Practice Problems 7-5
Answers at End of Chapter

Refer to Fig. 7-7.
a. How much is the bleeder current I_B through R_1, R_2, and R_3?
b. How much is the voltage for load E at tap E to ground?
c. How much is V_2 across R_2?

Summary

1. In a series circuit V_T is divided into IR voltage drops proportional to the resistances. Each $V_R = (R/R_T) \times V_T$, for any number of series resistances. The largest series R has the largest voltage drop.

2. In a parallel circuit, I_T is divided into branch currents. Each I is inversely proportional to the branch R. The inverse division of branch currents is given by Formula (7-2), for two resistances only. The smaller branch R has the larger branch current.

3. For any number of parallel branches, I_T is divided into branch currents directly proportional to each conductance G. Each $I = (G/G_T) \times I_T$.

4. A series voltage divider is often tapped for a parallel load, as in Fig. 7-6. Then the voltage at the tap is reduced because of the load current.

5. The design of a loaded voltage divider, as in Fig. 7-7, involves calculating each R. Find the I and potential difference V for each R. Then $R = V/I$.

Self-Examination
Answers at Back of Book

Answer True or False.

1. In a series voltage divider, each IR voltage is proportional to its R.
2. With parallel branches, each branch I is inversely proportional to its R.
3. With parallel branches, each branch I is directly proportional to its G.
4. Formula (7-2) for parallel current dividers can be used for three or more resistances.
5. Formula (7-3) for parallel current dividers can be used for five or more branch conductances.
6. In the series voltage divider of Fig. 7-1, V_1 is 2.5 times V_3 because R_1 is 2.5 times R_3.
7. In the parallel current divider of Fig. 7-3, I_1 is double I_2 because R_1 is one-half R_2.
8. In the parallel current divider of Fig. 7-5, I_2 is five times I_1 because G_2 is five times G_1.
9. In Fig. 7-6b, the branch current I_L flows through R_L, R_2, and R_1.
10. In Fig. 7-7, the bleeder current I_B flows through R_1, R_2, and R_3.

Essay Questions

1. Define a series voltage divider.
2. Define a parallel current divider.
3. Give two differences between a series voltage divider and a parallel current divider.
4. Give three differences between Formula (7-2) for branch resistances and Formula (7-3) for branch conductances.
5. Define bleeder current.
6. What is the main difference between the circuits in Fig. 7-6a and b?
7. Referring to Fig. 7-1, why is V_1 series-aiding with V_2 and V_3 but in series opposition to V_T? Show polarity of each IR drop.

8. Show the algebra for deriving Formula (7-2) for each branch current in a parallel bank of two resistances. [Hint: The voltage across the bank is $I_T \times R_T$ and R_T is $R_1R_2/(R_1 + R_2)$.]

Problems

Answers to Odd-Numbered Problems at Back of Book

1. A 200-Ω R_1 is in series with a 400-Ω R_2 and a 2-kΩ R_3. The applied voltage is 52 V. Calculate V_1, V_2, and V_3.
2. Find R_1 and R_2 for a voltage divider that takes 10 mA from a 20-V source, with 5 V across R_2. There are no load-current branches.
3. How much is the bleeder current through R_1 and R_2 in Fig. 7-6b?
4. If I_T is 14 mA for two branches, R_1 is 20 kΩ, and R_2 is 56 kΩ, find I_1 and I_2 in this current-divider circuit.
5. Three parallel branches have $G_1 = 1000$ μS, $G_2 = 2000$ μS, $G_3 = 10,000$ μS, and I_T is 39 mA. Find I_1, I_2, and I_3.
6. Referring to Fig. 7-3, find R_T for the two branch resistances and calculate the voltage across the bank as I_TR_T.
7. For the voltage divider in Fig. 7-7, how much is the equivalent resistance for load D, load E, and load F?
8. Referring to the voltage divider in Fig. 7-7, calculate the power dissipated in R_1, R_2, and R_3.
9. Design a voltage divider similar to that in Fig. 7-7 with R_1, R_2, and R_3 across a 48-V source and the following loads: 48 V at 800 mA, 28 V at 300 mA, and 9 V at 100 mA. Use the bleeder current I_B of 120 mA.
10. Find the value of V_6 in Fig. 7-8.
11. Find the value of I_2 in Fig. 7-9.

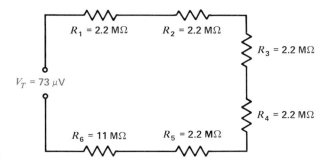

Fig. 7-8 For Prob. 10.

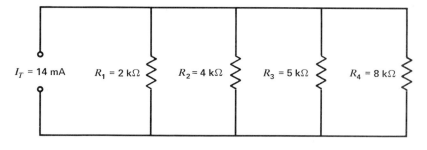

Fig. 7-9 For Prob. 11.

Answers to Practice Problems

7-1 a. $R_T = 100$ kΩ
 b. $V_3 = (^2\!/_{10}) \times V_T$

7-2 a. 2 to 1
 b. 1 to 2

7-3 a. 2 to 1
 b. 2 to 1

7-4 a. $^1\!/_3$
 b. $^1\!/_5$

7-5 a. $I_B = 30$ mA
 b. $V_{EG} = 40$ V
 c. $V_2 = 22$ V

Chapter 8
Direct-Current Meters

Voltage, current, and resistance measurements are generally made with a combination volt-ohm-milliammeter (VOM). To measure voltage, the voltmeter test leads are connected across the two points of potential difference. Similarly, when measuring resistance, the two leads are con-

nected across the resistance to be measured, but the power must be turned off. No power is needed in the circuit being tested because the ohmmeter has its own internal battery. To measure current, the meter is connected as a series component in the circuit. A combination meter with all three functions is generally used as a multitester to check V, I, and R when troubleshooting electronic circuits.

Important terms in this chapter are:

Ayrton shunt

back-off ohmmeter scale

clamp probe

continuity testing

D'Arsonval meter

decibel scale

digital multimeter (DMM)

galvanometer

loading effect of voltmeter

meter shunt

multiplier resistor

ohmmeter

ohms-per-volt sensitivity

taut-band meter

volt-ohm-milliammeter (VOM)

zero-ohms adjustment

Details of meter construction and how to make meter measurements are explained in the following sections:

8-1　Moving-Coil Meter

8-2　Measurement of Current

8-3　Meter Shunts

8-4　The Ayrton or Universal Shunt

8-5　Voltmeters

8-6　Loading Effect of a Voltmeter

8-7　Ohmmeters

8-8　Multimeters

8-9　Digital Multimeter (DMM)

8-10 Meter Applications

8-11 Checking Continuity with the Ohmmeter

8-1   Moving-Coil Meter

A typical volt-ohm-milliammeter is shown in Fig. 8-1. It can be used to measure voltage, as in Fig. 8-1a, or resistance, as in Fig. 8-1b. The typical VOM can also measure dc current.

A moving coil meter movement, shown in Fig. 8-2, is generally used in a VOM. The construction consists essentially of a coil of fine wire on a drum mounted between the poles of a permanent magnet. When direct current flows in the coil, the magnetic field of the current reacts with the field of the magnet. The resultant force turns the drum with its pointer. The amount of deflection indicates the amount of current in the coil. Correct polarity allows the pointer to read up-scale, to the right; the opposite polarity forces the pointer off-scale, to the left.

The pointer deflection is directly proportional to the amount of current in the coil. If 100 μA is the current needed for full-scale deflection, 50 μA in the coil will produce a half-scale deflection. The accuracy of the moving-coil meter mechanism is 0.1 to 2 percent.

The moving-coil principle is applied in several meter types which have different names. A *galvanometer* is an extremely sensitive instrument for measuring very small values of current. Laboratory-type galvanometers, which include a suspended moving coil with an optical system to magnify small deflection, can measure a small fraction of one microampere. A *ballistic galvanometer* is used for reading the value of a small momentary current, to measure electric charge. The suspended moving-coil arrangement of a galvanometer is often called a *D'Arsonval movement,* after its inventor, who patented this meter movement in 1881. The practical, commercial moving-coil meter in Fig. 8-2 is a *Weston movement.*

Values of I_M   The full-scale deflection current I_M is the amount needed to deflect the pointer all the way to the right to the last mark on the printed scale. Typical values of I_M are from about 10 μA to 30 mA for Weston movements. In a VOM, the I_M is typically either 50 μA or 1 mA.

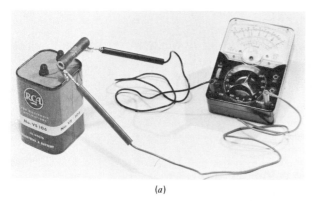

(a)

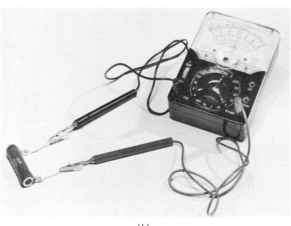

(b)

Fig. 8-1 Using a VOM for voltage and resistance measurements. (a) To read voltage, connect the voltmeter test leads across the potential difference being measured. Observe polarity for dc voltage. (b) To read resistance, connect the ohmmeter test leads across R, but with the power off. Polarity of meter leads does not matter for resistance.

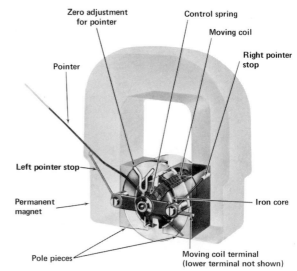

Fig. 8-2 Construction of moving-coil meter. (*Weston Instrument Corp.*)

As an example, I_M is 50 μA for the microammeter shown in Fig. 8-3. Notice the mirror along the scale to eliminate parallax error. You read the meter where the pointer and its mirror reflection are one. This eliminates the optical error of parallax when you look at the meter from the side. The schematic symbol for a current meter is a circle, as in Fig. 8-3b.

Values of r_M   This is the internal resistance of the wire of the moving coil. Typical values range from 1.2 Ω for a 30-mA movement to 2000 Ω for a 50-μA movement. A movement with a smaller I_M has a higher r_M because many turns of fine wire are needed. An average value of r_M for a 1-mA movement is about 120 Ω.

Taut-Band Meters   The meter movement can be constructed with the moving coil and pointer suspended by a metal band, instead of the pivot and jewel design with a restoring spring. Both types of movements have similar operating characteristics. However, taut-band meters generally have lower values of r_M because a smaller coil can be used to force the pointer up-scale.

Practice Problems 8-1
Answers at End of Chapter

a. Is a voltmeter connected in parallel or in series?
b. Is a milliammeter connected in parallel or in series?

8-2   Measurement of Current

Whether we are measuring amperes, milliamperes, or microamperes, two important facts to remember are:

1. The current meter must be in series in the circuit where the current is to be measured. The amount of deflection depends on the current through the meter. In a series circuit, the current is the same through all series components. Therefore, the current to be measured must be made to flow through the meter as a series component in the circuit.
2. A dc meter must be connected in the correct polarity

(a)

(b)

Fig. 8-3 Microammeter with mirror on scale to minimize parallax error. (a) Photo of meter. (b) Schematic symbol. (*Weston Instrument Corp.*)

for the meter to read up-scale. Reversed polarity makes the meter read down-scale, forcing the pointer against the stop at the left, which can bend the pointer.

How to Connect a Current Meter in Series
As illustrated in Fig. 8-4, the circuit must be opened at one point in order to insert the current meter in series in the circuit. Since R_1, R_2, R_3, and the meter are all in series, the current is the same in each and the meter reads the current in any part of the series circuit.

In Fig. 8-4, V_T is 150 V with a total series resistance of 1500 Ω for the current of 0.1 A, or 100 mA. This value is the current in R_1, R_2, R_3, and the battery, as shown in Fig. 8-4a. Note that in Fig. 8-4b, the circuit is opened at the junction of R_1 and R_2 for insertion of the meter. In Fig. 8-4c, the meter completes the series circuit to read the current of 100 mA. The meter inserted in series at any point in the circuit would read the same current. Turn off the power to connect the current meter and then put the power back on to read the current.

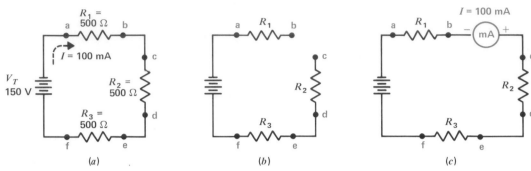

Fig. 8-4 Inserting a current meter in series. (*a*) Circuit without meter. (*b*) Circuit opened between points b and c for meter. (*c*) Meter connected between R_1 and R_2 in series with the circuit.

How to Connect a DC Meter in the Correct Polarity

A dc meter has its terminals marked for polarity, either with + and − signs or red for plus and black for minus. Electrons must flow into the negative side through the movement and out from the positive side for the meter to read up-scale.

To have the meter polarity correct, always connect the negative terminal to the point in the circuit that has a path back to the negative side of the voltage source, *without going through the meter*. Similarly, the positive terminal of the meter returns to the positive terminal of the voltage source (Fig. 8-5). Here the negative terminal of the meter is joined to R_2 because this path with R_1 connects to the negative terminal of the battery. The positive meter terminal is connected to R_3. Electrons in the circuit will flow through R_1 and R_2 into the negative side of the meter, through the movement, and out from the meter and return through R_3 to the positive battery terminal.

A Current Meter Should Have Very Low Resistance

Referring back to Fig. 8-4a, the milliammeter reads 100 mA because its resistance is negligible compared with the total series R of 1500 Ω. Then I is the same with or without the meter.

In general, a current meter should have very low R compared with the circuit where the current is being measured. We take an arbitrary figure of $\frac{1}{100}$. For the circuit in Fig. 8-4, then, the meter resistance should be less than $^{1500}\!/_{100} = 15$ Ω. Actually, a meter for 100 mA would have an internal R of about 1 Ω or less because of its internal shunt resistor. The higher the current range of the meter, the smaller its resistance.

An extreme case of a current meter with too much R is shown in Fig. 8-6. Here the series R_T is doubled when the meter is inserted in the circuit. The result is one-half the actual I in the circuit without the meter.

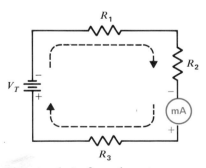

Fig. 8-5 Correct polarity for a dc meter.

Practice Problems 8-2
Answers at End of Chapter

a. In Fig. 8-4, how much will the milliammeter read when inserted at point a?

b. In Fig. 8-5, which R is connected to the positive side of the meter to make it read up-scale?

c. Should a current meter have very low or very high resistance?

d. Should a voltmeter have very high or very low resistance?

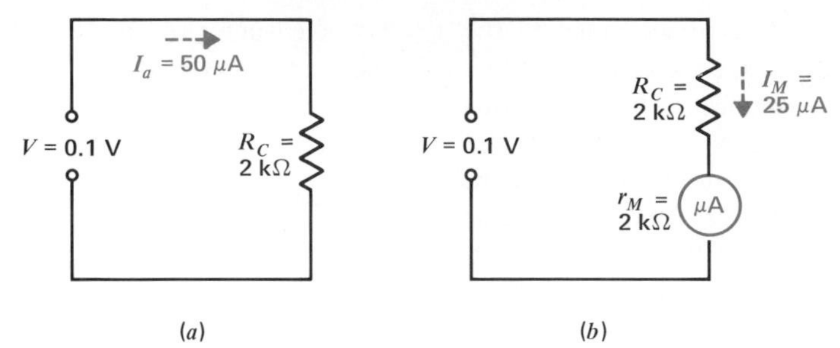

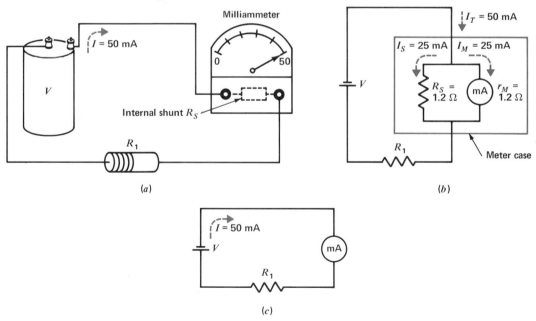

Fig. 8-6 Example of a current meter having too high a resistance. (a) Circuit without the meter has an I of 50 μA. (b) Meter resistance reduces I to 25 μA.

8-3   Meter Shunts

A meter shunt is a precision resistor connected across the meter movement for the purpose of shunting, or bypassing, a specific fraction of the circuit's current around the meter movement. The combination then provides a current meter with an extended range. The shunts are usually inside the meter case. However, the schematic symbol for the current meter usually does not show the shunt.

In current measurements, the parallel bank of the movement with its shunt is still connected as a current meter in series in the circuit (Fig. 8-7). It should be noted that a meter with an internal shunt has the scale calibrated to take into account the current through both the shunt and the meter movement. Therefore, the scale reads total circuit current.

Resistance of the Meter Shunt In Fig. 8-7b, the 25-mA movement has a resistance of 1.2 Ω, which is the resistance of the moving coil r_M. To double the

Fig. 8-7 Effect of a shunt in bypassing current around the meter movement to extend its range from 25 to 50 mA. (a) Wiring diagram. (b) Schematic showing effect of shunt R_S. With $R_S = r_M$ the current range is doubled. (c) Circuit with 50-mA meter.

range, the shunt resistance R_S is made equal to the 1.2 Ω of the movement. When the meter is connected in series in a circuit where the current is 50 mA, this total current into one terminal of the meter divides equally between the shunt and the meter movement. At the opposite meter terminal, these two branch currents combine to provide the 50 mA of the circuit current.

Inside the meter, the current is 25 mA through the shunt and 25 mA through the moving coil. Since it is a 25-mA movement, this current produces full-scale deflection. The scale is doubled, however, reading 50 mA, to account for the additional 25 mA through the shunt. Therefore, the scale reading indicates total current at the meter terminals, not just coil current. The movement with its shunt, then, is a 50-mA meter. Its internal resistance is 1.2 × ½ = 0.6 Ω.

Another example is shown in Fig. 8-8. In general, the shunt resistance for any range can be calculated with Ohm's law from the formula

$$R_S = \frac{V_M}{I_S} \tag{8-1}$$

where R_S is the resistance of the shunt and I_S is the current through it.

Voltage V_M is equal to $I_M \times r_M$. This is the voltage across both the shunt and the meter movement, which are in parallel.

Calculating I_S   This current through the shunt alone is the difference between the total current I_T through the meter and the divided current I_M through the movement. Or

$$I_S = I_T - I_M \tag{8-2}$$

Use the values of current for full-scale deflection, as these are known. In Fig. 8-8, $I_S = 50 - 10 = 40$ mA, or 0.04 A.

Calculating R_S   The complete procedure for using the formula $R_S = V_M/I_S$ can be as follows:

1. Find V_M. Calculate this for full-scale deflection as $I_M \times r_M$. In Fig. 8-8, with a 10-mA full-scale current through the 8-Ω movement, V_M is 0.01 × 8 = 0.08 V.
2. Find I_S. For the values that are shown in Fig. 8-8, $I_S = 50 - 10 = 40$ mA = 0.04 A.
3. Divide V_M by I_S to find R_S. For the final result, $R_S = 0.08/0.04 = 2$ Ω.

This shunt enables the 10-mA movement to be used for the extended range of 0 to 50 mA.

Note that R_S and r_M are inversely proportional to their full-scale currents. The 2 Ω for R_S equals one-fourth the 8 Ω of r_M because the shunt current of 40 mA is four times the 10 mA through the movement for full-scale deflection.

Example 1   A shunt extends the range of a 50-μA meter movement to 1 mA. How much is the current through the shunt of full-scale deflection?

Answer   All the currents must be in the same units for Formula (8-2). To avoid fractions, use 1000 μA for the 1-mA I_T. Then

$$I_S = I_T - I_M$$
$$= 1000 \ \mu A - 50 \ \mu A$$
$$I_S = 950 \ \mu A$$

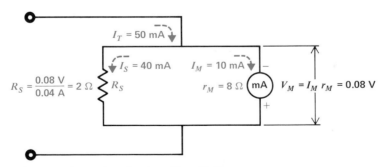

Fig. 8-8 Calculating the resistance of a meter shunt as $R_S = V_M/I_S$.

Example 2 A 50-μA meter movement has r_M of 1000 Ω. What R_S is needed to extend the range to 500 μA?

Answer The shunt current I_S is 500 − 50, or 450 μA. Then

$$R_S = \frac{V_M}{I_S}$$

$$= \frac{50 \times 10^{-6} \text{ A} \times 10^3 \text{ }\Omega}{450 \times 10^{-6} \text{ A}}$$

$$= \frac{50{,}000}{450} = \frac{1000}{9}$$

$$R_S = 111.1 \text{ }\Omega$$

The shunts usually are precision wire-wound resistors. For very low values, a short wire of precise size can be used.

Practice Problems 8-3
Answers at End of Chapter

A 50-μA movement with a 900-Ω r_M has a shunt R_S for the range of 500 μA.
a. How much is I_S?
b. How much is V_M?

8-4 The Ayrton or Universal Shunt

In Fig. 8-9, R_1, R_2, and R_3 are used in series-parallel combinations with the meter movements for different current ranges. The circuit is called an *Ayrton shunt* or *universal shunt*. This method is generally used for multiple current ranges in a VOM because the series-parallel circuit provides a safe method of switching between current ranges without danger of excessive current through the meter movement.

The wide contact on the switch arm in Fig. 8-9a indicates that it makes the next connection before breaking the old contact. This short-circuiting type of switch protects the meter movement by providing a shunt at all times during the switching to change ranges.

The universal shunt consists of R_1, R_2, and R_3 in Fig. 8-9. How they are connected as a shunt is determined by the switch S for the different current ranges. Their total resistance R_{ST} is 40 + 9 + 1 = 50 Ω. This resistance is used as a shunt in parallel with r_M for the 2-mA range in Fig. 8-9b. For the higher current ranges in Fig. 8-9c and d, part of R_{ST} is connected in series with r_M while the remainder of R_{ST} is in parallel as a shunt path.

The values in Fig. 8-9 are calculated as follows: Since the 2-mA range in Fig. 8-9b is double the 1-mA current rating of the meter movement, the shunt resistance must equal the r_M of 50 Ω so that 1 mA can flow in each of the two parallel paths. Therefore, R_{ST} is equal to the 50 Ω of r_M.

For the 10-mA range in Fig. 8-9c, 9 mA must flow through the shunt path and 1 mA through the meter path. Now r_M has R_1 in series with it, in the path bad. The shunt now includes R_2 in series with R_3, in the path bcd. Remember that the voltage is the same across the two parallel paths bad and bcd. The current is 1 mA in one path and 9 mA in the other path. To calculate R_1 we can equate the voltage across the two paths:

$$1 \text{ mA} \times (R_1 + r_M) = 9 \text{ mA} \times (R_2 + R_3)$$

We know r_M is 50 Ω. We also know R_{ST} is 50 Ω. We do not know R_1, R_2, or R_3, but $(R_2 + R_3)$ must be 50 Ω minus R_1. Therefore,

$$1 \text{ mA} \times (R_1 + 50) = 9 \text{ mA} \times (50 - R_1)$$

Solving for R_1,

$$R_1 + 50 = 450 - 9R_1$$
$$10R_1 = 400$$
$$R_1 = 40 \text{ }\Omega$$

Not only do we know now that R_1 is 40 Ω, but we also know that $(R_2 + R_3)$ must be 10 Ω, as they all must add up to 50 Ω. This value of 10 Ω for $(R_2 + R_3)$ is used for the next step in the calculations.

For the 100-mA range in Fig. 8-9d, 1 mA flows through R_1, R_2, and r_M in the path cbad, and 99 mA flows through R_3 in the path cd. The voltage is the same across both paths. To calculate R_2,

$$1 \text{ mA} \times (R_1 + R_2 + r_M) = 9 \text{ mA} \times R_3$$

We know R_1 is 40 Ω. Then

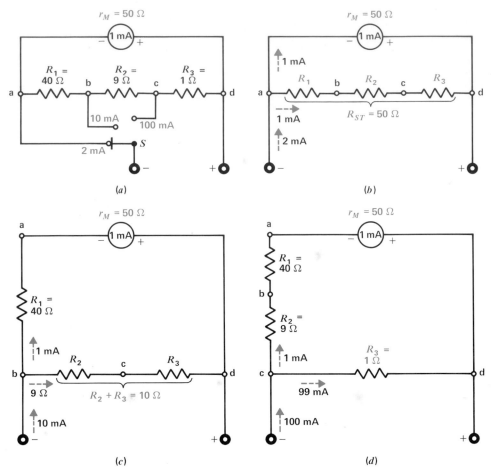

Fig. 8-9 Universal or Ayrton shunt for three current ranges. The I values are shown for full-scale deflection. (a) Actual circuit with switch S to choose different ranges. (b) Circuit for 2-mA range. (c) Circuit for 10-mA range. (d) Circuit for 100-mA range.

$$40 + R_2 + 50 = 99\,R_3$$

If $(R_2 + R_3)$ is 10 Ω, then R_3 must be $(10 - R_2)$. Substituting $(10 - R_2)$ for R_3, the equation becomes

$$40 + R_2 + 50 = 99 \times (10 - R_2)$$
$$R_2 + 90 = 990 - 99R_2$$
$$100R_2 = 900$$
$$R_2 = 9\ \Omega$$

Finally, R_3 must be 1 Ω.

The total of $R_1 + R_2 + R_3$ equals $40 + 9 + 1$, which equals the 50 Ω of R_{ST}.

As a proof of the resistance values, note that in Fig. 8-9b, 1 mA in each 50-Ω branch produces 50 mV

across both parallel branches. In Fig. 8-9c, 1 mA in the 90-Ω branch with the meter produces 90 mV from b to d, while 9 mA through the 10 Ω of $R_1 + R_2$ produces the same 90 mV. In Fig. 8-9d, 99 mA through the 1-Ω R_3 produces 99 mV, while 1 mA through the 99 Ω in path cbad produces the same 99 mV.

Practice Problems 8-4
Answers at End of Chapter

Refer to Fig. 8-9 and give the full-scale I through the meter movement on the

a. 2-mA range.

b. 100-mA range.

8-5   Voltmeters

Although a meter movement responds only to current in the moving coil, it is commonly used for measuring voltage by the addition of a high resistance in series with the movement (Fig. 8-10). The series resistance must be much higher than the coil resistance in order to limit the current through the coil. The combination of the meter movement with this added series resistance then forms a voltmeter. The series resistor, called a *multiplier,* is usually connected inside the voltmeter case.

Since a voltmeter has high resistance, it must be connected in parallel to measure the potential difference across two points in a circuit. Otherwise, the high-resistance multiplier would add so much series resistance that the current in the circuit would be reduced to a very low value. Connected in parallel, though, the high resistance of the voltmeter is an advantage. The higher the voltmeter resistance, the smaller the effect of its parallel connection on the circuit being tested.

The circuit is not opened to connect the voltmeter in parallel. Because of this convenience, it is common practice to make voltmeter tests in troubleshooting. The voltage measurements apply the same way to an *IR* drop or a generated emf.

The correct polarity must be observed in using a dc voltmeter. Connect the negative voltmeter lead to the negative side of the potential difference being measured and the positive lead to the positive side.

Multiplier Resistance   Figure 8-10 illustrates how the meter movement and its multiplier R_1 form a voltmeter. With 10 V applied by the battery in Fig. 8-10a, there must be 10,000 Ω of resistance to limit the current to 1 mA for full-scale deflection of the meter movement. Since the movement has a 50-Ω resistance, 9950 Ω is added in series, resulting in a 10,000-Ω total resistance. Then I is 10 V/10 kΩ = 1 mA.

With 1 mA in the movement, the full-scale deflection can be calibrated as 10 V on the meter scale, as long as the 9950-Ω multiplier is included in series with the movement. The multiplier can be connected on either side of the movement.

If the battery is taken away, as in Fig. 8-10b, the movement with its multiplier forms a voltmeter that can indicate a potential difference of 0 to 10 V applied across its terminals. When the voltmeter leads are connected across a potential difference of 10 V in a dc circuit, the resulting 1-mA current through the meter movement produces full-scale deflection, and the read-

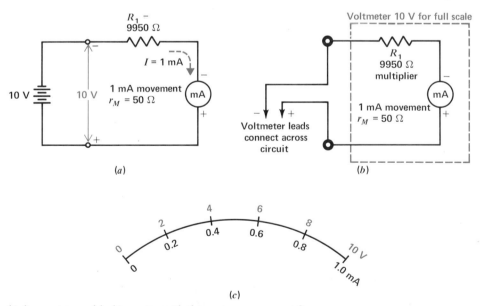

Fig. 8-10 A multiplier resistor added in series with the meter movement forms a voltmeter. (a) The multiplier R_1 allows a full-scale meter deflection with 10 V applied. (b) The voltmeter leads can be connected across a circuit to measure 0 to 10 V. (c) 10-V scale and the corresponding 1-mA scale.

ing is 10 V. In Fig. 8-10c the 10-V scale is shown corresponding to the 1-mA range of the movement.

If the voltmeter is connected across a 5-V potential difference, the current in the movement is ½ mA, the deflection is one-half of full scale, and the reading is 5 V. Zero voltage across the terminals means no current in the movement, and the voltmeter reads zero. In summary, then, any potential difference up to 10 V, whether an *IR* voltage drop or a generated emf, can be applied across the meter terminals. The meter will indicate less than 10 V in the same ratio that the meter current is less than 1 mA.

The resistance of a multiplier can be calculated from the formula

$$R_{mult} = \frac{\text{full-scale } V}{\text{full-scale } I} - r_M \qquad (8\text{-}3)$$

Applying this formula to the example of R_1 in Fig. 8-10 gives

$$R_{mult} = \frac{10 \text{ V}}{0.001 \text{ A}} - 50 \text{ }\Omega$$
$$= 10,000 - 50$$
$$R_{mult} = 9950 \text{ }\Omega$$

We can take another example for the same 10-V scale but with a 50-μA meter movement, which is commonly used. Now the multiplier resistance is much higher, though, because less *I* is needed for full-scale

deflection. Let the resistance of the 50-μA movement be 2000 Ω. Then

$$R_{mult} = \frac{10 \text{ V}}{0.000\ 050 \text{ A}} - 2000 \text{ }\Omega$$
$$= 200,000 - 2000$$
$$R_{mult} = 198,000 \text{ }\Omega$$

Multiple Voltmeter Ranges Voltmeters often have several multipliers which are used with one meter movement. A range switch selects one multiplier for the required scale. The higher the voltage range is, the higher the multiplier resistance, in essentially the same proportion as the ranges.

Figure 8-11 illustrates two ranges. When the switch is on the 10-V range, multiplier R_1 is connected in series with the 1-mA movement. Then you read the 10-V scale on the meter face. With the range switch on 25 V, R_2 is the multiplier, and the measured voltage is read on the 25-V scale.

Several examples of using these two scales are listed in Table 8-1. Note that voltages less than 10 V can be read on either scale. It is preferable, however, to have the pointer read on the middle third of the scale. That is why the scales are usually multiples of 10 and 2.5 or 3.

Range Switch With multiple ranges, the setting of the selector switch is the voltage that produces full-scale deflection (Fig. 8-12). One scale is generally used

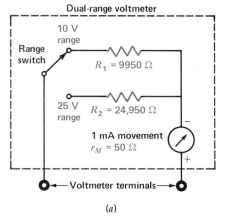
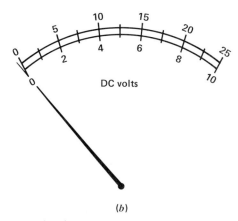

Fig. 8-11 Voltmeter with a range of either 10 or 25 V. (a) Range switch selects scale by connecting either R_1 and R_2 as the series multiplier. (b) Both voltage ranges on the face of the meter.

Table 8-1. Multiple Voltage-Scale Readings for Fig. 8-11

10-V scale, R_V* = 10,000 Ω			25-V scale, R_V* = 25,000 Ω		
Meter, mA	Deflection	Scale reading, V	Meter, mA	Deflection	Scale reading, V
0	0	0	0	0	0
0.5	½	5	0.2	²⁄₁₀	5
1.0	Full scale	10	0.4	⁴⁄₁₀	10
			0.5	½	12.5
			1.0	Full scale	25

*R_V is total voltmeter resistance of multiplier and meter movement.

for ranges that are multiples of 10. If the range switch is set for 250 V in Fig. 8-12, read the top scale as is. With the range switch at 25 V, however, the readings on the 250-V scale are divided by 10.

Similarly, the 100-V scale is used for the 100-V range and the 10-V range. In Fig. 8-12 the pointer indicates 30 V when the switch is on the 100-V range; this reading on the 10-V range is 3 V.

Typical Multiple Voltmeter Circuit

Another example of multiple voltage ranges is shown in Fig. 8-13, with a typical switching arrangement. Resistance R_1 is the series multiplier for the lowest voltage range of 2.5 V. When higher resistance is needed for the

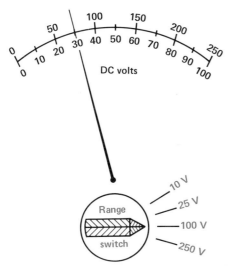

Fig. 8-12 The range switch selects the voltage that can produce full-scale deflection. The reading shown here is 30 V on the 100-V range.

higher ranges, the switch adds the required series resistors.

The meter in Fig. 8-13 requires 50 μA for full-scale deflection. For the 2.5-V range, a series resistance of $2.5/(50 \times 10^{-6})$, or 50,000 Ω, is needed. Since r_M is 2000 Ω, the value of R_1 is 50,000 − 2000, which equals 48,000 Ω or 48 kΩ.

For the 10-V range, a resistance of $10/(50 \times 10^{-6})$, or 200,000 Ω, is the value needed. Since $R_1 + r_M$ provides 50,000 Ω, R_2 is made 150,000 Ω, for a total of 200,000 Ω series resistance on the 10-V range. Similarly, additional resistors are switched in to increase the multiplier resistance for the higher voltage ranges. Note the separate jack and extra multiplier R_6 on the highest range for 5000 V. This method of adding series multipliers for higher voltage ranges is the circuit generally used in commercial multimeters.

Voltmeter Resistance

The high resistance of a voltmeter with a multiplier is essentially the value of the multiplier resistance. Since the multiplier is changed for each range, the voltmeter resistance changes.

Table 8-2 shows how the voltmeter resistance increases for the higher ranges. The middle column lists the total internal resistance R_V, including R_{mult} and r_M, for the voltmeter circuit in Fig. 8-13. With a 50-μA movement, R_V increases from 50 kΩ on the 2.5-V range to 20 MΩ on the 1000-V range. It should be noted that R_V has these values on each range whether you read full-scale or not.

Ohms-per-Volt Rating

To indicate the voltmeter's resistance independently of the range, voltmeters are generally rated in ohms of resistance needed for

Table 8-2. Characteristics of a Voltmeter Using a 50-μA Movement

Full-scale voltage V_F	$R_V = R_{mult} + r_M$	Ohms per volt = R_V/V_F
2.5	50 kΩ	20,000 Ω/V
10	200 kΩ	20,000 Ω/V
50	1 MΩ	20,000 Ω/V
250	5 MΩ	20,000 Ω/V
1000	20 MΩ	20,000 Ω/V

1 V of deflection. This value is the ohms-per-volt rating of the voltmeter. As an example, see the last column in Table 8-2. The values in the top row show that this meter needs 50,000 Ω R_V for 2.5 V of full-scale deflection. The resistance per 1 V of deflection then is 50,000/2.5, which equals 20,000 Ω/V.

The ohms-per-volt value is the same for all ranges. The reason is that this characteristic is determined by the full-scale current I_M of the meter movement. To calculate the ohms-per-volt rating, take the reciprocal of I_M in ampere units. For example, a 1-mA movement results in 1/0.001 or 1000 Ω/V; a 50-μA movement allows 20,000 Ω/V, and a 20-μA movement allows 50,000 Ω/V. The ohms-per-volt rating is also called the *sensitivity* of the voltmeter.

A high ohms-per-volt value means a high voltmeter resistance R_V. In fact R_V can be calculated as the product of the ohms-per-volt rating and the full-scale voltage of each range. For instance, across the second row in Table 8-2, on the 10-V range with a 20,000 Ω/V rating,

$$R_V = 10 \text{ V} \times \frac{20,000 \text{ }\Omega}{\text{volts}}$$

$$R_V = 200,000 \text{ }\Omega$$

These values are for dc volts only. The sensitivity for ac voltage is made lower, generally, to prevent erratic meter deflection produced by stray magnetic fields before the meter leads are connected into the circuit. Usually the ohms-per-volt rating of a voltmeter is printed on the meter face.

The sensitivity of 1000 Ω/V with a 1-mA movement used to be common for dc voltmeters, but 20,000 Ω/V with a 50-μA movement is generally used now. Higher sensitivity is an advantage, not only for less voltmeter loading, but because lower voltage ranges and higher ohmmeter ranges can be obtained.

Practice Problems 8-5
Answers at End of Chapter

Refer to Fig. 8-13 to calculate the voltmeter resistance R_V on the
a. 2.5-V range.
b. 50-V range.

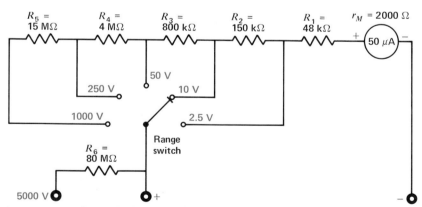

Fig. 8-13 Typical voltmeter circuit for multiple ranges (Simpson VOM Model 260).

8-6   Loading Effect of a Voltmeter

When the voltmeter resistance is not high enough, connecting it across a circuit can reduce the measured voltage, compared with the voltage present without the voltmeter. This effect is called *loading down* the circuit, since the measured voltage decreases because of the additional load current for the meter.

Loading Effect   Voltmeter loading can be appreciable in high-resistance circuits, as shown in Fig. 8-14. In Fig. 8-14*a*, without the voltmeter, R_1 and R_2 form a voltage divider across the applied voltage of 120 V. The two equal resistances of 100 kΩ each divide the applied voltage equally, with 60 V across each.

When the voltmeter in Fig. 8-14*b* is connected across R_2 to measure its potential difference, however, the voltage division changes. The voltmeter resistance R_V of 100 kΩ is the value for a 1000-ohms-per-volt meter on the 100-V range. Now the voltmeter in parallel with R_2 draws additional current and the equivalent resistance between the measured points 1 and 2 is reduced from 100,000 to 50,000 Ω. This resistance is one-third the total circuit resistance, and the measured voltage across points 1 and 2 drops to 40 V, as shown in Fig. 8-14*c*.

As additional current drawn by the voltmeter flows through the other series resistance R_1, this voltage goes up to 80 V.

Similarly, if the voltmeter were connected across R_1, this voltage would go down to 40 V, with the voltage across R_2 rising to 80 V. When the voltmeter is discon-

nected, the circuit returns to the condition in Fig. 8-14*a*, with 60 V across both R_1 and R_2.

The loading effect is minimized by using a voltmeter with a resistance much greater than the resistance across which the voltage is measured. As shown in Fig. 8-15, with a voltmeter resistance of 10 MΩ, the loading effect is negligible. Because R_V is so high, it does not change the voltage division in the circuit. The 10 MΩ of the meter in parallel with the 100,000 Ω for R_2 results in an equivalent resistance practically equal to 100,000 Ω.

With multiple ranges on a VOM, the voltmeter resistance changes with the range selected. Higher ranges require more multiplier resistance, increasing the voltmeter resistance for less loading. As examples, a 20,000-ohms-per-volt meter on the 250-V range has an internal resistance R_V of 20,000 × 250, or 5 MΩ. However, on the 2.5-V range the same meter has an R_V of 20,000 × 2.5, which is only 50,000 Ω.

On any one range, though, the voltmeter resistance is constant whether you read full-scale or less than full-scale deflection. The reason is that the multiplier resistance set by the range switch is the same for any reading on that range.

Correction for Loading Effect   The following formula can be used:

Actual reading + Correction

$$V = V_M + \frac{R_1 R_2}{R_V (R_1 + R_2)} V_M \qquad (8\text{-}4)$$

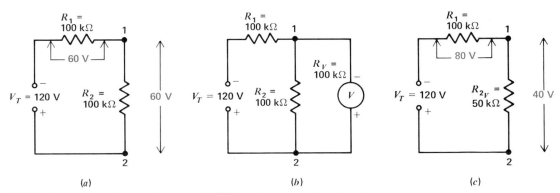

(a)   *(b)*   *(c)*

Fig. 8-14 Loading effect of a voltmeter. (*a*) High-resistance series circuit. (*b*) Voltmeter across one of the series resistances. (*c*) Reduced resistance and voltage between points 1 and 2 caused by the voltmeter resistance as a parallel branch. The R_{2V} includes R_2 and R_V.

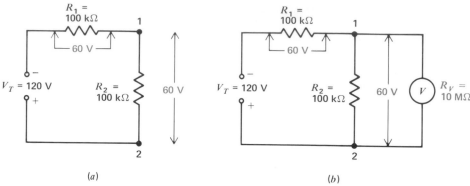

Fig. 8-15 Negligible loading with a high-resistance voltmeter. (a) High-resistance series circuit. (b) Same voltages in circuit with voltmeter connected.

Voltage V is the corrected reading the voltmeter would show if it had infinitely high resistance. Voltage V_M is the actual voltage reading. Resistances R_1 and R_2 are the voltage-dividing resistances in the circuit without the voltmeter resistance R_V. As an example, in Fig. 8-14,

$$V = 40 \text{ V} + \frac{100 \text{ k}\Omega \times 100 \text{ k}\Omega}{100 \text{ k}\Omega \times 200 \text{ k}\Omega} \times 40 \text{ V}$$

$$= 40 + \left(\frac{1}{2}\right) \times 40$$

$$= 40 + 20$$

$$V = 60 \text{ V}$$

The loading effect of a voltmeter reading too low because R_V is too low as a parallel resistance corresponds to the case of a current meter reading too low because r_M is too high as a series resistance. Both of these effects illustrate the general problem of trying to make any measurement without changing the circuit being measured.

Practice Problems 8-6
Answers at End of Chapter

With the voltmeter across R_2 in Fig. 8-14, what is the value for

a. V_1?

b. V_2?

8-7 Ohmmeters

Basically, an ohmmeter consists of an internal battery, the meter movement, and a current-limiting resistance, as illustrated in Fig. 8-16. For measuring resistance, the ohmmeter leads are connected across an external resistance to be measured, with power off in the circuit being tested. Then only the ohmmeter battery produces current for deflecting the meter movement. Since the amount of current through the meter depends on the external resistance, the scale can be calibrated in ohms.

The amount of deflection on the ohms scale indicates the measured resistance directly. The ohmmeter reads up-scale regardless of the polarity of the leads because the polarity of the internal battery determines the direction of current through the meter movement.

Series Ohmmeter Circuit In Fig. 8-16a, the circuit has 1500 Ω for ($R_1 + r_M$). Then the 1.5-V cell produces 1 mA, deflecting the moving coil full scale. When these components are enclosed in a case, as in Fig. 8-16b, the series circuit forms an ohmmeter.

If the leads are short-circuited together or connected across a short circuit, 1 mA flows. The meter movement is deflected full scale to the right. This ohmmeter reading is 0 Ω.

When the ohmmeter leads are open, not touching each other, the current is zero. The ohmmeter indicates infinitely high resistance or an open circuit across its terminals.

Therefore, the meter face can be marked zero ohms at the right for full-scale deflection and infinite ohms at

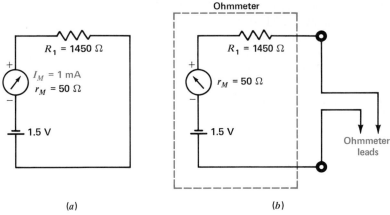

(a)   (b)

Fig. 8-16 Ohmmeter circuit. (*a*) Equivalent closed circuit when ohmmeter leads are short-circuited for zero ohms of external resistance. (*b*) Circuit with ohmmeter leads open.

the left for no deflection. In-between values of resistance result in less than 1 mA through the meter movement. The corresponding deflection on the ohms scale indicates how much resistance is across the ohmmeter terminals.

Back-off Ohmmeter Scale Table 8-3 and Fig. 8-17 illustrate the calibration of an ohmmeter scale in terms of meter current. The current equals V/R_T. Voltage V is the fixed applied voltage of 1.5 V supplied by the internal battery. Resistance R_T is the total resistance

Table 8-3. Calibration of Ohmmeter in Fig. 8-17

External R_x, Ω	Internal $R_i = R_1 + r_M$, Ω	$R_T = R_x + R_i$, Ω	$I = V/R_T$, mA	Deflection	Scale reading, Ω
0	1500	1500	1	Full scale	0
750	1500	2250	$\frac{2}{3} = 0.67$	$\frac{2}{3}$ scale	750
1500	1500	3000	$\frac{1}{2} = 0.5$	$\frac{1}{2}$ scale	1500
3000	1500	4500	$\frac{1}{3} = 0.33$	$\frac{1}{3}$ scale	3000
150,000	1500	151,500	0.01	$\frac{1}{100}$ scale	150,000
500,000	1500	501,500	0	None	∞

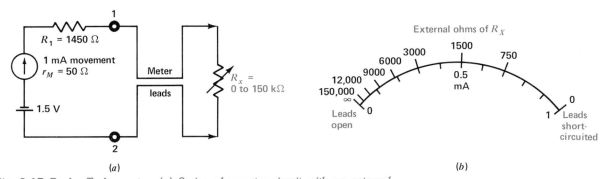

(a)   (b)

Fig. 8-17 Back-off ohmmeter. (*a*) Series ohmmeter circuit with an external resistor R_x. (*b*) Ohms scale reads higher resistances from right to left, as more R_x decreases I_M (see Table 8-3).

of R_x and the ohmmeter's internal resistance. Note that R_x is the external resistance to be measured.

The ohmmeter's internal resistance R_i is constant at 50 + 1450, or 1500 Ω here. If R_x also equals 1500 Ω, for example, R_T equals 3000 Ω. The current then is 1.5 V/3000 Ω, or 0.5 mA, resulting in half-scale deflection for the 1-mA movement. Therefore, the center of the ohms scale is marked for 1500 Ω. Similarly, the amount of current and meter deflection can be calculated for any value of the external resistance R_x.

Note that the ohms scale increases from right to left. This arrangement is called a *back-off scale*, with ohms values increasing to the left as the current backs off from full-scale deflection. The back-off scale is a characteristic of any ohmmeter where the internal battery is in series with the meter movement. Then more external R_x decreases the meter current.

A back-off ohmmeter scale is expanded at the right near zero ohms and crowded at the left near infinite ohms. This nonlinear scale results from the relation of $I = V/R$ with V constant at 1.5 V. Specifically, the back-off ohms scale represents the graph of a hyperbolic curve for the reciprocal function $y = 1/x$.

The highest resistance that can be indicated by the ohmmeter is about 100 times its total internal resistance. Therefore, the infinity mark on the ohms scale, or the "lazy eight" symbol ∞ for infinity, is only relative. It just means that the measured resistance is infinitely greater than the ohmmeter resistance.

For instance, if a 500,000-Ω resistor in good condition were measured with the ohmmeter in Fig. 8-17, it would indicate infinite resistance because this ohmmeter cannot measure as high as 500,000 Ω. To read higher values of resistance, the battery voltage can be increased to provide more current, or a more sensitive meter movement is necessary to provide deflection with less current.

Multiple Ohmmeter Ranges   Commercial multimeters provide for resistance measurements from less than 1 Ω up to many megohms, in several ranges. The range switch in Fig. 8-18 shows the multiplying factors for the ohms scale. On the $R \times 1$ range, for low-resistance measurements, read the ohms scale directly. In the example here, the pointer indicates 12 Ω. When the range switch is on $R \times 100$, multiply the scale reading

by 100; this reading would then be 12×100 or 1200 Ω. On the $R \times 10{,}000$ range, the pointer would indicate 120,000 Ω.

A multiplying factor, instead of full-scale resistance, is given for each ohms range because the highest resistance is infinite on all the ohms ranges. This method for ohms should not be confused with the full-scale values for voltage ranges. For the ohmmeter ranges, always multiply the scale reading by the $R \times$ factor. On voltage ranges, you may have to multiply or divide the scale reading to match the full-scale voltage with the value on the range switch.

Typical Ohmmeter Circuit   For high-ohms ranges a sensitive meter is necessary to read the low values of I with the high values of R_x. For the case of low ohms, however, less sensitivity is needed for the higher currents. These opposite requirements are solved by using a meter shunt across the meter movement and changing the shunt resistance for the multiple ohmmeter ranges. In Fig. 8-19, R_S is the meter shunt.

To analyze the ohmmeter circuit in Fig. 8-19, three conditions are shown. All are for the $R \times 1$ range with 12-Ω R_S. Figure 8-19a shows the internal circuit before the ohmmeter is adjusted for zero ohms. In Fig. 8-19b the test leads are short-circuited. Then there are two paths for branch current produced by the battery V. One

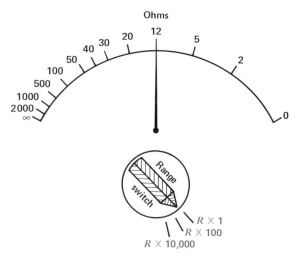

Fig. 8-18 Multiple ohmmeter ranges. Multiply ohms reading by factor set on range switch.

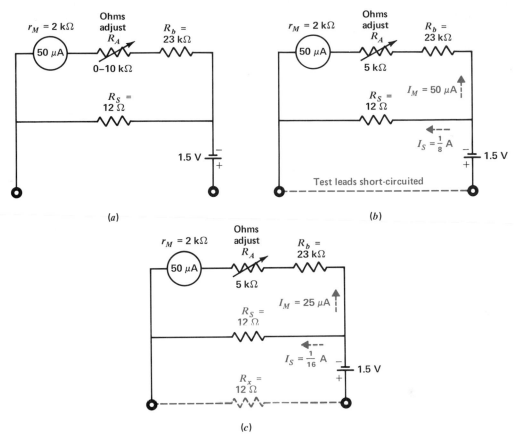

Fig. 8-19 Typical ohmmeter circuit for $R \times 1$ range. (a) Circuit before zero-ohms adjustment. (b) Test leads short-circuited to adjust for zero ohms. (c) Measuring external resistance R_x. The value of 12 Ω for R_x results in half-scale deflection, as shown on the ohms scale in Fig. 8-18.

branch is R_S. The other branch includes R_b, R_A, and the meter movement. The 1.5 V is across both branches.

To allow 50 μA through the meter, R_A is adjusted to 5000 Ω. Then the total resistance in this branch is 23 kΩ + 5 kΩ + 2 kΩ, which equals 30 kΩ. With 30 kΩ across 1.5 V, I_M equals 50 μA. Therefore, R_A is adjusted for full-scale deflection to read zero ohms with the test leads short-circuited.

In Fig. 8-19c assume that a resistance R_x being measured is 12 Ω, equal to R_S. Then the meter current is practically 25 μA for half-scale deflection. The center ohms reading on the $R \times 1$ scale, therefore, is 12 Ω. For higher values of R_x, the meter current decreases to indicate higher resistances on the back-off ohms scale.

For higher ohms ranges, the resistance of the R_S branch is increased. The half-scale ohms reading on each range is equal to the resistance of the R_S branch. A higher battery voltage can also be used for the highest ohms range.

On any range, R_A is adjusted for full-scale deflection to read zero ohms with the test leads short-circuited. This variable resistor is the *ohms adjust* or *zero-ohms* adjustment.

Zero-Ohms Adjustment   To compensate for lower voltage output as the internal battery ages, an ohmmeter includes a variable resistor such as R_A in Fig. 8-19, to calibrate the ohms scale. A back-off ohmmeter is always adjusted for zero ohms. With the test leads short-circuited, vary the ZERO OHMS control on the front panel of the meter until the pointer is exactly on zero at the right edge of the ohms scale. Then the ohms readings are correct for the entire scale.

This type of ohmmeter must be zeroed again every time you change the range. The reason is that the internal circuit is changed.

When the adjustment cannot deflect the pointer all the way to zero at the right edge, it usually means the battery voltage is too low and the internal dry cells must be replaced. Usually, this trouble shows up first on the $R \times 1$ range, which takes the most current from the battery. The ohmmeter battery in a typical VOM can be seen in Fig. 8-20.

Shunt-Ohmmeter Circuit   In this circuit, the internal battery, meter movement, and external R_x are in three parallel paths. The main advantage is a low-ohms scale that reads from left to right. However, the shunt

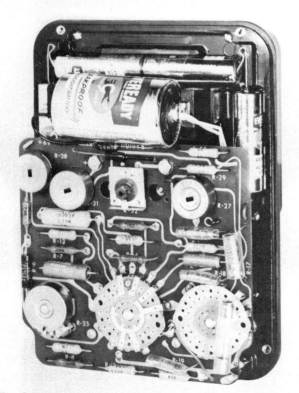

Fig. 8-20 Typical VOM with back cover off to show shunts, multipliers, and batteries for ohmmeter. Separate 1.5-V cell is for high ohms ranges. (*Triplett Corp.*)

ohmmeter circuit is seldom used because of constant current drain on the internal battery.

Practice Problems 8-7
Answers at End of Chapter

a. The ohmmeter reads 40 Ω on the $R \times 10$ range. How much is R_x?

b. A voltmeter reads 40 V on the 300-V scale, but with the range switch on 30 V. How much is the measured voltage?

8-8   Multimeters

These are also called *multitesters*, and they are used to measure voltage, current, or resistance. The main types are the volt-ohm-milliammeter (VOM), in Fig. 8-21, and the digital multimeter (DMM), in Fig. 8-22. Table 8-4 compares their features. The DMM is explained in more detail in the next section.

Because it is simple, compact, and portable, the VOM is probably more common. The cost of a basic type of VOM is less than for a DMM. Also, where a change in V or I must be checked, the VOM is more convenient.

Besides its digital readout, an advantage of the DMM is its high input resistance R_V, as a dc voltmeter. The R_V is usually 10 MΩ, the same on all ranges, which is high enough to prevent any loading effect by the voltmeter in most circuits. Some types have R_V of 22 MΩ.

Most multimeters measure ac voltage and current values. The ac variations are converted to dc voltage internally for the meter. However, the frequency response is generally limited to about 20 kHz or less. Also, the input impedance is lower as an ac voltmeter because of the rectifier circuit. In order to measure radio-frequency signal voltages, a special RF meter or an RF probe is necessary.

For either a VOM or a DMM, it is important to have a low-voltage dc scale with resolution good enough to read 0.2 V or less. The range of 0.2 to 0.6 V, or 200 to 600 mV, is needed for measuring dc bias voltages in transistor circuits.

Low-Power Ohms Another feature needed for transistor measurements is an ohmmeter that does not have enough battery voltage to bias a semiconductor junction into the ON or conducting condition. The limit is 0.2 V or less. The purpose is to prevent any parallel conduction paths in the transistor amplifier circuit, which can lower the ohmmeter reading.

For the VOM in Fig. 8-21, the open-circuit voltage is 0.1 V, or 100 mV, on the low-power ohms ranges. This voltage is too low to turn on a semiconductor junction. These ohms ranges are $R \times 1$ and $R \times 10$. The higher ohms ranges need more battery voltage.

Decibel Scale Most analog multimeters have an ac voltage scale calibrated in decibel (dB) units, for measuring ac signals. The decibel is a logarithmic unit used

Fig. 8-22 Portable digital multimeter (DMM). Power supplied by internal 9-V battery. (*Beckman Instruments Inc.*)

Fig. 8-21 Front view of typical VOM. Height is 7 in. Note switch positions at left side for clamp-on ammeter and at top for low-power ohms. (*Simpson Electric Company*)

Table 8-4. Comparison of VOM with DMM

VOM	DMM
Analog pointer reading	Digital readout
DC voltmeter R_V changes with range	R_V is 10 or 22 MΩ, the same on all ranges
Zero-ohms adjustment changed for each range	No zero-ohms adjustment
Ohms ranges up to $R \times 10,000$ Ω	Ohms ranges up to 20 MΩ

for comparisons of power levels or voltage levels. The mark of 0 dB on the scale indicates the reference level, which is usually 0.775 V for 1 mW across 600 Ω. Positive decibel values, above the zero mark, indicate ac voltages above the reference of 0.775 V; negative decibel values are less than the reference level.

Amp-Clamp Probe

The problem of opening a circuit to measure I can be eliminated by using a probe with a clamp that fits around the current-carrying wire. Its magnetic field is used to indicate the amount of current. The clamp in Fig. 8-23 is an accessory probe for the VOM in Fig. 8-21. This probe measures just ac amperes, generally for the 60-Hz power line.

High-Voltage Probe

The accessory probe in Fig. 8-24 can be used with a multimeter to measure dc voltages up to 30 kV. One application is measuring the high voltage of 20 to 30 kV at the anode of the color picture tube in a television receiver. The probe is basically just an external multiplier resistance for the dc voltmeter. The required R for a 30-kV probe is 580 MΩ with a 20-kΩ/V meter on the 1000-V range.

Practice Problems 8-8
Answers at End of Chapter

a. How much is R_V on the 1-V range for a VOM with a sensitivity of 20 kΩ/V?

b. If R_V is 10 MΩ for a DMM on the 100-V range, how much is R_V on the 200-mV range?

Fig. 8-24 High-voltage probe accessory for VOM. (*Triplett Corp.*)

8-9   Digital Multimeter (DMM)

This type has become very popular because the digital readout is displayed automatically with decimal point, polarity, and the unit, V, A, or Ω. Digital meters are generally easier to use because they eliminate the human error that often occurs in reading different scales on an analog meter.

Although the details of digital circuits are explained later, in Chap. 31, some of the main features of a typical DMM can be seen from the block diagram in Fig. 8-25. The input must be put into a form that fits the needs of the analog-to-digital (A/D) converter. As an example, the dc voltage range of −200 mV to +200 mV is needed here. The A/D converter accurately converts an unknown voltage to a digital equivalent, which is shown on the digital display. Either a light-emitting diode (LED) or liquid-crystal display (LCD) can be used. The LCD type requires less power.

When the input is ac voltage, it is converted to dc voltage for the A/D converter. For measuring resistance, a direct current is supplied that converts R to an IR voltage. To measure current, internal shunt resistors provide a proportional voltage to be measured. The DMM is connected in series in the circuit to measure current.

When the dc voltage is too high for the A/D converter, it is divided down to the range of 200 mV. When the dc voltage is too low, it is increased by a dc amplifier. The measured voltage can then be compared with a fixed reference voltage in the meter. The input resistance for the dc voltage ranges is 22 MΩ for this meter, but 10 MΩ is probably more common. The input resistance for the ac voltage ranges is much less because of the rectifier circuit needed to convert to dc voltage.

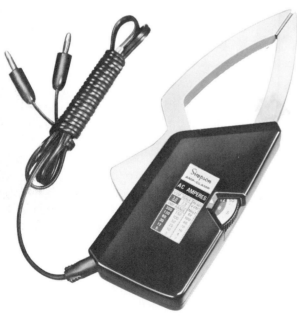

Fig. 8-23 Amp-clamp accessory for VOM in Fig. 8-21. (*Simpson Electric Company*)

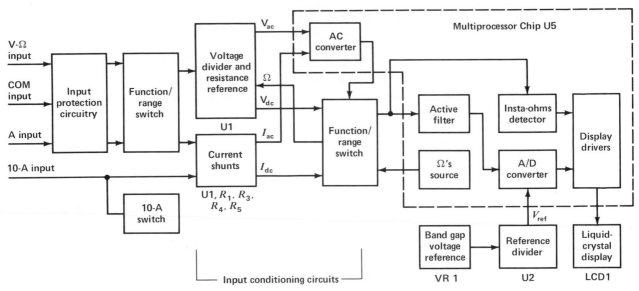

Fig. 8-25 Block diagram of circuits for DMM shown in Fig. 8-22. (*Beckman Instruments Inc.*)

The resolution of the meter reading is indicated by how many places can be used to display the digits 0 to 9, regardless of the decimal point. For instance, 9.99 V is a three-digit display. Also, 9.999 V would be a four-digit display. Many portable units, however, compromise with a 3½-digit display. This means that either the first of the four digits is a 1 or nothing is shown. A 3½-digit display can show 19.99 V on the 20-V range, but 7.999 V would be read as 8.00 V, as examples. This factor is the reason why the different ranges are multiples of 2.

Another useful feature of the DMM in Fig. 8-22 is indicated by the arrow for the position at the right on the ohms ranges. The arrow is the schematic symbol for a semiconductor diode, or a semiconductor junction in general. On this position, the meter supplies a fixed current in the forward direction and the voltage across the junction is shown on the display. A reading of 0.6 V is a normal value for a silicon junction, either for a diode or in NPN and PNP transistors. This method is much more accurate than checking junction resistance with an ohmmeter.

It should be noted that many technicians prefer an analog meter in some applications where the voltage or current must be adjusted. For instance, in adjusting for a null reading of zero, the moving needle of a VOM is probably easier to interpret.

Practice Problems 8-9
Answers at End of Chapter

Answer True or False.

a. A DMM has high input resistance of 10 to 22 MΩ when used as a dc voltmeter.

b. Ohms readings cannot be made accurately with a DMM.

c. A 3½-digit display can read 2.999 V.

8-10   Meter Applications

Table 8-5 summarizes the main points to remember in using a voltmeter, ohmmeter, or milliammeter. These rules apply whether the meter is a single unit or one function on a multimeter. Also, the voltage and current tests apply to either dc or ac circuits.

To avoid excessive current through the meter, it is good practice to start on a high range when measuring an unknown value of voltage or current. It is very important not to make the mistake of connecting a current meter in parallel, because usually this mistake ruins the meter. The mistake of connecting a voltmeter in series

Table 8-5. Direct-Current Meters

Voltmeter	Milliammeter or ammeter	Ohmmeter
Power on in circuit	Power on in circuit	Power off in circuit
Connect in parallel	Connect in series	Connect in parallel
High internal R	Low internal R	Has internal battery
Has internal series multipliers; higher R for higher ranges	Has internal shunts; lower resistance for higher current ranges	Higher battery voltage and more sensitive meter for higher ohms ranges

does not damage the meter, but the reading will be wrong.

If the ohmmeter is connected to a circuit where power is on, the meter can be damaged, besides giving the wrong reading. An ohmmeter has its own internal battery, and the power must be off in the circuit being tested.

Connecting a Current Meter in the Circuit
In a series-parallel circuit, the current meter must be inserted in a branch to read branch current. In the main line, the meter reads the total current. These different connections are illustrated in Fig. 8-26. The meters are shown by dashed lines to illustrate the different points at which a meter could be connected to read the respective currents.

If the circuit is opened at point a to insert the meter in series in the main line here, the meter will read total

line current I_T through R_1. A meter at b or c will read the same line current.

In order to read the branch current through R_2, this R must be disconnected from its junction with the main line at either end. A meter inserted at d or e, therefore, will read the R_2 branch current I_2. Similarly, a meter at f or g will read the R_3 branch current I_3.

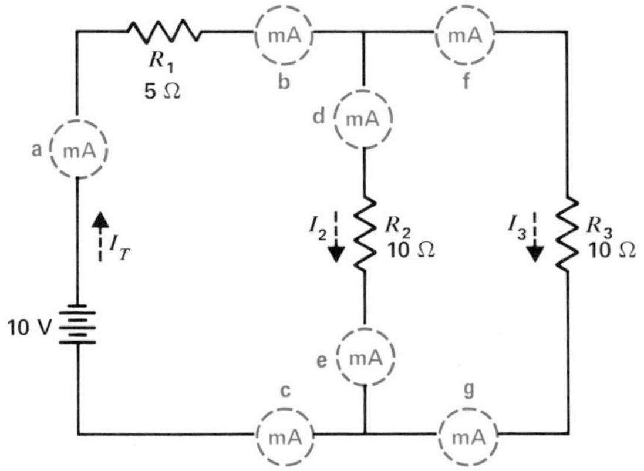

Calculating I from Measured Voltage
The inconvenience of opening the circuit to measure current can often be eliminated by the use of Ohm's law. The voltage and resistance can be measured without opening the circuit and the current calculated as V/R. In the example in Fig. 8-27, when the voltage across R_2 is 15 V and its resistance is 15 Ω, the current through R_2 must be 1 A. When values are checked during troubleshooting, if the voltage and resistance are normal, so is the current.

This technique can also be convenient for determining I in low-resistance circuits where the resistance of a microammeter may be too high. Instead of measuring I, measure V and R and calculate I as V/R.

Fig. 8-26 Inserting a current meter in a series-parallel circuit. At a, b, or c, meter reads I_T; at d or e, meter reads I_2; at f or g, meter reads I_3.

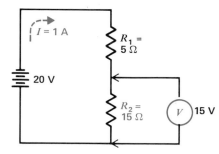

Fig. 8-27 With 15 V measured across 15 Ω, the current I is $V/R = 1$ A.

Furthermore, if necessary, we can insert a known resistance R_S in series in the circuit, temporarily, just to measure V_S. Then I is calculated as V_S/R_S. The resistance of R_S, however, must be small enough to have little effect on R_T and I in the series circuit.

This technique is often used with oscilloscopes to produce a voltage waveform of IR which has the same waveform as the current in a resistor. The oscilloscope must be connected as a voltmeter because of its high input resistance.

Checking Fuses Turn the power off or remove the fuse from the circuit to check with an ohmmeter. A good fuse reads 0 Ω. A blown fuse is open, which reads infinity on the ohmmeter.

A fuse can also be checked with the power on in the circuit by using a voltmeter. Connect the voltmeter across the two terminals of the fuse. A good fuse reads 0 V because there is practically no IR drop. With an open fuse, though, the voltmeter reading is equal to the full value of the applied voltage. Having the full applied voltage seems to be a good idea, but it should not be across the fuse.

Voltage Tests for an Open Circuit Figure 8-28 shows four equal resistors in series with a 100-V source. A ground return is shown here because voltage measurements are usually made to chassis ground. Normally, each resistor would have an IR drop of 25 V. Then, at point B the voltmeter to ground should read $100 - 25 = 75$ V. Also, the voltage at C should be 50 V, with 25 V at D, as shown in Fig. 8-28a.

However, the circuit in Fig. 8-28b has an open in R_3, toward the end of the series string of voltages to ground. Now when you measure at B, the reading is 100 V, equal to the applied voltage. This full voltage at B shows that the series circuit is open, without any IR drop across R_1. The question is, however, which R has the open? Continue the voltage measurements to ground until you find 0 V. In this example, the open is in R_3, between the 100 V at C and 0 V at D.

The points that read the full applied voltage have a path back to the source of voltage. The first point that reads 0 V has no path back to the high side of the source. Therefore, the open circuit must be between points C and D in Fig. 8-28b.

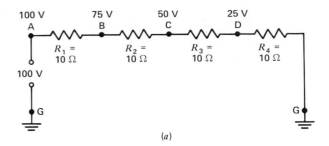

(a)

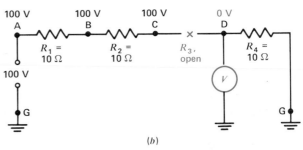

(b)

Fig. 8-28 Voltage tests to localize an open circuit. (a) Normal circuit with voltages to chassis ground. (b) Reading of 0 V at point D shows R_3 is open.

Practice Problems 8-10
Answers at End of Chapter

a. Which type of meter requires an internal battery?
b. How much is the normal voltage across a good fuse?
c. How much is the voltage across R_1 in Fig. 8-28a?
d. How much is the voltage across R_1 in Fig. 8-28b?

8-11 Checking Continuity with the Ohmmeter

A wire conductor that is continuous without a break has practically zero ohms of resistance. Therefore, the ohmmeter can be useful in testing for the continuity. This test should be done on the lowest ohms range. There are many applications. A wire conductor can have an internal break which is not visible because of the insulated cover, or the wire can have a bad connection at the terminal. Checking for zero ohms between any two points along the conductor tests continuity. A

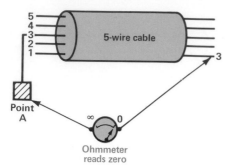

Fig. 8-29 Continuity from A to 3 shows that this wire is connected to terminal A.

break in the conducting path is evident from a reading of infinite resistance, showing an open circuit.

As another application of checking continuity, suppose there is a cable of wires harnessed together as illustrated in Fig. 8-29, where the individual wires cannot be seen, but it is desired to find the conductor that connects to terminal A. This is done by checking continuity for each conductor to point A. The wire that has zero ohms to A is the one connected to this terminal. Often the individual wires are color-coded, but it may be necessary to check the continuity of each lead.

An additional technique that can be helpful is illustrated in Fig. 8-30. Here it is desired to check the conti-

nuity of the two-wire line, but its ends are too far apart for the ohmmeter leads to reach. The two conductors are temporarily short-circuited at one end, however, so that the continuity of both wires can be checked at the other end.

In summary, then, the ohmmeter is helpful in checking the continuity of any wire conductor. This includes resistance-wire heating elements, like the wires in a toaster or the filament of an incandescent bulb. Their cold resistance is normally just a few ohms. Infinite resistance means that the wire element is open. Similarly, a good fuse has practically zero resistance; a burned-out fuse has infinite resistance, meaning it is open. Also, any coil for a transformer, solenoid, or motor will have infinite resistance if the winding is open.

Practice Problems 8-11
Answers at End of Chapter

a. On a back-off ohmmeter, is zero ohms at the left or the right end of the scale?

b. What is the ohmmeter reading for an open circuit?

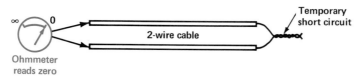

Fig. 8-30 Temporary short circuit at one end of a long two-wire line to check continuity from opposite end.

Summary

1. Direct current in a moving-coil meter deflects the coil in proportion to the amount of current.
2. A current meter is a low-resistance meter connected in series to read the amount of current in the circuit.
3. A meter shunt R_S in parallel with the meter movement extends the range of a current meter [see Formula (8-1)].
4. A voltmeter consists of the meter movement in series with a high-resistance multiplier. The voltmeter with its multiplier is connected across two points to measure their potential difference in volts. The multiplier R can be calculated from Formula (8-3).

5. The ohms-per-volt rating of a voltmeter with series multipliers specifies the sensitivity on all voltage ranges. It equals the reciprocal of the full-scale deflection current of the meter. A typical value is 20,000 Ω/V for a voltmeter using a 50-μA movement. The higher the ohms-per-volt rating, the better.

6. Voltmeter resistance R_V is higher for higher ranges because of higher-resistance multipliers. Multiply the ohms-per-volt rating by the voltage range to calculate the R_V for each range.

7. An ohmmeter consists of an internal battery in series with the meter movement. Power must be off in the circuit being checked with an ohmmeter. The series ohmmeter has a back-off scale with zero ohms at the right edge and infinity at the left. Adjust for zero ohms with the leads short-circuited each time the ohms range is changed.

8. The VOM is a portable multimeter that measures volts, ohms, and milliamperes.

9. The digital multimeter generally has an input resistance of 10 MΩ on all dc voltage ranges.

10. In checking wire conductors, the ohmmeter reads 0 Ω or very low R for normal continuity and infinite ohms for an open.

Self-Examination
Answers at Back of Book

Choose (a), (b), (c), or (d).

1. To connect a current meter in series: (a) open the circuit at one point and use the meter to complete the circuit; (b) open the circuit at the positive and negative terminals of the voltage source; (c) short-circuit the resistance to be checked and connect the meter across it; (d) open the circuit at one point and connect the meter to one end.

2. To connect a voltmeter in parallel to read an IR drop: (a) open the circuit at one end and use the meter to complete the circuit; (b) open the circuit at two points and connect the meter across both points; (c) allow the circuit to remain as is and connect the meter across the resistance; (d) allow the circuit to remain closed but disconnect the voltage source.

3. A shunt for a milliammeter (a) extends the range and reduces the meter resistance; (b) extends the range and increases the meter resistance; (c) decreases the range and the meter resistance; (d) decreases the range but increases the meter resistance.

4. For a 50-μA movement with 2000-Ω r_M, the voltage V_M at full-scale deflection is (a) 0.1 V; (b) 0.2 V; (c) 0.5 V; (d) 250 μV.

5. A voltmeter using a 20-μA meter movement has a sensitivity of (a) 1000 Ω/V; (b) 20,000 Ω/V; (c) 50,000 Ω/V; (d) 11 MΩ/V.

6. When using an ohmmeter, disconnect the applied voltage from the circuit being checked because: (a) the voltage source will increase the resistance; (b) the current will decrease the resistance; (c) the ohmmeter has its own internal battery; (d) no current is needed for the meter movement.

7. A multiplier for a voltmeter is (a) a high resistance in series with the meter movement; (b) a high resistance in parallel with the meter movement; (c) usu-

ally less than 1 Ω in series with the meter movement; (d) usually less than 1 Ω in parallel with the meter movement.

8. To double the current range of a 50-μA 2000-Ω meter movement, the shunt resistance is (a) 40 Ω; (b) 50 Ω; (c) 2000 Ω; (d) 18,000 Ω.

9. With a 50-μA movement, a VOM has an input resistance of 6 MΩ on the dc voltage range of (a) 3; (b) 12; (c) 60; (d) 300.

10. For a 1-V range, a 50-μA movement with an internal R of 2000 Ω needs a multiplier resistance of (a) 1 kΩ; (b) 3 kΩ; (c) 18 kΩ; (d) 50 kΩ.

Essay Questions

1. (a) Why is a milliammeter connected in series in a circuit? (b) Why should the milliammeter have low resistance?

2. (a) Why is a voltmeter connected in parallel in a circuit? (b) Why should the voltmeter have high resistance?

3. A circuit has a battery across two resistances in series. (a) Draw a diagram showing how to connect a milliammeter in the correct polarity to read current through the junction of the two resistances. (b) Draw a diagram showing how to connect a voltmeter in the correct polarity to read the voltage across one resistance.

4. Explain briefly why a meter shunt equal to the resistance of the moving coil doubles the current range.

5. Describe how to adjust the ZERO OHMS control on a back-off ohmmeter.

6. What is meant by a 3½-digit display on a DMM?

7. Give two advantages of the DMM in Fig. 8-22 compared with the conventional VOM in Fig. 8-21.

8. What is the function of the ZERO OHMS control in the circuit of a back-off ohmmeter?

9. State two precautions to be observed when you use a milliammeter.

10. State two precautions to be observed when you use an ohmmeter.

11. The resistance of a voltmeter R_V is 300 kΩ on the 300-V range when measuring 300 V. Why is R_V still 300 kΩ when measuring 250 V on the same range?

12. Redraw the schematic diagram in Fig. 5-1b, in Chap. 5, showing a milliammeter to read line current through R_1 and R_2, a meter for R_3 branch current, and a meter for R_4 branch current. Label polarities on each meter.

Problems
Answers to Odd-Numbered Problems at Back of Book

1. Calculate the shunt resistance needed to extend the range of a 50-Ω 1-mA movement to (a) 2 mA; (b) 10 mA; (c) 100 mA. (d) In each case, how much current is indicated by half-scale deflection?

2. With a 50-Ω 1-mA movement, calculate the multiplier resistances needed for ranges of (a) 10 V; (b) 30 V; (c) 100 V; (d) 300 V. How much voltage is indicated by half-scale deflection for each range?

3. A voltmeter reads 30 V across a 100-Ω resistance. (a) How much is the current

in the resistor? (**b**) If the current through the same resistance were doubled, how much would its *IR* voltage be?

4. A voltmeter has a sensitivity of 10,000 Ω/V on all ranges. (**a**) How much is the total voltmeter resistance on the 5-V range? (**b**) On the 50-V range? (**c**) On the 500-V range? (**d**) How much is the voltmeter resistance for a reading of 225 V on the 500-V range?

5. A 50-μA meter movement has an internal resistance of 1000 Ω. (**a**) Calculate the multiplier resistance needed for voltmeter ranges of 10, 30, and 500 V. (**b**) How much is the ohms-per-volt sensitivity rating on all ranges? (**c**) How much is the voltmeter resistance on the 500-V range?

6. For the same meter movement as in Prob. 5, calculate the shunt resistances needed for current ranges of 10, 30, and 500 mA. How much is the resistance of the meter with its shunt on each range? (Note: 1 mA = 1000 μA.)

7. Referring to the universal shunt in Fig. 8-9, calculate the required values of R_1, R_2, and R_3 for a 50-μA 2000-Ω movement to provide current ranges of 1.2, 12, and 120-mA.

8. Referring to the voltmeter loading problem in Fig. 8-14, exactly how much voltage would be indicated by a 20,000-Ω/V meter on its 100-V range?

9. Refer to the ohmmeter in Fig. 8-17. Assume that the movement is shunted to become a 10-mA meter. (**a**) Calculate the value of R_1 that would be required for full-scale deflection with the ohmmeter leads short-circuited. (**b**) How much would the half-scale reading be on the ohms scale?

10. Refer to Fig. 8-13. (**a**) How much is the total voltmeter resistance using the 5000-V jack with the range switch on the 1000-V position? (**b**) How much is the ohms-per-volt sensitivity? (**c**) Why must the range switch be on the 1000-V position?

11. In Fig. 8-14, if the voltmeter is connected across R_1 instead of R_2, what will the values be for V_1 and V_2?

Answers to Practice Problems

8-1 a. Parallel
   b. Series

8-2 a. I = 100 mA
   b. R_3
   c. Low
   d. High

8-3 a. I_S = 450 μA
   b. V_M = 0.045 V

8-4 a. I_M = 1 mA
   b. I_M = 1 mA

8-5 a. R_V = 50 kΩ
   b. R_V = 1 MΩ

8-6 a. V_1 = 80 V
   b. V_2 = 40 V

8-7 a. R_x = 400 Ω
   b. V = 4 V

8-8 a. 20 kΩ
   b. 10 MΩ

8-9 a. T
   b. F
   c. F

8-10 a. Ohmmeter
   b. 0 V
   c. 25 V
   d. 0 V

8-11 a. Right edge
   b. ∞ ohms

Review
Chapters
7 and 8

Summary

1. In a series voltage divider the IR drop across each resistance is proportional to its R. A larger R has a larger voltage drop. Each $V = (R/R_T) \times V_T$. In this way, the series voltage drops can be calculated from V_T without I.

2. In a parallel current divider, each branch current is inversely proportional to its R. A smaller R has more branch current. For only two resistances, we can use the inverse relation $I_1 = [R_2/(R_1 + R_2)] \times I_T$. In this way, the parallel branch currents can be calculated from I_T without V.

3. In a parallel current divider, each branch current is directly proportional to its conductance G. A larger G has more branch current. For any number of parallel resistances, each branch $I = (G/G_T) \times I_T$.

4. A milliammeter or ammeter is a low-resistance meter connected in series in a circuit to measure current.

5. Different current ranges are obtained by meter shunts in parallel with the meter.

6. A voltmeter is a high-resistance meter connected across the voltage to be measured.

7. Different voltage ranges are obtained by multipliers in series with the meter.

8. An ohmmeter has an internal battery to indicate the resistance of a component across its two terminals, with external power off.

9. In making resistance tests, remember that $R = 0\ \Omega$ for continuity or a short circuit, but the resistance of an open circuit is infinite.

Review Self-Examination
Answers at Back of Book

Answer True or False.

1. The internal R of a milliammeter must be low to have minimum effect on I in the circuit.

2. The internal R of a voltmeter must be high to have minimum current through the meter.
3. Power must be off when checking resistance in a circuit because the ohmmeter has its own internal battery.
4. In the series voltage divider in Fig. 8-28, the normal voltage from point B to ground is 75 V.
5. In Fig. 8-28, the normal voltage across R_1, between A and B, is 75 V.
6. The highest ohms range is best for checking continuity with an ohmmeter.
7. With four equal resistors in a series voltage divider with V_T of 44.4 V, each IR drop is 11.1 V.
8. With four equal resistors in parallel with I_T of 44.4 mA, each branch current is 11.1 mA.
9. Series voltage drops divide V_T in direct proportion to each series R.
10. Parallel currents divide I_T in direct proportion to each branch R.

References
Additional References at Back of Book

Gilmore, C. M.: *Instruments and Measurements,* McGraw-Hill Book Company, New York.

Prensky, S. D., and R. L. Castellucis: *Electronic Instrumentation,* Prentice-Hall, Inc., Englewood Cliffs, N.J.

Zbar, P. B.: *Basic Electricity,* McGraw-Hill Book Company, New York.

Chapter 9
Kirchhoff's Laws

Many types of circuits have components that are not in series, in parallel, or in series-parallel. For example, a circuit may have two voltages applied in different branches. Another example is an unbalanced bridge circuit. Where the rules of series and parallel circuits cannot be applied, more general methods of analysis become necessary. These methods include the application of Kirchhoff's laws, as described here, and the network theorems explained in Chap. 10.

Any circuit can be solved by Kirchhoff's laws because they do not depend on series or parallel connections. Stated in 1847 by the German physicist Gustav R. Kirchhoff, the two basic rules for voltage and current are:

1. *The algebraic sum of the voltage sources and IR voltage drops in any closed path must total zero.*
2. *At any point in a circuit the algebraic sum of the currents directed in and out must total zero.*

Specific methods for applying these basic rules in dc circuits are explained in this chapter. Applications for ac circuits are in Chap. 26.

Important terms in this chapter are:

current law	node voltage
loop	simultaneous equations
mesh current	ΣV
network	voltage law

More details are explained in the following sections:

9-1 Kirchhoff's Current Law
9-2 Kirchhoff's Voltage Law
9-3 Method of Branch Currents
9-4 Node-Voltage Analysis
9-5 Method of Mesh Currents

9-1   Kirchhoff's Current Law

The algebraic sum of the currents entering and leaving any point in a circuit must equal zero. Or stated another way: *The algebraic sum of the currents into any point of the circuit must equal the algebraic sum of the currents out of that point.* Otherwise, charge would accumulate at the point, instead of having a conducting path. An algebraic sum means combining positive and negative values.

Algebraic Signs   In using Kirchhoff's laws to solve circuits it is necessary to adopt conventions that determine the algebraic signs for current and voltage terms. A convenient system for currents is: *Consider all currents into a branch point as positive and all currents directed away from that point as negative.*

As an example, in Fig. 9-1 we can write the currents as

$$I_A + I_B - I_C = 0$$

or

$$5 \text{ A} + 3 \text{ A} - 8 \text{ A} = 0$$

Currents I_A and I_B are positive terms because these currents flow into P, but I_C, directed out, is negative.

Current Equations   For a circuit application, refer to point c at the top of the diagram in Fig. 9-2. The 6-A I_T into point c divides into the 2-A I_3 and 4-A I_{4-5}, both directed out. Note that I_{4-5} is the current through R_4 and R_5. The algebraic equation is

$$I_T - I_3 - I_{4-5} = 0$$

Substituting the values for these currents,

$$6 \text{ A} - 2 \text{ A} - 4 \text{ A} = 0$$

For the opposite directions, refer to point d at the bottom of Fig. 9-2. Here the branch currents into d combine to equal the main-line current I_T returning to the voltage source. Now I_T is directed out from d, with I_3 and I_{4-5} directed in. The algebraic equation is

$$-I_T + I_3 + I_{4-5} = 0$$
$$-6 \text{ A} + 2 \text{ A} + 4 \text{ A} = 0$$

The $I_{in} = I_{out}$ Note that at either point c or point d in Fig. 9-2, the sum of the 2-A and 4-A branch currents must equal the 6-A total line current. Therefore, Kirchhoff's current law can also be stated as: $I_{in} = I_{out}$. For Fig. 9-2, the equations of current can be written:

At point c: $6 \text{ A} = 2 \text{ A} + 4 \text{ A}$
At point d: $2 \text{ A} + 4 \text{ A} = 6 \text{ A}$

Kirchhoff's current law is really the basis for the practical rule in parallel circuits that the total line current must equal the sum of the branch currents.

Practice Problems 9-1
Answers at End of Chapter

a. With a 1-A I_1, 2-A I_2, and 3-A I_3 into a point, how much is I out?
b. If I_1 into a point is 3 A and I out of that point is 7 A, how much is I_2 into that point?

9-2   Kirchhoff's Voltage Law

The algebraic sum of the voltages around any closed path is zero. If you start from any point at one potential and come back to the same point and the same potential, the difference of potential must be zero.

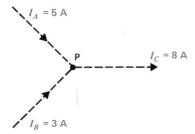

Fig. 9-1 Current I_C out from P equals 5 A + 3 A into P.

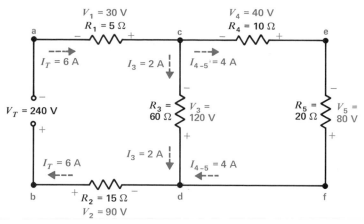

Fig. 9-2 Series-parallel circuit to illustrate Kirchhoff's laws. See text for current and voltage equations.

Algebraic Signs   In determining the algebraic signs for voltage terms, first mark the polarity of each voltage, as shown in Fig. 9-2. A convenient system then is: *Go around any closed path and consider any voltage whose plus terminal is reached first as positive, and vice versa.* This method applies to voltage drops and voltage sources. Also, the direction can be clockwise or counterclockwise. In any case, if you come back to the starting point, the algebraic sum of all the voltage terms must be zero.

If you do not come back to the start, then the algebraic sum is the voltage between the start and finish points.

You can follow any closed path. The reason is that the net voltage between any two points in a circuit is the same regardless of the path used in determining the potential difference.

Loop Equations   Any closed path is called a *loop*. A loop equation specifies the voltages around the loop.

Figure 9-2 has three loops. The outside loop, starting from point a at the top, through cefdb, and back to a, includes the voltage drops V_1, V_4, V_5, and V_2 and the source V_T.

The inside loop acdba includes V_1, V_3, V_2, and V_T. The other inside loop, cefdc with V_4, V_5, and V_3, does not include the voltage source.

Consider the voltage equation for the inside loop with V_T. In clockwise direction, starting from point a, the algebraic sum of the voltages is

$$-V_1 - V_3 - V_2 + V_T = 0$$

or

$$-30 \text{ V} - 120 \text{ V} - 90 \text{ V} + 240 \text{ V} = 0$$

Voltages V_1, V_3, and V_2 have the negative sign, because for each of these voltages the negative terminal is reached first. However, the source V_T is a positive term because its plus terminal is reached first, going in the same direction.

For the opposite direction, going counterclockwise in the same loop from point b at the bottom, V_2, V_3, and V_1 have positive values and V_T is negative. Then

$$V_2 + V_3 + V_1 - V_T = 0$$

or

$$90 \text{ V} + 120 \text{ V} + 30 \text{ V} - 240 \text{ V} = 0$$

When we transpose the negative term of -240 V, the equation becomes

$$90 \text{ V} + 120 \text{ V} + 30 \text{ V} = 240 \text{ V}$$

This equation states that the sum of the voltage drops equals the applied voltage.

$\Sigma V = V_T$ The Greek letter Σ means "sum of." In either direction, for any loop, the sum of the IR voltage drops must equal the applied voltage V_T. In Fig. 9-2, for the inside loop with the source V_T, going counterclockwise from point b:

$$90 \text{ V} + 120 \text{ V} + 30 \text{ V} = 240 \text{ V}$$

This system does not contradict the rule for algebraic signs. If 240 V were on the left side of the equation, this term would have a negative sign.

Stating a loop equation as $\Sigma V = V_T$ eliminates the step of transposing the negative terms from one side to the other to make them positive. In this form, the loop equations show that Kirchhoff's voltage law is really the basis for the practical rule in series circuits that the sum of the voltage drops must equal the applied voltage.

When a loop does not have any voltage source, the algebraic sum of the IR voltage drops alone must total zero. For instance, in Fig. 9-2, for the loop cefdc without the source V_T, going clockwise from point c, the loop equation of voltages is

$$-V_4 - V_5 + V_3 = 0$$
$$-40 \text{ V} - 80 \text{ V} + 120 \text{ V} = 0$$
$$0 = 0$$

Notice that V_3 is positive now, because its plus terminal is reached first by going clockwise from d to c in this loop.

Practice Problems 9-2
Answers at End of Chapter

Refer to Fig. 9-2.
a. For partial loop cefd, what is the total voltage across cd with -40 V for V_4 and -80 V for V_5?
b. For loop cefdc, what is the total voltage with -40 V for V_4, -80 V for V_5, and including 120 V for V_3?

9-3   Method of Branch Currents

Now we can use Kirchhoff's laws to analyze the circuit in Fig. 9-3. The problem is to find the currents and voltages for the three resistors.

First, indicate current directions and mark the volt-

age polarity across each resistor consistent with the assumed current. Remember that electron flow in a resistor produces negative polarity where the current enters. In Fig. 9-3, we assume that the source V_1 produces electron flow from left to right through R_1, while V_2 produces electron flow from right to left through R_2.

The three different currents in R_1, R_2, and R_3 are indicated as I_1, I_2, and I_3. However, three unknowns would require three equations for the solution. From Kirchhoff's current law, $I_3 = I_1 + I_2$, as the current out of point c must equal the current in. The current through R_3, therefore, can be specified as $I_1 + I_2$.

With two unknowns, two independent equations are needed to solve for I_1 and I_2. These equations are obtained by writing two Kirchhoff's voltage law equations around two loops. There are three loops in Fig. 9-3, the outside loop and two inside loops, but we need only two. The inside loops are used for the solution here.

Writing the Loop Equations   For the loop with V_1, start at point b, at the bottom left, and go clockwise through V_1, V_{R_1}, and V_{R_3}. This equation for loop 1 is

$$84 - V_{R_1} - V_{R_3} = 0$$

For the loop with V_2, start at point f, at the lower right, and go counterclockwise through V_2, V_{R_2}, and V_{R_3}. This equation for loop 2 is

$$21 - V_{R_2} - V_{R_3} = 0$$

Using the known values of R_1, R_2, and R_3 to specify the IR voltage drops,

$$V_{R_1} = I_1 R_1 = I_1 \times 12 = 12I_1$$
$$V_{R_2} = I_2 R_2 = I_2 \times 3 = 3I_2$$
$$V_{R_3} = (I_1 + I_2)R_3 = 6(I_1 + I_2)$$

Substituting these values in the voltage equation for loop 1,

$$84 - 12I_1 - 6(I_1 + I_2) = 0$$

Also, in loop 2,

$$21 - 3I_2 - 6(I_1 + I_2) = 0$$

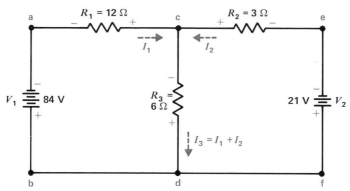

Fig. 9-3 Application of Kirchhoff's laws to a circuit with two sources in different branches. See text for solution by finding the branch currents.

Multiplying $(I_1 + I_2)$ by 6 and combining terms and transposing, the two equations are

$$-18I_1 - 6I_2 = -84$$
$$-6I_1 - 9I_2 = -21$$

Divide the top equation by -6 and the bottom equation by -3 to make the coefficients smaller and to have all positive terms. The two equations in their simplest form then become

$$3I_1 + I_2 = 14$$
$$2I_1 + 3I_2 = 7$$

Solving for the Currents These two equations in the two unknowns I_1 and I_2 contain the solution of the network. It should be noted that the equations include every resistance in the circuit. Currents I_1 and I_2 can be calculated by any of the methods for the solution of simultaneous equations. Using the method of elimination, multiply the top equation by 3 to make the I_2 terms the same in both equations. Then

$$9I_1 + 3I_2 = 42$$
$$2I_1 + 3I_2 = 7$$

Subtract the bottom equation from the top equation, term by term, to eliminate I_2. Then, since the I_2 term becomes zero,

$$7I_1 = 35$$
$$I_1 = 5 \text{ A}$$

The 5-A I_1 is the current through R_1. Its direction is from a to c, as assumed, because the answer for I_1 is positive.

To calculate I_2, substitute 5 for I_1 in either of the two loop equations. Using the bottom equation for the substitution,

$$2(5) + 3I_2 = 7$$
$$3I_2 = 7 - 10$$
$$3I_2 = -3$$
$$I_2 = -1 \text{ A}$$

The negative sign for I_2 means this current is opposite to the assumed direction. Therefore, I_2 flows through R_2 from c to e instead of from e to c.

Why the Solution for I_2 is Negative In Fig. 9-3, I_2 was assumed to flow from point e to c through R_2 because V_2 produces electron flow in this direction. However, the other voltage source V_1 produces electron flow through R_2 in the opposite direction, from point c to e. This solution of -1 A for I_2 shows that the current through R_2 produced by V_1 is more than the current produced by V_2. The net result is 1 A through R_2 from c to e.

The actual direction of I_2 is shown in Fig. 9-4 with all the values for the solution of this circuit. Notice that the polarity of V_{R_2} is reversed from the assumed polarity in Fig. 9-3. Since the net electron flow through R_2 is actually from c to e, the end of R_2 at c is the negative end. However, the polarity of V_2 is the same in both diagrams because it is a voltage source, which generates its own polarity.

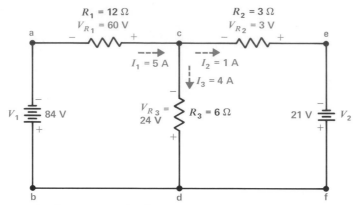

Fig. 9-4 Solution of Fig. 9-3 with all currents and voltages.

To calculate I_3 through R_3,

$I_3 = I_1 + I_2$

$\quad = 5 + (-1)$

$I_3 = 4$ A

The 4 A for I_3 is in the assumed direction from c to d. Although the negative sign for I_2 only means a reversed direction, its algebraic value of -1 must be used for substitution in the algebraic equations written for the assumed direction.

Calculating the Voltages   With all the currents known, the voltage across each resistor can be calculated as follows:

$V_{R_1} = I_1 R_1 = 5 \times 12 = 60$ V

$V_{R_2} = I_2 R_2 = 1 \times \ \ 3 = \ \ 3$ V

$V_{R_3} = I_3 R_3 = 4 \times \ \ 6 = 24$ V

All the currents are taken as positive, in the correct direction, to calculate the voltages. Then the polarity of each IR drop is determined from the actual direction of current, with electron flow into the negative end (see Fig. 9-4). Notice that V_{R_3} and V_{R_2} have opposing polarities in loop 2. Then the sum of $+3$ V and -24 V equals the -21 V of V_2.

Checking the Solution   As a summary of all the answers for this problem, Fig. 9-4 shows the network

with all the currents and voltages. The polarity of each V is marked from the known directions. In checking the answers, we can see whether Kirchhoff's current and voltage laws are satisfied:

At point c: 5 A = 4 A + 1 A
At point d: 4 A + 1 A = 5 A

Around the loop with V_1:

84 V $-$ 60 V $-$ 24 V $= 0$ clockwise from b

Around the loop with V_2:

21 V $+$ 3 V $-$ 24 V $= 0$ counterclockwise from f

It should be noted that the circuit has been solved using only the two Kirchhoff laws, without any of the special rules for series and parallel circuits. Any circuit can be solved just by applying Kirchhoff's laws for the voltages around a loop and the currents at a branch point.

Practice Problems 9-3
Answers at End of Chapter

Refer to Fig. 9-4.
a. How much is the voltage around partial loop cefd?
b. How much is the voltage around loop cefdc?

9-4 Node-Voltage Analysis

In the method of branch currents, these currents are used for specifying the voltage drops around the loops. Then loop equations are written to satisfy Kirchhoff's voltage law. Solving the loop equations, we can calculate the unknown branch currents.

Another method uses the voltage drops to specify the currents at a branch point, also called a *node*. Then node equations of currents are written to satisfy Kirchhoff's current law. Solving the node equations, we can calculate the unknown node voltages. This method of node-voltage analysis often is shorter than the method of branch currents.

A node is simply a common connection for two or more components. A *principal node* has three or more connections. In effect, a principal node is just a junction or branch point, where currents can divide or combine. Therefore, we can always write an equation of currents at a principal node. In Fig. 9-5, points N and G are principal nodes.

However, one node must be the reference for specifying the voltage at any other node. In Fig. 9-5, point G connected to chassis ground is the reference node. Therefore, we need only write one current equation for the other node N. In general, the number of current equations required to solve a circuit is one less than the number of principal nodes.

Writing the Node Equations The circuit of Fig. 9-3, earlier solved by the method of branch currents, is redrawn in Fig. 9-5 to be solved now by node-voltage analysis. The problem here is to find the node voltage V_N from N to G. Once this voltage is known, all the other voltages and currents can be determined.

The currents in and out of node N are specified as follows: I_1 is the only current through the 12-Ω R_1. Therefore, I_1 is V_{R_1}/R_1 or $V_{R_1}/12\ \Omega$. Similarly, I_2 is $V_{R_2}/3\ \Omega$. Finally, I_3 is $V_{R_3}/6\ \Omega$.

Note that V_{R_3} is the node voltage V_N that we are to calculate. Therefore, I_3 can also be stated as $V_N/6\ \Omega$. The equation of currents at node N is

$$I_1 + I_2 = I_3$$

or

$$\frac{V_{R_1}}{12} + \frac{V_{R_2}}{3} = \frac{V_N}{6}$$

There are three unknowns here, but V_{R_1} and V_{R_2} can be specified in terms of V_N and the known values of V_1 and V_2. We can use Kirchhoff's voltage law, because the applied voltage V must equal the algebraic sum of the voltage drops. For the loop with V_1 of 84 V,

$$V_{R_1} + V_N = 84 \qquad \text{or} \qquad V_{R_1} = 84 - V_N$$

For the loop with V_2 of 21 V,

$$V_{R_2} + V_N = 21 \qquad \text{or} \qquad V_{R_2} = 21 - V_N$$

Now substitute these values of V_{R_1} and V_{R_2} in the equation of currents:

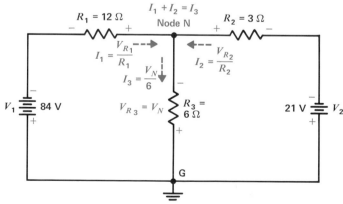

Fig. 9-5 Node-voltage analysis for the same circuit as in Fig. 9-3. See text for solution by finding V_N across R_3 from the principal node N to ground.

$$I_1 + I_2 = I_3$$

$$\frac{V_{R_1}}{R_1} + \frac{V_{R_2}}{R_2} = \frac{V_{R_3}}{R_3}$$

Using the value of each V in terms of V_N,

$$\frac{84 - V_N}{12} + \frac{21 - V_N}{3} = \frac{V_N}{6}$$

This equation has only the one unknown, V_N. Clearing fractions by multiplying each term by 12, the equation is

$$(84 - V_N) + 4(21 - V_N) = 2\,V_N$$
$$84 - V_N + 84 - 4\,V_N = 2\,V_N$$
$$-7\,V_N = -168$$
$$V_N = 24\text{ V}$$

This answer of 24 V for V_N is the same as that calculated for V_{R_3} by the method of branch currents. The positive value means the direction of I_3 is correct, making V_N negative at the top of R_3 in Fig. 9-5.

Calculating All Voltages and Currents The reason for finding the voltage at a node, rather than some other voltage, is the fact that a node voltage must be common to two loops. As a result, the node voltage can be used for calculating all the voltages in the loops.

In Fig. 9-5, with a V_N of 24 V, then V_{R_1} must be $84 - 24 = 60$ V. Also, I_1 is 60 V/12 Ω, which equals 5 A.

To find V_{R_2}, it must be $21 - 24$, which equals -3 V. The negative answer means that I_2 is opposite to the assumed direction and the polarity of V_{R_2} is the reverse of the signs shown across R_2 in Fig. 9-5. The correct directions are shown in the solution for the circuit in Fig. 9-4. The magnitude of I_2 is 3 V/3 Ω, which equals 1 A.

The following comparisons can be helpful in using node equations and loop equations. A node equation applies Kirchhoff's current law to the currents in and out of a node point. However, the currents are specified as V/R so that the equation of currents can be solved to find a node voltage.

A loop equation applies Kirchhoff's voltage law to the voltages around a closed path. However, the voltages are specified as IR so that the equation of voltages can be solved to find a loop current. This procedure with voltage equations is used for the method of branch currents explained before with Fig. 9-3 and for the method of mesh currents to be described next with Fig. 9-6.

Practice Problems 9-4
Answers at End of Chapter

a. Figure 9-5 has how many principal nodes?
b. How many node equations are necessary to solve a circuit with three principal nodes?

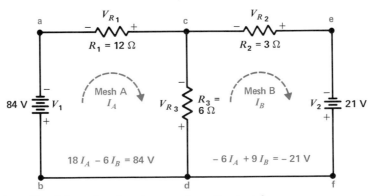

Fig. 9-6 The same circuit as Fig. 9-3 analyzed as two meshes. See text for solution by calculating the assumed mesh currents I_A and I_B.

9-5 Method of Mesh Currents

A mesh is the simplest possible closed path. The circuit in Fig. 9-6 has the two meshes acdba and cefdc. The outside path acefdba is a loop but not a mesh. Each mesh is like a single window frame. There is only one path without any branches.

A mesh current is assumed to flow around a mesh without dividing. In Fig. 9-6, the mesh current I_A flows through V_1, R_1, and R_3; mesh current I_B flows through V_2, R_2, and R_3. A resistance common to two meshes, such as R_3, has two mesh currents, which are I_A and I_B here.

The fact that a mesh current does not divide at a branch point is the difference between mesh currents and branch currents. A mesh current is an assumed current, while a branch current is the actual current. However, when the mesh currents are known, all the individual currents and voltages can be determined.

As an example, Fig. 9-6, which has the same circuit as Fig. 9-3, will now be solved by using the assumed mesh currents I_A and I_B. The mesh equations are

$$18I_A - 6I_B = 84 \text{ V} \qquad \text{in mesh A}$$
$$-6I_A + 9I_B = -21 \text{ V} \qquad \text{in mesh B}$$

Writing the Mesh Equations The number of meshes equals the number of mesh currents, which is the number of equations required. Here two equations are used for I_A and I_B in the two meshes.

The assumed current is usually taken in the same direction around each mesh, in order to be consistent. Generally, the clockwise direction is used, as shown for I_A and I_B in Fig. 9-6.

In each mesh equation, the algebraic sum of the voltage drops equals the applied voltage.

The voltage drops are added going around a mesh in the same direction as its mesh current. Any voltage drop in a mesh produced by its own mesh current is considered positive because it is added in the direction of the mesh current.

Since all the voltage drops of a mesh current in its own mesh must have the same positive sign, they can be written collectively as one voltage drop by adding all the resistances in the mesh. For instance, in the first equation, for mesh A, the total resistance equals 12 + 6, or 18 Ω. Therefore, the voltage drop for I_A is $18I_A$ in mesh A.

In the second equation, for mesh B, the total resistance is 3 + 6, or 9 Ω, making the total voltage drop $9I_B$ for I_B in mesh B. You can add all the resistances in a mesh for one R_T, because they can be considered in series for the assumed mesh current.

Any resistance common to two meshes has two opposite mesh currents. In Fig. 9-6, I_A flows down while I_B is up through the common R_3, with both currents clockwise. As a result, a common resistance has two opposing voltage drops. One voltage is positive for the current of the mesh whose equation is being written. The opposing voltage is negative for the current of the adjacent mesh.

In mesh A, the common 6-Ω R_3 has the opposing voltages $6I_A$ and $-6I_B$. The $6I_A$ of R_3 adds to the $12I_A$ of R_1 for the total positive voltage drop of $18I_A$ in mesh A. With the opposing voltage of $-6I_B$, then the equation for mesh A is $18I_A - 6I_B = 84$ V.

The same idea applies to mesh B. However, now the voltage $6I_B$ is positive because the equation is for mesh B. The $-6I_A$ voltage is negative here because I_A is for the adjacent mesh. The $6I_B$ adds to the $3I_B$ of R_2 for the total positive voltage drop of $9I_B$ in mesh B. With the opposing voltage of $-6I_A$, the equation for mesh B then is $-6I_A + 9I_B = -21$ V.

The algebraic sign of the source voltage in a mesh depends on its polarity. When the assumed mesh current flows into the positive terminal, as for V_1 in Fig. 9-6, it is considered positive for the right-hand side of the mesh equation. This direction of electron flow produces voltage drops that must add to equal the applied voltage.

With the mesh current into the negative terminal, as for V_2 in Fig. 9-5, it is considered negative. This is why V_2 is -21 V in the equation for mesh B. Then V_2 is actually a load for the larger applied voltage of V_1, instead of V_2 being the source. When a mesh has no source voltage, the algebraic sum of the voltage drops must equal zero.

Solving the Mesh Equations to Find the Mesh Currents The two equations for the two meshes in Fig. 9-6 are

$$18I_A - 6I_B = 84$$
$$-6I_A + 9I_B = -21$$

These equations have the same coefficients as in the voltage equations written for the branch currents, but the signs are different. The reason is that the directions of the assumed mesh currents are not the same as those of the branch currents.

The solution will give the same answers for either method, but you must be consistent in algebraic signs. Use either the rules for meshes with mesh currents or the rules for loops with branch currents, but do not mix the two methods.

For smaller coefficients, divide the first equation by 2 and the second equation by 3. Then

$$9I_A - 3I_B = 42$$
$$-2I_A + 3I_B = -7$$

Add the equations, term by term, to eliminate I_B. Then

$$7I_A = 35$$
$$I_A = 5 \text{ A}$$

To calculate I_B, substitute 5 for I_A in the second equation:

$$-2(5) + 3I_B = -7$$
$$3I_B = -7 + 10 = 3$$
$$I_B = 1 \text{ A}$$

The positive solutions mean that the electron flow for both I_A and I_B is actually clockwise, as assumed. When the solution for a mesh current is negative, its direction is opposite to the assumed direction.

Finding the Branch Currents and Voltage Drops
Referring to Fig. 9-6, the 5-A I_A is the only current through R_1. Therefore, I_A and I_1 are the same. Then V_{R_1} across the 12-Ω R_1 is 5×12, or 60 V. The polarity of V_{R_1} is marked negative at the left, with the electron flow into this side.

Similarly, the 1-A I_B is the only current through R_2. The direction of this electron flow through R_2 is from left to right. Note that this value of 1 A for I_B clockwise is the same as -1 A for I_2, assumed in the opposite direction in Fig. 9-3. Then V_{R_2} across the 3-Ω R_2 is 1×3 or 3 V, with the left side negative.

The current I_3 through R_3, common to both meshes, consists of I_A and I_B. Then I_3 is $5 - 1$ or 4 A. The

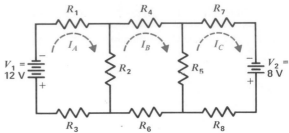

Fig. 9-7 A circuit with three meshes. Each R is 2 Ω. See text for mesh equations.

currents are subtracted because I_A and I_B are in opposing directions through R_3. When all the mesh currents are taken one way, they will always be in opposite directions through any resistance common to two meshes.

The direction of the net 4-A I_3 through R_3 is downward, the same as I_A, because it is larger than I_B. Then, V_{R_3} across the 6-Ω R_3 is $4 \times 6 = 24$ V, with the top negative.

The Set of Mesh Equations
The system for algebraic signs of the voltages in the mesh equations is different from the method used with branch currents, but the end result is the same. The advantage of mesh currents is the pattern of algebraic signs for the voltages, without the need for tracing any branch currents. This feature is especially helpful in a more elaborate circuit, such as the one in Fig. 9-7 that has three meshes. We can use Fig. 9-7 for more practice in writing mesh equations, without doing the numerical work of solving a set[1] of three equations. Each R is 2 Ω.

In mesh A:	$6I_A - 2I_B + 0 = 12$
In mesh B:	$-2I_A + 8I_B - 2I_C = 0$
In mesh C:	$0 - 2I_B + 6I_C = -8$

The zero term in equations A and C represents a missing mesh current. Only mesh B has all three mesh currents. However, note that mesh B has a zero term for the voltage source because it is the only mesh with all IR drops.

---

[1] A set with any number of simultaneous linear equations, for any number of meshes, can be solved by determinants. This procedure is shown in mathematics textbooks (see the bibliography at the end of this book).

In summary, the only positive *IR* voltage in a mesh is for the R_T of each mesh current in its own mesh. All other voltage drops for any adjacent mesh current across a common resistance are always negative. This procedure for assigning algebraic signs to the voltage drops is the same whether the source voltage in the mesh is positive or negative. It also applies even if there is no voltage source in the mesh.

Practice Problems 9-5
Answers at End of Chapter

Answer True or False.
a. A network with four mesh currents needs four mesh equations for a solution.
b. An *R* for two meshes has opposing mesh currents.

Summary

1. Kirchhoff's voltage law states that the algebraic sum of all voltages around any closed path must equal zero. Or the sum of the voltage drops equals the applied voltage.
2. Kirchhoff's current law states that the algebraic sum of all currents directed in and out at any point in a closed path must equal zero. Or the current in equals the current out.
3. A closed path is a loop. The method of using algebraic equations for the voltages around the loops to calculate the branch currents is illustrated in Fig. 9-3.
4. A principal node is a branch point where currents divide or combine. The method of using algebraic equations for the currents at a node to calculate each node voltage is illustrated in Fig. 9-5.
5. A mesh is the simplest possible loop. A mesh current is assumed to flow around the mesh without branching. The method of using algebraic equations for the voltages around the meshes to calculate the mesh currents is illustrated in Fig. 9-6.

Self-Examination
Answers at Back of Book

Answer True or False.

1. The algebraic sum of all voltages around any mesh or any loop must equal zero.
2. A mesh with two resistors has two mesh currents.
3. With $I_1 = 3$ A and $I_2 = 2$ A directed into a node, the current I_3 directed out must equal 5 A.
4. In a loop without any voltage source, the algebraic sum of the voltage drops must equal zero.
5. The algebraic sum of +40 V and −10 V equals +30 V.
6. A principal node is a junction where branch currents can divide or combine.
7. In the node-voltage method, the number of equations of current equals the number of principal nodes.
8. In the mesh-current method, the number of equations of voltage equals the number of meshes.

9. When all mesh currents are clockwise or all counterclockwise, any resistor common to two meshes has two currents in opposite directions.
10. The rules of series voltages and parallel currents are based on Kirchhoff's laws.

Essay Questions

1. State Kirchhoff's current law in two ways.
2. State Kirchhoff's voltage law in two ways.
3. What is the difference between a loop and a mesh?
4. What is the difference between a branch current and a mesh current?
5. Define a principal node.
6. Define a node voltage.
7. Use the values in Fig. 9-4 to show that the algebraic sum is zero for all voltages around the outside loop acefdba.
8. Use the values in Fig. 9-4 to show that the algebraic sum is zero for all the currents in and out at node c and at node d.

Problems

Answers to Odd-Numbered Problems at Back of Book

1. Find the current I_1 through R_1 in Fig. 9-8 by the method of mesh currents.
2. Find the voltage V_2 across R_2 in Fig. 9-8 by the method of node-voltage analysis.
3. Find all the currents and voltages in Fig. 9-8 by the method of branch currents.
4. Check your answers for Prob. 3 by showing that the algebraic sum is zero for the voltages in the three paths.
5. Reverse the polarity of V_2 in Fig. 9-8 and calculate the new I_1, compared with Prob. 1. (Hint: Use mesh currents to eliminate the need for tracing the branch currents.)
6. Write the mesh equations for the circuit in Fig. 9-9. Each R is 5 Ω. No solution is necessary.

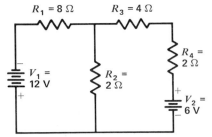

Fig. 9-8 For Probs. 1, 2, 3, 4, and 5.

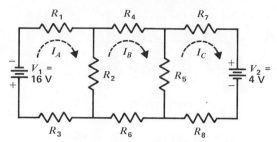

Fig. 9-9 For Prob. 6.

Answers to Practice Problems

9-1 a. 6 A
    b. 4 A

9-2 a. 120 V
    b. 0 V

9-3 a. 24 V
    b. 0 V

9-4 a. Two
    b. Two

9-5 a. T
    b. T

Chapter 10
Network Theorems

A network is just a combination of components, such as resistances, interconnected in any way. However, networks generally need more than the rules of series and parallel circuits for analysis. Kirchhoff's laws can always be applied for any circuit connections. The network theorems, though, usually provide shorter methods of solving the circuit. The reason is that the theorems enable us to convert the network into a simpler circuit, equivalent to the original. Then the equivalent circuit can be solved by the rules of series and parallel circuits.

Only the applications are given here, although all the network theorems can be derived from Kirchhoff's laws. It should also be noted that resistance networks with batteries are shown as examples, but the theorems can also be applied to ac networks, as explained in Chap. 26.

Important terms in this chapter are:

active components	Norton's theorem
bilateral components	passive components
current source	pi (π) network
delta network	superposition
Δ-Y transformations	T network
equivalent circuit	Thevenin-Norton conversions
equivalent current source	Thevenin's theorem
linear components	thevenizing a circuit
Millman's theorem	voltage source
nortonizing a circuit	Y network

More details are explained in the following sections:

10-1 Superposition

This theorem is very useful because it extends the use of Ohm's law to circuits that have more than one source. In brief, we can calculate the effect of one source at a time and then superimpose the results of all the sources. As a definition, the superposition theorem states that: *In a network with two or more sources, the current or voltage for any component is the algebraic sum of the effects produced by each source acting separately.*

In order to use one source at a time, all other sources are "killed" temporarily. This means disabling the source so that it cannot generate voltage or current, without changing the resistance of the circuit. A voltage source such as a battery is killed by assuming a

short circuit across its potential difference. The internal resistance remains.

Voltage Divider with Two Sources   The problem in Fig. 10-1 is to find the voltage at P to chassis ground for the circuit in Fig. 10-1a. The method is to calculate the voltage at P contributed by each source separately, as in Fig. 10-1b and c, and then superimpose these voltages.

To find the effect of V_1 first, short-circuit V_2 as shown in Fig. 10-1b. Note that the bottom of R_1 then becomes connected to chassis ground because of the short circuit across V_2. As a result, R_2 and R_1 form a series voltage divider for the V_1 source.

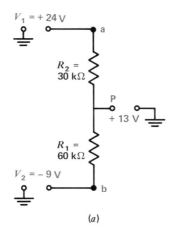

(a)

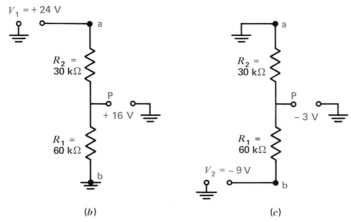

(b)   (c)

Fig. 10-1 Superposition applied to a voltage divider with two sources. (a) Actual circuit with +13 V between P and chassis ground. (b) Voltage V_1 producing +16 V at P. (c) Voltage V_2 producing −3 V at P.

Furthermore, the voltage across R_1 becomes the same as the voltage from P to ground. To find this V_{R_1} across R_1 as the contribution of the V_1 source, we use the voltage divider formula:

$$V_{R_1} = \frac{R_1}{R_1 + R_2} \times V_1$$

$$= \frac{60 \text{ k}\Omega}{30 \text{ k}\Omega + 60 \text{ k}\Omega} \times 24 \text{ V}$$

$$= \frac{2}{3} \times 24 \text{ V}$$

$$V_{R_1} = 16 \text{ V}$$

Next find the effect of V_2 alone, with V_1 short-circuited as shown in Fig. 10-1c. Then point a at the top of R_2 becomes grounded. R_1 and R_2 form a series voltage divider again, but here the R_2 voltage is the voltage at P to ground.

With one side of R_2 grounded and the other side to point P, V_{R_2} is the voltage to calculate. Again we have a series divider, but this time for the negative voltage V_2. Using the voltage divider formula for V_{R_2} as the contribution of V_2 to the voltage at P,

$$V_{R_2} = \frac{R_2}{R_1 + R_2} \times V_2$$

$$= \frac{30 \text{ k}\Omega}{30 \text{ k}\Omega + 60 \text{ k}\Omega} \times -9 \text{ V}$$

$$= \frac{1}{3} \times -9 \text{ V}$$

$$V_{R_2} = -3 \text{ V}$$

This voltage is negative at P because V_2 is negative.

Finally, the total voltage at P is

$$V_P = V_1 + V_2$$

$$= 16 - 3$$

$$V_P = 13 \text{ V}$$

This algebraic sum is positive for the net V_P because the positive V_1 is larger than the negative V_2.

By means of superposition, therefore, this problem was reduced to two series voltage dividers. The same procedure can be used with more than two sources. Also, each voltage divider can have any number of series resistances.

Requirements for Superposition All the components must be linear and bilateral in order to superimpose currents and voltages. *Linear* means that the current is proportional to the applied voltage. Then the currents calculated for different source voltages can be superimposed.

Bilateral means that the current is the same amount for opposite polarities of the source voltage. Then the values for opposite directions of current can be combined algebraically. Networks with resistors, capacitors, and air-core inductors are generally linear and bilateral. These are also *passive components,* meaning they do not amplify or rectify. *Active components,* such as transistors, semiconductor diodes, and electron tubes, are never bilateral and often are not linear.

Practice Problems 10-1
Answers at End of Chapter

a. In Fig. 10-1b, which R is shown grounded at one end?

b. In Fig. 10-1c, which R is shown grounded at one end?

10-2 Thevenin's Theorem

Named after M. L. Thevenin, a French engineer, this theorem is very useful in simplifying the voltages in a network. By Thevenin's theorem, many sources and components, no matter how they are interconnected, can be represented by an equivalent series circuit with respect to any pair of terminals in the network (see Fig. 10-2). Imagine that the block at the left contains a network connected to terminals a and b. Thevenin's theorem states that *the entire* network connected to a and b can be replaced by a single voltage source V_{Th} in series

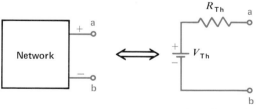

Fig. 10-2 Any network in the block at the left can be reduced to the Thevenin equivalent circuit at the right.

with a single resistance R_{Th}, connected to the same two terminals.

Voltage V_{Th} is the open-circuit voltage across terminals a and b. This means, find the voltage that the network produces across the two terminals with an open circuit between a and b. The polarity of V_{Th} is such that it will produce current from a to b in the same direction as in the original network.

Resistance R_{Th} is the open-circuit resistance across terminals a and b, but with all the sources killed. This means, find the resistance looking back into the network from terminals a and b. Although the terminals are open, an ohmmeter across ab would read the value of R_{Th} as the resistance of the remaining paths in the network, without any sources operating.

Thevenizing a Circuit As an example, refer to Fig. 10-3a, where we want to find the voltage V_L across the 2-Ω R_L and its current I_L. To use Thevenin's theorem, mentally disconnect R_L. The two open ends then become terminals a and b. Now we find the Thevenin equivalent of the remainder of the circuit that is still connected to a and b. In general, open the part of the circuit to be analyzed and ''thevenize'' the remainder of the circuit connected to the two open terminals.

Our only problem now is to find the value of the open-circuit voltage V_{Th} across ab and the equivalent resistance R_{Th}. The Thevenin equivalent always consists of a single voltage source in series with a single resistance, as in Fig. 10-3d.

The effect of opening R_L is shown in Fig. 10-3b. As a result, the 3-Ω R_1 and 6-Ω R_2 form a series voltage divider, without R_L.

Furthermore, the voltage across R_2 now is the same as the open-circuit voltage across terminals a and b. Therefore V_{R_2} with R_L open is V_{ab}. This is the V_{Th} we need for the Thevenin equivalent circuit. Using the voltage divider formula,

$$V_{R_2} = \% \times 36 \text{ V} = 24 \text{ V}$$

$$V_{R_2} = V_{ab} = V_{Th} = 24 \text{ V}$$

This voltage is positive at terminal a.

To find R_{Th}, the 2-Ω R_L is still disconnected. However, now the source V is short-circuited. So the circuit looks like Fig. 10-3c. The 3-Ω R_1 is now in parallel with the 6-Ω R_2, as they are both connected across the same two points. This combined resistance is

$$R_{Th} = {}^{18}\!/_9 = 2 \text{ } \Omega$$

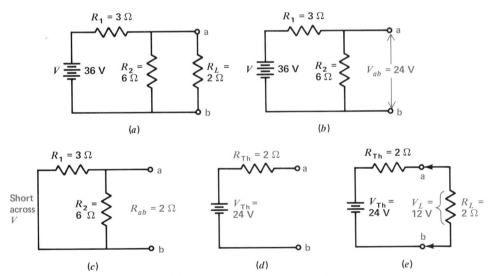

Fig. 10-3 Application of Thevenin's theorem. (a) Original circuit with terminals a and b across R_L. (b) Disconnect R_L to find V_{ab} is 24 V. (c) Short-circuit V to find R_{ab} is 2 Ω. (d) Thevenin equivalent. (e) Reconnect R_L at terminals a and b to find V_L is 12 V.

As shown in Fig. 10-3d, the Thevenin circuit to the left of terminals a and b then consists of the equivalent voltage V_{Th}, equal to 24 V, in series with the equivalent series resistance R_{Th}, equal to 2 Ω. This Thevenin equivalent applies for any value of R_L because R_L was disconnected. We are actually thevenizing the circuit that feeds the open ab terminals.

To find V_L and I_L, we can finally reconnect R_L to terminals a and b of the Thevenin equivalent circuit, as shown in Fig. 10-3e. Then R_L is in series with R_{Th} and V_{Th}. Using the voltage divider formula for the 2-Ω R_{Th} and 2-Ω R_L, $V_L = \frac{1}{2} \times 24$ V $= 12$ V. To find I_L as V_L/R_L, the value is 12 V/2 Ω, which equals 6 A.

These answers of 6 A for I_L and 12 V for V_L apply to R_L in both the original circuit in Fig. 10-3a and the equivalent circuit in Fig. 10-3e. Note that the 6-A I_L also flows through R_{Th}.

The same answers could be obtained by solving the series-parallel circuit in Fig. 10-3a, using Ohm's law. However, the advantage of thevenizing the circuit is that the effect of different values of R_L can be calculated easily. Suppose that R_L were changed to 4 Ω. In the Thevenin circuit, the new value of V_L would be $\frac{4}{6} \times 24$ V $= 16$ V. The new I_L would be 16 V/4 Ω, which equals 4 A. In the original circuit, a complete new solution would be required each time R_L was changed.

Looking Back from Terminals a and b

Which way to look at the resistance of a series-parallel circuit depends on where the source is connected. In general, we calculate the total resistance from the outside terminals of the circuit in toward the source, as the reference.

When the source is short-circuited for thevenizing a circuit, terminals a and b become the reference. Looking back from a and b to calculate R_{Th}, the situation becomes reversed from the way the circuit was viewed to determine V_{Th}.

For R_{Th}, imagine that a source could be connected across ab, and calculate the total resistance working from the outside in toward terminals a and b. Actually an ohmmeter placed across terminals a and b would read this resistance.

This idea of reversing the reference is illustrated in Fig. 10-4. The circuit in Fig. 10-4a has terminals a and b open, ready to be thevenized. This circuit is similar to Fig. 10-3 but with the 4-Ω R_3 inserted between R_2 and terminal a. The interesting point is that R_3 does not change the value of V_{ab} produced by the source V, but R_3 does increase the value of R_{Th}. When we look back from terminals a and b, the 4 Ω of R_3 is in series with 2 Ω to make R_{Th} 6 Ω, as shown for R_{ab} in Fig. 10-4b and R_{Th} in Fig. 10-4c.

Let us consider why V_{ab} is the same 24 V with or without R_3. Since R_3 is connected to the open terminal a, the source V cannot produce current in R_3. Therefore, R_3 has no IR drop. A voltmeter would read the same 24 V across R_2 and from a to b. Since V_{ab} equals 24 V, this is the value of V_{Th}.

Now consider why R_3 does change the value of R_{Th}. Remember that we must work from the outside in toward ab to calculate the total resistance. Then the 3-Ω R_1 and 6-Ω R_2 are in parallel, for a combined resistance of 2 Ω. Furthermore, this 2 Ω is in series with the 4-Ω R_3 because R_3 is in the main line from terminals a and b. Then R_{Th} is $2 + 4 = 6$ Ω. As shown in Fig. 10-4c, the Thevenin equivalent circuit consists of $V_{Th} = 24$ V and $R_{Th} = 6$ Ω.

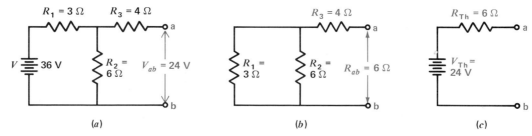

Fig. 10-4 Thevenizing the circuit of Fig. 10-3b, but with a 4-Ω R_3 in series with terminal a. (a) Voltage V_{ab} is still 24 V. (b) Resistance R_{ab} is $2 + 4 = 6$ Ω. (c) Thevenin equivalent.

Practice Problems 10-2
Answers at End of Chapter

Answer True or False. For a Thevenin equivalent circuit,
a. Terminals a and b are open to find both V_{Th} and R_{Th}.
b. The source voltage is short-circuited only to find R_{Th}.

10-3 Thevenizing a Circuit with Two Voltage Sources

The circuit in Fig. 10-5 has already been solved by Kirchhoff's laws, but we can use Thevenin's theorem to find the current I_3 through the middle resistance R_3. As shown in Fig. 10-5a, first mark the terminals a and b across R_3. In Fig. 10-5b, R_3 is disconnected. To calculate V_{Th}, find V_{ab} across the open terminals.

Superposition Method With two sources we can use superposition to calculate V_{ab}. First short-circuit V_2. Then the 84 V of V_1 is divided between R_1 and R_2. The voltage across R_2 is between terminals a and b. To calculate this divided voltage across R_2,

$$V_{R_2} = \frac{3}{15} \times V_1$$
$$= \frac{1}{5} \times (-84)$$
$$V_{R_2} = -16.8 \text{ V}$$

This is only the contribution of V_1 to V_{ab}. The polarity is negative at terminal a.

To find the voltage that V_2 produces between a and b, short-circuit V_1. Then the voltage across R_1 is connected from a to b. To calculate this divided voltage across R_1,

$$V_{R_1} = \frac{12}{15} \times V_2$$
$$= \frac{4}{5} \times (-21)$$
$$V_{R_1} = -16.8 \text{ V}$$

Both V_1 and V_2 produce -16.8 V across the ab terminals with the same polarity. Therefore, they are added.

The resultant value of $V_{ab} = -33.6$ V, shown in Fig. 10-5b, is the value of V_{Th}. The negative polarity means that terminal a is negative with respect to b.

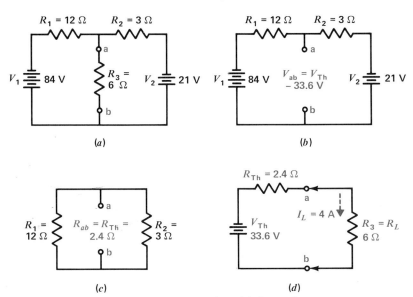

Fig. 10-5 Thevenizing a circuit with two voltage sources V_1 and V_2. (a) Original circuit with terminals a and b across the middle resistor R_3. (b) Disconnect R_3 to find V_{ab} is -33.6 V. (c) Short-circuit V_1 and V_2 to find $R_{ab} = 2.4$ Ω. (d) Thevenin equivalent with R_L reconnected to terminals a and b.

To calculate R_{Th}, short-circuit the sources V_1 and V_2, as shown in Fig. 10-5c. Then the 12-Ω R_1 and 3-Ω R_2 are in parallel across terminals a and b. Their combined resistance is $^{36}/_{15}$, or 2.4 Ω, which is the value of R_{Th}.

The final result is the Thevenin equivalent in Fig. 10-5d with an R_{Th} of 2.4 Ω and a V_{Th} of 33.6 V, negative toward terminal a.

In order to find the current through R_3, it is reconnected as a load resistance across terminals a and b. Then V_{Th} produces current through the total resistance of 2.4 Ω for R_{Th} and 6 Ω for R_3:

$$I_3 = \frac{V_{Th}}{R_{Th} + R_3} = \frac{33.6}{2.4 + 6} = \frac{33.6}{8.4} = 4 \text{ A}$$

This answer of 4 A for I_3 is the same value calculated before, using Kirchhoff's laws, in Fig. 9-4.

It should be noted that this circuit can be solved by superposition alone, without using Thevenin's theorem, if R_3 is not disconnected. However, opening terminals a and b for the Thevenin equivalent simplifies the superposition, as the circuit then has only series voltage dividers without any parallel current paths. In general, a circuit can often be simplified by disconnecting a component to open terminals a and b for Thevenin's theorem.

Short-cut Method The circuit in Fig. 10-5b with two voltage sources feeding terminals a and b can be thevenized more quickly by using the following formulas for V_{Th} and R_{Th}:

$$V_{Th} = \frac{V_1 R_2 + V_2 R_1}{R_1 + R_2}$$

$$= \frac{(-84)(3) + (-21)(12)}{12 + 3}$$

$$= \frac{-252 - 252}{15}$$

$$V_{Th} = \frac{-504}{15} = -33.6 \text{ V}$$

Voltages V_1 and V_2 are considered negative because the top is negative, compared with our reference at the bottom of the diagram.

To find R_{Th} the two resistances in series with the sources are combined in parallel:

$$R_{Th} = \frac{12 \times 3}{12 + 3} = \frac{36}{15} = 2.4 \ \Omega$$

Practice Problems 10-3
Answers at End of Chapter

In the Thevenin equivalent circuit in Fig. 10-5d,
a. How much is R_T?
b. How much is V_{R_L}?

10-4   Thevenizing a Bridge Circuit

As another example of Thevenin's theorem, we can find the current through the 2-Ω R_L at the center of the bridge circuit in Fig. 10-6a. When R_L is disconnected to open terminals a and b, the result is as shown in Fig. 10-6b. Notice how the circuit has become simpler because of the open. Instead of the unbalanced bridge in Fig. 10-6a which would require Kirchhoff's laws for a solution, the Thevenin equivalent in Fig. 10-6b consists of just two voltage dividers. Both the $R_3 R_4$ divider and the $R_1 R_2$ divider are across the same 30-V source.

Since the open terminal a is at the junction of R_3 and R_4, this divider can be used to find the potential at point a. Similarly the potential at terminal b can be found from the $R_1 R_2$ divider. Then V_{ab} is the difference between the potentials at terminals a and b.

Note the voltages for the two dividers. In the divider with the 3-Ω R_3 and 6-Ω R_4, the bottom voltage V_{R_4} is $^6/_9 \times 30 = 20$ V. Then V_{R_3} at the top is 10 V because both must add up to equal the 30-V source. The polarities are marked negative at the top, the same as the source voltage V.

Similarly, in the divider with the 6-Ω R_1 and 4-Ω R_2, the bottom voltage V_{R_2} is $^4/_{10} \times 30 = 12$ V. Then V_{R_1} at the top is 18 V, as the two must add up to equal the 30-V source. The polarities are also negative at the top, the same as V.

Now we can determine the potentials at terminals a and b, with respect to a common reference in order to find V_{ab}. Imagine that the positive side of the source V is connected to a chassis ground. Then we would use the bottom line in the diagram as our reference for voltages. Note that V_{R_4} at the bottom of the $R_3 R_4$ divider is the same as the potential of terminal a, with respect to ground. This value is -20 V, with terminal a negative.

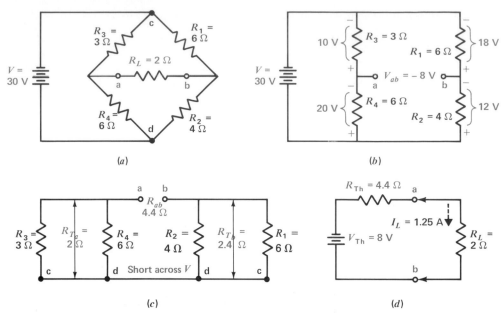

Fig. 10-6 Thevenizing a bridge circuit. (*a*) Original circuit with terminals a and b across the middle resistor R_L. (*b*) Disconnect R_L to find V_{ab} of -8 V. (*c*) With *V* short-circuited, R_{ab} is $2 + 2.4 = 4.4$ Ω. (*d*) Thevenin equivalent with R_L reconnected to terminals a and b.

Similarly, V_{R_2} in the R_1R_2 divider is the potential at b with respect to ground. This value is -12 V, with terminal b negative. As a result, V_{ab} is the difference between the -20 V at a and the -12 V at b, both with respect to the common ground reference.

The potential difference V_{ab} then equals

$$-20 - (-12) \quad \text{or} \quad -20 + 12 = -8 \text{ V}$$

Terminal a is 8 V more negative than b. Therefore, V_{Th} is 8 V, with the negative side toward terminal a as shown in the Thevenin equivalent in Fig. 10-6*d*.

The potential difference V_{ab} can also be found as the difference between V_{R_3} and V_{R_1} in Fig. 10-6*b*. In this case V_{R_3} is 10 V and V_{R_1} is 18 V, both positive with respect to the top line connected to the negative side of the source *V*. The potential difference between terminals a and b then is $10 - 18$, which also equals -8 V. Note that V_{ab} must have the same value no matter which path is used to determine the voltage.

To find R_{Th}, the 30-V source is short-circuited while terminals a and b are still open. Then the circuit looks like Fig. 10-6*c*. Looking back from terminals a and b, the 3-Ω R_3 and 6-Ω R_4 are in parallel, for a combined resistance R_{T_a} of $^{18}\!/_9$ or 2 Ω. The reason is that R_3 and

R_4 are joined at terminal a, while their opposite ends are connected by the short circuit across the source *V*. Similarly, the 6-Ω R_1 and 4-Ω R_2 are in parallel, for a combined resistance R_{T_b} of $^{24}\!/_{10} = 2.4$ Ω. Furthermore, the short circuit across the source now provides a path that connects R_{T_a} and R_{T_b} in series. The entire resistance is $2 + 2.4 = 4.4$ Ω for R_{ab} or R_{Th}.

The Thevenin equivalent in Fig. 10-6*d* represents the bridge circuit feeding the open terminals a and b, with 8 V for V_{Th} and 4.4 Ω for R_{Th}. Now connect the 2-Ω R_L to terminals a and b in order to calculate I_L. The current is

$$I_L = \frac{V_{\text{Th}}}{R_{\text{Th}} + R_L}$$

$$= \frac{8}{4.4 + 2} = \frac{8}{6.4}$$

$$I_L = 1.25 \text{ A}$$

This 1.25 A is the current through the 2-Ω R_L at the center of the unbalanced bridge in Fig. 10-6*a*. Furthermore, the amount of I_L for any value of R_L in Fig. 10-6*a* can be calculated from the equivalent circuit in Fig. 10-6*d*.

Practice Problems 10-4
Answers at End of Chapter

In the Thevenin equivalent circuit in Fig. 10-6d,
a. How much is R_T?
b. How much is V_{R_L}?

10-5   Norton's Theorem

Named after E. L. Norton, a scientist with Bell Telephone Laboratories, this theorem is used for simplifying a network in terms of currents instead of voltages. In many cases, analyzing the division of currents may be easier than voltage analysis. For current analysis, therefore, Norton's theorem can be used to reduce a network to a simple parallel circuit, with a current source. The idea of a current source is that it supplies a total line current to be divided among parallel branches, corresponding to a voltage source applying a total voltage to be divided among series components. This comparison is illustrated in Fig. 10-7.

Example of a Current Source   A source of electric energy supplying voltage is often shown with a series resistance which represents the internal resistance of the source, as in Fig. 10-7a. This method corresponds to showing an actual voltage source, such as a battery for dc circuits. However, the source may be represented also as a current source with a parallel resistance, as in Fig. 10-7b. Just as a voltage source is rated at, say, 10 V, a current source may be rated at 2 A. For the purpose of analyzing parallel branches,

the concept of a current source may be more convenient than a voltage source.

If the current I in Fig. 10-7b is a 2-A source, it supplies 2 A no matter what is connected across the output terminals a and b. Without anything connected across a and b, all the 2 A flows through the shunt R. When a load resistance R_L is connected across a and b, then the 2-A I divides according to the current division rules for parallel branches.

Remember that parallel currents divide inversely to branch resistances but directly with conductances. For this reason it may be preferable to consider the current source shunted by the conductance G, as shown in Fig. 10-7c. We can always convert between resistance and conductance, because $1/R$ in ohms is equal to G in siemens.

The symbol for a current source is a circle with an arrow inside, as shown in Fig. 10-7b and c, to show the direction of current. This direction must be the same as the current produced by the polarity of the corresponding voltage source. Remember that a source produces electron flow out from the negative terminal.

An important difference between voltage and current sources is that a current source is killed by making it open, compared with short-circuiting a voltage source. Opening a current source kills its ability to supply current without affecting any parallel branches. A voltage source is short-circuited to kill its ability to supply voltage without affecting any series components.

The Norton Equivalent Circuit   As illustrated in Fig. 10-8, Norton's theorem states that the entire network connected to terminals a and b can be replaced by a single current source I_N in parallel with a single resistance R_N. The value of I_N is equal to the short-cir-

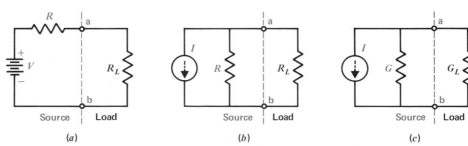

Fig. 10-7 General forms for a voltage or current source connected to a load R_L across terminals a and b. (a) Voltage source V with series R. (b) Current source I with parallel R. (c) Current source I with parallel conductance G.

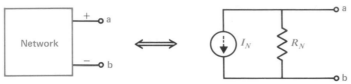

Fig. 10-8 Any network in the block at the left can be reduced to the Norton equivalent circuit at the right.

cuit current through the ab terminals. This means, find the current that the network would produce through a and b with a short circuit across these two terminals.

The value of R_N is the resistance looking back from open terminals a and b. These terminals are not short-circuited for R_N but are open, as in calculating R_{Th} for Thevenin's theorem. Actually, the single resistor is the same for both the Norton and Thevenin equivalent circuits. In the Norton case, this value of R_{ab} is R_N in parallel with the current source; in the Thevenin case, it is R_{Th} in series with the voltage source.

Nortonizing a Circuit   As an example, let us re-calculate the current I_L in Fig. 10-9a, which was solved

before by Thevenin's theorem. The first step in applying Norton's theorem is to imagine a short circuit across terminals a and b, as shown in Fig. 10-9b. How much current is flowing in the short circuit? Note that a short circuit across ab short-circuits R_L and the parallel R_2. Then the only resistance in the circuit is the 3-Ω R_1 in series with the 36-V source, as shown in Fig. 10-9c. The short-circuit current, therefore, is

$$I_N = \frac{36 \text{ V}}{3 \text{ Ω}} = 12 \text{ A}$$

This 12-A I_N is the total current available from the current source in the Norton equivalent in Fig. 10-9e.

To find R_N, remove the short circuit across a and b and consider the terminals open, without R_L. Now the

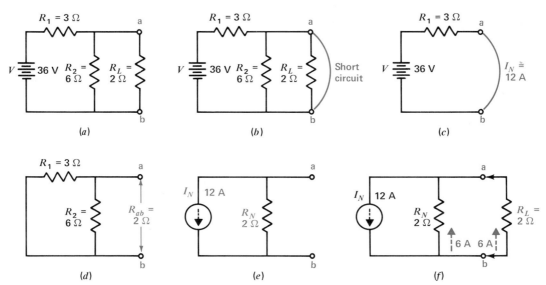

Fig. 10-9 Same circuit as shown in Fig. 10-3, but solved by Norton's theorem. (a) Original circuit. (b) Short circuit across terminals a and b. (c) Short-circuit I_N is 36/3 = 12 A. (d) Open terminals a and b, but short-circuit V to find R_{ab} = 2 Ω, the same as R_{Th}. (e) Norton equivalent circuit. (f) Resistance R_L reconnected to terminals a and b to find I_L is 6 A.

source V is considered to be short-circuited. As shown in Fig. 10-9d, the resistance seen looking back from terminals a and b is 6 Ω in parallel with 3 Ω, which equals 2 Ω for the value of R_N.

The resultant Norton equivalent is shown in Fig. 10-9e. It consists of a 12-A current source I_N shunted by the 2-Ω R_N. The arrow on the current source shows the direction of electron flow from terminal b to terminal a, as in the original circuit.

Finally, to calculate I_L, replace the 2-Ω R_L between terminals a and b, as shown in Fig. 10-9f. The current source still delivers 12 A, but now that current divides between the two branches of R_N and R_L. Since these two resistances are equal, the 12-A I_N divides into 6 A for each branch, and I_L is equal to 6 A. This value is the same current we calculated in Fig. 10-3, by Thevenin's theorem. Also, V_L can be calculated as $I_L R_L$, or 6 A × 2 Ω, which equals 12 V.

Looking at the Short-Circuit Current

In some cases, there may be a question of which current is I_N when terminals a and b are short-circuited. Imagine that a wire jumper is connected between a and b to short-circuit these terminals. Then I_N must be the current that flows in this wire between terminals a and b.

Remember that any components directly across these two terminals are also short-circuited by the wire jumper. Then these parallel paths have no effect. However, any components in series with terminal a or terminal b are in series with the wire jumper. Therefore, the short-circuit current I_N also flows through the series components.

An example of a resistor in series with the short circuit across terminals a and b is shown in Fig. 10-10.

The idea here is that the short-circuit I_N is a branch current, not the main-line current. Refer to Fig. 10-10a. Here the short circuit connects R_3 across R_2. Also, the short-circuit current I_N is now the same as the current I_3 through R_3. Note that I_3 is only a branch current.

To calculate I_3, the circuit is solved by Ohm's law. The parallel combination of R_2 with R_3 equals $^{72}/_{18}$ or 4 Ω. The R_T is 4 + 4 = 8 Ω. As a result, the I_T from the source is 48 V/8 Ω = 6 A.

This I_T of 6 A in the main line divides into 4 A for R_2 and 2 A for R_3. The 2-A I_3 for R_3 flows through short-circuited terminals a and b. Therefore, this current of 2 A is the value of I_N.

To find R_N in Fig. 10-10b, the short circuit is removed from terminals a and b. Now the source V is short-circuited. Looking back from open terminals a and b, the 4-Ω R_1 is in parallel with the 6-Ω R_2. This combination is $^{24}/_{10}$ = 2.4 Ω. The 2.4 Ω is in series with the 12-Ω R_3 to make R_{ab} = 2.4 + 12 = 14.4 Ω.

The final Norton equivalent is shown in Fig. 10-10c. Current I_N is 2 A because this branch current in the original circuit is the current that flows through R_3 and short-circuited terminals a and b. Resistance R_N is 14.4 Ω looking back from open terminals a and b with the source V short-circuited the same way as for R_{Th}.

Practice Problems 10-5
Answers at End of Chapter

Answer True or False. For a Norton equivalent circuit,
a. Terminals a and b are short-circuited to find I_N.
b. Terminals a and b are open to find R_N.

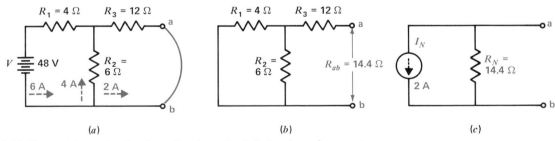

Fig. 10-10 Nortonizing a circuit where the short-circuit I_N is a branch current. (a) I_N is 2 A through short-circuited terminals a and b and R_3. (b) $R_N = R_{ab} =$ 14.4Ω. (c) Norton equivalent.

10-6   Thevenin-Norton Conversions

Thevenin's theorem says that any network can be represented by a voltage source and series resistance, while Norton's theorem says that the same network can be represented by a current source and shunt resistance. It must be possible, therefore, to convert directly from a Thevenin form to a Norton form and vice versa. Such conversions are often useful.

Norton from Thevenin   Consider the Thevenin equivalent circuit in Fig. 10-11a. What is its Norton equivalent? Just apply Norton's theorem, the same as for any other circuit. The short-circuit current through terminals a and b is

$$I_N = \frac{V_{Th}}{R_{Th}} = \frac{15 \text{ V}}{3 \text{ }\Omega} = 5 \text{ A}$$

The resistance, looking back from open terminals a and b with the source V_{Th} short-circuited, is equal to the 3 Ω of R_{Th}. Therefore, the Norton equivalent consists of a current source that supplies the short-circuit current of 5 A, shunted by the same 3-Ω resistance that is in series in the Thevenin circuit. The results are shown in Fig. 10-11b.

Thevenin from Norton   For the opposite conversion, we can start with the Norton circuit of Fig. 10-11b and get back to the original Thevenin circuit. To do this, apply Thevenin's theorem, the same as for any other circuit. First, we find the Thevenin resistance by looking back from open terminals a and b. An important principle here, though, is that while a voltage

source is short-circuited to find R_{Th}, a current source is an open circuit. In general, a current source is killed by opening the path between its terminals. Therefore, we have just the 3-Ω R_N, in parallel with the infinite resistance of the open current source. The combined resistance then is 3 Ω.

In general, the resistance R_N always has the same value as R_{Th}. The only difference is that R_N is connected in parallel with I_N, but R_{Th} is in series with V_{Th}.

Now all that is required is to calculate the open-circuit voltage in Fig. 10-11b to find the equivalent V_{Th}. Note that with terminals a and b open, all the current of the current source flows through the 3-Ω R_N. Then the open-circuit voltage across the terminals a and b is

$$I_N R_N = 5 \text{ A} \times 3 \text{ }\Omega = 15 \text{ V} = V_{Th}$$

As a result, we have the original Thevenin circuit, which consists of the 15-V source V_{Th} in series with the 3-Ω R_{Th}.

Conversion Formulas   In summary, the following formulas can be used for these conversions:

Thevenin from Norton

$$R_{Th} = R_N$$
$$V_{Th} = I_N \times R_N$$

Norton from Thevenin

$$R_N = R_{Th}$$
$$I_N = V_{Th} \div R_{Th}$$

Another example of these conversions is shown in Fig. 10-12.

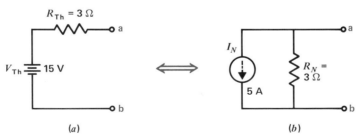

(a) (b)

Fig. 10-11 Thevenin equivalent circuit in (a) corresponds to the Norton equivalent in (b).

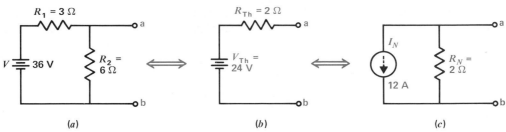

Fig. 10-12 Example of Thevenin-Norton conversions. (*a*) Original circuit, the same as Figs. 10-3*a* and 10-9*a*. (*b*) Thevenin equivalent. (*c*) Norton equivalent.

Practice Problems 10-6
Answers at End of Chapter

Answer True or False. In Thevenin-Norton conversions,
a. Resistances R_N and R_{Th} have the same value.
b. Current I_N is V_{Th}/R_{Th}.
c. Voltage V_{Th} is $I_N \times R_N$.

10-7   Conversion of Voltage and Current Sources

Norton conversion is a specific example of the general principle that any voltage source with its series resistance can be converted to an equivalent current source with the same resistance in parallel. In Fig. 10-13, the voltage source in Fig. 10-13*a* is equivalent to the current source in Fig. 10-13*b*. Just divide the source V by its series R to calculate the value of I for the equivalent current source shunted by the same R. Either source

will supply the same current and voltage for any components connected across terminals a and b.

Conversion of voltage and current sources can often simplify circuits, especially those with two or more sources. Current sources are easier for parallel connections, where we can add or divide currents. Voltage sources are easier for series connections, where we can add or divide voltages.

Two Sources in Parallel Branches   Referring to Fig. 10-14*a*, assume that the problem is to find I_3 through the middle resistor R_3. Note that V_1 with R_1 and V_2 with R_2 are branches in parallel with R_3. All three branches are connected across terminals a and b.

When we convert V_1 and V_2 to current sources in Fig. 10-14*b*, the circuit has all parallel branches. Current I_1 is $^{84}/_{12}$ or 7 A, while I_2 is $^{21}/_3$, which also happens to be 7 A. Current I_1 has its parallel R of 12 Ω, while I_2 has its parallel R of 3 Ω.

Furthermore, I_1 and I_2 can be combined for the one equivalent current source I_T in Fig. 10-14*c*. Since both sources produce current in the same direction through R_L, they are added for $I_T = 7 + 7 = 14$ A.

The shunt R for the 14-A combined source is the combined resistance of the 12-Ω R_1 and the 3-Ω R_2 in parallel. This R equals $^{36}/_{15}$ or 2.4 Ω, as shown in Fig. 10-14*c*.

To find I_L, we can use the current divider formula for the 6- and 2.4-Ω branches, into which the 14-A I_T from the current source was split. Then

$$I_L = \frac{2.4}{2.4 + 6} \times 14 = \frac{33.6}{8.4} = 4 \text{ A}$$

The voltage V_{R_3} across terminals a and b is $I_L R_L$, which equals $4 \times 6 = 24$ V. These are the same values

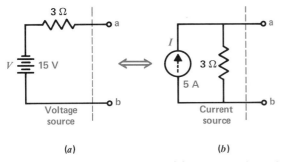

Fig. 10-13 The voltage source in (*a*) corresponds to the current source in (*b*).

—⋀⋀—⊣⎮⊢——————————————————————

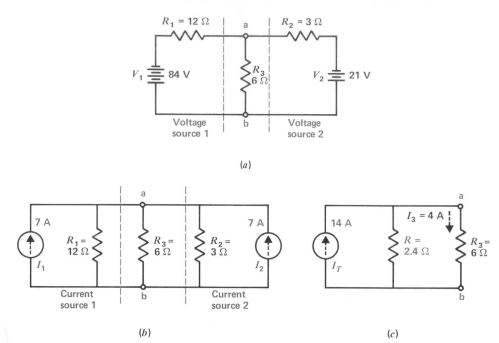

Fig. 10-14 Converting voltage sources in parallel branches to current sources that can be combined. (a) Original circuit. (b) Voltage sources V_1 and V_2 converted to parallel current sources I_1 and I_2. (c) Circuit with one combined current source I_T.

calculated for V_{R_3} and I_3 by Kirchhoff's laws in Fig. 9-4 and by Thevenin's theorem in Fig. 10-5.

Two Sources in Series   Referring to Fig. 10-15, assume that the problem is to find the current I_L through the load resistance R_L between terminals a and b. This circuit has the two current sources I_1 and I_2 in series with each other.

The problem here can be simplified by converting I_1 and I_2 to the series voltage sources V_1 and V_2 shown in Fig. 10-15b. The 2-A I_1 with its shunt 4-Ω R_1 is equivalent to 4 × 2 or 8 V for V_1 with a 4-Ω series resistance. Similarly, the 5-A I_2 with its shunt 2-Ω R_2 is equivalent to 5 × 2, or 10 V, for V_2 with a 2-Ω series resistance. The polarities of V_1 and V_2 produce electron flow in the same direction as I_1 and I_2.

The series voltages can now be combined as in Fig. 10-15c. The 8 V of V_1 and 10 V of V_2 are added because they are series-aiding, resulting in the total V_T of 18 V. And, the resistances of 4 Ω for R_1 and 2 Ω for R_2 are added, for a combined R of 6 Ω. This is the series

resistance of the 18-V source V_T connected across terminals a and b.

The total resistance of the circuit in Fig. 10-15c is R plus R_L, or 6 + 3 = 9 Ω. With 18 V applied, I_L = $^{18}\!/_{9}$ = 2 A through R_L between terminals a and b.

Practice Problems 10-7
Answers at End of Chapter

A voltage source has 21 V in series with 3 Ω. For the equivalent current source,
a. How much is I?
b. How much is the shunt R?

10-8   Millman's Theorem

This theorem provides a shortcut for finding the common voltage across any number of parallel branches with different voltage sources. A typical example is

—⋀⋀—⊣⎮⊢——————————————————————

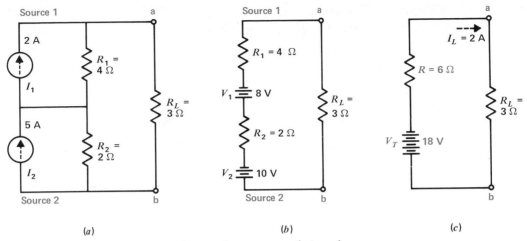

Fig. 10-15 Converting current sources in series to voltage sources that can be combined. (*a*) Original circuit. (*b*) Current sources I_1 and I_2 converted to series V_1 and V_2. (*c*) Circuit with one combined voltage source V_T.

shown in Fig. 10-16. For all the branches, the ends at point y are connected to chassis ground. Furthermore, the opposite ends of all the branches are also connected to the common point x. The voltage V_{xy}, therefore, is the common voltage across all the branches.

Finding the value of V_{xy} gives the net effect of all the sources in determining the voltage at x with respect to chassis ground. To calculate this voltage

$$V_{xy} = \frac{V_1/R_1 + V_2/R_2 + V_3/R_3}{1/R_1 + 1/R_2 + 1/R_3} \cdots \text{ etc.} \qquad (10\text{-}1)$$

This formula is derived from converting the voltage sources to current sources and combining the results.

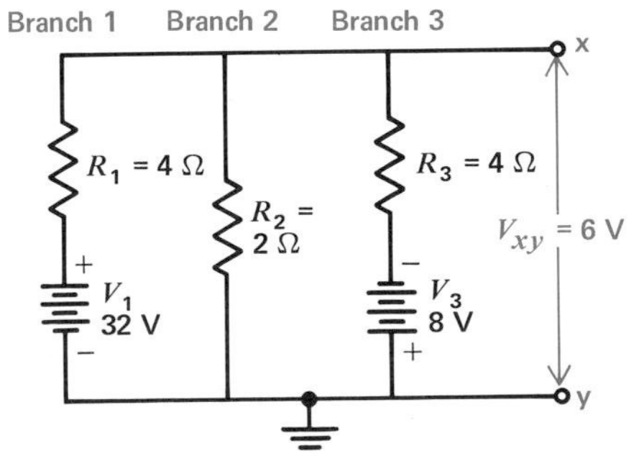

The numerator with V/R terms is the sum of the parallel current sources. The denominator with $1/R$ terms is the sum of the parallel conductances. The net V_{xy} then is the form of I/G or $I \times R$, which is in units of voltage.

Calculating V_{xy} For the values in Fig. 10-16,

$$V_{xy} = \frac{32/4 + 0/2 - 8/4}{1/4 + 1/2 + 1/4}$$

$$= \frac{8 + 0 - 2}{1}$$

$$V_{xy} = 6 \text{ V}$$

Note that in branch 3, V_3 is considered negative because it would make point x negative. However, all the resistances are positive. The positive answer for V_{xy} means that point x is positive with respect to y.

In branch 2, V_2 is zero because this branch has no voltage source. However, R_2 is still used in the denominator.

This method can be used for any number of branches, but they must all be in parallel, without any series resistances between the branches. In a branch with several resistances, they can be combined as one R_T. When a branch has more than one voltage source, they can be combined algebraically for one V_T.

Fig. 10-16 Example of Millman's theorem to find V_{xy}, the common voltage across branches with separate voltage sources.

Applications of Millman's Theorem

In many cases, a circuit can be redrawn to show the parallel branches and their common voltage V_{xy}. Then with V_{xy} known the entire circuit can be analyzed quickly. For instance, Fig. 10-17 has been solved before by other methods. For Millman's theorem the common voltage V_{xy} across all the branches is the same as V_3 across R_3. This voltage is calculated with Formula (10-1), as follows:

$$V_{xy} = \frac{-84/12 + 0/6 - 21/3}{1/12 + 1/6 + 1/3}$$

$$= \frac{-7 + 0 - 7}{7/12}$$

$$= \frac{-14}{7/12} = -14 \times \frac{12}{7}$$

$$V_{xy} = -24 \text{ V} = V_3$$

The negative sign means that point x is the negative side of V_{xy}.

With V_3 known to be 24 V across the 6-Ω R_3, then I_3 must be $^{24}/_6 = 4$ A. Similarly, all the voltages and currents in this circuit can then be calculated. (See Fig. 9-4 in Chap. 9, page 174.)

As another application, the example of superposition in Fig. 10-1 has been redrawn in Fig. 10-18 to show the parallel branches with a common voltage V_{xy} to be calculated by Millman's theorem. Then

$$V_{xy} = \frac{24 \text{ V}/30 \text{ k}\Omega - 9 \text{ V}/60 \text{ k}\Omega}{1/(30 \text{ k}\Omega) + 1/(60 \text{ k}\Omega)}$$

$$= \frac{0.8 \text{ mA} - 0.15 \text{ mA}}{3/(60 \text{ k}\Omega)}$$

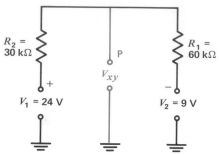

Fig. 10-18 Same circuit as in Fig. 10-1 for superposition, but shown with parallel branches to calculate V_{xy} by Millman's theorem.

$$= 0.65 \times \frac{60}{3} = \frac{39}{3}$$

$$V_{xy} = 13 \text{ V} = V_P$$

The answer of 13 V from point P to ground, using Millman's theorem, is the same value calculated before by superposition.

Practice Problems 10-8
Answers at End of Chapter

For the example of Millman's theorem in Fig. 10-16,
a. How much is V_{R_2}?
b. How much is V_{R_3}?

10-9 T and π Networks

The network in Fig. 10-19 is called a T (tee) or Y (wye) network, as suggested by its shape. T and Y are different names for the same network, the only difference being that the R_2 and R_3 arms are at an angle in the Y.

The network in Fig. 10-20 is called a π (pi) or Δ (delta) network, as the shape is similar to these Greek

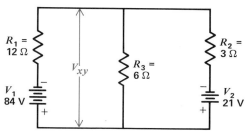

Fig. 10-17 The same circuit as in Fig. 9-4 for Kirchhoff's laws, but shown with parallel branches to calculate V_{xy} by Millman's theorem.

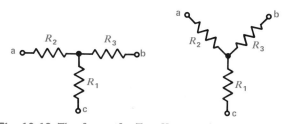

Fig. 10-19 The form of a T or Y network.

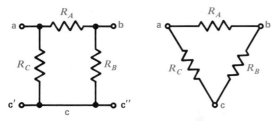

Fig. 10-20 The form of a π or Δ network.

letters. Actually, the network can have R_A at either the top or the bottom, between R_C and R_B. Note that when the single point c of the delta is separated into two points c$'$ and c$''$ for the π network, the connections are not really changed. The π and Δ are different names for the same network.

Conversion Formulas   In the analysis of networks, it is often helpful to convert a Δ to Y or vice versa. Either it may be impossible to solve the circuit without the conversion, or the conversion makes the solution simpler. The formulas for these transformations are given here. All are derived from Kirchhoff's laws. Note that letters are used as subscripts for R_A, R_B, and R_C in the Δ, while the resistances are numbered R_1, R_2, and R_3 in the Y.

Conversions of Y to Δ, or T to π

$$R_A = \frac{R_1R_2 + R_2R_3 + R_3R_1}{R_1}$$
$$R_B = \frac{R_1R_2 + R_2R_3 + R_3R_1}{R_2}$$
$$R_C = \frac{R_1R_2 + R_2R_3 + R_3R_1}{R_3}$$

(10-2)

or

$$R_\Delta = \frac{\Sigma \text{ all cross products in Y}}{\text{opposite } R \text{ in Y}}$$

These formulas can be used to convert a Y network to an equivalent Δ, or a T network to π. Both networks will have the same resistance across any pair of terminals.

The three formulas have the same general form, indi-

cated at the bottom as one basic rule. The symbol Σ is the Greek capital letter sigma, meaning "sum of."

For the opposite conversion:

Conversion of Δ to Y or π to T

$$R_1 = \frac{R_BR_C}{R_A + R_B + R_C}$$
$$R_2 = \frac{R_CR_A}{R_A + R_B + R_C}$$
$$R_3 = \frac{R_AR_B}{R_A + R_B + R_C}$$

(10-3)

or

$$R_Y = \frac{\text{product of two adjacent } R \text{ in } \Delta}{\Sigma \text{ all } R \text{ in } \Delta}$$

As an aid in using these formulas, the following scheme is useful. Place the Y inside the Δ, as shown in Fig. 10-21. Notice that the Δ has three closed sides, while the Y has three open arms. Also note how resistors can be considered opposite each other in the two networks. For instance, the open arm R_1 is opposite the closed side R_A; R_2 is opposite R_B; and R_3 is opposite R_C.

Furthermore, each resistor in an open arm has two adjacent resistors in the closed sides. For R_1, its adjacent resistors are R_B and R_C; also R_C and R_A are adjacent to R_2, while R_A and R_B are adjacent to R_3.

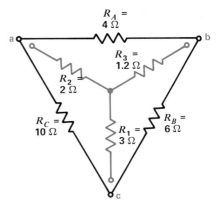

Fig. 10-21 Conversion between Y and Δ networks. See text for conversion formulas.

In the formulas for the Y-to-Δ conversion, each side of the delta is found by first taking all possible cross products of the arms of the wye, using two arms at a time. There are three such cross products. The sum of the three cross products is then divided by the opposite arm to find the value of each side of the delta. Note that the numerator remains the same, the sum of the three cross products. However, each side of the delta is calculated by dividing this sum by the opposite arm.

For the case of the Δ-to-Y conversion, each arm of the wye is found by taking the product of the two adjacent sides in the delta and dividing by the sum of the three sides of the delta. The product of two adjacent resistors excludes the opposite resistor. The denominator for the sum of the three sides remains the same in the three formulas. However, each arm is calculated by dividing the sum into each cross product.

An Example of Conversion   The values shown

for the equivalent Y and Δ in Fig. 10-21 are calculated as follows: Starting with 4, 6, and 10 Ω for sides R_A, R_B, and R_C, respectively, in the delta, the corresponding arms in the wye are:

$$R_1 = \frac{R_B R_C}{R_A + R_B + R_C}$$

$$= \frac{6 \times 10}{4 + 6 + 10} = \frac{60}{20} = 3\ \Omega$$

$$R_2 = \frac{R_C R_A}{20}$$

$$= \frac{10 \times 4}{20} = \frac{40}{20} = 2\ \Omega$$

$$R_3 = \frac{R_A R_B}{20}$$

$$= \frac{4 \times 6}{20} = \frac{24}{20} = 1.2\ \Omega$$

As a check on these values, we can calculate the equivalent delta for this wye. Starting with values of 3, 2, and 1.2 Ω for R_1, R_2, and R_3, respectively, in the wye, the corresponding values in the delta are:

$$R_A = \frac{R_1 R_2 + R_2 R_3 + R_3 R_1}{R_1} = \frac{6 + 2.4 + 3.6}{3}$$

$$= \frac{12}{3} = 4\ \Omega$$

$$R_B = \frac{12}{R_2} = \frac{12}{2} = 6\ \Omega$$

$$R_C = \frac{12}{R_3} = \frac{12}{1.2} = 10\ \Omega$$

These results show that the Y and Δ networks in Fig. 10-21 are equivalent to each other when they have the values obtained with the conversion formulas.

Simplifying a Bridge Circuit   As an example of

the use of such transformations, consider the bridge circuit of Fig. 10-22. The total current I_T from the battery is desired. Therefore, we must find the total resistance R_T.

One approach is to note that the bridge in Fig. 10-22a consists of two deltas connected between terminals P_1 and P_2. One of them can be replaced by an equivalent wye. We use the bottom delta with R_A across the top, in the same form as Fig. 10-21. We then replace this delta $R_A R_B R_C$ by an equivalent wye $R_1 R_2 R_3$, as shown in Fig. 10-22b. Using the conversion formulas,

$$R_1 = \frac{R_B R_C}{R_A + R_B + R_C} = \frac{24}{12} = 2\ \Omega$$

$$R_2 = \frac{R_C R_A}{12} = \frac{12}{12} = 1\ \Omega$$

$$R_3 = \frac{R_A R_B}{12} = \frac{8}{12} = \frac{2}{3}\ \Omega$$

We next use these values for R_1, R_2, and R_3 in an equivalent wye to replace the original delta. Then the resistances form the series-parallel circuit shown in Fig. 10-22c. The combined resistance of the two parallel branches here is $4 \times 6\frac{2}{3}$ divided by $10\frac{2}{3}$, which equals 2.5 Ω for R. Adding this 2.5 Ω to the series R_1 of 2 Ω, the total resistance is 4.5 Ω in Fig. 10-22d.

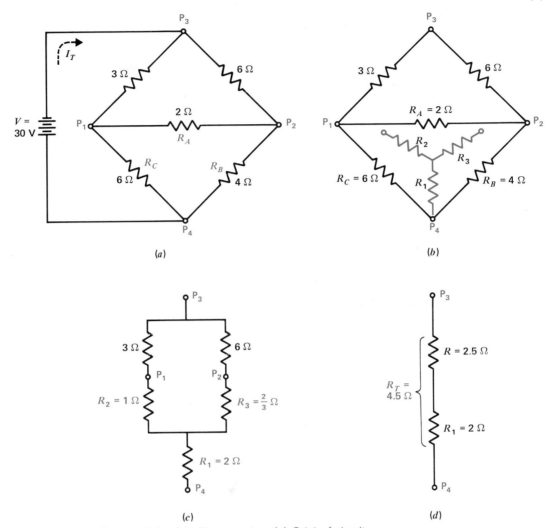

Fig. 10-22 Solving a bridge circuit by Δ-to-Y conversion. (*a*) Original circuit. (*b*) How Y of $R_1R_2R_3$ corresponds to Δ of $R_AR_BR_C$. (*c*) The Y substituted for the Δ network. Result is a series-parallel circuit with same R_T as the original bridge circuit. (*d*) Resistance R_T is 4.5 Ω between points P_3 and P_4.

This 4.5 Ω is the R_T for the entire bridge circuit between P_3 and P_4 connected to source V. Then I_T is 30 V/4.5 Ω, which equals 6⅔ A supplied by the source.

Another approach to finding R_T for the bridge circuit in Fig. 10-22*a* is to recognize that the bridge also consists of two T or Y networks between terminals P_3 and P_4. One of them can be transformed into an equivalent delta. The result is another series-parallel circuit but with the same R_T of 4.5 Ω.

Practice Problems 10-9
Answers at End of Chapter

In the standard form for conversion,
a. Which resistor R in the Y is opposite resistor R_A in the Δ?
b. Which two resistors in the Δ are adjacent to R_1 in the Y?

—⋀⋀⋀—|I|⊢—

Summary

1. *Superposition theorem*. In a linear, bilateral network having more than one source, the current and voltage in any part of the network can be found by adding algebraically the effect of each source separately. All other sources are temporarily killed by short-circuiting voltage sources and opening current sources.

2. *Thevenin's theorem*. Any network with two open terminals a and b can be replaced by a single voltage source V_{Th} in series with a single resistance R_{Th} driving terminals a and b. Voltage V_{Th} is the voltage produced by the network across terminals a and b. Resistance R_{Th} is the resistance across open terminals a and b with all sources short-circuited.

3. *Norton's theorem*. Any two-terminal network can be replaced by a single current source I_N in parallel with a single resistance R_N. The value of I_N is the current produced by the network through the short-circuited terminals. R_N is the resistance across the open terminals with all sources short-circuited.

4. *Millman's theorem*. The common voltage across parallel branches with different V sources can be determined with Formula (10-1).

5. A voltage source V with its series R can be converted to an equivalent current source I with parallel R, or vice versa. The value of I is V/R, or V is $I \times R$. The value of R is the same for both sources. However, R is in series with V but in parallel with I.

6. The comparison between delta and wye networks is illustrated in Fig. 10-21. To convert from one network to the other, Formula (10-2) or (10-3) is used.

Self-Examination
Answers at Back of Book

Answer True or False.

1. Voltage V_{Th} is an open-circuit voltage.
2. Current I_N is a short-circuit current.
3. Resistances R_{Th} and R_N have the same value.
4. A voltage source has series resistance.
5. A current source has parallel resistance.
6. A voltage source is killed by short-circuiting the terminals.
7. A current source is killed by opening the source.
8. A π network is the same as a T network.
9. Millman's theorem is useful for parallel branches with different voltage sources.
10. A 10-V source has a 2-Ω series R. Its equivalent current source is 2 A in parallel with 10 Ω.

Essay Questions

1. State the superposition theorem.
2. In applying the superposition theorem, how do we kill or disable voltage sources and current sources?

—⋀⋀⋀—|I|⊢—

3. State the method of calculating V_{Th} and R_{Th} for a Thevenin equivalent circuit.
4. State the method of calculating I_N and R_N for a Norton equivalent circuit.
5. How is a voltage source converted to a current source, and vice versa?
6. For what type of circuit is Millman's theorem used?
7. Draw a delta network and a wye network and give the six formulas needed to convert from one to the other.
8. Give two differences between a dc source and an ac source.

Problems
Answers to Odd-Numbered Problems at Back of Book

1. Refer to Fig. 10-23. Show the Thevenin equivalent and calculate V_L.
2. Show the Norton equivalent of Fig. 10-23 and calculate I_L.
3. In Fig. 10-23, convert V and R_1 to a current source and calculate I_L.
4. Use Ohm's law to solve Fig. 10-23 as a series-parallel circuit in order to calculate V_L and I_L. (Note: R_L is not opened for Ohm's law.)
5. Why is the value of V_L across terminals a and b in Prob. 4 not the same as V_{ab} for the Thevenin equivalent circuit in Prob. 1?

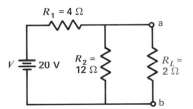

Fig. 10-23 For Probs. 1, 2, 3, 4, and 5.

6. Refer to Fig. 10-24. Determine V_P by superposition.
7. Redraw Fig. 10-24 as two parallel branches to calculate V_P by Millman's theorem.

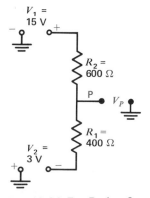

Fig. 10-24 For Probs. 6 and 7.

8. Refer to Fig. 10-25. Calculate V_L across R_L by Millman's theorem and also by superposition.
9. Show the Thevenin equivalent of Fig. 10-26, where terminals a and b are across the middle resistor R_2. Then calculate V_{R_2}.
10. In Fig. 10-26, solve for V_{R_2} by superposition.

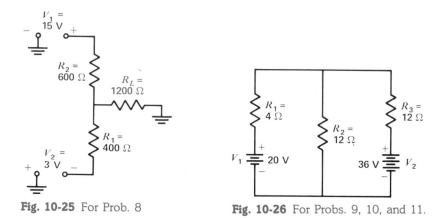

Fig. 10-25 For Prob. 8   **Fig. 10-26** For Probs. 9, 10, and 11.

11. In Fig. 10-26, solve for V_{R_2} by Millman's theorem.
12. In Fig. 10-27, solve for all the currents by Kirchhoff's laws.
13. In Fig. 10-27, find V_3 by Millman's theorem.
14. Convert the T network in Fig. 10-28 to an equivalent π network.

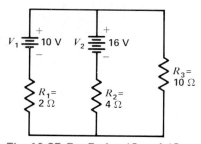

Fig. 10-27 For Probs. 12 and 13.

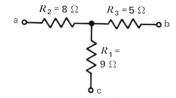

Fig. 10-28 For Prob. 14.

15. Convert the π network in Fig. 10-29 to an equivalent T network.
16. Show the Thevenin and Norton equivalent circuits for the diagram in Fig. 10-30.

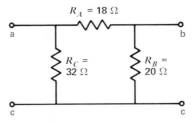

Fig. 10-29 For Prob. 15.

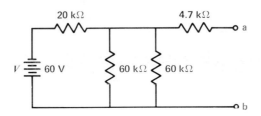

Fig. 10-30 For Prob. 16.

Answers to Practice Problems

10-1 a. R_1
  b. R_2

10-2 a. T
  b. T

10-3 a. 8.4 Ω
  b. 24 V

10-4 a. 6.4 Ω
  b. 2.5 V

10-5 a. T
  b. T

10-6 a. T
  b. T
  c. T

10-7 a. 7 A
  b. 3 Ω

10-8 a. 6 V
  b. 14 V

10-9 a. R_1
  b. R_B and R_C

Review Chapters 9 and 10

Summary

1. Methods of applying Kirchhoff's laws include (**a**) equations of voltages using the branch currents in the loops to specify the voltages; (**b**) equations of currents at a node using the node voltages to specify the node currents; (**c**) equations of voltages using assumed mesh currents to specify the voltages.
2. Methods of reducing a network to a simple equivalent circuit include (**a**) the superposition theorem using one source at a time; (**b**) Thevenin's theorem to convert the network to a series circuit with one source; (**c**) Norton's theorem to convert the network to a parallel circuit with one source; (**d**) Millman's theorem to find the common voltage across parallel branches with different sources; (**e**) delta-wye conversions to transform a network into a series-parallel circuit.

Review Self-Examination
Answers at Back of Book

Answer True or False.

1. In Fig. 9-3, V_3 can be found by using Kirchhoff's laws with either branch currents or mesh currents.
2. In Fig. 9-3, V_3 can be found by superposition, thevenizing, or using Millman's theorem.
3. In Fig. 10-6, I_L cannot be found by delta-wye conversion because R_L disappears in the transformation.
4. In Fig. 10-6, I_L can be calculated with Kirchhoff's laws, using mesh currents for three meshes.
5. With superposition, we can use Ohm's law for circuits that have more than one source.
6. A Thevenin equivalent is a parallel circuit.
7. A Norton equivalent is a series circuit.

8. Either a Thevenin or a Norton equivalent of a network will produce the same current in any load across terminals a and b.

9. A Thevenin-Norton conversion means converting a voltage source to a current source.

10. The units are volts for (volts/ohms) ÷ siemens.

11. A node voltage is a voltage between current nodes.

12. A π network can be converted to an equivalent T network.

13. A 10-V source with 10-Ω series R will supply 5 V to a 10-Ω load R_L.

14. A 10-A source with 10-Ω parallel R will supply 5 A to a 10-Ω load R_L.

15. Current sources in parallel can be added when they supply current in the same direction through R_L.

References
Additional References at Back of Book

Hayt and Kemmerling: *Engineering Circuit Analysis*, McGraw-Hill Book Company, New York.

Jackson, H. W.: *Introduction to Electric Circuits*, Prentice-Hall, Inc., Englewood Cliffs, N.J.

Ridsdale, R. E.: *Electric Circuits*, McGraw-Hill Book Company, New York.

Romanowitz, A. H.: *Electric Fundamentals and Circuit Analysis*, John Wiley & Sons, New York.

Chapter 11
Conductors and Insulators

Conductors have very low resistance. Less than 1 Ω for 10 ft of copper wire is a typical value. The function of the wire conductor is to connect a source of applied voltage to a load resistance with minimum *IR* volt-

age drop in the conductor. Then all the applied voltage can produce current in the load resistance.

At the opposite extreme, materials having a very high resistance of many megohms are insulators. Some common examples are air, paper, mica, glass, plastics, rubber, cotton, and shellac or varnish.

Between the extremes of conductors and insulators are semiconductor materials such as carbon, silicon, and germanium. Carbon is used in the manufacture of resistors. Silicon and germanium are used for transistors.

Important terms in this chapter are:

circuit breaker	hole current	resistance wire
circular mil	hot resistance	specific resistance
corona effect	ion current	superconductivity
covalent bond	printed wiring	switch types
F connector	RCA plug	temperature coefficient of R
fuse types	relay switch	wire gage sizes

More details are explained in the following sections:

11-1   Function of the Conductor
11-2   Standard Wire Gage Sizes
11-3   Types of Wire Conductors
11-4   Printed Wiring
11-5   Switches
11-6   Fuses
11-7   Wire Resistance
11-8   Temperature Coefficient of Resistance
11-9   Ion Current in Liquids and Gases
11-10 Electrons and Hole Charges in Semiconductors
11-11 Insulators
11-12 Troubleshooting Hints

11-1 Function of the Conductor

In Fig. 11-1, the resistance of the two 10-ft lengths of copper-wire conductors is approximately 0.6 Ω. This is negligibly small compared with the 144-Ω resistance for the tungsten filament in the bulb. When the current of approximately 0.9 A flows in the bulb and the series conductors, the IR voltage drop across the conductors is 0.54 V, with 119.5 V across the bulb. Practically all the applied voltage is across the filament of the bulb. Since the bulb then has its rated voltage of 120 V, approximately, it will dissipate its rated power of 100 W and light with full brilliance.

The current in the wire conductors and the bulb is the same, since they are in series. However, the IR voltage drop in the conductor is practically zero because its R is almost zero.

Also, the I^2R power dissipated in the conductor is negligibly small, allowing the conductor to operate without becoming hot. Therefore, the conductor delivers energy from the source to the load with minimum loss, by means of electron flow in the copper wires.

Although the resistance of wire conductors is very small, for some cases of excessive current the resultant IR drop can be appreciable. The complaint that the size of a television picture shrinks at night is one example. When many lights and possibly other appliances are on, the high value of current can produce too much voltage drop in the power line. A 30-V IR drop results in only 90 V at the load, which is low enough to reduce the picture size. As additional examples, excessive IR drop in the line and low voltage at the load can be the cause of a toaster not heating quickly or an electric motor not starting properly.

Practice Problems 11-1
Answers at End of Chapter

Refer to Fig. 11-1.
a. How much is R for the 20 ft of copper wire?
b. How much is the IR voltage drop for the wire conductors?

11-2 Standard Wire Gage Sizes

Table 11-1 lists the standard wire sizes in the system known as the American Wire Gage (AWG), or Brown and Sharpe (B&S) gage. The gage numbers specify the size of round wire in terms of its diameter and cross-sectional circular area. Note the following:

1. As the gage numbers increase from 1 to 40, the diameter and circular area decrease. Higher gage numbers indicate thinner wire sizes.
2. The circular area doubles for every three gage sizes. For example, No. 10 wire has approximately twice the area of No. 13 wire.
3. The higher the gage number and the thinner the wire, the greater the resistance of the wire for any given length.

In typical applications, hookup wire for electronic circuits with current in the order of milliamperes is generally about No. 22 gage. For this size, 0.5 to 1 A is the maximum current the wire can carry without heating.

House wiring for circuits where the current is 5 to 15 A is about No. 14 gage. Minimum sizes for house

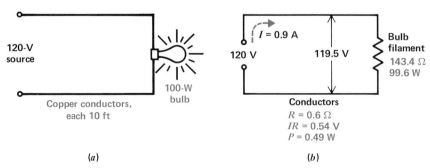

Fig. 11-1 The conductors should have minimum resistance to light the bulb with full brilliance. (*a*) Wiring diagram. (*b*) Schematic diagram.

Table 11-1. Copper-Wire Table

Gage no.	Diameter, mil	Circular-mil area	Ohms per 1000 ft of copper wire at 25°C*	Gage no.	Diameter, mil	Circular-mil area	Ohms per 1000 ft of copper wire at 25°C*
1	289.3	83,690	0.1264	21	28.46	810.1	13.05
2	257.6	66,370	0.1593	22	25.35	642.4	16.46
3	229.4	52,640	0.2009	23	22.57	509.5	20.76
4	204.3	41,740	0.2533	24	20.10	404.0	26.17
5	181.9	33,100	0.3195	25	17.90	320.4	33.00
6	162.0	26,250	0.4028	26	15.94	254.1	41.62
7	144.3	20,820	0.5080	27	14.20	201.5	52.48
8	128.5	16,510	0.6405	28	12.64	159.8	66.17
9	114.4	13,090	0.8077	29	11.26	126.7	83.44
10	101.9	10,380	1.018	30	10.03	100.5	105.2
11	90.74	8234	1.284	31	8.928	79.70	132.7
12	80.81	6530	1.619	32	7.950	63.21	167.3
13	71.96	5178	2.042	33	7.080	50.13	211.0
14	64.08	4107	2.575	34	6.305	39.75	266.0
15	57.07	3257	3.247	35	5.615	31.52	335.0
16	50.82	2583	4.094	36	5.000	25.00	423.0
17	45.26	2048	5.163	37	4.453	19.83	533.4
18	40.30	1624	6.510	38	3.965	15.72	672.6
19	35.89	1288	8.210	39	3.531	12.47	848.1
20	31.96	1022	10.35	40	3.145	9.88	1069

*20 to 25°C or 68 to 77°F is considered average room temperature.

wiring are set by local electrical codes, which are usually guided by the National Electrical Code published by the National Fire Protection Association. A gauge for measuring wire size is shown in Fig. 11-2.

Circular Mils The cross-sectional area of round wire is measured in circular mils, abbreviated cmil. A mil is one-thousandth of an inch, or 0.001 in. One circular mil is the cross-sectional area of a wire with a diameter of 1 mil. The number of circular mils in any circular area is equal to the square of the diameter in mils.

Example 1 What is the area in circular mils of a wire with a diameter of 0.005 in.?

Answer We must convert the diameter to mils. Since 0.005 in. equals 5 mil,

$$\text{Circular mil area} = (5 \text{ mil})^2$$
$$\text{Area} = 25 \text{ cmil}$$

Note that the circular mil is a unit of area, obtained by squaring the diameter, while the mil is a linear unit of length, equal to one-thousandth of an inch. Therefore, the circular-mil area increases as the square of the diameter. As illustrated in Fig. 11-3, doubling the diameter quadruples the area. Circular mils are convenient for round wire because the cross section is specified without using the formula πr^2 or $\pi d^2/4$ for the area of a circle.

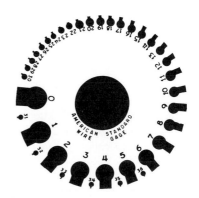

Fig. 11-2 Wire gage.

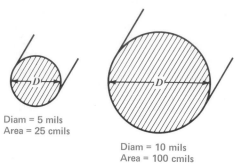

Fig. 11-3 Cross-sectional area for round wire. Double the diameter equals four times the circular area.

Diam = 5 mils
Area = 25 cmils

Diam = 10 mils
Area = 100 cmils

Practice Problems 11-2
Answers at End of Chapter

a. How much is R for 1 ft of No. 22 wire?
b. What is the cross-sectional area in circular mils for wire with a diameter of 0.025 in.?

11-3   Types of Wire Conductors

Most wire conductors are copper, although aluminum and silver are also used. Generally the copper is tinned with a thin coating of solder,[1] which gives it a silvery appearance. The wire can be solid or stranded, as shown in Fig. 11-4a and b.

Stranded wire is flexible and less likely to break open. Sizes for stranded wire are equivalent to the sum of the areas for the individual strands. For instance, two strands of No. 30 wire are equivalent to solid No. 27 wire.

Very thin wire, such as No. 30, often has an insulating coating of enamel or shellac. It may look like copper, but the coating must be scraped off at the ends to make a good connection to the wire.

Heavier wires generally are in an insulating sleeve, which may be rubber, cotton, or one of many plastics. General-purpose wire for connecting electronic components is generally plastic-coated hookup wire of No. 20 gage. Hookup wire that is bare should be enclosed in a hollow insulating sleeve called *spaghetti*.

---

[1]See App. H for more information about solder.

Wire Cable   Two or more conductors in a common covering form a cable. Each wire is insulated from the others. The two-conductor line in Fig. 11-4d is called *coaxial cable*. Its metallic braid is one conductor, which is connected to ground to shield the inner conductor against interference.

Constant spacing between two conductors provides a *transmission line*. Examples are the coaxial cable in Fig. 11-4d and the twin lead in Fig. 11-4e. Coaxial cable with an outside diameter of ¼ in. is generally used for cable television. Twin lead with a width of ⅝ in. is commonly used in television for connecting the antenna to the receiver. Wire of No. 20 gage is typical for the conductors.

Connectors   Refer to Fig. 11-5. The spade lug in Fig. 11-5a is often used for screw-type terminals. In Fig. 11-5b, the alligator clip is convenient for temporary connections. The banana pins in Fig. 11-5c have spring-type sides that make a tight connection. In Fig. 11-5d, the terminal strip provides a block for multiple soldered connections. The plugs in Fig. 11-5e, f, and g are convenient for cable connections. The F connector in Fig. 11-5g is popular because the center conductor of the cable serves as the center pin of the plug, without the need for any soldering. These are *male*-type plugs with pins that connect to a *female* type of socket.

The multiple connector in Fig. 11-5h is shown with a ribbon cable. It has eight wires in a flat, plastic ribbon to save space. This plug is a female type that mates with a male socket.

In Fig. 11-5i, the grabber has a small metal hook that is very convenient in crowded circuits. The hook is

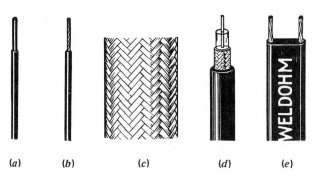

(a)   (b)   (c)   (d)   (e)

Fig. 11-4 Types of wire conductors (a) Solid wire. (b) Stranded wire. (c) Braided conductor for very low resistance. (d) Coaxial cable. (e) Twin-lead cable.

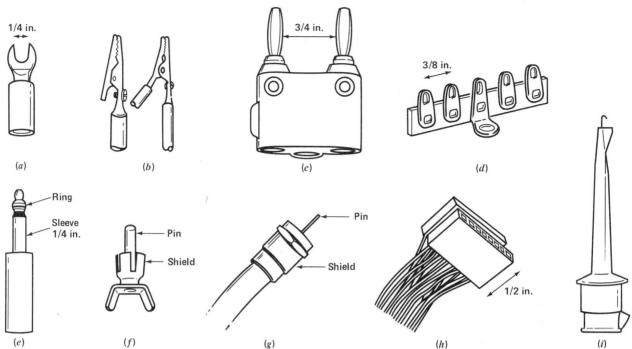

1/4 in.

3/4 in.

3/8 in.

(a)   (b)   (c)   (d)

Ring

Sleeve
1/4 in.

Pin

Shield

Pin

Shield

1/2 in.

(e)   (f)   (g)   (h)   (i)

Fig. 11-5 Common types of connectors for wire conductors. (a) Spade lug. (b) Alligator clips. (c) Double banana-pin plug. (d) Terminal strip. (e) Phone plug. (f) RCA plug for audio cables. (g) F-type plug for cable TV. (h) Multiple-pin connectors. (i) Spring-loaded metal hook as grabber for temporary connection.

spring-loaded to make a tight connection. Different sizes are available from about 1 to 3 in.

Practice Problems 11-3
Answers at End of Chapter

Answer True or False.

a. The plastic coating on wire conductors has very high resistance.

b. Coaxial cable is a shielded transmission line.

c. The F-type connector is used for coaxial cable.

11-4   Printed Wiring

Most electronic circuits are mounted on a plastic insulating board with printed wiring, as shown in Fig. 11-6. This is a printed-circuit (PC) or printed-wiring (PW) board. One side has the components, such as resistors, capacitors, coils, tubes, transistors, diodes, and inte-

grated-circuit (IC) units. The other side has the conducting paths printed with silver or copper on the board, instead of using wires. Sockets, small metal eyelets, or just holes in the board are used to connect the components to the wiring. With a bright light on one side, you can see through to the opposite side to trace the connections. However, the circuit is usually drawn on the PC board.

It is important not to use too much heat in soldering[1] or desoldering. Otherwise the printed wiring can be lifted off the board. Use a small iron of about 25 to 35 W rating. When soldering semiconductor diodes and transistors, hold the lead with pliers or connect an alligator clip as a heat sink to conduct heat away from the semiconductor junction.

In some cases, defective *R, L,* and *C* components can be replaced without disturbing the printed wiring. Just break the old component in the middle with diago-

---

[1]More details of solder, soldering, and desoldering are described in App. H.

Resistors

Transistors

(a)

Edge connectors

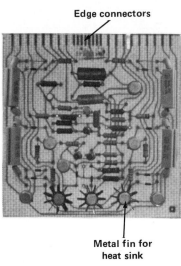

Metal fin for
heat sink

(b)

Fig. 11-6 Printed-wiring board. Size is 4 in. square. (a) Component side. (b) Wiring side.

nal cutting pliers. Then solder the new component to the old leads. However, the best way is to desolder the leads and solder the new component into the printed wiring.

For desoldering, use a solder-sucker tool, with a soldering iron, to clean each terminal. Another method is to use wire braid. Put the braid on the joint and heat it until the solder runs up into the braid. The terminal must be clean enough to lift out the component easily without damaging the PC board.

A small crack in the printed wiring can be repaired by soldering a short length of bare wire over the open circuit. If a larger section of printed wiring is open, or if the board is cracked, you can bridge the open circuit with a length of hookup wire soldered at two convenient end terminals of the printed wiring. Pins at end terminals, usually with multiple connections, are called *stakes*.

Practice Problems 11-4
Answers at End of Chapter

a. Which is the best size iron to use on a PC board, 25, 100, or 150 W?

b. How much is the resistance of a printed-wire conductor with a break in the middle?

11-5 Switches

As shown in Fig. 11-7, switches are commonly used to open or close a circuit. Closed is the ON, or *make,* position; open is the OFF, or *break,* position.

The switch is in series with the voltage source and its load. In the ON position, the closed switch has very little resistance. Then maximum current can flow in the load, with practically zero voltage drop across the switch. Open, the switch has infinite resistance, and no current flows in the circuit.

Note that the switch is in just one side of the line, but the entire series circuit is open when the switch is turned off. In the open position, the applied voltage is across the switch contacts. Therefore, the insulation must be good enough to withstand this amount of voltage without arcing.

The switch in Fig. 11-7 is a single-pole single-throw (SPST) switch. It provides an ON or OFF position for one circuit. Two connections are necessary.

Figure 11-8 shows double-throw switches for two circuits. Switch S_1 in Fig. 11-8a is single-pole double-throw (SPDT) to switch one side of the circuit. This switching can be done because R_1 and R_2 both have a common line. Three connections are necessary, one for the common line and one for each of the circuits to be switched.

In Fig. 11-8b, S_2 is double-pole double-throw (DPDT) to switch both sides of two circuits. This

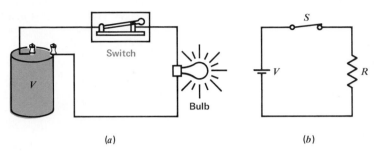

(a) (b)

Fig. 11-7 Single-pole, single-throw (SPST) switch to open or close one circuit. (a) Wiring diagram with knife switch. (b) Schematic diagram with general symbol for switch S.

switching is done because there is no common return line for the two separate antennas. Six connections are necessary here, two for each of the circuits to be switched and two for the center contacts.

Switch Types   Figure 11-9 illustrates a toggle switch, and Fig. 11-10 shows the construction of a rotary switch. Additional types include the knife switch, push-button switch, and rocker switch. The DIP switch in Fig. 11-11 has eight miniature rocker switches. Each switch can be set separately. The term DIP indicates a double inline package for the pin connections that fit an IC socket.

A spring switch that is normally closed is indicated as *NC*. When you activate the switch, its contacts open.

A normally open switch is indicated as *NO*. When you activate the switch, its contacts close.

Relay Switches   A relay is an automatic switch with contacts that can be closed or opened by current in

the relay coil. See Fig. 11-12. In some cases, the coil is also called a *solenoid*. However, a solenoid is usually for one contact, while relays can activate multiple contacts.

In operation, the relay is activated by current in the coil. Its magnetic field attracts the iron armature, which has the movable switch contacts. The construction can allow either dc or ac operation. Common coil voltages are 12 V, 24 V, and 120 V. The dc resistance of the coil may be 2000 Ω for a 120-V relay with a 2-W rating.

Practice Problems 11-5
Answers at End of Chapter

a. How much is the *IR* voltage drop across a closed switch?

b. How many connections are needed for a SPDT switch?

c. Are contacts that close when a relay switch is activated indicated as NO or NC?

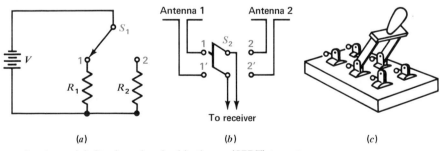

(a) (b) (c)

Fig. 11-8 Switch applications. (a) Single-pole, double-throw (SPDT) type to switch one connection for either of two circuits with R_1 and R_2. (b) Double-pole, double-throw (DPDT) switch to make two connections for two circuits. (c) Construction of DPDT knife switch.

Fig. 11-9 DPDT toggle switch. Length is 1 in. Note six soldering lugs.

Fig. 11-11 DIP switch with eight SPST rocker switches. (*EECO Inc.*)

11-6 Fuses

Many circuits have a fuse in series as a protection against an overload resulting from a short circuit. Excessive current melts the fuse element, blowing the fuse and opening the series circuit. The purpose is to let the fuse blow before the components are damaged. The blown fuse can easily be replaced by a new one, after the overload has been eliminated. A glass-cartridge fuse with holders is shown in Fig. 11-13. This is a type 3AG fuse, with a diameter of ¼ in. and length of 1¼ in. (Table 11-2). AG is an abbreviation of "automobile glass," since that was one of the first applications of fuses in a glass holder to make the wire link visible.

The metal fuse element may be made of aluminum, tin-coated copper, or nickel. Fuses are available with current ratings from ¹⁄₅₀₀ A to hundreds of amperes. The thinner the wire element in the fuse, the smaller its current rating. For example, a 2-in. length of No. 28 wire can serve as a 2-A fuse. As typical applications, the rating for plug fuses in each branch of house wiring is often 15 A; the high-voltage circuit in a television receiver is usually protected by a glass-cartridge ¼-A fuse. For automobile fuses, the ratings are generally 10 to 30 A because of the higher currents needed with a 12-V source for a given amount of power.

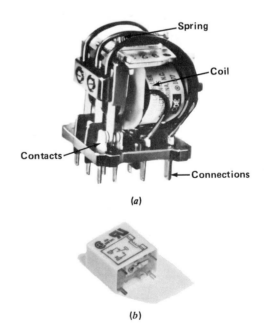

(a)

(b)

Fig. 11-12 (*a*) Construction of open type of relay. (*b*) Cover with schematic of switch contacts. (*Cornell-Dubilier*).

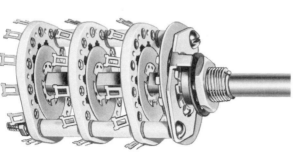

Fig. 11-10 Rotary switch with three wafers or decks on a common shaft. Diameter is 1½ in.

(a)   (b)   (c)

Fig. 11-13 (a) Glass cartridge fuse. Length of 3AG type is 1¼ in. (b) Fuse. (c) Panel-mounted fuse holder.

Slow-blow Fuses These have a coiled construction. They are designed to open only on a continued overload, such as a short circuit. The purpose of coiled construction is to prevent the fuse from blowing on just a temporary current surge. As an example, a slow-blow fuse will hold a 400 percent overload in current for up to 2 s. Typical ratings are shown by the curves in Fig. 11-14. Circuits with an electric motor use slow-blow fuses because the starting current is much more than the running current.

Circuit Breakers These have a thermal element in the form of a spring. The spring expands with heat and trips open the circuit. The circuit breaker can be reset for normal operation, however, after the short has been eliminated and the thermal element cools down.

Wire Links A short length of bare wire is often used as a fuse in television receivers. For instance, a 2-in. length of No. 24 gage wire can hold a 500-mA current but burn open with an overload. The wire link can be mounted between two terminal strips on the chassis. Or, the wire link may be wrapped over a small insulator to make a separate component.

Testing Fuses With glass fuses, you can usually see if the wire element inside is burned open. When measured with an ohmmeter, a good fuse has practically zero resistance. An open fuse reads infinite ohms. Power must be off or the fuse must be out of the circuit to test a fuse with an ohmmeter.

When you test with a voltmeter, a good fuse has zero volts across its two terminals (Fig. 11-15a). If you read appreciable voltage across the fuse, this means it is open. In fact, the full applied voltage is across the open fuse in a series circuit, as shown in Fig. 11-15b. This is why fuses also have a voltage rating, which gives the maximum voltage without arcing in the open fuse.

Referring to Fig. 11-15, notice the results when measuring the voltages to ground at the two fuse terminals. In Fig. 11-15a, the voltage is the same 120 V at both ends because there is no voltage drop across the good fuse. In Fig. 11-15b, however, terminal B reads 0 V, as this end is disconnected from V_T because of the open fuse. These tests apply to either dc or ac voltages.

Practice Problems 11-6
Answers at End of Chapter

a. How much is the resistance of a good fuse?
b. How much is the *IR* voltage drop across a good fuse?

11-7   Wire Resistance

The longer a wire, the higher its resistance. More work must be done to make electrons drift from one end to the other. However, the thicker the wire, the less the resistance, since there are more free electrons in the cross-sectional area. As a formula,

Table 11-2. Sizes for Type AG Fuses

Size	Diameter, in.	Length, in.
1AG	¼	⅝
3AG	¼	1¼
4AG	9/32	1¼
5AG	13/32	1½
7AG	¼	⅞
8AG	¼	1
9AG	¼	1 7/16

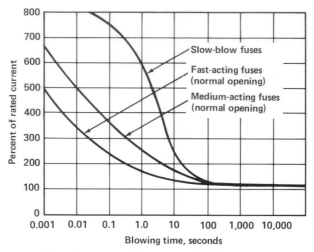

Fig. 11-14 Chart showing percentage of rated current to blowing time for fuses. (*Littelfuse Inc.*)

$$R = \rho \frac{l}{A} \qquad\qquad (11\text{-}1)$$

where R is the total resistance, l the length, A the cross-sectional area, and ρ the specific resistance or *resistivity*. The factor ρ then enables the resistance of different materials to be compared according to their nature without regard to different lengths or areas. Higher values of ρ mean more resistance. Note that ρ is the Greek letter *rho,* corresponding to *r*.

Specific Resistance Table 11-3 lists resistance values for different metals having the standard wire size of a 1-ft length with a cross-sectional area of 1 cmil. This rating is the *specific resistance* of the metal, in circular-mil ohms per foot. Since silver, copper, gold, and aluminum are the best conductors, they have the lowest values of specific resistance. Tungsten and iron have a much higher resistance.

Example 2 How much is the resistance of 100 ft of No. 20 copper wire?

Answer Note that from Table 11-1, the cross-sectional area for No. 20 wire is 1022 cmil; from Table 11-3, the ρ for copper is 10.4. Using Formula (11-1) gives

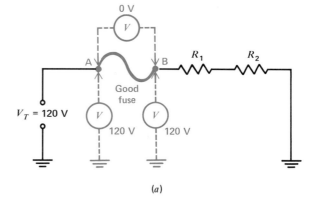

(a)

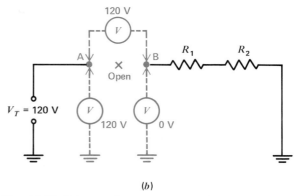

(b)

Fig. 11-15 When a fuse opens, the applied voltage is across the open fuse terminals. (*a*) Circuit closed with good fuse. Note schematic symbol. (*b*) Fuse open. Voltage readings explained in text.

Table 11-3. Properties of Conducting Materials*

Material	Description and symbol	ρ = specific resistance at 20°C, cmil · Ω/ft	Temperature coefficient per °C, α	Melting point, °C
Aluminum	Element (Al)	17	0.004	660
Carbon	Element (C)	†	−0.0003	3000
Constantan	Alloy, 55% Cu, 45% Ni	295	0 (average)	1210
Copper	Element (Cu)	10.4	0.004	1083
Gold	Element (Au)	14	0.004	1063
Iron	Element (Fe)	58	0.006	1535
Manganin	Alloy, 84% Cu, 12% Mn, 4% Ni	270	0 (average)	910
Nichrome	Alloy 65% Ni, 23% Fe, 12% Cr	676	0.0002	1350
Nickel	Element (Ni)	52	0.005	1452
Silver	Element (Ag)	9.8	0.004	961
Steel	Alloy, 99.5% Fe, 0.5% C	100	0.003	1480
Tungsten	Element (W)	33.8	0.005	3370

*Listings approximate only, since precise values depend on exact composition of material.
†Carbon has about 2500 to 7500 times the resistance of copper. Graphite is a form of carbon.

$$R = \rho \frac{l}{A}$$

$$= 10.4 \frac{\text{cmil} \cdot \Omega}{\text{ft}} \times \frac{100 \text{ ft}}{1022 \text{ cmil}}$$

$$R = 1.02 \ \Omega$$

All the units cancel except the ohms for R. Note that 1.02 Ω for 100 ft is approximately $\frac{1}{10}$ the resistance of 10.35 Ω for 1000 ft of No. 20 copper wire listed in Table 11-1, showing that the resistance is proportional to length.

Example 3   How much is the resistance of a 100-ft length of No. 23 copper wire?

Answer $R = \rho \dfrac{l}{A}$

$$= 10.4 \frac{\text{cmil} \cdot \Omega}{\text{ft}} \times \frac{100 \text{ ft}}{509.5 \text{ cmil}}$$

$$R = 2.04 \ \Omega$$

Note that a wire three gage sizes higher has half the circular area and double the resistance for the same wire length.

Units of Ohm-Centimeters for ρ

Except for wire conductors, specific resistances are usually compared for the standard size of a 1-cm cube. Then ρ is specified in Ω·cm for the unit cross-sectional area of 1 cm².

As an example, pure germanium has $\rho = 55$ Ω·cm, as listed in Table 11-4. This value means that R is 55 Ω for a cube with a cross-sectional area of 1 cm² and length of 1 cm.

For other sizes, use Formula (11-1) with l in cm and A in cm². Then all the units of size cancel to give R in ohms.

Example 4   How much is the resistance for a slab of germanium 0.2 cm long with a cross-sectional area of 1 cm²?

Answer $R = \rho \dfrac{l}{A}$

$$= 55 \ \Omega\cdot\text{cm} \times \dfrac{0.2 \ \text{cm}}{1 \ \text{cm}^2}$$

$$R = 11 \ \Omega$$

The same size slab of silicon would have R of 11,000 Ω. Note from Table 11-4 that ρ for silicon is 1000 times more for silicon than for germanium.

Types of Resistance Wire For applications in heating elements, such as a toaster, an incandescent light bulb, or a heater, it is necessary to use wire that has more resistance than good conductors like silver, copper, or aluminum. Higher resistance is preferable so that the required amount of I^2R power dissipated as heat in the wire can be obtained without excessive current. Typical materials for resistance wire are the elements tungsten, nickel, or iron and alloys[1] such as manganin, Nichrome, and constantan. These types are generally called resistance wire because R is greater than for copper wire, for the same length.

Practice Problems 11-7
Answers at End of Chapter

a. Does Nichrome wire have less or more resistance than copper wire?
b. For 100 ft of No. 14 copper wire, R is 0.26 Ω. How much is R for 1000 ft?

11-8 Temperature Coefficient of Resistance

This factor with the symbol alpha (α) states how much the resistance changes for a change in temperature. A positive value for α means R increases with temperature; with a negative α, R decreases; zero for α means

---

[1]An *alloy* is a fusion of elements, without chemical action between them. Metals are commonly alloyed to alter their physical characteristics.

Table 11-4. Comparison of Specific Resistances

Material	ρ, $\Omega\cdot$cm, at 25°C	Description
Silver	1.6×10^{-6}	Conductor
Germanium	55	Semiconductor
Silicon	55,000	Semiconductor
Mica	2×10^{12}	Insulator

R is constant. Some typical values of α, for metals and for carbon, are listed in Table 11-3 in the fourth column.

Positive α All metals in their pure form, such as copper and tungsten, have a positive temperature coefficient. The α for tungsten, for example, is 0.005. Although α is not exactly constant, an increase in wire resistance caused by a rise in temperature can be calculated approximately from the formula

$$R_t = R_0 + R_0(\alpha\Delta t) \tag{11-2}$$

where R_0 is the resistance at 20°C, R_t is the higher resistance at the higher temperature, and Δt is the temperature rise above 20°C.

Example 5 A tungsten wire has a 14-Ω R at 20°C. Calculate its resistance at 120°C.

Answer The temperature rise Δt here is 100°C; α is 0.005. Substituting in Formula (11-2),

$$R_t = 14 + 14(0.005 \times 100)$$
$$= 14 + 7$$
$$R_t = 21 \ \Omega$$

The added resistance of 7 Ω increases the wire resistance by 50 percent because of the 100°C rise in temperature.

In practical terms, a positive α means that heat increases R in wire conductors. Then I is reduced, with a specified applied voltage.

Negative α Note that carbon has a negative temperature coefficient. In general, α is negative for all semiconductors, including germanium and silicon. Also, all electrolyte solutions, such as sulfuric acid and water, have a negative α.

A negative value of α means less resistance at higher temperatures. The resistance of semiconductor diodes and transistors, therefore, can be reduced appreciably when they become hot with normal load current.

The negative α has a practical application in the use of carbon *thermistors*. A thermistor can be connected as a series component to decrease its resistance to compensate for the increased hot resistance of wire conductors.

Zero α This means R is constant with changes in temperature. The metal alloys constantan and manganin, for example, have the value of zero for α. They can be used for precision wire-wound resistors, which do not change resistance when the temperature increases.

Hot Resistance With resistance wire made of tungsten, Nichrome, iron, or nickel, there is usually a big difference in the amount of resistance the wire has when hot in normal operation and when cold without its normal load current. The reason is that the resistance increases with higher temperatures, since these materials have a positive temperature coefficient, as shown in Table 11-3.

As an example, the tungsten filament of a 100-W 120-V incandescent bulb has a current of 0.833 A when the bulb lights with normal brilliance at its rated power, since $I = P/V$. By Ohm's law, the hot resistance is V/I, or 120 V/0.833 A, which equals 144 Ω. If, however, the filament resistance is measured with an ohmmeter when the bulb is not lit, the cold resistance is only about 10 Ω.

The Nichrome heater elements in appliances and the tungsten heaters in vacuum tubes also become several hundred degrees hotter in normal operation. In these cases, only the cold resistance can be measured with an ohmmeter. The hot resistance must be calculated from voltage and current measurements with the normal value of load current. As a practical rule, the cold resistance is generally about one-tenth the hot resistance. In troubleshooting, however, the problem is usually just to check if the heater element is open. Then it reads infinite ohms on the ohmmeter.

Superconductivity The effect opposite to hot resistance is to cool a metal down to very low temperatures to reduce its resistance. Near absolute zero, 0 K or $-273°C$, some metals abruptly lose practically all their resistance. As an example, when cooled by liquid helium, the metal tin becomes superconductive at 3.7 K. Tremendous currents can be produced, resulting in very strong electromagnetic fields. Such work at very low temperatures,[1] near absolute zero, is called *cryogenics*.

Practice Problems 11-8
Answers at End of Chapter

Answer True or False.
a. Metal conductors have more R at higher temperatures.
b. A thermistor has a negative temperature coefficient.

11-9 Ion Current in Liquids and Gases

We usually think of metal wire for a conductor, but there are other possibilities. Liquids such as salt water or dilute sulfuric acid can also allow the movement of electric charges. For gases, consider the neon glow lamp, where neon serves as a conductor.

The mechanism may be different for conduction in metal wire, liquids, or gases, but in any case the current is a motion of charges. Furthermore, either positive or negative charges can be the carriers that provide electric current. The amount of current is Q/T. For one coulomb of charge per second, the current is one ampere.

In solid materials like the metals, the atoms are not free to move among each other. Therefore, conduction of electricity must take place by the drift of free electrons. Each atom remains neutral, neither gaining nor

---

[1]See App. D, "Physics Units," for a description of different temperature scales.

⊣|⊢

losing charge, but the metals are good conductors because they have plenty of free electrons that can be forced to drift through the solid substance.

In liquids and gases, however, each atom is able to move freely among all the other atoms because the substance is not solid. As a result, the atoms can easily take on electrons or lose electrons, particularly the valence electrons in the outside shell. The result is an atom that is no longer electrically neutral. Adding one or more electrons produces a negative charge; the loss of one or more electrons results in a positive charge. The charged atoms are called *ions*. Such charged particles are commonly formed in liquids and gases.

The Ion An ion is an atom that has a net electric charge, either positive or negative, resulting from a loss or gain of electrons. In Fig. 11-16a, the sodium atom is neutral, with 11 positive charges in the nucleus balanced by 11 electrons in the outside shells. This atom has only 1 electron in the shell farthest from the nucleus. When the sodium is in a liquid solution, this 1 electron can easily leave the atom. The reason may be another atom close by that needs 1 electron in order to have a stable ring of 8 electrons in its outside shell. Notice that if the sodium atom loses 1 valence electron, the atom will still have an outside ring of 8 electrons, as shown in Fig. 11-16b. This sodium atom now is a positive ion, with a charge equal to 1 proton. An ion still has the characteristics of the element because the nucleus is not changed.

Current of Ions Just as in electron flow, opposite ion charges are attracted to each other, while like

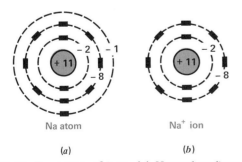

Na atom   Na⁺ ion

(a)   (b)

Fig. 11-16 Formation of ions. (a) Normal sodium (Na) atoms. (b) Positively charged ion indicated as Na⁺, missing one valence electron.

charges repel. The resultant motion of ions provides electric current. In liquids and gases, therefore, conduction of electricity results mainly from the movement of ions. This motion of ion charges is called *ionization current*. Since an ion includes the nucleus of the atom, the ion charge is much heavier than an electron charge and moves with less velocity. We can say that ion charges are less mobile than electron charges.

The direction of ionization current can be the same as electron flow or the opposite. When negative ions move, they are attracted to the positive terminal of an applied voltage, in the same direction as electron flow. However, when positive ions move, this ionization current is in the opposite direction, toward the negative terminal of an applied voltage.

For either direction, though, the amount of ionization current is determined by the rate at which the charge moves. If 3 C of positive ion charges move past a given point per second, the current is 3 A, the same as 3 C of negative ions or 3 C of electron charges.

Ionization in Liquids Ions are usually formed in liquids when salts or acids are dissolved in water. Salt water is a good conductor because of ionization, but pure distilled water is an insulator. In addition, metals immersed in acids or alkaline solutions produce ionization. Liquids that are good conductors because of ionization are called *electrolytes*. In general, electrolytes have a negative value of α, as more ionization at higher temperatures lowers the resistance.

Ionization in Gases Gases have a minimum striking or ionization potential, which is the lowest applied voltage that will ionize the gas. Before ionization the gas is an insulator, but the ionization current makes the ionized gas a low resistance. An ionized gas usually glows. Argon, for instance, emits blue light when the gas is ionized. Ionized neon gas glows red. The amount of voltage needed to reach the striking potential varies with different gases and depends on the gas pressure. For example, a neon glow lamp for use as a night light ionizes at approximately 70 V.

Ionic Bonds The sodium ion in Fig. 11-17 has a charge of +1 because it is missing 1 electron. If such positive ions are placed near negative ions with a

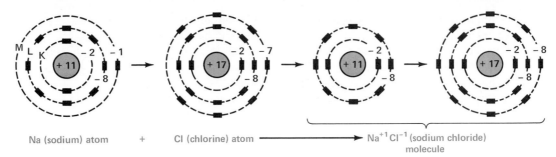

Fig. 11-17 Ionic bond between atoms of sodium (Na) and chlorine (Cl) to form a molecule of sodium chloride (NaCl).

charge of -1, there will be an electrical attraction to form an ionic bond.

A common example is the combination of sodium (Na) ions and chlorine (Cl) ions to form table salt (NaCl), as shown in Fig. 11-17. Notice that the 1 outer electron of the Na atom can fit into the 7-electron shell of the Cl atom. When these two elements are combined, the Na atom gives up 1 electron to form a positive ion, with a stable L shell having 8 electrons; also, the Cl atom adds this 1 electron to form a negative ion, with a stable M shell having 8 electrons. The two opposite types of ions are bound in NaCl because of the strong attractive force between opposite charges close together.

The ions in NaCl can separate in water to make salt water a conductor of electricity, while pure water is not. When current flows in salt water, then, the moving charges must be ions, as another example of ionization current.

Practice Problems 11-9
Answers at End of Chapter

a. How much is I for 2 C/s of positive ion charges?
b. Which have the greatest mobility, positive ions, negative ions, or electrons?
c. A dielectric material is a good conductor of electricity. True or False?

11-10 Electrons and Hole Charges in Semiconductors

The semiconductor materials like germanium and silicon are in a class by themselves as conductors, because the charge carriers for current flow are neither ions nor free valence electrons. With a valence of ± 4 for these elements, the tendency to gain or lose electrons to form a stable 8 shell is the same either way. As a result, these elements tend to share their outer electrons in pairs of atoms.

An example is illustrated in Fig. 11-18, for two silicon (Si) atoms, each sharing its 4 valence electrons with the other atom to form one Si_2 molecule. This type of combination of atoms sharing their outer electrons to form a stable molecule is called a *covalent bond*.

The covalent-bond structure in germanium and silicon is the basis for their use in transistors. This is because, although the covalent-bond structure is electrically neutral, it permits charges to be added by *doping* the semiconductor with a small amount of impurity atoms.

As a specific example, silicon, with a valence of 4, is combined with phosphorus, with a valence of 5. Then the doped germanium has covalent bonds with an excess of 1 electron for each impurity atom of phosphorus. The result is a negative, or N-type, semiconductor.

For the opposite case, silicon can be doped with aluminum, which has a valence of 3. Then covalent bonds

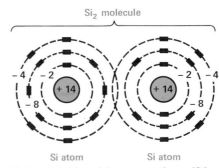

Fig. 11-18 Covalent bond between silicon (Si) atoms.

formed with the impurity atoms have 7 outer electrons, instead of the 8 with a pair of silicon atoms.

The 1 missing electron for each covalent bond with an impurity atom corresponds to a positive charge called a *hole*. The amount of charge for each hole is 0.16×10^{-18} C, the same as for an electron, but of opposite polarity. This type of doping results in a P-type semiconductor with positive hole charges.

For either N- or P-type semiconductors the charges can be made to move by an applied voltage that produces current. When electrons move, the current direction is the same as for electron flow. When the positive hole charges move, the direction is opposite from electron current. For either electrons or hole charges, when 1 C moves past a given point in 1 s, the amount of current is 1 A. However, electrons have greater mobility than hole charges.

For semiconductor diodes, the P and N types are combined in a PN junction. It should be noted that this junction is just an electrical boundary between opposite types, but without any physical separation. In a PNP transistor, the N type is between two P types. The opposite case, a P type between two N types, results in the NPN transistor. The NPN and PNP types are junction transistors, with two junctions. More details of semiconductors in general, as important solid-state components, and transistors in particular, as commonly used for amplifier circuits, are given in Chaps. 29 and 30, which explain the operation of semiconductor diodes and transistors. Solid-state integrated circuits are discussed in Chap. 32.

Practice Problems 11-10
Answers at End of Chapter

a. What is the polarity of the hole charges in P-type doped semiconductors?

b. What is the electron valence of silicon and germanium?

c. What are the charge carriers in N-type semiconductors?

11-11 Insulators

Substances that have very high resistance, of the order of many megohms, are classed as insulators. With such high resistance, an insulator cannot conduct apprecia-

ble current when voltage is applied. As a result, insulators can have either of two functions. One is to isolate conductors to eliminate conduction between them. The other is to store an electric charge when voltage is applied.

An insulator maintains its charge because electrons cannot flow to neutralize the charge. The insulators are commonly called *dielectric materials,* therefore, meaning that they can store a charge.

Among the best insulators, or dielectrics, are air, vacuum, rubber, wax, shellac, glass, mica, porcelain, oil, dry paper, textile fibers, and plastics such as Bakelite, formica, and polystyrene. Pure water is a good insulator, but salt water is not. Moist earth is a fairly good conductor, while dry, sandy earth is an insulator.

For any insulator, a high enough voltage can be applied to break down the internal structure of the material, forcing the dielectric to conduct. This dielectric breakdown is usually the result of an arc, which ruptures the physical structure of the material, making it useless as an insulator. Table 11-5 compares several insulators in terms of dielectric strength, which is the voltage breakdown rating. The higher the dielectric strength, the better the insulator, since it is less likely to break down with a high value of applied voltage. The breakdown voltages in Table 11-5 are approximate values for the standard thickness of 1 mil, or 0.001 in. More thickness allows a higher breakdown-voltage rating. Note that the value of 20 V/mil for air or vacuum is the same as 20 kV/in.

Insulator Discharge Current An insulator in contact with a voltage source stores charge, producing a potential on the insulator. The charge tends to remain on the insulator, but it can be discharged by one of the following methods:

1. Conduction through a conducting path. For instance, a wire across the charged insulator provides a discharge path. Then the discharged dielectric has no potential.

2. Brush discharge. As an example, high voltage on a sharp pointed wire can discharge through the surrounding atmosphere by ionization of the air molecules. This may be visible in the dark as a bluish or reddish glow, called the *corona effect.*

3. Spark discharge. This is a result of breakdown in the insulator because of a high potential difference

Table 11-5. Voltage Breakdown of Insulators

Material	Dielectric strength, V/mil	Material	Dielectric strength, V/mil
Air or vacuum	20	Paraffin wax	200–300
Bakelite	300–550	Phenol, molded	300–700
Fiber	150–180	Polystyrene	500–760
Glass	335–2000	Porcelain	40–150
Mica	600–1500	Rubber, hard	450
Paper	1250	Shellac	900
Paraffin oil	380		

that ruptures the dielectric. The current that flows across the insulator at the instant of breakdown causes the spark.

Corona is undesirable, as it reduces the potential by brush discharge into the surrounding air. In addition, the corona often indicates the beginning of a spark discharge. A potential of the order of kilovolts is usually necessary for corona, as the breakdown voltage for air is approximately 20 kV/in. To reduce the corona effect, conductors that have high voltage should be smooth, rounded, and thick. This equalizes the potential difference from all points on the conductor to the surrounding air. Any sharp point can have a more intense field, making it more susceptible to corona and eventual spark discharge.

Practice Problems 11-11
Answers at End of Chapter

a. Which has a higher voltage breakdown rating, air or mica?
b. Can 30 kV arc across an air gap of 1 in.?

11-12   Troubleshooting Hints

For all types of electronic equipment, a common problem is an open circuit in the wire conductors, the connectors, and the switch contacts.

For conductors, both wires and printed wiring, you can check the continuity with an ohmmeter. A good conductor reads 0 Ω for continuity. An open reads infinite ohms.

Connectors can also be checked for continuity between the wire and the connector itself. Also, the connector may be tarnished or rusted. Then it must be cleaned with either fine sandpaper or emery cloth.

With a plug connector for cable, make sure the wires have continuity to the plug. Except for the F-type connector, most plugs require careful soldering to the center pin.

For relays, the coil can be open. Then the armature cannot be activated. Although the operation of a relay is electromagnetic, the wire in the coil can be checked for continuity with an ohmmeter.

A common problem with relays is dirty contacts. Arcing across the switch contacts forms a carbon coating with constant use. Then they require cleaning, or else the relay must be replaced. A common indication of dirty contacts is *relay chatter* as the contacts vibrate between the open and closed positions.

Practice Problems 11-12
Answers at End of Chapter

Answer True or False.
a. Printed wiring cannot be checked for continuity with an ohmmeter.
b. An open relay coil will read infinite ohms when checked with an ohmmeter.

Summary

1. A conductor has very low resistance. All the metals are good conductors, the best being silver, copper, and aluminum. Copper is generally used for wire conductors.
2. The sizes for copper wire are specified by the American Wire Gage. Higher gage numbers mean thinner wire. Typical sizes are No. 22 gage hookup wire for electronic circuits and No. 12 and No. 14 for house wiring.
3. The cross-sectional area of round wire is measured in circular mils. One mil is 0.001 in. The area in circular mils equals the diameter in mils squared.
4. $R = \rho(l/A)$. The factor ρ is specific resistance. Wire resistance increases directly with length l, but decreases inversely with the cross-sectional area A, or the square of the diameter.
5. A switch inserted in one side of a circuit opens the entire series circuit. When open, the switch has the applied voltage across it.
6. A fuse protects the circuit components against overload, as excessive current melts the fuse element to open the entire series circuit. A good fuse has very low resistance and practically zero voltage across it.
7. Ionization in liquids and gases produces atoms that are not electrically neutral. These are ions. Negative ions have an excess of electrons; positive ions have a deficiency of electrons. In liquids and gases, electric current is a result of movement of the ions.
8. In the semiconductors, such as germanium and silicon, the charge carriers are electrons in N type and positive hole charges in P type. One hole charge is 0.16×10^{-18} C, the same as one electron, but with positive polarity.
9. The resistance of pure metals increases with temperature. For semiconductors and liquid electrolytes, the resistance decreases at higher temperatures.
10. An insulator has very high resistance. Common insulating materials are air, vacuum, rubber, paper, glass, porcelain, shellac, and plastics. Insulators are also called dielectrics.

Self-Examination
Answers at Back of Book

Choose (a), (b), (c), or (d).

1. A 10-ft length of copper-wire conductor of No. 20 gage has a total resistance of (a) less than 1 Ω; (b) 5 Ω; (c) 10.4 Ω; (d) approximately 1 MΩ.
2. A copper-wire conductor with 0.2 in. diameter has an area of (a) 200 cmil; (b) 400 cmil; (c) 20,000 cmil; (d) 40,000 cmil.
3. If a wire conductor of 0.1 Ω resistance is doubled in length, its resistance becomes (a) 0.01 Ω; (b) 0.02 Ω; (c) 0.05 Ω; (d) 0.2 Ω.
4. If two wire conductors are tied in parallel, their total resistance is (a) double the resistance of one wire; (b) one-half the resistance of one wire; (c) the same as the resistance of one wire; (d) two-thirds the resistance of one wire.
5. The hot resistance of the tungsten filament in a bulb is higher than its cold resistance because the filament's temperature coefficient is (a) negative; (b) positive; (c) zero; (d) about 10 Ω per degree.

6. A closed switch has a resistance of (a) zero; (b) infinity; (c) about 100 Ω at room temperature; (d) at least 1000 Ω.

7. An open fuse has a resistance of (a) zero; (b) infinity; (c) about 100 Ω at room temperature; (d) at least 1000 Ω.

8. Insulating materials have the function of (a) conducting very large currents; (b) preventing an open circuit between the voltage source and the load; (c) preventing a short circuit between conducting wires; (d) storing very high currents.

9. An ion is (a) a free electron; (b) a proton; (c) an atom with unbalanced charges; (d) a nucleus without protons.

10. Ionization current in liquids and gases results from a flow of (a) free electrons; (b) protons; (c) positive or negative ions; (d) ions that are lighter in weight than electrons.

Essay Questions

1. Name three good metal conductors in order of resistance. Give one application.
2. Name four insulators. Give one application.
3. Name two semiconductors. Give one application.
4. Name two types of resistance wire. Give one application.
5. What is meant by the dielectric strength of an insulator?
6. Why does ionization occur more readily in liquids and gases, compared with the solid metals? Give an example of ionization current.
7. Define the following: ion, ionic bond, covalent bond, molecule.
8. Draw a circuit with two bulbs, a battery, and an SPDT switch that determines which bulb lights.
9. Why is it not possible to measure the hot resistance of a filament with an ohmmeter?
10. Give one way in which negative ion charges are similar to electron charges and one way in which they are different.
11. Define the following abbreviations for switches: SPST, SPDT, DPST, DPDT, NO, and NC.
12. Give two common troubles with conductors and connector plugs.

Problems
Answers to Odd-Numbered Problems at Back of Book

1. A copper wire has a diameter of 0.032 in. (a) How much is its circular-mil area? (b) What is its AWG size? (c) How much is the resistance of a 100-ft length?

2. Draw the schematic diagram of a resistance in series with an open SPST switch and a 100-V source. (a) With the switch open, how much is the voltage across the resistance? Across the open switch? (b) With the switch closed, how much is the voltage across the switch and across the resistance? (c) Do the voltage drops around the series circuit add to equal the applied voltage in both cases?

3. Draw the schematic diagram of a fuse in series with the resistance of a 100-W 120-V bulb connected to a 120-V source. (a) What size fuse can be used? (b) How much is the voltage across the good fuse? (c) How much is the voltage across the fuse if it is open?

4. Compare the resistance of two conductors: 100 ft of No. 10 gage copper wire and 200 ft of No. 7 gage copper wire.

5. How much is the hot resistance of a 150-W 120-V bulb operating with normal load current?

6. How much is the resistance of a slab of silicon 0.1 cm long with a cross-sectional area of 1 cm^2?

7. A cable with two lengths of No. 10 copper wire is short-circuited at one end. The resistance reading at the open end is 10.35 Ω. What is the cable length in feet? (Temperature is 25°C.)

8. (a) How many hole charges are needed to equal 1 C? (b) How many electrons? (c) How many ions with a negative charge of 1 electron?

9. (a) If a copper wire has a resistance of 4 Ω at 25°C, how much is its resistance at 75°C? (b) If the wire is No. 10 gage, what is its length in feet?

10. A coil is wound with 1500 turns of No. 20 copper wire. If the average amount of wire in a turn is 4 in., how much is the total resistance of the coil? What will be its resistance if No. 30 wire is used instead? (Temperature is 25°C.)

11. Calculate the voltage drop across 1000 ft of No. 10 gage wire connected to a 3-A load.

12. What is the smallest size of copper wire that will limit the line drop to 5 V, with 120 V applied and a 6-A load? The total line length is 200 ft.

13. Refer to Fig. 11-19. Calculate the load current I for the IR drop of 24.6 V that reduces V_R to 95.4 V with the 120-V supply.

14. From Fig. 11-19, calculate the value of R_L.

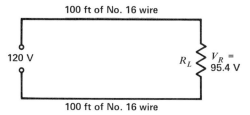

Fig. 11-19 For Probs. 13 and 14.

Answers to Practice Problems

11-1 a. $R = 0.6\ \Omega$
    b. $IR = 0.54$ V

11-2 a. $0.01646\ \Omega$
    b. 625 cmil

11-3 a. T
    b. T
    c. T

11-4 a. 25 W
    b. Infinite ohms

11-5 a. Zero
    b. Three
    c. *NO*

11-6 a. Zero
    b. Zero

11-7 a. More
    b. $2.6\ \Omega$

11-8 a. T
    b. T

11-9 a. $I = 2$ A
    b. Electrons
    c. F

11-10 a. Positive
    b. Four
    c. Electrons

11-11 a. Mica
    b. Yes

11-12 a. F
    b. T

Chapter 12
Batteries

A battery is a group of cells that generate electric energy from their internal chemical reaction. The cell itself consists of two different conducting materials as the electrodes immersed in an electrolyte. The chemical reaction results in a separation of electric charges, in the form of ions and free electrons. As a result, the two electrodes have a difference of potential that provides voltage output from the cell.

The main types are the carbon–zinc dry cell with an output of 1.5 V and the lead–sulfuric acid wet cell with an output of 2.1 V. The common 9-V flat battery for transistor radios has six cells connected in series internally, for an output of $6 \times 1.5 = 9$ V. Similarly, the 12-V

automotive battery has six lead-acid cells in series, for a nominal output of 12 V.

The function of a battery is to provide a source of steady dc voltage of fixed polarity. Furthermore, a battery is a good example of a generator with an internal resistance that affects the output voltage.

Important terms in this chapter are:

alkaline cell	lithium cell
ampere-hour capacity	load matching
carbon–zinc cell	mercury cell
constant-current generator	nickel–cadmium cell
constant-voltage generator	open-circuit voltage
galvanic cell	primary cell
hydrometer	secondary cell
internal resistance	solar cell
lead-acid cell	specific gravity
Leclanché cell	storage cell

More details are explained in the following sections:

12-1   General Features of Batteries

A battery is a combination of cells. The chemical battery has always been important as a dc voltage source for the operation of radio and electronic equipment. The reason is that a transistor amplifier needs dc operating voltages in order to conduct current. With current in the amplifier, the circuit can be used to amplify an ac signal. Originally, all radio receivers used batteries. Then rectifier power supplies were developed to convert the ac power-line voltage to dc output, eliminating the need for batteries. However, now batteries are used more than ever for transistorized portable equipment, which can operate without being connected to the ac power line.

From the old days of radio, dry batteries are still called A, B, and C batteries, according to their original functions in vacuum-tube operation. The A battery was used to supply enough current to heat the filament for thermionic emission of electrons from a heated cathode. A typical rating is 4.5 V or 6 V with a load current of 150 mA or more. The C battery was used for a small negative dc bias voltage at the control grid, typically 1.5 V, with practically no current drain.

The A battery is seldom used any more, although a 6-V lantern battery has ratings like an A battery. However, the function of a B battery with medium voltage and current ratings is the same now as it always was. Transistors need a steady dc voltage for the collector electrode, which receives charges from the emitter through the base electrode. For an NPN transistor, pos-

itive voltage is needed at the collector or negative voltage at the emitter. With a PNP transistor, the polarities are reversed. The positive dc supply voltage is called B^+ or V^+. In a small transistor radio with a 9-V battery, as an example, the battery is the V^+ supply. The same requirements for dc supply voltage apply to integrated circuits, as the IC chip contains transistor amplifiers.

Primary Cells   This type cannot be recharged. After it has delivered its rated capacity, the primary cell must be discarded. The reason is that the internal chemical reaction cannot be restored. All the cells and batteries shown in Fig. 12-1, which are mainly carbon–zinc, are of the primary type. In addition, the list in Table 12-1 indicates which are primary cells.

Secondary Cells   This type can be recharged because the chemical action is reversible. When it supplies current to a load resistance, the cell is *discharging*, as the current tends to neutralize the separated charges at the electrodes. For the opposite case, the current can be reversed to re-form the electrodes as the chemical action is reversed. This action is *charging* the cell. The charging current must be supplied by an external dc voltage source, with the cell serving just as a load resistance. The discharging and recharging is called *cycling* of the cell. Since a secondary cell can be recharged, it is also called a *storage cell*. The most common type is the lead-acid cell generally used in

Fig. 12-1 Typical cells and dry batteries. These are primary cells and batteries which cannot be recharged. (*Eveready-Union Carbide Corp.*)

Table 12-1. Cell Types and Open-Circuit Voltage

Cell Name	Type	Nominal Open-Circuit* Voltage (V_{dc})
Carbon-zinc	Primary	1.5
Zinc chloride	Primary	1.5
Manganese dioxide (alkaline)	Primary or secondary	1.5
Mercuric oxide	Primary	1.35
Silver oxide	Primary	1.5
Lead-acid	Secondary	2.1
Nickel-cadmium	Secondary	1.25
Nickel-iron (Edison cell)	Secondary	1.2
Silver-zinc	Secondary	1.5
Silver-cadmium	Secondary	1.1

*Open-circuit V is the terminal voltage without a load.

automotive batteries (Fig. 12-2). In addition, the list in Table 12-1 indicates which are secondary cells.

Dry Cells   What we call a ''dry cell'' really has moist electrolyte. However, the electrolyte cannot be spilled and the cell can operate in any position.

Sealed Rechargeable Cells   This type is a secondary cell that can be recharged, but it has a sealed electrolyte that cannot be refilled. These cells are capable of charge and discharge in any position.

Fig. 12-2 Typical 12-V automotive battery using six lead–sulfuric acid cells in series. This secondary type is rechargeable. (*Exide Corp.*)

Practice Problems 12-1
Answers at End of Chapter

a. How much is the output voltage of a carbon–zinc cell?
b. How much is the output voltage of a lead-acid cell?
c. Which type can be recharged, a primary or a secondary cell?

12-2   The Voltaic Cell

When two different conducting materials are dissolved in an electrolyte, as illustrated in Fig. 12-3*a*, the chemical action of forming a new solution results in the separation of charges. This method for converting chemical energy into electric energy is a voltaic cell. It is also called a *galvanic cell,* named after Luigi Galvani (1737–1798).

Referring to Fig. 12-3*a*, the charged conductors in the electrolyte are the electrodes or plates of the cell. They serve as the terminals for connecting the voltage output to an external circuit, as shown in Fig. 12-3*b*. Then the potential difference resulting from the separated charges enables the cell to function as a source of applied voltage. The voltage across the cell's terminals forces current to flow in the circuit to light the bulb.

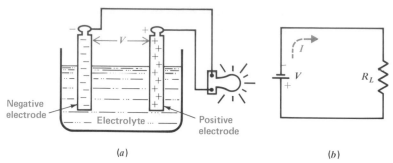

Fig. 12-3 Voltaic cell converts chemical energy into electric energy. (*a*) Electrodes or plates in liquid electrolyte solution. (*b*) Schematic diagram of cell as a dc voltage source *V* to produce current in the load resistance. The R_L is the resistance of bulb.

Current Outside the Cell Electrons from the negative terminal of the cell flow through the external circuit with R_L and return to the positive terminal. The chemical action in the cell separates charges continuously to maintain the terminal voltage that produces current in the circuit.

The current tends to neutralize the charges generated in the cell. For this reason, the process of producing load current is considered discharging of the cell. However, the internal chemical reaction continues to maintain the separation of charges that produces the output voltage.

Current Inside the Cell The current through the electrolyte is a motion of ion charges. Notice in Fig. 12-3*b* that the current inside the cell flows from the positive terminal to the negative terminal. This action represents the work being done by the chemical reaction to generate the voltage across the output terminals.

The negative terminal in Fig. 12-3*a* is considered to be the anode of the cell because it forms positive ions for the electrolyte. The opposite terminal of the cell is its cathode.

Internal Resistance Any generator has internal resistance, indicated as r_i, which limits the current that can be produced. For a chemical cell, as in Fig. 12-3, the r_i is mainly the resistance of the electrolyte. For a good cell, r_i is very low, with typical values less than 1 Ω. As the cell deteriorates, though, r_i increases, preventing the cell from producing its normal terminal voltage when load current is flowing. The reason is that the internal voltage drop across r_i opposes the output

terminal voltage. This factor is why you can often measure normal voltage on a dry cell with a voltmeter, which drains very little current, but the terminal voltage drops when the load is connected.

The voltage output of a cell depends on the elements used for the electrodes and the electrolyte. The current rating depends mostly on the physical size. Larger batteries can supply more current. Dry cells are generally rated up to 250 mA, while the lead-acid wet cell can supply current up to 300 A or more. A smaller r_i allows a higher current rating.

Electromotive Series The fact that the voltage output of a cell depends on its elements can be seen from Table 12-2. This list, called the *electrochemical series* or *electromotive series,* gives the relative activity

Table 12-2. Electromotive Series of Elements

Element	Potential, V
Lithium	−2.96
Magnesium	−2.40
Aluminum	−1.70
Zinc	−0.76
Cadmium	−0.40
Nickel	−0.23
Lead	−0.13
Hydrogen (reference)	0.00
Copper	+0.35
Mercury	+0.80
Silver	+0.80
Gold	+1.36

in forming ion charges for some of the chemical elements. The potential for each element is the voltage with respect to hydrogen as a zero reference. The difference between the potentials for two different elements indicates the voltage of an ideal cell using these electrodes. It should be noted, though, that other factors, such as the electrolyte, cost, stability, and long life, are important for the construction of commercial batteries.

Practice Problems 12-2
Answers at End of Chapter

Answer True or False.
a. The negative terminal of a chemical cell has a charge of excess electrons.
b. The internal resistance of a cell limits the amount of output current.
c. Two electrodes of the same metal provide the highest voltage output.

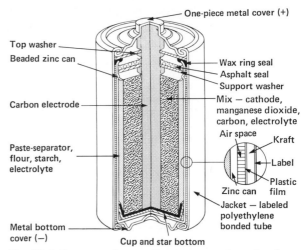

Fig. 12-4 Cutaway view to show construction of carbon–zinc dry cell, size D, with height of 2¼ in. (*Eveready-Union Carbide Corp.*)

12-3 Carbon–Zinc Dry Cell

This is probably the most common type of dry cell. It is also called the *Leclanché cell,* named after the inventor. Examples are shown in Fig. 12-1, while Fig. 12-4 illustrates internal construction for the D-size round cell. Voltage output for the carbon–zinc cell is 1.4 to 1.6 V, with a nominal value of 1.5 V. Suggested current range is up to 150 mA for the D size, which has a height of 2¼ in. and volume of 3.18 in.³. The C, A, AA, and AAA sizes are smaller, with lower current ratings. The larger No. 6 cell has a height of 6 in., a diameter of 2½ in., and a suggested current range of up to 1500 mA.

The electrochemical system consists of a zinc anode and a manganese–dioxide cathode in a moist electrolyte. The electrolyte is a combination of ammonium chloride and zinc chloride dissolved in water. For the round-cell construction, a carbon rod is used down the center, as shown in Fig. 12-4. The rod is chemically inert. However, it serves as a current collector for the positive terminal at the top. The path for current inside the cell includes the carbon rod as the positive terminal, the manganese dioxide, the electrolyte, and the zinc can for the negative electrode. As additional functions

of the carbon rod, it prevents leakage of the electrolyte but is porous to allow the escape of gases which accumulate in the cell.

In operation of the cell, the ammonia (NH_4) releases hydrogen gas. This hydrogen collects around the carbon electrode; this is called *polarization,* and it can reduce the voltage output. However, the manganese dioxide releases oxygen, which combines with the hydrogen to form water. The manganese dioxide functions as a *depolarizer.* Powdered carbon is also added to the depolarizer to improve conductivity and retain moisture.

The chemical efficiency of the carbon–zinc cell increases with less current drain. Stated another way, the application should allow for the largest battery possible, within practical limits. In addition, performance of the cell is generally better with intermittent operation. The reason is that the cell can recuperate between discharges, probably by the effect of depolarization.

Carbon–zinc dry cells are generally designed for an operating temperature of 70°F. Higher temperatures will enable the cell to provide greater output. However, temperatures of 125°F or more will cause rapid deterioration of the cell.

Shelf Life A dry cell loses its ability to produce output voltage even when it is out of use and stored on a shelf. The reasons are self-discharge within the cell and loss of moisture in the electrolyte. Therefore, dry cells should be used fresh from manufacture. The shelf life is shorter for smaller cells and for used cells.

It should be noted that shelf life can be extended by storing the cell at low temperatures, 40 to 50°F. Even temperatures below freezing will not harm the carbon–zinc cell. However, the cell should be allowed to reach normal room temperature before being used, preferably in its original packaging, in order to avoid condensation.

Practice Problems 12-3
Answers at End of Chapter

Answer True or False.
a. A size D cell and a larger No. 6 cell have the same voltage output of 1.5 V.
b. The zinc can of a carbon–zinc cell is the negative terminal.
c. Polarization at the carbon rod increases the voltage output.

12-4 Alkaline Cell

Another popular cell is the alkaline manganese–zinc type, shown in Fig. 12-5. It is available as a primary or secondary cell in flat, miniature button, or round types of construction. The alkaline cell was developed to answer the need for a high discharge rate in popular dry-cell sizes. The electrochemical system consists of a powdered zinc anode and a manganese–dioxide cathode in an alkaline electrolyte. The electrolyte is potassium hydroxide, which is the main difference between the alkaline and Leclanché cells. Hydroxide compounds are alkaline with negative hydroxyl (OH) ions, while an acid electrolyte has positive hydrogen (H) ions. Voltage output from the alkaline cell is 1.5 V.

The alkaline cell has many applications because of its ability to work with high efficiency with continuous high discharge rates. Depending on the application, an alkaline cell can provide up to seven times the service of a Leclanché cell. As examples, in a transistor radio an alkaline cell will normally have twice the service life of a general-purpose carbon–zinc cell; in toys the alkaline cell typically provides seven times more service.

The outstanding performance of the alkaline cell is due to its low internal resistance. Its r_i is low because of the dense cathode material, the large surface area of the anode in contact with the electrolyte, and the high conductivity of the electrolyte. In addition, alkaline cells will perform satisfactorily at low temperatures. Furthermore, the shelf life is better than that of the Leclanché cell.

Practice Problems 12-4
Answers at End of Chapter

Answer True or False.
a. Alkaline cells are available in both primary and secondary types.
b. Alkaline cells are used primarily for their ability to deliver a higher discharge rate than carbon–zinc cells.

12-5 Additional Types of Primary Cells

The miniature button construction shown in Fig. 12-6 is often used for the mercury cell and the silver–oxide cell. Diameter of the cell is ⅜ to 1 in.

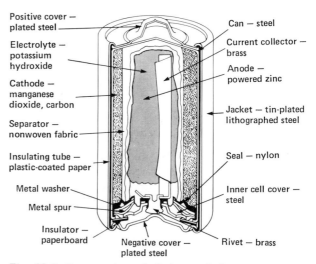

Fig. 12-5 Construction of alkaline cell. (*Eveready-Union Carbide Corp.*)

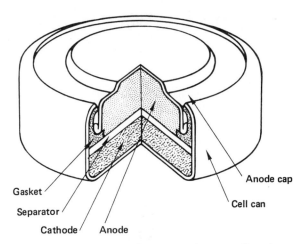

Gasket
Separator
Cathode   Anode
Anode cap
Cell can

● Anodes are a gelled mixture of amalgamated zinc powder and electrolyte.
● Cathodes
Silver cells: AgO_2, MnO_2, and conductor
Mercury cells: HgO and conductor (may contain MnO_2)
Manganese dioxide cells: MnO_2 and conductor

Fig. 12-6 Construction for miniature button type of primary cell. Diameter is ⅜ to 1 in. Note that AgO_2 is silver oxide, HgO is mercuric oxide, and MnO_2 is manganese dioxide. (*Eveready-Union Carbide Corp.*)

Mercury Cell

The electrochemical system consists of a zinc anode, a mercury compound for the cathode, and an electrolyte of potassium or sodium hydroxide. Mercury cells are available in flat, round cylinder, and miniature button shapes. It should be noted, though, that some round mercury cells have the top button as the negative terminal and the bottom terminal positive. The open-circuit voltage is 1.35 V when the cathode is mercuric oxide (HgO) and 1.4 V or more with mercuric oxide/manganese dioxide.

The mercury cell is used where a relatively flat discharge characteristic is required with high current density. Its internal resistance is low and essentially constant. These cells perform well at elevated temperatures, up to 130°F continuous or 200°F for short periods. One drawback of the mercury cell is its relatively high cost compared with a carbon–zinc cell.

Silver–Oxide Cell

The electrochemical system consists of a zinc anode, a cathode of silver oxide (AgO_2) with small amounts of manganese dioxide, and an electrolyte of potassium or sodium hydroxide. It is commonly available in the miniature button shape

shown in Fig. 12-6. The open-circuit voltage is 1.6 V, but the nominal output with a load is considered to be 1.5 V. Typical applications include hearing aids, cameras, and electronic watches.

Zinc–Chloride Cells

This type is actually a modified carbon–zinc cell with the construction illustrated in Fig. 12-4. However, the electrolyte contains only zinc chloride. The zinc–chloride cell is often referred to as the "heavy duty" type. It can normally deliver more current over a longer period of time than the Leclanché cell. Another difference is that the chemical reaction in the zinc–chloride cell consumes water along with the chemically active materials, so that the cell is nearly dry at the end of its useful life. As a result, liquid leakage is not a problem. Shelf life is approximately equal to that of the Leclanché cell.

Practice Problems 12-5
Answers at End of Chapter

Answer True or False.
a. The mercury cell has a minimum output of 2.6 V.
b. Most miniature button cells can be recharged.

12-6 Lithium Cell

This type is a relatively new primary cell. However, its high output voltage, long shelf life, low weight, and small volume make the lithium cell an excellent choice for special applications. The open-circuit output voltage is either 2.9 V or 3.7 V, depending on the electrolyte. Note the high potential of lithium in the electromotive list of elements shown before in Table 12-2. Examples of the lithium cell in the form of a small round disc are shown in Fig. 12-7.

A lithium cell can provide at least ten times more energy than the equivalent carbon–zinc cell with five times the service life. Up until now, lithium cells have not been widely used, except in military applications, because of safety concerns. Lithium is a very active chemical element. One problem is a possible explosive reaction that can occur without apparent warning during use or storage. However, many of the problems

Fig. 12-7 Lithium cells. (*Altus Corp.*)

associated with lithium cells have been solved, especially for small cells delivering low current. One interesting application is a lithium cell as the dc power source for a cardiac pacemaker. The long service life is important for this use.

Two forms of lithium cells have obtained widespread use. These are the lithium–sulfur dioxide ($LiSO_2$) type and the lithium–thionyl chloride type. Output of the $LiSO_2$ type is 2.9 V and 3.7 V from the other type.

In the $LiSO_2$ cell, the sulfur dioxide is kept in a liquid state using a high-pressure container and an organic liquid solvent, usually methyl cyanide. One problem is safe encapsulation of toxic vapor if the container should be punctured or cracked. This problem can be significant for safe disposal of the cells when they are discarded after use.

The shelf life for the lithium cell, 10 years or more, is much longer that that of other types. Now that safety problems have been solved, the use of small lithium cells is expanding rapidly.

Practice Problems 12-6
Answers at End of Chapter

Answer True or False.
a. Shelf life for the lithium cell is much longer than for a carbon–zinc cell.
b. Output voltage for a lithium cell is higher than for a silver–oxide cell.

12-7   Lead-Acid Wet Cell

Where high values of load current are necessary, the lead-acid cell is the type most commonly used. The electrolyte is a dilute solution of sulfuric acid (H_2SO_4). In the application of battery power to start the engine in an automobile, for example, the load current to the starter motor is typically 200 to 400 A. One cell has a nominal output of 2.1 V, but lead-acid cells are often used in a series combination of three for a 6-V battery and six for a 12-V battery. Examples are shown in Figs. 12-8 and 12-2.

The lead-acid type is a secondary cell or storage cell, which can be recharged. The charge and discharge cycle can be repeated many times to restore the output voltage, as long as the cell is in good physical condition. However, heat with excessive charge and discharge currents shortens the useful life to about 3 to 5 years for an automobile battery. Of the different types of secondary cells, the lead-acid type has the highest output voltage, which allows less cells for a specified battery voltage.

Construction   Inside a lead-acid battery, the positive and negative electrodes consist of a group of plates welded to a connecting strap. The plates are immersed in the electrolyte, consisting of 8 parts of water to 3 parts of concentrated sulfuric acid. Each plate is a grid or framework, made of a lead–antimony alloy. This construction enables the active material, which is lead oxide, to be pasted into the grid. In manufacture of the cell, a forming charge produces the positive and negative electrodes. In the forming process, the active material in the positive plate is changed to lead peroxide (PbO_2). The negative electrode is spongy lead (Pb).

Automobile batteries are usually shipped dry from the manufacturer. The electrolyte is put in at the time of installation, then the battery is charged to form the plates. With maintenance-free batteries, little or no water need be added in normal service. Some types are sealed, except for a pressure vent, without provision for adding water.

Chemical Action   Sulfuric acid is a combination of hydrogen and sulfate ions. When the cell discharges, lead peroxide from the positive electrode combines

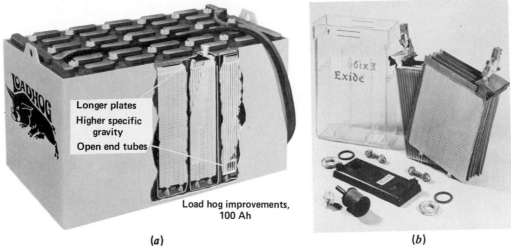

Longer plates
Higher specific
  gravity
Open end tubes

Load hog improvements,
100 Ah

(a)

(b)

Fig. 12-8 Lead-acid batteries. (a) A 12-V storage battery. (b) A 6-V storage battery. (*Exide Corp.*)

with hydrogen ions to form water and with sulfate ions to form lead sulfate. The lead sulfate is also produced by combining lead on the negative plate with sulfate ions. Therefore, the net result of discharge is to produce more water, which dilutes the electrolyte, and to form lead sulfate on the plates.

As discharge continues, the sulfate fills the pores of the grids, retarding circulation of acid in the active material. Lead sulfate is the powder often seen on the outside terminals of old batteries. When the combination of weak electrolyte and sulfation on the plate lowers the output of the battery, charging is necessary.

On charge, the external dc source reverses the current in the battery. The reversed direction of ions flowing in the electrolyte results in a reversal of the chemical reactions. Now the lead sulfate on the positive plate reacts with the water and sulfate ions to produce lead peroxide and sulfuric acid. This action re-forms the positive plate and makes the electrolyte stronger by adding sulfuric acid. At the same time, charging enables the lead sulfate on the negative plate to react with hydrogen ions; this also forms sulfuric acid while re-forming lead on the negative electrode.

As a result, the charging current can restore the cell to full output, with lead peroxide on the positive plates, spongy lead on the negative plate, and the required concentration of sulfuric acid in the electrolyte. The chemical equation for the lead-acid cell is

$$Pb + PbO_2 + 2H_2SO_4 \xmsup{\text{Charge}}{\text{Discharge}} 2PbSO_4 + 2H_2O$$

On discharge, the Pb and PbO_2 combine with the SO_4 ions at the left side of the equation to form lead sulfate ($PbSO_4$) and water (H_2O) at the right side of the equation.

On charge, with reverse current through the electrolyte, the chemical action is reversed. Then the Pb ions from the lead sulfate on the right side of the equation re-form the lead and lead peroxide electrodes. Also the SO_4 ions combine with H_2 ions from the water to produce more sulfuric acid at the left side of the equation.

Current Ratings   Lead-acid batteries are generally rated in terms of how much discharge current they can supply for a specified period of time. The output voltage must be maintained above a minimum level, which is 1.5 to 1.8 V per cell. A common rating is ampere-hours (A·h) based on a specific discharge time, which is often 8 h. Typical values for automobile batteries are 100 to 300 A·h.

As an example, a 200 A·h battery can supply a load current of $^{200}/_8$ or 25 A, based on 8 h discharge. The battery can supply less current for a longer time or more current for a shorter time. Automobile batteries may be rated for "cold cranking power," which is related to the job of starting the engine. A typical rating is 450 A for 30 s at a temperature of 0°F.

Note that the ampere-hour unit specifies coulombs of charge. For instance, 200 A·h corresponds to 200 A × 3600 s for one hour. This product equals 720,000 A·s, or coulombs. One ampere-second is equal to one coulomb. Then the charge equals 720,000 or 7.2×10^5 C. To put this much charge back into the battery would require 20 h with a charging current of 10 A.

The ratings for lead-acid batteries are for a temperature of 77 to 80°F. Higher temperatures increase the chemical reaction, but operation above 110°F shortens the battery life.

Low temperatures reduce the current capacity and voltage output. The ampere-hour capacity is reduced approximately 0.75 percent for each decrease of 1°F. At 0°F the available output is only 60 percent of the ampere-hour battery rating. In cold weather, therefore, it is very important to have an automobile battery up to full charge. In addition, the electrolyte freezes more easily when diluted by water in the discharged condition.

Specific Gravity   The state of discharge for a lead-acid cell is generally checked by measuring the specific gravity of the electrolyte. Specific gravity is a ratio comparing the weight of a substance with the weight of water. For instance, concentrated sulfuric acid is 1.835 times as heavy as water for the same volume. Therefore, its specific gravity equals 1.835. The specific gravity of water is 1, since it is the reference.

In a fully charged automotive cell, the mixture of sulfuric acid and water results in a specific gravity of 1.280 at room temperatures of 70 to 80°F. As the cell discharges, more water is formed, lowering the specific gravity. When it is down to about 1.150, the cell is completely discharged.

Specific-gravity readings are taken with a battery hydrometer, such as the one in Fig. 12-9. Note that the calibrated float with the specific gravity marks will rest higher in an electrolyte of higher specific gravity. The decimal point is often omitted for convenience. For example, the value of 1.220 in Fig. 12-9c is simply read "twelve twenty." A hydrometer reading of 1260 to 1280 indicates full charge, approximately 1250 is half charge, and 1150 to 1200 indicates complete discharge.

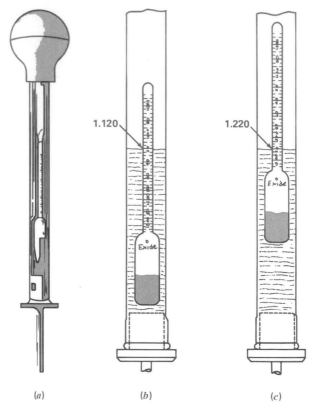

(a)   (b)   (c)

Fig. 12-9 Hydrometer to check specific gravity of lead-acid battery. (a) Syringe to suck up electrolyte. (b) Float at specific gravity of 1.120. (c) Float is higher for higher reading of 1.220. (*Exide Corp.*)

The importance of the specific gravity can be seen from the fact that the open-circuit voltage of the lead-acid cell is approximately equal to

$$V = \text{specific gravity} + 0.84$$

For the specific gravity of 1.280, the voltage is 1.280 + 0.84 = 2.12 V, as an example. These values are for a fully charged battery.

Charging the Lead-Acid Battery   The requirements are illustrated in Fig. 12-10. An external dc voltage source is necessary to produce current in one direction. Also, the charging voltage must be more than the battery emf. Approximately 2.5 V per cell is enough to overcome the cell emf so that the charging voltage can produce current opposite to the direction of discharge current.

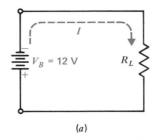
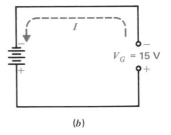

$V_B = 12$ V R_L

$V_G = 15$ V

(a)   (b)

Fig. 12-10 Reversed directions of charge and discharge currents of battery. (a) Voltage V_B discharges to supply load current for R_L. (b) Battery is load resistance for source of charging voltage V_G.

Note that the reversal of current is obtained just by connecting the battery V_B and charging source V_G with + to + and − to −, as shown in Fig. 12-10b. The charging current is reversed because the battery effectively becomes a load resistance for V_G when it is higher than V_B. In this example, the net voltage available to produce charging current is $15 − 12 = 3$ V.

A commercial charger for automobile batteries is shown in Fig. 12-11. The charger is essentially a dc power supply, rectifying input from the ac power line to provide dc output for charging batteries.

Float charging refers to a method in which the charger and the battery are always connected to each other for supplying current to the load. In Fig. 12-12, the charger provides current for the load and the current necessary to keep the battery fully charged. The battery here is an auxiliary source for dc power.

It may be of interest to note that an automobile battery is in a floating-charge circuit. The battery charger is an ac alternator with rectifier diodes, driven by a belt from the engine. When you start the car, the battery supplies the cranking power. Once the engine is running, the alternator charges the battery. It is not necessary for the car to be moving. A voltage regulator is used in this system to maintain the output at approximately 13 to 15 V.

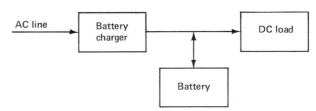

Fig. 12-12 Circuit for battery in float-charge application.

Fig. 12-11 Charger for automobile batteries. (*Exide Corp.*)

Practice Problems 12-7
Answers at End of Chapter

a. How many lead-acid cells in series are needed for a 12-V battery?

b. A battery is rated for 120 A·h at the 8-h rate at 77°F. How much discharge current can it supply for 8 h?

c. Which of the following is the specific gravity reading for a good lead-acid cell: 1070, 1170, or 1270?

12-8 Additional Types of Secondary Cells

A secondary cell is a storage cell that can be recharged by reversing the internal chemical reaction. A primary cell must be discarded after it has been completely discharged. The lead-acid cell is the most common type of storage cell. However, other types of secondary cells are available. Some of these are described next.

Nickel–Cadmium (NiCd) Cell This type is popular because of its ability to deliver high current and to be cycled many times for recharging. Also, the cell can be stored for a long time, even when discharged, without any damage. The NiCd cell is available in both sealed and nonsealed designs, but the sealed construction shown in Fig. 12-13 is common. Nominal output voltage is 1.25 V per cell. Applications include portable power tools, alarm systems, and portable radio or television equipment.

The chemical equation for the NiCd cell can be written as follows:

$$2\,Ni(OH)_3 + Cd \underset{\text{Discharge}}{\overset{\text{Charge}}{\rightleftharpoons}} 2\,Ni(OH)_2 + Cd(OH)_2$$

The electrolyte is potassium hydroxide (KOH), but it does not appear in the chemical equation. The reason is that the function of this electrolyte is just to act as a conductor for the transfer of hydroxyl (OH) ions. Therefore, unlike the lead-acid cell, the specific gravity of the NiCd cell does not change with the state of charge.

The NiCd cell is a true storage cell with a reversible chemical reaction for recharging that can be cycled up to 1000 times. Maximum charging current is equal to the 10-h discharge rate. It should be noted that a new NiCd battery may need charging before use.

Nickel–Iron (Edison) Cell Developed by Thomas Edison, this cell was once used extensively in industrial truck and railway applications. However, it has been replaced almost entirely by the lead-acid battery. New methods of construction for less weight, though, are making this cell a possible alternative in some applications.

The Edison cell has a positive plate of nickel oxide, a negative plate of iron, and an electrolyte of potassium hydroxide in water with a small amount of lithium hydroxide added. The chemical reaction is reversible for recharging. Nominal output is 1.2 V per cell.

Nickel–Zinc Cell This type has been used in limited railway applications. There has been renewed interest in it for use in electric cars, because of its high energy density. However, one drawback is its limited cycle life for recharging. Nominal output is 1.6 V per cell.

Zinc–Chlorine (Hydrate) Cell This cell has been under development in recent years for use in electric vehicles. It is sometimes considered a zinc–chloride cell. This type has high energy density with a good cycle life. Nominal output is 2.1 V per cell.

Lithium–Iron Sulfide Cell This type is under development for commercial energy applications. Nominal output is 1.6 V per cell. The normal operating temperature is 800 to 900°F, which is high compared with the more popular types of cells.

Sodium–Sulfur Cell This is another type of cell being developed for electric vehicle applications. It has the potential of long life at low cost with high efficiency. The cell is designed to operate at temperatures

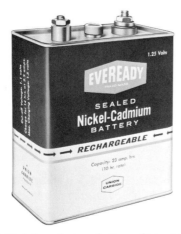

Fig. 12-13 Nickel–cadmium battery. Output is 1.25 V.

between 550 and 650°F. Its most interesting feature is the use of a ceramic electrolyte.

Plastic Cells

A recent development in battery technology is the rechargeable plastic cell made from a conductive polymer, which is a combination of organic chemical compounds. These cells could have ten times the power of the lead-acid type with one-tenth the weight and one-third the volume. In addition, the plastic cell does not require maintenance. One significant application could be for electric vehicles.

A plastic cell consists of an electrolyte between two polymer electrodes. Operation is similar to that of a capacitor. During charge, electrons are transferred from the positive electrode to the negative electrode by a dc source. On discharge, the stored electrons are driven through the external circuit to provide current in the load.

Solar Cells

This type converts the sun's light energy directly into electric energy. The cells are made of semiconductor materials, which generate voltage output with light input. Silicon, with an output of 0.5 V per cell, is mainly used now. Research is continuing, however, on other materials, such as cadmium sulfide and gallium arsenide, that might provide more output. In practice, the cells are arranged in modules that are assembled into a large solar array for the required power.

In most applications, the solar cells are used in combination with a lead-acid cell specifically designed for this use. When there is sunlight, the solar cells charge the battery, as well as supplying power to the load. When there is no light, the battery supplies the required power.

Practice Problems 12-8
Answers at End of Chapter

Answer True or False.
a. The NiCd cell is a primary type.
b. Output of the NiCd cell is 1.25 V.
c. The Edison cell is a storage type.
d. Output of a solar cell is typically 0.5 V.

12-9   Series and Parallel Cells

An applied voltage higher than the emf of one cell can be obtained by connecting cells in series. The total voltage available across the battery of cells is equal to the sum of the individual values for each cell. Parallel cells have the same voltage as one cell but have more current capacity. The combination of cells is a *battery*.

Figure 12-14 shows series-aiding connections for three dry cells. Here the three 1.5-V cells in series provide a total battery voltage of 4.5 V. Notice that the two end terminals A and B are left open to serve as the plus and minus terminals of the battery. These terminals are used to connect the battery to the load circuit, as shown in Fig. 12-14c.

In the lead-acid battery in Fig. 12-2, short, heavy metal straps connect the cells in series. The current capacity of a battery with cells in series is the same as for one cell because the same current flows through all the series cells.

Parallel Connections

For more current capacity, the battery has cells in parallel, as shown in Fig. 12-15. All the positive terminals are strapped together, as are

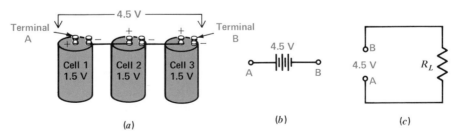

Fig. 12-14 Cells in series to add voltages. Current rating is the same as for one cell. (a) Wiring. (b) Schematic symbol. (c) Battery connected to load resistance R_L.

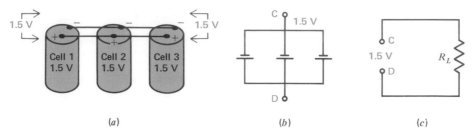

Fig. 12-15 Cells in parallel to increase rating for maximum current. Voltage is the same as for one cell. (*a*) Wiring. (*b*) Schematic symbol. (*c*) Battery connected to load resistance R_L.

all the negative terminals. Any point on the positive side can be the plus terminal of the battery, and any point on the negative side can be the negative terminal.

The parallel connection is equivalent to increasing the size of the electrodes and electrolyte, which increases the current capacity. The voltage output of the battery, however, is the same as for one cell.

Identical cells in parallel supply equal parts of the load current. For example, with three identical parallel cells producing a load current of 300 mA, each cell has

a drain of 100 mA. Bad cells should not be connected in parallel with good cells, however, since the cells in good condition will supply more current, which may overload the good cells. In addition, a cell with lower output voltage will act as a load resistance, draining excessive current from the cells that have higher output voltage.

Series-Parallel Connections   In order to provide higher output and more current capacity, cells can be connected in series-parallel combinations. Figure 12-16 shows four No. 6 cells connected in series-parallel to form a battery that has a 3-V output with a current capacity of ½ A. Two of the 1.5-V cells in series provide 3 V total output voltage. This series string has a current capacity of ¼ A, however, assuming this current rating for one cell.

To double the current capacity, another string is connected in parallel. The two strings in parallel have the same 3-V output as one string, but with a current capacity of ½ A instead of the ¼ A for one string.

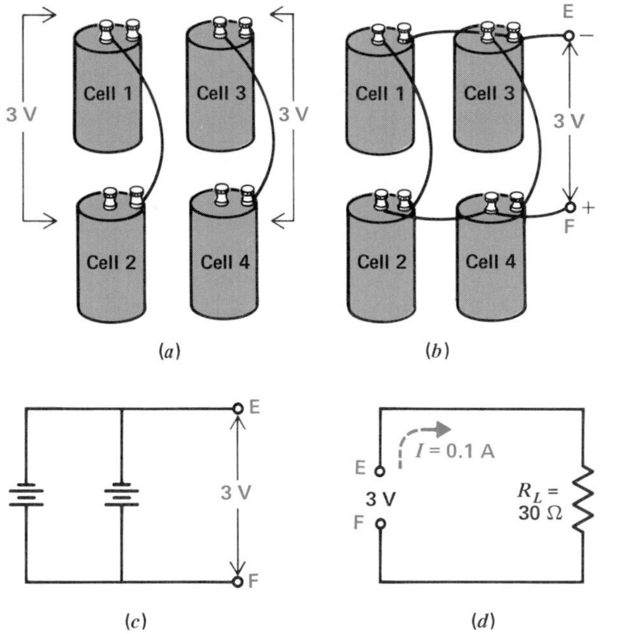

Fig. 12-16 Cells in series-parallel combinations. (*a*) Wiring two 3-V strings, each with two 1.5-V cells in series. (*b*) Wiring the two 3-V strings in parallel. (*c*) Schematic symbol for the battery in (*b*). (*d*) Battery connected to load resistance R_L.

Practice Problems 12-9
Answers at End of Chapter

a. How many carbon–zinc cells in series are required to obtain a 9-V dc output? How many lead-acid cells are required to obtain 12 V dc?

b. How many identical cells in parallel would be required to double the current rating of a single cell?

c. How many cells rated 1.5 V dc 300 mA would be required in a series-parallel combination which would provide a rating of 900 mA at 6 V dc?

12-10   Current Drain Depends on Load Resistance

It is important to note that the current rating of batteries, or any voltage source, is only a guide to typical values permissible for normal service life. The actual amount of current produced when the battery is connected to a load resistance is equal to $I = V/R$, by Ohm's law.

Figure 12-17 illustrates three different cases of using the applied voltage of 1.5 V from a dry cell. In Fig. 12-17a, the load resistance R_1 is 7.5 Ω. Then I is 1.5/7.5 = ⅕ A or 200 mA.

A No. 6 carbon–zinc cell used as the voltage source could supply this load of 200 mA continuously for about 74 h at a temperature of 70°F before dropping to an end voltage of 1.2 V. If an end voltage of 1.0 V could be used, the same load would be served for approximately 170 h.

In Fig. 12-17b, a larger load resistance R_2 is used. The value of 150 Ω limits the current to 1.5/150 = 0.01 A or 10 mA. Again using the No. 6 carbon–zinc cell at 70°F, the load could be served continuously for 4300 h with an end voltage of 1.2 V. The two principles here are:

1. The cell delivers less current for a higher resistance in the load circuit.
2. The cell can deliver a smaller load current for a longer time.

In Fig. 12-17c, the load resistance R_3 is reduced to 2.5 Ω. Then I is 1.5/2.5 = 0.6 A or 600 mA. The No. 6 cell could serve this load continuously for only 9 h for an end voltage of 1.2 V. The cell could deliver even more load current, but for a shorter time. The relationship between current and time is not linear. For any one example, though, the amount of current is determined by the circuit, not by the current rating of the battery.

Practice Problems 12-10
Answers at End of Chapter

Answer True or False.
a. A cell rated at 250 mA will produce this current for any value of R_L.
b. A higher value of R_L allows the cell to operate with normal voltage for a longer time.

12-11   Internal Resistance of a Generator

Any source that produces voltage output continuously is a generator. It may be a cell separating charges by chemical action or a rotary generator converting motion and magnetism into voltage output, for common examples. In any case, all generators have internal resistance, which is labeled r_i in Fig. 12-18.

The internal resistance r_i is important when a generator supplies load current because its internal Ir_i voltage drop subtracts from the generated emf, resulting in lower voltage across the output terminals. Physically, r_i may be the resistance of the wire in a rotary generator, or in a chemical cell r_i is the resistance of the electrolyte between electrodes. More generally, the internal resistance r_i is the opposition to load current inside the generator.

Since any current in the generator must flow through the internal resistance, r_i is in series with the generated

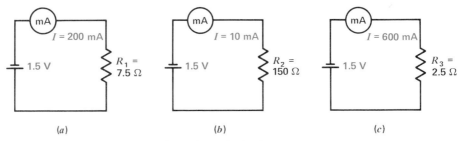

(a)   (b)   (c)

Fig. 12-17 Current drain from a voltage source depends on the load resistance. Different values of I shown for the same V of 1.5 V. (a) V/R_1 equals I of 200 mA. (b) V/R_2 equals I of 10 mA. (c) V/R_3 equals I of 600 mA.

⊣|⊢

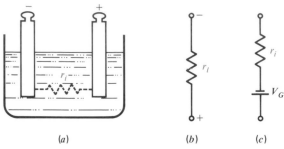

Fig. 12-18 Internal resistance r_i is in series with the generator voltage V_G. (a) Physical arrangement for a voltaic cell. (b) Schematic symbol for r_i. (c) Equivalent circuit of r_i in series with V_G.

voltage, as shown in Fig. 12-18c. It may be of interest to note that, with just one load resistance connected across a generator, they are in series with each other because R_L is in series with r_i.

If there is a short circuit across the generator, its r_i prevents the current from becoming infinitely high. As an example, if a 1.5-V cell is temporarily short-circuited, the short-circuit current I_{sc} could be about 15 A. Then r_i is V/I_{sc}, which equals 1.5/15, or 0.1 Ω for the internal resistance. These are typical values for a carbon–zinc D size cell.

Practice Problems 12-11
Answers at End of Chapter

Answer True or False.
a. The r_i is in series with the load.
b. More load current produces a larger voltage drop across r_i.

12-12   Why the Terminal Voltage Drops with More Load Current

Figure 12-19 illustrates how the output of a 100-V source can drop to 90 V because of the internal 10-V drop across r_i. In Fig. 12-19a, the voltage across the output terminals is equal to the 100 V of V_G because there is no load current on an open circuit. With no current, the voltage drop across r_i is zero. Then the full generated voltage is available across the output terminals. This value is the generated emf, *open-circuit voltage*, or *no-load voltage*.

We cannot connect the test leads inside the source to measure V_G. However, measuring this no-load voltage without any load current provides a method of determining the internally generated emf. We can assume the voltmeter draws practically no current because of its very high resistance.

In Fig. 12-19b with a load, however, current of 0.1 A flows, to produce a drop of 10 V across the 100 Ω of r_i. Note that R_T is $900 + 100 = 1000$ Ω. Then I_L equals $^{100}/_{1000}$, which is 0.1 A.

As a result, the voltage output V_L equals $100 - 10 = 90$ V. This terminal voltage or load voltage is available across the output terminals when the generator is in a closed circuit with load current. The 10-V internal drop is subtracted from V_G because they are series-opposing voltages.

The graph in Fig. 12-20 shows how the terminal voltage V_L drops with increasing load current I_L. The reason is the greater internal voltage drop across r_i, as shown by the calculated values listed in Table 12-3. For this example, V_G is 100 V and r_i is 100 Ω.

Across the top row, infinite ohms for R_L means an open circuit. Then I_L is zero, there is no internal drop V_i, and V_L is the same 100 V as V_G.

Table 12-3. For Fig. 12-19. How V_L Drops with more I_L

V_G, V	r_i, Ω	R_L, Ω	$R_T = R_L + r_i$, Ω	$I_L = V_G/R_T$, A	$V_i = I_L r_i$, V	$V_L = V_G - V_i$, V
100	100	∞	∞	0	0	100
100	100	900	1000	0.1	10	90
100	100	600	700	0.143	14.3	85.7
100	100	300	400	0.25	25	75
100	100	100	200	0.5	50	50
100	100	0	100	1.0	100	0

⊣|⊢

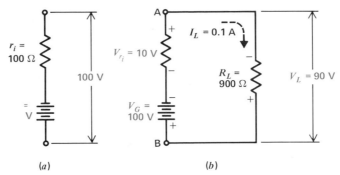

Fig. 12-19 Internal voltage drop decreases voltage at output terminals of generator. (*a*) Open-circuit voltage equals V_G of 100 V because there is no load current. (*b*) Terminal voltage V_L between A and B is reduced to 90 V because of 10-V drop across 100-Ω r_i with 0.1 A of I_L.

Across the bottom row, zero ohms for R_L means a short circuit. Then the short-circuit current of 1 A results in zero output voltage because the entire generator voltage is dropped across the internal resistance. Or, we can say that with a short circuit of zero ohms across the load, the current is limited to V_G/r_i.

The lower the internal resistance of a generator, the better it is in terms of being able to produce full output voltage when supplying current for a load. For example, the very low r_i, about 0.01 Ω, for a 12-V lead-acid battery is the reason it can supply high values of load current and maintain its output voltage.

For the opposite case, a higher r_i means that the terminal voltage of a generator is much less with load current. As an example, an old dry battery with r_i of 500 Ω would appear normal when measured by a voltmeter but be useless because of low voltage when normal load current flows in an actual circuit.

How to Measure r_i   The internal resistance of any generator can be measured indirectly by determining how much the output voltage drops for a specified amount of load current. The difference between the no-load voltage and the load voltage is the amount of internal voltage drop $I_L r_i$. Dividing by I_L gives the value of r_i. As a formula,

$$r_i = \frac{V_{NL} - V_L}{I_L} \qquad \text{(12-1)}$$

Example 1   Calculate r_i if the output of a generator drops from 100 V with zero load current to 80 V with 2 A for I_L.

Answer $r_i = \dfrac{100 - 80}{2}$

$= \dfrac{20}{2}$

$r_i = 10\ \Omega$

A convenient technique for measuring r_i is to use a variable load resistance R_L. Vary R_L until the load voltage is one-half the no-load voltage. This value of R_L is

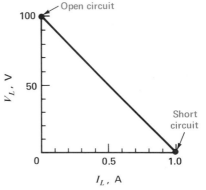

Fig. 12-20 How terminal voltage V_L drops with more load current I_L. Graph is plotted for values in Table 12-3.

also the value of r_i, since they must be equal to divide the generator voltage equally. For the same 100-V generator with the 10-Ω r_i used in Example 1, if a 10-Ω R_L were used, the load voltage would be 50 V, equal to one-half the no-load voltage.

You can solve this circuit by Ohm's law to see that I_L is 5 A with 20 Ω for the combined R_T. Then the two voltage drops of 50 V each add to equal the 100 V of the generator.

Practice Problems 12-12
Answers at End of Chapter

Answer True or False. For Eq. (12-1),
a. V_L must be more than V_{NL}.
b. When V_L is one-half V_{NL}, the r_i is equal to R_L.

12-13 Constant-Voltage and Constant-Current Sources

A generator with very low internal resistance is considered to be a constant-voltage source. Then the output voltage remains essentially the same when the load current changes. This idea is illustrated in Fig. 12-21a for a 6-V lead-acid battery with an r_i of 0.005 Ω. If the load current varies over the wide range of 1 to 100 A, for any of these values, the internal Ir_i drop across 0.005 Ω is less than 0.5 V.

Constant-Current Generator   It has very high resistance, compared with the external load resistance, resulting in constant current, although the output voltage varies.

The constant-current generator shown in Fig. 12-22 has such high resistance, with an r_i of 0.9 MΩ, that it is the main factor determining how much current can be produced by V_G. Here R_L varies in a 3:1 range from 50 to 150 kΩ. Since the current is determined by the total resistance of R_L and r_i in series, however, I is essentially constant at 1.05 to 0.95 mA, or approximately 1 mA. This relatively constant I_L is shown by the graph in Fig. 12-22b.

Note that the terminal voltage V_L varies in approximately the same 3:1 range as R_L. Also, the output voltage is much less than the generator voltage because of the high internal resistance compared with R_L. This is a necessary condition, however, in a circuit with a constant-current generator.

A common example is to insert a series resistance to keep the current constant, as shown in Fig. 12-23a. Resistance R_1 must be very high compared with R_L. In this example, I_L is 50 μA with 50 V applied, and R_T is practically equal to the 1 MΩ of R_1. The value of R_L can vary over a range as great as 10:1 without changing R_T or I_L appreciably.

The circuit with an equivalent constant-current source is shown in Fig. 12-23b. Note the arrow symbol for a current source. As far as R_L is concerned, its terminals A and B can be considered as receiving either 50 V in series with 1 MΩ or 50 μA in shunt with 1 MΩ.

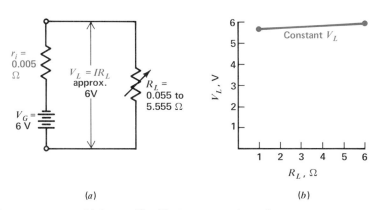

(a)   (b)

Fig. 12-21 Constant-voltage generator with low r_i. The V_L stays approximately the same 6 V as I varies with R_L. (a) Circuit. (b) Graph.

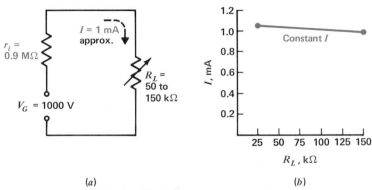

(a)   (b)

Fig. 12-22 Constant-current generator with high r_i. The I_L stays approximately the same 1 mA as V_L varies with R_L. (a) Circuit. (b) Graph.

Practice Problems 12-13
Answers at End of Chapter

Is the internal resistance high or low for:
a. A constant-voltage source?
b. A constant-current source?

12-14   Matching a Load Resistance to the Generator

In the diagram in Fig. 12-24, when R_L equals r_i the load and generator are matched. The matching is significant because the generator then produces maximum power in R_L, as verified by the values listed in Table 12-4.

Maximum Power in R_L   When R_L is 100 Ω to match the 100 Ω of r_i, maximum power is transferred from the generator to the load. With higher resistance for R_L, the output voltage V_L is higher, but the current is reduced. Lower resistance for R_L allows more current, but V_L is less. When r_i and R_L both equal 100 Ω, this combination of current and voltage produces the maximum power of 100 W across R_L.

With generators that have very low resistance, however, matching is often impractical. For example, if a 6-V lead-acid battery with a 0.003-Ω internal resistance were connected to a 0.003-Ω load resistance, the battery could be damaged by excessive current as high as 1000 A.

Maximum Voltage across R_L   If maximum voltage, rather than power, is desired, the load should have as high a resistance as possible. Note that R_L and

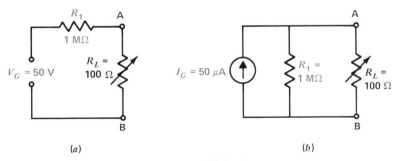

(a)   (b)

Fig. 12-23 Voltage source in (a) equivalent to current source in (b) for R_L across terminals A and B.

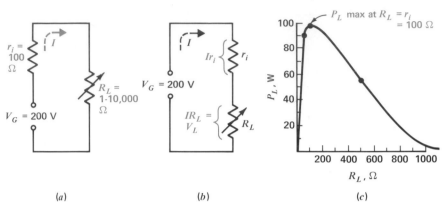

Fig. 12-24 Circuit for varying R_L to match r_i. (a) Schematic diagram. (b) Equivalent voltage divider for voltage output across R_L. (c) Graph of power output P_L for different R_L values. Values from Table 12-4.

r_i form a voltage divider for the generator voltage, as illustrated in Fig. 12-24b. The values for IR_L listed in Table 12-4 show how the output voltage V_L increases with higher values of R_L.

Maximum Efficiency   Note also that the efficiency increases as R_L increases because there is less current, resulting in less power lost in r_i. When R_L equals r_i, the efficiency is only 50 percent, since one-half the total generated power is dissipated in r_i, the internal resistance of the generator. In conclusion,

then, matching the load and generator resistances is desirable when the load requires maximum power rather than maximum voltage or efficiency, assuming that the match does not result in excessive current.

Practice Problems 12-14
Answers at End of Chapter

Answer True or False.
a. When $R_L = r_i$, the P_L is maximum.
b. The V_L is maximum when R_L is maximum.

Table 12-4. Effect of Load Resistance on Generator Output*

	R_L, Ω	$I = V_G/R_T$, A	Ir_i, V	IR_L, V	P_L, W	P_i, W	P_T, W	Efficiency = P_L/P_T, %
	1	1.98	198	2	4	392	396	1
	50	1.33	133	67	89	178	267	33
$R_L = r_i \rightarrow$	100	1	100	100	100	100	200	50
	500	0.33	33	167	55	11	66	83
	1,000	0.18	18	180	32	3.24	35.24	91
	10,000	0.02	2	198	4	0.04	4.04	99

*Values calculated approximately for circuit in Fig. 12-24, with $V_G = 200$ V and $r_i = 100$ Ω.

Summary

1. A voltaic cell consists of two different conductors as electrodes immersed in an electrolyte. The voltage output depends only on the chemicals in the cell. The current capacity increases with larger sizes. A primary cell cannot be recharged. A secondary or storage cell can be recharged.

2. A battery is a group of cells in series or in parallel. With cells in series, the voltages add, but the current capacity is the same as that of one cell. With cells in parallel, the voltage output is the same as that of one cell, but the total current capacity is the sum of the individual values.

3. The zinc–carbon–sal ammoniac dry cell is the most common type of primary cell. Zinc is the negative electrode; carbon is the positive electrode. Its output voltage is approximately 1.5 V.

4. The lead-acid cell is the most common form of storage battery. The positive electrode is lead peroxide; spongy lead is the negative electrode. Both are in a dilute solution of sulfuric acid for the electrolyte. The voltage output is approximately 2.1 V per cell.

5. To charge a lead-acid battery, connect it to a dc voltage equal to approximately 2.5 V per cell. Connecting the positive terminal of the battery to the positive side of the charging source and the negative terminal to the negative side results in charging current through the battery.

6. The mercury cell is a primary cell with an output of 1.35 or 1.4 V.

7. The nickel–cadmium cell is a dry cell that is rechargeable, with an output of 1.25 V.

8. A constant-voltage generator has very low internal resistance. Output voltage is relatively constant with changing values of load because of the small internal voltage drop.

9. A constant-current generator has a very high internal resistance. This determines the constant value of current in the generator circuit relatively independent of the load resistance.

10. Any generator has an internal resistance r_i. With load current I_L, the internal $I_L r_i$ drop reduces the voltage across the output terminals. When I_L makes the terminal voltage drop to one-half the zero load voltage, the external R_L equals the internal r_i.

11. Matching a load to a generator means making the R_L equal to the generator's r_i. The result is maximum power delivered to the load from the generator.

Self-Examination
Answers at Back of Book

Choose (a), (b), (c), or (d).

1. Which of the following is false? (a) A lead-acid cell can be recharged. (b) A primary cell has an irreversible chemical reaction. (c) A storage cell has a reversible chemical reaction. (d) A carbon–zinc cell has unlimited shelf life.

2. The output of a lead-acid cell is (a) 1.25 V; (b) 1.35 V; (c) 2.1 V; (d) 6 V.

3. The current in a chemical cell is a movement of (a) positive hole charges; (b) positive and negative ions; (c) positive ions only; (d) negative ions only.

4. Cells are connected in series to (a) increase the voltage output; (b) decrease the voltage output; (c) decrease the internal resistance; (d) increase the current capacity.

5. Cells are connected in parallel to (a) increase the voltage output; (b) increase the internal resistance; (c) decrease the current capacity; (d) increase the current capacity.

6. Which of the following is a dry storage cell? (a) Edison cell; (b) carbon–zinc cell; (c) mercury cell; (d) nickel–cadmium cell.

7. When R_L equals the generator r_i, which of the following is maximum? (a) Power in R_L; (b) current; (c) voltage across R_L; (d) efficiency of the circuit.

8. Five carbon–zinc cells in series have an output of (a) 1.5 V; (b) 5.0 V; (c) 7.5 V; (d) 11.0 V.

9. A constant-voltage generator has (a) low internal resistance; (b) high internal resistance; (c) minimum efficiency; (d) minimum current capacity.

10. A generator has an output of 10 V on open circuit, which drops to 5 V with a load current of 50 mA and an R_L of 1000 Ω. The internal resistance r_i equals (a) 25 Ω; (b) 50 Ω; (c) 100 Ω; (d) 1000 Ω.

Essay Questions

1. Draw a sketch illustrating the construction of a carbon–zinc dry cell. Indicate the negative and positive electrodes and the electrolyte.

2. Draw a sketch illustrating construction of the lead-acid cell. Indicate the negative and positive electrodes and the electrolyte.

3. Show the wiring for the following batteries: (a) six lead-acid cells for a voltage output of approximately 12 V; (b) six standard No. 6 dry cells for a voltage output of 4.5 V with a current capacity of ½ A. Assume a current capacity of ¼ A for one cell.

4. (a) What is the advantage of connecting cells in series? (b) What is connected to the end terminals of the series cells?

5. (a) What is the advantage of connecting cells in parallel? (b) Why can the load be connected across any one of the parallel cells?

6. How many cells are necessary in a battery to double the voltage and current ratings of a single cell? Show the wiring diagram.

7. Draw a diagram showing two 12-V lead-acid batteries being charged by a 15-V source.

8. Why is a generator with very low internal resistance called a constant-voltage source?

9. Why does discharge current lower the specific gravity in a lead-acid cell?

10. Would you consider the lead-acid battery a constant-current source or a constant-voltage source? Why?

11. List five types of chemical cells, giving two features of each.

12. Referring to Fig. 12-21b, draw the corresponding graph that shows how I varies with R_L.

13. Referring to Fig. 12-22b, draw the corresponding graph that shows how V_L varies with R_L.

14. Referring to Fig. 12-24c, draw the corresponding graph that shows how V_L varies with R_L.

Problems

Answers to Odd-Numbered Problems at Back of Book

1. A 1.5-V No. 6 carbon–zinc dry cell is connected across an R_L of 1000 Ω. How much current flows in the circuit?

2. Draw the wiring diagram for six No. 6 cells providing a 3-V output with a current capacity of ¾ A. Draw the schematic diagram of this battery connected across a 10-Ω resistance. (a) How much current flows in the circuit? (b) How much power is dissipated in the resistance? (c) How much power is supplied by the battery? Assume a current capacity of ¼ A for one cell.

3. A 6-V lead-acid battery has an internal resistance of 0.01 Ω. How much current will flow if the battery has a short circuit?

4. How much is the specific gravity of a solution with equal parts of sulfuric acid and water?

5. A lead-acid battery discharges at the rate of 8 A for 10 h. (a) How many coulombs of charge must be put back into the battery to restore the original charge, assuming 100 percent efficiency? (b) How long will this recharging take, with a charging current of 2 A?

6. The output voltage of a battery drops from 90 V at zero load to 60 V with a load current of 50 mA. (a) How much is the internal r_i of the battery? (b) How much is R_L for this load current? (c) How much R_L reduces the load voltage to one-half the no-load voltage?

7. The output voltage of a source reads 60 V with a DMM. When a meter with 1000 Ω/V sensitivity is used, the reading is 50 V on the 100-V range. How much is the internal resistance of the source?

8. A 100-V source with an internal resistance of 10 kΩ is connected to a variable load resistance R_L. Tabulate I, V_L, and power in R_L for values of 1 kΩ, 5 kΩ, 10 kΩ, 15 kΩ, and 20 kΩ.

9. A generator has an open-circuit emf of 18 V. Its terminal voltage drops to 15 V with an R_L of 30 Ω. Calculate r_i.

10. Referring to Fig. 12-24, calculate P_L when R_L is 200 Ω. Compare this value with the maximum P_L at $R_L = r_i = 100$ Ω.

Answers to Practice Problems

12-1	**a.** 1.5 V	**12-6**	**a.** T	**12-10**	**a.** F	
	b. 2.1 V		**b.** T		**b.** T	
	c. Secondary	**12-7**	**a.** Six	**12-11**	**a.** T	
12-2	**a.** T		**b.** 15 A		**b.** T	
	b. T		**c.** 1270	**12-12**	**a.** F	
	c. F	**12-8**	**a.** F		**b.** T	
12-3	**a.** T		**b.** T	**12-13**	**a.** Low	
	b. T		**c.** T		**b.** High	
	c. F		**d.** T	**12-14**	**a.** T	
12-4	**a.** T	**12-9**	**a.** Six		**b.** T	
	b. T		**b.** Two			
12-5	**a.** F		**c.** Twelve			
	b. F					

Review

Chapters 11 and 12

Summary

1. A conductor has very low resistance. Silver and copper are the best conductors, with copper generally used for wire.
2. The gage numbers for copper wire are listed in Table 11-1.
3. Resistance wire for use in heating elements and filaments has a much higher R when hot, compared with its cold resistance. The hot resistance, equal to V/I, cannot be measured with an ohmmeter.
4. An applied voltage can produce charged ions in liquids and gases. The ions may be either positive or negative.
5. Insulators or dielectrics have very high resistance. Examples are air, vacuum, paper, glass, rubber, shellac, wood, and plastics.
6. The main types of cells used for batteries are the carbon–zinc dry cell, the lead-acid wet cell, and the nickel–cadmium dry cell.
7. The carbon–zinc type is a primary cell, which cannot be recharged. The lead-acid and nickel–cadmium types are storage cells or secondary cells, which can be recharged. Additional types of cells are listed in Table 12-1.
8. With cells or batteries connected in series, the total voltage equals the sum of the individual values. The connections are series-aiding, with the + terminal of one connected to the − terminal of the next. The current rating for the series combination is the same as that for each cell.
9. For cells or batteries in parallel, the voltage across all is the same as that across one. However, the current rating of the combination equals the sum of the individual values.

Review Self-Examination
Answers at Back of Book

Choose (a), (b), (c), or (d).

1. Which of the following is the best conductor? (a) Carbon; (b) silicon; (c) rubber; (d) copper.
2. A 1-W 1-kΩ carbon resistor with current of 2 mA dissipates a power of (a) 4 W; (b) 4 mW; (c) 2 W; (d) 2 mW.
3. Which of the following cells has a reversible chemical reaction? (a) Carbon–zinc; (b) lead-acid; (c) silver–oxide; (d) mercury–oxide.
4. A tungsten filament measures 10 Ω with an ohmmeter. In a circuit with 100 V applied, 2 A flows. The hot resistance of the filament equals (a) 2 Ω; (b) 10 Ω; (c) 50 Ω; (d) 100 Ω.
5. Three resistors R_1, R_2, and R_3 are in series across a 100-V source. If R_2 opens: (a) the voltage across R_2 is zero; (b) the total resistance of R_1, R_2, and R_3 decreases; (c) the voltage across R_1 is 100 V; (d) the voltage across R_2 is 100 V.
6. The current flowing between electrodes inside a lead-acid battery is (a) electron current; (b) proton current; (c) ionization current; (d) polarization current.
7. Thirty zinc–carbon dry cells are connected in series. The total voltage output is (a) 1.5 V; (b) 30 V; (c) 45 V; (d) 60 V.
8. A 45-V source with an internal resistance of 2 Ω is connected across a wire-wound resistor. Maximum power will be dissipated in the resistor when its R is (a) zero; (b) 2 Ω; (c) 45 Ω; (d) infinity.

References
Additional References at Back of Book

Adams, J. A., and G. Rockmaker: *Industrial Electricity—Principles and Practices*, McGraw-Hill Book Company, New York.

Eveready–Union Carbide Corp.: *Battery Applications and Engineering Data*.

Fowler, R. J.: *Electricity—Principles and Applications*, McGraw-Hill Book Company, New York.

Richter, H. P.: *Practical Electrical Wiring*, McGraw-Hill Book Company, New York.

Slurzberg, M., and W. Osterheld: *Essentials of Electricity and Electronics*, McGraw-Hill Book Company, New York.

Marcus, A., and C. Thompson: *Electricity for Technicians*, Prentice-Hall, Englewood Cliffs, N.J.

Chapter 13
Magnetism

Electrical effects exist in two forms, voltage and current. In terms of voltage, separated electric charges have the potential to do mechanical work in attracting or repelling charges. Similarly, any electric current has an associated magnetic field that can do the work of attraction or repulsion. Materials made of iron, nickel, and cobalt, particularly, concentrate their magnetic effects at opposite ends, where the magnetic material meets a nonmagnetic medium such as air.

The points of concentrated magnetic strength are called north and south poles. The opposite magnetic poles correspond to the idea of opposite polarities of electric charges. The name *magnetism* is derived

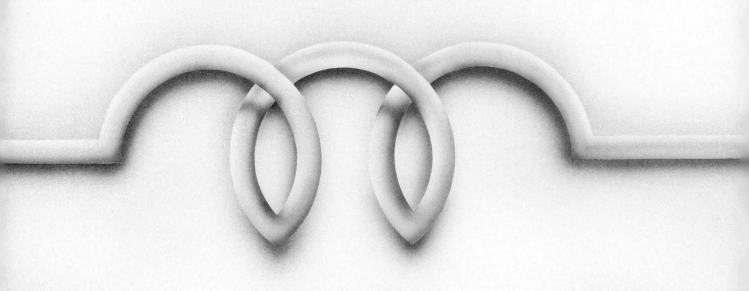

from the iron oxide mineral *magnetite*. Ferromagnetism refers specifically to the magnetic properties of iron.

Important terms in this chapter are:

air gap	magnetic induction
diamagnetic	magnetic poles
electromagnet	maxwell unit
ferrites	paramagnetic
ferromagnetic	permanent magnet
flux	relative permeability
flux density	shielding
gauss unit	tesla unit
Hall effect	toroid
keeper	weber unit

More details are explained in the following sections:

13-1 The Magnetic Field
13-2 Magnetic Flux ϕ
13-3 Flux Density B
13-4 Induction by the Magnetic Field
13-5 Air Gap of a Magnet
13-6 Types of Magnets
13-7 Ferrites
13-8 Magnetic Shielding
13-9 The Hall Effect

13-1 The Magnetic Field

As shown in Fig. 13-1, the north and south poles of a magnet are the points of concentration of magnetic strength. The practical effects of this ferromagnetism result from the magnetic field of force between the two poles at opposite ends of the magnet. Although the magnetic field is invisible, evidence of its force can be seen when small iron filings are sprinkled on a glass or paper sheet placed over a bar magnet (Fig. 13-2a). Each iron filing becomes a small bar magnet. If the sheet is tapped gently to overcome friction so that the filings can move, they become aligned by the magnetic field.

Many filings cling to the ends of the magnet, showing that the magnetic field is strongest at the poles. The field exists in all directions but decreases in strength inversely as the square of the distance from the poles of the magnet.

Field Lines In order to visualize the magnetic field without iron filings, we show the field as lines of force, as in Fig. 13-2b. The direction of the lines outside the magnet shows the path a north pole would follow in the field, repelled away from the north pole of the magnet and attracted to its south pole. Although we cannot actually have a unit north pole by itself, the field can be explored by noting how the north pole on a small compass needle moves.

The magnet can be considered as the generator for an external magnetic field, provided by the two opposite magnetic poles at the ends. This idea corresponds to the two opposite terminals on a battery as the source for an external electric field provided by opposite charges.

Magnetic field lines are unaffected by nonmagnetic materials such as air, vacuum, paper, glass, wood, or

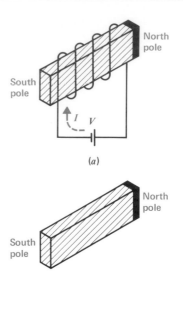

(a)

(b)

Fig. 13-1 Poles of a magnet. (a) Electromagnet (EM) produced by current from battery. (b) Permanent magnet (PM) without any external current source.

plastics. When these materials are placed in the magnetic field of a magnet, the field lines are the same as though the material were not there.

However, the magnetic field lines become concentrated when a magnetic substance like iron is placed in the field. Inside the iron, the field lines are more dense, compared with the field in air.

North and South Magnetic Poles The earth itself is a huge natural magnet, with its greatest strength at the north and south poles. Because of the earth's magnetic poles, if a small bar magnet is suspended so

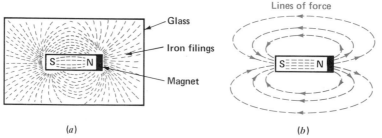

(a)

(b)

Fig. 13-2 Magnetic field of force around a bar magnet. (a) Field outlined by iron filings. (b) Field indicated by lines of force.

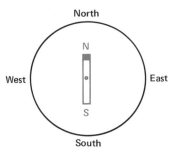

Fig. 13-3 Definition of north and south poles of bar magnet.

that it can turn easily, one end will always point north. This end of the bar magnet is defined as the *north-seeking pole,* as shown in Fig. 13-3. The opposite end is the *south-seeking pole.* When polarity is indicated on a magnet, the north-seeking end is the north pole (N) and the opposite end is the south pole (S). It should be noted that the magnetic north pole differs by about 15° from true geographic north on the axis of the earth's rotation.

Similar to the force between electric charges is a force between magnetic poles causing attraction of opposite poles and repulsion between similar poles:

1. A north pole (N) and a south pole (S) tend to attract each other.

2. A north pole (N) tends to repel another north pole (N), while a south pole (S) tends to repel another south pole (S).

These forces are illustrated by the field of iron filings between opposite poles in Fig. 13-4a and between similar poles in Fig. 13-4b.

Practice Problems 13-1
Answers at End of Chapter

Answer True or False.
a. On a magnet, the north-seeking pole is labeled N.
b. Like poles have a force of repulsion.

13-2   Magnetic Flux φ

The entire group of magnetic field lines, which can be considered to flow outward from the north pole of a magnet, is called *magnetic flux.* Its symbol is the Greek letter ϕ (phi). A strong magnetic field has more lines of force and more flux than a weak magnetic field.

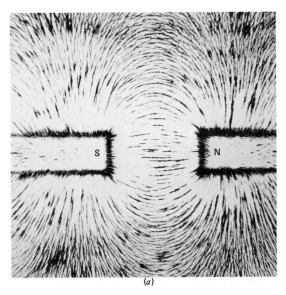

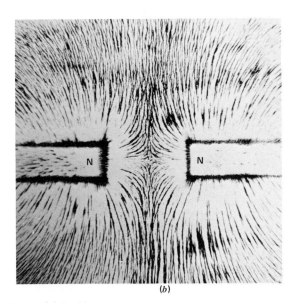

Fig. 13-4 Photos of magnetic field patterns produced by iron filings. (a) Field between opposite poles. The north and south poles could be reversed. (b) Field between similar poles. The two north poles could be south poles.

The Maxwell One maxwell (Mx) unit equals one magnetic field line. In Fig. 13-5, as an example, the flux illustrated is 6 Mx because there are 6 field lines flowing in or out for each pole. A 1-lb magnet can provide a magnetic flux ϕ of about 5000 Mx. This unit is named for James Clerk Maxwell (1831–1879), an important Scottish mathematical physicist who contributed much to electrical and field theory.

The Weber This is a larger unit of magnetic flux. One weber (Wb) equals 1×10^8 lines or maxwells. Since the weber is a large unit for typical fields, the microweber unit can be used. Then $1 \ \mu\text{Wb} = 10^{-6}$ Wb. This unit is named for Wilhelm Weber (1804–1890), a German physicist.

To convert microwebers to lines, multiply by the conversion factor 10^8 lines per weber, as follows:

$$1 \ \mu\text{Wb} = 1 \times 10^{-6} \ \text{Wb} \times 10^8 \ \frac{\text{lines}}{\text{Wb}}$$

$$= 1 \times 10^2 \ \text{lines}$$

$$1 \ \mu\text{Wb} = 100 \ \text{lines or Mx}$$

Note that the conversion is arranged to make the weber units cancel, since we want maxwell units in the answer. For the same 1-lb magnet producing the magnetic flux of 5000 Mx, this flux corresponds to 50 μWb.

Systems of Magnetic Units As explained in App. D, "Physics Units," units can be defined basically in two ways. The centimeter-gram-second (cgs) system defines small units. The meter-kilogram-second (mks) system is for larger units of a more practical size. Furthermore, the Système International (SI) units are in mks dimensions; these provide an international standard of practical units based on the ampere of current. More details of the SI units for electricity and magnetism are described on page 281. For the units of magnetic flux ϕ, the maxwell is a cgs unit, while the weber is an mks or SI unit.

Practice Problems 13-2
Answers at End of Chapter

The magnetic flux ϕ of 2000 lines is
a. How many Mx?
b. How many μWb?

13-3 Flux Density B

As shown in Fig. 13-5, the flux density is the number of magnetic field lines per unit area of a section perpendicular to the direction of flux. As a formula,

$$B = \frac{\phi}{A} \tag{13-1}$$

where ϕ is the flux through an area A, and the flux density is B.

The Gauss In the cgs system, this unit is one line per square centimeter, or 1 Mx/cm^2. As an example, in Fig. 13-5, the total flux ϕ is 6 lines, or 6 Mx. At point P in this field, however, the flux density B is 2 G because there are 2 lines per cm^2. The flux density has a higher value close to the poles, where the flux lines are more crowded.

As an example of flux density, B for a 1-lb magnet would be 1000 G at the poles. This unit is named for Karl F. Gauss (1777–1855), a German mathematician.

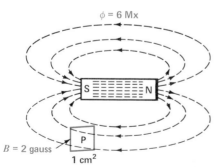

Fig. 13-5 Total flux ϕ is 6 lines or Mx. Flux density B at point P is 2 lines per square centimeter or 2 G.

> **Example 1** With a flux of 10,000 Mx through a perpendicular area of 5 cm^2, what is the flux density in gauss?
>
> **Answer** $\quad B = \dfrac{\phi}{A} = \dfrac{10{,}000 \ \text{Mx}}{5 \ \text{cm}^2} = 2000 \ \dfrac{\text{Mx}}{\text{cm}^2}$
>
> $\quad B = 2000 \ \text{G}$

As typical values, B for the earth's magnetic field can be about 0.2 G; a large laboratory magnet produces B of 50,000 G. Since the gauss is so small, it is often used in kilogauss units, where 1 kG = 10^3 G.

The Tesla   In SI, the unit of flux density B is webers per square meter (Wb/m^2). One weber per square meter is called a *tesla*, abbreviated T. This unit is named for Nikola Tesla (1857–1943), a Yugoslav-born American inventor in electricity and magnetism.

When converting between cgs and mks units, note that

$$1 \text{ m} = 100 \text{ cm}$$
$$1 \text{ m}^2 = 10,000 \text{ or } 10^4 \text{ cm}^2$$
$$1 \text{ cm}^2 = 0.0001 \text{ or } 10^{-4} \text{ m}^2$$

As an example, 5 cm^2 is 0.0005 m^2.

Example 2   With a flux of 400 μWb through an area of 0.0005 m^2, what is the flux density in tesla units?

Answer $B = \dfrac{\phi}{A} = \dfrac{400 \times 10^{-6} \text{ Wb}}{5 \times 10^{-4} \text{ m}^2}$

$\qquad\qquad = \dfrac{400}{5} \times 10^{-2} = 80 \times 10^{-2} \dfrac{\text{Wb}}{\text{m}^2}$

$\qquad B = 0.80 \text{ T}$

The tesla is a larger unit than the gauss, as 1 T = 1×10^4 G. For example, the flux density of 20,000 G is equal to 2 T.

It should be noted that all these units for flux or flux density apply to the magnetic field produced either by a permanent magnet (PM) or by an electromagnet (EM).

Comparison of Flux and Flux Density   Remember that the flux ϕ includes total area, while the flux density B is for a specified unit area. The difference between ϕ and B is illustrated in Fig. 13-6 with cgs units. The total area A here is 9 cm^2, equal to 3 cm × 3 cm. For one unit box of 1 cm^2, 16 lines are shown. Therefore, the flux density B is 16 lines or maxwells per cm^2, which equals 16 G. The total area includes nine of these boxes. Therefore, the total flux ϕ is 144 lines or maxwells, equal to 9 × 16 for $B \times A$.

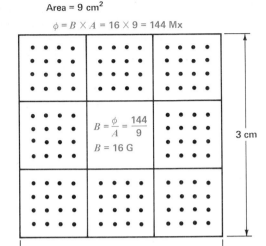

Fig. 13-6 Comparison of total flux ϕ and flux density B. Total area of 9 cm^2 has 144 lines or Mx. For 1 cm^2 the flux density is 144 Mx/9 cm^2 = 16 G.

For the opposite case, if the total flux ϕ is given as 144 lines or maxwells, the flux density is found by dividing 144 by 9 cm^2. This division of $^{144}/_9$ equals 16 lines or maxwells per cm^2, which is 16 G.

Practice Problems 13-3
Answers at End of Chapter

a. The ϕ is 9000 Mx through 3 cm^2. How much is B in gauss units?

b. How much is B in tesla units for ϕ of 90 μWb through 0.0003 m^2?

13-4   Induction by the Magnetic Field

The electrical effect of one body on another without any physical contact between them is called *induction*. For instance, a permanent magnet can induce an unmagnetized iron bar to become a magnet, without the two touching. The iron bar then becomes a magnet, as shown in Fig. 13-7. What happens is that the magnetic lines of force generated by the permanent magnet make the internal molecular magnets in the iron bar line up in

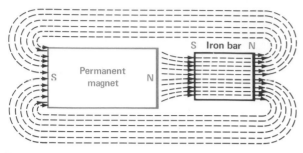

Fig. 13-7 Magnetizing an iron bar by induction.

the same direction, instead of the random directions in unmagnetized iron. The magnetized iron bar then has magnetic poles at the ends, as a result of the magnetic induction.

Note that the induced poles in the iron have opposite polarity from the poles of the magnet. Since opposite poles attract, the iron bar will be attracted. Any magnet attracts to itself all magnetic materials by induction.

Although the two bars in Fig. 13-7 are not touching, the iron bar is in the magnetic flux of the permanent magnet. It is the invisible magnetic field that links the two magnets, enabling one to affect the other. Actually, this idea of magnetic flux extending outward from the magnetic poles is the basis for many inductive effects in ac circuits. More generally, the magnetic field between magnetic poles and the electric field between electric charges form the basis for wireless radio transmission and reception.

Polarity of Induced Poles   Note that the north pole of the permanent magnet in Fig. 13-7 induces an opposite south pole at this end of the iron bar. If the permanent magnet were reversed, its south pole would induce a north pole. The closest induced pole will always be of opposite polarity. This is the reason why either end of a magnet can attract another magnetic material to itself. No matter which pole is used, it will induce an opposite pole, and the opposite poles are attracted.

Relative Permeability   Soft iron, as an example, is very effective in concentrating magnetic field lines, by induction in the iron. This ability to concentrate magnetic flux is called *permeability*. Any material that is easily magnetized has high permeability, therefore, as the field lines are concentrated because of induction.

Numerical values of permeability for different materials compared with air or vacuum can be assigned. For example, if the flux density in air is 1 G but an iron core in the same position in the same field has a flux density of 200 G, the relative permeability of the iron core equals $^{200}/_1$, or 200.

The symbol for relative permeability is μ_r (mu), where the subscript r indicates relative permeability. Typical values for μ_r are 100 to 9000 for iron and steel. There are no units, because μ_r is a comparison of two flux densities and the units cancel. The symbol K_m may also be used for relative permeability, to indicate this characteristic of a material for a magnetic field, corresponding to K_ϵ for an electric field.

Practice Problems 13-4
Answers at End of Chapter

Answer True or False.
a. Induced poles always have opposite polarity from the inducing poles.
b. The relative permeability of air or vacuum is approximately 300.

13-5   Air Gap of a Magnet

As shown in Fig. 13-8, the air space between poles of a magnet is its air gap. The shorter the air gap, the stronger the field in the gap for a given pole strength. Since air is not magnetic and cannot concentrate magnetic lines, a larger air gap only provides additional space for the magnetic lines to spread out.

Referring to Fig. 13-8a, note that the horseshoe magnet has more crowded magnetic lines in the air gap, compared with the widely separated lines around the bar magnet in Fig. 13-8b. Actually, the horseshoe magnet can be considered as a bar magnet bent around to place the opposite poles closer. Then the magnetic lines of the poles reinforce each other in the air gap. The purpose of a short air gap is to concentrate the magnetic field outside the magnet, for maximum induction in a magnetic material placed in the gap.

Ring Magnet without Air Gap   When it is desired to concentrate magnetic lines within a magnet, however, the magnet can be formed as a closed mag-

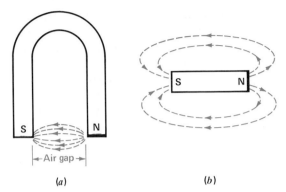

Fig. 13-8 The horseshoe magnet in (a) has a smaller air gap than the bar magnet in (b).

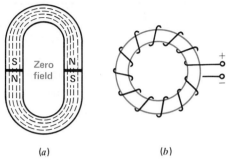

Fig. 13-9 Examples of a closed magnetic ring without any air gap. (a) Two PM horseshoe magnets with opposite poles touching. (b) Toroid electromagnet.

netic loop. This method is illustrated in Fig. 13-9a by the two permanent horseshoe magnets placed in a closed loop with opposite poles touching. Since the loop has no open ends, there can be no air gap and no poles. The north and south poles of each magnet cancel as opposite poles touch.

Each magnet has its magnetic lines inside, plus the magnetic lines of the other magnet, but outside the magnets the lines cancel because they are in opposite directions. The effect of the closed magnetic loop, therefore, is maximum concentration of magnetic lines in the magnet with minimum lines outside.

The same effect of a closed magnetic loop is obtained with the *toroid* or ring magnet in Fig. 13-9b, made in the form of a doughnut. Iron is often used for the core. This type of electromagnet has maximum strength in the iron ring, with little flux outside. As a result, the toroidal magnet is less sensitive to induction from external magnetic fields and, conversely, has little magnetic effect outside the coil.

It should be noted that, even if the winding is over only a small part of the ring, practically all the flux is in

the iron core because its permeability is so much greater than that of air. The small part of the field in the air is called *leakage flux*.

Keeper for a Magnet   The principle of the closed magnetic ring is used to protect permanent magnets in storage. In Fig. 13-10a, four PM bar magnets are in a closed loop, while Fig. 13-10b shows a stacked pair. Additional even pairs can be stacked this way, with opposite poles touching. The closed loop in Fig. 13-10c shows one permanent horseshoe magnet with a soft-iron *keeper* across the air gap. The keeper maintains the strength of the permanent magnet as it becomes magnetized by induction to form a closed loop. Then any external magnetic field is concentrated in the closed loop without inducing opposite poles in the permanent magnet. If permanent magnets are not stored this way, the polarity can be reversed with induced poles produced by a strong external field from a dc source; an alternating field can demagnetize the magnet.

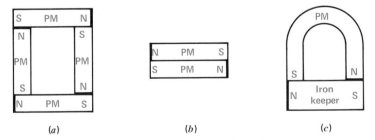

Fig. 13-10 Storing permanent magnets in a closed loop, with opposite poles touching. (a) Four bar magnets. (b) Two bar magnets. (c) Horseshoe magnet with iron keeper across air gap.

Practice Problems 13-5
Answers at End of Chapter

Answer True or False.
a. A short air gap has a stronger field than a large air gap, for the same magnetizing force.
b. A toroid magnet has no air gap.

13-6   Types of Magnets

The two broad classes are permanent magnets and electromagnets. An electromagnet needs current from an external source to maintain its magnetic field. With a permanent magnet, not only is its magnetic field present without any external current, but the magnet can maintain its strength indefinitely. Mechanical force has no effect. However, extreme heat can cause demagnetization.

Electromagnets   Current in a wire conductor has an associated magnetic field. If the wire is wrapped in the form of a coil, as in Fig. 13-11, the current and its magnetic field become concentrated in a smaller space, resulting in a stronger field. With the length much greater than its width, the coil is called a *solenoid*. It acts like a bar magnet, with opposite poles at the ends.

More current and more turns make a stronger magnetic field. Also, the iron core concentrates magnetic lines inside the coil. Soft iron is generally used for the core because it is easily magnetized and demagnetized.

The coil in Fig. 13-11, with the switch closed and current in the coil, is an electromagnet that can pick up the steel nail shown. If the switch is opened, the magnetic field is reduced to zero, and the nail will drop off. This ability of an electromagnet to provide a strong magnetic force of attraction that can be turned on or off easily has many applications in lifting magnets, buzzers, bells or chimes, and relays. A relay is a switch with contacts that are opened or closed by an electromagnet.

Another common application is magnetic tape recording. The tape is coated with fine particles of iron oxide. The recording head is a coil that produces a magnetic field in proportion to the current. As the tape passes through the air gap of the head, small areas of the coating become magnetized by induction. On playback, the moving magnetic tape produces variations in electric current.

Permanent Magnets   These are made of hard magnetic materials, such as cobalt steel, magnetized by induction in the manufacturing process. A very strong field is needed for induction in these materials. When the magnetizing field is removed, however, residual induction makes the material a permanent magnet. A common PM material is *alnico,* a commercial alloy of aluminum, nickel, and iron, with cobalt, copper, and titanium added to produce about 12 grades. The Alnico V grade is often used for PM loudspeakers (Fig. 13-12). In this application, a typical size of PM slug for a steady magnetic field is a few ounces to about 5 lb, with a flux B of 500 to 25,000 lines or maxwells. One advantage of a PM loudspeaker is that only two connecting leads are needed for the voice coil, as the steady magnetic field of the PM slug is obtained without any field-coil winding.

Commercial permanent magnets will last indefinitely if they are not subjected to high temperatures, to physical shock, or to a strong demagnetizing field. If the magnet becomes hot, however, the molecular structure can be rearranged, resulting in loss of magnetism that is not recovered after cooling. The point at which a magnetic material loses its ferromagnetic properties is the *Curie temperature*. For iron, this temperature is about 800°C, when the relative permeability drops to unity. A permanent magnet does not become exhausted with use, as its magnetic properties are determined by the structure of the internal atoms and molecules.

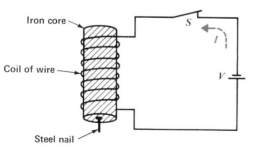

Fig. 13-11 Electromagnet holding nail when switch S is closed for current in coil.

Iron core

Coil of wire

Steel nail

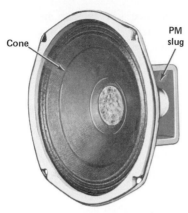

Fig. 13-12 Example of a PM loudspeaker.

Classification of Magnetic Materials

When we consider materials simply as either magnetic or nonmagnetic, this division is really based on the strong magnetic properties of iron. However, weak magnetic materials can be important in some applications. For this reason, a more exact classification includes the following three groups:

1. *Ferromagnetic materials.* These include iron, steel, nickel, cobalt, and commercial alloys such as alnico and Permalloy. They become strongly magnetized in the same direction as the magnetizing field, with high values of permeability from 50 to 5000. Permalloy has μ_r of 100,000 but is easily saturated at relatively low values of flux density.
2. *Paramagnetic materials.* These include aluminum, platinum, manganese, and chromium. The permeability is slightly more than 1. They become weakly magnetized in the same direction as the magnetizing field.
3. *Diamagnetic materials.* These include bismuth, antimony, copper, zinc, mercury, gold, and silver. The permeability is less than 1. They become weakly magnetized, but in the opposite direction from the magnetizing field.

The basis of all magnetic effects is the magnetic field associated with electric charges in motion. Within the atom, the motion of its orbital electrons generates a magnetic field. There are two kinds of electron motion in the atom. First is the electron revolving in its orbit. This motion provides a diamagnetic effect. However, this magnetic effect is weak because thermal agitation at normal room temperature results in random directions of motion that neutralize each other.

More effective is the magnetic effect from the motion of each electron spinning on its own axis. The spinning electron serves as a tiny permanent magnet. Opposite spins provide opposite polarities. Two electrons spinning in opposite directions form a pair, neutralizing the magnetic fields. In the atoms of ferromagnetic materials, however, there are many unpaired electrons with spins in the same direction, resulting in a strong magnetic effect.

In terms of molecular structure, iron atoms are grouped in microscopically small arrangements called *domains.* Each domain is an elementary *dipole magnet,* with two opposite poles. In crystal form, the iron atoms have domains that are parallel to the axes of the crystal. Still, the domains can point in different directions, because of the different axes. When the material becomes magnetized by an external magnetic field, though, the domains become aligned in the same direction. With PM materials, the alignment remains after the external field is removed.

Practice Problems 13-6
Answers at End of Chapter

Answer True or False.
a. An electromagnet needs current to maintain its magnetic field.
b. A relay coil is an electromagnet.
c. Steel is a diamagnetic material.

13-7   Ferrites

This is the name for recently developed nonmetallic materials that have the ferromagnetic properties of iron. The ferrites have very high permeability, like iron. However, a ferrite is a ceramic material, while iron is a conductor. The permeability of ferrites is in the range of 50 to 3000. The specific resistance is 10^5 $\Omega \cdot$cm, which makes the ferrite an insulator.

A common application is a ferrite core, usually adjustable, in the coils for RF transformers. The ferrite core is much more efficient than iron when the current alternates at a high frequency. The reason is that less I^2R power is lost by eddy currents in the core because of its very high resistance.

A ferrite core is used in small coils and transformers for signal frequencies up to 20 MHz, approximately. The high permeability means the transformer can be very small. However, the ferrites are easily saturated at low values of magnetizing current. This disadvantage means the ferrites are not used for power transformers.

Another application is in ferrite beads (Fig. 13-13). A bare wire is used as a string for one or more beads. The bead concentrates the magnetic field of the current in the wire. This construction serves as a simple, economical RF choke, instead of a coil. The purpose of the choke is to reduce the current just for an undesired radio frequency.

Practice Problems 13-7
Answers at End of Chapter

a. Which has more R, the ferrites or soft iron?
b. Which has more I^2R losses, an insulator or a conductor?

13-8   Magnetic Shielding

The idea of preventing one component from affecting another through their common electric or magnetic field is called *shielding*. Examples are the braided copper-wire shield around the inner conductor of a coaxial cable, a metal shield can that encloses an RF coil, or a shield of magnetic material enclosing a cathode-ray tube.

The problem in shielding is to prevent one component from inducing an effect in the shielded component. The shielding materials are always metals, but there is a difference between using good conductors

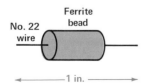

No. 22 wire — Ferrite bead

|← 1 in. →|

Fig. 13-13 Ferrite bead equivalent to coil with 20 μH of inductance at 10 MHz.

with low resistance like copper and aluminum and using good magnetic materials like soft iron.

A good conductor is best for two shielding functions. One is to prevent induction of static electric charges. The other is to shield against the induction of a varying magnetic field. For static charges, the shield provides opposite induced charges, which prevent induction inside the shield. For a varying magnetic field, the shield has induced currents that oppose the inducing field. Then there is little net field strength to produce induction inside the shield.

The best shield for a steady magnetic field is a good magnetic material of high permeability. A steady field is produced by a permanent magnet, a coil with steady direct current, or the earth's magnetic field. A magnetic shield of high permeability concentrates the magnetic flux. Then there is little flux to induce poles in a component inside the shield. The shield can be considered as a short circuit for the lines of magnetic flux.

Practice Problems 13-8
Answers at End of Chapter

Answer True or False
a. Magnetic material with high permeability is a good shield for a steady magnetic field.
b. A conductor is a good shield against a varying magnetic field.

13-9   The Hall Effect

In 1879, E. H. Hall observed that a small voltage is generated across a conductor carrying current in an external magnetic field. The Hall voltage was very small with typical conductors, and little was made of this effect. However, with the development of semiconductors, larger values of Hall voltage can be generated. The semiconductor material indium arsenide (InAs) is generally used. As illustrated in Fig. 13-14, the InAs element inserted in the magnetic field can generate 60 mV with B equal to 10 kG and an I of 100 mA. The applied flux must be perpendicular to the direction of current. With current in the direction of the length of conductor, the generated voltage is developed across the width.

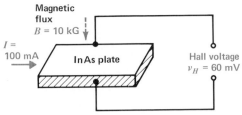

Fig. **13-14** The Hall effect. The voltage v_H generated across the element is proportional to the perpendicular flux density B.

Fig. **13-15** Gaussmeter to measure flux density, with probe containing indium arsenide element. (*RFL Industries Inc.*)

The amount of Hall voltage v_H is directly proportional to the value of flux density B. This means that values of B can be measured by means of v_H. As an example, the gaussmeter in Fig. 13-15 uses an InAs probe in the magnetic field to generate a proportional Hall voltage v_H. This value of v_H is then read by the meter, which is calibrated in gauss. The original calibration is made in terms of a reference magnet with a specified flux density.

Practice Problems 13-9
Answers at End of Chapter

a. In Fig. 13-14, how much is the generated Hall voltage?
b. Does the gaussmeter in Fig. 13-15 measure flux or flux density?

Summary

1. Iron, nickel, and cobalt are common examples of magnetic materials. Air, paper, wood, and plastics are nonmagnetic, meaning there is no effect with a magnetic field.
2. The pole of a magnet that seeks the magnetic north pole of the earth is called a north pole; the opposite pole is a south pole.
3. Opposite magnetic poles have a force of attraction; similar poles repel.
4. An electromagnetic needs current from an external source to provide a magnetic field. Permanent magnets retain their magnetism indefinitely.
5. Any magnet has an invisible field of force outside the magnet, indicated by magnetic field lines. Their direction is from the north to the south pole outside the magnet.
6. The open ends of a magnet where it meets a nonmagnetic material provide magnetic poles. At opposite open ends, the poles have opposite polarity.
7. A magnet with an air gap has opposite poles with magnetic lines of force across the gap. A closed magnetic ring has no poles. Practically all the magnetic lines are in the ring.
8. Magnetic induction enables the field of a magnet to induce magnetic poles in a magnetic material without touching.
9. Permeability is the ability to concentrate magnetic flux. A good magnetic material has high permeability, similar to the idea of high conductance for a good conductor of electricity.
10. Magnetic shielding means isolating a component from a magnetic field. The best shield against a steady magnetic field is a material with high permeability.

Table 13-1. Magnetic Flux ϕ and Flux Density B

Name	Symbol	CGS units	MKS or SI units
Flux, or total lines	$\phi = B \times$ area	1 maxwell (Mx) = 1 line	1 weber (Wb) = 10^8 Mx
Flux density, or lines per unit area	$B = \dfrac{\phi}{\text{area}}$	1 gauss (G) = $\dfrac{1 \text{ Mx}}{\text{cm}^2}$	1 tesla (T) = $\dfrac{1 \text{ Wb}}{\text{m}^2}$

11. The Hall voltage is a small emf generated across the width of a conductor carrying current through its length, when magnetic flux is applied perpendicular to the current. This effect is generally used in the gaussmeter to measure flux density.

12. Table 13-1 summarizes the units of magnetic flux ϕ and flux density B.

Self-Examination
Answers at Back of Book

Answer True or False.

1. Iron and steel are ferromagnetic materials with high permeability.
2. Ferrites are ferromagnetic but have high electrical resistance.
3. Air, vacuum, wood, paper, and plastics have practically no effect on magnetic flux.
4. Aluminum is ferromagnetic.
5. Magnetic poles exist on opposite sides of an air gap.
6. A closed magnetic ring has no poles and no air gap.
7. A magnet can pick up a steel nail by magnetic induction.
8. Induced poles are always opposite from the original field poles.
9. Soft iron concentrates magnetic flux by means of induction.
10. Without current, an electromagnet has practically no magnetic field.
11. The total flux ϕ of 5000 lines equals 5 Mx.
12. A flux ϕ of 5000 Mx through a cross-sectional area of 5 cm^2 has a flux density B of 1000 G or 1 kG.
13. The flux density B of 1000 G equals 1000 lines per cm^2.
14. A magnetic pole is a terminal where a magnetic material meets a nonmagnetic material.
15. High permeability for magnetic flux corresponds to high resistance for a conductor of current.

Essay Questions

1. Name two magnetic materials and three nonmagnetic materials.
2. Explain briefly the difference between a permanent magnet and an electromagnet.
3. Draw a horseshoe magnet, with its magnetic field. Label the magnetic poles, indicate the air gap, and show the direction of flux.

4. Define the following: relative permeability, shielding, induction, Hall voltage.
5. Give the symbol, cgs unit, and SI unit for magnetic flux and for flux density.
6. How would you determine the north and south poles of a bar magnet, using a magnetic compass?
7. Referring to Fig. 13-11, why can either end of the magnet pick up the nail?
8. What is the difference between flux ϕ and flux density B?

Problems
Answers to Odd-Numbered Problems at Back of Book

1. A magnetic pole produces 5000 field lines. How much is the flux ϕ in maxwells and webers?
2. If the area of the pole in Prob. 1 is 5 cm^2, calculate the flux density B in gauss units.
3. Calculate B in tesla units for a 200-μWb flux through an area of 5×10^{-4} m^2.
4. Convert 3000 G to tesla units.
5. For a flux density B of 3 kG at a pole with a cross-sectional area of 8 cm^2, how much is the total flux ϕ in maxwell units?
6. Convert 17,000 Mx to weber units.
7. The flux density is 0.002 T in the air core of an electromagnet. When an iron core is inserted, the flux density in the core is 0.6 T. How much is the relative permeability μ_r of the iron core?
8. Draw the diagram of an electromagnet operated from a 12-V battery, in series with a switch. (a) If the coil resistance is 60 Ω, how much is the current in the coil with the switch closed? (b) Why is the magnetic field reduced to zero when the switch is opened?
9. Derive the conversion of 1 μWb = 100 Mx from the fact that 1 μWb = 10^{-6} Wb and 1 Wb = 10^8 Mx.
10. Derive the relation 1 T = 10^4 G. (Note: 1 m^2 = 10,000 cm^2.)

Answers to Practice Problems

13-1 a. T
    b. T

13-2 a. 2000 Mx
    b. 20 μWb

13-3 a. 3000 G
    b. 0.3 T

13-4 a. T
    b. F

13-5 a. T
    b. T

13-6 a. T
    b. T
    c. F

13-7 a. Ferrites
    b. Conductor

13-8 a. T
    b. T

13-9 a. 60 mV
    b. Flux density

Chapter 14
Magnetic Units

A magnetic field is always associated with charges in motion. Therefore, the magnetic units can be derived from the current that produces the field.

The current in a conductor and its magnetic flux through the medium outside the conductor are related as follows: (1) The current I supplies a magnetizing force, or magnetomotive force (mmf), that increases with

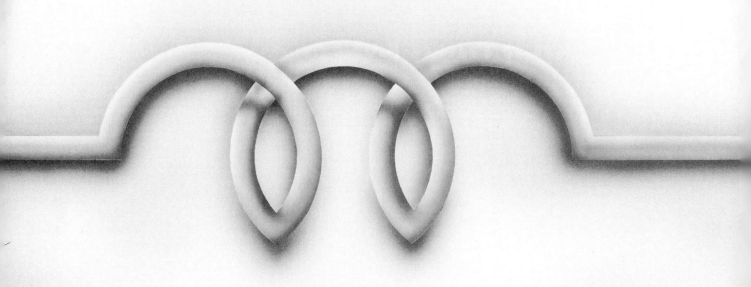

the amount of I. (2) The mmf results in a magnetic field intensity H that decreases with the length of conductor, as the field is less concentrated with more length. (3) The field intensity H produces a flux density B that increases with the permeability of the medium.

Important terms in this chapter are:

ampere-turns	International System of Units (SI)
ampere-turns/meter to oersteds	magnetomotive force (mmf)
Coulomb's law	permeability (μ or mu)
degaussing	reluctance
demagnetization	saturation
hysteresis	tesla

More details of magnetic units based on the practical ampere unit of current are explained in the following sections:

14-1 Ampere-turns (NI)
14-2 Field Intensity (H)
14-3 Permeability (μ)
14-4 B-H Magnetization Curve
14-5 Magnetic Hysteresis
14-6 Ohm's Law for Magnetic Circuits
14-7 Relations between Magnetic Units
14-8 Comparison of Magnetic and Electric Fields

14-1 Ampere-Turns (*NI*)

With a coil magnet, the strength of the magnetic field depends on how much current flows in the turns of the coil. The more current, the stronger the magnetic field. Also, more turns in a specific length concentrate the field. The coil serves as a bar magnet, with opposite poles at the ends, providing a magnetic field proportional to the ampere-turns. As a formula,

$$\text{Ampere-turns} = NI \qquad (14\text{-}1)$$

where *N* is the number of turns, multiplied by the current *I* in amperes. The quantity *NI* specifies the amount of *magnetizing force, magnetic potential,* or *magnetomotive force (mmf).*

The practical unit for *NI* is the ampere-turn. The SI abbreviation for ampere-turn is A, the same as for the ampere, since the number of turns in a coil usually is constant but the current can be varied. However, for clarity we shall use the abbreviation A·t.

As shown in Fig. 14-1, a solenoid with 5 turns and 2 amperes has the same magnetizing force as one with 10 turns and 1 ampere, as the product of the amperes and turns is 10 for both cases. With thinner wire, more turns can be used in a given space. The amount of current is determined by the resistance of the wire and the source voltage. How many ampere-turns are necessary depends on the required magnetic field strength.

> **Example 1** Calculate the ampere-turns for a coil with 2000 turns and a 5-mA current.

Answer $NI = 2000 \times 5 \times 10^{-3}$
$NI = 10$ A·t

Example 2 A coil with 4 amperes is to provide the magnetizing force of 600 A·t. How many turns are necessary?

Answer $N = \dfrac{NI}{I} = \dfrac{600}{4}$
$N = 150$ turns

Example 3 A coil with 400 turns must provide 800 A·t of magnetizing force. How much current is necessary?

Answer $I = \dfrac{NI}{N} = \dfrac{800}{400}$
$I = 2$ A

Example 4 The wire in a solenoid of 250 turns has a resistance of 3 Ω. (a) How much is the current with the coil connected to a 6-V battery? (b) Calculate the ampere-turns.

Answer

a. $I = \dfrac{V}{R} = \dfrac{6}{3}$
$I = 2$ A
b. $NI = 250 \times 2$
$NI = 500$ A·t

The ampere-turn is an SI unit. The cgs unit of mmf is the *gilbert*,[1] abbreviated Gb. One ampere-turn equals 1.26 Gb. The number 1.26 is approximately $4\pi/10$, derived from the surface area of a sphere, which is $4\pi r^2$.

To convert *NI* to gilberts, multiply the ampere-turns by the constant conversion factor 1.26 Gb/1 A·t. As an

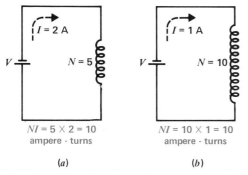

$NI = 5 \times 2 = 10$
ampere - turns

$NI = 10 \times 1 = 10$
ampere - turns

(a)   (b)

Fig. 14-1 Two examples of equal ampere-turns: (*a*) *NI* is $5 \times 2 = 10$. (*b*) *NI* is $10 \times 1 = 10$.

[1]William Gilbert (1540–1603) was an English scientist who investigated the magnetism of the earth.

example, 1000 A·t is the same mmf as 1260 Gb. The calculations are

$$1000 \text{ A·t} \times 1.26 \ \frac{\text{Gb}}{1 \text{ A·t}} = 1260 \text{ Gb}$$

Note that the units of A·t cancel in the conversion.

Practice Problems 14-1
Answers at End of Chapter

a. If *NI* is 243 A·t, and *I* is doubled from 2 to 4 A with the same number of turns, how much is the new *NI*?

b. Convert 500 A·t to gilberts.

14-2   Field Intensity (*H*)

The ampere-turns of mmf specify the magnetizing force, but the intensity of the magnetic field depends on how long the coil is. At any point in space, a specific value of ampere-turns must produce less field intensity for a long coil than for a short coil that concentrates the same *NI*. Specifically, the field intensity *H* in mks units is

$$H = \frac{NI \text{ ampere-turns}}{l \text{ meters}} \tag{14-2}$$

This formula is for a solenoid. *H* is the intensity at the center of an air core. With an iron core, *H* is the intensity through the entire core. By means of units for *H*, the magnetic field intensity can be specified for either electromagnets or permanent magnets, since both provide the same kind of magnetic field.

The length in Formula (14-2) is between poles. In Fig. 14-2*a*, the length is 1 m between the poles at the ends of the coil. In Fig. 14-2*b*, also, *l* is 1 m between the ends of the iron core. In Fig. 14-2*c*, though, *l* is 2 m between the poles at the ends of the iron core, although the winding is only 1 m long.

The examples in Fig. 14-2 illustrate the following comparisons:

1. In all three cases, the mmf equal to *NI* is 1000 A·t.

2. In Fig. 14-2*a* and *b*, *H* equals 1000 A·t/m. In *a*, this *H* is the intensity at the center of the air core; in *b*, this *H* is the intensity through the entire iron core.

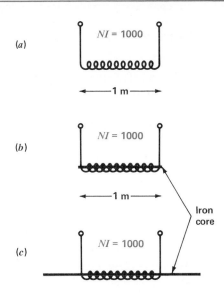

Fig. 14-2 Relation between ampere-turns of mmf and resultant field intensity H for different cores. Note that H = mmf/length. (a) Intensity H is 1000 A·t/m in air. (b) Intensity H is 1000 A·t/m in the iron core. (c) Intensity H is 1000/2 = 500 A·t/m in the longer iron core.

3. In Fig. 14-2*c*, because *l* is 2 m, *H* is 1000/2, or 500 A·t/m. This *H* is the intensity in the entire iron core.

Units for *H*   The field intensity is basically mmf per unit of length. In practical units, *H* is ampere-turns per meter. The cgs unit for *H* is the *oersted*,[1] abbreviated Oe, which equals one gilbert of mmf per centimeter.

Conversion of Units   To convert SI units of A·t/m to cgs units of Oe, multiply by the conversion factor 0.0126 Oe per 1 A·t/m. As an example, 1000 A·t/m is the same *H* as 12.6 Oe. The calculations are

$$1000 \ \frac{\text{A·t}}{\text{m}} \times 0.0126 \ \frac{\text{Oe}}{1 \text{ A·t/m}} = 12.6 \text{ Oe}$$

Note that the units of A·t and m cancel. The m in the conversion factor becomes inverted to the numerator.

---

[1]H. C. Oersted (1777–1851), a Danish physicist, discovered electromagnetism.

Practice Problems 14-2
Answers at End of Chapter

a. Suppose that H is 250 A·t/m. The length is doubled from 0.2 to 0.4 m for the same NI. How much is the new H?

b. Convert 500 A·t/m to oersted units.

14-3   Permeability (μ)

Whether we say H is 1000 A·t/m or 12.6 Oe, these units specify how much field intensity is available to produce magnetic flux. However, the amount of flux actually produced by H depends on the material in the field. A good magnetic material with high relative permeability can concentrate flux and produce a large value of flux density B for a specified H. These factors are related by the formula:

$$B = \mu \times H \qquad \text{(14-3)}$$

or

$$\mu = \frac{B}{H} \qquad \text{(14-4)}$$

Using SI units, B is the flux density in webers per square meter, or teslas; H is the field intensity in ampere-turns per meter. In the cgs system the units are gauss for B and oersted for H. The factor μ is the absolute permeability, not referred to any other material, in units of B/H.

In the cgs system the units of gauss for B and oersteds for H have been defined to give μ the value of 1 G/Oe, for vacuum, air, or space. This simplification means that B and H have the same numerical values in air and in vacuum. For instance, the field intensity H of 12.6 Oe produces the flux density of 12.6 G, in air.

Furthermore, the values of relative permeability μ_r are the same as those for absolute permeability in B/H units in the cgs system. The reason is that μ is 1 for air or vacuum, used as the reference for the comparison. As an example, if μ_r for an iron sample is 600, the absolute μ is also 600 G/Oe.

In SI, however, the permeability of air or vacuum is not 1. Specifically, this value is $4\pi \times 10^{-7}$, or

1.26×10^{-6}, with the symbol μ_0. Therefore, values of relative permeability μ_r must be multiplied by 1.26×10^{-6} for μ_0 to calculate μ as B/H in SI units.

For an example of $\mu_r = 100$, the SI value of μ can be calculated as follows:

$$\mu = \mu_r \times \mu_0$$
$$= 100 \times 1.26 \times 10^{-6} \frac{\text{T}}{\text{A·t/m}}$$
$$\mu = 126 \times 10^{-6} \frac{\text{T}}{\text{A·t/m}}$$

Example 5   A magnetic material has a μ_r of 500. Calculate the absolute μ as B/H (**a**) in cgs units, and (**b**) in SI units.

Answer

a. $\mu = \mu_r \times \mu_0$ in cgs units. Then

$$\mu = 500 \times 1 \frac{\text{G}}{\text{Oe}}$$
$$\mu = 500 \frac{\text{G}}{\text{Oe}}$$

b. $\mu = \mu_r \times \mu_0$ in SI units. Then

$$\mu = 500 \times 1.26 \times 10^{-6} \frac{\text{T}}{\text{A·t/m}}$$
$$\mu = 630 \times 10^{-6} \frac{\text{T}}{\text{A·t/m}}$$

Example 6   For this example of $\mu = 630 \times 10^{-6}$ in SI units, calculate the flux density B that will be produced by the field intensity H equal to 1000 A·t/m.

Answer

$$B = \mu H = \left(630 \times 10^{-6} \frac{\text{T}}{\text{A·t/m}}\right)\left(1000 \frac{\text{A·t}}{\text{m}}\right)$$
$$= 630 \times 10^{-3} \text{ T}$$
$$B = 0.63 \text{ T}$$

Note that the ampere-turns and meter units cancel, leaving only the tesla unit for the flux density B.

Practice Problems 14-3
Answers at End of Chapter

a. What is the value of μ_r for air, vacuum, or space?
b. An iron core has 200 times more flux density than air for the same field intensity H. How much is μ_r?
c. An iron core produces 200 G of flux density for 1 Oe of field intensity H. How much is μ?

14-4   B-H Magnetization Curve

The B-H curve in Fig. 14-3 is often used to show how much flux density B results from increasing the amount of field intensity H. This curve is for soft iron, plotted for the values in Table 14-1, but similar curves can be obtained for all magnetic materials.

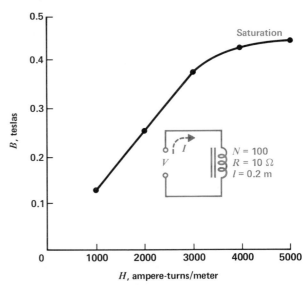

Fig. 14-3 B-H magnetization curve for soft iron. No values shown near zero, where μ may vary with previous magnetization.

Calculating H and B   The values in Table 14-1 are calculated as follows:

1. The current I in the coil equals V/R. For a 10-Ω coil resistance with 20 V applied, I is 2 A, as listed in the top row of Table 14-1. Increasing values of V produce more current in the coil.
2. The ampere-turns NI of magnetizing force increase with more current. Since the turns are constant at 100, the values of NI increase from 200 for 2 A in the top row to 1000 for 10 A in the bottom row.
3. The field intensity H increases with higher NI. The values of H are in mks units of ampere-turns per meter. These values equal $NI/0.2$, as the length is 0.2 m. Therefore, each NI is just divided by 0.2, or multiplied by 5, for the corresponding values of H. Since H increases in the same proportion as I, sometimes the horizontal axis on a B-H curve is calibrated only in amperes, instead of in H units.

4. The flux density B depends on the field intensity H and permeability of the iron. The values of B in the last column are obtained by multiplying $\mu \times H$. However, with SI units the values of μ_r listed must be multiplied by 1.26×10^{-6} to obtain $\mu \times H$ in teslas.

Saturation   Note that the permeability decreases for the highest values of H. With less μ, the iron core cannot provide proportional increases in B for increasing values of H. In the graph, for values of H above 4000 A·t/m, approximately, the values of B increase at a much slower rate, making the curve relatively flat at the top. The effect of little change in flux density when the field intensity increases is called *saturation*.

The reason is that the iron becomes saturated with magnetic lines of induction. After most of the molecu-

Table 14-1. B-H Values for Fig. 14-3

V, volts	R, ohms	I = V/R, amperes	N, turns	NI, A·t	l, m	H, A·t/m	μ_r	B = $\mu \times H$, T
20	10	2	100	200	0.2	1000	100	0.126
40	10	4	100	400	0.2	2000	100	0.252
60	10	6	100	600	0.2	3000	100	0.378
80	10	8	100	800	0.2	4000	85	0.428
100	10	10	100	1000	0.2	5000	70	0.441

lar dipoles and the magnetic domains are aligned by the magnetizing force, very little additional induction can be produced. When the value of μ is specified for a magnetic material, it is usually the highest value before saturation.

Practice Problems 14-4
Answers at End of Chapter

Refer to Fig. 14-3.
a. How much is B, in tesla units, for 1500 A·t/m?
b. What value of H starts to produce saturation?

14-5 Magnetic Hysteresis

Hysteresis means "a lagging behind." With respect to the magnetic flux in an iron core of an electromagnet, the flux lags the increases or decreases in magnetizing force. The hysteresis results from the fact that the magnetic dipoles are not perfectly elastic. Once aligned by an external magnetizing force, the dipoles do not return exactly to their original positions when the force is removed. The effect is the same as if the dipoles were forced to move against an internal friction between molecules. Furthermore, if the magnetizing force is reversed in direction by reversal of the current in an electromagnet, the flux produced in the opposite direction lags behind the reversed magnetizing force.

Hysteresis Loss When the magnetizing force reverses thousands or millions of times per second, as with rapidly reversing alternating current, the hysteresis can cause a considerable loss of energy. A large part of the magnetizing force is then used just to overcome the internal friction of the molecular dipoles. The work done by the magnetizing force against this internal friction produces heat. This energy wasted in heat as the molecular dipoles lag the magnetizing force is called *hysteresis loss*. For steel and other hard magnetic materials, the hysteresis losses are much higher than in soft magnetic materials like iron.

When the magnetizing force varies at a slow rate, the hysteresis losses can be considered negligible. An example is an electromagnet with direct current that is

simply turned on and off, or the magnetizing force of an alternating current that reverses 60 times per second or less. The faster the magnetizing force changes, however, the greater the hysteresis effect.

Hysteresis Loop To show the hysteresis characteristics of a magnetic material, its values of flux density B are plotted for a periodically reversing magnetizing force. See Fig. 14-4. This curve is the hysteresis loop of the material. The larger the area enclosed by the curve, the greater the hysteresis loss. The hysteresis loop is actually a *B-H* curve with an ac magnetizing force.

On the vertical axis, values of flux density B are indicated. The units can be gauss or teslas.

The horizontal axis indicates values of field intensity H. On this axis the units can be oersteds, ampere-turns per meter, ampere-turns, or just magnetizing current, as all factors are constant except I.

Opposite directions of current result in the opposite directions of $+H$ and $-H$ for the field lines. Similarly, opposite polarities are indicated for flux density as $+B$ or $-B$.

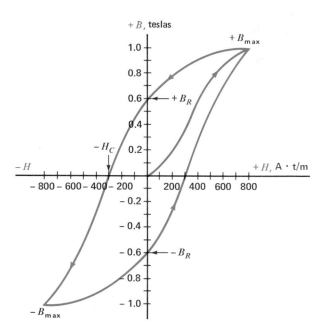

Fig. 14-4 Hysteresis loop for magnetic materials. This graph is a *B-H* curve like Fig. 14-3, but H alternates in polarity with alternating current.

The current starts from zero at the center, when the material is unmagnetized. Then positive H values increase B to saturation at $+B_{\max}$. Next H decreases to zero, but B drops to the value B_R, instead of to zero, because of hysteresis. When H becomes negative, B drops to zero and continues to $-B_{\max}$, which is saturation in the opposite direction from $+B_{\max}$ because of the reversed magnetizing current.

Then, as the $-H$ values decrease, the flux density is reduced to $-B_R$. Finally, the loop is completed, with positive values of H producing saturation at $B_{\max}$ again. The curve does not return to the zero origin at the center because of hysteresis. As the magnetizing force periodically reverses, the values of flux density are repeated to trace out the hysteresis loop.

The value of either $+B_R$ or $-B_R$, which is the flux density remaining after the magnetizing force has been reduced to zero, is the *residual induction* of a magnetic material, also called its *retentivity*. In Fig. 14-4, the residual induction is 0.6 T, in either the positive or negative direction.

The value of $-H_C$, which equals the magnetizing force that must be applied in the reverse direction to reduce the flux density to zero, is the *coercive force* of the material. In Fig. 14-4, the coercive force $-H_C$ is 300 A·t/m.

Demagnetization In order to demagnetize a magnetic material completely, the residual induction B_R must be reduced to zero. This usually cannot be accomplished by a reversed dc magnetizing force, because the material then would just become magnetized with opposite polarity. The practical way is to magnetize and demagnetize the material with a continuously decreasing hysteresis loop. This can be done with a magnetic field produced by alternating current. Then as the magnetic field and the material are moved away from each other, or the current amplitude is reduced, the hysteresis loop becomes smaller and smaller. Finally, with the weakest field, the loop collapses practically to zero, resulting in zero residual induction.

This method of demagnetization is also called *degaussing*. One application is degaussing the metal electrodes in a color picture tube, with a deguassing coil providing alternating current from the power line. Another example is erasing the recorded signal on magnetic tape by demagnetizing with an ac bias current.

The average level of the erase current is zero, and its frequency is much higher than the recorded signal.

Practice Problems 14-5
Answers at End of Chapter

Answer True or False.
a. Hysteresis loss increases with higher frequencies.
b. Degaussing is done with alternating current.

14-6 Ohm's Law for Magnetic Circuits

In comparison with electric circuits, the magnetic flux ϕ corresponds to current. The flux ϕ is produced by ampere-turns NI of magnetomotive force. Therefore, the mmf corresponds to voltage.

Opposition to the production of flux in a material is called its *reluctance*, comparable with resistance. The symbol for reluctance is $\mathcal{R}$. Reluctance is inversely proportional to permeability. Iron has high permeability and low reluctance. Air or vacuum has low permeability and high reluctance.

In Fig. 14-5, the ampere-turns of the coil produce magnetic flux throughout the magnetic path. The reluctance is the total opposition to the flux ϕ. In Fig. 14-5a, there is little reluctance in the closed iron path, and few ampere-turns are necessary. In Fig. 14-5b,

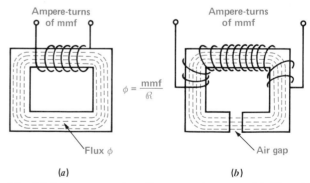

$$\phi = \frac{\text{mmf}}{\mathcal{R}}$$

(a) *(b)*

Fig. 14-5 Two examples of a magnetic circuit. (*a*) Closed iron path with low reluctance that requires little mmf. (*b*) Higher-reluctance path with air gap that requires more mmf.

however, the air gap has high reluctance, which requires many more ampere-turns for the same flux as in Fig. 14-5a.

The three factors—flux, ampere-turns, and reluctance—are related as follows:

$$\phi = \frac{\text{mmf}}{\mathcal{R}} \qquad (14\text{-}5)$$

which is known as Ohm's law for magnetic circuits, corresponding to $I = V/R$. The mmf is considered to produce flux ϕ in a magnetic material against the opposition of its reluctance $\mathcal{R}$. This relationship corresponds to emf or voltage producing current in a conducting material against the opposition of its resistance.

Remember that the units for the flux ϕ are maxwells and webers. These units measure total lines, as distinguished from flux density B, which equals lines per unit area.

There are no specific units for reluctance, but it can be considered as an mmf/ϕ ratio, just as resistance is a V/I ratio. Then $\mathcal{R}$ is ampere-turns per weber in SI units, or gilberts per maxwell in the cgs system.

The units for mmf are either gilberts in the cgs system or ampere-turns in SI.

Example 7   A coil has an mmf of 600 A·t and reluctance of 2×10^6 A·t/Wb. Calculate the total flux ϕ in microwebers.

Answer $\phi = \dfrac{\text{mmf}}{\mathcal{R}} = \dfrac{600 \text{ A·t}}{2 \times 10^6 \text{ A·t/Wb}}$

$\phi = 300 \times 10^{-6}$ Wb = 300 μWb

Example 8   A magnetic material has a total flux ϕ of 80 μWb with an mmf of 160 A·t. Calculate the reluctance in ampere-turns per weber.

Answer $\mathcal{R} = \dfrac{\text{mmf}}{\phi} = \dfrac{160 \text{ A·t}}{80 \times 10^{-6} \text{ Wb}}$

$\mathcal{R} = 2 \times 10^6 \dfrac{\text{A·t}}{\text{Wb}}$

Practice Problems 14-6
Answers at End of Chapter

Answer True or False.
a. Air has higher reluctance than soft iron.
b. More reluctance means more flux for a specified mmf.

14-7   Relations between Magnetic Units

The following examples show how the values of NI, H, ϕ, B, and $\mathcal{R}$ depend on each other. These calculations are in SI units, which are generally used for magnetic circuits.

Example 9   For a coil having 50 turns and 2 A, how much is the mmf?

Answer   mmf = NI = 50 $\times$ 2
            mmf = 100 A·t

The value of 100 A·t for NI is the mmf producing the magnetic field, with either an air core or an iron core.

Example 10   If this coil is on an iron core with a length of 0.2 m, how much is the field intensity H throughout the iron?

Answer $H = \dfrac{\text{mmf}}{l} = \dfrac{100 \text{ A·t}}{0.2 \text{ m}}$

$H = 500$ A·t/m

This is an example of calculating the field intensity of the external magnetic field from the mmf of the current in the coil.

Example 11   If this iron core with an H of 500 A·t/m has a relative permeability μ_r of 200, calculate the flux density B in teslas.

Answer

$$B = \mu H = \mu_r \times 1.26 \times 10^{-6} \times H$$

$$= 200 \times 1.26 \times 10^{-6} \frac{T}{A \cdot t/m} \times \frac{500\ A \cdot t}{m}$$

$$B = 0.126\ T$$

Example 12   For this iron core with a flux density B of 0.126 T, if its cross-sectional area is 2×10^{-4} m², calculate the amount of flux ϕ in the core.

Answer   Use the relations between flux and density: $\phi = B \times$ Area. Since $B = 0.126$ T, or 0.126 Wb/m², then

$$\phi = B \times \text{Area}$$

$$= 0.126\ \frac{Wb}{m^2} \times 2 \times 10^{-4}\ m^2$$

$$= 0.252 \times 10^{-4}\ Wb = 25.2 \times 10^{-6}\ Wb$$

$$\phi = 25.2\ \mu Wb$$

Example 13   With the mmf of 100 A·t for the coil in Fig. 14-5a and a value for ϕ of approximately 25×10^{-6} Wb in the iron core, calculate its reluctance $\mathcal{R}$.

Answer   Using Ohm's law for magnetic circuits,

$$\mathcal{R} = \frac{\text{mmf}}{\phi}$$

$$= \frac{100\ A \cdot t}{25 \times 10^{-6}\ Wb}$$

$$\mathcal{R} = 4 \times 10^6\ A \cdot t/Wb$$

Example 14   Refer to Fig. 14-5b. If the reluctance of the path (including the air gap) were 400×10^6 A·t/Wb, how much mmf would be required for the same flux of 25 μWb?

Answer

$$\text{mmf} = \phi \times \mathcal{R}$$

$$= 25 \times 10^{-6}\ Wb \times 400 \times 10^6\ \frac{A \cdot t}{Wb}$$

$$\text{mmf} = 10,000\ A \cdot t$$

Notice that the 10,000 A·t of mmf here is 100 times more than the 100 A·t in Example 13, because of the higher reluctance with an air gap. This idea corresponds to the higher voltage needed to produce the same current in a higher resistance.

Practice Problems 14-7
Answers at End of Chapter

Answer True or False.
a. More I in a coil produces more mmf for a specified number of turns.
b. More length for the coil produces more field intensity H for a specified mmf.
c. Higher permeability in the core produces more flux density B for a specified H.

14-8   Comparison of Magnetic and Electric Fields

As shown in Fig. 14-6a, there is an external field of lines of force between two electric charges, similar to the magnetic field between the magnetic poles in Fig. 14-6b. We cannot see the force of attraction and repulsion, just as the force of gravity is invisible, but the force is evident in the work it can do. For both fields, the force tends to make opposite polarities attract and similar polarities repel.

The electric lines show the path an electron would follow in the field; the magnetic lines show how a north pole would move. The entire group of electric lines of force of the static charges is called *electrostatic flux*. Its symbol is the Greek letter ψ (psi), corresponding to ϕ for magnetic flux.

In general, magnetic flux is associated with moving charges, or current, while electrostatic flux is associated with the voltage between static charges. For elec-

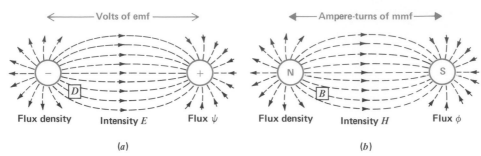

Fig. 14-6 Comparison of electric and magnetic fields. (*a*) Attraction between static charges of opposite polarity. (*b*) Attraction between opposite magnetic poles.

tric circuits, the application of magnetic flux is often a coil of wire, which is the construction of an inductor. With current, the wire has a magnetic field. As a coil, the wire's magnetic flux is concentrated in the coil. Furthermore, when the magnetic field varies, the change in magnetic flux produces an induced voltage, as explained in Chap. 15, "Electromagnetic Induction," and Chap. 17, "Inductance."

For the case of an electric field, the application is often an insulator between two conducting plates, which is the construction of a capacitor. With voltage across the insulator, it has an electric field. As a capacitor, the insulator's electric field is concentrated between the plates. Furthermore, when the electric field varies, the result is induced current through any conducting path connected to the capacitor. More details are explained in Chap. 20, "Capacitance."

Coulomb's Law
The electric lines of force in Fig. 14-6*a* illustrate the force on an electron in the field. The amount of force between two charges is given by Coulomb's law:

$$F = 9 \times 10^9 \times \frac{q_1 q_2}{r^2} \qquad (14\text{-}6)$$

where q_1 and q_2 are in coulomb units, F is in newtons, and r is the distance in meters between the charges. The constant factor 9×10^9 converts the values to SI units of newtons for the force in air or vacuum.

Coulomb's law states that the force increases with the amount of charge, but decreases as the square of the distance between charges. Typical values of q are in microcoulombs, since the coulomb is a very large unit of charge.

International System of Units
In order to provide a closer relation between practical units for electricity and magnetism, these mks units were standardized in 1960 by international agreement. The abbreviation is *SI*, for *Système International*. Table 14-2 lists the magnetic SI units. The corresponding electrical SI units include the coulomb, which is used for both electric flux and charge, the ampere for current, the volt for potential, and the ohm for resistance. The henry unit for inductance and the farad unit for capacitance are also SI units.

In Table 14-2, note that the reciprocal of reluctance is *permeance,* corresponding to conductance as the reciprocal of resistance. The SI unit for conductance is the siemens (S), equal to $1/\Omega$. This unit is named after Ernst von Siemens, a European inventor.

As another comparison, the permeability μ of a magnetic material with magnetic flux corresponds to the electric *permittivity* ϵ of an insulator with electric flux. Just as permeability is the ability of a magnetic material to concentrate magnetic flux, permittivity is the ability of an insulator to concentrate electric flux. The symbol K_ϵ is used for relative permittivity, corresponding to K_m for relative permeability.

Practice Problems 14-8
Answers at End of Chapter

Give the SI units for the following:
a. Voltage potential.
b. Magnetic potential.
c. Electric current.
d. Magnetic flux.

Summary

Table 14-2 summarizes the magnetic units and their definitions.

Table 14-2. International System of Mks Units (SI) for Magnetism

Quantity	Symbol	Unit
Flux	ϕ	Weber (Wb)
Flux density	B	Wb/m^2 = tesla (T)
Potential	mmf	Ampere-turn (A·t)
Field intensity	H	Ampere-turn per meter (A·t/m)
Reluctance	$\mathcal{R}$	Ampere-turn per weber (A·t/Wb)
Permeance	$\rho = \dfrac{1}{\mathcal{R}}$	Weber per ampere-turn (Wb/A·t)
Relative μ	μ_r or K_m	None, pure number
Permeability	$\mu = \mu_r \times 1.26 \times 10^{-6}$	$\dfrac{B}{H} = \dfrac{\text{tesla (T)}}{\text{ampere-turn per meter (A·t/m)}}$

Self-Examination
Answers at Back of Book

Answer True or False.

1. A current of 4 A through 200 turns provides an mmf of $NI = 200$ A·t.
2. For the mmf of 200 NI with 100 turns, a current of 2 A is necessary.
3. An mmf of 200 A·t across a flux path of 0.1 m provides a field intensity H of 2000 A·t/m.
4. A magnetic material with relative permeability μ_r of 100 has an absolute permeability μ of 126 G/Oe in cgs units.
5. There are no units for relative permeability μ_r.
6. Hysteresis losses are greater in soft iron than in air.
7. Magnetic saturation means that flux density B does not increase in proportion to increases in field intensity H.
8. The units for a B-H curve can be teslas plotted against ampere-turns.
9. In Ohm's law for magnetic circuits, reluctance $\mathcal{R}$ is the opposition to flux ϕ.
10. Ampere-turns of mmf between magnetic poles do not depend on the length of the coil.
11. The ampere-turn is a unit of magnetomotive force.
12. The tesla is an SI unit.

Essay Questions

1. In Ohm's law for magnetic circuits, what magnetic quantities correspond to V, I, and R?

2. Why can reluctance and permeability be considered opposite characteristics?

3. Give the SI magnetic unit and symbol for each of the following: (**a**) flux; (**b**) flux density; (**c**) field intensity; (**d**) absolute permeability.

4. What cgs units correspond to the following SI units? (**a**) weber; (**b**) tesla; (**c**) ampere-turn; (**d**) ampere-turn per meter.

5. Define the following: (**a**) saturation; (**b**) relative permeability; (**c**) relative permittivity.

6. Explain briefly how to demagnetize a metal object that has become temporarily magnetized.

7. Draw a *B-H* curve with μ, *N*, *L*, and *V* the same as in Fig. 14-3, but with a coil resistance of 5 Ω.

8. Give the formula for Coulomb's law for the force between electrostatic charges, with SI units.

9. What is meant by magnetic hysteresis? Why can hysteresis cause losses in an iron core with a reversing field produced by alternating current?

Problems
Answers to Odd-Numbered Problems at Back of Book

1. A coil of 2000 turns with a 100-mA current has a length of 0.4 m. (**a**) Calculate the mmf in ampere-turns. (**b**) Calculate the field intensity *H* in ampere-turns per meter.

2. If the current in the coil of Prob. 1 is increased to 400 mA, calculate the increased values of mmf and *H*.

3. An iron core has a flux density *B* of 3600 G with an *H* of 12 Oe. Calculate (**a**) the permeability μ in cgs units; (**b**) the permeability μ in SI units; (**c**) the relative permeability μ_r of the iron core.

4. A coil of 250 turns with a 400-mA current is 0.2 m long with an iron core of the same length. Calculate the following in mks units: (**a**) mmf; (**b**) *H*; (**c**) *B* in the iron core with a μ_r of 200; (**d**) *B* with an air core instead of the iron core.

5. Referring to the *B-H* curve in Fig. 14-3, calculate the μ in SI units for the iron core at an *H* of: (**a**) 3000 A·t/m; (**b**) 5000 A·t/m.

6. Referring to the hysteresis loop in Fig. 14-4, give the values of (**a**) residual induction B_R, and (**b**) coercive force $-H_C$.

7. A battery is connected across a coil of 100 turns and a 20-Ω *R*, with an iron core 0.2 m long. (**a**) Draw the circuit diagram. (**b**) How much battery voltage is needed for 200 A·t? (**c**) Calculate *H* in the iron core in ampere-turns per meter. (**d**) Calculate *B* in teslas in the iron core if its μ_r is 300. (**e**) Calculate ϕ in webers at each pole with an area of 8×10^{-4} m². (**f**) How much is the reluctance $\mathcal{R}$ of the iron core, in ampere-turns per weber?

8. In cgs units, how much is the flux density *B* in gauss, for a field intensity *H* of 24 Oe, with μ of 500?

9. Calculate the force, in newtons, between two 4-μC charges separated by 0.1 m in air or vacuum.

Answers to Practice Problems

14-1 **a.** 486
b. 630 Gb

14-2 **a.** 125
b. 6.3 Oe

14-3 **a.** 1
b. 200
c. 200 G/Oe

14-4 **a.** 0.2 T
b. 4000 A·t/m

14-5 **a.** T
b. T

14-6 **a.** T
b. F

14-7 **a.** T
b. F
c. T

14-8 **a.** Volt
b. Ampere-turn
c. Ampere
d. Weber

Chapter 15

Electromagnetic Induction

The link between electricity and magnetism was discovered in 1824 by Oersted, who found that current in a wire could move a magnetic compass needle. A few years later the opposite effect was discovered: A magnetic field in motion forces electrons to move, producing current. This important effect was studied by the English physicist and pioneer in electromagnetism, Michael Faraday (1791–1867); the American physicist, Joseph Henry (1797–1878); and the Russian physicist, H. F. E. Lenz (1804–1865).

Electrons in motion have an associated magnetic field; a moving magnetic field forces electrons to move, producing current. These electromagnetic effects have many practical applications that are the basis for motors and generators.

Important terms in this chapter are:

attraction	magnetic flux
Faraday's law of induced voltage	magnetic polarity
field intensity	motor action
generator action	repulsion
induced current and voltage	solenoid
left-hand rule	time rate of change
Lenz' law	torque

More details are explained in the following sections:

15-1 Magnetic Field Around an Electric Current
15-2 Magnetic Polarity of a Coil
15-3 Motor Action between Two Magnetic Fields
15-4 Induced Current
15-5 Lenz' Law
15-6 Generating an Induced Voltage
15-7 Faraday's Law of Induced Voltage

15-1   Magnetic Field Around an Electric Current

In Fig. 15-1, the iron filings aligned in concentric rings around the conductor show the magnetic field of the current in the wire. The iron filings are dense next to the conductor, showing that the field is strongest at this point. Furthermore, the field strength decreases inversely as the square of the distance from the conductor. It is important to note the following two factors about the magnetic lines of force:

1. The magnetic lines are circular, as the field is symmetrical with respect to the wire in the center.
2. The magnetic field with circular lines of force is in a plane perpendicular to the current in the wire.

From points c to d in the wire, its circular magnetic field is in the horizontal plane because the wire is vertical. Also, the vertical conductor between points ef and ab has the associated magnetic field in the horizontal plane. Where the conductor is horizontal, as from b to c and d to e, the magnetic field is in a vertical plane.

These two requirements of a circular magnetic field in a perpendicular plane apply to any charge in motion. Whether electron flow or a motion of positive charges is considered, the associated magnetic field must be at right angles to the direction of current.

In addition, the current need not be in a wire conductor. As an example, the beam of moving electrons in the vacuum of a cathode-ray tube has an associated magnetic field. In all cases, the magnetic field has circular lines of force in a plane perpendicular to the direction of motion of the electric charges.

Clockwise and Counterclockwise Fields

With circular lines of force, the magnetic field would tend to move a magnetic pole in a circular path. Therefore, the direction of the lines must be considered as either clockwise or counterclockwise. This idea is illustrated in Fig. 15-2, showing how a north pole would move in the circular field.

The directions are tested with a magnetic compass needle. When the compass is in front of the wire, the north pole on the needle points up. On the opposite side, the compass points down. If the compass were placed at the top, its needle would point toward the back of the wire; below the wire, the compass would point forward.

When all these directions are combined, the result is the circular magnetic field shown, with counterclockwise lines of force. This field has the magnetic lines upward at the front of the conductor and downward at the back.

Instead of testing every conductor with a magnetic compass, however, we can use the following rule to determine the circular direction of the magnetic field: *If you look along the wire in the direction of electron flow, the magnetic field is counterclockwise.* In Fig. 15-2, the line of electron flow is from left to right. Facing this way, you can assume the circular magnetic flux in a perpendicular plane has lines of force in the counterclockwise direction.

The opposite direction of current produces a reversed field. Then the magnetic lines of force have clockwise rotation. If the charges were moving from right to left in Fig. 15-2, the associated magnetic field would be in the opposite direction, with clockwise lines of force.

Fields Aiding or Canceling

When the magnetic lines of two fields are in the same direction, the lines of force aid each other, making the field stronger. With magnetic lines in opposite directions, the fields cancel.

In Fig. 15-3 the fields are shown for two conductors with opposite directions of current. The dot in the middle of the field at the left indicates the tip of an arrowhead to show current up from the paper. The cross symbolizes the back of an arrow to indicate current into the paper.

Notice that the magnetic lines *between the conductors* are in the same direction, although one field is clockwise and the other counterclockwise. Therefore, the fields aid here, making a stronger total field. On

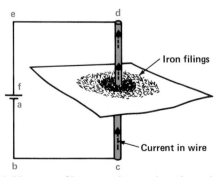

Fig. 15-1 How iron filings can be used to show the invisible magnetic field around current in a wire conductor.

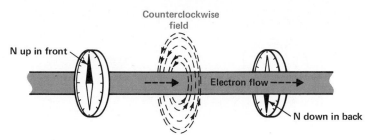

Fig. 15-2 Rule for determining direction of circular field around straight conductor. Field is counterclockwise for direction of electron flow shown here. Opposite electron flow would provide clockwise field.

either side of the conductors, the two fields are opposite in direction and tend to cancel each other. The net result, then, is to strengthen the field in the space between the conductors.

Practice Problems 15-1
Answers at End of Chapter

Answer True or False.
a. Magnetic field lines around a conductor are circular in a perpendicular plane.
b. In Fig. 15-3, the field is strongest between the conductors.

15-2 Magnetic Polarity of a Coil

Bending a straight conductor into the form of a loop, as shown in Fig. 15-4, has two effects. First, the magnetic field lines are more dense inside the loop. The total number of lines is the same as for the straight conductor, but inside the loop the lines are concentrated in a smaller space. Furthermore, all the lines inside the loop are aiding in the same direction. This makes the loop field effectively the same as a bar magnet with opposite poles at opposite faces of the loop.

Solenoid as a Bar Magnet A coil of wire conductor with more than one turn is generally called a *solenoid*. An ideal solenoid, however, has a length much greater than its diameter. Like a single loop, the solenoid concentrates the magnetic field inside the coil and provides opposite magnetic poles at the ends. These effects are multiplied, however, by the number of turns as the magnetic field lines aid each other in the same direction inside the coil. Outside the coil, the field corresponds to a bar magnet with north and south poles at opposite ends, as illustrated in Fig. 15-5.

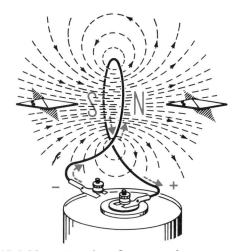

Fig. 15-4 Magnetic poles of a current loop.

Fig. 15-3 Magnetic fields aiding between parallel conductors with opposite directions of current.

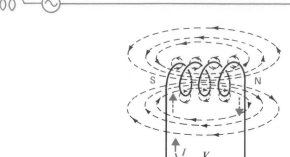 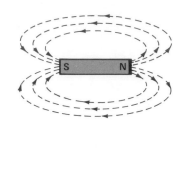

Fig. 15-5 Magnetic poles of a solenoid. (a) Coil winding. (b) Equivalent bar magnet.

Magnetic Polarity   To determine the magnetic polarity, use the *left-hand rule* illustrated in Fig. 15-6: *If the coil is grasped with the fingers of the left hand curled around the coil in the direction of electron flow, the thumb points to the north pole of the coil.* The left hand is used here because the current is electron flow.

The solenoid acts like a bar magnet whether it has an iron core or not. Adding an iron core increases the flux density inside the coil. In addition, the field strength then is uniform for the entire length of the core. The polarity is the same, however, for air-core and iron-core coils.

The magnetic polarity depends on the direction of current flow and the direction of winding. The current is determined by the connections to the voltage source. Electron flow is from the negative side of the voltage source, through the coil, and back to the positive terminal.

The direction of winding can be over and under,

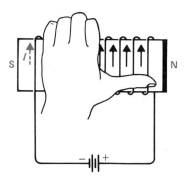

Fig. 15-6 Left-hand rule for north pole of a coil with current in it. Current *I* is electron flow.

starting from one end of the coil, or under and over with respect to the same starting point. Reversing either the direction of winding or the direction of current reverses the magnetic poles of the solenoid. See Fig. 15-7. With both reversed, though, the polarity is the same.

Practice Problems 15-2
Answers at End of Chapter

a. In Fig. 15-5, if the battery is reversed, will the north pole be at the left or the right?
b. If one end of a solenoid is a north pole, is the opposite end a north or a south pole?

15-3   Motor Action between Two Magnetic Fields

The physical motion resulting from the forces of magnetic fields is called *motor action*. One example is the simple attraction or repulsion between bar magnets.

We know that like poles repel and unlike poles attract. It can also be considered that fields in the same direction repel and opposite fields attract.

Consider the repulsion between two north poles, illustrated in Fig. 15-8. Similar poles have fields in the same direction. Therefore, the similar fields of the two like poles repel each other.

A more fundamental reason for motor action, however, is the fact that the force in a magnetic field tends

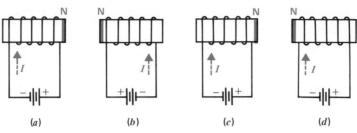

(a)	(b)	(c)	(d)

Fig. 15-7 Examples for determining the magnetic polarity of a coil. The polarities are opposite in (a) and (b) because the battery is reversed to reverse the direction of current. Also, (d) is the opposite of (c) because of the reversed winding.

to produce motion from a stronger field toward a weaker field. In Fig. 15-8, note that the field intensity is greatest in the space between the two north poles. Here the field lines of similar poles in both magnets reinforce in the same direction. Farther away the field intensity is less, for essentially one magnet only. As a result there is a difference in field strength, providing a net force that tends to produce motion. The direction of motion is always toward the weaker field.

To remember the directions, we can consider that the stronger field moves to the weaker field, tending to equalize the field intensity. Otherwise, the motion would make the strong field stronger and the weak field weaker. This must be impossible, because then the magnetic field would multiply its own strength without any work being added.

Force on a Straight Conductor in a Magnetic Field Current in a conductor has its associated magnetic field. When this conductor is placed in another magnetic field from a separate source, the two fields can react to produce motor action. The conductor must be perpendicular to the magnetic field, however, as illustrated in Fig. 15-9. This way, the perpendicular magnetic field of the current then is in the same plane as the external magnetic field.

Unless the two fields are in the same plane, they cannot affect each other. In the same plane, however, lines of force in the same direction reinforce to make a stronger field, while lines in the opposite direction cancel and result in a weaker field.

To summarize these directions:

1. With the conductor at 90°, or perpendicular to the external field, the reaction between the two magnetic fields is maximum.

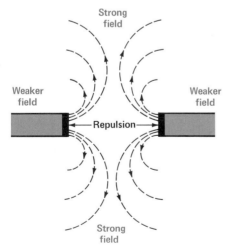

Fig. 15-8 Repulsion between similar poles of two bar magnets. The motion is from stronger field to weaker field.

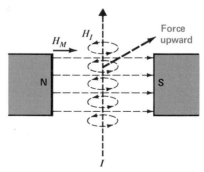

Fig. 15-9 Motor action of current in a straight conductor when it is in an external magnetic field. The H_I is the circular field of the current. The H_M indicates field lines between north and south poles of the external magnet.

2. With the conductor at 0°, or parallel to the external field, there is no effect between them.

3. When the conductor is at an angle between 0 and 90°, only the perpendicular component is effective.

In Fig. 15-9, electrons flow in the wire conductor in the plane of the paper, from the bottom to the top of the page. This flow provides the counterclockwise field H_I around the wire, in a perpendicular plane cutting through the paper. The external field H_M has lines of force from left to right in the plane of the paper. Then lines of force in the two fields are parallel above and below the wire.

Below the conductor, its field lines are left to right in the same direction as the external field. Therefore, these lines reinforce to produce a stronger field. Above the conductor the lines of the two fields are in opposite directions, causing a weaker field. As a result, the net force of the stronger field makes the conductor move upward out of the page, toward the weaker field.

If electrons flow in the reverse direction in the conductor, or if the external field is reversed, the motor action will be in the opposite direction. Reversing both the field and the current, however, results in the same direction of motion.

Rotation of a Current Loop in a Magnetic Field

With a loop of wire in the magnetic field, opposite sides of the loop have current in opposite directions. Then the associated magnetic fields are opposite. The resulting forces are upward on one side of the loop and downward on the other side, making it rotate. This effect of a force in producing rotation is called *torque*.

The principle of motor action between magnetic fields producing rotational torque is the basis of all electric motors. Also, the moving-coil meter described in Sec. 8-1 is a similar application. Since the torque is proportional to current, the amount of rotation indicates how much current flows through the coil.

Practice Problems 15-3
Answers at End of Chapter

Answer True or False.

a. In Fig. 15-8, the field is strongest between the two north poles.

b. In Fig. 15-9, if both the magnetic field and the current are reversed, the motion will still be upward.

15-4   Induced Current

Just as electrons in motion provide an associated magnetic field, when magnetic flux moves, the motion of magnetic lines cutting across a conductor forces free electrons in the conductor to move, producing current. This action is called *induction* because there is no physical connection between the magnet and the conductor. The induced current is a result of generator action as the mechanical work put into moving the magnetic field is converted into electric energy when current flows in the conductor.

Referring to Fig. 15-10, let the conductor AB be placed at right angles to the flux in the air gap of the horseshoe magnet. Then, when the magnet is moved up or down, its flux cuts across the conductor. The action of magnetic flux cutting across the conductor generates current. The fact that current flows is indicated by the microammeter.

When the magnet is moved downward, current flows in the direction shown. If the magnet is moved upward, current will flow in the opposite direction. Without motion, there is no current.

Direction of Motion   Motion is necessary in order to have the flux lines of the magnetic field cut across the conductor. This cutting can be accomplished by motion of either the field or the conductor. When the conductor is moved upward or downward, it cuts across the flux. The generator action is the same as moving the field, except that the relative motion is opposite. Moving the conductor upward, for instance, corresponds to moving the magnet downward.

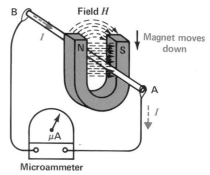

Fig. 15-10 Induced current produced by magnetic flux cutting across a conductor. Current I is electron flow.

Conductor Perpendicular to External Flux

In order to have electromagnetic induction, the conductor and the magnetic lines of flux must be perpendicular to each other. Then the motion makes the flux cut through the cross-sectional area of the conductor. As shown in Fig. 15-10, the conductor is at right angles to the lines of force in the field H.

The reason the conductor must be perpendicular is to make its induced current have an associated magnetic field in the same plane as the external flux. If the field of the induced current does not react with the external field, there can be no induced current.

How Induced Current Is Generated

The induced current can be considered the result of motor action between the external field H and the magnetic field of free electrons in every cross-sectional area of the wire. Without an external field, the free electrons move at random without any specific direction, and they have no net magnetic field. When the conductor is in the magnetic field H, there still is no induction without relative motion, since the magnetic fields for the free electrons are not disturbed. When the field or conductor moves, however, there must be a reaction opposing the motion. The reaction is a flow of free electrons resulting from motor action on the electrons.

Referring to Fig. 15-10, for example, the induced current must flow in the direction shown because the field is moved downward, pulling the magnet away from the conductor. The induced current of electrons then has a clockwise field, with lines of force aiding H above the conductor and canceling H below. With motor action between the two magnetic fields tending to move the conductor toward the weaker field, the conductor will be forced downward, staying with the magnet to oppose the work of pulling the magnet away from the conductor.

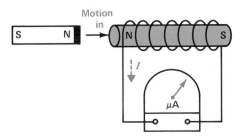

Fig. 15-11 Induced current produced by magnetic flux cutting across turns of wire in a coil. Current I is electron flow.

The effect of electromagnetic induction is increased where a coil is used for the conductor. Then the turns concentrate more conductor length in a smaller area. As illustrated in Fig. 15-11, moving the magnet into the coil enables the flux to cut across many turns of conductors.

Practice Problems 15-4
Answers at End of Chapter

Answer True or False. Refer to Fig. 15-10.
a. If the conductor is moved up, instead of the magnet down, the induced current will flow in the same direction.
b. The electron flow through the meter is from terminal A to B.

15-5   Lenz' Law

This basic principle is used to determine the direction of an induced voltage or current. Based on the principle of conservation of energy, Lenz' law simply states that the direction of the induced current must be such that its own magnetic field will oppose the action that produced the induced current.

In Fig. 15-11, for example, the induced current has the direction that produces a north pole at the left to oppose the motion by repulsion of the north pole being moved in. This is why it takes some work to push the permanent magnet into the coil. The work expended in moving the permanent magnet is the source of energy for the current induced in the coil.

Using Lenz' law, we can start with the fact that the left end of the coil in Fig. 15-11 must be a north pole to oppose the motion. Then the direction of the induced current is determined by the left-hand rule for electron flow. If the fingers coil around the direction of electron flow shown, under and over the winding, the thumb will point to the left for the north pole.

For the opposite case, suppose that the north pole of the permanent magnet in Fig. 15-11 is moved away from the coil. Then the induced pole at the left end of the coil must be a south pole, by Lenz' law. The induced south pole will attract the north pole to oppose the motion of the magnet being moved away. For a south pole at the left end of the coil, then, the electron flow will be reversed from the direction shown in Fig. 15-11. We could actually generate an alternating current in the coil by moving the magnet periodically in and out.

Practice Problems 15-5
Answers at End of Chapter

Refer to Fig. 15-11
a. If the north end of the magnet is moved away from the coil, will its left side be north or south?
b. If the south end of the magnet is moved in, will the left end of the coil be north or south?

15-6   Generating an Induced Voltage

Consider the case of magnetic flux cutting a conductor that is not in a closed circuit, as shown in Fig. 15-12. The motion of flux across the conductor forces free electrons to move, but with an open circuit, the displaced electrons produce opposite electric charges at the two open ends.

For the directions shown, free electrons in the conductor are forced to move to point A. Since the end is open, electrons accumulate here. Point A then develops a negative potential.

At the same time, point B loses electrons and becomes charged positive. The result is a potential difference across the two ends, provided by the separation of electric charges in the conductor.

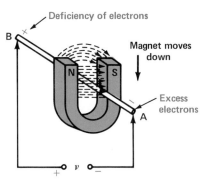

Fig. 15-12 Voltage induced across open ends of conductor cut by magnetic flux.

The potential difference is an electromotive force (emf), generated by the work of cutting across the flux. You can measure this potential difference with a voltmeter. However, a conductor cannot store electric charge. Therefore, the voltage is present only while the motion of flux cutting across the conductor is producing the induced voltage.

Induced Voltage Across a Coil   With a coil, as in Fig. 15-13a, the induced emf is increased by the number of turns. Each turn cut by flux adds to the induced voltage, since they all force free electrons to accumulate at the negative end of the coil, with a deficiency of electrons at the positive end.

The polarity of the induced voltage follows from the direction of induced current. The end of the conductor to which the electrons go and at which they accumulate is the negative side of the induced voltage. The opposite end, with a deficiency of electrons, is the positive side. The total emf across the coil is the sum of the induced voltages, since all the turns are in series.

Furthermore, the total induced voltage acts in series with the coil, as illustrated by the equivalent circuit in Fig. 15-13b, showing the induced voltage as a separate generator. This generator represents a voltage source with a potential difference resulting from the separation of charges produced by electromagnetic induction. The source v then can produce current in an external load circuit connected across the negative and positive terminals, as shown in Fig. 15-13c.

The induced voltage is in series with the coil because current produced by the generated emf must flow through all the turns. An induced voltage of 10 V, for example, with R_L equal to 5 Ω, results in a current of 2 A, which flows through the coil, the equivalent generator v, and the load resistance R_L.

The direction of current in Fig. 15-13c shows electron flow around the circuit. Outside the source v, the electrons move from its negative terminal, through R_L, and back to the positive terminal of v because of its potential difference.

Inside the generator, however, the electron flow is from the + terminal to the − terminal. This direction of electron flow results from the fact that the left end of the coil in Fig. 15-13a must be a north pole by Lenz' law, to oppose the north pole being moved in.

Notice how motors and generators are similar in using the motion of a magnetic field, but with opposite

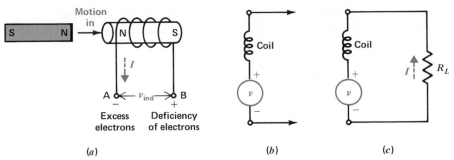

Fig. 15-13 Voltage induced across coil cut by magnetic flux. (a) Motion of flux generating voltage across coil. (b) Induced voltage acts in series with coil. (c) Induced voltage is a source that can produce current in an external load circuit connected across coil.

applications. In a motor, current is supplied so that an associated magnetic field can react with the external flux to produce motion of the conductor. In a generator, motion must be supplied so that the flux and conductor can cut across each other to induce voltage across the ends of the conductor.

Practice Problems 15-6
Answers at End of Chapter

Refer to Fig. 15-13.
a. Is terminal A or B the negative side of the induced voltage?
b. Is the bottom of R_L the negative side of V_{R_L}?

15-7 Faraday's Law of Induced Voltage

The voltage induced by magnetic flux cutting the turns of a coil depends upon the number of turns and how fast the flux moves across the conductor. Either the flux or the conductor can move. Specifically, the amount of induced voltage is determined by the following three factors:

1. *Amount of flux.* The more magnetic lines of force that cut across the conductor, the higher the amount of induced voltage.
2. *Number of turns.* The more turns in a coil, the higher the induced voltage. The v_{ind} is the sum of all the individual voltages generated in each turn in series.

3. *Time rate of cutting.* The faster the flux cuts a conductor, the higher the induced voltage. Then more lines of force cut the conductor within a specific period of time.

These factors are of fundamental importance in many applications. Any conductor with current will have voltage induced in it by a change in current and its associated magnetic flux.

The amount of induced voltage can be calculated by Faraday's law:

$$v_{ind} = N \frac{d\phi \text{ (webers)}}{dt \text{ (seconds)}} \qquad (15\text{-}1)$$

where N is the number of turns and $d\phi/dt$ specifies how fast the flux ϕ cuts across the conductor. With $d\phi/dt$ in webers per second, the induced voltage is in volts.

As an example, suppose that magnetic flux cuts across 300 turns at the rate of 2 Wb/s. To calculate the induced voltage,

$$V_{ind} = N \frac{d\phi}{dt}$$
$$= 300 \times 2$$
$$V_{ind} = 600 \text{ V}$$

It is assumed that all the flux links all the turns, which is true with an iron core.

Time Rate of Change The symbol d in $d\phi$ and dt is an abbreviation for delta (Δ), which means a *change*. The $d\phi$ means a change in the flux ϕ, while dt means a change in time. In calculus, *dt* represents an infinitesimally small change in time, but we are using the symbol *d* for rate of change in general.

As an example, if the flux ϕ is 4 Wb one time but then changes to 6 Wb, the change in flux $d\phi$ is 2 Wb. The same idea applies to a decrease as well as an increase. If the flux changed from 6 to 4 Wb, $d\phi$ would still be 2 Wb. However, an increase is usually considered a change in the positive direction, with an upward slope, while a decrease has a negative slope downward.

Similarly, *dt* means a change in time. If we consider the flux at a time 2 s after the start and at a later time 3 s after the start, the change in time is 3 − 2, or 1 s for *dt*. Time always increases in the positive direction.

Combining the two factors of $d\phi$ and *dt*, we can say that for magnetic flux increasing by 2 Wb in 1 s, $d\phi/dt$ equals ²⁄₁, or 2 Wb/s. This states the time rate of change of the magnetic flux.

As another example, suppose that the flux increases by 2 Wb in the time of ½ or 0.5 s. Then

$$\frac{d\phi}{dt} = \frac{2\ \text{Wb}}{0.5\ \text{s}} = 4\ \text{Wb/s}$$

Analysis of Induced Voltage as $N(d\phi/dt)$

This fundamental concept of voltage induced by a change in flux is illustrated by the graphs in Fig. 15-14, for the values listed in Table 15-1. The linear rise in Fig. 15-14*a* shows values of flux ϕ increasing at a uniform rate. In this case, the curve goes up 2 Wb for every 1-s interval of time. The slope of this curve, then, equal to $d\phi/dt$, is 2 Wb/s. Note that, although ϕ increases, the rate of change is constant because the linear rise has a constant slope.

For induced voltage, only the $d\phi/dt$ factor is important, not the actual value of flux. To emphasize this basic concept, the graph in Fig. 15-14*b* shows the $d\phi/dt$ values alone. This graph is just a straight horizontal line for the constant value of 2 Wb/s.

The induced-voltage graph in Fig. 15-14*c* is also a straight horizontal line. Since $v_{\text{ind}} = N(d\phi/dt)$, the graph of induced voltage is just the $d\phi/dt$ values multiplied by the number of turns. The result is a constant 600 V, with 300 turns cut by flux changing at the constant rate of 2 Wb/s.

The example illustrated here can be different in several ways without changing the basic fact that the induced voltage is equal to $N(d\phi/dt)$. First, the number of turns or the $d\phi/dt$ values can be greater than the values assumed here, or less. More turns will provide more induced voltage, while fewer turns mean less voltage. Similarly, a higher value for $d\phi/dt$ results in more induced voltage.

Note that two factors are included in $d\phi/dt$. Its value can be increased by a higher value of $d\phi$ or a smaller value of *dt*. As an example, the value of 2 Wb/s for $d\phi/dt$ can be doubled by either increasing $d\phi$ to 4 Wb or reducing *dt* to ½ s. Then $d\phi/dt$ is ⁴⁄₁ or ²⁄₀.₅, which equals 4 Wb/s in either case. The same flux changing within a shorter time means a faster rate of flux cutting

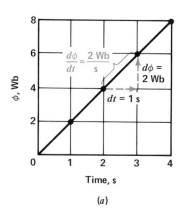

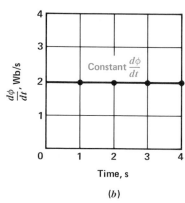

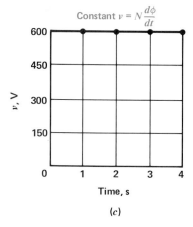

Fig. 15-14 Graphs of induced voltage produced by flux changes in a coil. (*a*) Linear increase of flux ϕ. (*b*) Constant rate of change for $d\phi/dt$ at 2 Wb/s. (*c*) Constant induced voltage of 600 V, for a coil with 300 turns.

Table 15-1. Induced-Voltage Calculations for Fig. 15-14

ϕ, Wb	$d\phi$, Wb	t, s	dt, s	$d\phi/dt$, Wb/s	N, turns	$N(d\phi/dt)$, V
2	2	1	1	2	300	600
4	2	2	1	2	300	600
6	2	3	1	2	300	600
8	2	4	1	2	300	600

for a higher value of $d\phi/dt$ and more induced voltage.

For the opposite case, a smaller value of $d\phi/dt$, with less flux or a slower rate of change, results in a smaller value of induced voltage. When $d\phi/dt$ decreases, the induced voltage has opposite polarity, compared with an increase.

Finally, it should be noted that the $d\phi/dt$ graph in Fig. 15-14b has the constant value of 2 Wb/s because the flux is increasing at a linear rate. However, the flux need not have a uniform rate of change. Then the $d\phi/dt$ values will not be constant. In any case, though, the values of $d\phi/dt$ at all instants of time will determine the values of the induced voltage equal to $N(d\phi/dt)$.

Polarity of the Induced Voltage The polarity is determined by Lenz' law. Any induced voltage has the polarity that opposes the change causing the induction. Sometimes this fact is indicated by using a negative sign for v_{ind} in Formula (15-1). However, the absolute polarity depends on whether the flux is increasing or decreasing, the method of winding, and which end of the coil is the reference.

When all these factors are considered, v_{ind} has the polarity such that the current it produces and the associated magnetic field will oppose the change in flux producing the induced voltage. If the external flux increases, the magnetic field of the induced current will be in the opposite direction. If the external field decreases, the magnetic field of the induced current will be in the same direction as the external field to oppose the change by sustaining the flux. In short, the induced voltage has the polarity that opposes the change.

Practice Problems 15-7
Answers at End of Chapter

a. The magnetic flux of 8 Wb changes to 10 Wb in 1 s. How much is $d\phi/dt$?

b. The flux of 8 μWb changes to 10 μWb in 1 μs. How much is $d\phi/dt$?

Summary

1. Current in a straight conductor has an associated magnetic field with circular lines of force in a plane perpendicular to the conductor. The direction of the circular field is counterclockwise when you look along the conductor in the direction of electron flow.
2. With two fields in the same plane, produced by either current or a permanent magnet, lines of force in the same direction aid each other to provide a stronger field. Lines of force in opposite directions cancel, resulting in a weaker field.
3. A solenoid is a long, narrow coil of wire which concentrates the conductor and its associated magnetic field. Because the fields for all turns aid inside the coil and cancel outside, a solenoid has a resultant electromagnetic field like a bar magnet with north and south poles at opposite ends.
4. The left-hand rule for polarity of an electromagnet says that when your fingers curl around the turns in the direction of electron flow, the thumb points to the north pole.

5. Motor action is the motion that results from the net force of two fields that can aid or cancel each other. The direction of the resultant force is always from the stronger field to the weaker field.

6. Generator action refers to induced voltage. For N turns, $v_{ind} = N(d\phi/dt)$, with $d\phi/dt$ in webers per second. There must be a change in the flux to produce induced voltage.

7. Lenz' law states that the polarity of the induced voltage will oppose the change in magnetic flux causing the induction.

8. The faster the flux changes, the higher the induced voltage.

9. When the flux changes at a constant rate, the induced voltage is constant.

Self-Examination
Answers at Back of Book

Answer True or False.

1. A vertical wire with electron flow downward through this page has an associated magnetic field counterclockwise in the plane of the paper.

2. Lines of force of two magnetic fields in the same direction aid each other to produce a stronger resultant field.

3. Motor action always tends to produce motion toward the weaker field.

4. In Fig. 15-6, if the battery connections are reversed, the magnetic poles of the coil will be reversed.

5. A solenoid is a coil that acts as a bar magnet only when current flows.

6. A torque is a force tending to cause rotation.

7. In Fig. 15-9, if external poles are reversed, motor action will be downward.

8. In Fig. 15-10, if the conductor is moved down, instead of the magnet, the induced current flows in the opposite direction.

9. An induced voltage is present only while the flux is changing.

10. Faraday's law determines the amount of induced voltage.

11. Lenz' law determines the polarity of an induced voltage.

12. Induced voltage increases with a faster rate of flux cutting.

13. An induced voltage is effectively in series with the turns of the coil in which the voltage is produced.

14. A decrease in flux will induce a voltage of opposite polarity from an increase in flux, with the same direction of field lines in both cases.

15. The flux of 1000 lines increasing to 1001 lines in 1 s produces a flux change $d\phi/dt$ of 1 line per s.

Essay Questions

1. Draw a diagram showing two conductors connecting a battery to a load resistance through a closed switch. (a) Show the magnetic field of the current in the negative side of the line and in the positive side. (b) Where do the two fields aid? Where do they oppose?

2. State the rule for determining the magnetic polarity of a solenoid. (a) How can the polarity be reversed? (b) Why are there no magnetic poles when the current through the coil is zero?

3. Why does the motor action between two magnetic fields result in motion toward the weaker field?

4. Why does current in a conductor perpendicular to this page have a magnetic field in the plane of the paper?

5. Why must the conductor and the external field be perpendicular to each other in order to have motor action or to generate induced voltage?

6. Explain briefly how either motor action or generator action can be obtained with the same conductor in a magnetic field.

7. Assume that a conductor being cut by the flux of an expanding magnetic field has 10 V induced with the top end positive. Now analyze the effect of the following changes: (a) The magnetic flux continues to expand, but at a slower rate. How does this affect the amount of induced voltage and its polarity? (b) The magnetic flux is constant, neither increasing nor decreasing. How much is the induced voltage? (c) The magnetic flux contracts, cutting across the conductor with the opposite direction of motion. How does this affect the polarity of the induced voltage?

8. Redraw the graph in Fig. 15-14c for 500 turns, with all other factors the same.

9. Redraw the circuit with the coil and battery in Fig. 15-6, showing two different ways to reverse the magnetic polarity.

10. Referring to Fig. 15-14, suppose that the flux decreases from 8 Wb to zero at the same rate as the increase. Tabulate all the values as in Table 15-1 and draw the three graphs corresponding to those in Fig. 15-14.

Problems

Answers to Odd-Numbered Problems at Back of Book

1. A magnetic flux of 800 Mx cuts across a coil of 1000 turns in 1 μs. How much is the voltage induced in the coil? [1 Mx = 10^{-8} Wb]

2. Refer to Fig. 15-13. (a) Show the induced voltage here connected to a load resistance R_L of 50 Ω. (b) If the induced voltage is 100 V, how much current flows in R_L? (c) Give one way to reverse the polarity of the induced voltage.

3. Calculate the rate of flux change $d\phi/dt$ in webers per second for the following: (a) 6 Wb increasing to 8 Wb in 1 s; (b) 8 Wb decreasing to 6 Wb in 1 s.

4. Calculate the voltage induced in 400 turns by each of the flux changes in Prob. 3.

5. Draw a circuit with a 20-V battery connected to a 100-Ω coil of 400 turns with an iron core 0.2 m long. Using SI magnetic units, calculate (a) I; (b) NI; (c) the field intensity H; (d) the flux density B in a core with a μ_r of 500; (e) the total flux ϕ at each pole with an area of 6×10^{-4} m^2. (f) Show the direction of winding and magnetic polarity of the coil.

6. For the coil in Prob. 5: (a) If the iron core is removed, how much will the flux be in the air-core coil? (b) How much induced voltage would be produced by this change in flux while the core is being moved out in 1 s? (c) How much is the induced voltage after the core is removed?

Answers to Practice Problems

15-1 a. T	**15-3 a.** T	**15-5 a.** South	**15-7 a.** 2 Wb/s
b. T	**b.** T	**b.** South	**b.** 2 Wb/s
15-2 a. Left	**15-4 a.** T	**15-6 a.** A	
b. South pole	**b.** T	**b.** Yes	

Chapter 16

Alternating Voltage and Current

This chapter begins the analysis of ac voltage, as used for the 120-V ac power line, and the alternating current it produces in an ac circuit. Alternating current and voltage reverses between positive and negative polarities and varies in amplitude with time. One cycle includes a complete set of variations, with two alternations in polarity. The number of cycles per second is the frequency. The ac waveform is important be-

cause, in addition to the ac power-line voltage, audio, video, and radio signals are all examples of ac voltages.

We can use the principles for dc circuits to analyze ac circuits also. All the rules for series and parallel circuits and Ohm's law calculations still apply. However, the new factor to consider with an ac source is that the voltage alternately reverses in polarity, producing current that reverses in direction. Most important, between the polarity reversals, the voltage and current values are always changing instead of remaining at a steady value. A specific value in the cycle can be used, however, to indicate the amplitude.

Important terms in this chapter are:

alternator	delta connections	peak value	square wave
armature	effective value	phase angle	three-phase power
average value	field winding	rms value	wavelength
brushes	frequency	sawtooth wave	wye connections
commutator	harmonic	sine wave	
cycle	octave	slip rings	

More details are explained in the following sections:

16-1   AC Applications

Figure 16-1 shows the reversals between positive and negative polarities and the variations in amplitude for an ac waveform. The characteristic of varying values is the reason why ac circuits have so many uses. For instance, a transformer can operate only with alternating current, to step up or step down an ac voltage. The reason is that the changing current produces changes in its associated magnetic field. This application is just an example of inductance L in ac circuits, where the changing magnetic flux of a varying current can produce induced voltage. The details of inductance are explained in Chaps. 17, 18, and 19.

A similar but opposite effect in ac circuits is capacitance C. The capacitance is important with the changing electric field of a varying voltage. Just as L has a big effect with alternating current, C has an effect which depends on alternating voltage. The details of capacitance are explained in Chaps. 20, 21, and 22.

The L and C are additional factors, besides resistance R, in the operation of ac circuits. It should be noted that R is the same for either a dc or an ac circuit. However, the effects of L and C depend on having an ac source. The rate at which the ac variations occur, which determines the frequency, allows a greater or smaller reaction by L and C. Therefore, the effect is different for different frequencies. One important application is a resonant circuit with L and C which is tuned to a particular frequency. All examples of tuning in radio and television are applications of resonance in an LC circuit.

In general, electronic circuits are combinations of R, L, and C, with both direct current and alternating current. The audio, video, and radio signals are ac voltages and currents. However, the amplifiers that use transistors need dc voltages in order to conduct any current at all. The resulting output of an amplifier circuit, therefore, consists of direct current with a superimposed ac signal.

Practice Problems 16-1
Answers at End of Chapter

Answer True or False.
a. An ac voltage varies in magnitude and reverses in polarity.
b. A transformer can operate with either ac or steady dc input.
c. Inductance L and capacitance C are important factors in ac circuits.

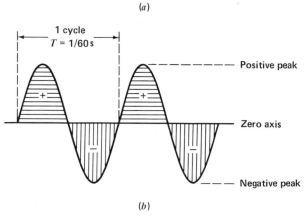

(a)

(b)

Fig. 16-1 Waveform of ac power-line voltage with frequency of 60 Hz. Two cycles shown. (a) Oscilloscope photograph. (b) Details of waveform.

16-2   Alternating-Voltage Generator

We can define an ac voltage as one that continuously varies in magnitude and periodically reverses in polarity. In Fig. 16-1, the variations up and down on the waveform show the changes in magnitude. The zero axis is a horizontal line across the center. Then voltages above the center have positive polarity, while the values below center are negative.

Figure 16-2 illustrates how such a voltage waveform is produced by a rotary generator. The conductor loop

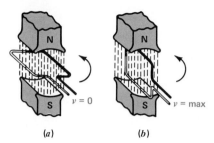

$v = 0$ $v = \text{max}$

(a)   (b)

Fig. 16-2 Loop rotating in magnetic field to produce induced voltage v with alternating polarity. Arrow indicates rotation. (a) Loop conductors moving parallel to field lines results in zero voltage. (b) Loop conductors cutting across field produce maximum induced voltage.

rotates through the magnetic field to generate the induced ac voltage across its open terminals. The magnetic flux shown here is vertical, with lines of force down in the plane of the paper.

In Fig. 16-2a the loop is in its horizontal starting position in a plane perpendicular to the paper. When the loop rotates counterclockwise, the two longer conductors move around a circle. Note that in the flat position shown, the two long conductors of the loop move vertically up or down through the paper but parallel to the vertical flux lines. In this position, motion of the loop does not induce a voltage because the conductors are not cutting across the flux.

When the loop rotates through the upright position in Fig. 16-2b, however, the conductors cut across the flux, producing maximum induced voltage. The shorter connecting wires in the loop do not have any appreciable voltage induced in them.

Each of the longer conductors has opposite polarity of induced voltage because the conductor at the top is moving to the left while the bottom conductor is moving to the right. The amount of voltage varies from zero to maximum as the loop moves from a flat position to upright, where it can cut across the flux. Also, the polarity at the terminals of the loop reverses as the motion of each conductor reverses during each half-revolution.

With one revolution of the loop in a complete circle back to the starting position, therefore, the induced voltage provides a potential difference v across the loop, varying in the same way as the wave of voltage shown in Fig. 16-1. If the loop rotates at the speed of 60 revolutions per second, the ac voltage will have the frequency of 60 Hz.

The Cycle   One complete revolution of the loop around the circle is a *cycle*. In Fig. 16-3, the generator loop is shown in its position at each quarter-turn during one complete cycle. The corresponding wave of induced voltage also goes through one cycle. Although not shown, the magnetic field is from top to bottom of the page as in Fig. 16-2.

At position A in Fig. 16-3, the loop is flat and moves parallel to the magnetic field, so that the induced voltage is zero. Counterclockwise rotation of the loop moves the dark conductor to the top at position B, where it cuts across the field to produce maximum induced voltage. The polarity of the induced voltage here makes the open end of the dark conductor positive. This conductor at the top is cutting across the flux from right to left. At the same time, the opposite conductor below is moving from left to right, causing its induced voltage to have opposite polarity. Therefore, maximum induced voltage is produced at this time across the two open ends of the loop. Now the top conductor is positive with respect to the bottom conductor.

In the graph of induced voltage values below the loop in Fig. 16-3, the polarity of the dark conductor is shown with respect to the other conductor. Positive voltage is shown above the zero axis in the graph. As the dark conductor rotates from its starting position parallel to the flux toward the top position, where it cuts maximum flux, more and more induced voltage is produced, with positive polarity.

When the loop rotates through the next quarter-turn, it returns to the flat position shown in C, where it cannot cut across flux. Therefore, the induced voltage values shown in the graph decrease from the maximum value to zero at the half-turn, just as the voltage was zero at the start. The half-cycle of revolution is called an *alternation*.

The next quarter-turn of the loop moves it to the position shown at D in Fig. 16-3, where the loop cuts across the flux again for maximum induced voltage. Note, however, that here the dark conductor is moving left to right at the bottom of the loop. This motion is reversed from the direction it had when it was at the top, moving right to left. Because the direction of motion is reversed during the second half-revolution, the induced voltage has opposite polarity, with the dark conductor negative. This polarity is shown as negative voltage, below the zero axis. The maximum value of induced voltage at the third quarter-turn is the same as at the first quarter-turn but with opposite polarity.

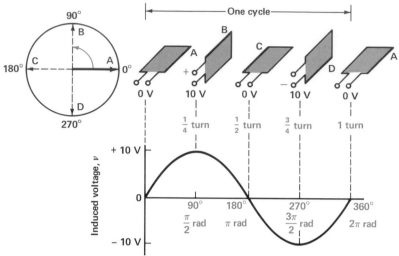

Fig. 16-3 One cycle of alternating voltage generated by rotating loop. Magnetic field, not shown here, is from top to bottom, as in Fig. 16-2.

When the loop completes the last quarter-turn in the cycle, the induced voltage returns to zero as the loop returns to its flat position at A, the same as at the start. This cycle of values of induced voltage is repeated as the loop continues to rotate, with one complete cycle of voltage values, as shown, for each circle of revolution.

Note that zero at the start and zero after the half-turn of an alternation are not the same. At the start, the voltage is zero because the loop is flat, but the dark conductor is moving upward in the direction that produces positive voltage. After one half-cycle, the voltage is zero with the loop flat, but the dark conductor is moving downward in the direction that produces negative voltage. After one complete cycle, the loop and its corresponding waveform of induced voltage are the same as at the start. *A cycle can be defined, therefore, as including the variations between two successive points having the same value and varying in the same direction.*

Angular Measure   Because the cycle of voltage in Fig. 16-3 corresponds to rotation of the loop around a circle, it is convenient to consider parts of the cycle in angles. The complete circle includes 360°. One half-cycle, or one alternation, is 180° of revolution. A quarter-turn is 90°. The circle next to the loop positions in Fig. 16-3 illustrates the angular rotation of the dark conductor as it rotates counterclockwise from 0 to 90 to

180° for one half-cycle, then to 270° and returning to 360° to complete the cycle. Therefore, one cycle corresponds to 360°.

Radian Measure   In angular measure it is convenient to use a specific unit angle called the *radian* (abbreviated rad), which is an angle equal to 57.3°. Its convenience is due to the fact that a radian is the angular part of the circle that includes an arc equal to the radius r of the circle, as shown in Fig. 16-4. The circumference around the circle equals $2\pi r$. A circle includes 2π rad, then, as each radian angle includes one length r of the circumference. Therefore, one cycle equals 2π rad.

As shown in the graph in Fig. 16-3, divisions of the cycle can be indicated by angles in either degrees or radians. The comparison between degrees and radians can be summarized as follows:

Zero degrees is also zero radians

360° is 2π rad

180° is $\frac{1}{2} \times 2\pi$ rad = π rad

90° is $\frac{1}{2} \times \pi$ rad = $\pi/2$ rad

Also, 270° is 180° + 90° or π rad + $\pi/2$ rad, which equals $3\pi/2$ rad.

The constant 2π in circular measure is numerically equal to 6.2832. This is double the value of 3.1416 for

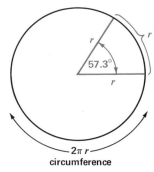

Fig. 16-4 One radian is the angle equal to 57.3°. The complete circle with 360° has 2π rad.

π. The π is a symbol for the ratio of the circumference to the diameter for any circle, which always has the numerical value of 3.1416. The fact that 2π rad is 360° can be shown as $2 \times 3.1416 \times 57.3° = 360°$ for a complete cycle.

Practice Problems 16-2
Answers at End of Chapter

Refer to Fig. 16-3.
a. How much is the induced voltage at $\pi/2$ rad?
b. How many degrees are in a complete cycle?

16-3 The Sine Wave

The voltage waveform in Figs. 16-1 and 16-3 is called a *sine wave, sinusoidal wave,* or *sinusoid* because the amount of induced voltage is proportional to the sine of the angle of rotation in the circular motion producing the voltage. The sine is a trigonometric function[1] of an angle; it is equal to the ratio of the opposite side to the hypotenuse in a right triangle. This numerical ratio increases from zero for 0° to a maximum value of 1 for 90° as the side opposite the angle becomes larger.

The voltage waveform produced by the circular motion of the loop is a sine wave, because the induced voltage increases to a maximum at 90°, when the loop is vertical, in the same way that the sine of the angle of rotation increases to a maximum at 90°. The induced voltage and sine of the angle correspond for the full 360° of the cycle. Table 16-1 lists the numerical values of the sine for several important angles, to illustrate the specific characteristics of a sine wave.

Notice that the sine wave reaches ½ its maximum value in 30°, which is only ⅓ of 90°. This fact means that the sine wave has a sharper slope of changing values when the wave is near the zero axis, compared with the more gradual changes near the maximum value.

---

[1]See App. E for an explanation of the sine, cosine, and tangent functions of an angle. More details are given in B. Grob, *Mathematics Outline and Review Problems for Basic Electronics*, McGraw-Hill Book Company, New York.

Table 16-1. Values in a Sine Wave

Angle θ		Sin θ	Loop voltage
Degrees	Radians		
0	0	0	Zero
30	$\dfrac{\pi}{6}$	0.500	50% of maximum
45	$\dfrac{\pi}{4}$	0.707	70.7% of maximum
60	$\dfrac{\pi}{3}$	0.866	86.6% of maximum
90	$\dfrac{\pi}{2}$	1.000	Positive maximum value
180	π	0	Zero
270	$\dfrac{3\pi}{2}$	-1.000	Negative maximum value
360	2π	0	Zero

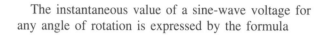

The instantaneous value of a sine-wave voltage for any angle of rotation is expressed by the formula

$$v = V_M \sin \theta \qquad (16\text{-}1)$$

where θ (Greek letter *theta*) is the angle, sin is the abbreviation for its sine, V_M is the maximum voltage value, and v is the instantaneous value for any angle.

Example 1   A sine wave of voltage varies from zero to a maximum of 100 V. How much is the voltage at the instant of 30° of the cycle? 45°? 90°? 270°?

Answer $v = V_M \sin \theta = 100 \sin \theta$

At 30°: $v = V_M \sin 30° = 100 \times 0.5$
          $v = 50$ V

At 45°: $v = V_M \sin 45° = 100 \times 0.707$
          $v = 70.7$ V

At 90°: $v = V_M \sin 90° = 100 \times 1$
          $v = 100$ V

At 270° $v = V_M \sin 270° = 100 \times -1$
          $v = -100$ V

The value of -100 V at 270° is the same as that at 90° but with opposite polarity.

Between zero at 0° and maximum at 90° the amplitudes of a sine wave increase exactly as the sine value for the angle of rotation. These values are for the first quadrant in the circle. From 90 to 180°, in the second quadrant, the values decrease as a mirror image of the first 90°. The values in the third and fourth quadrants, from 180 to 360°, are exactly the same as 0 to 180° but with opposite sign. At 360° the waveform is back to 0° to repeat its values every 360°.

In summary, the characteristics of the sine-wave ac waveform are:

1. The cycle includes 360° or 2π rad.
2. The polarity reverses each half-cycle.
3. The maximum values are at 90 and 270°.
4. The zero values are at 0 and 180°.
5. The waveform changes its values the fastest when it crosses the zero axis.
6. The waveform changes its values the slowest when

it is at its maximum value. The values must stop increasing before they can decrease.

A perfect example of the sine-wave ac waveform is the 60-Hz power-line voltage in Fig. 16-1.

Practice Problems 16-3
Answers at End of Chapter

A sine-wave voltage has a peak value of 170 V. What is its value at
a. 30°?
b. 45°?
c. 90°?

16-4   Alternating Current

When a sine wave of alternating voltage is connected across a load resistance, the current that flows in the circuit is also a sine wave. In Fig. 16-5, let the sine-wave voltage at the left in the diagram be applied across R of 100 Ω. The resulting sine wave of alternating current is shown at the right in the diagram. Note that the frequency is the same for v and i.

During the first half-cycle of v in Fig. 16-5, terminal 1 is positive with respect to terminal 2. Since the direction of electron flow is from the negative side of v, through R, and back to the positive side of v, current flows in the direction indicated by arrow a for the first half-cycle. This direction is taken as the positive direction of current in the graph for i, corresponding to positive values of v.

The amount of current is equal to v/R. If several instantaneous values are taken, when v is zero, i is zero; when v is 50 V, i equals 50 V/100, or 0.5 A; when v is 100 V, i equals 100 V/100, or 1 A. For all values of applied voltage with positive polarity, therefore, the current is in one direction, increasing to its maximum value and decreasing to zero, just like the voltage.

On the next half-cycle, the polarity of the alternating voltage reverses. Then terminal 1 is negative with respect to terminal 2. With reversed voltage polarity, current flows in the opposite direction. Electron flow is from terminal 1 of the voltage source, which is now the negative side, through R, and back to terminal 2. This direction of current, as indicated by arrow b in Fig. 16-5, is negative.

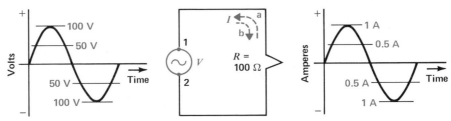

Fig. 16-5 A sine wave of alternating voltage applied across R produces a sine wave of alternating current in the circuit. (*a*) Waveform of applied voltage. (*b*) AC circuit. (*c*) Waveform of current in the circuit.

The negative values of i in the graph have the same numerical values as the positive values in the first half-cycle, corresponding to the reversed values of applied voltage. As a result, the alternating current in the circuit has sine-wave variations corresponding exactly to the sine-wave alternating voltage.

Only the waveforms for v and i can be compared. There is no comparison between relative values, because the current and voltage are different quantities.

It is important to note that the negative half-cycle of applied voltage is just as useful as the positive half-cycle in producing current. The only difference is that the reversed polarity of voltage produces the opposite direction of current.

Furthermore, the negative half-cycle of current is just as effective as the positive values when heating the filament to light a bulb. With positive values, electrons flow through the filament in one direction. Negative values produce electron flow in the opposite direction. In both cases, electrons flow from the negative side of the voltage source, through the filament, and return to the positive side of the source. For either direction, the current heats the filament. The direction does not matter, since it is just the motion of electrons against resist-

ance that produces power dissipation. In short, resistance R has the same effect in reducing I for either direct current or alternating current.

Practice Problems 16-4
Answers at End of Chapter

Refer to Fig. 16-5.
a. When v is 70.7 V, how much is i?
b. How much is i at 30°?

16-5   Voltage and Current Values for a Sine Wave

Since an alternating sine wave of voltage or current has many instantaneous values through the cycle, it is convenient to define specific magnitudes for comparing one wave with another. The peak, average, or root-mean-square (rms) value can be specified, as indicated in Fig. 16-6. These values can be used for either current or voltage.

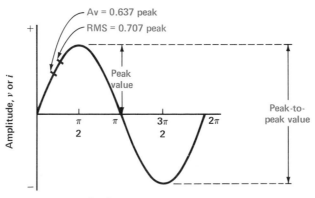

Fig. 16-6 Amplitude values for a sine wave of voltage or current.

Peak Value   This is the maximum value V_M or I_M. For example, specifying that a sine wave has a peak value of 170 V states how much it is, since all other values during the cycle follow a sine wave. The peak value applies to either the positive or the negative peak.

In order to include both peak amplitudes, the *peak-to-peak* (p-p) *value* may be specified. For the same example, the peak-to-peak value is 340 V, double the peak value of 170 V, since the positive and negative peaks are symmetrical. It should be noted, though, that the two opposite peak values cannot occur at the same time. Furthermore, in some waveforms the two peaks are not equal.

Average Value   This is an arithmetic average of all the values in a sine wave for one alternation, or half-cycle. The half-cycle is used for the average because over a full cycle the average value is zero, which is useless for comparison purposes. If the sine values for all angles up to 180°, for one alternation, are added and then divided by the number of values, this average equals 0.637. These calculations are shown in Table 16-2.

Since the peak value of the sine is 1 and the average equals 0.637, then

Average value = 0.637 × peak value   (16-2)

With a peak of 170 V, for example, the average value is 0.637 × 170 V, which equals approximately 108 V.

Root-Mean-Square, or Effective, Value   The most common method of specifying the amount of a sine wave of voltage or current is by stating its value at 45°, which is 70.7 percent of the peak. This is its *root-mean-square* value, abbreviated rms. Therefore,

rms value = 0.707 × peak value   (16-3)

or

$$V_{\text{rms}} = 0.707 V_{\text{max}} \qquad \text{and} \qquad I_{\text{rms}} = 0.707 I_{\text{max}}$$

With a peak of 170 V, for example, the rms value is 0.707 × 170, or 120 V, approximately. This is the voltage of the commercial ac power line, which is always given in rms value.

It is often necessary to convert from rms to peak value. This can be done by inverting Formula (16-3), as follows:

Table 16-2. Derivation of Average and RMS Values for a Sine-Wave Alternation

Interval	Angle θ	Sin θ	$(\text{Sin } \theta)^2$
1	15°	0.26	0.07
2	30°	0.50	0.25
3	45°	0.71	0.50
4	60°	0.87	0.75
5	75°	0.97	0.93
6	90°	1.00	1.00
7*	105°	0.97	0.93
8	120°	0.87	0.75
9	135°	0.71	0.50
10	150°	0.50	0.25
11	165°	0.26	0.07
12	180°	0.00	0.00
	Total	7.62	6.00
	Average →	$\dfrac{7.62}{12} = 0.635$†	$\sqrt{\dfrac{6}{12}} = \sqrt{0.5} = 0.707$

*For angles between 90 and 180°, sin θ = sin (180° − θ).
†More intervals and precise values are needed to get the exact average of 0.637.

$$\text{Peak} = \frac{1}{0.707} \times \text{rms} = 1.414 \times \text{rms} \qquad \textbf{(16-4)}$$

or

$$V_{\max} = 1.414 V_{\text{rms}} \qquad \text{and} \qquad I_{\max} = 1.414 I_{\text{rms}}$$

Dividing by 0.707 is the same as multiplying by 1.414.

For example, the commercial power-line voltage with an rms value of 120 V has a peak value of 120 × 1.414, which equals 170 V, approximately. Its peak-to-peak value is 2 × 170, or 340 V, which is double the peak value. As a formula,

$$\text{Peak-to-peak value} = 2.828 \times \text{rms value} \qquad \textbf{(16-5)}$$

The factor 0.707 for rms value is derived as the square root of the average (mean) of all the squares of the sine values. If we take the sine for each angle in the cycle, square each value, add all the squares, divide by the number of values added to obtain the average square, and then take the square root of this mean value, the answer is 0.707. These calculations are shown in Table 16-2 for one alternation from 0 to 180°. The results are the same for the opposite alternation.

The advantage of the rms value derived in terms of the squares of the voltage or current values is that it provides a measure based on the ability of the sine wave to produce power, which is I^2R or V^2/R. As a result, the rms value of an alternating sine wave corresponds to the same amount of direct current or voltage in heating power. An alternating voltage with an rms value of 120 V, for instance, is just as effective in heating the filament of a light bulb as 120 V from a steady dc voltage source. For this reason, the rms value is also the *effective* value.

Unless indicated otherwise, all sine-wave ac measurements are in rms values. The capital letters V and I are used, corresponding to the symbols for dc values. As an example, $V = 120$ V for the ac power-line voltage.

The ratio of the rms to average values is the *form factor*. For a sine wave, this ratio is 0.707/0.637 = 1.11.

Note that sine waves can have different amplitudes but still follow the sinusoidal waveform. Figure 16-7 compares a low-amplitude voltage with a high-amplitude voltage. Although different in amplitude, they are

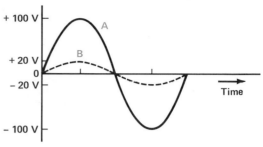

Fig. 16-7 Waveforms A and B have different amplitudes, but both are sine waves.

both sine waves. In each wave, the rms value is 0.707 of the peak value.

Practice Problems 16-5
Answers at End of Chapter

a. Convert 170 V peak to rms value.
b. Convert 10 V rms to peak value.

16-6   Frequency

The number of cycles per second is the *frequency,* with the symbol f. In Fig. 16-3, if the loop rotates through 60 complete revolutions, or cycles, during 1 s, the frequency of the generated voltage is 60 cps, or 60 Hz. You see only one cycle of the sine waveform, instead of 60 cycles, because the time interval shown here is $\frac{1}{60}$ s. Note that the factor of time is involved. More cycles per second means a higher frequency and less time for one cycle, as illustrated in Fig. 16-8. Then the changes in values are faster for higher frequencies.

A complete cycle is measured between two successive points that have the same value and direction. In Fig. 16-8 the cycle is between successive points where the waveform is zero and ready to increase in the positive direction. Or the cycle can be measured between successive peaks.

On the time scale of 1 s, waveform a goes through one cycle; waveform b has much faster variations, with four complete cycles during 1 s. Both waveforms are sine waves, even though each has a different frequency.

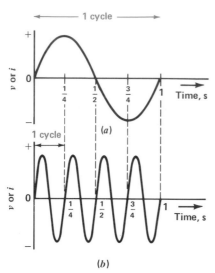

Fig. 16-8 Number of cycles per second is the frequency in hertz (Hz) units. (*a*) The *f* = 1 Hz. (*b*) The *f* = 4 Hz.

In comparing sine waves, the amplitude has no relation to frequency. Two waveforms can have the same frequency with different amplitudes (Fig. 16-7), the same amplitude but different frequencies (Fig. 16-8), or different amplitudes and frequencies. The amplitude indicates how much the voltage or current is, while the frequency indicates the time rate of change of the amplitude variations, in cycles per second.

Frequency Units   The unit called the *hertz* (Hz), named after H. Hertz, is used for cycles per second. Then 60 cps = 60 Hz. All the metric prefixes can be used. As examples:

1 kilocycle per second = 1×10^3 Hz = 1 kHz

1 megacycle per second = 1×10^6 Hz = 1 MHz

1 gigacycle per second = 1×10^9 Hz = 1 GHz

Audio and Radio Frequencies   The entire frequency range of alternating voltage or current from 1 Hz to many megahertz can be considered in two broad groups: audio frequencies (AF) and radio frequencies (RF). *Audio* is a Latin word meaning "I hear." The audio range includes frequencies that can be heard in the form of sound waves by the human ear. This range of audible frequencies is approximately 16 to 16,000 Hz.

Table 16-3. Examples of Common Frequencies

Frequency	Use
60 Hz	AC power line
50–15,000 Hz	Audio equipment
535–1605 kHz	AM radio band
54–60 MHz	TV channel 2
88–108 MHz	FM radio band

The higher the frequency, the higher the pitch or tone of the sound. High audio frequencies, about 3000 Hz and above, can be considered to provide *treble* tone. Low audio frequencies, about 300 Hz and below, provide *bass* tone.

Loudness is determined by amplitude. The greater the amplitude of the AF variation, the louder is its corresponding sound.

Alternating current and voltage above the audio range provide RF variations, since electrical variations of high frequency can be transmitted by electromagnetic radio waves. The more common frequency bands for radio broadcasting are listed in App. B. Some applications are listed here, though, in Table 16-3.

Sonic and Supersonic Frequencies   These terms refer to sound waves, which are variations in pressure generated by mechanical vibrations, rather than electrical variations. The velocity of transmission for sound waves equals 1130 ft/s, through dry air at 20°C. Sound waves above the audible range of frequencies are called *supersonic* waves. The range of frequencies for supersonic applications, therefore, is from 16,000 Hz up to several megahertz. Sound waves in the audible range of frequencies below 16,000 Hz can be considered *sonic* or sound frequencies, reserving *audio* for electrical variations that can be heard when converted to sound waves.

Practice Problems 16-6
Answers at End of Chapter

a. What is the frequency of the bottom waveform in Fig. 16-8?

b. Convert 1605 kHz to megahertz.

16-7   Period

The amount of time for one cycle is the *period*. Its symbol is T for time. With a frequency of 60 Hz, as an example, the time for one cycle is $\frac{1}{60}$ s. Therefore, the period is $\frac{1}{60}$ s in this case. The frequency and period are reciprocals of each other:

$$T = \frac{1}{f} \quad \text{or} \quad f = \frac{1}{T} \tag{16-6}$$

The higher the frequency, the shorter the period. In Fig. 16-8*a*, the period for the wave, with a frequency of 1 Hz, is 1 s, while the higher-frequency wave of 4 Hz in Fig. 16-8*b* has the period of $\frac{1}{4}$ s for a complete cycle.

Units of Time   The second is the basic unit, but for higher frequencies and shorter periods, smaller units of time are convenient. Those used most often are:

$T = 1 \text{ millisecond} = 1 \text{ ms} = 1 \times 10^{-3}$ s

$T = 1 \text{ microsecond} = 1 \text{ }\mu s = 1 \times 10^{-6}$ s

$T = 1 \text{ nanosecond} = 1 \text{ ns} = 1 \times 10^{-9}$ s

These units of time for period are reciprocals of the corresponding units for frequency. The reciprocal of frequency in kilohertz gives the period T in milliseconds; the reciprocal of megahertz is microseconds; the reciprocal of gigahertz is nanoseconds.

Example 2   An alternating current varies through one complete cycle in $\frac{1}{1000}$ s. Calculate the period and frequency.

Answer $T = \dfrac{1}{1000}$ s

$$f = \frac{1}{T} = \frac{1}{\frac{1}{1000}}$$

$$= \frac{1000}{1} = 1000$$

$f = 1000$ Hz or 1 kHz

Example 3   Calculate the period for the two frequencies of 1 MHz and 2 MHz.

Answer

a.   For 1 MHz,

$$T = \frac{1}{f} = \frac{1}{1 \times 10^6}$$

$$= 1 \times 10^{-6} \text{ s}$$

$T = 1 \text{ }\mu s$

b.   For 2 MHz,

$$T = \frac{1}{f} = \frac{1}{2 \times 10^6}$$

$$= 0.5 \times 10^{-6} \text{ s}$$

$T = 0.5 \text{ }\mu s$

Practice Problems 16-7
Answers at End of Chapter

a. $T = \frac{1}{400}$ s. Calculate f.
b. $f = 400$ Hz. Calculate T.

16-8   Wavelength

When a periodic variation is considered with respect to distance, one cycle includes the *wavelength*, which is the length of one complete wave or cycle (Fig. 16-9). For example, when a radio wave is transmitted, variations in the electromagnetic field travel through space. Also, with sound waves, the variations in air pressure corresponding to the sound wave move through air. In these applications, the distance traveled by the wave in

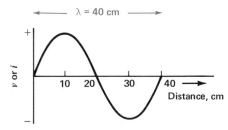

Fig. 16-9 Wavelength λ is the distance traveled by the wave in one cycle.

one cycle is the wavelength. The wavelength depends upon the frequency of the variation and its velocity of transmission:

$$\lambda = \frac{\text{velocity}}{\text{frequency}} \qquad (16\text{-}7)$$

where λ (lambda) is the symbol for one complete wavelength.

Wavelength of Radio Waves   For electromagnetic radio waves, the velocity in air or vacuum is 186,000 mi/s, or 3×10^{10} cm/s, which is the speed of light. Therefore,

$$\lambda \text{ (cm)} = \frac{3 \times 10^{10} \text{ cm/s}}{f(\text{Hz})} \qquad (16\text{-}8)$$

Note that the higher the frequency is, the shorter the wavelength. For instance, the short-wave radio broadcast band of 5.95 to 26.1 MHz includes higher frequencies than the standard radio broadcast band of 540 to 1620 kHz.

Example 4   Calculate λ for a radio wave with f of 30 GHz.

Answer $\lambda = \dfrac{3 \times 10^{10} \text{ cm/s}}{30 \times 10^9 \text{ Hz}} = \dfrac{3}{30} \times 10 \text{ cm}$

$$= 0.1 \times 10$$

$$\lambda = 1 \text{ cm}$$

Such short wavelengths are called *microwaves*. This range includes λ of 1 m or less, for frequencies of 300 MHz or more.

Example 5   The length of a TV antenna is $\lambda/2$ for radio waves with f of 60 MHz. What is the antenna length in centimeters and feet?

Answer

a. $\lambda = \dfrac{3 \times 10^{10} \text{ cm/s}}{60 \times 10^6 \text{ Hz}} = \dfrac{1}{20} \times 10^4 \text{ cm}$

$$= 0.05 \times 10^4$$

$$\lambda = 500 \text{ cm}$$

Then, $\lambda/2 = {}^{500}\!/_2 = 250$ cm.

b.   Since 2.54 cm = 1 in.,

$$\lambda/2 = \frac{250 \text{ cm}}{2.54 \text{ cm/in.}} = 98.4 \text{ in.}$$

$$\lambda/2 = \frac{98.4 \text{ in.}}{12 \text{ in./ft}} = 8.2 \text{ ft}$$

This half-wave dipole antenna is made with two quarter-wave poles, each about 4 ft long.

Example 6   For the 6-m band used in amateur radio, what is the corresponding frequency?

Answer   The formula $\lambda = v/f$ can be inverted to $f = v/\lambda$. Then

$$f = \frac{3 \times 10 \text{ cm/s}}{6 \text{ m}} = \frac{3 \times 10^{10} \text{ cm/s}}{6 \times 10^2 \text{ cm}}$$

$$= \frac{3}{6} \times 10^8 \times \frac{1}{\text{s}}$$

$$= 0.5 \times 10^8 \text{ Hz}$$

$$f = 50 \times 10^6 \text{ Hz} \qquad \text{or} \qquad 50 \text{ MHz}$$

Wavelength of Sound Waves   The velocity of sound waves is much lower, compared with that of radio waves, because sound waves result from mechanical vibrations rather than electrical variations. For average conditions the velocity of sound waves in air equals 1130 ft/s. To calculate the wavelength, therefore,

$$\lambda = \frac{1130 \text{ ft/s}}{f \text{ Hz}} \qquad (16\text{-}9)$$

This formula can also be used for supersonic waves. Although their frequencies are too high to be audible, supersonic waves are still sound waves rather than radio waves.

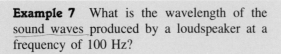

Example 7   What is the wavelength of the sound waves produced by a loudspeaker at a frequency of 100 Hz?

Answer $\lambda = \dfrac{1130 \text{ ft/s}}{100 \text{ Hz}}$

$\lambda = 11.3 \text{ ft}$

Example 8   For supersonic waves at a frequency of 34.44 kHz, calculate the wavelength in feet and in centimeters.

Answer $\lambda = \dfrac{1130}{34.44 \times 10^3}$

$= 32.8 \times 10^{-3} \text{ ft}$

$\lambda = 0.0328 \text{ ft}$

To convert to inches:

$0.0328 \text{ ft} \times 12 = 0.3936 \text{ in.}$

To convert to centimeters:

$0.3936 \text{ in.} \times 2.54 = 1 \text{ cm}$ approximately

Note that for sound waves with a frequency of 34.44 kHz in this example, the wavelength is the same 1 cm as radio waves with the much higher frequency of 30 GHz, as calculated in Example 4. The reason is that radio waves have a much higher velocity.

Practice Problems 16-8
Answers at End of Chapter

Answer True or False.

a. The higher the frequency, the shorter the wavelength λ.

b. The higher the frequency, the longer the period T.

c. The velocity of propagation for radio waves in free space is 3×10^{10} cm/s.

16-9   Phase Angle

Referring back to Fig. 16-3, suppose that the generator started its cycle at point B, where maximum voltage output is produced, instead of starting at the point of zero output. If we compare the two cases, the two output voltage waves would be as in Fig. 16-10. Each is the same waveform of alternating voltage, but wave B starts at maximum, while wave A starts at zero. The complete cycle of wave B through 360° takes it back to the maximum value from which it started. Wave A starts and finishes its cycle at zero. With respect to time, therefore, wave B is ahead of wave A in its values of generated voltage. The amount it leads in time equals one quarter-revolution, which is 90°. This angular difference is the phase angle between waves B and A. Wave B leads wave A by the phase angle of 90°.

The 90° phase angle between waves B and A is maintained throughout the complete cycle and in all successive cycles, as long as they both have the same frequency. At any instant of time, wave B has the value that A will have 90° later. For instance, at 180° wave A is at zero, but B is already at its negative maximum value, where wave A will be later at 270°.

In order to compare the phase angle between two waves, they must have the same frequency. Otherwise, the relative phase keeps changing. Also, they must have sine-wave variations, as this is the only kind of waveform that is measured in angular units of time. The amplitudes can be different for the two waves, although they are shown the same here. We can compare the phase of two voltages, two currents, or a current with a voltage.

The 90° Phase Angle   The two waves in Fig. 16-10 represent a sine wave and a cosine wave 90° out of phase with each other. The 90° phase angle means that one has its maximum amplitude when the other is at zero value. Wave A starts at zero, corresponding to the sine of 0°, has its peak amplitude at 90 and 270°, and is back to zero after one cycle of 360°. Wave B starts at its peak value, corresponding to the cosine of 0°, has its zero value at 90 and 270°, and is back to the peak value after one cycle of 360°.

However, wave B can also be considered a sine wave that starts 90° before wave A in time. This phase angle of 90° for current and voltage waveforms has

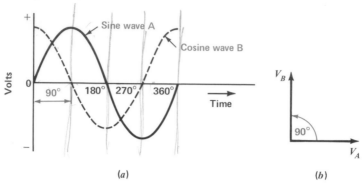

Fig. 16-10 Two voltage waveforms 90° out of phase. (a) Wave B leads wave A by 90°. (b) Corresponding phasors V_B and V_A for the two voltages with phase angle $\theta = 90°$. The right angle shows quadrature phase.

many applications in sine-wave ac circuits with inductance or capacitance.

The sine and cosine waveforms really have the same variations, but displaced by 90°. In fact, both waveforms are called *sinusoids*. The 90° angle is called *quadrature phase*.

Phase-Angle Diagrams   To compare phases of alternating currents and voltages, it is much more convenient to use phasor diagrams corresponding to the voltage and current waveforms, as shown in Fig. 16-10b. The arrows here represent the phasor quantities corresponding to the generator voltage.

A phasor is a quantity that has magnitude and direction. The length of the arrow indicates the magnitude of the alternating voltage, in rms, peak, or any ac value as long as the same measure is used for all the phasors. The angle of the arrow with respect to the horizontal axis indicates the phase angle.

The terms *phasor* and *vector* are used for a quantity that has direction, requiring an angle to specify the value completely. However, a vector quantity has direction in space, while a phasor quantity varies in time. As an example of a vector, a mechanical force can be represented by a vector arrow at a specific angle, with respect to either the horizontal or vertical direction.

For phasor arrows, the angles shown represent differences in time. One sinusoid is chosen as the reference. Then the timing of the variations in another sinusoid can be compared to the reference by means of the angle between the phasor arrows.

The phasor corresponds to the entire cycle of volt-

age, but is shown only at one angle, such as the starting point, since the complete cycle is known to be a sine wave. Without the extra details of a whole cycle, phasors represent the alternating voltage or current in a compact form that is easier for comparing phase angles.

In Fig. 16-10b, for instance, the phasor V_A represents the voltage wave A, with a phase angle of 0°. This angle can be considered as the plane of the loop in the rotary generator where it starts with zero output voltage. The phasor V_B is vertical to show the phase angle of 90° for this voltage wave, corresponding to the vertical generator loop at the start of its cycle. The angle between the two phasors is the phase angle.

The symbol for a phase angle is θ (theta). In Fig. 16-10, as an example, $\theta = 90°$.

Phase-Angle Reference   The phase angle of one wave can be specified only with respect to another as reference. How the phasors are drawn to show the phase angle depends on which phase is chosen as the reference. Generally, the reference phasor is horizontal, corresponding to 0°. Two possibilities are shown in Fig. 16-11. In Fig. 16-11a the voltage wave A or its phasor V_A is the reference. Then the phasor V_B is 90° counterclockwise. This method is standard practice, using counterclockwise rotation as the positive direction for angles. Also, a leading angle is positive. In this case, then, V_B is 90° counterclockwise from the reference V_A to show that wave B leads wave A by 90°.

However, wave B is shown as the reference in Fig. 16-11b. Now V_B is the horizontal phasor. In order to

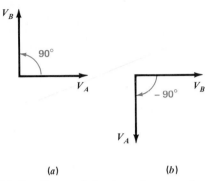

(a)   *(b)*

Fig. 16-11 Leading and lagging phase angles for 90°. (a) When phasor V_A is the horizontal reference, the phasor V_B leads by 90°. (b) When phasor V_B is the horizontal reference, the phasor V_A lags by −90°.

have the same phase angle, V_A must be 90° clockwise, or −90° from V_B. This arrangement shows that negative angles, clockwise from the 0° reference, are used to show a lagging phase angle. The reference determines whether the phase angle is considered leading or lagging in time.

The phase is not actually changed by the method of showing it. In Fig. 16-11, V_A and V_B are 90° out of phase, and V_B leads V_A by 90° in time. There is no fundamental difference whether we say V_B is ahead of V_A by +90° or V_A is behind V_B by −90°.

Two waves and their corresponding phasors can be out of phase by any angle, either less or more than 90°. For instance, a phase angle of 60° is shown in Fig. 16-12. For the waveforms in Fig. 16-12*a*, wave D is

behind C by 60° in time. For the phasors in Fig. 16-12*b* this lag is shown by the phase angle of −60°.

In-phase Waveforms A phase angle of 0° means the two waves are in phase (Fig. 16-13). Then the amplitudes add.

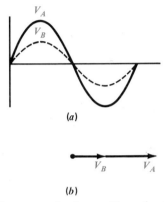

(a)

(b)

Fig. 16-13 Two waves in phase with angle of 0°. (a) Waveforms. (b) Phasor diagram.

Out-of-phase Waveforms An angle of 180° means opposite phase, or the two waveforms are exactly out of phase (Fig. 16-14). Then the amplitudes are opposing. Equal values of opposite phase cancel each other.

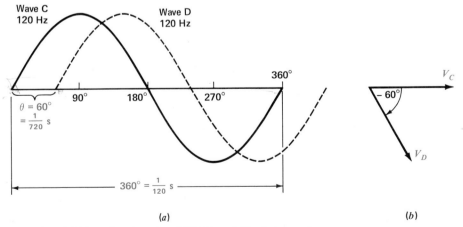

(a)   *(b)*

Fig. 16-12 Phase angle of 60° is the time for 60/360 or 1/6 of the cycle. (a) Waveforms. (b) Phasor diagram.

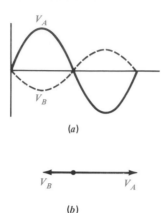

(a)

(b)

Fig. 16-14 Two waves out of phase with angle of 180°. *(a)* Waveforms. *(b)* Phasor diagram.

Practice Problems 16-9
Answers at End of Chapter

Give the phase angle in
a. Fig. 16-10.
b. Fig. 16-12.
c. Fig. 16-13.

16-10   The Time Factor in Frequency and Phase

It is important to remember that the waveforms we are showing are just graphs drawn on paper. The physical factors represented are variations in amplitude, usually on the vertical scale, with respect to equal intervals on the horizontal scale, which can represent either distance or time. To show wavelength, as in Fig. 16-9, the cycles of amplitude variations are plotted against distance or length units. To show frequency, the cycles of amplitude variations are shown with respect to time in angular measure. The angle of 360° represents the time for one cycle, or the period T.

As an example of how frequency involves time, a waveform with stable frequency is actually used in electronic equipment as a clock reference for very small units of time. Assume a voltage waveform with the frequency of 10 MHz. The period T is 0.1 μs. Every cycle is repeated at 0.1-μs intervals, therefore. When each cycle of voltage variations is used to indi-

cate time, then, the result is effectively a clock that measures 0.1-μs units. Even smaller units of time can be measured with higher frequencies. In everyday applications, an electric clock connected to the power line keeps correct time because it is controlled by the exact frequency of 60 Hz.

Furthermore, the phase angle between two waves of the same frequency indicates a specific difference in time. As an example, Fig. 16-12 shows a phase angle of 60°, with wave C leading wave D. They both have the same frequency of 120 Hz. The period T for each wave then is $1/120$ s. Since 60° is one-sixth of the complete cycle of 360°, this phase angle represents one-sixth of the complete period of $1/120$ s. Multiplying $1/6 \times 1/120$, the answer is $1/720$ s for the time corresponding to the phase angle of 60°. If we consider wave D lagging wave C by 60°, this lag is a time delay of $1/720$ s.

More generally, the time for a phase angle θ can be calculated as

$$t = \frac{\theta}{360} \times \frac{1}{f} \qquad (16\text{-}10)$$

With f in Hz and θ in degrees, then t is in seconds. The formula gives the time of the phase angle as its proportional part of the total period of one cycle. For the example of θ equal to 60° with f at 120 Hz,

$$t = \frac{\theta}{360} \times \frac{1}{f}$$

$$= \frac{60}{360} \times \frac{1}{120} = \frac{1}{6} \times \frac{1}{120}$$

$$t = \frac{1}{720} \text{ s}$$

Practice Problems 16-10
Answers at End of Chapter

a. In Fig. 16-12, how much time corresponds to 180°?
b. For two waves with the frequency of 1 MHz, how much time is the phase angle of 36°?

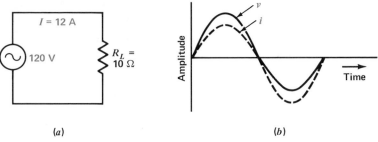

(a)   (b)

Fig. 16-15 An ac circuit with R alone. (a) Schematic diagram. (b) Waveforms showing I in phase with V.

16-11   AC Circuits with Resistance

An ac circuit has an ac voltage source. Note the symbol in Fig. 16-15 used for any source of sine-wave alternating voltage. This voltage connected across an external load resistance produces alternating current of the same waveform, frequency, and phase as the applied voltage.

The amount of current equals V/R by Ohm's law. When V is an rms value, I is also an rms value. For any instantaneous value of V during the cycle, the value of I is for the corresponding instant of time.

In an ac circuit with only resistance, the current variations are in phase with the applied voltage, as shown in Fig. 16-15b. This in-phase relationship between V and I means that such an ac circuit can be analyzed by the same methods used for dc circuits, since there is no phase angle to consider. Components that have R alone include resistors, the filaments for incandescent light bulbs, and vacuum-tube heaters.

The calculations in ac circuits are generally in rms values, unless noted otherwise. In Fig. 16-15a, for example, the 120 V applied across the 10-Ω R_L produces rms current of 12 A. The calculations are

$$I = \frac{V}{R_L} = \frac{120 \text{ V}}{10 \text{ }\Omega} = 12 \text{ A}$$

Furthermore, the rms power dissipation is I^2R, or

$$P = 144 \times 10 = 1440 \text{ W}$$

Series AC Circuit with R   In Fig. 16-16, R_T is 30 Ω, equal to the sum of 10 Ω for R_1 plus 20 Ω for R_2. The current in the series circuit is

$$I = \frac{V}{R_T} = \frac{120 \text{ V}}{30 \text{ }\Omega} = 4 \text{ A}$$

The 4-A current is the same in all parts of the series circuit. This principle applies for either an ac or a dc source and for R, L, and C components.

Next, we can calculate the series voltage drops in Fig. 16-16. With 4 A through the 10-Ω R_1, its IR voltage drop is

$$V_1 = I \times R_1 = 4 \text{ A} \times 10 \text{ }\Omega = 40 \text{ V}$$

The same 4 A through the 20-Ω R_2 produces an IR voltage drop of 80 V. The calculations are

$$V_2 = I \times R_2 = 4 \text{ A} \times 20 \text{ }\Omega = 80 \text{ V}$$

Note that the sum of 40 V for V_1 and 80 V for V_2 in series equals the 120 V applied.

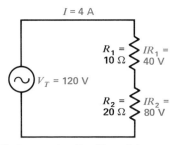

Fig. 16-16 Series ac circuit with resistance only.

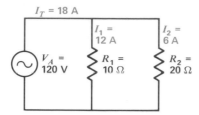

Fig. 16-17 Parallel ac circuit with resistance only.

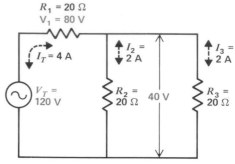

Fig. 16-18 Series-parallel ac circuit with resistance only.

Parallel AC Circuit with R   In Fig. 16-17, the 10-Ω R_1 and 20-Ω R_2 are in parallel across the 120-V ac source. Therefore, the voltage across the parallel branches is the same as the applied voltage.

Each branch current, then, is equal to 120 V divided by the branch resistance. The branch current for the 10-Ω R_1 is

$$I_1 = \frac{120 \text{ V}}{10 \text{ } \Omega} = 12 \text{ A}$$

The same 120 V is across the 20-Ω branch with R_2. Its branch current is

$$I_2 = \frac{120 \text{ V}}{20 \text{ } \Omega} = 6 \text{ A}$$

The total line current I_T is $12 + 6 = 18$ A, or the sum of the branch currents.

Series-Parallel AC Circuit with R   See Fig. 16-18. The 20-Ω R_2 and 20-Ω R_3 are in parallel, for an equivalent bank resistance of 20/2 or 10 Ω. This 10-Ω bank is in series with the 20-Ω R_1 in the main line, for a total of 30 Ω for R_T across the 120-V source. Therefore, the main line current produced by the 120-V source is

$$I_T = \frac{V}{R_T} = \frac{120 \text{ V}}{30 \text{ } \Omega} = 4 \text{ A}$$

The voltage drop across R_1 in the main line is calculated as

$$V_1 = I_T \times R_1 = 4 \text{ A} \times 20 \text{ } \Omega = 80 \text{ V}$$

Subtracting this 80-V drop from the 120 V of the source, the remaining 40 V is across the bank of R_2 and R_3 in parallel. Since the branch resistances are equal, the 4-A I_T divides equally, with 2 A in R_2 and 2 A in R_3. The branch currents can be calculated as

$$I_2 = \frac{40 \text{ V}}{20 \text{ } \Omega} = 2 \text{ A}$$
$$I_3 = \frac{40 \text{ V}}{20 \text{ } \Omega} = 2 \text{ A}$$

Note that the 2 A for I_1 and 2 A for I_2 in parallel branches add to equal the 4-A current in the main line.

AC Circuits with R and Reactance   The opposition of inductance and capacitance to sine-wave alternating current is called *reactance,* indicated as X. The symbols are X_L for inductive reactance and X_C for capacitive reactance. The opposition is measured in ohms, like resistance, but reactance has a phase angle of $\pm 90°$. The X_L is at $+90°$, while X_C is at $-90°$.

Resistance has a phase angle of $0°$. Reactance has a phase angle of $\pm 90°$. Therefore, when R and X are combined, the phase angle of the ac circuit is between 0 and $+90°$, or between 0 and $-90°$.

Furthermore, the ohms of resistance and reactance must be combined by phasor addition because of the $90°$ phase angle between the two. The resultant sum is called *impedance,* with the symbol Z, which is the total opposition of resistance and reactance to a sine-wave alternating current. The methods of combining R, X_L, and X_C to find the total Z and phase angle θ are explained in detail in Chap. 24, ''Alternating-Current Circuits.''

Practice Problems 16-11
Answers at End of Chapter

Calculate R_T in
a. Fig. 16-16.
b. Fig. 16-17.
c. Fig. 16-18.

16-12 Nonsinusoidal AC Waveforms

The sine wave is the basic waveform for ac variations for several reasons. This waveform is produced by a rotary generator, as the output is proportional to the angle of rotation. In addition, electronic oscillator circuits with inductance and capacitance naturally produce sine-wave variations.

Because of its derivation from circular motion, any sine wave can be analyzed in angular measure, either in degrees from 0 to 360° or in radians from 0 to 2π rad.

Another feature of a sine wave is its basic simplicity, as the rate of change for the amplitude variations corresponds to a cosine wave which is similar but 90° out of phase. The sine wave is the only waveform that has this characteristic of a rate of change with the same waveform as the original changes in amplitude.

In many electronic applications, however, other waveshapes are important. Any waveform that is not a sine or cosine wave is a *nonsinusoidal waveform*. Common examples are the square wave and sawtooth wave in Fig. 16-19.

With nonsinusoidal waveforms, for either voltage or current, there are important differences and similarities to consider. Note the following comparisons with sine waves.

1. In all cases, the cycle is measured between two points having the same amplitude and varying in the same direction. The period is the time for one cycle. In Fig. 16-19, T for any of the waveforms is 4 μs and the corresponding frequency is $1/T$, equal to ¼ MHz or 0.25 MHz.
2. Peak amplitude is measured from the zero axis to the maximum positive or negative value. However, peak-to-peak amplitude is better for measuring nonsinusoidal waveshapes because they can have unsymmetrical peaks, as in Fig. 16-19d. For all the

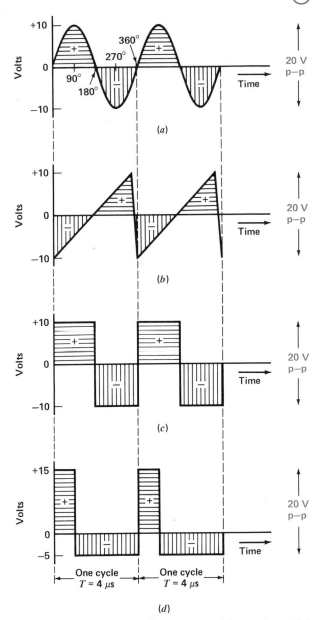

Fig. 16-19 Comparison of sine wave with nonsinusoidal ac waveforms. Two cycles shown. (a) Sine wave. (b) Sawtooth wave. (c) Symmetrical square wave. (d) Unsymmetrical rectangular wave or pulse waveform.

waveforms shown here, though, the peak-to-peak (p-p) amplitude is 20 V.
3. The rms value 0.707 of maximum applies only to sine waves, as this factor is derived from the sine values in the angular measure used only for the sine waveform.

4. Phase angles apply only to sine waves, as angular measure is used only for sine waves. Note that the horizontal axis for time is divided into angles for the sine wave in Fig. 16-19a, but there are no angles shown for the nonsinusoidal waveshapes.

5. All the waveforms represent ac voltages. Positive values are shown above the zero axis, with negative values below the axis.

The sawtooth wave in Fig. 16-19b represents a voltage that slowly increases, with a uniform or linear rate of change, to its peak value, and then drops sharply to its starting value. This waveform is also called a *ramp voltage*. It is also often referred to as a *time base* because of its constant rate of change.

Note that one complete cycle includes the slow rise and the fast drop in voltage. In this example, the period T for a complete cycle is 4 μs. Therefore, these sawtooth cycles are repeated at the frequency of ¼ MHz, which equals 0.25 MHz. The sawtooth waveform of voltage or current is often used for horizontal deflection of the electron beam in the cathode-ray tube (CRT) for oscilloscopes and TV receivers.

The square wave in Fig. 16-19c represents a switching voltage. First, the 10-V peak is instantaneously applied in positive polarity. This voltage remains on for 2 μs, which is one half-cycle. Then the voltage is instantaneously reduced to zero and applied in reverse polarity for another 2 μs. The complete cycle then takes 4 μs, and the frequency is ¼ MHz.

The rectangular waveshape in Fig. 16-19d is similar, but the positive and negative half-cycles are not symmetrical, either in amplitude or in time. However, the frequency is the same 0.25 MHz and the peak-to-peak amplitude is the same 20 V, as in all the waveshapes. This waveform shows pulses of voltage or current, repeated at a regular rate.

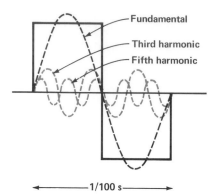

Fig. 16-20 Fundamental and harmonic frequencies for 100-Hz square wave.

16-13   Harmonic Frequencies

Consider a repetitive nonsinusoidal waveform, such as a 100-Hz square wave. Its fundamental rate of repetition is 100 Hz. Exact multiples of the fundamental frequency are called *harmonic frequencies*. The second harmonic is 200 Hz, the third harmonic is 300 Hz, etc. Even multiples are even harmonics, while odd multiples are odd harmonics.

Harmonics are useful in analyzing distorted sine waves or nonsinusoidal waveforms. Such waveforms consist of a pure sine wave at the fundamental frequency plus harmonic frequency components. For example, Fig. 16-20 illustrates how a square wave corresponds to a fundamental sine wave with odd harmonics. Typical audio waveforms include odd and even harmonics. It is the harmonic components that make one source of sound different from another with the same fundamental frequency.

Another unit for frequency multiples is the *octave,* which is a range of 2:1. Doubling the frequency range—from 100 to 200 Hz, from 200 to 400 Hz, and from 400 to 800 Hz, as examples—raises the frequency by one octave. The reason for this name is that an octave in music includes eight consecutive tones, for double the frequency. One-half the frequency is an octave lower.

Practice Problems 16-12
Answers at End of Chapter

a. In Fig. 16-19c, for how much time is the waveform at +10 V?

b. In Fig. 16-19d, what voltage is the positive peak amplitude?

Practice Problems 16-13
Answers at End of Chapter

a. What frequency is the fourth harmonic of 12 MHz?

b. Give the frequency one octave above 220 Hz.

16-14   The 60-Hz AC Power Line

Practically all homes in the United States are supplied alternating voltage at 115 to 125 V rms, with a frequency of exactly 60 Hz. This is a sine-wave voltage produced by a rotary generator. The electricity is distributed by power lines from the generating station to the main line in the home. Here the 120-V line is wired to all the wall outlets and electrical equipment in parallel. The 120-V source of commercial electricity is the *60-Hz power line* or the *mains,* indicating it is the main line for all the parallel branches.

Advantages   With an rms value of 120 V, the ac power is equivalent to 120-V dc power in heating effect. If the value were higher, there would be more danger of a fatal electric shock. Lower voltages would be less efficient in supplying power.

Higher voltage can supply electric power with less I^2R loss, since the same power is produced with less I. Note that the I^2R power loss increases as the square of the current. For industrial applications where large amounts of power are used, the main line is often 240 V single-phase or 208 V three-phase. Three-phase ac power is more efficient for the operation of large motors.

The advantage of ac over dc power is greater efficiency in distribution from the generating station. Ac voltages can easily be stepped up by means of a transformer, with very little loss, but a transformer cannot operate on direct current. The reason is that a transformer needs a varying current in the primary winding, with its varying magnetic field, to produce induced voltage in the secondary winding.

Therefore, the alternating voltage at the generating station can be stepped up to values as high as 80 kV for high-voltage distribution lines. These high-voltage lines supply large amounts of power with much less current and less I^2R loss, compared with a 120-V line. At the home, the lower voltage required is supplied by a step-down transformer. The step-up and step-down characteristics of a transformer refer to the ratio of voltages across the secondary winding compared with the primary winding.

The frequency of 60 Hz is convenient for commercial ac power. Much lower frequencies would require much bigger transformers because larger windings would be necessary. Also, too low a frequency for alternating current in a lamp could cause the light to flicker. For the opposite case, too high a frequency results in excessive iron-core heating in the transformer because of eddy currents and hysteresis losses. Based on these factors, 60 Hz is the frequency of the ac power line in the United States. It should be noted that the frequency of the ac power mains in England and most European countries is 50 Hz.

The 60-Hz Frequency Reference   All power companies in the United States, except those in Texas, are interconnected in a grid that maintains the ac power-line frequency between 59.98 and 60.02 Hz. The frequency is compared with the time standard provided by the Bureau of Standards radio station WWV at Fort Collins, Colorado. As a result the 60-Hz power-line frequency is maintained accurate to ± 0.033 percent. This accuracy makes the power-line voltage a good secondary standard for checking frequencies based on 60 Hz.

Residential Wiring   Most homes have at the electrical service entrance the three-wire power lines illustrated in Fig. 16-21. The three wires, including the grounded neutral, can be used for either 240 or 120 V single phase. They cannot be used for three-phase power.

Note the color coding for the wiring in Fig. 16-21. The grounded neutral is white, or bare wire is used. Each high side can use any color except white or green,

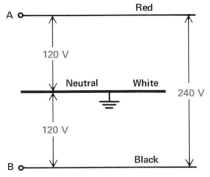

Fig. 16-21 Three-wire single-phase power lines for either 240 or 120 V.

but usually they use black[1] or red. White is reserved for the neutral wire, and green is reserved for grounding.

From either the red or black high side to the neutral, 120 V is available for separate branch circuits to the lights and outlets. Across the red and black wires, 240 V is available for high-power appliances. This three-wire service with a grounded neutral is called the *Edison system*.

The electrical service is commonly rated for 100 A. At 240 V, then, the power available is $100 \times 240 = 24,000$ W, or 24 kW.

The main wires to the service entrance, where the power enters the house, are generally No. 4 to 8 gage. Sizes 6 and heavier are always stranded wire. The 120-V branch circuits, usually rated at 15 A, use No. 8 to 14 gage wire. Each branch has its own fuse or circuit breaker. A main switch should be included to cut off all power from the service entrance.

The neutral wire is grounded at the service entrance to a water pipe or a metal rod driven into the earth. All 120-V branches must have one side connected to the grounded neutral. White wire is used for these connections. In addition, all the metal boxes for outlets, switches, and lights must have a continuous ground to each other and to the neutral. The wire cable usually has a bare wire for this grounding of the boxes.

Cables commonly used are armored sheath with the trade name BX and nonmetallic flexible cable with the trade name Romex. Each has two or more wires for the neutral, high-side connections, and grounding. In addition, the metal sheath is used for continuous grounding

---

[1]It should be noted that in electronic equipment black is the color-coded wiring used for chassis-ground returns. However, in electric power work, black wire is used for high-side connections.

between boxes. The flexible cable usually has an extra bare wire for grounding.

The purpose of grounding is safety against electric shock. Switches and fuses are never in the ground side of the line, in order to maintain the ground connections.

Rules for grounding and wiring are specified by local electrical codes. The National Electrical Code standards are available from the National Fire Protection Association.

Three-Phase Power   In an alternator with three generator windings equally spaced around the circle, they will produce output voltages 120° out of phase with each other. The three-phase output is illustrated by the sine-wave voltages in Fig. 16-22a and the corresponding phasors in Fig. 16-22b. The advantage of three-phase ac voltage is more efficient distribution of power. Also, ac induction motors are self-starting with three-phase alternating current. Finally, the ac ripple is easier to filter in the rectified output of a dc power supply.

In Fig. 16-23a, the three windings are in the form of a Y, also called *Wye* or *star* connections. All three coils are joined at one end, with the opposite ends for the output terminals A, B, and C. Note that any pair of terminals is across two coils in series. Each coil has 120 V. The voltage output across any two output terminals is $120 \times 1.73 = 208$ V, because of the 120° phase angle.

In Fig. 16-23b, the three windings are connected in the form of a *delta* (Δ). Any pair of terminals is across one generator winding. However, the other coils are in a parallel branch. Therefore, the current capacity to the line is increased by the factor 1.73.

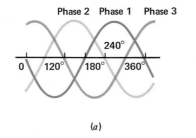

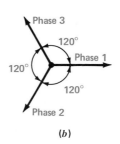

(a)   (b)

Fig. 16-22 Three-phase alternating voltage or current with 120° between each phase. (a) Sine waves. (b) Phasor diagram.

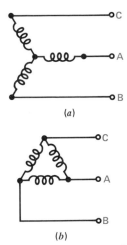

Fig. 16-23 The types of connections for three-phase ac power. (a) Y or wye. (b) Delta or Δ.

Practice Problems 16-14
Answers at End of Chapter

Answer True or False.
a. The 120 V of the ac power line is a peak-to-peak value.
b. The frequency of the ac power-line voltage is 60 Hz ± 0.033 percent.
c. In Fig. 16-21 the voltage between black and white wires is 120 V.
d. In Fig. 16-23a the voltage between terminals A and B is 208 V.

16-15 Motors and Generators

A generator converts mechanical energy into electric energy; a motor does the opposite, converting electricity into rotary motion. The main parts in the assembly of motors and generators are essentially the same (Fig. 16-24).

Armature In a generator, the armature connects to the external circuit to provide the generator output voltage. In a motor, it connects to the electrical source that drives the motor. The armature is often constructed in the form of a drum, using many conductor loops for increased output. In Fig. 16-24 the rotating armature is the *rotor* part of the assembly.

Field Winding This electromagnet provides the flux cut by the rotor. In a motor, current for the field is produced by the same source that supplies the armature. In a generator, the field current may be obtained from a separate exciter source, or from its own armature output. Residual magnetism in the iron yoke of the field allows this *self-excited generator* to start.

The field coil may be connected in series with the armature, in parallel, or in a series-parallel *compound winding*. When the field winding is stationary, it is the *stator* part of the assembly.

Slip Rings In an ac machine, two or more slip rings or *collector rings* enable the rotating loop to be connected to the stationary wire leads for the external circuit.

Brushes These graphite connectors are spring-mounted to brush against the spinning rings on the rotor. The stationary external leads are connected to the brushes for connection to the rotating loop. Constant rubbing slowly wears down the brushes, and they must be replaced after they are worn.

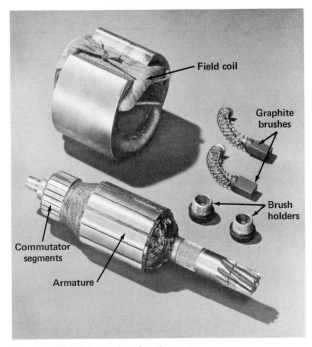

Fig. 16-24 Main parts of a dc motor.

Commutator A dc machine has a commutator ring instead of the slip rings. As shown in Fig. 16-24, the commutator ring has segments, with one pair for each loop in the armature. Each of the commutator segments is insulated from the others by mica.

The commutator converts the ac machine to dc operation. In a generator, the commutator segments reverse the loop connections to the brushes every half-cycle to maintain a constant polarity of output voltage. For a dc motor, the commutator segments allow the dc source to produce torque in one direction.

Brushes are necessary with a commutator ring. The two stationary brushes contact opposite segments on the rotating commutator. Graphite brushes are used for very low resistance.

AC Induction Motor This type, for alternating current (ac) only, does not have any brushes. The stator is connected directly to the ac source. Then alternating current in the stator winding induces current in the rotor without any physical connection between them. The magnetic field of the current induced in the rotor reacts with the stator field to produce rotation. Alternating-current induction motors are economical and rugged, without any troublesome brush arcing.

With a single-phase source, however, a starting torque must be provided for an ac induction motor. One method uses a starting capacitor in series with a separate starting coil. The capacitor supplies an out-of-phase current just for starting, and then is switched out. Another method of starting uses shaded poles. A solid copper ring on the main field pole makes the magnetic field unsymmetrical to allow starting.

The rotor of an ac induction motor may be wire-wound or the squirrel-cage type. This rotor is constructed with a frame of metal bars.

Universal Motor This type operates on either alternating or direct current because the field and armature are in series. Its construction is like that of a dc motor, with the rotating armature connected to a commutator and brushes. The universal motor is commonly used for small machines such as portable drills and food mixers.

Alternators Ac generators are alternators. For large power requirements, the alternator usually has a rotor field, while the armature is the stator. This method eliminates slip-ring connections, with their arcing problems, in the high-voltage output.

Practice Problems 16-15
Answers at End of Chapter

Answer True or False.
a. In Fig. 16-24 the commutator segments are on the armature.
b. Motor brushes are made of graphite for very low resistance.
c. A starting capacitor is used with dc motors that have small brushes.

Summary

1. Alternating voltage continuously varies in magnitude and reverses in polarity. When alternating voltage is applied across a load resistance, the result is alternating current in the circuit.
2. A complete set of values repeated periodically is one cycle of the ac waveform. The cycle can be measured from any one point on the wave to the next successive point having the same value and varying in the same direction. One cycle includes 360° in angular measure, or 2π rad.
3. The rms value of a sine wave is $0.707 \times$ peak value.
4. The peak amplitude, at 90 and 270° in the cycle, is $1.414 \times$ rms value.

5. The peak-to-peak value is double the peak amplitude, or 2.828 × rms for a symmetrical ac waveform.

6. The average value is 0.637 × peak value.

7. The frequency equals the number of cycles per second. One cps is 1 Hz. The audio-frequency (AF) range is 16 to 16,000 Hz. Higher frequencies up to 300,000 MHz are radio frequencies (RF).

8. The amount of time for one cycle is the period T. The period and frequency are reciprocals: $T = 1/f$, or $f = 1/T$. The higher the frequency, the shorter the period.

9. Wavelength λ is the distance a wave travels in one cycle. The higher the frequency, the shorter the wavelength. The wavelength also depends on the velocity at which the wave travels: $\lambda = $ velocity/frequency.

10. Phase angle is the angular difference in time between corresponding values in the cycles for two waveforms of the same frequency.

11. When one sine wave has its maximum value while the other is at zero, the two waves are 90° out of phase. Two waveforms with zero phase angle between them are in phase; a 180° phase angle means opposite phase.

12. The length of a phasor arrow indicates amplitude, while the angle corresponds to the phase. Leading phase is shown by counterclockwise angles.

13. Sine-wave alternating voltage V applied across a load resistance R produces alternating current I in the circuit. The current has the same waveform, frequency, and phase as the applied voltage because of the resistive load. The amount of $I = V/R$.

14. The sawtooth wave and square wave are two common examples of nonsinusoidal waveforms. The amplitudes of these waves are usually measured in peak-to-peak value.

15. Harmonic frequencies are exact multiples of the fundamental frequency.

16. The ac power line has a nominal voltage of 115 to 125 V rms and the exact frequency of 60 Hz.

17. For residential wiring, the three-wire single-phase Edison system shown in Fig. 16-21 is used to provide either 120 or 240 V.

18. In three-phase power, each phase angle is 120°. For the Y connections in Fig. 16-23a, any pair of output terminals has output of 120 × 1.73 = 208 V.

19. In a motor, the rotating armature connects to the power line. The stator field coils provide the magnetic flux cut by the armature as it is forced to rotate. A generator has the opposite effect; it converts mechanical energy into electrical output.

20. A dc motor has commutator segments contacted by graphite brushes for the external connections to the power source. An ac induction motor does not have brushes.

Self-Examination
Answers at Back of Book

Answer True or False.

1. An ac voltage varies in magnitude and reverses in polarity.
2. A dc voltage always has one polarity.

3. Sine-wave alternating current flows in a load resistor with sine-wave voltage applied.

4. When two waves are 90° out of phase, one has its peak value when the other is at zero.

5. When two waves are in phase, they have their peak values at the same time.

6. The positive peak of a sine wave cannot occur at the same time as the negative peak.

7. The angle of 90° is the same as π rad.

8. A period of 2 μs corresponds to a higher frequency than T of 1 μs.

9. A wavelength of 2 ft corresponds to a lower frequency than a wavelength of 1 ft.

10. When we compare the phase between two waveforms, they must have the same frequency.

Fill in the missing answers.

11. For the rms voltage of 10 V, the p-p value is _____ V.
12. With 120 V rms across 100 Ω R_L, the rms current equals _____ A.
13. For a peak value of 100 V, the rms value is _____ V.
14. The wavelength of a 1000-kHz radio wave is _____ cm.
15. The period of a 1000-kHz voltage is _____ ms.
16. The period of $\frac{1}{60}$ s corresponds to a frequency of _____ Hz.
17. The frequency of 100 MHz corresponds to a period of _____ μs.
18. The square wave in Fig. 16-19c has the frequency of _____ MHz.
19. The rms voltage for the sine wave in Fig. 16-19a is _____ V.
20. The ac voltage across R_2 in Fig. 16-18 is _____ V.
21. For an audio signal with a T of 0.001 s, its frequency is _____ Hz.
22. For the 60-Hz ac power-line voltage, the third harmonic is _____ Hz.
23. For a 10-V average value, the rms value is _____ V.
24. For a 340-V p-p value, the rms value is _____ V.
25. An audio signal that produces four cycles in the time it takes for one cycle of ac voltage from the power line has the frequency of _____ Hz.
26. In Fig. 16-21, the voltage between the red and black wires is _____ V.
27. In Fig. 16-22, the angle between the three phases is _____ degrees.
28. In Fig. 16-23a, the voltage between terminals B and C is _____ V.

Essay Questions

1. (a) Define an alternating voltage. (b) Define an alternating current. (c) Why does ac voltage applied across a load resistance produce alternating current in the circuit?

2. (a) State two characteristics of a sine wave of voltage. (b) Why does the rms value of 0.707 × peak value apply just to sine waves?

3. Draw two cycles of an ac sawtooth voltage waveform with a peak-to-peak amplitude of 40 V. Do the same for a square wave.

4. Give the angle, in degrees and radians, for each of the following: one cycle, one half-cycle, one quarter-cycle, three quarter-cycles.

5. The peak value of a sine wave is 1 V. How much is its average value? Rms value? Effective value? Peak-to-peak value?

6. State the following ranges in Hz: (a) audio frequencies; (b) radio frequencies; (c) standard AM radio broadcast band; (d) FM broadcast band; (e) VHF band; (f) microwave band. (Hint: See App. B.)

7. Make a graph with two waves, one with a frequency of 500 kHz and the other with 1000 kHz. Mark the horizontal axis in time, and label each wave.

8. Draw the sine waves and phasor diagrams to show (a) two waves 180° out of phase; (b) two waves 90° out of phase.

9. Give the voltage value for the 60-Hz ac line voltage with an rms value of 120 V at each of the following times in a cycle: 0°, 30°, 45°, 90°, 180°, 270°, 360°.

10. (a) The phase angle of 90° equals how many radians? (b) For two sine waves 90° out of phase with each other, compare their amplitudes at 0°, 90°, 180°, 270°, and 360°.

11. Tabulate the sine and cosine values every 30° from 0 to 360° and draw the corresponding sine wave and cosine wave.

12. Draw a graph of the values for $(\sin \theta)^2$ plotted against θ for every 30° from 0 to 360°.

13. Why is the wavelength of a supersonic wave at 34.44 kHz the same 1 cm as for the much higher-frequency radio wave at 30 GHz?

14. Draw the sine waves and phasors to show wave V_1 leading wave V_2 by 45°.

15. Why are amplitudes for nonsinusoidal waveforms generally measured in peak-to-peak values, rather than rms or average value?

16. Define harmonic frequencies, giving numerical values.

17. Define one octave, with an example of numerical values.

18. Which do you consider more important for applications of alternating current—the polarity reversals or the variations in value?

19. Define the following parts in the assembly of motors: (a) armature rotor; (b) field stator; (c) collector rings; (d) commutator segments.

20. Show diagrams of Y and Δ connections for three-phase ac power.

Problems
Answers to Odd-Numbered Problems at Back of Book

1. The 60-Hz power-line voltage of 120 V is applied across a resistance of 20 Ω. (a) How much is the rms current in the circuit? (b) What is the frequency of the current? (c) What is the phase angle between the current and the voltage? (d) How much dc applied voltage would be necessary for the same heating effect in the resistance?

2. What is the frequency for the following ac variations? (a) 10 cycles in 1 s; (b) 1 cycle in ¹⁄₁₀ s; (c) 50 cycles in 1 s; (d) 50 cycles in ½ s; (e) 50 cycles in 5 s.

3. Calculate the time delay for a phase angle of 45° at the frequency of (a) 500 Hz; (b) 2 MHz.

4. Calculate the period T for the following frequencies: (a) 500 Hz; (b) 5 MHz; (c) 5 GHz.

5. Calculate the frequency for the following periods: (a) 0.05 s; (b) 5 ms; (c) 5 µs; (d) 5 ns.

6. Referring to Fig. 16-18, calculate the I^2R power dissipated in R_1, R_2, and R_3.

7. Give the plus and minus peak values for each wave in Fig. 16-19a to d.
8. An ac circuit has a 5-MΩ resistor R_1 in series with a 10-MΩ resistor R_2 across a 200-V source. Calculate I, V_1, V_2, P_1, and P_2.
9. The two resistors in Prob. 8 are in parallel. Calculate I_1, I_2, V_1, V_2, P_1, and P_2.
10. A series-parallel ac circuit has two branches across the 60-Hz 120-V power line. One branch has a 10-Ω R_1 in series with a 20-Ω R_2. The other branch has a 10-MΩ R_3 in series with a 20-MΩ R_4. Find V_1, V_2, V_3, and V_4.
11. How much rms I does a 300-W 120-V bulb take from a 120-V 60-Hz line?
12. In Fig. 16-25, calculate V_{rms}, period T, and frequency f.

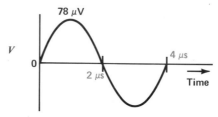

Fig. 16-25 For Prob. 12.

13. A sine-wave ac voltage has an rms value of 19.2 V. (**a**) Find the peak value. (**b**) What is the instantaneous value at 50° of the cycle?
14. In Fig. 16-26, calculate I, V_1, V_2, and V_3.
15. In Fig. 16-27, calculate I_1, I_2, I_3, and I_T.

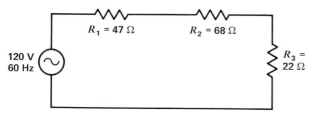

Fig. 16-26 For Prob. 14.

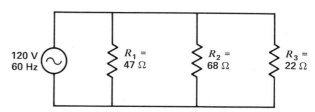

Fig. 16-27 For Prob. 15.

Answers to Practice Problems

16-1 a. T	**16-5 a.** 120 V rms	**16-9 a.** 90°	**16-12 a.** 2 μs
b. F	**b.** 14.14 V peak	**b.** 60°	**b.** 15 V
c. T	**16-6 a.** 4 Hz	**c.** 0°	**16-13 a.** 48 MHz
16-2 a. 10 V	**b.** 1.605 MHz	**16-10 a.** $\frac{1}{240}$ s	**b.** 440 Hz
b. 360°	**16-7 a.** 400 Hz	**b.** 0.1 μs	**16-14 a.** F
16-3 a. 85 V	**b.** $\frac{1}{400}$ s	**16-11 a.** 30 Ω	**b.** T
b. 120 V	**16-8 a.** T	**b.** 6.67 Ω	**c.** T
c. 170 V	**b.** F	**c.** 30 Ω	**d.** T
16-4 a. 0.707 A	**c.** T		**16-15 a.** T
b. 0.5 A			**b.** T
			c. F

Review

Chapters 13 to 16

Summary

1. Iron, nickel, and cobalt are magnetic materials. Magnets have a north pole and a south pole at opposite ends. Opposite poles attract; like poles repel.

2. A magnet has an invisible, external magnetic field. This magnetic flux is indicated by field lines. The direction of field lines outside the magnet is from north pole to south pole.

3. A permanent magnet is made of a hard magnetic material, such as alnico, to retain its magnetism indefinitely. Iron is a soft magnetic material which can be magnetized temporarily.

4. An electromagnet has an iron core that becomes magnetized when current flows in the coil winding.

5. Magnetic units are defined in Tables 13-1 and 14-2.

6. Continuous magnetization and demagnetization of an iron core by means of alternating current causes hysteresis losses, which increase with higher frequencies.

7. Ferrites are ceramic magnetic materials that are insulators.

8. Current in a conductor has an associated magnetic field with circular lines of force in a plane perpendicular to the wire. Their direction is counterclockwise when you look along the conductor in the direction of electron flow.

9. Motor action results from the net force of two fields that can aid or cancel. The direction of the resultant force is from the stronger field to the weaker.

10. The motion of magnetic flux cutting across a perpendicular conductor generates an induced emf. The amount of induced voltage increases with higher frequencies, more flux, and more turns of conductor.

11. Faraday's law of induced voltage is $v = N \, d\phi/dt$, where N is the turns and $d\phi/dt$ is the change in flux in webers per second.

12. Lenz' law states that an induced voltage must have the polarity that opposes the change causing the induction.

13. Alternating voltage varies in magnitude and reverses in direction. An ac voltage source produces alternating current.

14. One cycle includes the values between points having the same value and varying in the same direction. The cycle includes 360°, or 2π rad.
15. Frequency f equals the cycles per second (cps). One cps = 1 Hz.
16. Period T is the time for one cycle. It equals $1/f$. When f is in cycles per second, T is in seconds.
17. Wavelength λ is the distance a wave travels in one cycle. $\lambda = v/f$.
18. The rms, or effective, value of a sine wave equals $0.707 \times$ peak value. Or the peak value equals $1.414 \times$ rms value. The average value equals $0.637 \times$ peak value.
19. Phase angle θ is the angular difference in time between corresponding values in the cycles for two sine waves of the same frequency.
20. Phasors, similar to vectors, indicate the amplitude and phase angle of alternating voltage or current. The length of the phasor is the amplitude, while the angle is the phase.
21. The square wave and sawtooth wave are common examples of nonsinusoidal waveforms.
22. Harmonic frequencies are exact multiples of the fundamental frequency.
23. Dc motors generally use commutator segments with graphite brushes. Ac motors are usually the induction type without brushes.
24. Residential wiring generally uses three-wire single-phase power with the exact frequency of 60 Hz. Voltages available are 120 V to the grounded neutral or 240 V across the two high sides.
25. Three-phase ac power has three legs 120° out of phase. With Y connections, 208 V is available across any two legs.

Review Self-Examination
Answers at Back of Book

Choose (a), (b), (c), or (d).

1. Which of the following statements is true? (a) Alnico is commonly used for electromagnets. (b) Paper cannot affect magnetic flux because it is not a magnetic material. (c) Iron is generally used for permanent magnets. (d) Ferrites have lower permeability than air or vacuum.
2. Hysteresis losses (a) are caused by high-frequency alternating current in a coil with an iron core; (b) generally increase with direct current in a coil; (c) are especially important with permanent magnets that have a steady magnetic field; (d) cannot be produced in an iron core, because it is a conductor.
3. A magnetic flux of 25,000 lines through an area of 5 cm² results in (a) 5 lines of flux; (b) 5000 Mx of flux; (c) flux density of 5000 G; (d) flux density corresponding to 25,000 A.
4. If 10 V is applied across a relay coil with 100 turns having 2 Ω of resistance, the total force producing magnetic flux in the circuit is (a) 10 Mx; (b) 50 G; (c) 100 Oe; (d) 500 A·t.
5. The ac power-line voltage of 120 V rms has a peak value of (a) 100 V; (b) 170 V; (c) 240 V; (d) 338 V.
6. Which of the following can produce the most induced voltage? (a) 1-A direct

current; (b) 50-A direct current; (c) 1-A 60-Hz alternating current; (d) 1-A 400-Hz alternating current.

7. Which of the following has the highest frequency? (a) $T = \frac{1}{1000}$ s; (b) $T = \frac{1}{60}$ s; (c) $T = 1$ s; (d) $T = 2$ s.

8. Two waves of the same frequency have opposite phase when the phase angle between them is (a) 0°; (b) 90°; (c) 360°; (d) π rad.

9. The 120-V 60-Hz power-line voltage is applied across a 120-Ω resistor. The current equals (a) 1 A, peak value; (b) 120 A, peak value; (c) 1 A, rms value; (d) 5 A, rms value.

10. When an alternating voltage reverses in polarity, the current it produces (a) reverses in direction; (b) has a steady dc value; (c) has a phase angle of 180°; (d) alternates at 1.4 times the frequency of the applied voltage.

11. In Fig. 16-21, the voltage from either A or B to the grounded neutral is (a) 240; (b) 208; (c) 170; (d) 120.

12. In Fig. 16-23, the voltage across any one coil is (a) 120 V, single-phase; (b) 120 V, three-phase; (c) 208 V, single-phase; (d) 208 V, three-phase.

References
Additional References at Back of Book

Books

Adams, J. E., and G. Rockmaker: *Industrial Electricity—Principles and Practices,* McGraw-Hill Book Company, New York.

Duff, J. R., and M. Kaufman: *Alternating Current Fundamentals,* Delmar Publishers, Albany, New York.

NFPA: *National Electrical Code Handbook,* National Fire Protection Association, Quincy, Mass.

Richter, H. P., and W. P. Schwan: *Practical Electrical Wiring,* McGraw-Hill Book Company, New York.

Tocci, R. J.: *Introduction to Electric Circuit Analysis,* Charles E. Merrill Publishing Company, Columbus, Ohio.

Chapter 17
Inductance

Inductance is the ability of a conductor to produce induced voltage when the current varies. A long wire has more inductance than a short wire, since more conductor length cut by magnetic flux produces more induced voltage. Similarly, a coil has more inductance than the equivalent length of straight wire because the coil concentrates magnetic flux. Components manufactured to have a definite value of inductance are just coils of wire, therefore, called *inductors*. Coils can be wound around hollow forms so that air is part of the magnetic circuit. Other coils are wound around iron cores. At the radio frequency range air-core inductors are used to reduce RF current. Iron-core inductors are used in the audio frequency range and lower frequencies in general.

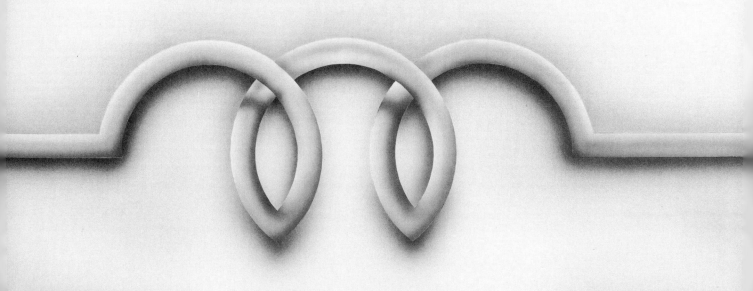

Important terms in this chapter are:

air-core   leakage flux
autotransformer   Lenz' law
coupling coefficient   magnetic coupling
eddy current   mutual inductance
efficiency   self-inductance
ferrite   stray inductance
henry unit   transformer
hysteresis   turns ratio
iron-core   variac

The construction, operation, and uses of inductors are explained in the following sections:

17-1   Induction by Alternating Current

Induced voltage is the result of flux cutting across a conductor. This action can be produced by physical motion of either the magnetic field or the conductor. When the current in a conductor varies in amplitude, however, the variations of current and its associated magnetic field are equivalent to motion of the flux. As the current increases in value, the magnetic field expands outward from the conductor. When the current decreases, the field collapses into the conductor. As the field expands and collapses with changes of current, the flux is effectively in motion. Therefore, a varying current can produce induced voltage without the need for motion of the conductor.

Figure 17-1 illustrates the changes in magnetic field associated with a sine wave of alternating current. Since the alternating current varies in amplitude and reverses in direction, its associated magnetic field has the same variations. At point A, the current is zero and there is no flux. At B, the positive direction of current provides some field lines taken here in the counterclockwise direction. Point C has maximum current and maximum counterclockwise flux.

At D there is less flux than at C. Now the field is collapsing because of the reduced current. At E, with zero current, there is no magnetic flux. The field can be considered as having collapsed into the wire.

The next half-cycle of current allows the field to expand and collapse again, but the directions are reversed. When the flux expands at points F and G, the field lines are clockwise, corresponding to current in the negative direction. From G to H and I, this clockwise field collapses into the wire.

The result of an expanding and collapsing field, then, is the same as that of a field in motion. This moving flux cuts across the conductor that is providing the current, producing induced voltage in the wire itself. Furthermore, any other conductor in the field, whether carrying current or not, also is cut by the varying flux and has induced voltage.

It is important to note that induction by a varying current results from the change in current, not the current value itself. The current must change to provide motion of the flux. A steady direct current of 1000 A, as an example of a large current, cannot produce any induced voltage as long as the current value is constant. A current of 1 μA changing to 2 μA, however, does induce voltage. Also, the faster the current changes, the higher the induced voltage because when the flux moves at a higher speed, it can induce more voltage.

Since inductance is a measure of induced voltage, the amount of inductance has an important effect in any circuit in which the current changes. The inductance is an additional characteristic of the circuit besides its resistance. The characteristics of inductance are important in:

1. *AC circuits*. Here the current is continuously changing and producing induced voltage. Lower frequencies of alternating current require more inductance to produce the same amount of induced voltage as a higher-frequency current. The current can have any waveform, as long as the amplitude is changing.

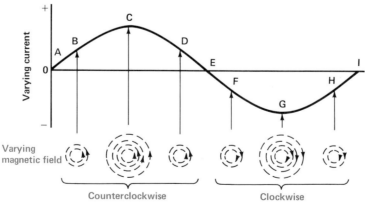

Fig. 17-1 Magnetic field of an alternating current is effectively in motion as it expands and contracts with the current variations.

2. *DC circuits in which the current changes in value*. It is not necessary for the current to reverse direction. One example is a dc circuit being turned on or off. When the direct current is changing between zero and its steady value, the inductance affects the circuit at the time of switching. This effect with a sudden change is called the *transient response*. A steady direct current that does not change in value is not affected by inductance, however, because there can be no induced voltage without a change in current.

Practice Problems 17-1
Answers at End of Chapter

a. Which has more inductance, a coil with an iron core or one without an iron core?
b. In Fig. 17-1, are the changes of current faster at time B or C?

17-2   Self-inductance *L*

The ability of a conductor to induce voltage in itself when the current changes is its *self-inductance* or simply *inductance*. The symbol for inductance is *L*, for linkages of the magnetic flux, and its unit is the *henry* (H). This unit is named after Joseph Henry (1797–1878).

Definition of the Henry Unit
As illustrated in Fig. 17-2, one henry is the amount of inductance that allows one volt to be induced when the current changes at the rate of one ampere per second. The formula is

$$L = \frac{v_L}{di/dt} \qquad\qquad (17\text{-}1)$$

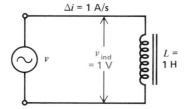

Fig. 17-2 When a change of 1 A/s induces 1 V across *L*, its inductance equals 1 H.

where v_L is in volts and di/dt is the current change in amperes per second.

Again the symbol *d* is used for *delta* (Δ) to indicate a small change. The factor di/dt for the current variation with respect to time really specifies how fast the current's associated magnetic flux is cutting the conductor to produce v_L.

Example 1 The current in an inductor changes from 12 to 16 A in 1 s. How much is the di/dt rate of current change in amperes per second?

Answer The *di* is the difference between 16 and 12, or 4 A in 1 s. Then

$$\frac{di}{dt} = \frac{4\text{ A}}{\text{s}}$$

Example 2 The current in an inductor changes by 50 mA in 2 μs. How much is the di/dt rate of current change in amperes per second?

Answer $$\frac{di}{dt} = \frac{50 \times 10^{-3}}{2 \times 10^{-6}}$$
$$= 25 \times 10^{3}$$
$$\frac{di}{dt} = 25{,}000\ \frac{\text{A}}{\text{s}}$$

Example 3 How much is the inductance of a coil that induces 40 V when its current changes at the rate of 4 A/s?

Answer $$L = \frac{v_L}{di/dt}$$
$$= \frac{40}{4}$$
$$L = 10\text{ H}$$

Example 4 How much is the inductance of a coil that induces 1000 V when its current changes at the rate of 50 mA in 2 μs?

Answer For this example, the $1/dt$ factor in the denominator of Formula (17-1) can be inverted to the numerator.

$$L = \frac{v_L}{di/dt} = \frac{v_L \times dt}{di}$$

$$= \frac{1 \times 10^3 \times 2 \times 10^{-6}}{50 \times 10^{-3}}$$

$$= \frac{2 \times 10^{-3}}{50 \times 10^{-3}} = \frac{2}{50}$$

$$L = 0.04 \text{ H}$$

Notice that the smaller inductance in Example 4 produces much more v_L than the inductance in Example 3. The very fast current change in Example 4 is equivalent to 25,000 A/s.

Inductance of Coils

In terms of physical construction, the inductance depends on how a coil is wound.[1] Note the following factors.

1. A greater number N of turns increases L because more voltage can be induced. Actually L increases in proportion to N^2. Double the number of turns in the same area and length increases the inductance four times.
2. More area A for each turn increases L. This means a coil with larger turns has more inductance. The L increases in direct proportion to A and as the square of the diameter of each turn.
3. The L increases with the permeability of the core. For an air core μ_r is 1. With a magnetic core, L is increased by the μ_r factor as the magnetic flux is concentrated in the coil.
4. The L decreases with more length for the same number of turns, as the magnetic field then is less concentrated.

These physical characteristics of a coil are illustrated in Fig. 17-3. For a long coil, where the length is at least ten times the diameter, the inductance can be calculated from the formula

$$L = \mu_r \times \frac{N^2 \times A}{l} \times 1.26 \times 10^{-6} \qquad \text{H} \qquad \textbf{(17-2)}$$

where l is in meters and A is in square meters. The constant factor 1.26×10^{-6} is the absolute permeabil-

---

[1]Methods of winding coils for a specific L are described in the *A.R.R.L. Handbook* and in *Bulletin* 74 of the National Bureau of Standards.

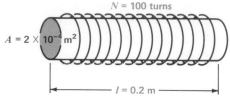

Fig. 17-3 Physical factors for inductance of a coil. See text for calculating L.

ity of air or vacuum, in SI units, to calculate L in henrys.

For the air-core coil in Fig. 17-3,

$$L = 1 \times \frac{10^4 \times 2 \times 10^{-4}}{0.2} \times 1.26 \times 10^{-6}$$

$$= 12.6 \times 10^{-6} \text{ H}$$

$$L = 12.6 \ \mu\text{H}$$

This value means that the coil can produce a self-induced voltage of 12.6 μV when its current changes at the rate of 1 A/s, as $v_L = L(di/dt)$. Furthermore, if the coil has an iron core with $\mu_r = 100$, then L will be 100 times greater.

Typical Coil Inductance Values

Air-core coils for RF applications have L values in millihenrys (mH) and microhenrys (μH). A typical air-core RF inductor (called a *choke*) is shown with its schematic symbol in Fig. 17-4a. Note that

$$1 \text{ mH} = 1 \times 10^{-3} \text{ H}$$

$$1 \ \mu\text{H} = 1 \times 10^{-6} \text{ H}$$

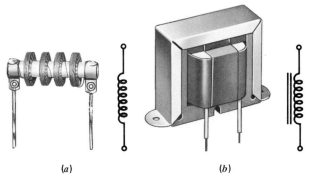

(a) (b)

Fig. 17-4 Typical inductors with their schematic symbols. (a) Air-core coil used as RF choke. Length is 2 in. (b) Iron-core coil used for 60 Hz. Height is 2 in.

For example, an RF coil for the radio broadcast band of 535 to 1605 kHz may have an inductance L of 250 μH or 0.250 mH.

Iron-core inductors for the 60-Hz power line and for audio frequencies have inductance values of about 1 to 25 H. A typical iron-core choke is shown in Fig. 17-4b.

Practice Problems 17-2
Answers at End of Chapter

a. A coil induces 2 V with di/dt of 1 A/s. How much is L?

b. A coil has L of 8 mH with 125 turns. If the number of turns is doubled, how much will L be?

17-3   Self-induced Voltage v_L

The self-induced voltage across an inductance L produced by a change in current di/dt can be stated as

$$v_L = L \frac{di}{dt} \tag{17-3}$$

where v_L is in volts, L in henrys, and di/dt in amperes per second. This formula is just an inverted version of $L = v_L/(di/dt)$, giving the definition of inductance.

Actually both versions are based on Formula (15-1): $v = N(d\phi/dt)$ for magnetism. This gives the voltage in terms of how much magnetic flux is cut per second. When the magnetic flux associated with the current varies the same as i, then Formula (17-3) gives the same results for calculating induced voltage. Remember also that the induced voltage across the coil is actually the result of inducing electrons to move in the conductor, so that there is also an induced current. In using Formula (17-3) to calculate v_L, just multiply L by the di/dt factor.

Example 5 How much is the self-induced voltage across a 4-H inductance produced by a current change of 12 A/s?

Answer $v_L = L \dfrac{di}{dt}$

$= 4 \times 12$

$v_L = 48$ V

Example 6 The current through a 200-mH L changes from 0 to 100 mA in 2 μs. How much is v_L?

Answer $v_L = L \dfrac{di}{dt}$

$= 200 \times 10^{-3} \times \dfrac{100 \times 10^{-3}}{2 \times 10^{-6}}$

$v_L = 10{,}000$ V

Note the high voltage induced in the 200-mH inductance because of the fast change in current.

The induced voltage is an actual voltage that can be measured, although v_L is produced only while the current is changing. When di/dt is present for only a short time, v_L is in the form of a voltage pulse. With a sine-wave current, which is always changing, v_L is a sinusoidal voltage 90° out of phase with i_L.

Practice Problems 17-3
Answers at End of Chapter

a. If L is 2 H and di/dt is 1 A/s, how much is v_L?

b. For the same coil, the di/dt is increased to 100 A/s. How much is v_L?

17-4   How v_L Opposes a Change in Current

By Lenz' law, the induced voltage must oppose the change of current that induces v_L. The polarity of v_L, therefore, depends on the direction of the current variation di. When di increases, v_L has the polarity that opposes the increase of current; when di decreases, v_L has the opposite polarity to oppose the decrease of current.

In both cases, the change of current is opposed by the induced voltage. Otherwise, v_L could increase to an unlimited amount without the need for adding any work. *Inductance, therefore, is the characteristic that opposes any change in current.* This is the reason why an induced voltage is often called a *counter emf* or *back emf*.

More details of applying Lenz' law to determine the polarity of v_L in a circuit are illustrated in Fig. 17-5. Note the directions carefully. In Fig. 17-5a, the electron flow is into the top of the coil. This current is increasing. By Lenz' law, v_L must have the polarity needed to oppose the increase. The induced voltage shown with the top side negative opposes the increase in current. The reason is that this polarity of v_L can produce current in the opposite direction, from minus to plus in the external circuit. Note that for this opposing current, v_L is the generator. This action tends to keep the current from increasing.

In Fig. 17-5b, the source is still producing electron flow into the top of the coil, but i is decreasing, because the source voltage is decreasing. By Lenz' law, v_L must have the polarity needed to oppose the decrease in current. The induced voltage shown with the top side positive now opposes the decrease. The reason is that this polarity of v_L can produce current in the same direction, tending to keep the current from decreasing.

In Fig. 17-5c, the voltage source reverses polarity to produce current in the opposite direction, with electron flow into the bottom of the coil. This reversed direction of current is now increasing. The polarity of v_L must oppose the increase. As shown, now the bottom of the coil is made negative by v_L to produce current opposing the source current. Finally, in Fig. 17-5d the reversed current is decreasing. This decrease is opposed by the polarity shown for v_L to keep the current flowing in the same direction as the source current.

Notice that the polarity of v_L reverses for either a reversal of direction for i or a reversal of change in di between increasing or decreasing values. When both the direction of the current and the direction of change are reversed, as in a comparison of Fig. 17-5a and d, the polarity of v_L is the same.

Sometimes the formulas for induced voltage are written with a minus sign, in order to indicate the fact that v_L opposes the change, as specified by Lenz' law. However, the negative sign is omitted here so that the actual polarity of the self-induced voltage can be determined in typical circuits.

In summary, Lenz' law states that the reaction v_L opposes its cause, which is the change in i. When i is increasing, v_L produces an opposing current. For the opposite case when i is decreasing, v_L produces an aiding current.

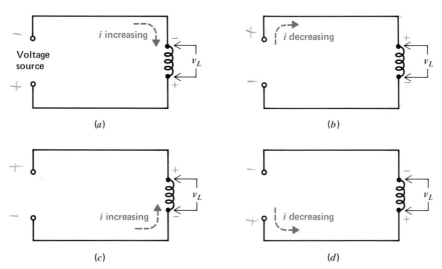

Fig. 17-5 Determining the polarity of v_L that opposes the change in i. (a) Amount of i is increasing, and v_L produces an opposing current. (b) Amount of i is decreasing, and v_L produces an aiding current. (c) The i is increasing but in the opposite direction. (d) Same direction of i as in (c), but decreasing values.

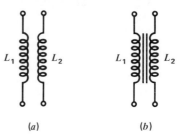

Fig. 17-7 Schematic symbols for two coils with mutual inductance. (*a*) Air core. (*b*) Iron core.

Practice Problems 17-4
Answers at End of Chapter

Answer True or False.
a. In Fig. 17-5*a* and *b* the v_L has opposite polarities.
b. In Fig. 17-5*b* and *c* the polarity of v_L is the same.

17-5 Mutual Inductance L_M

When the current in an inductor changes, the varying flux can cut across any other inductor nearby, producing induced voltage in both inductors. In Fig. 17-6, the coil L_1 is connected to a generator that produces varying current in the turns. The winding L_2 is not connected to L_1, but the turns are linked by the magnetic field. A varying current in L_1, therefore, induces voltage across L_1 and across L_2. If all the flux of the current in L_1 links all the turns of the coil L_2, each turn in L_2 will have the same amount of induced voltage as each turn in L_1. Furthermore, the induced voltage v_{L_2} can produce current in a load resistance connected across L_2.

When the induced voltage produces current in L_2, its varying magnetic field induces voltage in L_1. The two coils L_1 and L_2 have mutual inductance, therefore, because current in one can induce voltage in the other.

The unit of mutual inductance is the henry, and the symbol is L_M. *Two coils have L_M of one henry when a current change of one ampere per second in one coil induces one volt in the other coil.*

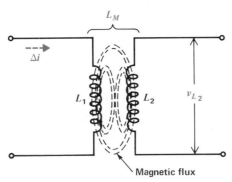

Fig. 17-6 Mutual inductance L_M between L_1 and L_2 linked by magnetic flux.

The schematic symbol for two coils with mutual inductance is shown in Fig. 17-7*a* for an air core, and for an iron core in Fig. 17-7*b*. Iron increases the mutual inductance, since it concentrates magnetic flux. Any magnetic lines that do not link the two coils result in *leakage flux*.

Coefficient of Coupling The fraction of total flux from one coil linking another coil is the coefficient of coupling k between the two coils. As examples, if all the flux of L_1 in Fig. 17-6 links L_2, then k equals 1, or unity coupling; if half the flux of one coil links the other, k equals 0.5. Specifically, the coefficient of coupling is

$$k = \frac{\text{flux linkages between } L_1 \text{ and } L_2}{\text{flux produced by } L_1}$$

There are no units for k, as it is just a ratio of two values of magnetic flux. The value of k is generally stated as a decimal fraction, like 0.5, rather than as a percent.

The coefficient of coupling is increased by placing the coils close together, possibly with one wound on top of the other, by placing them parallel rather than perpendicular to each other, or by winding the coils on a common iron core. Several examples are shown in Fig. 17-8.

A high value of k, called *tight coupling*, allows the current in one coil to induce more voltage in the other coil. *Loose coupling*, with a low value of k, has the opposite effect. In the extreme case of zero coefficient of coupling, there is no mutual inductance. Two coils may be placed perpendicular to each other and far apart for essentially zero coupling when it is desired to minimize interaction between the coils.

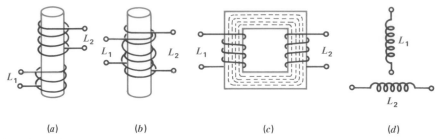

Fig. 17-8 Examples of coupling between two coils linked by L_M. (a) Inductances L_1 and L_2 on paper or plastic form with air core; k is 0.1. (b) Inductance L_1 wound over L_2 for tighter coupling; k is 0.3. (c) Inductances L_1 and L_2 on the same iron core; k is 1. (d) Zero coupling between perpendicular air-core coils.

Air-core coils wound on one form have values of k equal to 0.05 to 0.3, approximately, corresponding to 5 to 30 percent linkage. Coils on a common iron core can be considered to have practically unity coupling, with k equal to 1. As shown in Fig. 17-8c, for both windings L_1 and L_2 practically all the magnetic flux is in the common iron core.

Example 7 A coil L_1 produces 80 μWb of magnetic flux. Of this total flux, 60 μWb are linked with L_2. How much is k between L_1 and L_2?

Answer $k = \dfrac{60\ \mu\text{Wb}}{80\ \mu\text{Wb}}$

$k = 0.75$

Example 8 A 10-H inductance L_1 on an iron core produces 4 Wb of magnetic flux. Another coil L_2 is on the same core. How much is k between L_1 and L_2?

Answer Unity or 1. All coils on a common iron core have practically perfect coupling.

Calculating L_M The mutual inductance increases with higher values for the primary and secondary inductances and tighter coupling:

$$L_M = k \sqrt{L_1 \times L_2} \qquad \text{H} \qquad \textbf{(17-4)}$$

where L_1 and L_2 are the self-inductance values of the

two coils, k is the coefficient of coupling, and L_M is the mutual inductance linking L_1 and L_2, in the same units as L_1 and L_2. The k factor is needed to indicate the flux linkages between the two coils.

As an example, suppose that $L_1 = 2$ H and $L_2 = 8$ H, with both coils on an iron core for unity coupling. Then the mutual inductance is

$$L_M = 1 \sqrt{2 \times 8} = \sqrt{16} = 4 \text{ H}$$

The value of 4 H for L_M in this example means that when the current changes at the rate of 1 A/s in either coil, it will induce 4 V in the other coil.

Example 9 Two 400-mH coils L_1 and L_2 have a coefficient of coupling k equal to 0.2. Calculate L_M.

Answer

$L_M = k \sqrt{L_1 \times L_2}$
$\quad = 0.2 \sqrt{400 \times 10^{-3} \times 400 \times 10^{-3}}$
$\quad = 0.2 \times 400 \times 10^{-3}$
$\quad = 80 \times 10^{-3}$
$L_M = 80 \text{ mH}$

Example 10 If the two coils in Example 9 had a mutual inductance L_M of 40 mH, how much would k be? (Note: Invert Formula (17-4) to find k.)

Answer $k = \dfrac{L_M}{\sqrt{L_1 \times L_2}}$

$\qquad\quad = \dfrac{40 \times 10^{-3}}{\sqrt{400 \times 10^{-3} \times 400 \times 10^{-3}}}$

$\qquad\quad = \dfrac{40 \times 10^{-3}}{400 \times 10^{-3}}$

$\qquad k = 0.1$

Notice that the same two coils have one-half the mutual inductance L_M, because the coefficient of coupling k is 0.1 instead of 0.2.

Practice Problems 17-5
Answers at End of Chapter

a. All the flux from the current in L_1 links L_2. How much is the coefficient of coupling k?
b. Mutual inductance L_M is 9 mH with k of 0.2. If k is doubled to 0.4, how much will L_M be?

17-6 Transformers

The transformer is an important application of mutual inductance. As shown in Fig. 17-9, a transformer has the primary winding L_P connected to a voltage source that produces alternating current, while the secondary winding L_S is connected across the load resistance R_L. The purpose of the transformer is to transfer power from the primary, where the generator is connected, to the secondary, where the induced secondary voltage can produce current in the load resistance that is connected across L_S.

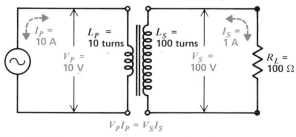

Fig. 17-9 Iron-core transformer with 1:10 turns ratio. Primary current I_P induces secondary voltage V_S that produces current in the secondary load R_L.

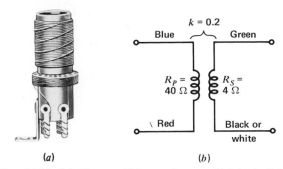

Fig. 17-10 (a) Air-core RF transformer. Height is 2 in. (b) Color code and dc resistance of windings.

Although the primary and secondary are not connected to each other, power in the primary is coupled into the secondary by the magnetic field linking the two windings. The transformer is used to provide power for the load resistance R_L, instead of connecting R_L directly across the generator, whenever the load requires an ac voltage higher or lower than the generator voltage. By having more or fewer turns in L_S, compared with L_P, the transformer can step up or step down the generator voltage to provide the required amount of secondary voltage. Typical transformers are shown in Figs. 17-10 and 17-11. It should be noted that a steady dc voltage cannot be stepped up or down by a transformer, because a steady current cannot produce induced voltage.

Turns Ratio   The ratio of the number of turns in the primary to the number in the secondary is the turns ratio of the transformer:

$$\text{Turns ratio} = \frac{N_P}{N_S} \qquad (17\text{-}5)$$

For example, 500 turns in the primary and 50 turns in the secondary provide a turns ratio of $^{500}\!/_{50}$, or 10:1.

Voltage Ratio   With unity coupling between primary and secondary, the voltage induced in each turn of the secondary is the same as the self-induced voltage of each turn in the primary. Therefore, the voltage ratio is in the same proportion as the turns ratio:

$$\frac{V_P}{V_S} = \frac{N_P}{N_S} \qquad (17\text{-}6)$$

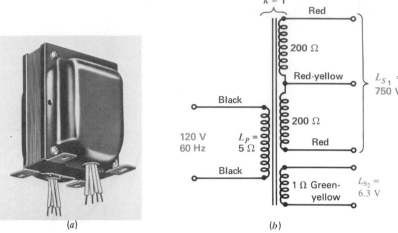

Fig. 17-11 (a) Iron-core power transformer. Height is 4 in. (b) Color code and dc resistance of windings.

When the secondary has more turns, the secondary voltage is higher and the primary voltage is stepped up. This principle is illustrated in Fig. 17-9 with a step-up ratio of 10 to 100, or 1:10. When the secondary has fewer turns, the voltage is stepped down.

In either case, the ratio is in terms of the primary voltage, which may be stepped up or down in the secondary winding.

These calculations apply only to iron-core transformers with unity coupling. Air-core transformers for RF circuits are generally tuned to resonance. In this case, the resonance factor is considered instead of the turns ratio.

Example 11 A power transformer has 100 turns for L_P and 600 turns for L_S. What is the turns ratio? How much is the secondary voltage V_S with a primary voltage V_P of 120 V?

Answer The turns ratio is $^{100}/_{600}$, or 1:6. Therefore, V_P is stepped up by the factor 6, making V_S equal to 6×120, or 720 V.

Example 12 A power transformer has 100 turns for L_P and 5 turns for L_S. What is the turns ratio? How much is the secondary voltage V_S with a primary voltage of 120 V?

Answer The turns ratio is $^{100}/_5$, or 20:1. Secondary voltage is stepped down by a factor of $^1/_{20}$, making V_S equal to $^{120}/_{20}$, or 6 V.

Secondary Current By Ohm's law, the amount of secondary current equals the secondary voltage divided by the resistance in the secondary circuit. In Fig. 17-9, with a value of 100 Ω for R_L and negligible coil resistance assumed,

$$I_S = \frac{V_S}{R_L} = \frac{100 \text{ V}}{100 \text{ }\Omega} = 1 \text{ A}$$

Power in the Secondary The power dissipated by R_L in the secondary is $I_S^2 \times R_L$ or $V_S \times I_S$, which equals 100 W in this example. The calculations are

$$P = I_S^2 \times R_L = 1 \times 100 = 100 \text{ W}$$
$$P = V_S \times I_S = 100 \times 1 = 100 \text{ W}$$

It is important to note that power used by the secondary load, such as R_L in Fig. 17-9, is supplied by the generator in the primary. How the load in the secondary draws power from the generator in the primary can be explained as follows.

With current in the secondary winding, its magnetic field opposes the varying flux of the primary current.

The generator must then produce more primary current to maintain the self-induced voltage across L_P and the secondary voltage developed in L_S by mutual induction. If the secondary current doubles, for instance, because the load resistance is reduced one-half, the primary current will also double in value to provide the required power for the secondary. Therefore, the effect of the secondary-load power on the generator is the same as though R_L were in the primary, except that in the secondary the voltage for R_L is stepped up or down by the turns ratio.

Current Ratio With zero losses assumed for the transformer, the power in the secondary equals the power in the primary:

$$V_S I_S = V_P I_P \qquad (17\text{-}7)$$

or

$$\frac{I_S}{I_P} = \frac{V_P}{V_S} \qquad (17\text{-}8)$$

The current ratio is the inverse of the voltage ratio; that is, voltage step-up in the secondary means current step-down, and vice versa. The secondary does not generate power but only takes it from the primary. Therefore, the current step-up or step-down is in terms of the secondary current I_S, which is determined by the load resistance across the secondary voltage. These points are illustrated by the following two examples.

Example 13 A transformer with a 1:6 turns ratio has 720 V across 7200 Ω in the secondary. **(a)** How much is I_S? **(b)** Calculate the value of I_P.

Answer

a. $I_S = \dfrac{V_S}{R_L} = \dfrac{720 \text{ V}}{7200 \text{ }\Omega}$

$I_S = 0.1$ A

b. With a turns ratio of 1:6, the current ratio is 6:1. Therefore,

---

$I_P = 6 \times I_S$

$ = 6 \times 0.1$

$I_P = 0.6$ A

Example 14 A transformer with a 20:1 voltage step-down ratio has 6 V across 0.6 Ω in the secondary. **(a)** How much is I_S? **(b)** How much is I_P?

Answer

a. $I_S = \dfrac{V_S}{R_L} = \dfrac{6 \text{ V}}{0.6 \text{ }\Omega}$

$I_S = 10$ A

b. $I_P = \frac{1}{20} \times I_S$

$ = \frac{1}{20} \times 10$

$I_P = 0.5$ A

As an aid in these calculations, remember that the side with the higher voltage has the lower current. The primary and secondary V and I are in the same proportion as the number of turns in the primary and secondary.

Total Secondary Power Equals Primary Power Figure 17-12 illustrates a power transformer with two secondary windings L_1 and L_2. There can be one, two, or more secondary windings with unity coupling to the primary as long as all the windings are on the same iron core. Each secondary winding has induced voltage in proportion to its turns ratio with the primary winding, which is connected across the 120-V source.

The secondary winding L_1 has a voltage step-up of 6:1, providing 720 V. The 7200-Ω load resistance R_1, across L_1, allows the 720 V to produce 0.1 A for I_1 in this secondary circuit. The power here is 720 V × 0.1 A, therefore, which equals 72 W.

The other secondary winding L_2 provides voltage step-down, with the ratio 20:1, resulting in 6 V for R_2. The 0.6-Ω load resistance in this circuit allows 10 A for I_2. Therefore, the power here is 6 V × 10 A, or 60 W. Since the windings have separate connections,

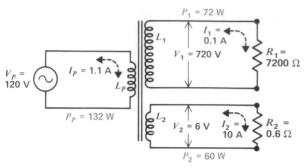

Fig. 17-12 Total power used by the two secondary loads R_1 and R_2 is equal to the power supplied by the source in the primary.

each can have its individual values of voltage and current.

All the power used in the secondary circuit is supplied by the primary, however. In this example, the total secondary power is 132 W, equal to 72 W for R_1 and 60 W for R_2. The power supplied by the 120-V source in the primary then is 72 + 60 = 132 W.

The primary current I_P equals the primary power P_P divided by the primary voltage V_P. This is 132 W divided by 120 V, which equals 1.1 A for the primary current. The same value can be calculated as the sum of 0.6 A of primary current providing power for L_1 plus 0.5 A of primary current for L_2, resulting in the total of 1.1 A as the value of I_P.

This example shows how to analyze a loaded power transformer. The main idea is that the primary current depends on the secondary load. The calculations can be summarized as follows:

1. Calculate V_S from the turns ratio and V_P.
2. Use V_S to calculate $I_S = V_S/R_L$.

3. Use I_S to calculate $P_S = V_S \times I_S$.
4. Use P_S to find $P_P = P_S$.
5. Finally, I_P can be calculated as P_P/V_P.

With more than one secondary, calculate each I_S and P_S. Then add for the total secondary power, which equals the primary power.

Autotransformers As illustrated in Fig. 17-13, an autotransformer consists of one continuous coil with a tapped connection such as terminal 2 between the ends at terminals 1 and 3. In Fig. 17-13a the autotransformer steps up the generator voltage. Voltage V_P between 1 and 2 is connected across part of the total turns, while V_S is induced across all the turns. With six times the turns for the secondary voltage, V_S also is six times V_P.

In Fig. 17-13b the autotransformer steps down the primary voltage connected across the entire coil. Then the secondary voltage is taken across less than the total turns.

The winding that connects to the voltage source to supply power is the primary, while the secondary is across the load resistance R_L. The turns ratio and voltage ratio apply the same way as in a conventional transformer having an isolated secondary winding.

Autotransformers are used often because they are compact, efficient, and usually cost less with only one winding. However, the same wire size must be suitable for both the primary and secondary. Note that the autotransformer in Fig. 17-13 has only three leads, compared with four leads for the transformer in Fig. 17-9 with an isolated secondary.

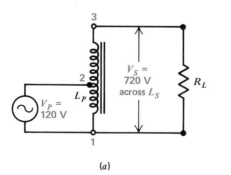

(a)

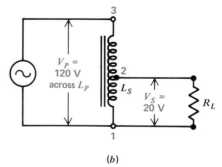

(b)

Fig. 17-13 Autotransformer with tap at terminal 2 for 10 turns of 60-turn winding. (a) Voltage V_P between terminals 1 and 2 stepped up across 1 and 3. (b) Voltage V_P between terminals 1 and 3 stepped down across 1 and 2.

Isolation of the Secondary In a transformer with a separate winding for L_S, as in Fig. 17-9, the secondary load is not connected directly to the ac power line in the primary. This isolation is an advantage in reducing the chance of electric shock. With an autotransformer, as in Fig. 17-13, the secondary is not isolated. Another advantage of an isolated secondary is the fact that any direct current in the primary is blocked from the secondary. Sometimes a transformer with a 1:1 turns ratio is used just for isolation from the ac power line.

Transformer Efficiency Efficiency is defined as the ratio of power out to power in. Stated as a formula,

$$\text{Efficiency} = \frac{P_{\text{out}}}{P_{\text{in}}} \times 100\% \qquad (17\text{-}9)$$

For example, when the power out in watts equals one-half the power in, the efficiency is one-half, which equals 0.5×100 percent, or 50 percent. In a transformer, power out is secondary power, while power in is primary power.

Assuming zero losses in the transformer, power out equals power in and the efficiency is 100 percent. Power transformers actually, however, have an efficiency slightly less than 100 percent. The efficiency is approximately 80 to 90 percent for power transformers in receivers, with a power rating of 50 to 300 W. Transformers for higher power are more efficient because they require heavier wire, which has less resistance. In a transformer that is less than 100 percent efficient, the primary supplies more than the secondary power. The primary power missing from the output is dissipated as heat in the transformer.

Transformer Color Codes The colors of the leads show the required connections in electronic circuits. For the RF transformer in Fig. 17-10, the leads are:

Blue—Output electrode of transistor amplifier

Red—Dc supply voltage for this electrode

Green—Input electrode of next amplifier

Black or white—Return line of secondary winding

This system applies to all coupling transformers be-

tween amplifier stages, including iron-core transformers for audio circuits.

For the power transformer in Fig. 17-11, the primary is connected to the ac power line. The leads are:

Black—Primary leads without tap

Black with yellow—Tap on primary

Red—High-voltage secondary to rectifier in power supply

Red with yellow—Tap on high-voltage secondary

Green with yellow—Low-voltage secondary

The R of the primary winding is generally about 10 Ω or less, for power transformers.

Practice Problems 17-6
Answers at End of Chapter

a. A transformer connected to the 120-V ac line has a turns ratio of 1:2. Calculate the stepped-up V_S.

b. This V_S is connected across a 2400-Ω R_L. Calculate I_S.

17-7 Core Losses

The fact that the magnetic core can become warm, or even hot, shows that some of the energy supplied to the coil is used up in the core as heat. The two main effects are eddy-current losses and hysteresis losses.

Eddy Currents In any inductance with an iron core, alternating current induces voltage in the core itself. Since it is a conductor, the iron core has current produced by the induced voltage. This current is called an *eddy current* because it flows in a circular path through the cross section of the core, as illustrated in Fig. 17-14.

The eddy currents represent wasted power dissipated as heat in the core, equal to I^2R, where R is the resistance of the core. Note in Fig. 17-14 that the eddy-current flux opposes the coil flux, so that more current is required in the coil to maintain its magnetic field. The higher the frequency of the alternating current in the inductance, the greater the eddy-current loss.

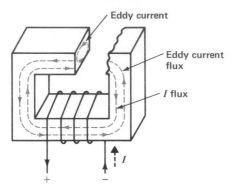

Fig. 17-14 Cross-sectional view of iron core showing eddy currents.

Eddy currents can be induced in any conductor near a coil with alternating current, not only in its core. For instance, a coil has eddy-current losses in a metal cover. In fact, the technique of induction heating is an application of heat resulting from induced eddy currents.

RF Shielding   The reason why a coil may have a metal cover, usually copper or aluminum, is to provide a shield against the varying flux of RF current. In this case, the shielding effect depends on using a good conductor for the eddy currents produced by the varying flux, rather than the magnetic materials used for shielding against static magnetic flux.

The shield cover not only isolates the coil from external varying magnetic fields, but also minimizes the effect of the coil's RF current for external circuits. The reason why the shield helps both ways is the same, as the induced eddy currents have a field that opposes the field that is inducing the current. It should be noted that the clearance between the sides of the coil and the metal should be equal to or greater than the coil radius, to minimize the effect of the shield in reducing the inductance.

Hysteresis Losses   Another factor with a magnetic core for RF coils is hysteresis losses, although these are not so great as eddy-current losses. The hysteresis losses result from the additional power needed to reverse the magnetic field in magnetic materials with RF alternating current.

Air-Core Coils   It should be noted that air has practically no losses from eddy currents or hysteresis. However, the inductance for small coils with an air core is limited to low values in the microhenry or millihenry range.

Practice Problems 17-7
Answers at End of Chapter

a. Which has greater eddy-current losses, an iron core or an air core?
b. Which produces more hysteresis losses, 60 Hz or 60 MHz?

17-8   Types of Cores

In order to minimize losses while maintaining high flux density, the core can be made of laminated sheets insulated from each other, or insulated powdered-iron granules and ferrite materials can be used. These core types are illustrated in Figs. 17-15 and 17-16. The purpose is to reduce the amount of eddy currents.

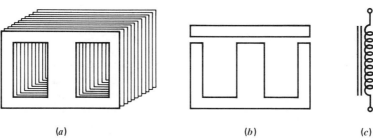

(a)   (b)   (c)

Fig. 17-15 Laminated iron core. (a) Shell-type construction. (b) E- and I-shaped laminations. (c) Symbol for iron core.

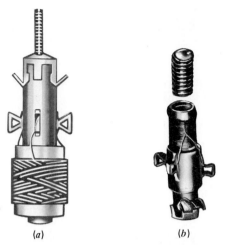

(a)   *(b)*

Fig. 17-16 RF coils with ferrite core. Width of coil is ½ in. (*a*) Variable *L* from 1 to 3 mH. (*b*) Tuning coil for 40 MHz.

Laminated Core Figure 17-15*a* shows a shell-type core formed with a group of individual laminations. Each laminated section is insulated by a very thin coating of iron oxide and varnish. The insulating material increases the resistance in the cross section of the core to reduce the eddy currents, but allows a low-reluctance path for high flux density around the core. Transformers for audio frequencies and 60-Hz power are generally made with a laminated iron core.

Powdered-Iron Core To reduce eddy currents in the iron core of an inductance for radio frequencies, powdered iron is generally used. It consists of individual insulated granules pressed into one solid form called a *slug*.

Ferrite Core The ferrites are synthetic ceramic materials that are ferromagnetic. They provide high values of flux density, like iron, but have the advantage of being insulators. Therefore, a ferrite core can be used for high frequencies with minimum eddy-current losses.

This core is usually a slug that can move in or out of the coil to vary *L*. In Fig. 17-16*a*, the screw at the top moves the core; in Fig. 17-16*b*, the core has a hole to fit a plastic alignment tool for tuning the coil. Maximum *L* results with the slug in the coil.

Practice Problems 17-8
Answers at End of Chapter

Answer True or False.

a. An iron core provides a coefficient of coupling *k* of unity or 1.
b. A laminated iron core reduces eddy-current losses.
c. The ferrites have less eddy-current losses than iron.

17-9 Variable Inductance

The inductance of a coil can be varied by one of the methods illustrated in Fig. 17-17. In Fig. 17-17*a*, more or fewer turns can be used by connection to one of the taps on the coil. In Fig. 17-17*b*, a slider contacts the coil to vary the number of turns used. These methods are for large coils. Note that the unused turns are short-circuited to prevent the tapped coil from acting as an autotransformer. The reason is that stepped-up voltage could cause arcing across the turns.

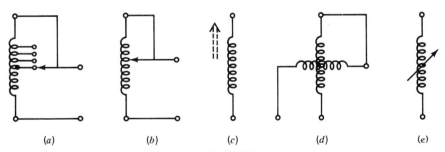

(a)   *(b)*   *(c)*   *(d)*   *(e)*

Fig. 17-17 Methods of varying inductance. (*a*) Tapped coil. (*b*) Slider contact. (*c*) Adjustable slug. (*d*) Variometer. (*e*) Symbol for variable *L*.

Figure 17-17c shows the schematic symbol for a coil with a slug of powdered iron or ferrite. The dotted lines indicate that the core is not solid iron. The arrow shows that the slug is variable. Usually, an arrow at the top means the adjustment is at the top of the coil. An arrow at the bottom, pointing down, shows the adjustment is at the bottom.

The symbol in Fig. 17-17d is a *variometer,* which is an arrangement for varying the position of one coil within the other. The total inductance of the series-aiding coils is minimum when they are perpendicular.

For any method of varying L, the coil with an arrow in Fig. 17-17e can be used. However, an adjustable slug is usually shown as in Fig. 17-17c.

A practical application of variable inductance is the *Variac* in Fig. 17-18. This unit is an autotransformer with a variable tap to change the turns ratio. The output voltage in the secondary can be varied from 0 to 140 V, with input from the 120-V 60-Hz power line. One use is to test equipment with voltage above or below the normal line voltage.

The Variac is plugged into the power line, and the equipment to be tested is plugged into the Variac. Note that the power rating of the Variac should be equal to or more than the power used by the equipment being tested.

Fig. 17-18 Variac rated at 300 W. Length is 5 in. (*General Radio Corp.*)

17-10   Inductances in Series or Parallel

As shown in Fig. 17-19, the total inductance of coils connected in series is the sum of the individual L values, as for series R. Since the series coils have the same current, the total induced voltage is a result of the total number of turns. Therefore, in series,

$$L_T = L_1 + L_2 + L_3 + \cdots + \text{etc.} \qquad (17\text{-}10)$$

where L_T is in the same units of inductance as L_1, L_2, and L_3. This formula assumes no mutual induction between the coils.

> **Example 15**   Inductance L_1 in Fig. 17-19 is 5 mH and L_2 is 10 mH. How much is L_T?
>
> **Answer** $L_T = 5 \text{ mH} + 10 \text{ mH} = 15 \text{ mH}$

With coils connected in parallel, the total inductance is calculated from the reciprocal formula

$$\frac{1}{L_T} = \frac{1}{L_1} + \frac{1}{L_2} + \frac{1}{L_3} + \cdots + \text{etc.} \qquad (17\text{-}11)$$

Again, no mutual induction is assumed, as illustrated in Fig. 17-20.

> **Example 16**   Inductances L_1 and L_2 in Fig. 17-20 are each 8 mH. How much is L_T?
>
> **Answer** $\dfrac{1}{L_T} = \dfrac{1}{8} + \dfrac{1}{8} = \dfrac{2}{8}$
>
> $$L_T = \frac{8}{2} = 4 \text{ mH}$$

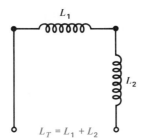

Fig. 17-19 Inductances in series without mutual coupling.

$$L_T = L_1 + L_2$$

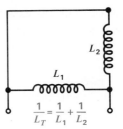

$$\frac{1}{L_T} = \frac{1}{L_1} + \frac{1}{L_2}$$

Fig. 17-20 Inductances in parallel without mutual coupling.

All the shortcuts for calculating parallel R can be used with parallel L, since both are based on the reciprocal formula. In this example L_T is $\frac{1}{2} \times 8 = 4$ mH.

Series Coils with L_M This case depends on the amount of mutual coupling and on whether the coils are connected series-aiding or series-opposing. *Series-aiding* means that the common current produces the same direction of magnetic field for the two coils. The *series-opposing* connection results in opposite fields.

The coupling depends on the coil connections and direction of winding. Reversing either one reverses the field. Inductances L_1 and L_2 with the same direction of winding are connected series-aiding in Fig. 17-21a. However, they are series-opposing in Fig. 17-21b because L_1 is connected to the opposite end of L_2.

To calculate the total inductance of two coils that are series-connected and have mutual inductance,

$$L_T = L_1 + L_2 \pm 2L_M \qquad \text{(17-12)}$$

The mutual inductance L_M is plus, increasing the total inductance, when the coils are series-aiding, or minus when they are series-opposing to reduce the total inductance.

Note the large dots just above the coils in Fig. 17-21. This method is generally used to indicate the sense of the windings without the need for showing the actual physical construction. Coils with dots at the same end have the same direction of winding. When current enters the dotted ends for two coils, their fields are aiding and L_M has the same sense as L.

How to Measure L_M Formula (17-12) provides a method of determining the mutual inductance between two coils L_1 and L_2 of known inductance. First, the total inductance is measured for the series-aiding connection. Let this be L_{T_a}. Then the connections to one coil are reversed to measure the total inductance for the series-opposing coils. Let this be L_{T_o}. Then

$$L_M = \frac{L_{T_a} - L_{T_o}}{4} \qquad \text{(17-13)}$$

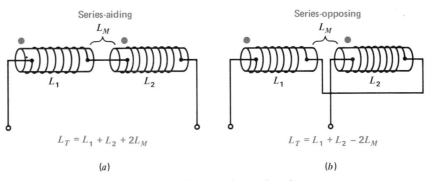

Series-aiding	Series-opposing
L_M	L_M
L_1 L_2	L_1 L_2
$L_T = L_1 + L_2 + 2L_M$	$L_T = L_1 + L_2 - 2L_M$
(a)	(b)

Fig. 17-21 Inductances L_1 and L_2 in series but with mutual coupling L_M. (a) Aiding magnetic fields. (b) Opposing magnetic fields.

When the mutual inductance is known, the coefficient of coupling k can be calculated from the fact that $L_M = k \sqrt{L_1 L_2}$.

Example 17   Two series coils, each with an L of 250 μH, have a total inductance of 550 μH connected series-aiding and 450 μH series-opposing. **(a)** How much is the mutual inductance L_M between the two coils? **(b)** How much is the coupling coefficient k?

Answer

a. $L_M = \dfrac{L_{T_a} - L_{T_o}}{4}$

$= \dfrac{550 - 450}{4} = \dfrac{100}{4}$

$L_M = 25 \ \mu$H

b. $L_M = k \sqrt{L_1 L_2}$, or

$k = \dfrac{L_M}{\sqrt{L_1 L_2}} = \dfrac{25}{\sqrt{250 \times 250}}$

$= \dfrac{25}{250} = \dfrac{1}{10}$

$k = 0.1$

Coils may also be in parallel with mutual coupling. However, the inverse relations with parallel connections and the question of aiding or opposing fields make this case complicated. Actually, it would hardly ever be used.

Practice Problems 17-10
Answers at End of Chapter

a. A 500-μH coil and a 1-mH coil are in series without L_M. Calculate L_T.
b. The same coils are in parallel without L_M. Calculate L_T.

17-11   Stray Inductance

Although practical inductors are generally made as coils, all conductors have inductance. The amount of L is $v_L/(di/dt)$, as with any inductance producing induced voltage when the current changes. The inductance of any wiring not included in the conventional inductors can be considered stray inductance. In most cases, the stray inductance is very small, typical values being less than 1 μH. For high radio frequencies, though, even a small L can have an appreciable inductive effect.

One source of stray inductance is the connecting leads. A wire of 0.04 in. diameter and 4 in. long has an L of approximately 0.1 μH. At low frequencies, this inductance is negligible. However, consider the case of RF current, where i varies from a 0- to a 20-mA peak value in the short time of 0.025 μs for a quarter-cycle of a 10-MHz sine wave. Then v_L equals 80 mV, which is an appreciable inductive effect. This is one reason why the connecting leads must be very short in RF circuits.

As another example, wire-wound resistors can have appreciable inductance when wound as a straight coil. This is why carbon resistors are preferred for minimum stray inductance in RF circuits. However, noninductive wire-wound resistors can also be used. These are wound in such a way that adjacent turns have current in opposite directions, so that the magnetic fields oppose each other to cancel the inductance. Another application of this technique is twisting a pair of connecting leads to reduce the inductive effect.

Practice Problems 17-11
Answers at End of Chapter

Answer True or False.
a. A straight wire 1 ft long can have L less than 1 μH.
b. Carbon resistors have less L than wire-wound resistors.

17-12   Energy in Magnetic Field of Inductance

Magnetic flux associated with current in an inductance has electric energy supplied by the voltage source producing the current. The energy is stored in the field, since it can do the work of producing induced voltage when the flux moves. The amount of electric energy stored is

$$\text{Energy} = \mathcal{E} = \tfrac{1}{2} LI^2 \quad \text{J} \qquad (17\text{-}14)$$

The factor of ½ gives the average result of I in producing energy. With L in henrys and I in amperes, the energy is in watt-seconds, or *joules*. For a 10-H L with a 3-A I, the electric energy stored in the magnetic field equals

$$\text{Energy} = \frac{1}{2}\, LI^2 = \frac{10 \times 9}{2} = 45 \text{ J}$$

This 45 J of energy is supplied by the voltage source that produces 3 A in the inductance. When the circuit is opened, the magnetic field collapses. The energy in the collapsing magnetic field is returned to the circuit in the form of induced voltage, which tends to keep the current flowing.

The entire 45 J is available for the work of inducing voltage, since no energy is dissipated by the magnetic field. With resistance in the circuit, however, the I^2R loss with induced current dissipates all the energy after a period of time.

Practice Problems 17-12
Answers at End of Chapter

a. What is the unit of energy?
b. Does a 4-H coil store more or less energy than a 2-H coil?

17-13   Troubles in Coils

The most common trouble in coils is an open winding. As illustrated in Fig. 17-22, an ohmmeter connected across the coil reads infinite resistance for the open circuit. It does not matter whether the coil has an air core or an iron core. Since the coil is open, it cannot conduct current and therefore has no inductance, because it cannot produce induced voltage. When the resistance is checked, the coil should be disconnected from the external circuit to eliminate any parallel paths that could affect the resistance readings.

DC Resistance of a Coil   A coil has dc resistance equal to the resistance of the wire used in the winding. The amount of resistance is less with heavier wire and fewer turns. For RF coils with inductance values up to several millihenrys, requiring 10 to 100 turns of fine wire, the dc resistance is 1 to 20 Ω, approximately. Inductors for 60 Hz and audio frequencies with several hundred turns may have resistance values of 10 to 500 Ω, depending on the wire size.

As shown in Fig. 17-23, the dc resistance and inductance of a coil are in series, since the same current that induces voltage in the turns must overcome the resistance of the wire. Although resistance has no function in producing induced voltage, it is useful to know the dc coil resistance because if it is normal, usually the inductance can also be assumed to have its normal value.

Open Coil   An open winding has infinite resistance, as indicated by an ohmmeter reading. With a transformer that has four leads or more, check the resistance across the two leads for the primary, across the two leads for the secondary, and across any other pairs of leads for additional secondary windings. For an autotransformer with three leads, check the resistance from one lead to each of the other two.

When the open circuit is inside the winding, it is usually not practical to repair the coil, and the entire

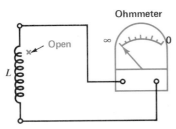

Fig. 17-22 An open coil reads infinite ohms when its continuity is checked with an ohmmeter.

Fig. 17-23 The internal dc resistance r_i is in series with the inductance of the coil.

unit is replaced. In some cases, an open connection at the terminals can be resoldered.

Open Primary Winding

When the primary of a transformer is open, no primary current can flow and there is no voltage induced in any of the secondary windings.

Open Secondary Winding

When the secondary of a transformer is open, it cannot supply power to any load resistance across the open winding. Furthermore, with no current in the secondary, the primary current is also practically zero, as though the primary winding were open. The only primary current needed is the small magnetizing current to sustain the field producing induced voltage across the secondary without any load. If the transformer has several secondary windings, however, an open winding in one secondary does not affect transformer operation for the secondary circuits that are normal.

Short across Secondary Winding

In this case excessive primary current flows, as though it were short-circuited, often burning out the primary winding. The reason is that the large secondary current has a strong field that opposes the flux of the self-induced voltage across the primary, making it draw more current from the generator.

Practice Problems 17-13
Answers at End of Chapter

a. The normal R of a coil is 18 Ω. How much will an ohmmeter read if the coil is open?
b. The primary of a 1:3 step-up autotransformer is connected to the 120-V ac power line. How much will the secondary voltage be if the primary is open?

Summary

1. Varying current induces voltage in a conductor, since the expanding and collapsing field of the current is equivalent to flux in motion.
2. Lenz' law states that the induced voltage opposes the change in current causing the induction. Inductance, therefore, tends to keep the current from changing.
3. The ability of a conductor to produce induced voltage across itself when the current varies is its self-inductance, or inductance. The symbol is L, and the unit of inductance is the henry. One henry of inductance allows 1 V to be induced when the current changes at the rate of 1 A/s. For smaller units, 1 mH = 1×10^{-3} H and 1 μH = 1×10^{-6} H.
4. To calculate the self-induced voltage, $v_L = L(di/dt)$, with v in volts, L in henrys, and di/dt in amperes per second.
5. Mutual inductance is the ability of varying current in one conductor to induce voltage in another conductor nearby. Its symbol is L_M, measured in henrys. $L_M = k \sqrt{L_1 L_2}$, where k is the coefficient of coupling between conductors.
6. A transformer consists of two or more windings with mutual inductance. The primary winding connects to the source voltage; the load resistance is connected across the secondary winding. A separate winding is an isolated secondary. The transformer is used to step up or down ac voltage.
7. An autotransformer is a tapped coil, used to step up or step down the primary voltage. There are three leads with one connection common to both the primary and secondary.
8. A transformer with an iron core has essentially unity coupling. Therefore, the voltage ratio is the same as the turns ratio: $V_P/V_S = N_P/N_S$.
9. Assuming 100 percent efficiency for an iron-core power transformer, the power supplied to the primary equals the power used in the secondary.

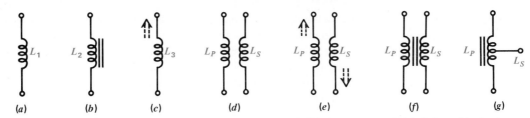

Fig. 17-24 Summary of types of inductors. (*a*) Air-core coil. (*b*) Iron-core coil. (*c*) Adjustable ferrite core. (*d*) Air-core transformer. (*e*) Variable L_P and L_S. (*f*) Iron-core transformer. (*g*) Autotransformer.

10. Eddy currents are induced in the iron core of an inductance, causing wasted power that heats the core. Eddy-current losses increase with higher frequencies of alternating current. To reduce eddy currents, the iron core is laminated with insulated sections. Powdered-iron and ferrite cores have minimum eddy-current losses for radio frequencies. Hysteresis losses also cause wasted power.

11. Assuming no mutual coupling, series inductances are added like series resistances. For parallel inductances, the total inductance is calculated by the reciprocal formula, as for parallel resistances.

12. The magnetic field of an inductance has stored energy $\frac{1}{2} LI^2$. With I in amperes and L in henrys, the energy is in joules.

13. In addition to its inductance, a coil has dc resistance equal to the resistance of the wire in the coil. An open coil has infinitely high resistance.

14. An open primary in a transformer results in no induced voltage in any of the secondary windings.

15. Figure 17-24 summarizes the main types of inductors, or coils, with their schematic symbols.

Self-Examination
Answers at Back of Book

Choose (a), (b), (c), or (d).

1. Alternating current can induce voltage because alternating current has a (a) high peak value; (b) varying magnetic field; (c) stronger magnetic field than direct current; (d) constant magnetic field.

2. When current in a conductor increases, Lenz' law says that the self-induced voltage will (a) tend to increase the amount of current; (b) aid the applied voltage; (c) produce current opposite to the increasing current; (d) aid the increasing current.

3. A 1:5 voltage step-up transformer has 120 V across the primary and a 600-Ω resistance across the secondary. Assuming 100 percent efficiency, the primary current equals (a) $\frac{1}{5}$ A; (b) 600 mA; (c) 5 A; (d) 10 A.

4. An iron-core transformer with an 1:8 step-up ratio has 120 V applied across the primary. The voltage across the secondary equals (a) 15 V; (b) 120 V; (c) 180 V; (d) 960 V.

5. With double the number of turns but the same length and area, the inductance is (a) the same; (b) double; (c) quadruple; (d) one-quarter.

6. Current changing from 4 to 6 A in 1 s induces 40 V in a coil. Its inductance equals (a) 40 mH; (b) 4 H; (c) 6 H; (d) 20 H.

7. A laminated iron core has reduced eddy-current losses because (a) the laminations are stacked vertically; (b) the laminations are insulated from each other; (c) the magnetic flux is concentrated in the air gap of the core; (d) more wire can be used with less dc resistance in the coil.

8. Two 250-μH coils in series without mutual coupling have a total inductance of (a) 125 μH; (b) 250 μH; (c) 400 μH; (d) 500 μH.

9. The dc resistance of a coil made with 100 ft of No. 30 gage copper wire is approximately (a) less than 1 Ω; (b) 10.5 Ω; (c) 104 Ω; (d) more than 1 MΩ.

10. An open coil has (a) infinite resistance and zero inductance; (b) zero resistance and high inductance; (c) infinite resistance and normal inductance; (d) zero resistance and inductance.

Essay Questions

1. Define 1 H of self-inductance and 1 H of mutual inductance.

2. State Lenz' law in terms of induced voltage produced by varying current.

3. Refer to Fig. 17-5. Explain why the polarity of v_L is the same for the examples in Fig. 17-5a and d.

4. Make a schematic diagram showing primary and secondary for an iron-core transformer with a 1:6 voltage step-up ratio: (a) using an autotransformer; (b) using a transformer with isolated secondary winding. Then (c) with 100 turns in the primary, how many turns are in the secondary for both cases?

5. Define the following: coefficient of coupling, transformer efficiency, stray inductance, and eddy-current losses.

6. Why are eddy-current losses reduced with the following cores: (a) laminated; (b) powdered iron; (c) ferrite?

7. Why is a good conductor used for an RF shield?

8. Show two methods of providing a variable inductance.

9. (a) Why will the primary of a power transformer have excessive current if the secondary is short-circuited? (b) Why is there no voltage across the secondary if the primary is open?

10. (a) Describe briefly how to check a coil for an open winding with an ohmmeter. What ohmmeter range should be used? (b) What leads will be checked on an autotransformer with one secondary and a transformer with two isolated secondary windings?

11. Derive the formula $L_M = (L_{T_a} - L_{T_o})/4$ from the fact that $L_{T_a} = L_1 + L_2 + 2L_M$ while $L_{T_o} = L_1 + L_2 - 2L_M$.

Problems

Answers to Odd-Numbered Problems at Back of Book

1. Convert the following current changes to amperes per second: (a) zero to 3 A in 2 s; (b) zero to 50 mA in 5 μs; (c) 100 to 150 mA in 5 μs; (d) 150 to 100 mA in 5 μs.

2. Convert into henrys using powers of 10: (a) 250 μH; (b) 40 μH; (c) 40 mH; (d) 7 mH; (e) 0.005 H.

3. Calculate the values of v_L across a 5-mH inductance for each of the current variations in Prob. 1.

4. A coil produces a self-induced voltage of 42 mV when i varies at the rate of 19 mA/ms. How much is L?

5. A power transformer with a 1:8 turns ratio has 60 Hz 120 V across the primary. (a) What is the frequency of the secondary voltage? (b) How much is the secondary voltage? (c) With a load resistance of 10,000 Ω across the secondary, how much is the secondary current? Draw the schematic diagram showing primary and secondary circuits. (d) How much is the primary current? Assume 100 percent efficiency. (Note: 1:8 is the ratio of L_P to L_S.)

6. How much would the primary current be in a power transformer having a primary resistance of 5 Ω if it were connected by mistake to a 120-V dc line instead of the 120-V ac line?

7. For a 100-μH inductance L_1 and a 200-μH inductance L_2, calculate the following: (a) the total inductance L_T of L_1 and L_2 in series without mutual coupling; (b) the combined inductance of L_1 and L_2 in parallel without mutual coupling; (c) the L_T of L_1 and L_2 series-aiding, and series-opposing, with 10-μH mutual inductance; (d) the value of the coupling factor k.

8. Calculate the inductance L for the following long coils: (a) Air core, 20 turns, area 3.14 cm^2, length 25 cm; (b) same coil as (a) with ferrite core having a μ of 5000; (c) air core, 200 turns, area 3.14 cm^2, length 25 cm; (d) air core, 20 turns, area 3.14 cm^2, length 50 cm; (e) air core, 20 turns, diameter 4 cm, length 50 cm. (Note: 1 cm = 10^{-2} m, and 1 cm^2 = 10^{-4} m^2.)

9. Calculate the resistance of the following coil, using Table 11-1 on page 211: 400 turns, each using 3 in. of No. 30 gage wire.

10. (a) Calculate the period T for one cycle of a 5-MHz sine wave. (b) How much is the time for one quarter-cycle? (c) If i increases from 0 to 20 mA in this time, how much is v_L across a 0.1-μH inductance?

11. Calculate the energy in joules stored in the magnetic field of a 60-mH L with a 90-mA I.

12. For a power transformer connected to the 120-V ac line, calculate the turns ratio needed for each of the following secondary voltages: (a) 5 V; (b) 9 V; (c) 24 V; (d) 30 V; (e) 120 V.

13. (a) A transformer delivers 400 W out with 500 W in. (a) Calculate the efficiency in percent. (b) A transformer with 80 percent efficiency delivers 400 W total secondary power. Calculate the primary power.

14. A 20-mH L and a 40-mH L are connected series-aiding, with $k = 0.4$. Calculate L_T.

15. Calculate the inductance of the coil in Fig. 17-3 with $\mu r = 100$.

Answers to Practice Problems

17-1 a. Coil with an iron core
    b. Time B
17-2 a. $L = 2$ H
    b. $L = 32$ mH
17-3 a. $v_L = 2$ V
    b. $v_L = 200$ V
17-4 a. T
    b. T

17-5 a. $k = 1$
    b. $L_M = 18$ mH
17-6 a. $V_S = 240$ V
    b. $I_S = 0.1$ A
17-7 a. Iron core
    b. 60 MHz

17-8 a. T
    b. T
    c. T
17-9 a. T
    b. T
17-10 a. $L_T = 1.5$ mH
    b. $L_T = 0.33$ mH

17-11 a. T
    b. T
17-12 a. Joule
    b. More
17-13 a. Infinite ohms
    b. 0 V

Chapter 18
Inductive Reactance

When alternating current flows in an inductance L, the amount of current is much less than the resistance alone would allow. The reason is that the current variations induce a voltage across L that opposes the applied voltage. This additional opposition of an inductance to sine-wave alternating current is specified by the amount of its inductive

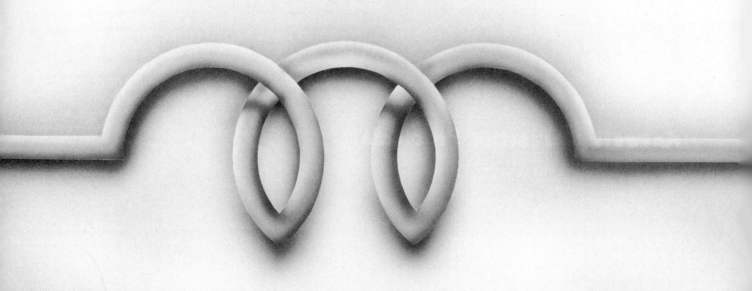

reactance X_L. The X indicates reactance. It is an opposition to current, measured in ohms. The X_L is the ohms of opposition, therefore, that an inductance L has for sine-wave current.

The amount of X_L equals $2\pi fL$ ohms, with f in hertz and L in henrys. Note that the opposition in ohms of X_L increases for higher frequencies and more inductance. The constant factor 2π indicates sine-wave variations.

The requirements for having X_L correspond to what is needed for producing induced voltage. There must be variations in current and its associated magnetic flux. For a steady direct current without any changes in current, the X_L is zero. However, with sine-wave alternating current, the X_L is the best way to analyze the effect of L.

Important terms in this chapter are:

cosine wave   90° phase angle
inductive reactance X_L quadrature angle

This important factor of X_L in sine-wave ac circuits is explained in the following sections:

18-1 How X_L Reduces the Amount of I
18-2 $X_L = 2\pi fL$
18-3 Series or Parallel Inductive Reactances
18-4 Ohm's Law Applied to X_L
18-5 Applications of X_L for Different Frequencies
18-6 Waveshape of v_L Induced by Sine-Wave Current

18-1   How X_L Reduces the Amount of I

Figure 18-1 illustrates the effect of X_L in reducing the alternating current for a light bulb. The more ohms of X_L, the less current flows. When X_L reduces I to a very small value, the bulb cannot light.

In Fig. 18-1a, there is no inductance, and the ac voltage source produces a 2.4-A current to light the bulb with full brilliance. This 2.4-A I results from 120 V applied across the 50-Ω R of the bulb's filament.

In Fig. 18-1b, however, a coil is connected in series with the bulb. The coil has a dc resistance of only 1 Ω, which is negligible, but the reactance of the inductance is 1000 Ω. This X_L is a measure of the coil's reaction to sine-wave current in producing a self-induced voltage that opposes the applied voltage and reduces the current. Now I is 120 V/1000 Ω, approximately, which equals 0.12 A. This I is not enough to light the bulb.

Although the dc resistance is only 1 Ω, the X_L of 1000 Ω for the coil limits the amount of alternating current to such a low value that the bulb cannot light. This X_L of 1000 Ω for a 60-Hz current can be obtained with an inductance L of approximately 2.65 H.

In Fig. 18-1c, the coil is also in series with the bulb, but the applied battery voltage produces a steady value of direct current. Without any current variations, the coil cannot induce any voltage and, therefore, it has no reactance. The amount of direct current, then, is practically the same as though the dc voltage source were connected directly across the bulb, and it lights with

full brilliance. In this case, the coil is only a length of wire, as there is no induced voltage without current variations. The dc resistance is the resistance of the wire in the coil.

In summary, we can make the following conclusions:

1. An inductance can have appreciable X_L in ac circuits, to reduce the amount of current. Furthermore, the higher the frequency of the alternating current, and the greater the inductance, the higher is the X_L opposition.
2. There is no X_L for steady direct current. In this case, the coil is just a resistance equal to the resistance of the wire.

These effects have almost unlimited applications in practical circuits. Consider how useful ohms of X_L can be for different kinds of current, compared with resistance, which always has the same ohms of opposition. One example is to use X_L where it is desired to have high ohms of opposition to alternating current but little opposition to direct current. Another example is to use X_L for more opposition to a high-frequency alternating current, compared with lower frequencies.

X_L Is an Inductive Effect   The reason why an inductance can have X_L to reduce the amount of alternating current is the fact that self-induced voltage is produced to oppose the applied voltage. In Fig. 18-2, V_L is the voltage across L, induced by the variations in sine-wave current produced by the applied voltage V_A.

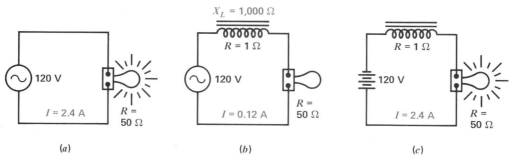

Fig. 18-1 Illustrating the effect of X_L in reducing the amount of alternating current. (a) Bulb lights with 2.4 A. (b) Inserting an X_L of 1000 Ω reduces I to 0.12 A, and the bulb cannot light. (c) With direct current, the coil has no inductive reactance, and the bulb lights.

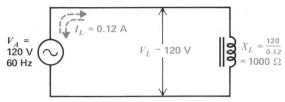

Fig. 18-2 The inductive reactance X_L equals the V_L/I_L ratio in ohms.

The two voltages V_A and V_L are the same because they are in parallel. However, the current I_L is the amount that allows the self-induced voltage V_L to be equal to V_A. In this example, I is 0.12 A. This value of a 60-Hz current in the inductance produces a V_L of 120 V.

The Reactance Is a *V/I* Ratio   When we consider the V/I ratio for the ohms of opposition to the sine-wave current, this value is 120/0.12, which equals 1000 Ω. This 1000 Ω is what we call X_L, to indicate how much current can be produced by sine-wave voltage across an inductance. The ohms of X_L can be almost any amount, but the 1000 Ω here is a typical example.

The Effect of *L* and *f* on X_L   The X_L value depends on the amount of inductance and the frequency of the alternating current. If L in Fig. 18-2 were increased, it could induce the same 120 V for V_L with less current. Then the ratio of V_L/I_L would be greater, meaning more X_L for more inductance.

Also, if the frequency were increased in Fig. 18-2, the current variations would be faster with a higher frequency. Then the same L could produce the 120 V for V_L with less current. For this condition also, the V_L/I_L ratio would be greater because of the smaller current, indicating more X_L for a higher frequency.

Practice Problems 18-1
Answers at End of Chapter

a. For the dc circuit in Fig. 18-1*c*, how much is X_L?
b. For the ac circuit in Fig. 18-1*b*, how much is the V/I ratio for the ohms of X_L?

18-2 $X_L = 2\pi fL$

This formula includes the effects of frequency and inductance for calculating the reactance. The frequency is in hertz and L is in henrys for an X_L in ohms. As an example, we can calculate X_L for a 2.65-h L at the frequency of 60 Hz:

$$X_L = 2\pi fL \tag{18-1}$$
$$= 6.28 \times 60 \times 2.65$$
$$X_L = 1000 \ \Omega$$

Note the following factors in the formula $X_L = 2\pi fL$.

1. The constant factor 2π is always $2 \times 3.14 = 6.28$. It indicates the circular motion from which a sine wave is derived. Therefore, this formula applies only to sine-wave ac circuits. The 2π is actually 2π rad or 360° for a complete circle or cycle. Furthermore, $2\pi \times f$ is the angular velocity, in radians per second, for a rotating phasor corresponding to the sine-wave V or I of that particular frequency.
2. The frequency f is a time element. Higher frequency means that the current varies at a faster rate. A faster current change can produce more self-induced voltage across a given amount of inductance.
3. The inductance L indicates the physical factors of the coil that determine how much voltage it can induce for a given current change.
4. Inductive reactance X_L is in ohms, corresponding to a V_L/I_L ratio for sine-wave ac circuits, to determine how much current L allows for a given applied voltage.

Stating X_L as V_L/I_L and as $2\pi fL$ are two ways of specifying the same value of ohms. The $2\pi fL$ formula gives the effect of L and f on the X_L. The V_L/I_L ratio gives the result of $2\pi fL$ in reducing the amount of I.

The formula $2\pi fL$ shows that X_L is proportional to frequency. When f is doubled, for instance, X_L is doubled. This linear increase of inductive reactance with frequency is illustrated in Fig. 18-3.

The reactance formula also shows that X_L is proportional to the inductance. When the value of henrys for L is doubled, the ohms of X_L is also doubled. This linear increase of inductive reactance with frequency is illustrated in Fig. 18-4.

X_L increases with higher f

Frequency, Hz	$X_L = 2\pi f L$, Ω
0	0
100	200
200	400
300	600
400	800

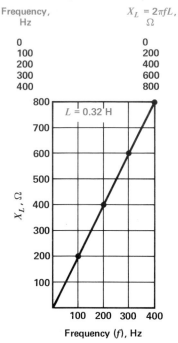

Fig. 18-3 Linear increase of X_L with higher frequencies. The L has constant value of 0.32 H.

X_L increases with higher L

Inductance, H	$X_L = 2\pi f L$, Ω
0	0
0.32	200
0.64	400
0.96	600
1.28	800

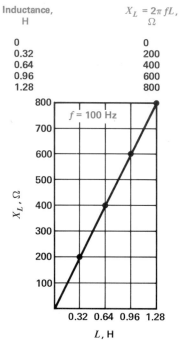

Fig. 18-4 Linear increase of X_L with higher values of inductance. The f is constant at 100 Hz.

Example 1   How much is X_L of a 6-mH L at 41.67 kHz?

Answer

$X_L = 2\pi f L$

$\quad = 6.28 \times 41.67 \times 10^3 \times 6 \times 10^{-3}$

$X_L = 1570\ \Omega$

Example 2   Calculate the X_L of (**a**) a 10-H L at 60 Hz and (**b**) a 5-H L at 60 Hz.

Answer

a.   For a 10-H L,

$X_L = 2\pi f L$

$\quad = 6.28 \times 60 \times 10$

$X_L = 3768\ \Omega$

b.   For a 5-H L,

$X_L = \frac{1}{2} \times 3768 = 1884\ \Omega$

Example 3   Calculate the X_L of a 250-μH coil at (**a**) 1 MHz and (**b**) 10 MHz.

Answer

a.   At 1 MHz,

$X_L = 2\pi f L$

$\quad = 6.28 \times 1 \times 10^6 \times 250 \times 10^{-6}$

$\quad = 6.28 \times 250$

$X_L = 1570\ \Omega$

b.   At 10 MHz,

$X_L = 10 \times 1570 = 15{,}700\ \Omega$

The last two examples illustrate the fact that X_L is proportional to frequency and inductance. In Example 2**b**, X_L is one-half the value in Example 2**a** because the inductance is one-half. In Example 3**b**, X_L is ten times more than in Example 3**a** because the frequency is ten times higher.

Finding L from X_L

Not only can X_L be calculated from f and L, but if any two factors are known, the third can be found. Very often X_L can be determined from voltage and current measurements. With the frequency known, L can be calculated as

$$L = \frac{X_L}{2\pi f} \qquad (18\text{-}2)$$

This formula just has the factors inverted from $X_L = 2\pi f L$. Use the basic units with ohms for X_L and hertz for f to calculate L in henrys.

Example 4   A coil with negligible resistance has 62.8 V across it with 0.01 A of current. How much is X_L?

Answer $X_L = \dfrac{V_L}{I_L}$

$$= \frac{62.8 \text{ V}}{0.01 \text{ A}}$$

$$X_L = 6280 \ \Omega$$

Example 5   Calculate the L of the coil in Example 4 when the frequency is 1000 Hz.

Answer $L = \dfrac{X_L}{2\pi f}$

$$= \frac{6280}{6.28 \times 1000} = \frac{6280}{6280}$$

$$L = 1 \text{ H}$$

Example 6   Calculate the inductance of a coil that has a 15,700-Ω X_L at 10 MHz.

Answer $L = \dfrac{X_L}{2\pi f} = \dfrac{15,700}{6.28 \times 10 \times 10^6}$

$$= \frac{15,700}{62.8} \times 10^{-6}$$

$$= 250 \times 10^{-6}$$

$$L = 250 \ \mu\text{H}$$

Find f from X_L

For the third and final version of the inductive reactance formula,

$$f = \frac{X_L}{2\pi L} \qquad (18\text{-}3)$$

Use the basic units of ohms for X_L and henrys for L to calculate the frequency in hertz.

Example 7   At what frequency will an inductance of 1 H have the reactance of 1000 Ω?

Answer $f = \dfrac{X_L}{2\pi L}$

$$= \frac{1000}{6.28 \times 1} = 0.159 \times 10^3$$

$$f = 159 \text{ Hz}$$

Practice Problems 18-2
Answers at End of Chapter

a. If L is 1 H and f is 100 Hz, how much is X_L?
b. If L is 0.5 H and f is 100 Hz, how much is X_L?
c. If L is 1 H and f is 1000 Hz, how much is X_L?

18-3   Series or Parallel Inductive Reactances

Since reactance is an opposition in ohms, inductive reactances in series or in parallel are combined the same way as ohms of resistance. With series reactances the total reactance is the sum of the individual values, as shown in Fig. 18-5a. For example, the series reactances of 100 and 200 Ω add to equal 300 Ω of X_L across both inductances. Therefore, in series,

$$X_{L_T} = X_{L_1} + X_{L_2} + X_{L_3} + \cdots + \text{etc.} \qquad (18\text{-}4)$$

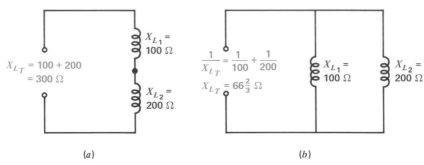

Fig. 18-5 Combining ohms of X_L for inductive reactances. (a) Reactances X_{L_1} and X_{L_2} in series. (b) Reactances X_{L_1} and X_{L_2} in parallel.

For the case of parallel reactances, the combined reactance is calculated by the reciprocal formula. As shown in Fig. 18-5b, in parallel

$$\frac{1}{X_{L_T}} = \frac{1}{X_{L_1}} + \frac{1}{X_{L_2}} + \frac{1}{X_{L_3}} + \cdots + \text{etc.} \qquad \textbf{(18-5)}$$

The combined parallel reactance will be less than the lowest branch reactance. Any short cuts for calculating parallel resistances also apply to the parallel reactances. For instance, the combined reactance of two equal reactances in parallel is one-half either reactance.

Practice Problems 18-3
Answers at End of Chapter

a. An X_L of 200 Ω is in series with a 300-Ω X_L. How much is the total X_{L_T}?

b. An X_L of 200 Ω is in parallel with a 300-Ω X_L. How much is the combined X_{L_T}?

18-4 Ohm's Law Applied to X_L

The amount of current in an ac circuit with just inductive reactance is equal to the applied voltage divided by X_L. Three examples are illustrated in Fig. 18-6. No dc resistance is indicated, since it is assumed to be practically zero for the coils shown. In Fig. 18-6a, there is just one reactance of 100 Ω. Then I equals V/X_L, or 100 V/100 Ω, which is 1 A.

In Fig. 18-6b, the total reactance is the sum of the two individual series reactances of 100 Ω each, for a total of 200 Ω. The current, calculated as V/X_{L_T}, then equals 100 V/200 Ω, which is ½ A or 0.5 A. This cur-

rent is the same in both series reactances. Therefore, the voltage across each reactance equals its IX_L product. This is 0.5 A × 100 Ω, or 50 V across each X_L.

In Fig. 18-6c, each parallel reactance has its individual branch current, equal to the applied voltage divided by the branch reactance. Then each branch current equals 100 V/100 Ω, which is 1 A. The voltage is the same across both reactances, equal to the generator voltage, since they are all in parallel.

The total line current of 2 A is the sum of the two individual 1-A branch currents. With rms value for the applied voltage, all the calculated values of currents and voltage drops in Fig. 18-6 are also rms values.

Practice Problems 18-4
Answers at End of Chapter

a. In Fig. 18-6b, how much is the I through both X_{L_1} and X_{L_2}?

b. In Fig. 18-6c, how much is the V across both X_{L_1} and X_{L_2}?

18-5 Applications of X_L for Different Frequencies

The general use of inductance is to provide minimum reactance for relatively low frequencies but more for higher frequencies. In this way, the current in an ac circuit can be reduced for higher frequencies because of more X_L. There are many circuits in which voltages of different frequencies are applied to produce current with different frequencies. Then, the general effect of X_L is to allow the most current for direct current and

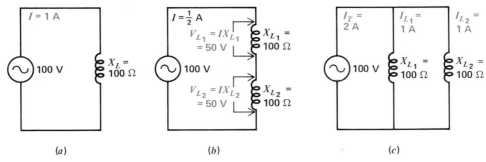

Fig. 18-6 Circuit calculations with V, I, and ohms of reactance X_L. (a) One reactance. (b) Two series reactances. (c) Two parallel reactances.

low frequencies, with less current for higher frequencies, as X_L increases.

Compare this frequency factor for ohms of X_L with ohms of resistance. The X_L increases with frequency, but R has the same effect in limiting direct current or alternating current of any frequency.

If 1000 Ω is taken as a suitable value of X_L for many applications, typical inductances can be calculated for different frequencies. These are listed in Table 18-1. At 60 Hz, for example, the inductance L in the top row of Table 18-1 is 2.65 H for 1000 Ω of X_L. The calculations are

$$L = \frac{X_L}{2\pi f} = \frac{1000}{2\pi \times 60}$$

$$= \frac{1000}{377}$$

$$L = 2.65 \text{ H}$$

For this case, the inductance has practically no reactance for direct current or for very low frequencies

below 60 Hz. However, above 60 Hz, the inductive reactance increases to more than 1000 Ω.

To summarize, the effects of increasing frequencies for this 2.65-H inductance are as follows:

Inductive reactance X_L is zero for 0 Hz which corresponds to a steady direct current.

Inductive reactance X_L is less than 1000 Ω for frequencies below 60 Hz.

Inductive reactance X_L equals 1000 Ω at 60 Hz.

Inductive reactance X_L is more than 1000 Ω for frequencies above 60 Hz.

Note that the smaller inductances at the bottom of the first column still have the same X_L of 1000 Ω as the frequency is increased. Typical RF coils, for instance, have an inductance value of the order of 100 to 300 μH. For the very high radio-frequency (VHF) range, only several microhenrys of inductance are needed for an X_L of 1000 Ω.

It is necessary to use smaller inductance values as the frequency is increased because a coil that is too large

Table 18-1. Values of Inductance L for X_L of 1000 Ω

L^* (approx.)	Frequency	Remarks
2.65 H	60 Hz	Power-line frequency and low audio frequency
160 mH	1000 Hz	Medium audio frequency
16 mH	10,000 Hz	High audio frequency
160 μH	1000 kHz (RF)	In radio broadcast band
16 μH	10 MHz (HF)	In short-wave radio band
1.6 μH	100 MHz (VHF)	In FM broadcast band

*Calculated as $L = 1000/(2\pi f)$.

can have excessive losses at high frequencies. With iron-core coils, particularly, the hysteresis and eddy-current losses increase with frequency.

Practice Problems 18-5
Answers at End of Chapter

Refer to Table 18-1.
a. Which frequency uses the smallest L for 1000 Ω of X_L?
b. How much would X_L be for the 1.6-μH L at 200 MHz?

18-6 Waveshape of v_L Induced by Sine-Wave Current

More details of inductive circuits can be analyzed by means of the waveshapes in Fig. 18-7, plotted for the calculated values in Table 18-2. The top curve shows a sine wave of current i_L flowing through a 6-mH inductance L. Since induced voltage depends on rate of change of current rather than the absolute value of i, the curve in Fig. 18-7b shows how much the current changes. In this curve the di/dt values are plotted for the current changes every 30° of the cycle. The bottom curve shows the actual induced voltage v_L. This v_L curve is similar to the di/dt curve because v_L equals the constant factor L multiplied by di/dt.

90° Phase Angle The v_L curve at the bottom of Fig. 18-7 has its zero values when the i_L curve at the top is at maximum. This comparison shows that the curves are 90° out of phase. The v_L is a cosine wave of voltage for the sine wave of current i_L.

The 90° phase difference results from the fact that v_L depends on the di/dt rate of change, rather than on i itself. More details of this 90° phase angle between v_L and i_L for inductance are explained in the next chapter.

Frequency For each of the curves, the period T is 24 μs. Therefore, the frequency is $1/T$ or $\frac{1}{24}$ μs, which equals 41.67 kHz. Each curve has the same frequency.

Ohms of X_L The ratio of v_L/i_L actually specifies the inductive reactance in ohms. For this comparison, we use the actual value of i_L, which has a peak value of 100 mA. The rate-of-change factor is included in the induced voltage v_L. Although the peak of v_L at 150 V is 90° before the peak of i_L at 100 mA, we can compare these two peak values. Then v_L/i_L is 150/0.1, which equals 1500 Ω.

This X_L is only an approximate value because v_L cannot be determined exactly for the large dt changes every 30°. If we used smaller intervals of time, the peak v_L would be 157 V. Then X_L would be 1570 Ω, the same as $2\pi fL$ Ω with a 6-mH L and a frequency of 41.67 kHz. This is the same X_L problem as Example 1 on page 358.

The Tabulated Values from 0 to 90° The numerical values in Table 18-2 are calculated as follows: The i curve is a sine wave. This means it rises to one-half its peak value in 30° and to 0.866 of the peak in 60°, and the peak value is at 90°.

In the di/dt curve the changes in i are plotted. For the first 30° the di is 50 mA; the dt change is 2 μs. Then di/dt is 25 mA/μs. This point is plotted between 0 and 30° to indicate that 25 mA/μs is the rate of change of current for the 2-μs interval between 0 and 30°. If smaller intervals were used, the di/dt values could be determined more accurately.

During the next 2-μs interval, from 30 to 60°, the current increases from 50 to 86.6 mA. The change of current during this time is $86.6 - 50$, which equals 36.6 mA. The time is the same 2 μs for all the intervals. Then di/dt for the next plotted point is 36.6/2, or 18.3.

For the final 2-μs change before i reaches its peak at 100 mA, the di value is $100 - 86.6$, or 13.4 mA, and the di/dt value is 6.7. All these values are listed in Table 18-2.

Notice that the di/dt curve in Fig. 18-7b has its peak at the zero value of the i curve, while the peak i values correspond to zero on the di/dt curves. These conditions result because the sine wave of i has its sharpest slope at the zero values. The rate of change is greatest when the i curve is going through the zero axis. The i curve flattens near the peaks and has zero rate of change exactly at the peak. The curve must stop going up before it can come down. In summary, then, the

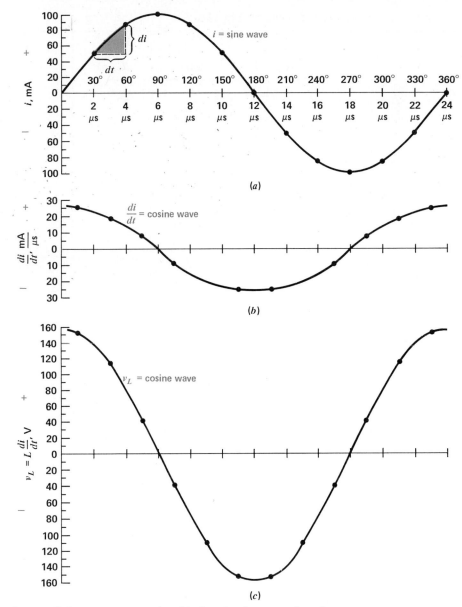

Fig. 18-7 Waveshapes of sine-wave current i and induced voltage v_L, plotted for values in Table 18-2.

di/dt curve and the i curve are 90° out of phase with each other.

The v_L curve follows the di/dt curve exactly, as $v_L = L(di/dt)$. The phase of the v_L curve is exactly the same as that of the di/dt curve, 90° out of phase with the i curve. For the first plotted point,

$$v_L = L\frac{di}{dt}$$

$$= 6 \times 10^{-3} \times \frac{50 \times 10^{-3}}{2 \times 10^{-6}}$$

$$v_L = 150 \text{ V}$$

The other v_L values are calculated the same way, multiplying the constant factor of 6 mH by the di/dt value for each 2-μs interval.

Table 18-2. Values for $v_L = L(di/dt)$ Curves in Fig. 18-7

Time		dt			di/dt,		
θ	μs	θ	μs	di, mA	mA/μs	L, mH	$v_L = L\,(di/dt)$, V
30°	2	30°	2	50	25	6	150
60°	4	30°	2	36.6	18.3	6	109.8
90°	6	30°	2	13.4	6.7	6	40.2
120°	8	30°	2	−13.4	−6.7	6	−40.2
150°	10	30°	2	−36.6	−18.3	6	−109.8
180°	12	30°	2	−50	−25	6	−150
210°	14	30°	2	−50	−25	6	−150
240°	16	30°	2	−36.6	−18.3	6	−109.8
270°	18	30°	2	−13.4	−6.7	6	−40.2
300°	20	30°	2	13.4	6.7	6	40.2
330°	22	30°	2	36.6	18.3	6	109.8
360°	24	30°	2	50	25	6	150

90 to 180°   In this quarter-cycle, the sine wave of i decreases from its peak of 100 mA at 90° to zero at 180°. This decrease is considered a negative value for di, as the slope is negative going downward. Physically, the decrease in current means its associated magnetic flux is collapsing, compared with the expanding flux as the current increases. The opposite motion of the collapsing flux must make v_L of opposite polarity, compared with the induced voltage polarity for increasing flux. This is why the di values are negative from 90 to 180°. The di/dt values are also negative, and the v_L values are negative.

180 to 270°   In this quarter-cycle, the current increases in the reverse direction. If the magnetic flux is considered counterclockwise around the conductor with $+i$ values, the flux is in the reversed clockwise direction with $-i$ values. Any induced voltage produced by expanding flux in one direction will have opposite polarity from voltage induced by expanding flux in the opposite direction. This is why the di values are considered negative from 180 to 270°, as in the second quarter-cycle, compared with the positive di values from 0 to 90°. Actually, increasing negative values and decreasing positive values are changing in the same direction. This is why v_L is negative for both the second and third quarter-cycles.

270 to 360°   In the last quarter-cycle, the negative i values are decreasing. Now the effect on polarity is like

two negatives making a positive. The current and its magnetic flux have the negative direction. But the flux is collapsing, which induces opposite voltage from increasing flux. Therefore, the di values from 270 to 360° are positive, as are the di/dt values and the induced voltages v_L.

The same action is repeated for each cycle of sine-wave current. Then the current i_L and the induced voltage v_L are 90° out of phase. The reason is that v_L depends on di/dt, not on i alone.

Application of the 90° Phase Angle in a Circuit   The phase angle of 90° between V_L and I will always apply for any L with sine-wave current. Remember, though, that the specific comparison is only between the induced voltage across any one coil and the current flowing in its turns. To emphasize this im-

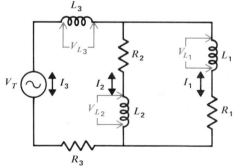

Fig. 18-8 How 90° phase angle for the V_L applies in a complex circuit with more than one inductance. Current I_1 lags V_{L_1} by 90°; I_2 lags V_{L_2} by 90°; I_3 lags V_{L_3} by 90°.

portant principle, Fig. 18-8 shows an ac circuit with a few coils and resistors. The details of this complex circuit are not to be analyzed now. However, for each L in the circuit, the V_L is 90° out of phase with its I. The I lags V_L by 90°, or V_L leads I. For the examples of three coils in Fig. 18-8:

Current I_1 lags V_{L_1} by 90°.

Current I_2 lags V_{L_2} by 90°.

Current I_3 lags V_{L_3} by 90°.

Note that I_3 is also I_T for the series-parallel circuit.

Practice Problems 18-6
Answers at End of Chapter

Refer to Fig. 18-7.
a. At what angle does i have its maximum positive value?
b. At what angle does v_L have its maximum positive value?
c. What is the phase angle difference between the waveforms for i and v_L?

Summary

1. Inductive reactance, indicated X_L, is the opposition of an inductance to the flow of sine-wave alternating current.
2. Reactance X_L is measured in ohms because it limits the current to the value $I = V/X_L$. With V in volts and X_L in ohms, I is in amperes.
3. $X_L = 2\pi fL$. With f in hertz and L in henrys, X_L is in ohms.
4. With one constant L, its X_L increases proportionately with higher frequencies.
5. At one frequency, X_L increases proportionately with higher inductances.
6. With X_L and f known, the inductance $L = X_L/(2\pi f)$.
7. With X_L and L known, the frequency $f = X_L/(2\pi L)$.
8. The total X_L of reactances in series is the sum of the individual values, as for series resistances. Series reactances have the same current. The voltage across each inductive reactance is IX_L.
9. With parallel reactances, the total reactance is calculated by the reciprocal formula, as for parallel resistances. Each branch current is V/X_L. The total line current is the sum of the individual branch currents.
10. Table 18-3 summarizes the differences between L and X_L.
11. Table 18-4 compares X_L and R.

Table 18-3. Comparison of Inductance and Inductive Reactance

Inductance	Inductive reactance
Symbol is L	Symbol is X_L
Measured in henry units	Measured in ohm units
Depends on construction of coil	Depends on frequency
$L = v_L/(di/dt)$, in H units	$X_L = v_L/i_L$ or $2\pi fL$, in Ω units

Table 18-4. Comparison of X_L and R

X_L	R
Ohm unit	Ohm unit
Increases for higher frequencies	Same for all frequencies
Phase angle is 90°	Phase angle is 0°

Self-Examination
Answers at Back of Book

Choose (a), (b), (c), or (d).

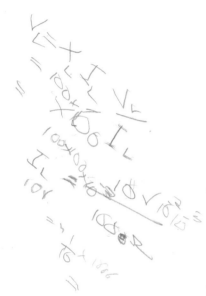

1. Inductive reactance is measured in ohms because it (a) reduces the amplitude of alternating current; (b) increases the amplitude of alternating current; (c) increases the amplitude of direct current; (d) has a back emf opposing a steady direct current.
2. Inductive reactance applies only to sine waves because it (a) increases with lower frequencies; (b) increases with lower inductance; (c) depends on the factor 2π; (d) decreases with higher frequencies.
3. An inductance has a reactance of 10,000 Ω at 10,000 Hz. At 20,000 Hz, its inductive reactance equals (a) 500 Ω; (b) 2000 Ω; (c) 20,000 Ω; (d) 32,000 Ω.
4. A 16-mH inductance has a reactance of 1000 Ω. If two of these are connected in series without any mutual coupling, their total reactance equals (a) 500 Ω; (b) 1000 Ω; (c) 1600 Ω; (d) 2000 Ω.
5. Two 5000-Ω inductive reactances in parallel have an equivalent reactance of (a) 2500 Ω; (b) 5000 Ω; (c) 10,000 Ω; (d) 50,000 Ω.
6. With 10 V applied across an inductive reactance of 100 Ω, the current equals (a) 10 μA; (b) 10 mA; (c) 100 mA; (d) 10 A.
7. A current of 100 mA through an inductive reactance of 100 Ω produces a voltage drop equal to (a) 1 V; (b) 6.28 V; (c) 10 V; (d) 100 V.
8. The inductance required for a 2000-Ω reactance at 20 MHz equals (a) 10 μH; (b) 15.9 μH; (c) 159 μH; (d) 320 μH.
9. A 160-μH inductance will have a 5000-Ω reactance at the frequency of (a) 5 kHz; (b) 200 kHz; (c) 1 MHz; (d) 5 MHz.
10. A coil has an inductive reactance of 1000 Ω. If its inductance is doubled and the frequency is doubled, then the inductive reactance will be (a) 1000 Ω; (b) 2000 Ω; (c) 4000 Ω; (d) 16,000 Ω.

Essay Questions

1. Explain briefly why X_L limits the amount of alternating current.
2. Give two differences and one similarity between X_L and R.
3. Explain briefly why X_L increases with higher frequencies and more inductance.
4. Give two differences between inductance L of a coil and its reactance X_L.
5. Why are the waves in Fig. 18-7a and b considered to be 90° out of phase, while the waves in Fig. 18-7b and c have the same phase?
6. Referring to Fig. 18-3, how does this graph show a linear proportion between X_L and frequency?
7. Referring to Fig. 18-4, how does this graph show a linear proportion between X_L and L?
8. Referring to Fig. 18-3, tabulate the values of L that would be needed for each frequency listed but for an X_L of 2000 Ω. (Do not include 0 Hz.)
9. (a) Draw the circuit for a 40-Ω R across a 120-V 60-Hz source. (b) Draw the circuit for a 40-Ω X_L across a 120-V 60-Hz source. (c) Why is I equal to 3 A for both circuits? (d) Give two differences between the circuits.
10. Why are coils for RF applications generally smaller than AF coils?

Problems

Answers to Odd-Numbered Problems at Back of Book

1. Calculate the X_L of a 0.5-H inductance at 100, 200, and 1000 Hz.
2. How much is the inductance for 628 Ω reactance at 100 Hz? 200 Hz? 1000 Hz? 500 kHz?
3. A coil with an X_L of 748 Ω is connected across a 16-V ac generator. (a) Draw the schematic diagram. (b) Calculate the current. (c) How much is the voltage across the coil?
4. A 20-H coil has 10 V applied, with a frequency of 60 Hz. (a) Draw the schematic diagram. (b) How much is the inductive reactance of the coil? (c) Calculate the current. (d) What is the frequency of the current?
5. How much is the inductance of a coil with negligible resistance if the current is 0.1 A when connected across the 60-Hz 120-V power line?
6. Referring to Fig. 18-6, how much is the inductance of L_T, L_1, and L_2 if the frequency of the source voltage is 400 Hz?
7. How much is the inductance of a coil that has a reactance of 1000 Ω at 1000 Hz? How much will the reactance be for the same coil at 10 kHz?
8. How much is the reactance of a 20-μH inductance at 40 MHz?
9. A 1000-Ω X_{L_1} and a 4000-Ω X_{L_2} are in series across a 10-V 60-Hz source. Draw the schematic diagram and calculate the following: (a) total X_L; (b) current in X_{L_1} and in X_{L_2}; (c) voltage across X_{L_1} and across X_{L_2}; (d) L_1 and L_2.
10. The same 1000-Ω X_{L_1} and 4000-Ω X_{L_2} are in parallel across the 10-V 60-Hz source. Draw the schematic diagram and calculate the following: branch currents in X_{L_1} and in X_{L_2}, total current in the generator, voltage across X_{L_1} and across X_{L_2}, inductance of L_1 and L_2.
11. At what frequencies will X_L be 2000 Ω for the following inductors: (a) 2 H; (b) 250 mH; (c) 800 μH; (d) 200 μH; (e) 20 μH?
12. A 6-mH L_1 is in series with an 8-mH L_2. The frequency is 40 kHz. (a) How much is L_T? (b) Calculate X_{L_T}. (c) Calculate X_{L_1} and X_{L_2}.
13. Calculate X_L of a 2.4-mH coil at 108 kHz.
14. Calculate X_L of a 40-μH coil at 3.2 MHz.
15. Calculate X_L of a 2-H coil at 60 Hz.
16. How much is I when the X_L of Prob. 15 is connected to the 120-V 60-Hz power line?
17. A 250-mH inductor with negligible resistance is connected across a 10-V source. Tabulate the values of X_L and current in the circuit for alternating current at (a) 20 Hz; (b) 60 Hz; (c) 100 Hz; (d) 500 Hz; (e) 5000 Hz; (f) 15,000 Hz.
18. Do the same as in Prob. 17 for an 8-H inductor.

Answers to Practice Problems

18-1 **a.** 0 Ω	**18-3** **a.** X_{L_T} = 500 Ω	**18-5** **a.** 100 MHz
b. 1000 Ω	**b.** X_{L_T} = 120 Ω	**b.** 2000 Ω
18-2 **a.** X_L = 628 Ω	**18-4** **a.** 0.5 A	**18-6** **a.** 90°
b. X_L = 314 Ω	**b.** 100 V	**b.** 0 or 360°
c. X_L = 6280 Ω		**c.** 90°

Chapter 19

Inductive Circuits

This unit analyzes circuits that combine inductive reactance X_L and resistance R. The main questions are: How do we combine the ohms of opposition, how much current flows, and what is the phase angle? In addition, the practical application of using a coil as a choke to reduce the current for a specific frequency is illustrated.

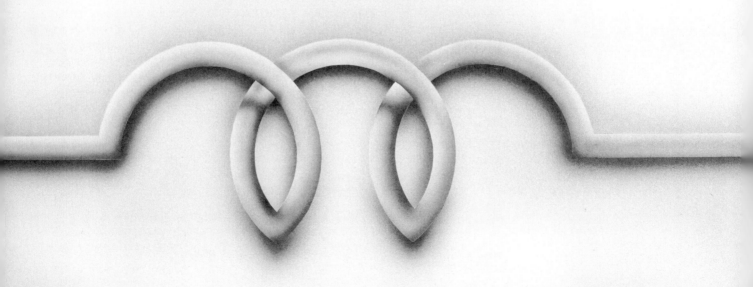

Finally, the general case of induced voltage across L is shown with nonsinusoidal current variations. Here, we compare the waveshapes of i_L and v_L instead of their phase. With nonsinusoidal waveforms, the circuit can be analyzed in terms of its L/R time constant instead of its reactance. Remember that X_L and its 90° phase angle apply only to sine waves.

Important terms in this chapter are:

choke	Q of a coil
impedance Z	sawtooth waveform
lagging current	steady-state value
L/R time constant	transient response

More details are explained in the following sections:

19-1 Sine-Wave i_L Lags v_L by 90°
19-2 X_L and R in Series
19-3 Impedance Z
19-4 X_L and R in Parallel
19-5 Q of a Coil
19-6 AF and RF Chokes
19-7 The General Case of Inductive Voltage
19-8 Calculating the L/R Time Constant

19-1   Sine-Wave i_L Lags v_L by 90°

With sine-wave variations of current producing an induced voltage, the current lags its induced voltage by exactly 90°, as shown in Fig. 19-1. The inductive circuit in Fig. 19-1a has the current and voltage waveshapes shown in Fig. 19-1b. The phasors in Fig. 19-1c show the 90° phase angle between i_L and v_L. Therefore, we can say that i_L lags v_L by 90°. Or, v_L leads i_L by 90°.

This 90° phase relationship between i_L and v_L is true in any sine-wave ac circuit, whether L is in series or parallel, and whether L is alone or combined with other components. We can always say that the voltage across any X_L is 90° out of phase with the current through it.

Why the Phase Angle is 90°   This results because v_L depends on the rate of change of i_L. As previously shown in Fig. 18-7 (see page 363) for a sine wave of i_L, the induced voltage is a cosine wave. In other words, v_L has the phase of di/dt, not the phase of i.

Why i_L Lags v_L   The 90° difference can be measured between any two points having the same value on the i_L and v_L waves. A convenient point is the positive peak value. Note that the i_L wave does not have its positive peak until 90° after the v_L wave. Therefore, i_L lags v_L by 90°. This 90° lag is in time. The time lag equals one quarter-cycle, which is one-quarter of the time for a complete cycle.

Inductive Current Is the Same in a Series Circuit   The time delay and resultant phase angle for the current in an inductance apply only with respect to the voltage across the inductance. This condition does not change the fact that the current is the same in all parts of a series circuit. In Fig. 19-1a, the current in the generator, the connecting wires, and L must be the same because they are in series. At any instant, whatever the current value is at that time, it is the same in all the series components. The time lag is between current and voltage.

Inductive Voltage Is the Same across Parallel Branches   In Fig. 19-1a, the voltage across the generator and the voltage across L are the same because they are in parallel. There cannot be any lag or lead in time between these two parallel voltages. At any instant, whatever the voltage value is across the generator at that time, the voltage across L is the same. Considering the parallel voltage v_A or v_L, it is 90° out of phase with the current.

In this circuit the voltage across L is determined by the applied voltage, since they must be the same. The inductive effect here is to make the current have the values that produce $L(di/dt)$ equal to the parallel voltage.

The Frequency Is the Same for i_L and v_L   Although i_L lags v_L by 90°, both waves have the same frequency. The i_L wave reaches it peak values 90° later than the v_L wave, but the complete cycles of variations are repeated at the same rate. As an example, if the frequency of the sine wave v_L in Fig. 19-1b is 100 Hz, this is also the frequency for i_L.

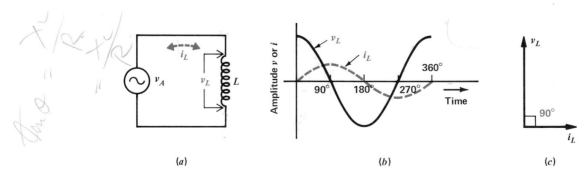

Fig. 19-1 (a) Circuit with inductance L. (b) Sine wave of i_L lags v_L by 90°. (c) Phasor diagram.

Refer to Fig. 19-1.
a. What is the phase between v_A and v_L?
b. What is the phase between v_L and i_L?
c. Does i_L lead or lag v_L?

is the angle between current through an inductance and its self-induced voltage.

4. The current I through R and its IR voltage drop have the same phase. There is no reactance to sine-wave current in any resistance. Therefore, I and IR have the same phase, or this phase angle is $0°$.

Resistance R can be either the internal resistance of the coil or an external series resistance. The I and V values may be rms, peak, or instantaneous, as long as the same measure is applied to all. Peak values are used here for convenience in comparing the waveforms.

19-2 X_L and R in Series

When a coil has series resistance, the current is limited by both X_L and R. This current I is the same in X_L and R, since they are in series. Each has its own series voltage drop, equal to IR for the resistance and IX_L for the reactance.

Note the following points about a circuit that combines series X_L and R, as in Fig. 19-2:

1. The current is labeled I, rather than I_L, because I flows through all the series components.
2. The voltage across X_L, labeled V_L, can be considered an IX_L voltage drop, just as we use V_R for an IR voltage drop.
3. The current I through X_L must lag V_L by $90°$, as this

Phase Comparisons Note the following:

1. Voltage V_L is $90°$ out of phase with I.
2. However, V_R has the same phase as I.
3. Therefore, V_L is also $90°$ out of phase with V_R.

Specifically, V_R lags V_L by $90°$, just as the current I lags V_L. These phase relations are shown by the waveforms in Fig. 19-2b and the phasors in Fig. 19-2c.

Combining V_R and V_L As shown in Fig. 19-2b, when the V_R voltage wave is combined with the V_L voltage wave, the result is the voltage wave for the

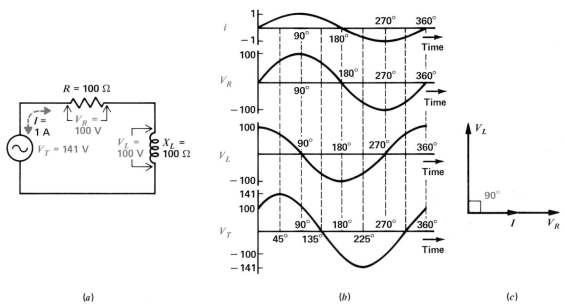

(a) (b) (c)

Fig. 19-2 Reactance X_L and R in series. (a) Circuit. (b) Waveforms of current and voltages. (c) Phasor diagram.

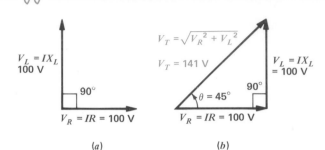

Fig. 19-3 Addition of two voltages 90° out of phase. (a) Phasors at 90°. (b) Resultant of the two phasors is the hypotenuse of the right triangle.

applied generator voltage V_T. The voltage drops must add to equal the applied voltage. The 100-V peak values for V_R and for V_L total 141 V, however, instead of 200 V, because of the 90° phase difference.

Consider some instantaneous values to see why the 100-V peak V_R and 100-V peak V_L cannot be added arithmetically. When V_R is at its maximum of 100 V, for instance, V_L is at zero. The total for V_T then is 100 V. Similarly, with V_L at its maximum of 100 V, then V_R is zero and the total V_T is also 100 V.

Actually, V_T has its maximum value of 141 V at the time when V_L and V_R are each 70.7 V. When series voltage drops that are out of phase are combined, therefore, they cannot be added without taking the phase difference into account.

Phasor-Voltage Triangle   Instead of combining waveforms that are out of phase, we can add them more quickly by using their equivalent phasors, as shown in Fig. 19-3. The phasors in Fig. 19-3a just show the 90° angle without any addition. The method in Fig. 19-3b is to add the tail of one phasor to the arrowhead of the other, using the angle required to show their relative phase. Voltages V_R and V_L are at right angles because they are 90° out of phase. The sum of the phasors is a resultant phasor from the start of one to the end of the other. Since the V_R and V_L phasors form a right angle, the resultant phasor is the hypotenuse of a right triangle. The hypotenuse is the side opposite the 90° angle.

From the geometry of a right triangle, the pythagorean theorem states that the hypotenuse is equal to the square root of the sum of the squares of the sides. For the voltage triangle in Fig. 19-3b, therefore, the resultant is

$$V_T = \sqrt{V_R^2 + V_L^2} \qquad (19\text{-}1)$$

where V_T is the phasor sum of the two voltages V_R and V_L 90° out of phase.

This formula is for V_R and V_L when they are in series, since then they are 90° out of phase. All the voltages must be in the same units. When V_A is an rms value, V_R and V_L are also rms values.

In calculating the value of V_T, note that the terms V_R and V_L must each be squared before they are added to find the square root. For the example in Fig. 19-3,

$$V_T = \sqrt{100^2 + 100^2}$$
$$= \sqrt{10{,}000 + 10{,}000}$$
$$= \sqrt{20{,}000}$$
$$V_T = 141 \text{ V}$$

Practice Problems 19-2
Answers at End of Chapter

a. In a series circuit with X_L and R, what is the phase angle between I and V_R?

b. What is the phase angle between V_R and V_L?

19-3   Impedance Z

A phasor triangle of R and X_L in series corresponds to the voltage triangle, as shown in Fig. 19-4. It is similar to the voltage triangle in Fig. 19-3, but the common factor I cancels because the current is the same in X_L and R. The resultant of the phasor addition of R and X_L is their total opposition in ohms, called *impedance*,

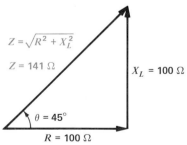

Fig. 19-4 Phasor addition of R and X_L 90° out of phase in series circuit, to find the resultant impedance Z.

with the symbol Z. The Z takes into account the 90° phase relation between R and X_L.

For the impedance triangle of a series circuit with reactance and resistance,

$$Z = \sqrt{R^2 + X_L^2} \qquad (19\text{-}2)$$

With R and X_L in ohms, Z is also in ohms. For the example in Fig. 19-4,

$$\begin{aligned} Z &= \sqrt{100^2 + 100^2} \\ &= \sqrt{10,000 + 10,000} \\ &= \sqrt{20,000} \\ Z &= 141 \ \Omega \end{aligned}$$

Note that the total impedance of 141 Ω divided into the applied voltage of 141 V results in 1 A of current in the series circuit. The IR voltage is 1×100, or 100 V; the IX_L voltage is also 1×100, or 100 V. The total of the series IR drops of 100 V each added by phasors equals the applied voltage of 141 V. Finally, the applied voltage equals IZ, or 1×141, which is 141 V.

To summarize the similar phasor triangles for volts and ohms in a series circuit:

1. The phasor for R, IR, or V_R is at 0°.
2. The phasor for X_L, IX_L, or V_L is at 90°.
3. The phasor for Z, IZ, or V_T has the phase angle θ of the complete circuit.

Phase Angle with Series X_L   The angle between the generator voltage and its current is the phase angle of the circuit. Its symbol is θ (theta). In Fig. 19-3, the phase angle between V_T and IR is 45°. Since IR and I have the same phase, the angle is also 45° between V_T and I.

In the corresponding impedance triangle in Fig. 19-4, the angle between Z and R is also equal to the phase angle. Therefore, the phase angle can be calculated from the impedance triangle of a series circuit by the formula

$$\tan \theta_Z = \frac{X_L}{R} \qquad (19\text{-}3)$$

The tangent (tan) is a trigonometric function[1] of any angle, equal to the ratio of the opposite side to the adjacent side of a triangle. In this impedance triangle, X_L is the opposite side and R is the adjacent side of the angle. We use the subscript $_Z$ for θ to show that θ_Z is found from the impedance triangle for a series circuit. To calculate this phase angle,

$$\tan \theta_Z = \frac{X_L}{R} = \frac{100}{100} = 1$$

From the trigonometric table of sines, cosines, and tangents in App. E, the angle that has the tangent equal to 1 is 45°. Therefore, the phase angle is 45° in this example. The numerical values of the trigonometric functions can be found from a table, slide rule, or scientific calculator.

Note that the phase angle of 45° is halfway to 90° because R and X_L are equal.

Example 1   If a 30-Ω R and a 40-Ω X_L are in series with 100 V applied, find the following: Z, I, V_R, V_L, and θ_Z. What is the phase of V_L and V_R with respect to the phase of I? Prove that the sum of the series voltage drops equals the applied voltage V_T.

Answer $Z = \sqrt{R^2 + X_L^2} = \sqrt{900 + 1600}$
$$\qquad\quad = \sqrt{2500}$$
$$\qquad Z = 50 \ \Omega$$

$$I = \frac{V_T}{Z} = \frac{100}{50} = 2 \text{ A}$$

$$V_R = IR = 2 \times 30 = 60 \text{ V}$$

$$V_L = IX_L = 2 \times 40 = 80 \text{ V}$$

$$\tan \theta_Z = \frac{X_L}{R} = \frac{40}{30} = \frac{4}{3} = 1.33$$

$$\theta_Z = 53°$$

---

[1]Numerical trigonometry using the sine, cosine, or tangent functions for any angle is explained in B. Grob, *Mathematics Outline and Review Problems for Basic Electronics*, McGraw-Hill Book Company, New York.

Therefore, I lags V_T by 53°. Furthermore, I and V_R have the same phase, and I lags V_L by 90°. Finally.

$$V_T = \sqrt{V_R{}^2 + V_L{}^2} = \sqrt{60^2 + 80^2}$$
$$= \sqrt{3600 + 6400} = \sqrt{10,000}$$
$$V_T = 100 \text{ V}$$

Therefore, the sum of the voltage drops equals the applied voltage.

Series Combinations of X_L and R

In a series circuit, the higher the value of X_L compared with R, the more inductive the circuit is. This means there is more voltage drop across the inductive reactance and the phase angle increases toward 90°. The series current lags the applied generator voltage. With all X_L and no R, the entire applied voltage is across X_L and θ_Z equals 90°.

Several combinations of X_L and R in series are listed in Table 19-1 with their resultant impedance and phase angle. Note that a ratio of 10:1 or more for X_L/R means that the circuit is practically all inductive. The phase angle of 84.3° is only slightly less than 90° for the ratio of 10:1, and the total impedance Z is approximately equal to X_L. The voltage drop across X_L in the series circuit will be practically equal to the applied voltage, with almost none across R.

At the opposite extreme, when R is ten times as large as X_L, the series circuit is mainly resistive. The phase angle of 5.7°, then, means the current has almost the same phase as the applied voltage, the total impedance Z is approximately equal to R, and the voltage drop across R is practically equal to the applied voltage, with almost none across X_L.

Table 19-1. Series R and X_L Combinations

R, Ω	X_L, Ω	Z, Ω (approx.)	Phase angle θ_Z
1	10	$\sqrt{101} = 10$	84.3°
10	10	$\sqrt{200} = 14$	45°
10	1	$\sqrt{101} = 10$	5.7°

Note: θ_Z is the angle of Z_T with respect to the reference I in a series circuit.

For the case when X_L and R equal each other, their resultant impedance Z is 1.41 times the value of either one. The phase angle then is 45°, halfway between 0° for resistance alone and 90° for inductive reactance alone.

Practice Problems 19-3
Answers at End of Chapter

a. How much is Z_T for a 20-Ω R in series with a 20-Ω X_L?
b. How much is V_T for 20 V across R and 20 V across X_L in series?
c. What is the phase angle of this circuit?

19-4 X_L and R in Parallel

For parallel circuits with X_L and R, the 90° phase angle must be considered for each of the branch currents, instead of voltage drops in a series circuit. Remember that any series circuit has different voltage drops but one common current. A parallel circuit has different branch currents but one common voltage.

In the parallel circuit in Fig. 19-5a, the applied voltage V_A is the same across X_L, R, and the generator, since they are all in parallel. There cannot be any phase difference between these voltages. Each branch, however, has its individual current. For the resistive branch, $I_R = V_A/R$; in the inductive branch, $I_L = V_A/X_L$.

The resistive branch current I_R has the same phase as the generator voltage V_A. The inductive branch current I_L lags V_A, however, because the current in an inductance lags the voltage across it by 90°.

The total line current, therefore, consists of I_R and I_L, which are 90° out of phase with each other. The phasor sum of I_R and I_L equals the total line current I_T. These phase relations are shown by the waveforms in Fig. 19-5b, with the phasors in Fig. 19-5c. Either way, the phasor sum of 10 A for I_R and 10 A for I_L is equal to 14.14 A for I_T.

Both methods illustrate the general principle that quadrature components must be combined by phasor addition. The branch currents are added by phasors here because they are the factors that are 90° out of phase in a parallel circuit. This method is similar to combining voltage drops 90° out of phase in a series circuit.

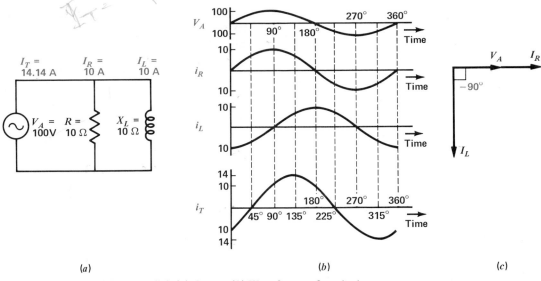

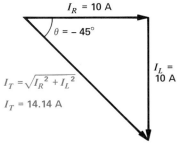

(a)   (b)   (c)

Fig. 19-5 Reactance X_L and R in parallel. (a) Circuit. (b) Waveforms of applied voltage and branch currents. (c) Phasor diagram.

Phasor Current Triangle   Note that the phasor diagram in Fig. 19-5c has the applied voltage V_A of the generator as the reference phase. The reason is that V_A is the same throughout the parallel circuit.

The phasor for I_L is down, as compared with up for an X_L phasor. Here the parallel branch current I_L lags the parallel voltage reference V_A. In a series circuit the X_L voltage leads the series current reference I. For this reason the I_L phasor is shown with a negative 90° angle. The −90° means the current I_L lags the reference phasor V_A.

The phasor addition of the branch currents in a parallel circuit can be calculated by the phasor triangle for currents shown in Fig. 19-6. Peak values are used for convenience in this example, but when the applied voltage is an rms value, the calculated currents are also in rms values. To calculate the total line current, we have

$$I_T = \sqrt{I_R^2 + I_L^2} \tag{19-4}$$

For the values in Fig. 19-6,

$$I_T = \sqrt{10^2 + 10^2}$$
$$= \sqrt{100 + 100}$$
$$= \sqrt{200}$$
$$I_T = 14.14 \text{ A}$$

Impedance of X_L and R in Parallel   A practical approach to the problem of calculating the total impedance of X_L and R in parallel is to calculate the total line current I_T and divide this into the applied voltage:

$$Z_T = \frac{V_A}{I_T} \tag{19-5}$$

For example, in Fig. 19-5, V_A is 100 V and the resultant I_T, obtained as the vector sum of the resistive and reactive branch currents, is equal to 14.14 A. Therefore, we calculate the impedance as

Fig. 19-6 Phasor triangle of branch currents 90° out of phase in parallel circuit to find I_T.

$$Z_T = \frac{V_A}{I_T}$$

$$= \frac{100 \text{ V}}{14.14 \text{ A}}$$

$$Z_T = 7.07 \ \Omega$$

This impedance is the combined opposition in ohms across the generator, equal to the resistance of 10 Ω in parallel with the reactance of 10 Ω.

Note that the impedance for equal values of R and X_L in parallel is not one-half but equals 70.7 percent of either one. Still, the combined value of ohms must be less than the lowest ohms value in the parallel branches.

For the general case of calculating the impedance of X_L and R in parallel, any number can be assumed for the applied voltage because in the calculations for Z in terms of the branch currents the value of V_A cancels. A good value to assume for V_A is the value of either R or X_L, whichever is the higher number. This way there are no fractions smaller than one in calculation of the branch currents.

Example 2   What is the total Z of a 600-Ω R in parallel with a 300-Ω X_L? Assume 600 V for the applied voltage. Then

Answer $I_R = \dfrac{600 \text{ V}}{600 \ \Omega} = 1 \text{ A}$

$I_L = \dfrac{600 \text{ V}}{300 \ \Omega} = 2 \text{ A}$

$I_T = \sqrt{I_R^2 + I_L^2}$

$\quad = \sqrt{1 + 4} = \sqrt{5}$

$I_T = 2.24 \text{ A}$

Then, dividing the total line current into the assumed value of 600 V for the applied voltage gives

$$Z_T = \frac{V_A}{I_T} = \frac{600 \text{ A}}{2.24 \text{ A}}$$

$$Z_T = 268 \ \Omega$$

The combined impedance of a 600-Ω R in parallel with a 300-Ω X_L is equal to 268 Ω, no matter how much the applied voltage is.

Phase Angle with Parallel X_L and R   In a parallel circuit, the phase angle is between the line current I_T and the common voltage V_A applied across all the branches. However, the resistive branch current I_R has the same phase as V_A. Therefore, the phase of I_R can be substituted for the phase of V_A. This is shown in Fig. 19-5c. The triangle of currents is in Fig. 19-6. To find θ_I from the branch currents, use the tangent formula

$$\tan \theta_I = -\frac{I_L}{I_R} \qquad \text{(19-6)}$$

We use the subscript $_I$ for θ to show that θ_I is found from the triangle of branch currents in a parallel circuit. In Fig. 19-6, θ_I is $-45°$ because I_L and I_R are equal. Then $\tan \theta_I = -1$.

The negative sign is used for this current ratio because I_L is lagging at $-90°$, compared with I_R. The phase angle of $-45°$ here means that I_T lags I_R and V_A by 45°.

Note that the phasor triangle of branch currents gives θ_I as the angle of I_T with respect to the generator voltage V_A. This phase angle for I_T is with respect to the applied voltage as the reference at 0°. For the phasor triangle of voltages in a series circuit, the phase angle θ_Z for Z_T and V_T is with respect to the series current as the reference at 0°.

Parallel Combinations of X_L and R   Several combinations of X_L and R in parallel are listed in Table 19-2. When X_L is ten times R, the parallel circuit is practically resistive because there is little inductive current in the line. The small value of I_L results from the high X_L. The total impedance of the parallel circuit is approximately equal to the resistance, then, since the high value of X_L in a parallel branch has little effect. The phase angle of $-5.7°$ is practically 0° because almost all the line current is resistive.

As X_L becomes smaller, it provides more inductive current in the main line. When X_L is $\frac{1}{10}$ R, practically all the line current is the I_L component. Then the parallel circuit is practically all inductive, with a total im-

Table 19-2. Parallel Resistance and Inductance Combinations*

R, Ω	X_L, Ω	I_R, A	I_L, A	I_T, A (approx.)	$Z_T = V_A/I_T$, Ω	Phase angle θ_I
1	10	10	1	$\sqrt{101} = 10$	1	$-5.7°$
10	10	1	1	$\sqrt{2} = 1.4$	7.07	$-45°$
10	1	1	10	$\sqrt{101} = 10$	1	$-84.3°$

*$V_A = 10$ V. Note that θ_I is the angle of I_T with respect to the reference V_A in parallel circuits.

pedance practically equal to X_L. The phase angle of $-84.3°$ is almost $-90°$ because the line current is mostly inductive. Note that these conditions are opposite from the case of X_L and R in series.

When X_L and R are equal, their branch currents are equal and the phase angle is $-45°$. All these phase angles are negative for parallel I_L and I_R.

As additional comparisons between series and parallel circuits, remember that

1. The series voltage drops V_R and V_L have individual values that are 90° out of phase. Therefore, V_R and V_L are added by phasors to equal the applied voltage V_T. The phase angle θ_Z is between V_T and the common series current I. More series X_L allows more V_L to make the circuit more inductive with a larger positive phase angle for V_T with respect to I.

2. The parallel branch currents I_R and I_L have individual values that are 90° out of phase. Therefore, I_R and I_L are added by phasors to equal I_T, which is the main-line current. The negative phase angle $-\theta_I$ is between the line current I_T and the common parallel voltage V_A. Less parallel X_L allows more I_L to make the circuit more inductive with a larger negative phase angle for I_T with respect to V_A.

Practice Problems 19-4
Answers at End of Chapter

a. How much is I_T for a branch current I_R of 2 A and I_L of 2 A?

b. Find the phase angle θ_I.

19-5   Q of a Coil

The ability of a coil to produce self-induced voltage is indicated by X_L, since it includes the factors of frequency and inductance. However, a coil has internal resistance equal to the resistance of the wire in the coil. This internal r_i of the coil reduces the current, which means less ability to produce induced voltage. Combining these two factors of X_L and r_i, the quality or merit of a coil is indicated by

$$Q = \frac{X_L}{r_i} = \frac{2\pi f L}{r_i} \tag{19-7}$$

As shown in Fig. 19-7, the internal r_i is in series with X_L.

As an example, a coil with X_L of 500 Ω and r_i of 5 Ω has a Q of $500/5 = 100$. The Q is a numerical value without any units, since the ohms cancel in the ratio of reactance to resistance. This Q of 100 means that the X_L of the coil is 100 times more than its r_i.

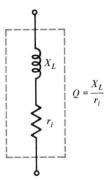

Fig. 19-7 The Q of a coil depends on its inductive reactance X_L and internal resistance r_i.

The Q of coils may range in value from less than 10 for a low-Q coil up to 1000 for a very high Q. Radio-frequency coils generally have a Q of about 30 to 300.

At low frequencies, r_i is just the dc resistance of the wire in the coil. However, for RF coils the losses increase with higher frequencies and the effective r_i increases. The increased resistance results from eddy currents and other losses.

Because of these losses, the Q of a coil does not increase without limit as X_L increases for higher frequencies. Generally, the Q can increase by a factor of about 2 for higher frequencies, within the range for which the coil is designed. The highest Q for RF coils generally results with an inductance value that provides an X_L of about 1000 Ω at the operating frequency.

More fundamentally, Q can be defined as the ratio of reactive power in the inductance to the real power dissipated in the resistance. Then

$$Q = \frac{P_L}{P_{r_i}} = \frac{I^2 X_L}{I^2 r_i} = \frac{X_L}{r_i} = \frac{2\pi f L}{r_i}$$

which is the same as Formula (19-7).

Skin Effect   Radio-frequency current tends to flow at the surface of a conductor, at very high frequencies, with little current in the solid core at the center. This skin effect results from the fact that current in the center of the wire encounters slightly more inductance because of the magnetic flux concentrated in the metal, compared with the edges, where part of the flux is in air. For this reason, conductors for VHF currents are often made of hollow tubing. The skin effect increases the effective resistance, as a smaller cross-sectional area is used for the current path in the conductor.

AC Effective Resistance   When the power and current applied to a coil are measured for RF applied voltage, the I^2R loss corresponds to a much higher resistance than the dc resistance measured with an ohmmeter. This higher resistance is the ac effective resistance R_e. Although a result of high-frequency alternating current, R_e is not a reactance; R_e is a resistive component because it draws in-phase current from the ac voltage source.

The factors that make the R_e of a coil more than its dc resistance include skin effect, eddy currents, and

hysteresis losses. Air-core coils have low losses but are limited to small values of inductance.

For a magnetic core in RF coils, a powdered-iron or ferrite slug is generally used. In a powdered-iron slug, the granules of iron are insulated from each other to reduce eddy currents. Ferrite materials have small eddy-current losses, as they are insulators, although magnetic. A ferrite core is easily saturated. Therefore, its use must be limited to coils with low values of current. A common application is the ferrite-core antenna coil in Fig. 19-8.

To reduce the R_e for small RF coils, stranded wire can be made with separate strands insulated from each other and braided so that each strand is as much on the outer surface as all the other strands. This is called *litzendraht* or *litz wire*.

As an example of the total effect of ac losses, assume that an air-core RF coil of 50-μH inductance has a resistance of 1 Ω with the dc measurement of the battery in an ohmmeter. However, in an ac circuit with a 2-MHz current, the effective coil resistance R_e can increase to 12 Ω. The increased resistance reduces the Q of the coil.

Actually, the Q can be used to determine the effective ac resistance. Since Q is X_L/R_e, then R_e equals X_L/Q. For this 50-μH L at 2 MHz, its X_L, equal to $2\pi f L$, is 628 Ω. The Q of the coil can be measured on a Q meter, which operates on the principle of resonance. Let the measured Q be 50. Then $R_e = {}^{628}\!/_{50}$, equal to 12.6 Ω.

Example 3   An air-core coil has an X_L of 700 Ω and an R_e of 2 Ω. Calculate Q.

Answer
$$Q = \frac{X_L}{R_e}$$
$$= \frac{700}{2}$$
$$Q = 350$$

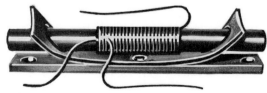

Fig. 19-8 Ferrite coil antenna for radio receiver. Length is 4 in.; inductance is 700 μH. (*J. W. Miller Co.*)

$Q = \dfrac{X_L}{R_e}$

Example 4 A 200-μH coil has a Q of 40 at 0.5 MHz. Find R_e.

Answer

$$R_e = \frac{X_L}{Q} = \frac{2\pi fL}{Q}$$

$$= \frac{2\pi \times 0.5 \times 10^6 \times 200 \times 10^{-6}}{40}$$

$$= \frac{628}{40}$$

$$R_e = 15.7 \ \Omega$$

In general, the lower the internal resistance for a coil, the higher is its Q.

Practice Problems 19-5
Answers at End of Chapter

a. A 200-μH coil with an 8-Ω internal R_e has an X_L of 600 Ω. Calculate the Q.
b. A coil with a Q of 50 has a 500-Ω X_L at 4 MHz. Calculate its internal R_e.

19-6   AF and RF Chokes

Inductance has the useful characteristic of providing more ohms of reactance at higher frequencies. Resistance has the same opposition at all frequencies and for direct current. These characteristics are applied to the circuit in Fig. 19-9, where X_L is much greater than R for the frequency of the ac source V_T. The result is that L has practically all the voltage drop in this series circuit with very little of the applied voltage across R.

The inductance L is used here as a *choke*. Therefore, a choke is an inductance in series with an external R to prevent the ac signal voltage from developing any appreciable output across R, at the frequency of the source.

The dividing line in calculations for a choke can be taken as X_L ten or more times the series R. Then the circuit is primarily inductive. Practically all the ac voltage drop is across L, with little across R. This case also results in θ of practically 90°, but the phase angle is not related to the action of X_L as a choke.

Fig. 19-9b illustrates how a choke is used to prevent ac voltage in the input from developing voltage in the output for the next circuit. Note that the output here is V_R from point A to chassis ground. Practically all the ac input voltage is across X_L between points B and C. However, this voltage is not coupled out because neither B nor C is grounded.

The desired output across R could be direct current from the input side without any ac component. Then X_L has no effect on the steady dc component. Practically all the dc voltage would be across R for the output, but the ac voltage would be just across X_L. The same idea applies to passing an AF signal through to R, while blocking an RF signal as IX_L across the choke because of more X_L at the higher frequency.

Calculations for a Choke Typical values for audio or radio frequencies can be calculated if we assume a series resistance of 100 Ω, as an example. Then

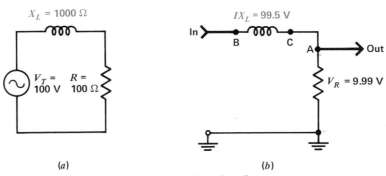

(a)   (b)

Fig. 19-9 Coil used as a choke with X_L at least $10 \times R$. Note that R is an external resistor. Voltage V_L is practically all the applied voltage, with almost none for V_R. (a) Circuit with X_L and R in series. (b) Input and output voltages.

Table 19-3. Typical Chokes for a Reactance of 1000 Ω*

F	L	Remarks
100 Hz	1.6 H	Low audio frequency
1000 Hz	0.16 H	Medium audio frequency
10 kHz	16 mH	High audio frequency
1000 kHz	0.16 mH	Radio frequency
100 MHz	1.6 μH	Very high radio frequency

*For an X_L ten times a series R of 100 Ω.

X_L must be at least 1000 Ω. As listed in Table 19-3, at 100 Hz the relatively large inductance of 1.6 H provides 1000 Ω of X_L. Higher frequencies allow a smaller value of L for a choke with the same reactance. At 100 MHz, in the VHF range, the choke is only 1.6 μH.

Some typical chokes are shown in Fig. 19-10. The iron-core choke in Fig. 19-10a is for audio frequencies. The air-core choke in Fig. 19-10b is for radio frequencies. The RF choke in Fig. 19-10c has color coding, which is often used for small coils. The color values are the same as for resistors, but with L in microhenrys. As an example, a coil with yellow, red, and black stripes or dots is 42 μH.

Choosing a Choke for a Circuit As an example of using these calculations, suppose that we have the problem of determining what kind of a coil to use as a choke for the following application. The L is to be an RF choke in series with an external R of 300 Ω, with a current of 90 mA and a frequency of 0.2 MHz. Then X_L must be at least $10 \times 300 = 3000$ Ω. At f of 0.2 MHz,

$$L = \frac{X_L}{2\pi f} = \frac{30{,}000}{2\pi \times 2 \times 10^6}$$

$$= \frac{3 \times 10^3}{1.256 \times 10^6}$$

$$= \frac{3}{1.256} \times 10^{-3}$$

$$L = 2.39 \text{ mH}$$

A typical commercial size easily available is 2.5 mH, with a current rating of 115 mA and an internal resistance of 20 Ω, similar to the RF choke in Fig. 19-10b. Note that the higher current rating is suitable. Also, the internal resistance is negligible compared with the external R. An inductance a little higher than the calculated value will provide more X_L, which is better for a choke.

Practice Problems 19-6
Answers at End of Chapter

a. How much is the minimum X_L for a choke in series with R of 80 Ω?
b. If X_L is 800 Ω at 3 MHz, how much will X_L be at 6 MHz for the same coil?

(a)

(b)

(c)

Fig. 19-10 Typical chokes. (a) Choke for 60 Hz with 8-H inductance and r_i of 350 Ω. Width is 2 in. (b) RF choke with 5 mH inductance and r_i of 50 Ω. Height is 1 in. (c) Small RF choke encapsulated in plastic with leads for printed-circuit board. Inductance L is 42 μH. Width is ¾ in.

19-7 The General Case of Inductive Voltage

The voltage across any inductance in any circuit is always equal to $L(di/dt)$. This formula gives the instantaneous values of v_L based on the self-induced voltage which is produced by a change in magnetic flux associated with a change in current.

A sine waveform of current i produces a cosine waveform for the induced voltage v_L, equal to $L(di/dt)$. This means v_L has the same waveform as i, but they are 90° out of phase for sine-wave variations.

The inductive voltage can be calculated as IX_L in sine-wave ac circuits. Since X_L is $2\pi fL$, the factors that determine the induced voltage are included in the frequency and inductance. Usually, it is more convenient to work with IX_L for the inductive voltage in sine-wave ac circuits, instead of $L(di/dt)$.

However, with a nonsinusoidal current waveform, the concept of reactance cannot be used. The X_L applies only to sine waves. Then v_L must be calculated as $L(di/dt)$, which applies for any inductive voltage.

An example is illustrated in Fig. 19-11a for sawtooth current. This waveform is often used in the deflection circuits for the picture tube in television receivers. The sawtooth rise is a uniform or linear increase of current from zero to 90 mA in this example. The sharp drop in current is from 90 mA to zero. Note that the rise is relatively slow; it takes 90 μs. This is nine times longer than the fast drop in 10 μs.

The complete period of one cycle of this sawtooth wave is 100 μs. A cycle includes the rise of i to the peak value and its drop back to the starting value.

The Slope of i The slope of any curve is a measure of how much it changes vertically for each horizontal unit. In Fig. 19-11a the increase of current has a constant slope. Here i increases 90 mA in 90 μs, or 10 mA for every 10 μs of time. Then di/dt is constant at 10 mA/10 μs for the entire rise time of the sawtooth waveform. Actually di/dt is the slope of the i curve. The constant di/dt is why the v_L waveform has a constant value of voltage during the linear rise of i. Remember that the amount of induced voltage depends on the change in current.

The drop in i is also linear but much faster. During this time, the slope is 90 mA/10 μs for di/dt.

The Polarity of v_L In Fig. 19-11, apply Lenz' law to indicate that v_L opposes the change in current. With electron flow into the top of L, the v_L is negative to oppose an increase of current. This polarity opposes the direction of electron flow shown for the current i produced by the source. For the rise time, then, the induced voltage here is labeled $-v_L$.

During the drop of current, the induced voltage has opposite polarity, which is labeled $+v_L$. These voltage polarities are for the top of L with respect to chassis ground.

Calculations for v_L The values of induced voltage across the 300-mH L are calculated as follows:

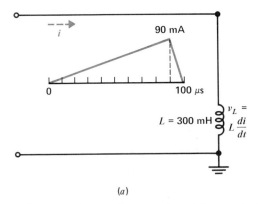

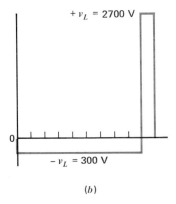

Fig. 19-11 Rectangular waveshape of v_L produced by sawtooth current. (a) Waveform of i for inductance L. (b) Induced voltage equal to $L(di/dt)$.

For the sawtooth rise:

$$-v_L = L \frac{di}{dt}$$

$$= 300 \times 10^{-3} \times \frac{10 \times 10^{-3}}{10 \times 10^{-6}}$$

$$-v_L = 300 \text{ V}$$

For the sawtooth drop:

$$+v_L = L \frac{di}{dt}$$

$$= 300 \times 10^{-3} \times \frac{90 \times 10^{-3}}{10 \times 10^{-6}}$$

$$+v_L = 2700 \text{ V}$$

The decrease in current produces nine times more voltage because the sharp drop in i is nine times faster than the relatively slow rise.

Remember that the di/dt factor can be very large, even with small currents, when the time is short. For instance, a current change of 1 mA in 1 μs is equivalent to the very high di/dt value of 1000 A/s.

In conclusion, it is important to note that v_L and i_L have different waveshapes with nonsinusoidal current. In this case, we compare the waveshapes instead of the phase angle in sine-wave circuits. Common examples of nonsinusoidal waveshapes for either v or i are the sawtooth waveform, square wave, and rectangular pulses.

19-8   Calculating the L/R Time Constant

With nonsinusoidal waveforms of i, the reaction of L to di/dt is the *transient response*, meaning a temporary result of a sudden change in i. The transient response of an inductive circuit is measured in terms of the ratio L/R, which is the time constant. In Fig. 19-12, L/R is the time for I to increase 63.2 percent when dc voltage is suddenly applied by the switch S.

To calculate the time constant,

$$T = \frac{L}{R} \qquad \text{s} \tag{19-8}$$

where T is the time constant in seconds, L the inductance in henrys, and R the series resistance in ohms. The R may be the internal coil resistance, an external resistance, or both in series. In Fig. 19-12,

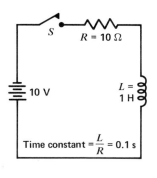

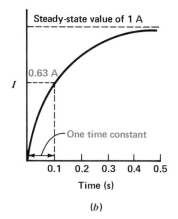

(a)   (b)

Fig. 19-12 Transient response of inductive circuit. (a) Circuit. (b) Time for a 63 percent change in i is equal to L/R.

$$T = \frac{L}{R} = \frac{1\ H}{10\ \Omega} = 0.1\ s$$

Remember that if L were not present, the current in Fig. 19-12 would rise to 10 V/10 Ω or 1 A instantly. Eventually, I will rise to 1 A, which is the steady-state value determined by V/R, and stay there. At the first instant when S is closed, though, the transient response of L opposes the increase in I from zero. The time constant of 0.1 s here specifies that I will rise 63 percent or to 0.63 A in 0.1 s. It should be noted that I is the same in R and L, because they are in series.

The current will rise practically all the way to its steady-state value of 1 A in approximately 0.5 s, which is equal to five time constants. In other words, five time constants is the time for the inductance to complete its transient response and allow the steady-state value determined by R.

In general, the time constant always measures a 63 percent change. When i is increasing, it rises by 63 percent in one time constant. When i is decreasing, it drops by 63 percent in one time constant. For any part of the transient response curve, during one time constant, the value changes by 63 percent.

On the rise, the curve goes from zero to the steady-state value at the top in five time constants. For a decrease in i, the values drop from the steady-state value to zero in five times constants. More details of the L/R time constant and a comparison with the RC time constant for capacitive circuits are explained in Chap. 23.

Practice Problems 19-8
Answers at End of Chapter

a. Calculate the time constant for L of 240 mH in series with 20 Ω.

b. If the time constant in Fig. 19-12 were 12 ms, how long would it take for I to rise to 0.63 A?

Summary

1. In a sine-wave ac circuit, the current through an inductance lags 90° behind the voltage across the inductance because $v_L = L(di/dt)$. This fundamental fact is the basis of all the following relations.
2. Therefore, inductive reactance X_L is a phasor quantity 90° out of phase with R. The phasor combination of X_L and R is their impedance Z.
3. These three types of opposition to current are compared in Table 19-4.
4. The phase angle θ is the angle between the applied voltage and its current.
5. The opposite characteristics for series and parallel circuits with X_L and R are summarized in Table 19-5 on the next page.
6. The Q of a coil is X_L/r_i, where r_i is its internal resistance.
7. A choke is an inductance with X_L greater than the series R by a factor of 10 or more.
8. In sine-wave circuits, calculate V_L as IX_L. Then V_L has a phase angle 90° different from the current.

Table 19-4. Comparison of R, X_L, and Z

R	$X_L = 2\pi fL$	$Z = \sqrt{R^2 + X_L{}^2}$
Ohms unit IR voltage same phase as I Same for all frequencies	Ohms unit IX_L voltage leads I by 90° Increases at higher frequencies	Ohms unit IZ is applied voltage Increases with X_L at higher frequencies

Table 19-5. Series and Parallel *RL* Circuits

X_L and R in series	X_L and R in parallel
I the same in X_L and R $V_T = \sqrt{V_R^2 + V_L^2}$	V_A the same across X_L and R $I_T = \sqrt{I_R^2 + I_L^2}$
$Z = \sqrt{R^2 + X_L^2}$	$Z = \dfrac{V_A}{I_T}$
V_L leads V_R by 90°	I_L lags I_R by 90°
$\tan \theta_Z = \dfrac{X_L}{R}$; θ increases as more X_L makes circuit inductive	$\tan \theta_I = -\dfrac{I_L}{I_R}$; $-\theta$ decreases as more X_L means less I_L

9. When the current is not a sine wave, figure $V_L = L(di/dt)$. Then the waveshape of V_L is different from the waveshape of current.
10. The time constant, equal to L/R, is the time in seconds for I to change by 63.2 percent, with L in henrys and R in ohms in series.

Self-Examination
Answers at Back of Book

Choose (a), (b), (c), or (d).

1. In a sine-wave ac circuit with inductive reactance, the (a) phase angle of the circuit is always 45°; (b) voltage across the inductance must be 90° out of phase with the applied voltage; (c) current through the inductance lags its induced voltage by 90°; (d) current through the inductance and voltage across it are 180° out of phase.
2. In a sine-wave ac circuit with X_L and R in series, the (a) voltages across R and X_L are in phase; (b) voltages across R and X_L are 180° out of phase; (c) voltage across R lags the voltage across X_L by 90°; (d) voltage across R leads the voltage across X_L by 90°.
3. In a sine-wave ac circuit with a 40-Ω R in series with a 30-Ω X_L, the total impedance Z equals (a) 30 Ω; (b) 40 Ω; (c) 50 Ω; (d) 70 Ω.
4. In a sine-wave ac circuit with a 90-Ω R in series with a 90-Ω X_L, phase angle θ equals (a) 0°; (b) 30°; (c) 45°; (d) 90°.
5. A 250-μH inductance is used as a choke at 10 MHz. At 12 MHz the choke (a) does not have enough inductance; (b) has more reactance; (c) has less reactance; (d) needs more turns.
6. The combined impedance of a 1000-Ω R in parallel with a 1000-Ω X_L equals (a) 500 Ω; (b) 707 Ω; (c) 1000 Ω; (d) 2000 Ω.
7. A coil with a 1000-Ω X_L at 3 MHz and 10 Ω internal resistance has a Q of (a) 3; (b) 10; (c) 100; (d) 1000.
8. In a sine-wave ac circuit with a resistive branch and an inductive branch in parallel, the (a) voltage across the inductance leads the voltage across the resist-

ance by 90°; (b) resistive branch current is 90° out of phase with the inductive branch current; (c) resistive and inductive branch currents have the same phase; (d) resistive and inductive branch currents are 180° out of phase.

9. With a 2-A I_R and a 2-A I_L in parallel branches, I_T is (a) 1 A; (b) 2 A; (c) 2.8 A; (d) 4 A.

10. In Fig. 19-11 the di/dt for the drop in sawtooth current is (a) 90 mA/s; (b) 100 mA/s; (c) 100 A/s; (d) 9000 A/s.

Essay Questions

1. What characteristic of the current in an inductance determines the amount of induced voltage? State briefly why.
2. Draw a schematic diagram showing an inductance connected across a sine-wave voltage source and indicate the current and voltage that are 90° out of phase.
3. Why does the voltage across a resistance have the same phase as the current through the resistance?
4. (a) Draw the sine waveforms for two voltages 90° out of phase, each with a peak value of 100 V. (b) Explain why their vector sum equals 141 V and not 200 V. (c) When will the sum of two 100-V drops in series equal 200 V?
5. (a) Define the phase angle of a sine-wave ac circuit. (b) State the formula for the phase angle in a circuit with X_L and R in series.
6. Define the following: (a) Q of a coil; (b) ac effective resistance; (c) RF choke; (d) sawtooth current.
7. Referring to Fig. 19-2b, why do the waveshapes shown all have the same frequency?
8. Describe how to check the trouble of an open choke with an ohmmeter.
9. Redraw the circuit and graph in Fig. 19-11 for a sawtooth current with a peak of 30 mA.
10. Why is the R_e of a coil considered resistance rather than reactance?
11. Define the time constant of an inductive circuit.
12. Why are RF chokes usually smaller than AF chokes?

Problems
Answers to Odd-Numbered Problems at Back of Book

1. Draw the schematic diagram of a circuit with X_L and R in series across a 100-V source. Calculate Z, I, IR, IX_L, and θ, approximately, for the following values: (a) 100-Ω R, 1-Ω X_L; (b) 1-Ω R, 100-Ω X_L; (c) 50-Ω R, 50-Ω X_L.
2. Draw the schematic diagram of a circuit with X_L and R in parallel across a 100-V source. Calculate I_R, I_L, I_T, and Z for the following values: (a) 100-Ω R, 1-Ω X_L; (b) 1-Ω R, 100-Ω X_L; (c) 50-Ω R, 50-Ω X_L.
3. A coil has an inductance of 1 H and a 100-Ω internal resistance. (a) Draw the equivalent circuit of the coil showing its internal resistance in series with its inductance. (b) How much is the coil's inductive reactance at 60 Hz? (c) How much is the total impedance of the coil at 60 Hz? (d) How much current will

flow when the coil is connected across a 120-V source with a frequency of 60 Hz? (e) How much is I with an f of 400 Hz?

4. Calculate the minimum inductance required for a choke in series with a resistance of 100 Ω when the frequency of the current is 5 kHz, 5 MHz, and 50 MHz. Do the same for the case where the series resistance is 10 Ω.

5. How much is the impedance Z of a coil that allows 0.3 A current when connected across a 120-V 60 Hz source? How much is the X_L of the coil if its resistance is 5 Ω? (Hint: $X_L^2 = Z^2 - R^2$.)

6. A 200-Ω R is in series with L across a 141-V 60-Hz generator V_T. The V_R is 100 V. Find L. (Hint: $V_L^2 = V_T^2 - V_R^2$.)

7. A 350-μH L has a Q of 35 at 1.5 MHz. Calculate the effective ac resistance R_e.

8. How much L is required to produce V_L equal to 9 kV when i_L drops from 300 mA to zero in 8 μs?

9. A 400-Ω R and 400-Ω X_L are in series with a 100-V 400-Hz source. Find Z, I, V_L, V_R, and θ_Z.

10. The same R and X_L of Prob. 9 are in parallel. Find I_R, I_L, I_T, Z, and θ_I.

11. The frequency is raised to 800 Hz for the parallel circuit in Prob. 10. Compare the values of I_R, I_L, and θ_I for the two frequencies of 400 and 800 Hz.

12. A 0.4-H L and a 180-Ω R are in series across a 120-V 60-Hz source. Find the current I and θ_Z.

13. An inductance L has 20 V across it at 40 mA. The frequency is 5 kHz. Calculate X_L in ohms and L in henrys.

14. A 500-Ω R is in series with 300-Ω X_L. Find Z_T, I, and θ_Z. $V_T = 120$ V.

15. A 300-Ω R is in series with a 500-Ω X_L. Find Z_T, I, and θ_Z. Compare θ_Z here with Prob. 14, with the same 120 V applied.

16. A 500-Ω R is in parallel with a 300-Ω X_L. Find I_T, Z_T, and θ_I. Compare θ_I here with θ_Z in Prob. 14 with the same 120 V applied.

17. The current shown in Fig. 19-13 flows through an 8-mH inductance. Show the corresponding waveform of induced voltage with values.

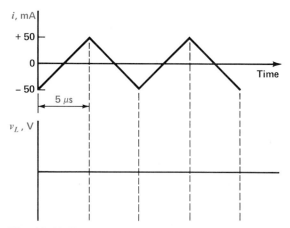

Fig. 19-13 For Prob. 17.

Answers to Practice Problems

19-1 **a.** $0°$
     b. $90°$
     c. lag

19-2 **a.** $0°$
     b. $90°$

19-3 **a.** $28.28 \ \Omega$
     b. $28.28 \ V$
     c. $\theta_Z = 45°$

19-4 **a.** $I_T = 2.828 \ A$
     b. $\theta_I = -45°$

19-5 **a.** $Q = 75$
     b. $R_e = 10 \ \Omega$

19-6 **a.** $X_L = 800 \ \Omega$
     b. $X_L = 1600 \ \Omega$

19-7 **a.** $di/dt = 1000 \ A/s$
     b. $di/dt = 9000 \ A/s$

19-8 **a.** $12 \ ms$
     b. $12 \ ms$

Review

Chapters 17 to 19

Summary

1. The ability of a conductor to produce induced voltage across itself when the current changes is its self-inductance, or inductance. The symbol is L, and the unit is the henry. One henry allows 1 V to be induced when the current changes at the rate of 1 A/s.

2. The polarity of the induced voltage always opposes the change in current that is causing the induced voltage. This is Lenz' law.

3. Mutual inductance is the ability of varying current in one coil to induce voltage in another coil nearby, without any connection between them. Its symbol is L_M, and the unit is also the henry.

4. A transformer consists of two or more windings with mutual inductance. The primary connects to the source voltage, the secondary to the load. With an iron core, the voltage ratio between primary and secondary equals the turns ratio.

5. Efficiency of a transformer equals the ratio of power output from the secondary to power input to the primary, $\times$ 100 percent.

6. Eddy currents are induced in the iron core of an inductance, causing I^2R losses that increase with higher frequencies. Laminated iron, powdered-iron, or ferrite cores have minimum eddy-current losses. Hysteresis also increases the losses.

7. Series inductances without mutual coupling add like series resistances. With parallel inductances, the combined inductance is calculated by the reciprocal formula, as with parallel resistances.

8. Inductive reactance X_L equals $2\pi fL$ Ω, where f is in hertz and L in henrys. Reactance X_L increases with more inductance and higher frequencies.

9. A common application of X_L is an AF or RF choke, which has high reactance for one group of frequencies but less reactance for lower frequencies.

10. Reactance X_L is a phasor quantity that has its current lagging 90° behind its induced voltage. In series circuits, R and X_L are added by phasors because their voltage drops are 90° out of phase. In parallel circuits, the resistive and inductive branch currents are 90° out of phase.

11. Impedance Z, in ohms, is the total opposition of an ac circuit with resistance and reactance. For series circuits, $Z = \sqrt{R^2 + X_L{}^2}$ and $I = V_T/Z$. For parallel circuits, $I_T = \sqrt{I_R{}^2 + I_L{}^2}$ and $Z = V_A/I_T$.

12. The Q of a coil is X_L/r_i.

13. Energy stored by an inductance is $\frac{1}{2} LI^2$. With I in amperes and L in henrys, the energy is in joules.

14. The quantity L/R is the time constant in seconds for I to change by 63.2 percent. The L is in henrys and the R in ohms.

15. The voltage across L is always equal to $L(di/dt)$ for any waveshape of current.

Review Self-Examination
Answers at Back of Book

Choose (a), (b), (c), or (d).

1. A coil induces 200 mV when the current changes at the rate of 1 A/s. The inductance L is: (a) 1 mH; (b) 2 mH; (c) 200 mH; (d) 100 mH.

2. Alternating current in an inductance produces maximum induced voltage when the current has its (a) maximum value; (b) maximum change in magnetic flux; (c) minimum change in magnetic flux; (d) rms value of $0.707 \times$ peak.

3. An iron-core transformer connected to the 120-V 60-Hz power line has a turns ratio of 1:20. The voltage across the secondary equals (a) 20 V; (b) 60 V; (c) 120 V; (d) 2400 V.

4. Two 250-mH chokes in series have a total inductance of (a) 60 mH; (b) 125 mH; (c) 250 mH; (d) 500 mH.

5. Which of the following will have minimum eddy-current losses? (a) Iron core; (b) laminated iron core; (c) powdered-iron core; (d) air core.

6. Which of the following will have maximum inductive reactance? (a) 2-H inductance at 60 Hz; (b) 2-mH inductance at 60 kHz; (c) 5-mH inductance at 60 kHz; (d) 5-mH inductance at 100 kHz.

7. A 100-Ω R is in series with 100 Ω of X_L. The total impedance Z equals (a) 70.7 Ω; (b) 100 Ω; (c) 141 Ω; (d) 200 Ω.

8. A 100-Ω R is in parallel with 100 Ω of X_L. The total impedance Z equals (a) 70.7 Ω; (b) 100 Ω; (c) 141 Ω; (d) 200 Ω.

9. If two waves have the frequency of 1000 Hz and one is at the maximum value when the other is at zero, the phase angle between them is (a) 0°; (b) 90°; (c) 180°; (d) 360°.

10. If an ohmmeter check on a 50-μH choke reads 3 Ω, the coil is probably (a) open; (b) defective; (c) normal; (d) partially open.

References
Additional References at Back of Book

Gilli, A. C.: *Electrical Principles for Electronics*, McGraw-Hill Book Company, New York.

Oppenheimer, S. L., F. R. Hess, and J. P. Borchers: *Direct and Alternating Currents*, McGraw-Hill Book Company, New York.

Ridsdale, R. E.: *Electric Circuits*, McGraw-Hill Book Company, New York.

Tocci, R. J.: *Introduction to Electric Circuit Analysis*, Charles E. Merrill Publishing Co., Columbus, Ohio.

Chapter 20
Capacitance

Just as inductance is important for variations of current in a wire, capacitance is a similar but opposite characteristic that is important when the voltage changes across an insulator or dielectric. Specifically, capacitance is the ability of a dielectric to store electric charge. The unit is the farad, named after Michael Faraday. A capacitor consists of an insulator between two conductors. Capacitors are manufactured for specific values of capacitance.

The types of capacitors are named according to the dielectric. Most common are air, paper, mica, ceramic, and electrolytic capacitors. This chapter explains how capacitance becomes charged by a voltage source, and how the capacitor discharges. Typical troubles of an open or short

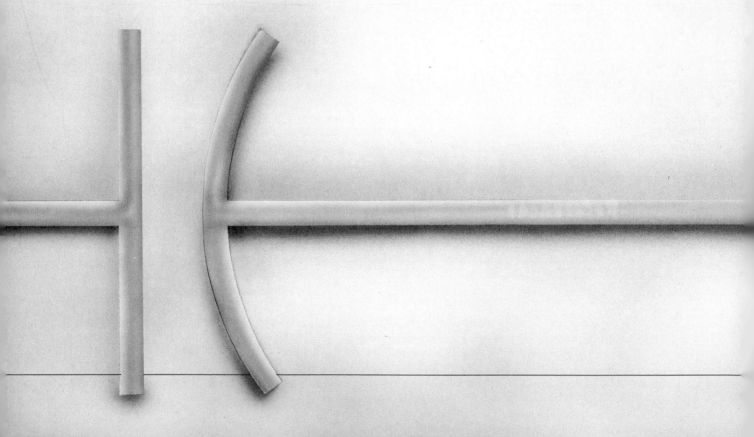

are explained, including the method of checking capacitors with an ohmmeter, even though a capacitor is actually an insulator.

Important terms in this chapter are:

aluminum capacitor
capacitor charge and discharge
ceramic capacitor
color coding of capacitors
dielectric constant
distributed capacitance
electrolytic capacitor
electrostatic induction
farad (F) unit
ganged capacitors

leakage resistance
mica capacitor
Mylar capacitor
paper capacitor
shelf life of capacitors
stored charge
stray capacitance
tantalum capacitor
testing of capacitors
tuning capacitor

More details are described in the following sections:

20-1 How Charge Is Stored in the Dielectric
20-2 Charging and Discharging a Capacitor
20-3 The Farad Unit of Capacitance
20-4 Typical Capacitors
20-5 Electrolytic Capacitors
20-6 Capacitor Color Coding
20-7 Parallel Capacitances
20-8 Series Capacitances
20-9 Stray Capacitive and Inductive Effects
20-10 Energy in Electrostatic Field of Capacitance
20-11 Troubles in Capacitors

20-1   How Charge Is Stored in the Dielectric

It is possible for dielectric materials such as air or paper to hold an electric charge because free electrons cannot flow through an insulator. However, the charge must be applied by some source. In Fig. 20-1a, the battery can charge the capacitor shown. With the dielectric contacting the two conductors connected to the potential difference V, electrons from the voltage source accumulate on the side of the capacitor connected to the negative terminal of V. The opposite side of the capacitor connected to the positive terminal of V loses electrons.

As a result, the excess of electrons produces a negative charge on one side of the capacitor, while the opposite side has a positive charge. As an example, if 6.25×10^{18} electrons are accumulated, the negative charge equals 1 C. The charge on only one plate need be considered, as the number of electrons accumulated on one plate is exactly the same as the number taken from the opposite plate.

What the voltage source does is simply redistribute some electrons from one side of the capacitor to the other side. This process is charging the capacitor. The charging continues until the potential difference across the capacitor is equal to the applied voltage. Without any series resistance, the charging is instantaneous. Practically, however, there is always some series resistance. This charging current is transient, or temporary, as it flows only until the capacitor is charged to the applied voltage. Then there is no current in the circuit.

The result is a device for storing charge in the dielectric. Storage means that the charge remains even after the voltage source is disconnected. The measure of how much charge can be stored is the capacitance C. More charge stored for a given amount of applied voltage means more capacitance. Components made to provide a specified amount of capacitance are called *capacitors,* or by their old name *condensers.*

Electrically, then, capacitance is the ability to store charge. Physically, a capacitance consists simply of two conductors separated by an insulator. For example, Fig. 20-1b shows a capacitor using air for the dielectric between the metal plates. There are many types with different dielectric materials, including paper, mica, and ceramics, but the schematic symbols shown in Fig. 20-1c apply to all capacitors.

Electric Field in the Dielectric   Any voltage has a field of electric lines of force between the opposite electric charges. The electric field corresponds to the magnetic lines of force of the magnetic field associated with electric current.[1] What a capacitor does is concentrate the electric field in the dielectric between the plates. This concentration corresponds to a magnetic field concentrated in the turns of a coil. The only function of the capacitor plates and wire conductors is to connect the voltage source V across the dielectric. Then the electric field is concentrated in the capacitor, instead of being spread out in all directions.

Electrostatic Induction   The capacitor has opposite charges because of electrostatic induction by the electric field. Electrons that accumulate on the negative

---

[1]Electric and magnetic fields are compared in Fig. 14-6, Chap. 14.

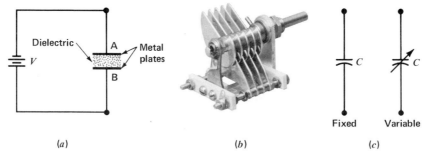

(a)   (b)   (c)

Fig. 20-1 Capacitance stores charge in the dielectric between two conductors. (a) Structure. (b) Air-dielectric variable capacitor. Length is 2 in. (c) Schematic symbols for fixed and variable capacitances.

side of the capacitor provide electric lines of force that repel electrons from the opposite side. When this side loses electrons, it becomes positively charged. The opposite charges induced by an electric field correspond to the idea of opposite poles induced in magnetic materials by a magnetic field.

Practice Problems 20-1
Answers at End of Chapter

a. In a capacitor, is the electric charge stored in the dielectric or in the metal plates?
b. What is the unit of capacitance?

20-2   Charging and Discharging a Capacitor

These are the two main effects with capacitors. Applied voltage puts charge in the capacitor. The accumulation of charge results in a buildup of potential difference across the capacitor plates. When the capacitor voltage equals the applied voltage, there is no more charging. The charge remains in the capacitor, with or without the applied voltage connected.

The capacitor discharges when a conducting path is provided between the plates, without any applied voltage. Actually, it is only necessary that the capacitor voltage be more than the applied voltage. Then the capacitor can serve as voltage source, temporarily, to produce discharge current in the discharge path. The capacitor discharge continues until the capacitor voltage drops to zero or is equal to the applied voltage.

Applying the Charge In Fig. 20-2a, the capacitor is neutral with no charge because it has not been connected to any source of applied voltage and there is no electrostatic field in the dielectric. Closing the switch in Fig. 20-2b, however, allows the negative battery terminal to repel free electrons in the conductor to plate A. At the same time, the positive terminal attracts free electrons from plate B. The side of the dielectric at plate A accumulates electrons because they cannot flow through the insulator, while plate B has an equal surplus of protons.

Remember that the opposite charges have an associated potential difference, which is the voltage across the capacitor. The charging process continues until the capacitor voltage equals the battery voltage, which is 10 V in this example. Then no further charging is possible because the applied voltage cannot make free electrons flow in the conductors.

Note that the potential difference across the charged capacitor is 10 V between plates A and B. There is no potential difference from each plate to its battery terminal, however, which is the reason why the capacitor stops charging.

Storing the Charge The negative and positive charges on opposite plates have an associated electric field through the dielectric, as shown by the dotted lines in Fig. 20-2b and c. The direction of these electric lines of force is shown repelling electrons from plate B, making this side positive. It is the effect of electric lines of force through the dielectric that results in storage of the charge. The electric field distorts the molecular structure so that the dielectric is no longer neutral. The dielectric is actually stressed by the invisible force

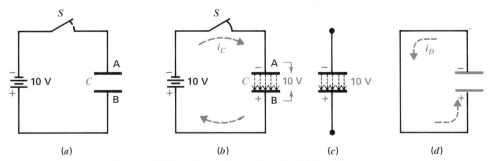

Fig. 20-2 Storing charge in a capacitance. (*a*) Capacitor with no charge. (*b*) Battery charges capacitor to the applied voltage. (*c*) Stored charge remains in capacitor, providing 10 V without the battery. (*d*) Discharging the capacitor.

⊣⊢

of the electric field. As evidence, the dielectric can be ruptured by a very intense field with high voltage across the capacitor.

The result of the electric field, then, is that the dielectric has charge supplied by the voltage source. Since the dielectric is an insulator that cannot conduct, the charge remains in the capacitor even after the voltage source is removed, as illustrated in Fig. 20-2c. You can now take this charged capacitor by itself out of the circuit, and it still has 10 V across the two terminals.

Discharging The action of neutralizing the charge by connecting a conducting path across the dielectric is discharging the capacitor. As shown in Fig. 20-2d, the wire between plates A and B is a low-resistance path for discharge current. With the stored charge in the dielectric providing the potential difference, 10 V is available to produce discharge current. The negative plate repels electrons, which are attracted to the positive plate through the wire, until the positive and negative charges are neutralized. Then there is no net charge, the capacitor is completely discharged, the voltage across it equals zero, and there is no discharge current. Now the capacitor is in the same uncharged condition as in Fig. 20-2a. It can be charged again, however, by a source of applied voltage.

Nature of the Capacitance A capacitor has the ability to store the amount of charge necessary to provide a potential difference equal to the charging voltage. If 100 V were applied in Fig. 20-2, the capacitor would charge to 100 V.

The capacitor charges to the applied voltage because, when the capacitor voltage is less, it takes on more charge. As soon as the capacitor voltage equals the applied voltage, no more charging current can flow. *Note that any charge or discharge current flows through the conducting wires to the plates but not through the dielectric.*

Charge and Discharge Currents In Fig. 20-2b, i_C is in the opposite direction from i_D in Fig. 20-2d. In both cases the current is electron flow. However, i_C is charging current to the capacitor and i_D is discharge current from the capacitor. The charge and discharge currents must always be in opposite direc-

tions. In Fig. 20-2b, the negative plate of C accumulates electrons from the voltage source. In Fig. 20-2d, the charged capacitor serves as a voltage source to produce electron flow around the discharge path.

Practice Problems 20-2
Answers at End of Chapter

Refer to Fig. 20-2.
a. If the applied voltage were 14.5 V, how much would the voltage be across C after it has charged?
b. How much is the voltage across C after it is completely discharged?
c. Can the capacitor be charged again after it is discharged?

20-3   The Farad Unit of Capacitance

With more charging voltage, the electric field is stronger and more charge is stored in the dielectric. The amount of charge Q stored in the capacitance is therefore proportional to the applied voltage. Also, a larger capacitance can store more charge. These relations are summarized by the formula

$$Q = CV \qquad \text{coulombs} \qquad (20\text{-}1)$$

where Q is the charge stored in the dielectric in coulombs (C), and V is the voltage across the plates of the capacitor.

The C is a physical constant, indicating the capacitance in terms of how much charge can be stored for a given amount of charging voltage. When one coulomb is stored in the dielectric with a potential difference of one volt, the capacitance is one *farad*.

Practical capacitors have sizes in millionths of a farad, or smaller. The reason is that typical capacitors store charge of microcoulombs or less. Therefore, the common units are

$$1 \text{ microfarad} = 1 \ \mu\text{F} = 1 \times 10^{-6} \text{ F}$$
$$1 \text{ picofarad} = 1 \text{ pF} = 1 \times 10^{-12} \text{ F}$$

⊣⊢

Example 1   How much charge is stored in a 2-μF capacitor with 50 V across it?

Answer $Q = CV$

$\qquad = 2 \times 10^{-6} \times 50$

$\qquad Q = 100 \times 10^{-6}$ coulomb

Example 2   How much charge is stored in a 40-μF capacitor with 50 V across it?

Answer $Q = CV$

$\qquad = 40 \times 10^{-6} \times 50$

$\qquad Q = 2000 \times 10^{-6}$ coulomb

Note that the larger capacitor stores more charge for the same voltage, in accordance with the definition of capacitance as the ability to store charge.

The factors in $Q = CV$ can be inverted to

$$C = \frac{Q}{V} \qquad\qquad (20\text{-}2)$$

or

$$V = \frac{Q}{C} \qquad\qquad (20\text{-}3)$$

For all three formulas, the basic units are volts for V, coulombs for Q, and farads for C. Note that the formula $C = Q/V$ actually defines one farad of capacitance as one coulomb of charge stored for one volt of potential difference. The symbol C (in italic, or slanted, type) is for capacitance. The same symbol C (in roman, or upright, type) is the coulomb unit of charge. This typographical distinction and the application indicate which we mean.

Example 3   A constant current of 2 μA charges a capacitor for 20 s. How much charge is stored? Remember $I = Q/t$ or $Q = I \times t$.

Answer $Q = I \times t$

$\qquad = 2 \times 10^{-6} \times 20$

$\qquad Q = 40$ microcoulombs (μC)

Example 4   The voltage across the charged capacitor in Example 3 is 20 V. Calculate C.

Answer $C = \dfrac{Q}{V}$

$\qquad = \dfrac{40 \times 10^{-6}}{20} = 2 \times 10^{-6}$

$\qquad C = 2\ \mu$F

Example 5   A constant current of 5 mA charges a 10-μF capacitor for 1 s. How much is the voltage across the capacitor?

Answer   Find the stored charge first:

$Q = I \times t = 5 \times 10^{-3} \times 1$

$\qquad = 5 \times 10^{-3}$ coulomb

$V = \dfrac{Q}{C} = \dfrac{5 \times 10^{-3}}{10 \times 10^{-6}}$

$\qquad = \dfrac{5}{10} \times 10^{3}$

$V = 500$ V

Larger Plate Area Increases Capacitance

As illustrated in Fig. 20-3, when the area of each plate is doubled, the capacitance in Fig. 20-3b stores twice the charge of Fig. 20-3a. The potential difference in both cases is still 10 V. This voltage produces a given strength of electric field. A larger plate area, however, means that more of the dielectric surface can contact each plate, allowing more lines of force through the dielectric between the plates and less flux leakage outside the dielectric. Then the field can store more charge in the dielectric. The result of larger plate area is more charge stored for the same applied voltage, which means the capacitance is larger.

Thinner Dielectric Increases Capacitance

As illustrated in Fig. 20-3c, when the distance between plates is reduced one-half, the capacitance stores twice the charge of Fig. 20-3a. The potential difference is still 10 V, but its electric field has greater flux density in the thinner dielectric. Then the field between oppo-

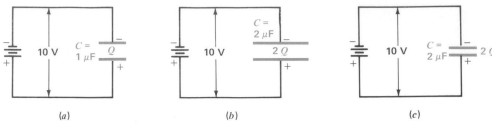

Fig. 20-3 Increasing stored charge and capacitance by increasing plate area and decreasing distance between plates. (*a*) Capacitance of 1 μF. (*b*) A 2-μF capacitance with twice the plate area and the same distance. (*c*) A 2-μF capacitance with one-half the distance and the same plate area.

site plates can store more charge in the dielectric. With less distance between the plates, the stored charge is greater for the same applied voltage, which means the capacitance is larger.

Dielectric Constant K_ϵ

This indicates the ability of an insulator to concentrate electric flux. Its numerical value is specified as the ratio of flux in the insulator compared with the flux in air or vacuum. The dielectric constant of air or vacuum is 1, since it is the reference.

Mica, for example, has an average dielectric constant of 6, meaning it can provide a density of electric flux six times as great as that of air or vacuum for the same applied voltage and equal physical size. Insulators generally have a dielectric constant K_ϵ greater than 1, as listed in Table 20-1. Higher values of K_ϵ allow greater values of capacitance.

It should be noted that the aluminum oxide and tantalum oxide listed in Table 20-1 are used for the dielectric in electrolytic capacitors. Also, the plastic film is used instead of paper for the rolled-foil type of capacitor.

The dielectric constant for an insulator is actually its *relative permittivity,* with the symbol ϵ_r or K_ϵ, indicating the ability to concentrate electric flux. This factor corresponds to relative permeability, with the symbol μ_r or K_m, for magnetic flux. Both ϵ_r and μ_r are pure numbers without units, as they are just ratios.[1]

Dielectric Strength

Table 20-1 also lists breakdown-voltage ratings for typical dielectrics. Dielectric strength is the ability of a dielectric to withstand a po-

tential difference without arcing across the insulator. This voltage rating is important because rupture of the insulator provides a conducting path through the dielectric. Then it cannot store charge, because the capacitor has been short-circuited. Since the breakdown voltage increases with greater thickness, capacitors for higher voltage ratings have more distance between the plates. This increased distance reduces the capacitance, however, all other factors remaining the same.

These physical factors for a parallel-plate capacitor are summarized by the formula

$$C = K_\epsilon \times \frac{A}{d} \times 8.85 \times 10^{-12} \text{ F} \qquad (20\text{-}4)$$

A is the area in square meters of either plate, d is the distance in meters between plates, and K_ϵ is the dielectric constant, or relative permittivity, as listed in Table 20-1. The constant factor 8.85×10^{-12} is the absolute permittivity of air or vacuum, in SI, to calculate C in farads, which is an SI unit.

Table 20-1. Dielectric Materials*

Material	Dielectric constant K_ϵ	Dielectric strength, V/mil
Air or vacuum	1	20
Aluminum oxide	7	
Ceramics	80–1200	600–1250
Glass	8	335–2000
Mica	3–8	600–1500
Oil	2–5	275
Paper	2–6	1250
Plastic film	2–3	
Tantalum oxide	25	

*Exact values depend on the specific composition of different types.

---

[1]The absolute permittivity ϵ_0 is 8.854×10^{-12} F/m, in SI units, for electric flux in air or vacuum. This value corresponds to an absolute permeability μ_0 of $4\pi \times 10^{-7}$ H/m, in SI units, for magnetic flux in air or vacuum.

Example 6 Calculate C for two plates each with an area 2 m², separated by 1 cm, or 10^{-2} m, with a dielectric of air.

Answer Substituting in Formula (20-4),

$$C = 1 \times \frac{2}{10^{-2}} \times 8.85 \times 10^{-12} \text{ F}$$
$$= 200 \times 8.85 \times 10^{-12}$$
$$= 1770 \times 10^{-12} \text{ F}$$
$$C = 1770 \text{ pF}$$

This value means the capacitor can store 1770×10^{-12} coulomb of charge with 1 V. Note the relatively small capacitance, in picofarad units, with the extremely large plates of 2 m², which is really the size of a table or a desk top.

If the dielectric used is paper with a dielectric constant of 6, then C will be six times greater. Also, if the spacing between plates is reduced by one-half to 0.5 cm, the capacitance will be doubled. It should be noted that practical capacitors for electronic circuits are much smaller than this parallel-plate capacitor. They use a very thin dielectric, with a high dielectric constant, and the plate area can be concentrated in a small space.

Practice Problems 20-3
Answers at End of Chapter

a. A capacitor charged to 100 V has 1000 microcoulombs of charge. How much is C?
b. A mica capacitor and ceramic capacitor have the same physical dimensions. Which has more C?

20-4   Typical Capacitors

Commercial capacitors are generally classified according to the dielectric. Most common are air, mica, paper, and ceramic capacitors, plus the electrolytic type. Electrolytic capacitors use a molecular-thin oxide film as the dielectric, resulting in large capacitance values in little space. These types are compared in Table 20-2 and discussed in the sections that follow.

There is no required polarity, since either side can be the more positive plate, except for electrolytic capacitors. These are marked to indicate which side must be positive to maintain the internal electrolytic action that produces the dielectric required to form the capacitance. *It should be noted that it is the polarity of the charging source that determines the polarity of the capacitor voltage.*

Mica Capacitors Thin mica sheets are stacked between tinfoil sections for the conducting plates to provide the required capacitance. Alternate strips of tinfoil are connected together and brought out as one terminal for one set of plates, while the opposite terminal connects to the other set of plates. The entire unit is generally in a molded Bakelite case. Mica capacitors are often used for small capacitance values of 50 to 500 pF; their length is ¾ in. or less with about ⅛-in. thickness. Typical mica capacitors are shown in Fig. 20-4.

Paper Capacitors In this construction, two rolls of tinfoil conductor separated by a tissue-paper insulator are rolled into a compact cylinder. Each outside lead connects to its roll of tinfoil as a plate. The entire

Table 20-2. Types of Capacitors

Dielectric	Construction	Capacitance	Breakdown, V
Air	Meshed plates	10–400 pF	400 (0.02-in. air gap)
Ceramic	Tubular	0.5–1600 pF	500–20,000
	Disk	0.002–0.1 μF	
Electrolytic	Aluminum	5–1000 μF	10–450
	Tantalum	0.01–300 μF	6–50
Mica	Stacked sheets	10–5000 pF	500–20,000
Paper or plastic film	Rolled foil	0.001–1 μF	200–1600

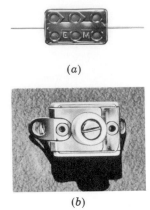

(a)

(b)

Fig. 20-4 Mica capacitors, about ⅝ in. wide. (a) Fixed value, color-coded in picofarads. (b) Variable trimmer of 5 to 30 pF.

cylinder is generally placed in a cardboard container coated with wax or encased in plastic. Paper capacitors are often used for medium capacitance values of 0.001 to 1.0 μF, approximately. The physical size for 0.05 μF is typically 1 in. long with ⅜-in. diameter. Paper capacitors are shown in Fig. 20-5.

A black band at one end of a paper capacitor indicates the lead connected to the outside foil. This lead should be used for the ground or low-potential side of the circuit to take advantage of shielding by the outside

(a)

(b)

Fig. 20-5 Paper capacitors. (a) Tubular type 1 in. long. Capacitance C is 0.068 μF. (b) Encapsulated type with leads for printed-circuit board; length is ¾ in. Capacitance C is 430 pF.

foil. There is no required polarity, however, since the capacitance is the same no matter which side is grounded. It should also be noted that in the schematic symbol for C the curved line usually indicates the low-potential side of the capacitor.

Many capacitors of foil construction use a plastic film instead of tissue paper. Two types are Teflon[1] and Mylar[1] plastic film. These feature very high insulation resistance, of over 1000 MΩ, low losses, and longer service life without voltage breakdown, compared with paper capacitors. The plastic capacitors are available in sizes of 0.001 to 1.0 μF, like paper capacitors.

Ceramic Capacitors The ceramic dielectric materials are made from earth fired under extreme heat. By use of titanium dioxide, or several types of silicates, very high values of dielectric constant K_ϵ can be obtained.

In the disk form, silver is fired onto both sides of the ceramic, to form the conductor plates. With a K_ϵ value of 1200, the disk ceramics feature capacitance values up to 0.01 μF in much less space than a paper capacitor.

For tubular ceramics, the hollow ceramic tube has a silver coating on the inside and outside surfaces. With values of 1 to 500 pF, these capacitors have the same applications as mica capacitors but are smaller. Typical ceramic capacitors are shown in Fig. 20-6.

Temperature Coefficient Ceramic capacitors are often used for temperature compensation, to increase or decrease capacitance with a rise in temperature. The temperature coefficient is given in parts per million (ppm) per degree Celsius, with a reference of 25°C. As an example, a negative 750 ppm unit is stated as N750. A positive temperature coefficient of the same value would be stated as P750. Units that do not change in capacitance are labeled NPO.

Variable Capacitors Figure 20-1b shows a variable air capacitor. In this construction, the fixed metal plates connected together form the *stator*. The movable

---

[1]Du Pont trademarks.

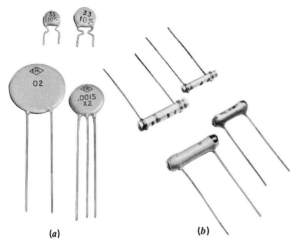

Fig. 20-6 Ceramic capacitors, shown actual size. (*a*) Disk type. (*b*) Tubular type. (*Centralab Division, Globe-Union Inc.*)

plates connected together on the shaft form the *rotor*. Capacitance is varied by rotating the shaft to make the rotor plates mesh with the stator plates. They do not touch, however, since air is the dielectric. Full mesh is maximum capacitance. Moving the rotor completely out of mesh provides minimum capacitance.

A common application is the tuning capacitor in radio receivers. When you tune to different stations, the capacitance varies as the rotor moves in or out of mesh. Combined with an inductance, the variable capacitance then tunes the receiver to a different resonant frequency for each station. Usually two or three capacitor sections are *ganged* on one common shaft.

Capacitance Tolerance

Ceramic disk capacitors for general applications usually have a tolerance of ±20 percent. Paper capacitors usually have a tolerance of ±10 percent. For closer tolerances, mica or ceramic tubular capacitors are used. These have tolerance values of ±2 to 20 percent. Silver-plated mica capacitors are available with a tolerance of ±1 percent.

The tolerance may be less on the minus side to make sure there is enough capacitance, particularly with electrolytic capacitors, which have a wide tolerance. For instance, a 20-μF electrolytic with a tolerance of −10 percent, +50 percent may have a capacitance of 18 to 30 μF. However, the exact capacitance value is not critical in most applications of capacitors for filtering, ac coupling, and bypassing.

Voltage Rating of Capacitors

This rating specifies the maximum potential difference that can be applied across the plates without puncturing the dielectric. Usually the voltage rating is for temperatures up to about 60°C. Higher temperatures result in a lower voltage rating. Voltage ratings for general-purpose paper, mica, and ceramic capacitors are typically 200 to 500 V. Ceramic capacitors with ratings of 1 to 5 kV are also available.

Electrolytic capacitors are commonly used in 25-, 150-, and 450-V ratings. In addition, 6- and 10-V electrolytic capacitors are often used in transistor circuits. For applications where a lower voltage rating is permissible, more capacitance can be obtained in a smaller physical size.

The potential difference across the capacitor depends upon the applied voltage and is not necessarily equal to the voltage rating. A voltage rating higher than the potential difference applied across the capacitor provides a safety factor for long life in service. With electrolytic capacitors, however, the actual capacitor voltage should be close to the rated voltage to produce the oxide film that provides the specified capacitance.

The voltage ratings are for dc voltage applied. The breakdown rating is lower for ac voltage because of the internal heat produced by continuous charge and discharge.

Capacitor Applications

In most electronic circuits, a capacitor has dc voltage applied, combined with a much smaller ac signal voltage. The usual function of the capacitor is to block the dc voltage but pass the ac signal voltage, by means of the charge and discharge current. These applications include coupling, bypassing, and filtering for ac signal.

Practice Problems 20-4
Answers at End of Chapter

Answer True or False.
a. An electrolytic capacitor must be connected in the correct polarity.
b. The potential difference across a capacitor is always equal to its maximum voltage rating.
c. Ceramic and paper capacitors generally have less C than electrolytic capacitors.
d. The letters NPO indicate zero temperature coefficient.

20-5   Electrolytic Capacitors

These capacitors are commonly used for C values of 5 to 2000 μF, approximately, because electrolytics provide the most capacitance in the smallest space with least cost.

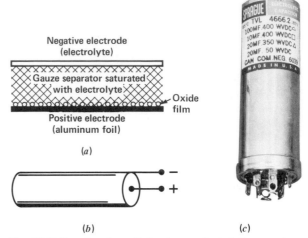

Negative electrode (electrolyte)

Gauze separator saturated with electrolyte

Oxide film

Positive electrode (aluminum foil)

(a)

(b)   (c)

Fig. 20-7 Construction of aluminum electrolytic capacitor. (a) Internal electrodes. (b) Foil rolled into cartridge. (c) Typical capacitor with multiple sections. Height is about 3 in. (*Sprague Electric Co.*)

Construction   Figure 20-7 shows the aluminum-foil type. The two aluminum electrodes are in an electrolyte of borax, phosphate, or carbonate. Between the two aluminum strips, absorbent gauze soaks up electrolyte to provide the required electrolysis that produces an oxide film. This type is considered a wet electrolytic, but it can be mounted in any position.

When dc voltage is applied to form the capacitance in manufacture, the electrolytic action accumulates a molecular-thin layer of aluminum oxide at the junction between the positive aluminum foil and the electrolyte. The oxide film is an insulator. As a result, capacitance is formed between the positive aluminum electrode and the electrolyte in the gauze separator. The negative aluminum electrode simply provides a connection to the electrolyte. Usually, the metal can itself is the negative terminal of the capacitor, as shown in Fig. 20-7c.

Because of the extremely thin dielectric film, very large C values can be obtained. The area is increased by using long strips of aluminum foil and gauze, which are rolled into a compact cylinder with very high capacitance. For example, an electrolytic capacitor the same size as a 0.1-μF paper capacitor, but rated at 10 V breakdown, may have 1000 μF of capacitance or more. Higher voltage ratings, up to 450 V, are used, with typical C values of 8 to 200 μF for electrolytic capacitors.

Polarity   Electrolytic capacitors are used in circuits that have a combination of dc voltage and ac voltage. The dc voltage maintains the required polarity across the electrolytic capacitor to form the oxide film. A common application is for electrolytic filter capacitors to eliminate 60-Hz ac ripple in a dc power supply. Another use is for audio coupling capacitors in transistor amplifiers. In both these applications, for filtering or coupling, electrolytics are needed for large C with a low-frequency ac component, while the circuit has a dc component for the required voltage polarity. Incidentally, the difference between filtering an ac component

out or coupling it into a circuit is only a question of parallel or series connections.

If the electrolytic is connected in opposite polarity, the reversed electrolysis forms gas in the capacitor. It becomes hot and may explode. This is a possibility only with electrolytic capacitors.

Leakage Current   The disadvantage of electrolytics, in addition to the required polarization, is their relatively high leakage current, since the oxide film is not a perfect insulator. Leakage current through the dielectric is about 0.1 to 0.5 mA/μF of capacitance for the aluminum-foil type. As an example, a 10-μF electrolytic capacitor can have a leakage current of 5 mA. For the opposite case, a mica capacitor has practically zero leakage current.

The problem with leakage current in a capacitor is that it allows part of the dc component to be coupled into the next circuit along with the ac component. However, electrolytics are generally used in low-resistance circuits where some leakage current is acceptable because of the small IR drop.

Nonpolarized Electrolytics   This type is available for applications in circuits without any dc polarizing voltage, as in the 60-Hz ac power line. One application is the starting capacitor for ac motors. A

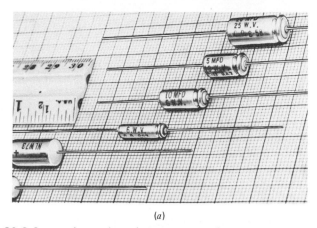

(a)

(b)

Fig. 20-8 Low-voltage electrolytic capacitors for transistor circuits. These are tantalum type, with C of 5 to 25 μF. (a) With axial leads. (b) Miniature type with radial leads for printed-circuit board.

nonpolarized electrolytic actually contains two capacitors, connected internally in series-opposing polarity. The capacitance is one-half either C, but the oxide film is maintained.

Tantalum Capacitors This is a new form of electrolytic capacitor, using tantalum (Ta) instead of aluminum. Titanium (Ti) is also used. Typical tantalum capacitors are shown in Fig. 20-8. They feature:

1. Larger C in a smaller size
2. Longer shelf life
3. Less leakage current

However, tantalum electrolytics cost more than the aluminum type. Methods of construction for tantalum capacitors include the wet-foil type and a solid chip or slug. The solid tantalum is processed in manufacture to have an oxide film as the dielectric. Referring back to Table 20-1, note that tantalum oxide has a dielectric constant of 25, compared with 7 for aluminum oxide.

Practice Problems 20-5
Answers at End of Chapter

Answer True or False.
a. The rating of 1000 μF at 25 V could be for an electrolytic capacitor.
b. Electrolytic capacitors have practically zero leakage current.
c. Tantalum capacitors have a longer shelf life than aluminum electrolytics.

20-6   Capacitor Color Coding

Mica and tubular ceramic capacitors are color-coded to indicate their capacitance value. Since coding is necessary only for very small sizes, the color-coded capacitance value is always in pF units. The colors used are the same as for resistor coding, from black for 0 up to white for 9.

Mica capacitors generally use the six-dot system shown in Fig. 20-9. Read the top row first from left to right, then the bottom row, in reverse order right to left. White for the first dot indicates the new EIA coding, but the capacitance value is read from the next three dots. As an example, if the colors are red, green, and brown for dots 2, 3, and 4, the capacitance is 250 pF. If the first dot is silver, it indicates a paper capacitor, but the capacitance is still read from dots 2, 3, and 4. Dot 5 specifies tolerance, while dot 6 gives

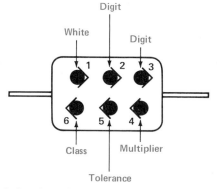

Fig. 20-9 Six-dot color code for mica capacitors.

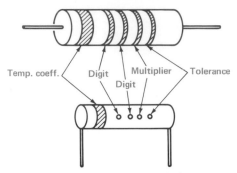

Fig. 20-10 Color code for ceramic tubular capacitors.

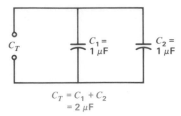

$$C_T = C_1 + C_2$$
$$= 2 \ \mu F$$

Fig. 20-11 Capacitances in parallel.

the EIA class. There are seven classes from A to G, specifying temperature coefficient, leakage resistance, and additional variable factors. Appendix G has more detailed information on the tolerance and class coding. Also listed are discontinued codes found on capacitors in old equipment.

For tubular ceramic capacitors, the system shown in Fig. 20-10 is used with color dots or bands. The wide color band specifying temperature coefficient indicates the left end, which is the side connected to the inner electrode. Capacitance is read from the next three colors, in either dots or stripes. For instance, brown, black and brown for bands or dots 2, 3, and 4 means 100 pF.

Gray and white are used as decimal multipliers for very small values, with gray for 0.01 and white for 0.1. For instance, green, black, and white in dots 2, 3, and 4 means 50×0.1, or 5 pF. The color codes for tolerance and temperature coefficient of ceramic capacitors are listed in App. G.

In reading the color-coded capacitance value, keep in mind that mica capacitors generally range from 10 to 5000 pF. The small tubular ceramic capacitors are usually 0.5 to 1000 pF. With paper and ceramic disc capacitors, the capacitance and voltage rating is generally printed on the case. Where no voltage rating is specified, it is usually about 200 to 600 V. Electrolytic capacitors have the capacitance, voltage rating, and polarity printed on the case.

Practice Problems 20-6
Answers at End of Chapter

a. How much is C for red, green, and black color stripes or dots?
b. What is the value of gray as a decimal multiplier?

20-7   Parallel Capacitances

Connecting capacitances in parallel is equivalent to adding the plate areas. Therefore, the total capacitance is the sum of the individual capacitances. As illustrated in Fig. 20-11,

$$C_T = C_1 + C_2 + \cdots + \text{etc.} \qquad (20\text{-}5)$$

A 10-μF capacitor in parallel with a 5-μF capacitor, for example, provides a 15-μF capacitance for the parallel combination. The voltage is the same across the parallel capacitors. Note that adding parallel capacitances is opposite to the case of inductances in parallel, and resistances in parallel.

Practice Problems 20-7
Answers at End of Chapter

a. How much is C_T for 0.01 μF in parallel with 0.02 μF?
b. What C must be connected in parallel with 100 pF to make C_T of 250 pF?

20-8   Series Capacitances

Connecting capacitances in series is equivalent to increasing the thickness of the dielectric. Therefore, the combined capacitance is less than the smallest individual value. As shown in Fig. 20-12, the combined equivalent capacitance is calculated by the reciprocal formula:

$$\frac{1}{C_T} = \frac{1}{C_1} + \frac{1}{C_2} + \cdots + \text{etc.} \qquad (20\text{-}6)$$

Any of the short-cut calculations for the reciprocal formula apply. For example, the combined capacitance of two equal capacitances of 10 μF in series is 5 μF.

Capacitors are used in series to provide a higher voltage breakdown rating for the combination. For instance, each of three equal capacitances in series has one-third the applied voltage.

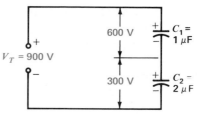

Fig. 20-13 With series capacitances the smaller capacitance has more voltage for the same charge.

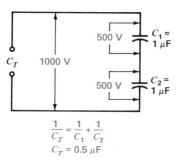

$$\frac{1}{C_T} = \frac{1}{C_1} + \frac{1}{C_2}$$
$$C_T = 0.5 \ \mu F$$

Fig. 20-12 Capacitances in series.

Division of Voltage across Unequal Capacitances In series, the voltage across each C is inversely proportional to its capacitance, as illustrated in Fig. 20-13. The smaller capacitance has the larger proportion of the applied voltage. The reason is that the series capacitances all have the same charge because they are in one current path. With equal charge, a smaller capacitance has a greater potential difference.

We can consider the amount of charge in the series capacitors in Fig. 20-13. Let the charging current be 600 μA flowing for 1 s. The charge Q equals $I \times t$, then, or 600 μC. Both C_1 and C_2 have Q equal to 600 μC, as they are in the same series path for charging current.

Although the charge is the same in C_1 and C_2, they have different voltages because of different capacitance values. For each capacitor $V = Q/C$. For the two capacitors in Fig. 20-13, then:

$$V_1 = \frac{Q}{C_1} = \frac{600 \ \mu C}{1 \ \mu F} = 600 \ V$$

$$V_2 = \frac{Q}{C_2} = \frac{600 \ \mu F}{2 \ \mu F} = 300 \ V$$

Charging Current for Series Capacitances
The charging current is the same in all parts of the series path, including the junction between C_1 and C_2, even though this point is separated from the source voltage by two insulators. At the junction, the current is the resultant of electrons repelled by the negative plate of C_2 and attracted by the positive plate of C_1. The amount of current is how much would be produced by one capacitance of $\frac{2}{3} \ \mu$F, which is the equivalent capacitance of C_1 and C_2 in series in Fig. 20-13.

Practice Problems 20-8
Answers at End of Chapter

a. How much is C_T for two 0.2-μF capacitors in series?
b. With 50 V applied across both, how much is V_C across each capacitor?
c. How much is C_T for 100 pF in series with 50 pF?

20-9   Stray Capacitive and Inductive Effects

These two important characteristics can be evident in all circuits with all types of components. A capacitor has a small amount of inductance in the conductors. A coil has some capacitance between windings. A resistor has a small amount of inductance and capacitance. After all, a capacitance physically is simply an insulator between two points having a difference of potential. An inductance is basically just a conductor carrying current.

Actually, though, these stray effects are usually quite small, compared with the concentrated or lumped

$\dashv\vdash$

values of capacitors or inductors. Typical values of stray capacitance may be 1 to 10 pF, while stray inductance is usually a fraction of 1 μH. For very high radio frequencies, however, when small values of L and C must be used, the stray effects become important. As another example, any wire cable has capacitance between the conductors.

Stray Circuit Capacitance   The wiring and the components in a circuit have capacitance to the metal chassis. This stray capacitance C_S is typically 5 to 10 pF. To reduce C_S, the wiring should be short, with the leads and components placed high off the chassis. Sometimes, for very high frequencies, the stray capacitance is included as part of the circuit design. Then changing the placement of components or wiring affects the circuit operation. Such critical *lead dress* is usually specified in the manufacturer's service notes.

Leakage Resistance of a Capacitor   Consider a capacitor charged by a dc voltage source. After the charging voltage is removed, a perfect capacitor would keep its charge indefinitely. After a long period of time, however, the charge will be neutralized by a small leakage current through the dielectric and across the insulated case between terminals, because there is no perfect insulator. For paper, ceramic, and mica capacitors, though, the leakage current is very slight or, inversely, the leakage resistance is very high. As shown in Fig. 20-14, the leakage resistance R_l is indicated by a high resistance in parallel with the capacitance C. For paper, ceramic, or mica capacitors R_l is 100 MΩ or more. However, electrolytic capacitors may have a leakage resistance of 0.5 MΩ or less.

Absorption Losses in Capacitors   With ac voltage applied to a capacitor, the continuous charge, discharge, and reverse charging action cannot be followed instantaneously in the dielectric. This corresponds to hysteresis in magnetic materials. With high-frequency charging voltage for a capacitor, there may be a difference between the amount of ac voltage applied and the ac voltage stored in the dielectric. The difference can be considered *absorption loss* in the dielectric. With higher frequencies, the losses increase. In Fig. 20-14, the small value of 0.5 Ω for R_d indicates a

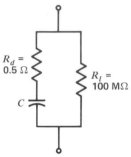

Fig. 20-14 Equivalent circuit of a capacitor. Resistance R_l is leakage resistance; R_d is absorption loss dissipated in dielectric.

typical value for paper capacitors. For ceramic and mica capacitors, the dielectric losses are even smaller. These losses need not be considered for electrolytic capacitors because they are generally not used for radio frequencies.

Power Factor of a Capacitor   The quality of a capacitor in terms of minimum loss is often indicated by its power factor, which states the fraction of input power dissipated as heat loss in the capacitor. The lower the numerical value of the power factor, the better is the quality of the capacitor. Since the losses are in the dielectric, the power factor of the capacitor is essentially the power factor of the dielectric, independent of capacitance value or voltage rating. At radio frequencies, approximate values of power factor are 0.000 for air or vacuum, 0.0004 for mica, about 0.01 for paper, and 0.0001 to 0.03 for ceramics.

The reciprocal of the power factor can be considered the Q of the capacitor, similar to the idea of Q of a coil. For instance, a power factor of 0.001 corresponds to a Q of 1000. A higher Q therefore means better quality for the capacitor.

Inductance of a Capacitor   Capacitors with a coiled construction, particularly paper and electrolytic capacitors, have some internal inductance. The larger the capacitor, the greater is its series inductance. Mica and ceramic capacitors have very little inductance, however, which is why they are generally used for radio frequencies.

For use above audio frequencies, the rolled-foil type of capacitor must have a noninductive construction.

$\dashv\vdash$

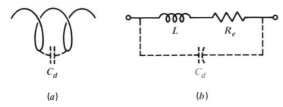

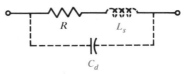

Fig. 20-15 Equivalent circuit of an RF coil. (*a*) Distributed capacitance C_d between turns of wire. (*b*) Equivalent circuit.

Fig. 20-16 High-frequency equivalent circuit of a resistor.

This means the start and finish of the foil winding must not be the terminals of the capacitor. Instead, the foil windings are offset. Then one terminal can contact all layers of one foil at one edge, while the opposite edge of the other foil contacts the second terminal. Most rolled-foil capacitors, including the paper and Mylar types, are constructed this way.

Distributed Capacitance of a Coil   As illustrated in Fig. 20-15, a coil has distributed capacitance C_d between turns. Note that each turn is a conductor separated from the next turn by an insulator, which is the definition of capacitance. Furthermore, the potential of each turn is different from the next, providing part of the total voltage as a potential difference to charge C_d. The result then is the equivalent circuit shown for an RF coil. The L is the inductance and R_e its internal effective ac resistance in series with L, while the total distributed capacitance C_d for all the turns is across the entire coil.

Special methods for minimum C_d include *space-wound* coils, where the turns are spaced far apart; the honeycomb or *universal* winding, with the turns crossing each other at right angles; and the *bank winding,* with separate sections called *pies.* These windings are for RF coils. In audio and power transformers, a grounded conductor shield, called a *Faraday screen,* is often placed between windings to reduce capacitive coupling.

Reactive Effects in Resistors   As illustrated by the high-frequency equivalent circuit in Fig. 20-16, a resistor can include a small amount of inductance and capacitance. For carbon-composition resistors, the inductance is usually negligible. However, approximately 0.5 pF of capacitance across the ends may have

an effect, particularly with large resistances used for high radio frequencies. Wire-wound resistors definitely have enough inductance to be evident at radio frequencies. However, special resistors are available with double windings in a noninductive method based on cancellation of opposing magnetic fields.

Capacitance of an Open Circuit   An open switch or a break in a conducting wire has capacitance C_O across the open. The reason is that the open consists of an insulator between two conductors. With a voltage source in the circuit, C_O charges to the applied voltage. Because of the small C_O, in the order of picofarads, the capacitance charges to the source voltage in a short time. This charging of C_O is the reason why an open series circuit has the applied voltage across the open terminals. After a momentary flow of charging current, C_O charges to the applied voltage and stores the charge needed to maintain this voltage.

Practice Problems 20-9
Answers at End of Chapter

Answer True or False.
a. A two-wire cable has distributed C between the conductors.
b. A coil has distributed C between the turns.
c. The leakage resistance of ceramic capacitors is very high.

20-10   Energy in Electrostatic Field of Capacitance

The electrostatic field of the charge stored in the dielectric has electric energy supplied by the voltage source that charges C. This energy is stored in the dielectric. The proof is the fact that the capacitance can produce discharge current when the voltage source is removed. The electric energy stored is

$$\text{Energy} = \mathscr{E} = \tfrac{1}{2}CV^2 \qquad \text{joules} \qquad (20\text{-}7)$$

where C is the capacitance in farads and V is the voltage across the capacitor. For example, a 1-μF capacitor charged to 400 V has stored energy equal to

$$\mathscr{E} = \tfrac{1}{2}\,CV^2 = \frac{1 \times 10^{-6} \times (4 \times 10^2)^2}{2}$$

$$= \frac{1 \times 10^{-6} \times (16 \times 10^4)}{2}$$

$$= 8 \times 10^{-2}$$

$$\mathscr{E} = 0.08 \text{ J}$$

This 0.08 J of energy is supplied by the voltage source that charges the capacitor to 400 V. When the charging circuit is opened, the stored energy remains as charge in the dielectric. With a closed path provided for discharge, the entire 0.08 J is available to produce discharge current. As the capacitor discharges, the energy is used in producing discharge current. When the capacitor is completely discharged, the stored energy is zero.

The stored energy is the reason why a charged capacitor can produce an electric shock, even when not connected into a circuit. When you touch the two leads of the charged capacitor, its voltage produces discharge current through your body. Stored energy greater than 1 J can be dangerous with a capacitor charged to a voltage high enough to produce an electric shock.

Practice Problems 20-10
Answers at End of Chapter

Answer True or False.
a. The stored energy in C increases with more V.
b. The stored energy decreases with less C.

20-11 Troubles in Capacitors

Capacitors can become open or short-circuited. In either case, the capacitor is useless because it cannot store charge. A leaky capacitor is equivalent to a partial

short circuit where the dielectric gradually loses its insulating properties under the stress of applied voltage, lowering its resistance. A good capacitor has very high resistance of the order of megohms; a short-circuited capacitor has zero ohms resistance, or continuity; the resistance of a leaky capacitor is lower than normal.

Checking Capacitors with an Ohmmeter A capacitor usually can be checked with an ohmmeter. The highest ohms range, such as $R \times 1$ MΩ, is preferable. Also, disconnect one side of the capacitor from the circuit to eliminate any parallel resistance paths that can lower the resistance. Keep your fingers off the connections, since the body resistance lowers the reading.

As illustrated in Fig. 20-17, the ohmmeter leads are connected across the capacitor. For a good capacitor, the meter pointer moves quickly toward the low-resistance side of the scale and then slowly recedes toward infinity. The reading when the pointer stops moving is the insulation resistance of the capacitor, which is normally very high. For paper, mica, and ceramic capacitors, the resistance can be 500 to 1000 MΩ, or more, which is practically infinite resistance. Electrolytic capacitors, however, have a lower normal resistance of about 0.5 MΩ. In all cases, discharge the capacitor before checking with the ohmmeter.

When the ohmmeter is initially connected, its battery charges the capacitor. This charging current is the reason the meter pointer moves away from infinity, since more current through the ohmmeter means less resist-

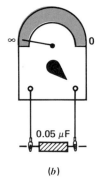

Fig. 20-17 Checking a capacitor with an ohmmeter. The R scale is shown right to left, as on a VOM. Use highest ohms range. (a) Capacitor action as needle is moved by charging current. (b) Practically infinite leakage-resistance reading after capacitor has charged.

ance. Maximum current flows at the first instant of charge. Then the charging current decreases as the capacitor voltage increases toward the applied voltage; therefore, the needle pointer slowly moves toward infinite resistance. Finally, the capacitor is completely charged to the ohmmeter battery voltage, the charging current is zero, and the ohmmeter reads just the small leakage current through the dielectric. This charging effect, called *capacitor action,* shows that the capacitor can store charge, indicating a normal capacitor. It should be noted that both the rise and fall of the meter readings are caused by charging. The capacitor discharges when the meter leads are reversed.

Ohmmeter Readings
Troubles in a capacitor are indicated as follows:

1. If an ohmmeter reading immediately goes practically to zero and stays there, the capacitor is short-circuited.
2. If the capacitor shows charging, but the final resistance reading is appreciably less than normal, the capacitor is leaky. Such capacitors are particularly troublesome in high-resistance circuits. When checking electrolytics, reverse the ohmmeter leads and take the higher of the two readings.
3. If the capacitor shows no charging action but just reads very high resistance, it may be open. Some precautions must be remembered, however, since very high resistance is a normal condition for capacitors. Reverse the ohmmeter leads to discharge the capacitor, and check it again. In addition, remember that capacitance values of 100 pF, or less, normally have very little charging current for the low battery voltage of the ohmmeter.

Short-circuited Capacitors
In normal service, capacitors can become short-circuited because the dielectric deteriorates with age, usually over a period of years under the stress of charging voltage, especially with higher temperatures. This effect is more common with paper and electrolytic capacitors. The capacitor may become leaky gradually, indicating a partial short circuit, or the dielectric may be punctured, causing a short circuit.

Open Capacitors
In addition to the possibility of an open connection in any type of capacitor, electrolytics develop high resistance in the electrolyte with age, particularly at high temperatures. After service of a few years, if the electrolyte dries up, the capacitor will be partially open. Much of the capacitor action is gone, and the capacitor should be replaced.

Leaky Capacitors
A leaky capacitor reads R less than normal with an ohmmeter. However, dc voltage tests are more definite. In a circuit, the dc voltage at one terminal of the capacitor should not affect the dc voltage at the other terminal.

Shelf Life
Except for electrolytics, capacitors do not deteriorate with age while stored, since there is no applied voltage. Electrolytic capacitors, however, like dry cells, should be used fresh from manufacture. The reason is the wet electrolyte.

Practice Problems 20-11
Answers at End of Chapter

a. What is the ohmmeter reading for a shorted capacitor?
b. Does capacitor action with an ohmmeter show the capacitor is good or bad?

Summary

1. A capacitor consists of two conductors separated by a dielectric insulator. Its ability to store charge is the capacitance C. Applying voltage to store charge is charging the capacitor; short-circuiting the two conductors of the capacitor to

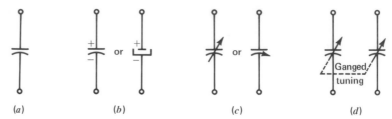

Fig. 20-18 Schematic symbols for types of *C*. (*a*) Fixed—air, paper, mica, or ceramic. (*b*) Electrolytic, with polarity. (*c*) Variable. (*d*) Ganged on one shaft.

neutralize the charge is discharging the capacitor. Schematic symbols for *C* are summarized in Fig. 20-18.

2. The unit of capacitance is the farad. One farad of capacitance stores one coulomb of charge with one volt applied. Practical capacitors have much smaller capacitance values, from 1 pF to 1000 μF. One pF is 1×10^{-12} F; one μF is 1×10^{-6} F.

3. $Q = CV$, where Q is the charge in coulombs, C the capacitance in farads, and V the potential difference across the capacitor in volts.

4. Capacitance increases with larger plate area and less distance between the plates.

5. The ratio of charge stored in different insulators to the charge stored in air is the dielectric constant K_ϵ of the material. Air or vacuum has a dielectric constant of 1.

6. The most common types of commercial capacitors are air, paper, mica, ceramic, and electrolytic. Electrolytics are the only capacitors with polarity. The different types are compared in Table 20-2.

7. Mica and tubular ceramic capacitors are color-coded as shown in Figs. 20-9 and 20-10.

8. For parallel capacitors, $C_T = C_1 + C_2 + C_3 + \cdots +$ etc.

9. For series capacitors, $1/C_T = 1/C_1 + 1/C_2 + 1/C_3 + \cdots +$ etc.

10. The electric field of a capacitance has stored energy $CV^2/2$. With V in volts and C in farads, the energy is in joules.

11. When checked with an ohmmeter, a good capacitor shows charging current, and then the ohmmeter reading steadies at the insulation resistance. All types except electrolytics normally have a very high insulation resistance of 500 to 1000 MΩ. Electrolytics have more leakage current, with a typical resistance of 0.5 MΩ.

12. The main comparisons between the opposite characteristics of capacitance and inductance are summarized in Table 20-3.

Self-Examination
Answers at Back of Book

Choose (a), (b), (c), or (d).

1. A capacitor consists of two (a) conductors separated by an insulator; (b) insulators separated by a conductor; (c) conductors alone; (d) insulators alone.

Table 20-3. Comparison of Capacitance and Inductance

Capacitance	Inductance
Symbol is C	Symbol is L
Farad unit	Henry unit
Stores charge Q	Conducts current I
Needs dielectric as insulator	Needs wire conductor
More plate area allows more C	More turns allow more L
Dielectric with higher K_ϵ or ϵ_r concentrates electric field for more C	Core with higher K_m or μ_r concentrates magnetic field for more L
$\dfrac{1}{C_T} = \dfrac{1}{C_1} + \dfrac{1}{C_2}$ in series	$L_T = L_1 + L_2$ in series
$C_T = C_1 + C_2$ in parallel	$\dfrac{1}{L_T} = \dfrac{1}{L_1} + \dfrac{1}{L_2}$ in parallel

2. A capacitance of 0.02 μF equals (a) 0.02×10^{-12} F; (b) 0.02×10^{-6} F; (c) 0.02×10^6 F; (d) 200×10^{-12} F.

3. A 10-μF capacitance charged to 10 V has a stored charge equal to (a) 10 μC; (b) 100 μC; (c) 200 μC; (d) 1 C.

4. Capacitance increases with (a) larger plate area and greater distance between plates; (b) smaller plate area and less distance between plates; (c) larger plate area and less distance between plates; (d) higher values of applied voltage.

5. Which of the following statements is correct? (a) Air capacitors have a black band to indicate the outside foil. (b) Mica capacitors are available in capacitance values of 1 to 10 μF. (c) Electrolytic capacitors must be connected in the correct polarity. (d) Ceramic capacitors must be connected in the correct polarity.

6. Voltage applied across a ceramic dielectric produces an electrostatic field 100 times greater than in air. The dielectric constant K_ϵ of the ceramic equals (a) 33⅓; (b) 50; (c) 100; (d) 10,000.

7. A six-dot mica capacitor color-coded white, red, green, brown, red, and yellow has the capacitance value of (a) 25 pF; (b) 124 pF; (c) 250 pF; (d) 925 pF.

8. The combination of two 0.02-μF 500-V capacitors in series has a capacitance and breakdown rating of (a) 0.01 μF, 500 V; (b) 0.01 μF, 1000 V; (c) 0.02 μF, 500 V; (d) 0.04 μF, 500 V.

9. The combination of two 0.02-μF 500-V capacitors in parallel has a capacitance and breakdown rating of (a) 0.01 μF, 1000 V; (b) 0.02 μF, 500 V; (c) 0.04 μF, 500 V; (d) 0.04 μF, 1000 V.

10. For a good 0.05-μF paper capacitor, the ohmmeter reading should (a) go quickly to 100 Ω, approximately, and remain there; (b) show low resistance momentarily and back off to a very high resistance; (c) show high resistance momentarily and then a very low resistance; (d) not move at all.

Essay Questions

1. Define capacitance with respect to physical structure and electrical function. Explain how a two-wire conductor has capacitance.

⊣⊢

2. **(a)** What is meant by a dielectric material? **(b)** Name five common dielectric materials. **(c)** Define dielectric flux.

3. Explain briefly how to charge a capacitor. How is a charged capacitor discharged?

4. Define 1 F of capacitance. Convert the following into farads using powers of 10: **(a)** 50 pF; **(b)** 0.001 μF; **(c)** 0.047 μF; **(d)** 0.01 μF; **(e)** 10 μF.

5. State the effect on capacitance of **(a)** larger plate area; **(b)** thinner dielectric; **(c)** higher value of dielectric constant.

6. Give one reason for your choice of the type of capacitor to be used in the following applications: **(a)** 80-μF capacitance for a circuit where one side is positive and the applied voltage never exceeds 150 V; **(b)** 1.5-pF capacitance for an RF circuit where the required voltage rating is less than 500 V; **(c)** 5-μF capacitance for an audio circuit where the required voltage rating is less than 25 V.

7. **(a)** Give the capacitance value of six-dot mica capacitors color-coded as follows: (1) Black, red, green, brown, black, black. (2) White, green, black, black, green, brown. (3) White, gray, red, brown, silver, black. **(b)** Give the capacitance value of the tubular ceramic capacitors color-coded as follows: (4) Black, brown, black, black, brown. (5) Brown, gray, black, gray, black.

8. Draw a diagram showing the least number of 400-V 2-μF capacitors needed for a combination rated at 800 V with 2 μF total capacitance.

9. Given two identical uncharged capacitors. One is charged to 50 V and connected across the uncharged capacitor. Why will the voltage across both capacitors then be 25 V?

10. Describe briefly how you would check a 0.05-μF capacitor with an ohmmeter. State the ohmmeter indications for the case of the capacitor being good, short-circuited, or open.

11. Define the following: **(a)** stray circuit capacitance; **(b)** distributed capacitance of a coil; **(c)** leakage resistance of a capacitor; **(d)** power factor and Q of a capacitor.

12. Give two comparisons between the electric field in a capacitor and the magnetic field in a coil.

13. Give three types of troubles in capacitors.

14. When a capacitor discharges, why is its discharge current in the opposite direction from the charging current?

15. Compare the features of aluminum and tantalum electrolytic capacitors.

16. Why can plastic film be used instead of paper for capacitors?

Problems

Answers to Odd-Numbered Problems at Back of Book

1. How much charge in coulombs is in a 4-μF capacitor charged to 100 V?

2. A 4-μF capacitor has 400 μC of charge. **(a)** How much voltage is across the capacitor? **(b)** How much is the voltage across an 8-μF capacitor with the same 400-μC charge?

3. A 2-μF capacitor is charged by a constant 3-μA charging current for 6 s. **(a)** How much charge is stored in the capacitor? **(b)** How much is the voltage across the capacitor?

⊣⊢

4. A 1-μF capacitor C_1 and a 10-μF capacitor C_2 are in series with a constant 2-mA charging current. **(a)** After 4 s, how much charge is in C_1 and in C_2? **(b)** How much is the voltage across C_1 and across C_2?

5. Calculate C for a mica capacitor, with $K_\epsilon = 8$, a thickness of 0.02 cm, plates of 6 cm^2, and five sectons in parallel. (Hint: 1 cm $= 10^{-2}$ m and 1 cm^2 $= 10^{-4}$ m^2.)

6. How much capacitance stores 6000 μC of charge with 150 V applied? The charge of how many electrons is stored? What type of capacitor is this most likely to be?

7. With 100 V across a capacitor, it stores 100 μC of charge. Then the applied voltage is doubled to 200 V. **(a)** How much is the voltage across the capacitor? **(b)** How much charge is stored? **(c)** How much is its capacitance?

8. Referring to the parallel capacitors in Fig. 20-11, calculate the charge Q_1 in C_1 and Q_2 in C_2 with 50 V. How much is the total charge Q_T in both capacitors? Calculate the total capacitance C_T as Q_T/V.

9. Calculate the energy in joules stored in **(a)** a 500-pF C charged to 10 kV; **(b)** a 1-μF C charged to 5 kV; **(c)** a 40-μF C charged to 400 V.

10. Three capacitors are in series. C_1 is 100 pF, C_2 is 100 pF, and C_3 is 50 pF. Calculate C_T.

11. Calculate C_T for the series-parallel combination of capacitors in Fig. 20-19a and b.

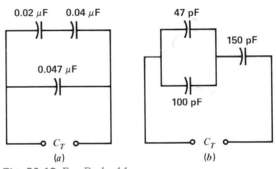

Fig. 20-19 For Prob. 11.

Answers to Practice Problems

20-1	**a.**	Dielectric	**20-5**	**a.**	T	**20-9**	**a.**	T
	b.	Farad		**b.**	F		**b.**	T
20-2	**a.**	14.5 V		**c.**	T		**c.**	T
	b.	0 V	**20-6**	**a.**	250 pF	**20-10**	**a.**	T
	c.	Yes		**b.**	0.01		**b.**	T
20-3	**a.**	10 μF	**20-7**	**a.**	0.03 μF	**20-11**	**a.**	0 Ω
	b.	Ceramic		**b.**	150 pF		**b.**	Good
20-4	**a.**	T	**20-8**	**a.**	0.1 μF			
	b.	F		**b.**	25 V			
	c.	T		**c.**	33.3 pF			
	d.	T						

Chapter 21
Capacitive Reactance

When a capacitor charges and discharges with varying voltage applied, alternating current can flow. Although there cannot be any current through the dielectric of the capacitor, its charge and discharge produces current in the circuit connected to the capacitor plates. How

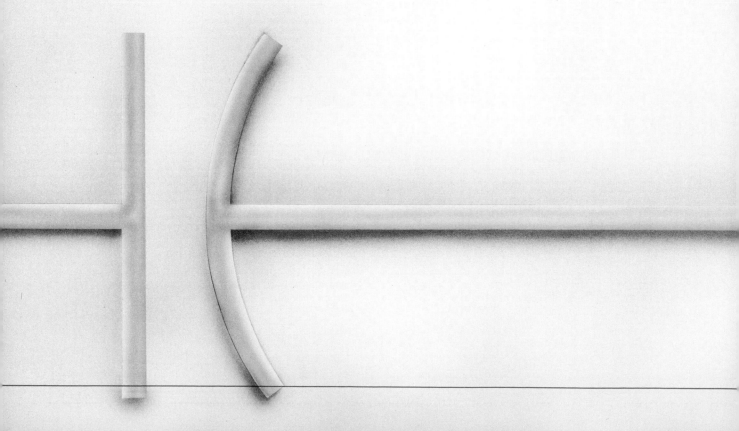

much alternating current flows with sine-wave voltage applied depends on the capacitive reactance X_C.

The amount of X_C is $1/(2\pi fC)$, with f in hertz and C in farads for X_C in ohms. The reactance of X_C is in ohm units, like X_L, but their effects are opposite in terms of frequency. While X_L is directly proportional to f, the X_C is inversely proportional to f. Because of the reciprocal relation in $X_C = 1/(2\pi fC)$, the ohms of X_C decrease for higher frequencies and more capacitance. The reason is more charge and discharge current.

Important terms in this chapter are:

capacitive reactance   inverse relation
charge current   parallel capacitance
dc blocking   phase angle
discharge current   series capacitance

More details are explained in the following sections:

21-1 Alternating Current in a Capacitive Circuit
21-2 $X_C = 1/(2\pi fC)$
21-3 Series or Parallel Capacitive Reactances
21-4 Ohm's Law Applied to X_C
21-5 Applications of Capacitive Reactance
21-6 Sine-Wave Charge and Discharge Current

21-1   Alternating Current in a Capacitive Circuit

The fact that current flows with ac voltage applied is demonstrated in Fig. 21-1, where the bulb lights in Fig. 21-1*a* and *b* because of the capacitor charge and discharge current. There is no current through the dielectric, which is an insulator. While the capacitor is being charged by increasing applied voltage, however, the charging current flows in one direction in the conductors to the plates. While the capacitor is discharging, when the applied voltage decreases, the discharge current flows in the reverse direction. With alternating voltage applied, the capacitor alternately charges and discharges.

First the capacitor is charged in one polarity, and then it discharges; next the capacitor is charged in the opposite polarity, and then it discharges again. The cycles of charge and discharge current provide alternating current in the circuit, at the same frequency as the applied voltage. This is the current that lights the bulb.

In Fig. 21-1*a*, the 4-μF capacitor provides enough alternating current to light the bulb brightly. In Fig. 21-1*b*, the 1-μF capacitor has less charge and discharge current because of the smaller capacitance, and the light is not so bright. Therefore, the smaller capacitor has more opposition to alternating current as less current flows with the same applied voltage; that is, it has more reactance for less capacitance.

In Fig. 21-1*c*, the steady dc voltage will charge the capacitor to 120 V. Because the applied voltage does not change, though, the capacitor will just stay charged. Since the potential difference of 120 V across the charged capacitor is a voltage drop opposing the applied voltage, no current can flow. Therefore, the bulb cannot light. The bulb may flicker on for an instant as charging current flows when voltage is applied, but this current is only temporary until the capacitor is charged. Then the capacitor has the applied voltage of 120 V, but there is zero voltage across the bulb.

As a result, the capacitor is said to *block* direct current or voltage. In other words, after the capacitor has been charged by a steady dc voltage, there is no current in the dc circuit. All the applied dc voltage is across the charged capacitor, with zero voltage across any series resistance.

In summary, then, this demonstration shows the following points:

1. Alternating current flows in a capacitive circuit with ac voltage applied.
2. A smaller capacitance allows less current, which means more X_C with more ohms of opposition.
3. Lower frequencies for the applied voltage result in less current and more X_C. With a steady dc voltage source, which corresponds to a frequency of zero, the opposition of the capacitor is infinite and there is no current. In this case the capacitor is effectively an open circuit.

These effects have almost unlimited applications in practical circuits because X_C depends on frequency. A very common use of a capacitor is to provide little opposition for ac voltage but to block any dc voltage. Another example is to use X_C for less opposition to a high-frequency alternating current, compared with lower frequencies.

Capacitive Current   The reason why a capacitor allows current to flow in an ac circuit is the alternate charge and discharge. If we insert an ammeter in the

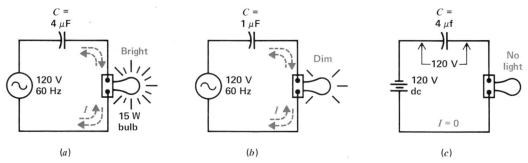

(*a*)   (*b*)   (*c*)

Fig. 21-1 Current in a capacitive circuit. (*a*) The 4-μF capacitor allows enough 60-Hz current to light the bulb brightly. (*b*) Less current with smaller capacitor causes dim light. (*c*) Bulb cannot light with dc voltage applied.

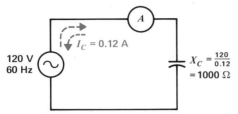

Fig. 21-2 Capacitive reactance X_C is the ratio of V_C/I_C.

circuit, as shown in Fig. 21-2, the ac meter will read the amount of charge and discharge current. In this example I_C is 0.12 A. This current is the same in the voltage source, the connecting leads, and the plates of the capacitor. However, there is no current through the insulator between the plates of the capacitor.

Values for X_C   When we consider the ratio of V_C/I_C for the ohms of opposition to the sine-wave current, this value is $120/0.12$, which equals 1000 Ω. This 1000 Ω is what we call X_C, to indicate how much current can be produced by sine-wave voltage applied to a capacitor. In terms of current, $X_C = V_C/I_C$. In terms of frequency and capacitance, $X_C = 1/(2\pi fC)$.

The X_C value depends on the amount of capacitance and the frequency of the applied voltage. If C in Fig. 21-2 were increased, it could take on more charge for more charging current and then produce more discharge current. Then X_C is less for more capacitance. Also, if the frequency in Fig. 21-2 were increased, the capacitor could charge and discharge faster to produce more current. This action also means V_C/I_C would be less, with more current for the same applied voltage. Therefore, X_C is less for higher frequencies. Reactance X_C can actually have almost any value, from practically zero to almost infinite ohms.

Practice Problems 21-1
Answers at End of Chapter

a. Which has more reactance, a 0.1- or a 0.5-μF capacitor, at the same frequency?
b. Which allows more charge and discharge current, a 0.1- or a 0.5-μF capacitor?

21-2 $X_C = 1/(2\pi fC)$

This formula includes the effects of frequency and capacitance for calculating the ohms of reactance. The frequency is in hertz and C is in farads for X_C in ohms. As an example, we can calculate X_C for 2.65 μF and 60 Hz.

$$X_C = \frac{1}{2\pi fC} \tag{21-1}$$

$$= \frac{1}{2\pi \times 60 \times 2.65 \times 10^{-6}}$$

$$= \frac{0.159 \times 10^6}{60 \times 2.65} = \frac{159,000}{159}$$

$$X_C = 1000 \ \Omega$$

The constant factor 2π, equal to 6.28, indicates the circular motion from which a sine wave is derived. Therefore, the formula applies only to sine-wave circuits. To simplify calculations of X_C, the constant reciprocal $1/6.28$ can be taken as 0.159, approximately. Then

$$X_C = \frac{0.159}{fC} \tag{21-2}$$

Remember that C must be in farads for X_C in ohms. Although C values are usually microfarads (10^{-6}) or picofarads (10^{-12}), substitute the value of C in farads with the required negative power of 10.

Example 1   How much is X_C for **(a)** a 0.1-μF C at 1000 Hz? **(b)** a 1-μF C at the same frequency?

Answer

a. $$X_C = \frac{0.159}{fC} = \frac{0.159 \times 10^6}{0.1 \times 1000}$$

$$= \frac{0.159 \times 10^3}{0.1}$$

$$X_C = 1590 \ \Omega$$

b.   At the same frequency, with ten times more C, the X_C is 1590/10, which equals 159 Ω.

Note that X_C in Example 1**b** is one-tenth the X_C in 1**a** because C is ten times larger.

Example 2   How much is the X_C of a 100-pF C at (**a**) 1 MHz? (**b**) 10 MHz?

Answer

a. $X_C = \dfrac{0.159}{fC} = \dfrac{0.159}{1 \times 10^6 \times 100 \times 10^{-12}}$

$ = \dfrac{0.159 \times 10^6}{100}$

$X_C = 1590 \ \Omega$

b. At ten times the frequency, X_C is 1590/10, which equals 159 Ω.

Note that X_C in Example 2**b** is one-tenth the X_C in 2**a** because f is ten times higher.

Example 3   How much is the X_C of a 240-pF C at 41.67 kHz?

Answer $X_C = \dfrac{0.159}{fC}$

$ = \dfrac{0.159}{41.67 \times 10^3 \times 240 \times 10^{-12}}$

$ = \dfrac{0.159 \times 10^9}{41.67 \times 240}$

$X_C = 15,900 \ \Omega$

X_C Is Inversely Proportional to Capacitance

This statement means that X_C increases as the capacitance is reduced. In Fig. 21-3, when C is reduced by the factor of $\frac{1}{10}$, from 1.0 to 0.1 μF, then X_C increases ten times, from 1000 to 10,000 Ω. Also, decreasing C one-half, from 0.2 to 0.1 μF, doubles X_C from 5000 to 10,000 Ω.

This inverse relation between C and X_C is illustrated by the graph in Fig. 21-3. Note that values of X_C increase downward on the graph, indicating negative reactance that is opposite from inductive reactance. With C increasing to the right, the decreasing values of X_C approach the zero axis of the graph.

X_C Is Inversely Proportional to Frequency

Figure 21-4 illustrates the inverse relation between X_C and f. With f increasing to the right in the graph from 0.1 to 1 MHz, the negative value of X_C for the 159-pF capacitor decreases from 10,000 to 1000 Ω as the X_C curve comes closer to the zero axis.

The graphs are nonlinear because of the inverse relation between X_C and f or C. At one end, the curves approach infinitely high reactance for zero capacitance or zero frequency. At the other end, the curves approach zero reactance for infinitely high capacitance or frequency.

In some applications, it is necessary to find the value of capacitance required for a desired value of X_C, at a specific frequency. For this case the reactance formula can be inverted:

$$C = \frac{0.159}{fX_C} \tag{21-3}$$

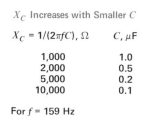

X_C Increases with Smaller C

$X_C = 1/(2\pi fC),\ \Omega$	C, μF
1,000	1.0
2,000	0.5
5,000	0.2
10,000	0.1

For f = 159 Hz

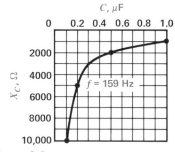

Fig. 21-3 Capacitive reactance X_C decreases with higher values of C.

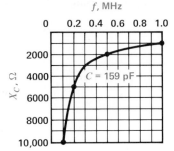

X_C Increases with Lower Frequencies

X_C* = 1/(2πfC), Ω   f, MHz

1,000	1.0
2,000	0.5
5,000	0.2
10,000	0.1

*For C = 159 pF

Fig. 21-4 Capacitive reactance X_C decreases with higher frequencies.

Example 4 What capacitance is needed to have a 100-Ω X_C at 1 MHz?

Answer
$$C = \frac{0.159}{fX_C} = \frac{0.159}{1 \times 10^6 \times 100}$$
$$= \frac{0.159 \times 10^{-6}}{1 \times 100}$$
$$= 0.001\,59 \times 10^{-6} \text{ F}$$
$$C = 0.001\,59 \text{ μF}$$

Or, to find the frequency at which a given capacitance has a specified X_C, the reactance formula can be inverted to the form

$$f = \frac{0.159}{CX_C} \tag{21-4}$$

Example 5 At what frequency will a 0.1-μF capacitor have an X_C equal to 1000 Ω?

Answer
$$f = \frac{0.159}{CX_C} = \frac{0.159}{0.1 \times 10^{-6} \times 1000}$$
$$= \frac{0.159}{0.1 \times 10^{-6} \times 10^3}$$
$$= 0.159 \times 10^4$$
$$f = 1590 \text{ Hz}$$

In summary, Formula (21-2) gives X_C in terms of f and C; with Formula (21-3) we can calculate C when X_C and f are known; Formula (21-4) is used to find f given the values of C and X_C. The value of X_C can be measured as V_C/I_C.

Practice Problems 21-2
Answers at End of Chapter

The X_C is 400 Ω for a capacitor at 8 MHz.
a. How much is X_C at 16 MHz?
b. How much is X_C at 4 MHz?

21-3   Series or Parallel Capacitive Reactances

Because capacitive reactance is an opposition in ohms, series or parallel reactances are combined in the same way as resistances. As shown in Fig. 21-5a, series reactances of 100 and 200 Ω add to equal 300 Ω of X_{C_T}. The formula is

$$X_{C_T} = X_{C_1} + X_{C_2} + \cdots + \text{etc.} \quad \text{in series} \tag{21-5}$$

For parallel reactances, the combined reactance is calculated by the reciprocal formula, as shown in Fig. 21-5b.

$$\frac{1}{X_{C_T}} = \frac{1}{X_{C_1}} + \frac{1}{X_{C_2}} + \cdots + \text{etc.} \quad \text{in parallel} \tag{21-6}$$

In Fig. 21-5b the parallel combination of 100 and 200 Ω is 66⅔ Ω for X_{C_T}. The combined parallel reactance is smaller than the lowest branch reactance. Any short cuts for combining parallel resistances also apply to parallel reactances.

Combining reactances is opposite to the way capacitances are combined. The two procedures are equivalent, however, because capacitive reactance is in-

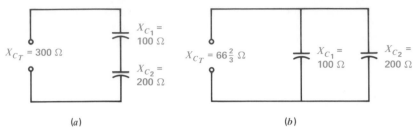

Fig. 21-5 Reactances combine like resistances. (*a*) Addition of series reactances. (*b*) Two reactances in parallel equal their product over their sum.

versely proportional to capacitance. The general case is that ohms of opposition add in series but combine by the reciprocal formula in parallel. This rule applies to resistances, to a combination of inductive reactances alone, or to capacitive reactances alone.

Practice Problems 21-3
Answers at End of Chapter

a. How much is X_{C_T} for a 200-Ω X_{C_1} in series with a 300-Ω X_{C_2}?
b. How much is X_{L_T} for a 200-Ω X_{L_1} in series with a 300-Ω X_{L_2}?

21-4   Ohm's Law Applied to X_C

The current in an ac circuit with X_C alone is equal to the applied voltage divided by the ohms of X_C. Three examples with X_C are illustrated in Fig. 21-6. In Fig. 21-6*a* there is just one reactance of 100 Ω. The current I then is equal to V/X_C, or 100 V/100 Ω, which is 1 A.

For the series circuit in Fig. 21-6*b*, the total reactance, equal to the sum of the series reactances, is 300 Ω. Then the current is 100 V/300 Ω, which equals $\frac{1}{3}$ A. Furthermore, the voltage across each reactance is equal to its IX_C product. The sum of these series voltage drops equals the applied voltage.

For the parallel circuit in Fig. 21-6*c*, each parallel reactance has its individual branch current, equal to the applied voltage divided by the branch reactance. The applied voltage is the same across both reactances, since they are all in parallel. In addition, the total line current of $1\frac{1}{2}$ A is equal to the sum of the individual branch currents of 1 and $\frac{1}{2}$ A each. With the applied voltage an rms value, all the calculated currents and voltage drops in Fig. 21-6 are also rms values.

Practice Problems 21-4
Answers at End of Chapter

a. In Fig. 21-6*b*, how much is X_{C_T}?
b. In Fig. 21-6*c*, how much is X_{C_T}?

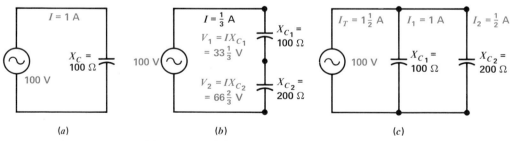

Fig. 21-6 Circuit calculations with X_C. (*a*) $I = V/X_C$. (*b*) Sum of series voltage drops equals the applied voltage V_T. Sum of parallel branch currents equals total line current I_T.

Table 21-1. Capacitance Values for a Reactance of 100 Ω

C (approx.)	Frequency	Remarks
27 μF	60 Hz	Power-line and low audio frequency
1.6 μF	1,000 Hz	Medium audio frequency
0.16 μF	10,000 Hz	High audio frequency
1600 pF	1,000 kHz (RF)	In AM radio broadcast band
160 pF	10 MHz (HF)	In short-wave radio band
16 pF	100 MHz (VHF)	In FM radio broadcast band

21-5 Applications of Capacitive Reactance

The general use of X_C is to block direct current but provide low reactance for alternating current. In this way, a varying ac component can be separated from a steady direct current. Furthermore, a capacitor can have less reactance for alternating current of high frequencies, compared with lower frequencies.

Note the following differences in ohms of R, X_L, and X_C. Ohms of R remain the same for dc circuits or ac circuits. Ohms of reactance, however, either X_L or X_C, depend on the frequency. The effects of X_L and X_C are opposite, since X_L increases with frequency and X_C decreases with frequency.

If 100 Ω is taken as a desired value of X_C, capacitor values can be calculated for different frequencies, as listed in Table 21-1. The C values indicate typical capacitor sizes for different frequency applications. Note that the required C becomes smaller for higher frequencies.

The 100 Ω of reactance for Table 21-1 is taken as a low X_C in common applications of C as a coupling capacitor, bypass capacitor, or filter capacitor for ac variations. For all these functions, the X_C must be low compared with the resistance in the circuit. Typical values of C, then, are 16 to 1600 pF for RF signals and 0.16 to 27 μF for AF signals. The power line frequency of 60 Hz, which is a low audio frequency, requires C values of about 27 μF or more.

Practice Problems 21-5
Answers at End of Chapter

A 20-μF C has 100 Ω of X_C at 60 Hz.
a. How much is X_C at 120-Hz?
b. How much is X_C at 6 Hz?

21-6 Sine-Wave Charge and Discharge Current

In Fig. 21-7 sine-wave voltage applied across a capacitor produces alternating charge and discharge current. The action is considered for each quarter-cycle. Note that the voltage v_C across the capacitor is the same as the applied voltage v_A at all times because they are in parallel. The values of current i, however, depend on

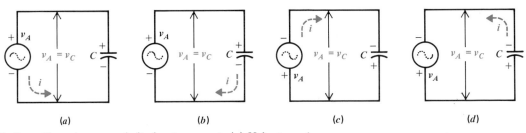

(a) *(b)* *(c)* *(d)*

Fig. 21-7 Capacitive charge and discharge current. (a) Voltage v_A increases positive to charge C. (b) Capacitance C discharges as v_A decreases. (c) Voltage v_A increases negative to charge C in opposite polarity. (d) Capacitance C discharges as reversed v_A decreases.

⊣⊢

the charge and discharge of C. When v_A is increasing, it charges C to keep v_C at the same voltage as v_A; when v_A is decreasing, C discharges to maintain v_C at the same voltage as v_A. When v_A is not changing, there is no charge or discharge current.

During the first quarter-cycle, in Fig. 21-7a, v_A is positive and increasing, charging C in the polarity shown. The electron flow is from the negative terminal of the source voltage, producing charging current in the direction indicated by the arrow for i. Next, when the applied voltage decreases during the second quarter-cycle, v_C also decreases by discharging. The discharge current is from the negative plate of C through the source, and back to the positive plate. Note that the discharge current in Fig. 21-7b has the opposite direction from the charge current in Fig. 21-7a.

For the third quarter-cycle, in Fig. 21-7c, the applied voltage v_A increases again but in the negative direction. Now C charges again but in reversed polarity. Here the charging current is in the opposite direction from the charge current in Fig. 21-7a but in the same direction as the discharge current in Fig. 21-7b. Finally, the negative applied voltage decreases during the final quarter-cycle in Fig. 21-7d. As a result, C discharges. This discharge current is opposite to the charge current in Fig. 21-7c, but in the same direction as the charge current in Fig. 21-7a.

For the sine wave of applied voltage, therefore, the capacitor provides a cycle of alternating charge and discharge current. Notice that capacitive current flows for either charge or discharge, whenever the voltage changes, for either an increase or decrease. Also, i and v have the same frequency.

Calculating the Values of i_C

The greater the voltage change, the greater is the amount of capacitive current. Furthermore, a larger capacitor can allow more charge current when the applied voltage increases and produce more discharge current. Because of these factors the amount of capacitive current can be calculated as

$$i_C = C\frac{dv}{dt} \qquad\qquad (21\text{-}7)$$

where i is in amperes, with C in farads and dv/dt in volts per second. As an example, suppose that the voltage across a 240-pF capacitor changes by 25 V in 1 μs. The amount of capacitive current then is

$$i_C = C\frac{dv}{dt} = 240 \times 10^{-12} \times \frac{25}{1 \times 10^{-6}}$$

$$= 240 \times 25 \times 10^{-6} = 6000 \times 10^{-6}$$

$$= 6 \times 10^{-3} \text{ A}$$

$$i_C = 6 \text{ mA}$$

Notice how Formula (21-7) is similar to the capacitor charge formula $Q = CV$. When the voltage changes, this dv/dt factor produces a change in the charge Q. When the charge moves, this dq/dt change is the current i_C. Therefore, dq/dt or i_C is proportional to dv/dt. With the constant factor C, then, i_C becomes equal to $C(dv/dt)$.

The formula for capacitive current $i_C = C(dv/dt)$ corresponds to the formula for induced voltage $v_L = L(di/dt)$. In both cases there must be a change to have an effect. For inductance, v_L is induced when the current changes. For capacitance, i_C results when the voltage changes.

These formulas give the fundamental definitions for the amount of reactive effect for inductance or capacitance. Just as one henry is defined as the amount of inductance that produces one volt of v_L when the current changes at the rate of one ampere per second, one farad can also be defined as the amount of capacitance that produces one ampere of i_C when the voltage changes at the rate of one volt per second.

By means of Formula (21-7), then, i_C can be calculated to find the instantaneous value of charge or discharge current when the voltage changes across a capacitor.

Example 6 Calculate the instantaneous value of charging current i_C produced by a 6-μF C when its potential difference is increased by 50 V in 1 s.

Answer $i_C = C\dfrac{dv}{dt} = 6 \times 10^{-6} \times \dfrac{50}{1}$

$\qquad\qquad i_C = 300 \ \mu$A

Example 7 Calculate i_C for the same C as in Example 6 where its potential difference is *decreased* by 50 V in 1 s.

⊣⊢

Answer For the same $C(dv/dt)$, i_C is the same 300 μA. However, this 300 μA is discharge current, which flows in the opposite direction from i_C on charge. If desired, the i_C for discharge current can be considered negative, or -300 μA.

Example 8 Calculate i_C produced by a 250-pF capacitor for a change of 50 V in 1 μs.

Answer $i_C = C \dfrac{dv}{dt}$

$$= 250 \times 10^{-12} \times \frac{50}{1 \times 10^{-6}}$$

$$= 12{,}500 \times 10^{-6}$$

$$i_C = 12{,}500 \ \mu\text{A}$$

Notice that more i_C is produced here, although C is smaller than in Example 6, because dv/dt is a much faster voltage change.

Waveshapes of v_C and i_C More details of capacitive circuits can be analyzed by means of the waveshapes in Fig. 21-8, plotted for the calculated values in Table 21-2. The top curve shows a sine wave of voltage v_C across a 240-pF capacitance C. Since the capacitive current i_C depends on the rate of change of voltage,

rather than the absolute value of v, the curve in Fig. 21-8b shows how much the voltage changes. In this curve, the dv/dt values are plotted for every 30° of the cycle.

The bottom curve shows the actual capacitive current i_C. This i_C curve is similar to the dv/dt curve because i_C equals the constant factor C multiplied by dv/dt.

All three curves are similar to the three curves shown before in Fig. 18-7 in Chap. 18 for inductive reactance, but with the voltage and current curves interchanged. Both examples illustrate the effects of the rate of change in a sine wave.

90° Phase Angle The i_C curve at the bottom in Fig. 21-8 has its zero values when the v_C curve at the top is at maximum. This comparison shows that the curves are 90° out of phase, as i_C is a cosine wave of current for the sine wave of voltage v_C. The 90° phase difference results from the fact that i_C depends on the dv/dt rate of change, rather than on v itself. More details of this 90° phase angle for capacitance are explained in the next chapter.

For each of the curves, the period T is 24 μs. Therefore, the frequency is $1/T$ or $1/24$ μs, which equals 41.67 kHz. Each curve has the same frequency, although there is a 90° phase difference between i and v.

Ohms of X_C The ratio of v_C/i_C actually specifies the capacitive reactance, in ohms. For this comparison, we

Table 21-2. Values for $i_C = C(dv/dt)$ Curves in Fig. 21-8 (on Page 422)

Time		dt		dv, V	dv/dt, V/μs	C, pF	$i_C = C(dv/dt)$, mA
θ	μs	θ	μs				
30°	2	30°	2	50	25	240	6
60°	4	30°	2	36.6	18.3	240	4.4
90°	6	30°	2	13.4	6.7	240	1.6
120°	8	30°	2	−13.4	−6.7	240	−1.6
150°	10	30°	2	−36.6	−18.3	240	−4.4
180°	12	30°	2	−50	−25	240	−6
210°	14	30°	2	−50	−25	240	−6
240°	16	30°	2	−36.6	−18.3	240	−4.4
270°	18	30°	2	−13.4	−6.7	240	−1.6
300°	20	30°	2	13.4	6.7	240	1.6
330°	22	30°	2	36.6	18.3	240	4.4
360°	24	30°	2	50	25	240	6

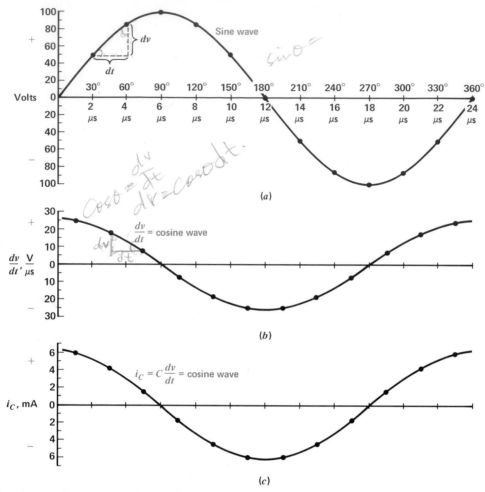

Fig. 21-8 Waveshapes of sine-wave voltage with charge and discharge current i_C, plotted for values in Table 21-2.

use the actual value of v_C, which has the peak of 100 V. The rate-of-change factor is included in i_C. Although the peak of i_C at 6 mA is 90° ahead of the peak of v_C at 100 V, we can compare these two peak values. Then v_C/i_C is $^{100}\!/_{.006}$, which equals 16,667 Ω.

This X_C is only an approximate value because i_C cannot be determined exactly for the large dt changes every 30°. If we used smaller intervals of time, the peak i_C would be 6.28 mA with X_C then 15,900 Ω, the same as $1/(2\pi fC)$ with a 240-pF C and a frequency of 41.67 kHz. This is the same X_C problem as Example 3 on page 416.

Practice Problems 21-6
Answers at End of Chapter

Refer to the curves in Fig. 21-8.

a. At what angle does v have its maximum positive value?

b. At what angle does dv/dt have its maximum positive value?

c. What is the phase angle difference between v_C and i_C?

Summary

1. Capacitive reactance, indicated by X_C, is the opposition of a capacitance to the flow of sine-wave alternating current.
2. Reactance X_C is measured in ohms because it limits the current to the value V/X_C. With V in volts and X_C in ohms, I is in amperes.
3. $X_C = 1/(2\pi fC)$. With f in hertz and C in farads, X_C is in ohms.
4. For one value of capacitance, X_C decreases with higher frequencies.
5. At one frequency, X_C decreases with higher values of capacitance.
6. With X_C and f known, the capacitance $C = 1/(2\pi fX_C)$.
7. With X_C and C known, the frequency $f = 1/(2\pi CX_C)$.
8. The total X_C of capacitive reactances in series equals the sum of the individual values, as for series resistances. The series reactances have the same current. The voltage across each reactance is IX_C.
9. With parallel capacitive reactances, the combined reactance is calculated by the reciprocal formula, as for parallel resistances. Each branch current is V/X_C. The total line current is the sum of the individual branch currents.
10. Table 21-3 summarizes the differences between C and X_C.
11. Table 21-4 compares the opposite reactances X_L and X_C.

Table 21-3. Comparison of Capacitance and Capacitive Reactance

Capacitance	Capacitive reactance
Symbol is C Measured in farad units Depends on construction of capacitor $C = i_C/(dv/dt)$ or Q/V	Symbol is X_C Measured in ohm units Depends on frequency of sine-wave voltage $X_C = v_C/i_C$ or $1/(2\pi fC)$

Table 21-4. Comparison of Inductive and Capacitive Reactances

X_L, Ω	X_C, Ω
Increases with more inductance Increases for higher frequencies Allows more current for lower frequencies; passes direct current	Decreases with more capacitance Decreases for higher frequencies Allows less current for lower frequencies; blocks direct current

Self-Examination
Answers at Back of Book

Choose (a), (b), (c), or (d).

1. Alternating current can flow in a capacitive circuit with ac voltage applied because (a) of the high peak value; (b) varying voltage produces charge and discharge current; (c) charging current flows when the voltage decreases; (d) discharge current flows when the voltage increases.

2. With higher frequencies, the amount of capacitive reactance (a) increases; (b) stays the same; (c) decreases; (d) increases only when the voltage increases.

3. At one frequency, larger capacitance results in (a) more reactance; (b) the same reactance; (c) less reactance; (d) less reactance if the voltage amplitude decreases.

4. The capacitive reactance of a 0.1-μF capacitor at 1000 Hz equals (a) 1000 Ω; (b) 1600 Ω; (c) 2000 Ω; (d) 3200 Ω.

5. Two 1000-Ω X_C values in series have a total reactance of (a) 500 Ω; (b) 1000 Ω; (c) 1414 Ω; (d) 2000 Ω.

6. Two 1000-Ω X_C values in parallel have a combined reactance of (a) 500 Ω; (b) 707 Ω; (c) 1000 Ω; (d) 2000 Ω.

7. With 50 V rms applied across a 100-Ω X_C, the rms current in the circuit equals (a) 0.5 A; (b) 0.637 A; (c) 0.707 A; (d) 1.414 A.

8. With steady dc voltage from a battery applied to a capacitance, after it charges to the battery voltage, the current in the circuit (a) depends on the current rating of the battery; (b) is greater for larger values of capacitance; (c) is smaller for larger values of capacitance; (d) is zero for any capacitance value.

9. The capacitance needed for a 1000-Ω reactance at 2 MHz is (a) 2 pF; (b) 80 pF; (c) 1000 pF; (d) 2000 pF.

10. A 0.2-μF capacitance will have a reactance of 1000 Ω at the frequency of (a) 800 Hz; (b) 1 kHz; (c) 1 MHz; (d) 8 MHz.

Essay Questions

1. Why is capacitive reactance measured in ohms? State two differences between capacitance and capacitive reactance.

2. Explain briefly why the bulb lights in Fig. 21-1a but not in Fig. 21-1c.

3. Explain briefly what is meant by two factors being inversely proportional. How does this apply to X_C and C? X_C and f?

4. In comparing X_L with X_C, give two differences and one similarity.

5. In comparing X_C and R, give two differences and one similarity.

6. Why are the waves in Fig. 21-8a and b considered to be 90° out of phase, while the waves in Fig. 21-8b and c have the same phase?

7. Referring to Fig. 21-3, how does this graph show an inverse relation between X_C and C?

8. Referring to Fig. 21-4, how does this graph show an inverse relation between X_C and f?

9. Referring to Fig. 21-8, draw three similar curves but for a sine wave of voltage with a period $T = 12$ μs for the full cycle. Use the same C of 240 pF. Compare the value of X_C obtained as $1/(2\pi fC)$ and v_C/i_C.

10. (a) What is the relation between charge q and current i? (b) How is this comparison similar to the relation between the two formulas $Q = CV$ and $i = C(dv/dt)$?

Problems
Answers to Odd-Numbered Problems at Back of Book

1. Referring to Fig. 21-4, give the values of C needed for 2000 Ω of X_C at the four frequencies listed.

2. What size capacitance is needed for 50-Ω reactance at 100 kHz?

3. A capacitor with an X_C of 2000 Ω is connected across a 9-V 1000-Hz source. (a) Draw the schematic diagram. (b) How much is the current in the circuit? (c) What is the frequency of the current?

4. How much is the capacitance of a capacitor that draws 0.1 A from the 60-Hz 120-V power line?

5. A 1000-Ω X_{C_1} and a 4000-Ω X_{C_2} are in series across a 10-V source. (a) Draw the schematic diagram. (b) Calculate the current in the series circuit. (c) How much is the voltage across X_{C_1}? (d) How much is the voltage across X_{C_2}?

6. The 1000-Ω X_{C_1} and 4000-Ω X_{C_2} in Prob. 5 are in parallel across the 10-V source. (a) Draw the schematic diagram. (b) Calculate the branch current in X_{C_1}. (c) Calculate the branch current in X_{C_2}. (d) Calculate the total line current. (e) How much is the voltage across both reactances?

7. At what frequency will a 0.01-μF capacitor have a reactance of 5000 Ω?

8. Four capacitive reactances of 100, 200, 300, and 400 Ω each are connected in series across a 40-V source. (a) Draw the schematic diagram. (b) How much is the total X_{C_T}? (c) Calculate I. (d) Calculate the voltages across each capacitance. (e) If the frequency of the applied voltage is 1600 kHz, calculate the required value of each capacitance.

9. Three equal capacitive reactances of 600 Ω each are in parallel. (a) How much is the equivalent combined reactance? (b) If the frequency of the applied voltage is 800 kHz, how much is the capacitance of each capacitor and how much is the equivalent combined capacitance of the three in parallel?

10. A 2-μF C is in series with a 4-μF C. The frequency is 5 kHz. (a) How much is C_T? (b) Calculate X_{C_T}. (c) Calculate X_{C_1} and X_{C_2} to see if their sum equals X_{C_T}.

11. A capacitor across the 120-V 60-Hz ac power line allows a 0.4-A current. (a) Calculate X_C and C. (b) What size C is needed to double the current?

12. A 0.01-μF capacitor is connected across a 10-V source. Tabulate the values of X_C and current in the circuit at 0 Hz (for steady dc voltage) and at 20 Hz, 60 Hz, 100 Hz, 500 Hz, 5 kHz, 10 kHz, and 455 kHz.

13. Calculate X_C for 470 pF at 1640 kHz.

14. What C is needed for the same X_C in Prob. 13 but at 500 Hz?

15. How much is I with 162 mV applied for the X_C in Probs. 13 and 14?

16. At what frequencies will X_C be 200 Ω for the following capacitors: (a) 2 μF; (b) 0.1 μF; (c) 0.05 μF; (d) 0.002 μF; (e) 250 pF; (f) 100 pF; (g) 47 pF?

17. What size C is needed to have X_C the same as the X_L of a 6-mH L at 100 kHz?

Answers to Practice Problems

21-1 a. 0.1 μF		**21-3 a.** 500 Ω		**21-5 a.** 50 Ω	
b. 0.5 μF		**b.** 500 Ω		**b.** 1000 Ω	
21-2 a. 200 Ω		**21-4 a.** 300 Ω		**21-6 a.** 90°	
b. 800 Ω		**b.** 66.7 Ω		**b.** 0 or 360°	
				c. 90°	

Chapter 22

Capacitive Circuits

This chapter analyzes circuits that combine capacitive reactance X_C and resistance R. The main questions are: How do we combine the ohms of opposition, how much current flows, and what is the phase angle? The method is similar to the procedures using impedance for inductive circuits, but remember that some important characteristics of X_C are op-

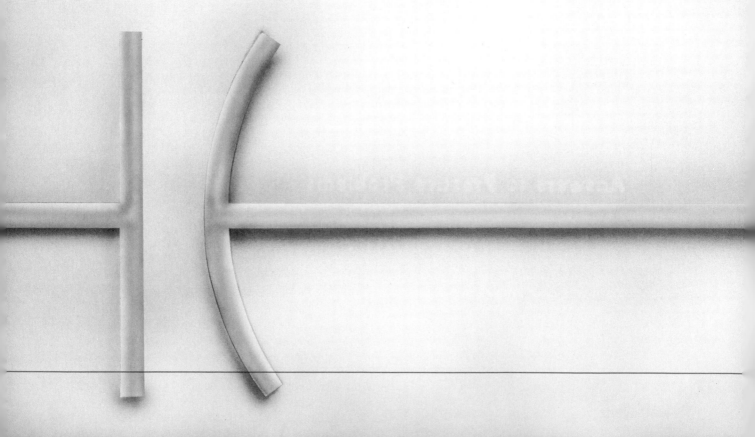

posite from X_L. In addition, methods of using series capacitances for a voltage divider are explained. Also, the practical application of a coupling capacitor shows how it is used to pass ac variations while blocking the steady dc value.

Finally, the general case of capacitive charge and discharge current is shown with nonsinusoidal voltage variations. With nonsinusoidal waveforms, the circuit can be analyzed in terms of its time constant, which is the product of $R \times C$. Remember that X_C and its 90° phase angle apply only to sine waves.

Important terms in this chapter are:

capacitive voltage divider   phasor triangle
coupling capacitor RC time constant
leading current   rectangular waveform
phase-shifter circuit   sawtooth waveform

More details are explained in the following sections:

22-1   Sine-Wave v_C Lags i_C by 90°

For a sine wave of applied voltage, the capacitor provides a cycle of alternating charge and discharge current, as shown in Fig. 22-1a. In Fig. 22-1b, the waveshape of this charge and discharge current i_C is compared with the voltage v_C.

Note that the instantaneous value of i_C is zero when v_C is at its maximum value. At either its positive or negative peak, v_C is not changing. For one instant at both peaks, therefore, the voltage must have a static value before changing its direction. Then v is not changing and C is not charging or discharging. The result is zero current at this time.

Also note that i_C is maximum when v_C is zero. When v_C crosses the zero axis, i_C has its maximum value because then the voltage is changing most rapidly.

Therefore, i_C and v_C are 90° out of phase, since the maximum value of one corresponds to the zero value of the other; i_C leads v_C because i_C has its maximum value a quarter-cycle before the time that v_C reaches its peak. The phasors in Fig. 22-1c show i_C leading v_C by the counterclockwise angle of 90°. Here v_C is the horizontal phasor for the reference angle of 0°. In Fig. 22-1d, however, the current i_C is the horizontal phasor for reference. Since i_C must be 90° leading, v_C is shown lagging by the clockwise angle of −90°. In series circuits, the current i_C is the reference and then the voltage v_C can be considered to lag i_C by 90°.

The 90° phase angle results because i_C depends on the rate of change of v_C. As shown previously in Fig. 21-8 for a sine wave of v_C, the capacitive charge and discharge current is a cosine wave. This 90° phase between v_C and i_C is true in any sine-wave ac circuit, whether C is in series or parallel and whether C is alone or combined with other components. We can always say that for any X_C its current and voltage are 90° out of phase.

Capacitive Current the Same in Series Circuit

The leading phase angle of capacitive current is only with respect to the voltage across the capacitor, which does not change the fact that the current is the same in all parts of a series circuit. In Fig. 22-1a, for instance, the current in the generator, the connecting wires, and both plates of the capacitor must be the same because they are all in the same path.

Capacitive Voltage the Same across Parallel Branches

In Fig. 22-1a, the voltage is the same across the generator and C because they are in parallel. There cannot be any lag or lead in time between these two parallel voltages. At any instant, whatever the voltage value is across the generator at that time, the voltage across C is the same. With respect to the series current, however, both v_A and v_C are 90° out of phase with i_C.

The Frequency Is the Same for v_C and i_C

Although v_C lags i_C by 90°, both waves have the same frequency. For example, if the frequency of the sine wave v_C in Fig. 22-1b is 100 Hz, this is also the frequency of i_C.

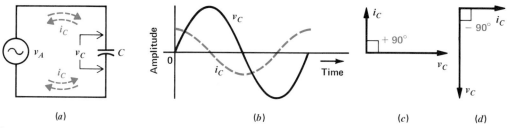

Fig. 22-1 Capacitive current i_C leads v_C by 90°. (a) Circuit with sine-wave v_A across C. (b) Waveshapes of i_C 90° ahead of v_C in phase. (c) Phasor diagram of i_C leading v_C by counterclockwise angle of 90°. (d) Phasor diagram with i_C as horizontal reference to show v_C lagging i_C by clockwise angle of −90°.

Practice Problems 22-1
Answers at End of Chapter

Refer to Fig. 22-1.
a. What is the phase between v_A and v_C?
b. What is the phase between v_C and i_C?
c. Does v_C lead or lag i_C?

22-2 X_C and R in Series

When resistance is in series with capacitive reactance (Fig. 22-2), both determine the current. Current I is the same in X_C and R since they are in series. Each has its own series voltage drop, equal to IR for the resistance and IX_C for the reactance.

If the capacitive reactance alone is considered, its voltage drop lags the series current I by 90°. The IR voltage has the same phase as I, however, because resistance provides no phase shift. Therefore, R and X_C combined in series must be added by phasors because they are 90° out of phase with each other.

Phasor Addition of V_C and V_R
In Fig. 22-2b, the current phasor is shown horizontal, as the reference phase, because I is the same throughout the series circuit. The resistive voltage drop IR has the same phase as I. The capacitor voltage IX_C must be 90° clockwise from I and IR, as the capacitive voltage lags.

Note that the IX_C phasor is downward, exactly opposite from an IX_L phasor, because of the opposite phase

angle. The phasor voltages V_R and V_C, being 90° out of phase, still form a right triangle. Therefore

$$V_T = \sqrt{V_R^2 + V_C^2} \qquad (22\text{-}1)$$

This formula applies just to series circuits because then V_C is 90° out of phase with V_R. All the voltages must be in the same units. When V_R and V_C are rms values, then V_T is an rms value.

In calculating the value of V_T, first square V_R and V_C, then add and take the square root. For the example in Fig. 22-2,

$$
\begin{aligned}
V_T &= \sqrt{100^2 + 100^2} \\
&= \sqrt{10,000 + 10,000} \\
&= \sqrt{20,000} \\
V_T &= 141 \text{ V}
\end{aligned}
$$

The two phasor voltages total 141 V instead of 200 V because the 90° phase means the peak value of one occurs when the other is at zero.

Phasor Addition of X_C and R
The voltage triangle in Fig. 22-2b corresponds to the impedance triangle in Fig. 22-2c because the common factor I can be canceled with the same current in X_C and R. Their phasor sum is the combined impedance

$$Z = \sqrt{R^2 + X_C^2} \qquad (22\text{-}2)$$

With R and X_C in ohms, Z is also in ohms. For the example in Fig. 22-2c, the values are

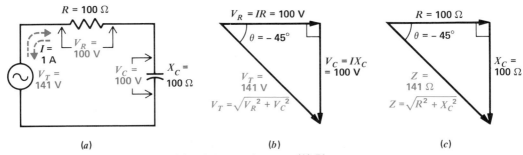

(a) (b) (c)

Fig. 22-2 Circuit with X_C and R in series. (a) Schematic diagram. (b) Phasor triangle of voltages, showing V_C lagging V_R by −90°. (c) Similar impedance triangle, showing X_C lagging R by −90°.

$$Z = \sqrt{100^2 + 100^2}$$
$$= \sqrt{10,000 + 10,000}$$
$$= \sqrt{20,000}$$
$$Z = 141 \ \Omega$$

Note that the total impedance of 141 Ω divided into the applied voltage of 141 V allows the current of 1 A in the series circuit. The IR voltage drop is 1×100, or 100 V; the IX_C voltage drop is also 1×100, or 100 V.

The phasor sum of the two series voltage drops of 100 V each equals the applied voltage of 141 V. Also, the applied voltage is equal to $I \times Z$, or 1×141, which is 141 V for V_T.

Phase Angle with Series X_C

As with inductive reactance, θ is the phase angle between the generator voltage and its series current. As shown in Fig. 22-2b and c, the θ can be calculated from the voltage or impedance triangle.

With series X_C, the phase angle is negative, clockwise from the zero reference angle of I, because the X_C voltage lags its current. To indicate the negative phase angle, therefore, this 90° phasor points downward from the horizontal reference, instead of upward as with series inductive reactance. To calculate the phase angle with series X_C and R,

$$\tan \theta = \frac{-X_C}{R} \qquad (22\text{-}3)$$

Using this formula for the circuit in Fig. 22-2c,

$$\tan \theta = \frac{-X_C}{R}$$
$$= -\frac{100}{100} = -1$$
$$\theta = -45°$$

The negative sign means the angle is clockwise from zero, to indicate that V_T lags behind the leading I.

Series Combinations of X_C and R

In series, the higher the X_C compared with R, the more capacitive the circuit. There is more voltage drop across the ca-

Table 22-1. Series R and X_C Combinations

R, Ω	X_C, Ω	Z, Ω (approx.)	Phase angle θ_Z
1	10	$\sqrt{101} = 10$	$-84.3°$
10	10	$\sqrt{200} = 14$	$-45°$
10	1	$\sqrt{101} = 10$	$-5.7°$

Note: θ_Z is angle of Z_T or V_T with respect to the reference I in series circuits.

pacitive reactance, and the phase angle increases toward $-90°$. The series X_C always makes the current lead the applied voltage. With all X_C and no R, the entire applied voltage is across X_C, and θ equals $-90°$.

Several combinations of X_C and R in series are listed in Table 22-1, with their resultant impedance values and phase angle. Note that a ratio of 10:1, or more, for X_C/R means the circuit is practically all capacitive. The phase angle of $-84.3°$ is almost $-90°$, and the total impedance Z is approximately equal to X_C. The voltage drop across X_C in the series circuit is then practically equal to the applied voltage, with almost none across the R.

At the opposite extreme, when R is ten times more than X_C, the series circuit is mainly resistive. The phase angle of $-5.7°$ then means the current has almost the same phase as the applied voltage; Z is approximately equal to R, and the voltage drop across R is practically equal to the applied voltage with almost none across the X_C.

For the case when X_C and R equal each other, the resultant impedance Z is 1.41 times either one. The phase angle then is $-45°$, halfway between 0° for resistance alone and $-90°$ for capacitive reactance alone.

Practice Problems 22-2
Answers at End of Chapter

a. How much is Z_T for a 20-Ω R in series with a 20-Ω X_C?

b. How much is V_T for 20 V across R and 20 V across X_C in series?

c. What is the phase angle of this circuit?

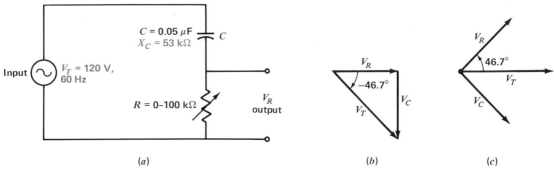

Fig. 22-3 The RC phase-shifter circuit. (a) Schematic diagram. (b) Phasor triangle with IR or V_R as the horizontal reference. Voltage V_R leads the applied V_T by $-46.7°$, with R set at 50 kΩ. (c) Phasors shown with V_T as the horizontal reference.

22-3   *RC* Phase-Shifter Circuit

Figure 22-3 shows an application of X_C and R in series for the purpose of providing a desired phase shift in the output V_R compared with the input V_T. The R can be varied up to 100 kΩ to change the phase angle. The C is 0.5 μF here for the 60-Hz ac power line voltage, but a smaller C would be used for a higher frequency. The capacitor must have an appreciable value of reactance for the phase shift.

For the circuit in Fig. 22-3a, assume that R is set for 50 kΩ, at its middle value. The reactance for the 0.05-μF capacitor at 60 Hz is approximately 53 kΩ. For these values of X_C and R, the phase angle of the circuit is $-46.7°$. This angle has the tangent of $-53/50 = -1.06$.

The phasor triangle in Fig. 22-3b shows that IR or V_R is out of phase with V_T by the leading angle or 46.7°. Note that V_C is always 90° lagging V_R in a series circuit. The angle between V_C and V_T then becomes $90° - 46.7° = 43.3°$.

The purpose of this circuit is to provide a phase-shifted voltage V_R in the output, with respect to the input. For this reason, the phasors are redrawn in Fig. 22-3c to show the voltages with the input V_T as the horizontal reference. The conclusion, then, is that the output voltage across R leads the input V_T by 46.7°.

Now let R be varied for a higher value at 90 kΩ, while X_C stays the same. The phase angle becomes $-30.5°$. This angle has the tangent $-53/90 = -0.59$. As a result, V_R leads V_T by 30.5°.

For the opposite case, let R be reduced to 10 kΩ. Then the phase angle becomes $-79.3°$. This angle has the tangent $-53/10 = -5.3$. Then V_R leads V_T by 79.3°. Notice that the phase angle becomes larger as the series circuit becomes more capacitive with less resistance.

A practical application for this circuit is to provide a voltage of variable phase to set the conduction time of semiconductors in power-control circuits. As R is varied, the phase of the output V_R is varied with respect to the power-line voltage V_T.

Practice Problems 22-3
Answers at End of Chapter

In Fig. 22-3, give the phase angle between
a. V_R and V_T.
b. V_R and V_C.
c. V_C and V_T.

22-4   *X$_C$* and *R* in Parallel

Now the 90° phase angle for X_C must be with respect to branch currents instead of voltage drops in a series circuit. In the parallel circuit in Fig. 22-4a, the voltage is the same across X_C, R, and the generator, since they are all in parallel. There cannot be any phase difference between the parallel voltages.

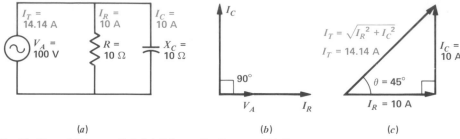

Fig. 22-4 Circuit with X_C and R in parallel. (a) Schematic diagram. (b) Current phasors showing I_C leading V_A by 90°. (c) Phasor triangle of branch circuits I_C and I_R to calculate total line current I_T.

Each branch, however, has its individual current. For the resistive branch, I_R is V_A/R; in the capacitive branch, $I_C = V_A/X_C$. These current phasors are shown in Fig. 22-4b.

Note that the phasor diagram has the generator voltage V_A as the reference phase because it is the same throughout the circuit. The resistive branch current I_R has the same phase as V_A, but the capacitive branch current I_C leads V_A by 90°.

The phasor for I_C is up, compared with down for an X_C phasor, because the parallel branch current I_C leads the reference V_A. This I_C phasor for a parallel branch current is opposite from an X_C phasor.

The total line current I_T then consists of I_R and I_C 90° out of phase with each other. The phasor sum of I_R and I_C equals I_T. As a result, the formula is

$$I_T = \sqrt{I_R^2 + I_C^2} \qquad (22\text{-}4)$$

In Fig. 22-2c, the phasor sum of 10 A for I_R and 10 A for I_C equals 14.14 A. The branch currents are added by phasors since they are the factors 90° out of phase in a parallel circuit, corresponding to the voltage drops 90° out of phase in a series circuit.

Impedance of X_C and R in Parallel   As usual, the impedance of a parallel circuit equals the applied voltage divided by the total line current: $Z = V_A/I_T$. In Fig. 22-4, for example,

$$Z = \frac{V_A}{I_T} = \frac{100}{14.14 \text{ A}} = 7.07 \text{ } \Omega$$

which is the opposition in ohms across the generator. This Z of 7.07 Ω is equal to the resistance of 10 Ω in

parallel with the reactance of 10 Ω. Notice that the impedance of equal values of R and X_C is not one-half but equals 70.7 percent of either one.

Phase Angle in Parallel Circuits
22-4c, the phase angle θ is 45° because R and X_C are equal, resulting in equal branch currents. The phase angle is between the total current I_T and the generator voltage V_A. However, the phase of V_A is the same as the phase of I_R. Therefore θ is also between I_T and I_R.

Using the tangent formula to find θ from the current triangle in Fig. 22-4c gives

$$\tan \theta_I = \frac{I_C}{I_R} \qquad (22\text{-}5)$$

The phase angle is positive because the I_C phasor is upward, leading V_A by 90°. This direction is opposite from the lagging phasor of series X_C. The effect of X_C is no different, however. Only the reference is changed for the phase angle.

Note that the phasor triangle of branch currents for parallel circuits gives θ_I as the angle of I_T with respect to the generator voltage V_A. This phase angle for I_T is labeled θ_I with respect to the applied voltage. For the phasor triangle of voltages in a series circuit, the phase angle for Z_T and V_T is labeled θ_Z with respect to the series current.

Parallel Combinations of X_C and R   In Table 22-2, when X_C is ten times R, the parallel circuit is practically resistive because there is little leading capacitive current in the main line. The small value of I_C results from the high reactance of shunt X_C. Then the

Table 22-2. Parallel Resistance and Capacitance Combinations*

R, Ω	X_C, Ω	I_R, A	I_C, A	I_T, A (approx.)	Z_T, Ω (approx.)	Phase angle θ_I
1	10	10	1	$\sqrt{101} = 10$	1	5.7°
10	10	1	1	$\sqrt{2} = 1.4$	7.07	45°
10	1	1	10	$\sqrt{101} = 10$	1	84.3°

*$V_A = 10$ V. Note that θ_I is angle of I_T with respect to the reference V_A in parallel circuits.

total impedance of the parallel circuit is approximately equal to the resistance, since the high value of X_C in a parallel branch has little effect. The phase angle of 5.7° is practically 0° because almost all the line current is resistive.

As X_C becomes smaller, it provides more leading capacitive current in the main line. When X_C is $\frac{1}{10}$ R, practically all the line current is the I_C component. Then, the parallel circuit is practically all capacitive, with a total impedance practically equal to X_C. The phase angle of 84.3° is almost 90° because the line current is mostly capacitive. Note that these conditions are opposite to the case of X_C and R in series. With X_C and R equal, their branch currents are equal and the phase angle is 45°.

As additional comparisons between series and parallel circuits, remember that

1. The series voltage drops V_R and V_C have individual values that are 90° out of phase. Therefore, V_R and V_C are added by phasors to equal the applied voltage V_T. The negative phase angle $-\theta_Z$ is between V_T and the common series current I. More series X_C allows more V_C to make the circuit more capacitive, with a larger negative phase angle for V_T with respect to I.

2. The parallel branch currents I_R and I_C have individual values that are 90° out of phase. Therefore, I_R and I_C are added by phasors to equal I_T, which is the main-line current. The positive phase angle θ_I is between the line current I_T and the common parallel voltage V_A. Less parallel X_C allows more I_C to make the circuit more capacitive, with a larger positive phase angle for I_T with respect to V_A.

Practice Problems 22-4
Answers at End of Chapter

a. How much is I_T for branch currents I_R of 2 A and I_C of 2 A?

b. Find the phase angle θ_I.

22-5   RF and AF Coupling Capacitors

In Fig. 22-5, C_C is used in the application of a coupling capacitor. Its low reactance allows practically all the ac signal voltage of the generator to be developed across R. Very little of the ac voltage is across C_C.

The coupling capacitor is used for this application because at lower frequencies it provides more reactance, resulting in less ac voltage coupled across R and more across C_C. For dc voltage, all the voltage is across C with none across R, since the capacitor blocks direct current. As a result, the output signal voltage across R includes the desired higher frequencies but not direct current or very low frequencies. This application of C_C, therefore, is called *ac coupling*.

The dividing line for C_C to be a coupling capacitor at a specific frequency can be taken as X_C one-tenth or less of the series R. Then the series RC circuit is pri-

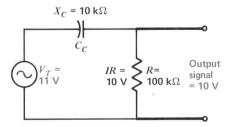

Fig. 22-5 Series circuit for *RC* coupling. Small X_C compared with R allows practically all the applied voltage across R for output, with little across C.

Table 22-3. Coupling Capacitors with a Reactance of 1600 Ω*

f	C_C	Remarks
100 Hz	1 μF	Low audio frequencies
1000 Hz	0.1 μF	Medium audio frequencies
10 kHz	0.01 μF	High audio frequencies
1000 kHz	100 pF	Radio frequencies
100 MHz	1 pF	Very high frequencies

*For an X_C one-tenth of a series R of 16,000 Ω.

marily resistive. Practically all the voltage drop of the ac generator is across R, with little across C. In addition, this case results in a phase angle of practically 0°.

Typical values of a coupling capacitor for audio or radio frequencies can be calculated if we assume a series resistance of 16,000 Ω. Then X_C must be 1600 Ω or less. Typical values for C_C are listed in Table 22-3. At 100 Hz, a coupling capacitor must be 1 μF to provide 1600 Ω of reactance. Higher frequencies allow a smaller value of C_C for a coupling capacitor having the same reactance. At 100 MHz in the VHF range the required capacitance is only 1 pF.

It should be noted that the C_C values are calculated for each frequency as a lower limit. At higher frequencies, the same size C_C will have less reactance than one-tenth of R, which improves the coupling.

Choosing a Coupling Capacitor for a Circuit

As an example of using these calculations, suppose that we have the problem of determining C_C for a transistorized audio amplifier. This application also illustrates the relatively large capacitance needed with low series resistance. The C is to be a coupling capacitor for audio frequencies of 50 Hz and up, with a series R of 4000 Ω. Then the required X_C is 4000/10, or 400 Ω. To find C at 50 Hz,

$$C = \frac{0.159}{f \times X_C} = \frac{0.159}{50 \times 400}$$
$$= \frac{159,000 \times 10^{-6}}{20 \times 10^3}$$
$$= 7.95 \times 10^{-6}$$
$$C = 7.95 \ \mu F$$

A typical commercial size of low-voltage electrolytic readily available is 10 μF. The slightly higher capacitance value is better for coupling. The voltage rating can be 3 to 10 V, depending on the circuit, with a typical transistor supply voltage of 9 V. Although electrolytic capacitors have relatively high leakage current, they can be used for coupling capacitors in this application because of the low series resistance.

Practice Problems 22-5
Answers at End of Chapter

a. The X_C of a coupling capacitor is 70 Ω at 200 Hz. How much is its X_C at 400 Hz?

b. From Table 22-3, what C would be needed for 1600 Ω of X_C at 50 MHz?

22-6   Capacitive Voltage Dividers

When capacitors are connected in series across a voltage source, the series capacitors serve as a voltage divider. Each capacitor has part of the applied voltage, and the sum of all the series voltage drops equals the source voltage.

The amount of voltage across each is inversely proportional to its capacitance. For instance, with 2 μF in series with 1 μF, the smaller capacitor has double the voltage of the larger capacitor. Assuming 120 V applied, one-third of this, or 40 V, is across the 2-μF capacitor, with two-thirds, or 80 V, across the 1-μF capacitor.

The two series voltage drops of 40 and 80 V add to equal the applied voltage of 120 V. The addition is just the arithmetic sum of the two voltages. It is only when voltages are out of phase with each other that phasor addition becomes necessary.

AC Divider   With sine-wave alternating current, the voltage division between series capacitors can be calculated on the basis of reactance. In Fig. 22-6a, the total reactance is 120 Ω across the 120-V source. The current in the series circuit then is 1 A. This current is the same for X_{C_1} and X_{C_2} in series. Therefore, the IX_C voltage across C_1 is 40 V, with 80 V across C_2.

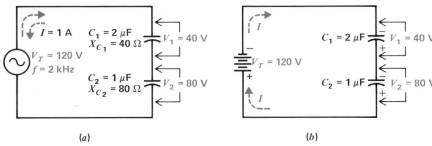

Fig. 22-6 Series capacitors divide V_T inversely proportional to each C. The smaller C has more V. (a) An ac divider. (b) A dc divider.

The voltage division is proportional to the series reactances, as it is to series resistances. However, reactance is inversely proportional to capacitance. As a result, the smaller capacitance has more reactance and a greater part of the applied voltage.

DC Divider   In Fig. 22-6b, both C_1 and C_2 will be charged by the battery. The voltage across the series combination of C_1 and C_2 must equal V_T. When charging current flows, electrons repelled from the negative battery terminal accumulate on the negative plate of C_1, repelling electrons from its positive plate. These electrons flow through the conductor to the negative plate of C_2. With the positive battery terminal attracting electrons, the charging current from the positive plate of C_2 returns to the positive side of the dc source. Then C_1 and C_2 become charged in the polarity shown.

Since C_1 and C_2 are in the same series path for charging current, both have the same amount of charge. However, the potential difference provided by the equal charges is inversely proportional to capacitance. The reason is that $Q = CV$, or $V = Q/C$. Therefore, the 1-μF capacitor has double the voltage of the 2-μF capacitor, with the same charge in both.

If you measure with a dc voltmeter across C_1, the meter reads 40 V. Across C_2 the dc voltage is 80 V. The measurement from the negative side of C_1 to the positive side of C_2 is the same as the applied battery voltage of 120 V.

If the meter is connected from the positive side of C_1 to the negative plate of C_2, however, the voltage is zero. These plates have the same potential because they are joined by a conductor of zero resistance.

The polarity marks at the junction between C_1 and C_2 indicate the voltage at this point with respect to the opposite plate of each capacitor. This junction is positive compared with the opposite plate of C_1 with a surplus of electrons. However, the same point is negative compared with the opposite plate of C_2, which has a deficiency of electrons.

In general, the following formula can be used for capacitances in series as a voltage divider:

$$V_C = \frac{C_T}{C} \times V_T \qquad (22\text{-}6)$$

Note that C_T is in the numerator, since it must be less than the smallest individual C with series capacitances. For the divider examples in Fig. 22-6a and b,

$$V_1 = \frac{C_T}{C_1} \times 120 = \frac{\frac{2}{3}}{2} \times 120 = 40 \text{ V}$$

$$V_2 = \frac{C_T}{C_2} \times 120 = \frac{\frac{2}{3}}{1} \times 120 = 80 \text{ V}$$

This method applies to series capacitances as a divider for either dc or ac voltage, as long as there is no series resistance.

Practice Problems 22-6
Answers at End of Chapter

a. Capacitance C_1 of 10 pF and C_2 of 90 pF are across 20 kV. Calculate V_1 and V_2.
b. In Fig. 22-6a, how much is X_{C_T}?

22-7   The General Case of Capacitive Current i_C

The capacitive charge and discharge current i_C is always equal to $C(dv/dt)$. A sine wave of voltage variations for v_C produces a cosine wave of current i. This means v_C and i_C have the same waveform, but they are 90° out of phase.

It is usually convenient to use X_C for calculations in sine-wave circuits. Since X_C is $1/(2\pi fC)$, the factors that determine the amount of charge and discharge current are included in f and C. Then I_C equals V_C/X_C. Or, if I_C is known, V_C can be calculated as $I_C \times X_C$.

With a nonsinusoidal waveform for voltage v_C, the concept of reactance cannot be used. Reactance X_C applies only to sine waves. Then i_C must be determined as $C(dv/dt)$. An example is illustrated in Fig. 22-7 to show the change of waveform here, instead of the change of phase angle in sine-wave circuits.

Note that the sawtooth waveform of voltage v_C corresponds to a rectangular waveform of current. The linear rise of the sawtooth wave produces a constant amount of charging current i_C because the rate of change is constant for the charging voltage. When the

capacitor discharges, v_C drops sharply. Then discharge current is in the opposite direction from charge current. Also, the discharge current has a much larger value because of the faster rate of change in v_C.

An interesting feature of these capacitive waveshapes is the fact that they are the same as the inductive waveshapes shown before in Fig. 19-11 on page 381, but with the current and voltage waveshapes interchanged. This comparison follows from the fact that both i_C and v_L depend on rate of change.

Practice Problems 22-7
Answers at End of Chapter

a. In Fig. 22-7a, how much is dv/dt in V/s for the sawtooth rise from 0 to 90 V in 90 μs?

b. How much is the charge current i_C, as $C(dv/dt)$ for this dv/dt?

22-8   Calculating the RC Time Constant

With nonsinusoidal waveforms, the transient response of C is measured by its RC time constant. As a formula

$$T = R \times C \qquad \text{s} \qquad (22\text{-}7)$$

where T is the time constant, C the capacitance in farads, and R the resistance in series with C for charge or discharge. As an example, for 1000 Ω of R in series with 4 μF of C,

$$T = 1000 \times 4 \times 10^{-6}$$
$$= 4000 \times 10^{-6}$$
$$= 4 \times 10^{-3}$$
$$T = 4 \text{ ms}$$

The time constant T is the time for the voltage across C to change by 63.2 percent. For instance, if the applied voltage is 100 V, then C will charge to 63.2 V in 4 ms for this example. The capacitor is completely charged in five time constants.

The RC time constant for capacitor voltage is similar to the L/R time constant for inductive current. In each case, the change is 63.2 percent in one time constant.

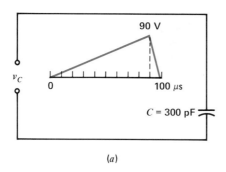

(a)

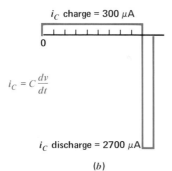

(b)

Fig. 22-7 Waveshape of i_C equal to $C(dv/dt)$. (a) Sawtooth waveform of v_C. (b) Rectangular current waveform of i_C.

Actually, the RC time constant is much more important in its applications than the L/R time constant. The reason is that capacitors and resistors are small and economical to provide almost any value of RC time constant, without any coil problems. The applications of the RC time constant are explained in more detail in the next chapter, along with a comparison to the L/R time constant.

Practice Problems 22-8
Answers at End of Chapter

a. How much is the RC time constant for 2 MΩ in series with 2 μF for charge?
b. How much is the RC time constant for 100 Ω in series with 2 μF for discharge?

Summary

1. In a sine-wave ac circuit, the voltage across a capacitance lags its charge and discharge current by 90°.
2. Therefore, capacitive reactance X_C is a phasor quantity out of phase with its series resistance by $-90°$ because $i_C = C(dv/dt)$. This fundamental fact is the basis of all the following relations.
3. The vector combination of X_C and R in series is their impedance Z. These three types of ohms of opposition to current are compared in Table 22-4.

Table 22-4. Comparison of R, X_C, and Z

R	$X_C = 1/(2\pi fC)$	$Z = \sqrt{R^2 + X_C^2}$
Ohms unit	Ohms unit	Ohms unit
IR voltage same phase as I	IX_C voltage lags I_C by 90°	IZ is the applied voltage
Same for all f	Decreases for higher f	Decreases with X_C

4. The opposite characteristics for series and parallel circuits with X_C and R are summarized in Table 22-5.

Table 22-5. Series and Parallel RC Circuits

X_C and R in series	X_C and R in parallel
I the same in X_C and R	V the same across X_C and R
$V_T = \sqrt{V_R^2 + V_C^2}$	$I_T = \sqrt{I_R^2 + I_C^2}$
$Z = \sqrt{R^2 + X_C^2}$	$Z = \dfrac{V}{I_T}$
V_C lags V_R by 90°	I_C leads I_R by 90°
$\tan \theta_Z = -\dfrac{X_C}{R}$; θ_Z increases as	$\tan \theta_I = \dfrac{I_C}{I_R}$; θ_I decreases as
more X_C means more V_C	more X_C means less I_C

5. In a comparison of capacitive and inductive circuits, I_L always lags V_L, but I_C leads V_C.

6. Two or more capacitors in series across a voltage source serve as a voltage divider. The smallest C has the largest part of the applied voltage.

7. A coupling capacitor has X_C less than its series resistance by the factor of $\frac{1}{10}$ or less, for the purpose of providing practically all the ac applied voltage across R with little across C.

8. In sine-wave circuits, $I_C = V_C/X_C$. Then I_C has a phase angle of 90° compared with V_C.

9. When the voltage is not a sine wave, $i_C = C(dv/dt)$. Then the waveshape of i_C is different from the voltage.

10. The RC time constant T in seconds is the product of C in farads and the series R in ohms. T is the time for a 63.2 percent change in V_C on charge or discharge.

Self-Examination
Answers at Back of Book

Choose (a), (b), (c), or (d).

1. In a capacitive circuit (a) a decrease in applied voltage makes a capacitor charge; (b) a steady value of applied voltage causes discharge; (c) an increase in applied voltage makes a capacitor discharge; (d) an increase in applied voltage makes a capacitor charge.

2. In a sine-wave ac circuit with X_C and R in series, the (a) phase angle of the circuit is 180° with high series resistance; (b) voltage across the capacitance must be 90° out of phase with its charge and discharge current; (c) voltage across the capacitance has the same phase as its charge and discharge current; (d) charge and discharge current of the capacitor must be 90° out of phase with the applied voltage.

3. When v_C across a 1-μF C drops from 43 to 42 V in 1 s, the discharge current i_C equals (a) 1 μA; (b) 42 μA; (c) 43 μA; (d) 43 A.

4. In a sine-wave ac circuit with R and C in parallel, (a) the voltage across C lags the voltage across R by 90°; (b) resistive I_R is 90° out of phase with I_C; (c) I_R and I_C have the same phase; (d) I_R and I_C are 180° out of phase.

5. In a sine-wave ac circuit with a 90-Ω R in series with a 90-Ω X_C, the phase angle equals (a) −90°; (b) −45°; (c) 0°; (d) 90°.

6. The combined impedance of a 1000-Ω R in parallel with a 1000-Ω X_C equals (a) 500 Ω; (b) 707 Ω; (c) 1000 Ω; (d) 2000 Ω.

7. With 100 V applied across two series capacitors of 5 μF each, the voltage across each capacitor will be (a) 5 V; (b) 33⅓ V; (c) 50 V; (d) 66⅔ V.

8. In a sine-wave ac circuit with X_C and R in series, the (a) voltages across R and X_C are in phase; (b) voltages across R and X_C are 180° out of phase; (c) voltage across R leads the voltage across X_C by 90°; (d) voltage across R lags the voltage across X_C by 90°.

9. A 0.01-μF capacitance in series with R is used as a coupling capacitor C_C for 1000 Hz. At 10,000 Hz: (a) C_C has too much reactance to be good for coupling; (b) C_C has less reactance, which improves the coupling; (c) C_C has the same reactance and coupling; (d) the voltage across R is reduced by one-tenth.

10. In an RC coupling circuit the phase angle is (a) 90°; (b) close to 0°; (c) −90°; (d) 180°.

Essay Questions

1. (a) Why does a capacitor charge when the applied voltage increases? (b) Why does the capacitor discharge when the applied voltage decreases?
2. A sine wave of voltage V is applied across a capacitor C. (a) Draw the schematic diagram. (b) Draw the sine waves of voltage and current out of phase by 90°. (c) Draw a vector diagram showing the phase angle of −90° between V and I.
3. Why will a circuit with R and X_C in series be less capacitive as the frequency of the applied voltage is increased?
4. Define the following: coupling capacitor, sawtooth voltage, capacitive voltage divider.
5. Give two comparisons between RC circuits with sine-wave voltage applied and nonsinusoidal voltage applied.
6. Give three differences between RC circuits and RL circuits.
7. Compare the functions of a coupling capacitor with a choke coil, with two differences in their operation.
8. State two troubles possible in coupling capacitors and describe briefly how you would check with an ohmmeter.
9. Define the time constant of a capacitive circuit.
10. Explain the function of R and C in an RC coupling circuit for ac signal from one transistor amplifier to the next stage.

Problems

Answers to Odd-Numbered Problems at Back of Book

1. A 40-Ω R is in series with a 30-Ω X_C across a 100-V sine-wave ac source. (a) Draw the schematic diagram. (b) Calculate Z. (c) Calculate I. (d) Calculate the voltages across R and C. (e) What is the phase angle of the circuit?
2. A 40-Ω R and a 30-Ω X_C are in parallel across a 100-V sine-wave ac source. (a) Draw the schematic diagram. (b) Calculate each branch current. (c) How much is I_T? (d) Calculate Z. (e) What is the phase angle of the circuit? (f) Compare the phase of the voltage across R and X_C.
3. Draw the schematic diagram of a capacitor in series with a 20-kΩ resistance across a 10-V ac source. What size C is needed for equal voltages across R and X_C at frequencies of 100 Hz and 100 kHz?
4. Draw the schematic diagram of two capacitors C_1 and C_2 in series across 10,000 V. The C_1 is 900 pF and has 9000 V across it. (a) How much is the voltage across C_2? (b) How much is the capacitance of C_2?
5. In Fig. 22-2a, how much is C for the X_C value of 100 Ω at frequencies of 60 Hz, 1000 Hz, and 1 MHz?

6. A 1500-Ω R is in series with a 0.01-μF C across a 30-V source with a frequency of 8 kHz. Calculate X_C, Z_T, θ_Z, I, V_R, and V_C.

7. The same R and C as in Prob. 6 are in parallel. Calculate I_C, I_R, I_T, θ_I, Z, V_R, and V_C.

8. A 0.05-μF capacitor is in series with a 50,000-Ω R and a 10-V source. Tabulate the values of X_C, I, V_R, and V_C at the frequencies of 0 (for steady dc voltage), 20, 60, 100, 500, 5000, and 15,000 Hz.

9. A capacitive voltage divider has C_1 of 1 μF, C_2 of 2 μF, and C_3 of 4 μF in series across a 700-V source V_T. (a) Calculate V_1, V_2, and V_3 for a steady dc source. (b) Calculate V_1, V_2, and V_3 for an ac source with a frequency of 400 Hz.

10. (a) A 40-Ω X_C and a 30-Ω R are in series across a 120-V source. Calculate Z_T, I, and θ_Z. (b) The same X_C and R are in parallel. Calculate I_T, Z, and θ_I. (c) A 40-Ω X_L and a 30-Ω R are in series across a 120-V source. Calculate Z_T, I, and θ_Z. (d) The same X_L and R are in parallel. Calculate I_T, Z, and θ_I. [Note that capacitive reactance X_C is used in (a) and (b) compared with inductive reactance X_L in (c) and (d).]

11. Calculate the values for L and C in Prob. 10, with a frequency of 60 Hz for the source voltage.

12. A 500-Ω R is in series with 300-Ω X_C. Find Z_T, I, and θ_Z. $V_T = 120$ V.

13. A 300-Ω R is in series with a 500-Ω X_C. Find Z_T, I, and θ_Z. Compare θ_Z here with Prob. 12, with the same 120 V applied.

14. A 500-Ω R is parallel with a 300-Ω X_C. Find I_T, Z_T, and θ_I. Compare θ_I here with θ_Z in Prob. 12, with the same 120 V applied.

15. For the waveshape of capacitor voltage v_C in Fig. 22-8, show the corresponding charge and discharge current i_C, with values for a 200-pF capacitance. Compare these waveshapes with Prob. 17 in Chap. 19.

16. Calculate the values needed in Fig. 22-6a for the same voltage division but with a frequency of 60 Hz for V_T.

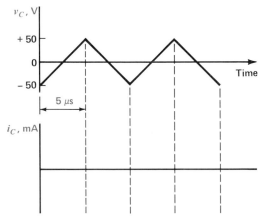

Fig. 22-8 For Prob. 15.

Answers to Practice Problems

22-1 **a.** 0°
    b. 90°
    c. Lag

22-2 **a.** 28.28 Ω
    b. 28.28 V
    c. $\theta_Z = -45°$

22-3 **a.** 46.7°
    b. 90°
    c. 43.3°

22-4 **a.** 2.828 A
    b. $\theta_I = 45°$

22-5 **a.** 35 Ω
    b. 2 pF

22-6 **a.** $V_1 = 18$ kV
      $V_2 = 2$ kV
    b. $X_{C_T} = 120$ Ω

22-7 **a.** $dv/dt = 1 \times 10^6$ V/s
    b. $i_C = 300$ μA

22-8 **a.** 4 s
    b. 200 μs

Chapter 23
RC and *L/R* Time Constants

Many applications of inductance are for sine-wave ac circuits, but any time the current changes, *L* has the effect of producing induced voltage. Examples of nonsinusoidal waveshapes include dc voltages that are switched on or off, square waves, sawtooth waves, and rectangular pulses. For capacitance, also, many applications are for sine waves, but any time the voltage changes, *C* produces charge or discharge current.

With nonsinusoidal voltage and current, the effect of *L* or *C* is to produce a change in waveshape. This effect can be analyzed by means

of the time constant for *RC* and *RL* circuits. The time constant is the time for a change of 63.2 percent in the current through *L* or the voltage across *C*.

Important terms in this chapter are:

collapsing magnetic field	rate of charge or discharge
decay	*RC* time constant
differentiation	short time constant
energy stored	steady-state *I* or *V*
integration	transient response
L/R time constant	voltage pulses
long time constant	waveshapes

More details are explained in the following sections:

23-1   Response of Resistance Alone

In order to emphasize the special features of *L* or *C*, the circuit in Fig. 23-1 illustrates how ordinary a resistive circuit is. When the switch is closed, the battery supplies 10 V across the 10-Ω *R* and the resultant *I* is 1 A. The graph in Fig. 23-1*b* shows that *I* changes from 0 to 1 A instantly when the switch is closed. If the applied voltage is changed to 5 V, the current will change instantly to 0.5 A. If the switch is opened, *I* will immediately drop to zero.

Resistance has only opposition to current; there is no reaction to a change. The reason is that *R* has no concentrated magnetic field to oppose a change in *I*, like inductance, and no electric field to store charge that opposes a change in *V*, like capacitance.

Practice Problems 23-1
Answers at End of Chapter

Answer True or False.
a. Resistance *R* does not produce induced voltage for a change in *I*.
b. Resistance *R* does not produce charge or discharge current for a change in *V*.

23-2   *L/R* Time Constant

Consider the circuit in Fig. 23-2 where *L* is in series with *R*. When *S* is closed, the current changes as *I* increases from zero. Eventually, *I* will have the steady value of 1 A, equal to the battery voltage of 10 V di-

vided by the circuit resistance of 10 Ω. While the current is building up from 0 to 1 A, however, *I* is changing and the inductance opposes the change. The action of the *RL* circuit during this time is its *transient response,* meaning a temporary condition existing only until the steady-state current of 1 A is reached. Similarly, when *S* is opened, the transient response of the *RL* circuit opposes the decay of current toward the steady-state value of zero.

The transient response is measured in terms of the ratio *L/R*, which is the time constant of an inductive circuit. To calculate the time constant

$$T = \frac{L}{R} \tag{23-1}$$

where *T* is the time constant in seconds and *L* the inductance in henrys. Resistance *R* is the ohms of resistance in series with *L*, being either the coil resistance, an external resistance, or both in series. In Fig. 23-2,

$$T = \frac{L}{R} = \frac{1}{10} = 0.1 \text{ s}$$

Specifically, the time constant is a measure of how long it takes the current to change by 63.2 percent, or approximately 63 percent. In Fig. 23-2, the current increases from 0 to 0.63 A, which is 63 percent of the steady-state value, in the period of 0.1 s, which is the time constant. In the period of five time constants, the current is practically equal to its steady-state value of 1 A.

If the switch is opened now so that the current can decay to zero, *I* will decrease to 36.8 percent, or approximately 37 percent, of the steady-state value in one time constant. For the example in Fig. 23-2, *I* will decay from 1 to 0.37 A in one time constant. Note that the decrease to 0.37 A from 1 A is a change of 63 percent. The current decays practically to zero in five time constants.

The reason why *L/R* equals time can be illustrated as follows: Since induced voltage *V* = *L* (*di/dt*), by transposing terms, *L* has the dimensions of *V* × *T/I*. Dividing *L* by *R* results in *V* × *T/IR*. As the *IR* and *V* factors cancel, *T* remains to indicate the dimension of time for the ratio *L/R*.

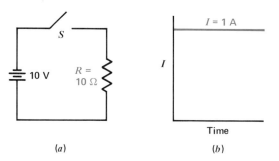

Fig. 23-1 Response of circuit with *R* alone. When switch is closed, *I* = 1 A. (*a*) Circuit. (*b*) Graph of steady *I*.

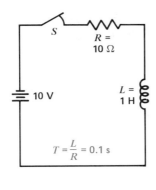

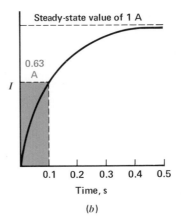

(a)

(b)

Fig. 23-2 Transient response of circuit with *R* and *L*. When switch is closed, *I* rises from zero to the steady-state 1 A. (a) Circuit with time constant of 0.1 s. (b) Graph of *I* during five time constants.

Example 1   What is the time constant of a 20-H coil having 100 Ω of series resistance?

Answer $T = \dfrac{L}{R} = \dfrac{20\ \text{H}}{100\ \Omega}$

$T = 0.2\ \text{s}$

Example 2   An applied dc voltage of 10 V will produce a steady-state current of 100 mA in the 100-Ω coil of Example 1. How much is the current after 0.2 s? After 1 s?

Answer   Since 0.2 s is one time constant, *I* then is 63 percent of 100 mA, which equals 63 mA. After five time constants, or 1 s, the current will reach its steady-state value of 100 mA and remain at this value as long as the applied voltage stays at 10 V.

Example 3   If a 1-MΩ *R* is added in series with the coil of Example 1, how much will the time constant be for the higher-resistance *RL* circuit?

Answer $T = \dfrac{L}{R} = \dfrac{20\ \text{H}}{1{,}000{,}000}$

$= 20 \times 10^{-6}\ \text{s}$

$T = 20\ \mu\text{s}$

The *L/R* time constant becomes longer with larger values of *L*. More series *R*, however, makes the time constant shorter. Then the circuit is less inductive, with more series resistance.

Practice Problems 23-2
Answers at End of Chapter

a. Calculate the time constant for 2 H in series with 100 Ω.
b. Calculate the time constant for 2 H in series with 4000 Ω.

23-3   High Voltage Produced by Opening *RL* Circuit

When an inductive circuit is opened, the time constant for current decay becomes very short because *L/R* becomes smaller with the high resistance of the open circuit. Then the current drops toward zero much faster than the rise of current when the switch is closed. The result is a high value of self-induced voltage V_L across a coil whenever an *RL* circuit is opened. This high voltage can be much greater than the applied voltage.

There is no gain in energy, though, because the high-voltage peak exists only for the short time the current is decreasing at a very fast rate at the start of the decay. Then, as *I* decays with a slower rate of change, the value of V_L is reduced. After the current has dropped to zero, there is no voltage across L.

This effect can be demonstrated by a neon bulb connected across the coil, as shown in Fig. 23-3. The neon bulb requires 90 V for ionization, at which time it glows. The source here is only 8 V, but when the switch is opened, the self-induced voltage is high enough to light the bulb for an instant. The sharp voltage pulse or spike is more than 90 V just after the switch is opened, when I drops very fast at the start of the decay in current.

Note that the 100-Ω R_1 is the internal resistance of the 2-H coil. This resistance is in series with L whether S is closed or open. The 4-kΩ R_2 across the switch is in the circuit only when S is opened, in order to have a specific resistance across the open switch. Since R_2 is much more than R_1, the L/R time constant is much shorter with the switch open.

Closing the Circuit
In Fig. 23-3a, the switch is closed to allow current in L and to store energy in the magnetic field. Since R_2 is short-circuited by the switch, the 100-Ω R_1 is the only resistance. The steady-state I is $V/R_1 = {}^8/_{100} = 0.08$ A. This value of I is reached after five time constants.

One time constant is $L/R = {}^2/_{100} = 0.02$ s. Five time constants equal $5 \times 0.02 = 0.1$ s. Therefore, I is 0.08 A after 0.1 s, or 100 ms. The energy stored in the magnetic field is 64×10^{-4} J, equal to $\frac{1}{2}LI^2$.

Opening the Circuit
When the switch is opened in Fig. 23-3b, R_2 is in series with L, making the total resistance 4100 Ω, or approximately 4 kΩ. The result

is a much shorter time constant for current decay. Then L/R is $^2/_{4000}$, or 0.5 ms. The current decays practically to zero in five time constants, or 2.5 ms.

This rapid drop in current results in a magnetic field collapsing at a fast rate, inducing a high voltage across L. The peak v_L in this example is 320 V. Then v_L serves as the voltage source for the bulb connected across the coil. As a result, the neon bulb becomes ionized, and it lights for an instant.

Calculating the Peak of v_L
The value of 320 V for the peak induced voltage when S is opened in Fig. 23-3 can be determined as follows: With the switch closed, I is 0.08 A in all parts of the series circuit. The instant S is opened, R_2 is added in series with L and R_1. The energy stored in the magnetic field maintains I at 0.08 A for an instant before the current decays. With 0.08 A in the 4-kΩ R_2 its potential difference is $0.08 \times 4000 = 320$ V. The collapsing magnetic field induces this 320-V pulse to allow an I of 0.08 A at the instant the switch is opened.

The di/dt for v_L
The required rate of change in current is 160 A/s for the v_L of 320 V induced by the L of 2 H. Since $v_L = L$ (di/dt), this formula can be transposed to specify di/dt as equal to v_L/L. Then di/dt corresponds to 320 V/2 H, or 160 A/s. This value is the actual di/dt at the start of the decay in current when the switch is opened in Fig. 23-3b, as a result of the short time constant.[1]

---

[1]The di/dt value can be calculated from the slope at the start of decay, shown by the dashed line for curve b in Fig. 23-9.

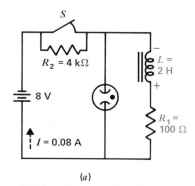

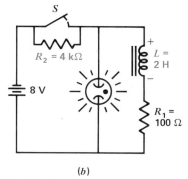

(a)   (b)

Fig. 23-3 Demonstration of high voltage produced by opening inductive circuit. (a) With switch closed, 8 V applied cannot light the 90-V neon bulb. (b) Short time constant when S is opened results in a high V_L that lights bulb.

Applications of Inductive Voltage Pulses

There are many uses of the high voltage generated by opening an inductive circuit. One example is the high voltage produced for the ignition system in an automobile. Here the circuit of the battery in series with a high-inductance spark coil is opened by the breaker points of the distributor to produce the high voltage needed for each spark plug. By opening an inductive circuit very rapidly, 10,000 V can easily be produced. Another important application is the high voltage of 10 to 30 kV for the anode of the picture tube in television receivers. One problem is that the high v_L produced when an inductive circuit is opened can cause arcing.

Practice Problems 23-3
Answers at End of Chapter

a. Is the L/R time constant longer or shorter in Fig. 23-3 when S is opened?
b. Which produces more v_L, a faster di/dt or a slower di/dt?

23-4   RC Time Constant

For capacitive circuits, the transient response is measured in terms of the product $R \times C$. To calculate the time constant

$$T = R \times C \tag{23-2}$$

with R in ohms and C in farads, T is in seconds. In Fig. 23-4, for example, with an R of 3 MΩ and a C of 1 μF,

$$T = 3 \times 10^6 \times 1 \times 10^{-6}$$
$$T = 3 \text{ s}$$

Note that the 10^6 for megohms and the 10^{-6} for microfarads cancel. Therefore, multiplying the units of MΩ × μF gives the RC product in seconds.

The reason why the RC product corresponds to time can be illustrated as follows: $C = Q/V$. The charge Q is the product of $I \times T$. The factor V is IR. Therefore, RC is equivalent to $(R \times Q)/V$, or $(R \times IT)/IR$. Since I and R cancel, T remains to indicate the dimension of time.

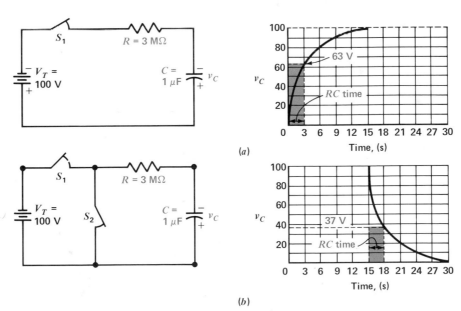

Fig. 23-4 The RC charge and discharge. (a) With S_1 closed, C charges through R to 63 percent of V_T in one time constant of 3 s. (b) With S_1 open and S_2 closed, C discharges through R. Then v_C drops to 37 percent of its initial voltage in one time constant of 3 s.

The Time Constant Indicates the Rate of Charge or Discharge

On charge, *RC* specifies the time it takes *C* to charge to 63 percent of the charging voltage. Similarly, on discharge, *RC* specifies the time it takes *C* to discharge 63 percent of the way down, to the value equal to 37 percent of the initial voltage across *C* at the start of discharge.

In Fig. 23-4*a*, for example, the time constant on charge is 3 s. Therefore, in 3 s, *C* charges to 63 percent of the 100 V applied, reaching 63 V in *RC* time. After five time constants, which is 15 s here, *C* is almost completely charged to the full 100 V applied. If *C* discharges after being charged to 100 V, then *C* will discharge down to 36.8 V or approximately 37 V in 3 s. After five time constants, *C* discharges down to zero.

A shorter time constant allows the capacitor to charge or discharge faster. If the *RC* product in Fig. 23-4 is 1 s, then *C* will charge to 63 V in 1 s instead of 3 s. Also, v_C will reach the full applied voltage of 100 V in 5 s instead of 15 s. Charging to the same voltage in less time means a faster charge.

On discharge also, the shorter time constant will allow *C* to discharge from 100 to 37 V in 1 s instead of 3 s. Also, v_C will be down to zero in the time of 5 s instead of 15 s.

For the opposite case, a longer time constant means slower charge or discharge of the capacitor. More *R* or *C* results in a longer time constant.

RC Applications

Several examples are given here to illustrate how the time constant can be applied to *RC* circuits.

Example 4 What is the time constant of a 0.01-μF capacitor in series with a 1-MΩ resistance?

Answer
$$T = R \times C$$
$$= 1 \times 10^6 \times 0.01 \times 10^{-6}$$
$$T = 0.01 \text{ s}$$

This is the time constant for charging or discharging, assuming the series resistance is the same for charge or discharge.

Example 5 With a dc voltage of 300 V applied, how much is the voltage across *C* in Example 4 after 0.01 s of charging? After 0.05 s? After 2 hours? After 2 days?

Answer Since 0.01 s is one time constant, the voltage across *C* then is 63 percent of 300 V, which equals 189 V. After five time constants, or 0.05 s, *C* will be charged practically to the applied voltage of 300 V. After 2 hours or 2 days *C* will still be charged to 300 V if the applied voltage is still connected.

Example 6 If the capacitor in Example 5 is allowed to charge to 300 V and then discharged, how much is the capacitor voltage 0.01 s after the start of discharge? The series resistance is the same on discharge as on charge.

Answer In one time constant *C* discharges to 37 percent of its initial voltage, or 0.37 × 300 V, which equals 111 V.

Example 7 If the capacitor in Example 5 is made to discharge after being charged to 200 V, how much will the voltage across *C* be 0.01 s later? The series resistance is the same on discharge as on charge.

Answer In one time constant *C* discharges to 37 percent of its initial voltage, or 0.37 × 200, which equals 74 V.

This example shows that the capacitor can charge or discharge from any voltage value, not just after one *RC* time constant or five *RC* time constants.

Example 8 If a 1-MΩ resistance is added in series with the capacitor in Example 4, how much will the time constant be?

Answer Now the series resistance is 2 MΩ. Therefore, *RC* is 2 × 0.01, or 0.02 s.

The *RC* time constant becomes longer with larger values of *R* and *C*. More capacitance means that the capacitor can store more charge. Therefore, it takes longer to store the charge needed to provide a potential difference equal to 63 percent of the applied voltage. More resistance reduces the charging current, requiring more time for charging the capacitor.

It should be noted that the *RC* time constant specifies just a rate. The actual amount of voltage across *C* depends upon the applied voltage as well as upon the *RC* time constant.

The capacitor takes on charge whenever its voltage is less than the applied voltage. The charging continues at the *RC* rate until either the capacitor is completely charged or the applied voltage decreases.

The capacitor discharges whenever its voltage is more than the applied voltage. The discharge continues at the *RC* rate until either the capacitor is completely discharged or the applied voltage increases.

To summarize these two important principles:

1. Capacitor *C* charges when the net charging voltage is more than v_C.
2. Capacitor *C* discharges when v_C is more than the net charging voltage.

The net charging voltage equals the difference between v_C and the applied voltage.

Practice Problems 23-4
Answers at End of Chapter

a. How much is the *RC* time constant for 470 pF in series with 2 MΩ on charge?
b. How much is the *RC* time constant for 470 pF in series with 1 kΩ on discharge?

23-5 *RC* Charge and Discharge Curves

In Fig. 23-4, the *RC* charge curve has the rise shown because the charging is fastest at the start, then tapers off as *C* takes on additional charge at a slower rate. As *C* charges, its potential difference increases. Then the difference in voltage between V_T and v_C is reduced. Less potential difference reduces the current that puts the charge in *C*. The more *C* charges, the more slowly it takes on additional charge.

Similarly, on discharge, *C* loses its charge at a slower rate. At first, v_C has its highest value and can produce maximum discharge current. With the discharge continuing, v_C goes down and there is less discharge current. The more *C* discharges, the more slowly it can lose the remainder of its charge.

Charge and Discharge Current There is often the question of how current can flow in a capacitive circuit with a battery as the dc source. The answer is that current flows any time there is a change in voltage. When V_T is connected, the applied voltage changes from zero. Then charging current flows to charge *C* to the applied voltage. After v_C equals V_T, there is no net charging voltage and *I* is zero.

Similarly, *C* can produce discharge current any time v_C is greater than *V*. When V_T is disconnected, v_C can discharge down to zero, producing discharge current in the opposite direction from the charging current. After v_C equals zero, there is no current.

Capacitance Opposes Voltage Changes Across Itself This ability corresponds to the ability of inductance to oppose a change of current. In terms of the *RC* circuit, when the applied voltage increases, the voltage across the capacitance cannot increase until the charging current has stored enough charge in *C*. The increase in applied voltage is present across the resistance in series with *C* until the capacitor has charged to the higher applied voltage. When the applied voltage decreases, the voltage across the capacitor cannot go down immediately because the series resistance limits the discharge current.

The voltage across the capacitance in an *RC* circuit, therefore, cannot follow instantaneously the changes in applied voltage. As a result, the capacitance is able to oppose changes in voltage across itself. The instantaneous variations in V_T are present across the series resistance, however, since the series voltage drops must add to equal the applied voltage at all times.

Practice Problems 23-5
Answers at End of Chapter

a. From the curve in Fig. 23-4*a*, how much is v_C after 3 s of charge?

b. From the curve in Fig. 23-4*b*, how much is v_C after 3 s of discharge?

23-6   High Current Produced by Short-Circuiting *RC* Circuit

Specifically, a capacitor can be charged slowly with a small charging current through a high resistance and then discharged fast through a low resistance to obtain a momentary surge, or pulse, of discharge current. This idea corresponds to the pulse of high voltage obtained by opening an inductive circuit.

The circuit in Fig. 23-5 illustrates the application of a battery-capacitor (BC) unit to fire a flash bulb for cameras. The flash bulb needs 5 A to ignite, but this is too much load current for the small 15-V battery that has a rating of 30 mA for normal load current. Instead of using the bulb as a load for the battery, though, the 100-μF capacitor is charged by the battery through the 3-kΩ *R* in Fig. 23-5*a*, and then the capacitor is discharged through the bulb in Fig. 23-5*b*.

Charging the Capacitor   In Fig. 23-5*a*, S_1 is closed to charge *C* through the 3-kΩ R_1 without the bulb. The time constant of the *RC* charging circuit is 0.3 s.

After five time constants, or 1.5 s, *C* is charged to the 15 V of the battery. The peak charging current, at the first instant of charge, is *V/R* or 15 V/3 kΩ, which equals 5 mA. This value is an easy load current for the battery.

Discharging the Capacitor   In Fig. 23-5*b*, v_C is 15 V without the battery. Now S_2 is closed, and *C* discharges through the 3-Ω resistance of the bulb. The time constant for discharge with the lower *r* of the bulb is $3 \times 100 \times 10^{-6}$, which equals 300 μs. At the first instant of discharge, when v_C is 15 V, the peak discharge current is 15/3, which equals 5 A. This current is enough to fire the bulb.

Energy Stored in *C*   When the 100-μF *C* is charged to 15 V by the battery, the energy stored in the electric field is $CV^2/2$, which equals 0.01 J, approximately. This energy is available to maintain v_C at 15 V for an instant when the switch is closed. The result is the 5-A *I* through the 3-Ω *r* of the bulb at the start of the decay. Then v_C and i_C drop to zero in five time constants.

The *dv/dt* for *i_C*   The required rate of change in voltage is 0.05×10^6 V/s for the discharge current i_C of 5 A produced by the *C* of 100 μF. Since $i_C = C(dv/dt)$, this formula can be transposed to specify *dv/dt* as equal to i_C/C. Then *dv/dt* corresponds to 5 A/100 μF, or 0.05×10^6 V/s. This value is the actual *dv/dt* at the start of discharge when the switch is

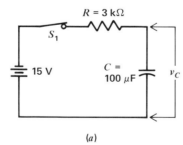

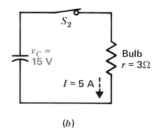

(*a*)   (*b*)

Fig. 23-5 High current produced by discharging a charged capacitor through a low resistance. (*a*) When S_1 is closed, *C* charges to 15 V through 3 kΩ. (*b*) When S_2 is closed, v_C produces the peak discharge current of 5 A through the 3 Ω of the bulb.

closed in Fig. 23-5*b*. The *dv/dt* is high because of the short *RC* time constant.[1]

Practice Problems 23-6
Answers at End of Chapter

a. Is the *RC* time constant longer or shorter in Fig. 23-5*b* compared with Fig. 23-5*a*?

b. Which produces more i_C, a faster *dv/dt* or a slower *dv/dt*?

23-7   *RC* Waveshapes

The voltage and current waveshapes in an *RC* circuit are shown in Fig. 23-6 for the case where a capacitor is allowed to charge through a resistance for *RC* time and then discharge through the same resistance for the same amount of time. It should be noted that this particular case is not typical of practical *RC* circuits, but the waveshapes show some useful details about the voltage and current for charging and discharging. The *RC* time constant here equals 0.1 s to simplify the calculations.

Square Wave of Applied Voltage
The idea of closing S_1 to apply 100 V and then opening it to disconnect V_T at a regular rate corresponds to a square wave of applied voltage, as shown by the waveform in Fig. 23-6*a*. When S_1 is closed for charge, S_2 is open; when S_1 is open, S_2 is closed for discharge. Here the voltage is on for the *RC* time of 0.1 s and off for the same time of 0.1 s. The period of the square wave is 0.2 s, and *f* is 1/0.2 s, which equals 5 Hz for the frequency.

Capacitor Voltage v_C
As shown in Fig. 23-6*b*, the capacitor charges to 63 V, equal to 63 percent of the charging voltage, in the *RC* time of 0.1 s. Then the capacitor discharges because the applied V_T drops to zero. As a result, v_C drops to 37 percent of 63 V, or 23.3 V in *RC* time.

The next charge cycle begins with v_C at 23.3 V. The

net charging voltage now is $100 - 23.3 = 76.7$ V. The capacitor voltage increases by 63 percent of 76.7 V, or 48.3 V. Adding 48.3 V to 23.3 V, then v_C rises to 71.6 V. On discharge, after 0.3 s, v_C drops to 37 percent of 71.6 V, or to 26.5 V.

Charge and Discharge Current
As shown in Fig. 23-6*c*, the current *i* has its positive peak at the start

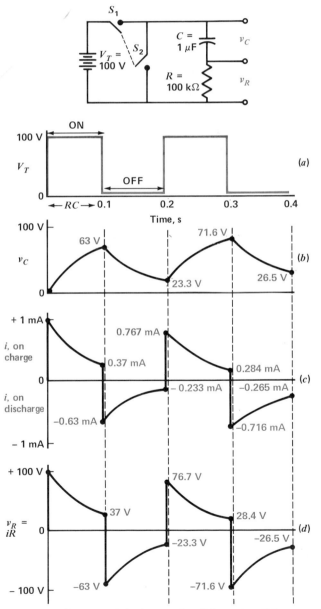

Fig. 23-6 Charge and discharge of an *RC* circuit in *RC* time.

---

[1]See footnote on p. 446.

of charge and its negative peak at the start of discharge. On charge, i is calculated as the net charging voltage, which is $(V_T - v_C)$, divided by R. On discharge, i always equals the value of v_C/R.

At the start of charge, i is maximum because the net charging voltage is maximum before C charges. Similarly, the peak i for discharge occurs at the start when v_C is maximum before C discharges.

Note that i is actually an ac waveform around the zero axis, since the charge and discharge currents are in opposite directions. We are arbitrarily taking the charging current as positive values for i.

Resistor Voltage v_R   This waveshape in Fig. 23-6d follows the waveshape of current, as v_R is $i \times R$. Because of the opposite directions of charge and discharge current, the iR waveshape is an ac voltage.

Note that on charge v_R must always be equal to $V_T - v_C$ because of the series circuit.

On discharge v_R has the same values as v_C because they are parallel, without V_T. Then S_2 is closed to connect R across C.

Why the i_C Waveshape Is Important   The v_C waveshape of capacitor voltage in Fig. 23-6 shows the charge and discharge directly, but the i_C waveshape is very interesting. First, the voltage waveshape across R is the same as the i_C waveshape. Also, whether C is charging or discharging, the i_C waveshape is really the same except for the reversed polarity. We can see the i_C waveshape as the voltage across R. It generally is better to connect an oscilloscope for voltage waveshapes across R, especially with one side grounded.

Finally, we can tell what v_C is from the v_R waveshape. The reason is that at any instant of time V_T must equal the sum of v_R and v_C. Therefore v_C is equal to $V_T - v_R$, when V_T is charging C. For the case when C is discharging, there is no V_T. Then v_R is the same as v_C.

Practice Problems 23-7
Answers at End of Chapter

Refer to Fig. 23-6.
a. When v_C is 63 V, how much is v_R?
b. When v_R is 76.7 V, how much is v_C?

23-8   Long and Short Time Constants

Useful waveshapes can be obtained by using RC circuits with the required time constant. In practical applications, RC circuits are used more than RL circuits because almost any value of an RC time constant can be obtained easily. With coils, the internal series resistance cannot be short-circuited and the distributed capacitance often causes resonance effects.

Long RC Time   Whether an RC time constant is long or short depends on the pulse width of the applied voltage. We can arbitrarily define a long time constant as at least five times longer than the time for applied voltage. As a result, C takes on very little charge. The time constant is too long for v_C to rise appreciably before the applied voltage drops to zero and C must discharge. On discharge also, with a long time constant, C discharges very little before the applied voltage rises to make C charge again.

Short RC Time   A short time constant can be defined as no more than one-fifth the time for applied voltage V_T. Then V_T is applied for a period of at least five time constants, allowing C to become completely charged. After C is charged, v_C remains at the value of V_T, while the voltage is applied. When V_T drops to zero, C discharges completely in five time constants and remains at zero while there is no applied voltage. On the next cycle, C charges and discharges completely again.

Differentiation   The voltage across R in an RC circuit is called differentiated output because v_R can change instantaneously. A short time constant is generally used for differentiating circuits to provide sharp pulses of v_R.

Integration   The voltage across C is called integrated output because it must accumulate over a period of time. A medium or long time constant is generally used for integrating circuits.

Practice Problems 23-8
Answers at End of Chapter

a. Voltage V_T is on for 0.4 s and off for 0.4 s. *RC* is 6 ms for charge and discharge. Is this a long or short *RC* time constant?

b. Voltage V_T is on for 2 μs and off for 2 μs. *RC* is 6 ms for charge and discharge. Is this a long or short *RC* time constant?

23-9 Charge and Discharge with Short *RC* Time Constant

Usually, the time constant is made much shorter or longer than the factor of 5, to obtain better waveshapes. In Fig. 23-7, *RC* is 0.1 ms. The frequency for the square wave is 25 Hz, with a period of 0.04 s, or 40 ms. One-half this period is the time V_T is applied. Therefore, the applied voltage is on for 20 ms and off for 20 ms. The *RC* time constant of 0.1 ms is shorter than the pulse width of 20 ms by a factor of $\frac{1}{200}$.

Note that the time axis of all the waveshapes is calibrated in seconds for the period of V_T, not in *RC* time constants.

Square Wave of V_T Is Across *C*
The waveshape of v_C in Fig. 23-7b is essentially the same as the square wave of applied voltage. The reason is that the short time constant allows *C* to charge or discharge completely very soon after V_T is applied or removed. The charge or discharge time of five time constants is much less than the pulse width.

Sharp Pulses of *i*
The waveshape of *i* shows sharp peaks for the charge or discharge current. Each current peak is $V_T/R = 1$ mA, decaying to zero in five *RC* time constants. These pulses coincide with the leading and trailing edges of the square wave of V_T.

Actually, the pulses are much sharper than shown. They are not to scale horizontally in order to indicate the charge and discharge action. Also, v_C is actually a square wave like the applied voltage but with slightly rounded corners for the charge and discharge.

Sharp Pulses of v_R
The waveshape of voltage across the resistor follows the current waveshape, as $v_R = iR$. Each current pulse of 1 mA across the 100-kΩ *R* results in a voltage pulse of 100 V.

More fundamentally, the peaks of v_R equal the applied voltage V_T before *C* charges. Then v_R drops to zero as v_C rises to the value of V_T.

On discharge, $v_R = v_C$, which is 100 V at the start of discharge. Then the pulse drops to zero in five time constants. The pulses of v_R in Fig. 23-7 are useful as timing pulses that match the edges of the square-wave applied voltage V_T. Either the positive or the negative pulses can be used.

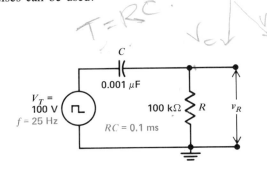

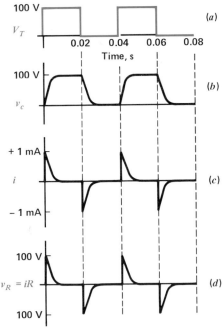

Fig. 23-7 Charge and discharge of an *RC* circuit with a short time constant. Note that the waveshape of v_R in (*d*) has sharp voltage peaks for the leading and trailing edges of the square wave of applied voltage.

Practice Problems 23-9
Answers at End of Chapter

Refer to Fig. 23-7.
a. Is the time constant here short or long?
b. Is the square wave of applied voltage across *C* or *R*?

Practice Problems 23-10
Answers at End of Chapter

Refer to Fig. 23-8.
a. Is the *RC* time constant here short or long?
b. Is the square wave of applied voltage across *R* or *C*?

23-10   Long Time Constant for *RC* Coupling Circuit

The *RC* circuit in Fig. 23-8 is the same as in Fig. 23-7, but now the *RC* time constant is long because of the higher frequency of the applied voltage. Specifically, the *RC* time of 0.1 ms is 200 times longer than the 0.5-μs pulse width of V_T with a frequency of 1 MHz. Note that the time axis is calibrated in microseconds for the period of V_T, not in *RC* time constants.

Very Little of V_T is Across *C*
The waveshape of v_C in Fig. 23-8b shows very little voltage rise because of the long time constant. During the 0.5 μs that V_T is applied, *C* charges to only $\frac{1}{200}$ of the charging voltage. On discharge, also, v_C drops very little.

Square Wave of i
The waveshape of i stays close to the 1-mA peak at the start of charge. The reason is that v_C does not increase much, allowing V_T to maintain the charging current. On discharge, the reverse i for discharge current is very small because v_C is low.

Square Wave of V_T Is Across *R*
The waveshape of v_R is the same square wave as i, as $v_R = iR$. Actually, the waveshapes of i and v_R are essentially the same as the square-wave V_T applied. They are not shown to scale vertically in order to indicate the slight charge and discharge action.

Eventually, v_C will climb to the average dc value of 50 V, i will vary ±0.5 mA above and below zero, while v_R will vary ±50 V above and below zero. This application is an *RC* coupling circuit to block the average value of the varying dc voltage V_T as the capacitive voltage v_C, while v_R provides an ac voltage output having the same variations as V_T.

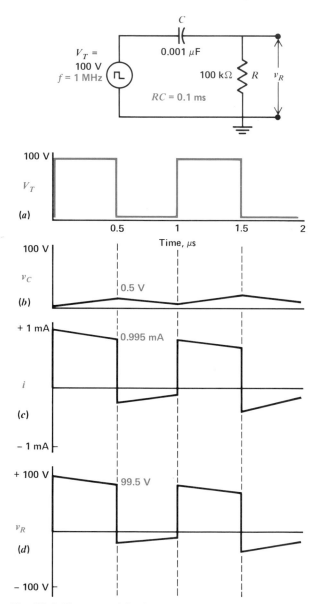

Fig. 23-8 Charge and discharge of an *RC* circuit with a long time constant. Note that the waveshape of v_R in (d) has essentially the same waveform as the applied voltage.

23-11   Universal Time Constant Graph

We can determine transient voltage and current values for any amount of time, with the curves in Fig. 23-9. The rising curve *a* shows how v_C builds up as *C* charges in an *RC* circuit; the same curve applies to i_L, increasing in the inductance for an *RL* circuit. The decreasing curve *b* shows how v_C drops as *C* discharges or the decay of i_L in an inductance.

Note that the horizontal axis is in units of time constants rather than absolute time. Suppose that the time constant of an *RC* circuit is 5 μs. Therefore, one *RC* time unit = 5 μs, two *RC* units = 10 μs, three *RC* units = 15 μs, four *RC* units = 20 μs, and five *RC* units = 25 μs.

As an example, to find v_C after 10 μs of charging, we can take the value of curve *a* in Fig. 23-9 at two *RC*. This point is at 86 percent amplitude. Therefore, we can say that in this *RC* circuit with a time constant of 5 μs, v_C charges to 86 percent of the applied V_T, after 10 μs. Similarly, some important values that can be read from the curve are listed in Table 23-1.

If we consider curve *a* in Fig. 23-9 as an *RC* charge curve, v_C adds 63 percent of the net charging voltage for each additional unit of one time constant, although

it may not appear so. For instance, in the second interval of *RC* time, v_C adds 63 percent of the net charging voltage, which is 0.37 V_T. Then 0.63×0.37 equals 0.23, which is added to 0.63 to give 0.86, or 86 percent, as the total charge from the start.

Slope at $t = 0$ The curves in Fig. 23-9 can be considered linear for the first 20 percent of change. In 0.1 time constant, for instance, the change in amplitude is 10 percent; in 0.2 time constant, the change is 20 percent. The dotted lines in Fig. 23-9 show that if this constant slope continued, the result would be 100 percent charge in one time constant. This does not happen,

Table 23-1. Time Constant Factors

Factor	Amplitude
0.2 time constant	20%
0.5 time constant	40%
0.7 time constant	50%
1 time constant	63%
2 time constants	86%
3 time constants	96%
4 time constants	98%
5 time constants	99%

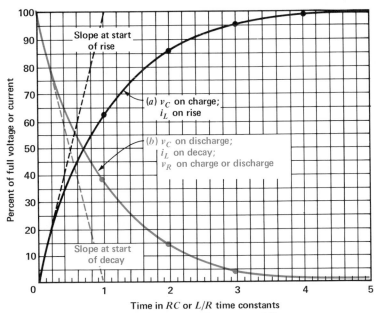

Fig. 23-9 Universal time-constant chart for *RC* or *RL* circuits. The rise or fall changes by 63 percent in one time constant.

though, because the change is opposed by the energy stored in L and C. However, at the first instant of rise or decay, at $t = 0$, the change in v_C or i_L can be calculated from the dotted slope line.

Equation of the Decay Curve

The rising curve a in Fig. 23-9 may seem more interesting because it describes the buildup of v_C or i_L, but the decaying curve b is more useful. For RC circuits, curve b can be applied to

1. v_C on discharge
2. i and v_R on charge or discharge

If we use curve b for the voltage in RC circuits, the equation of this decay curve can be written as

$$v = V \times \epsilon^{-t/RC} \tag{23-3}$$

where V is the voltage at the start of decay and v is the reduced voltage after the time t. Specifically, v can be v_R on charge and discharge, or v_C only on discharge.

The constant ϵ is the base 2.718 for natural logarithms. The negative exponent $-t/RC$ indicates a declining exponential or logarithmic curve. The value of t/RC is the ratio of actual time of decline t to the RC time constant.

This equation can be converted to common logarithms for easier calculations. Since the natural base ϵ is 2.718, its logarithm to base 10 equals 0.434. Therefore, the equation becomes

$$v = \text{antilog} \left(\log V - 0.434 \times \frac{t}{RC} \right) \tag{23-4}$$

Calculations for v_R

As an example, let us calculate v_R dropping from 100 V, after RC time. Then the factor t/RC is 1. Substituting these values,

$v_R = \text{antilog} (\log 100 - 0.434 \times 1)$

$\quad = \text{antilog} (2 - 0.434)$

$\quad = \text{antilog} 1.566$

$v_R = 37$ V

All these logs are to base 10. The antilog of 1.566 is 37.

We can also use v_R to find v_C, which is $V_T - v_R$. Then $100 - 37 = 63$ V for v_C. These answers agree with the fact that in one time constant v_R drops 63 percent while v_C rises 63 percent. However, the formula can be used to calculate any decaying value on curve b in Fig. 23-9.

Calculations for t

Furthermore, Formula (23-4) can be transposed to find the time t for a specific voltage decay. Then

$$t = 2.30 \, RC \, \log \frac{V}{v} \tag{23-5}$$

where V is the higher voltage at the start and v is the lower voltage at the finish. The factor 2.30 is 1/0.434.

As an example, let RC be 1 s. How long will it take for v_R to drop from 100 to 50 V? The required time for this decay is

$t = 2.30 \times 1 \times \log \dfrac{100}{50}$

$\quad = 2.30 \times 1 \times \log 2$

$\quad = 2.30 \times 1 \times 0.3$

$t = 0.7$ s   approximately

This answer agrees with the fact that the time for a drop of 50 percent takes 0.7 time constant. Formula (23-5) can also be used to calculate the time for any decay of v_C or v_R.

The formula cannot be used for a rise in v_C. However, if you convert this rise to an equivalent drop in v_R, the calculated time is the same for both cases.

Practice Problems 23-11
Answers at End of Chapter

Answer True or False for the universal curves in Fig. 23-9.
a. Curve a applies to v_C on charge.
b. Curve b applies to v_C on discharge.
c. Curve b applies to v_R when C charges or discharges.

23-12   Comparison of Reactance and Time Constant

The formula for capacitive reactance includes the factor of time in terms of frequency as $X_C = 1/(2\pi fC)$. Therefore, X_C and the *RC* time constant are both measures of the reaction of *C* to a change in voltage. The reactance X_C is a special case but a very important one that applies only to sine waves. The *RC* time constant can be applied to any waveshape.

Phase Angle of Reactance   The capacitive charge and discharge current i_C is always equal to $C \, (dv/dt)$. A sine wave of voltage variations for v_C produces a cosine wave of current i_C. This means v_C and i_C are both sinusoids, but 90° out of phase.

In this case, it is usually more convenient to use X_C for calculations in sine-wave ac circuits to determine *Z*, *I*, and the phase angle θ. Then $I_C = V_C/X_C$. Moreover, if I_C is known, $V_C = I_C \times X_C$. The phase angle of the circuit depends on the amount of X_C compared with the resistance *R*.

Changes in Waveshape   With nonsinusoidal voltage applied, X_C cannot be used. Then i_C must be calculated as $C \, (dv/dt)$. In this comparison of i_C and v_C, their waveshapes can be different, instead of the change in phase angle for sine waves. The waveshapes of v_C and i_C depend on the *RC* time constant.

Coupling Capacitors   If we consider the application of a coupling capacitor, X_C must be one-tenth or less of its series *R* at the desired frequency. This condition is equivalent to having an *RC* time constant that is long compared with the period of one cycle. In terms of X_C, the *C* has little IX_C voltage, with practically all the applied voltage across the series *R*. In terms of a long *RC* time constant, *C* cannot take on much charge. Practically all the applied voltage is developed as $v_R = iR$ across the series resistance by the charge and discharge current. These comparisons are summarized in Table 23-2.

Inductive Circuits   Similar comparisons can be made between $X_L = 2\pi fL$ for sine waves and the *L/R* time constant. The voltage across any inductance is $v_L = L \, (di/dt)$. Sine-wave variations for i_L produce a cosine wave of voltage v_L, 90° out of phase.

In this case X_L can be used to determine *Z*, *I*, and the phase angle θ. Then $I_L = V_L/X_L$. Furthermore, if I_L is known, $V_L = I_L \times X_L$. The phase angle of the circuit depends on the amount of X_L compared with *R*.

With nonsinusoidal voltage, however, X_L cannot be used. Then v_L must be calculated as $L \, (di/dt)$. In this comparison, i_L and v_L can have different waveshapes, depending on the *L/R* time constant.

Choke Coils   For this application, the idea is to have almost all the applied ac voltage across *L*. The condition of X_L being at least ten times *R* corresponds

Table 23-2. Comparison of Reactance X_C and *RC* Time Constant

Sine-wave voltage	Nonsinusoidal voltage
Examples are 60-Hz power line, AF signal voltage, RF signal voltage	Examples are dc circuit turned on and off, square waves, sawtooth waves
Reactance $X_C = \dfrac{1}{2\pi fC}$	Time constant $T = RC$
Larger *C* results in smaller reactance X_C	Larger *C* results in longer time constant
Higher frequency results in smaller X_C	Shorter pulse width corresponds to longer time constant
$I_C = \dfrac{V_C}{X_C}$	$i_C = C\dfrac{dv}{dt}$
X_C makes I_C and V_C 90° out of phase	Waveshape changes between i_C and v_C

to having a long time constant. The high value of X_L means practically all the applied ac voltage is across X_L as IX_L, with little IR voltage.

The long L/R time constant means i_L cannot rise appreciably, resulting in little v_R voltage across the resistor. The waveform for i_L and v_R in an inductive circuit corresponds to v_C in a capacitive circuit.

When Do We Use the Time Constant?

In electronic circuits, the time constant is useful in analyzing the effect of L or C on the waveshape of nonsinusoidal voltages, particularly rectangular pulses. Another application is the transient response when a dc voltage is turned on or off. The 63 percent change in one time constant is a natural characteristic of v or i, where the magnitude of one is proportional to the rate of change of the other.

When Do We Use Reactance?

The X_L and X_C are generally used for sine-wave V or I. We can determine Z, I, voltage drops, and phase angles. The phase angle of 90° is a natural characteristic of a cosine wave where its magnitude is proportional to the rate of change in a sine wave.

Practice Problems 23-12
Answers at End of Chapter

a. Does an RC coupling circuit have a small or large X_C compared with R?

b. Does an RC coupling circuit have a long or short time constant for the frequency of applied voltage?

Summary

1. The transient response of an inductive circuit with nonsinusoidal current is indicated by the time constant L/R. With L in henrys and R in ohms, T is the time in seconds for the current i_L to change by 63 percent. In five time constants, i_L reaches the steady value of V_T/R.

2. At the instant an inductive circuit is opened, high voltage is generated across L because of the fast current decay with a short time constant. The induced voltage $v_L = L\,(di/dt)$. The di is the change in i_L.

3. The transient response of a capacitive circuit with nonsinusoidal voltage is indicated by the time constant RC. With C in farads and R in ohms, T is the time in seconds for the voltage across the capacitor v_C to change by 63 percent. In five time constants, v_C reaches the steady value of V_T.

4. At the instant a charged capacitor is discharged through a low resistance, a high value of discharge current can be produced. The discharge current $i_C = C\,(dv/dt)$ can be large because of the fast discharge with a short time constant. The dv is the change in v_C.

5. The waveshapes of v_C and i_L correspond, as both rise relatively slowly to the steady-state value. This is integrated output.

6. Also i_C and v_L correspond, as they are the waveforms that can change instantaneously. This is differentiated output.

7. For both RC and RL circuits the resistor voltage $v_R = iR$.

8. A short time constant is one-fifth or less of the time for applied voltage.

9. A long time constant is greater than the time for applied voltage by a factor of 5 or more.

10. An RC circuit with a short time constant produces sharp voltage spikes for v_R at the leading and trailing edges of a square-wave applied voltage. The waveshape of applied voltage V_T is across the capacitor as v_C. See Fig. 23-7.

11. An *RC* circuit with a long time constant allows v_R to be essentially the same as the variations in applied voltage V_T, while the average dc value of V_T is blocked as v_C. See Fig. 23-8.

12. The universal rise and decay curves in Fig. 23-9 can be used for current or voltage in *RC* and *RL* circuits for any time up to five time constants.

13. The concept of reactance is useful for sine-wave ac circuits with *L* and *C*.

14. The time constant method is used with *L* or *C* to analyze nonsinusoidal waveforms.

Self-Examination
Answers at Back of Book

Choose (a), (b), (c), or (d).

1. A 250-μH *L* is in series with a 50-Ω *R*. The time constant is (a) 5 μs; (b) 25 μs; (c) 50 μs; (d) 250 μs.

2. If V_T is 500 mV in the preceding circuit, after 5 μs *I* rises to the value of (a) 3.7 mA; (b) 5 mA; (c) 6.3 mA; (d) 10 mA.

3. In the preceding circuit, *I* will have the steady-state value of 10 mA after (a) 5 μs; (b) 6.3 μs; (c) 10 μs; (d) 25 μs.

4. The arc across a switch when it opens an *RL* circuit is a result of the (a) long time constant; (b) large self-induced voltage across *L*; (c) low resistance of the open switch; (d) surge of resistance.

5. A 250-pF *C* is in series with a 1-MΩ *R*. The time constant is (a) 63 μs; (b) 100 μs; (c) 200 μs; (d) 250 μs.

6. If V_T is 100 V in the preceding circuit, after 250 μs, v_C rises to the value of (a) 37 V; (b) 50 V; (c) 63 V; (d) 100 V.

7. In the preceding circuit, v_C will have the steady-state value of 100 V after (a) 250 μs; (b) 630 μs; (c) 1000 μs or 1 ms; (d) 1.25 ms.

8. In the preceding circuit, after 3 hours v_C will be (a) zero; (b) 63 V; (c) 100 V; (d) 200 V.

9. For a square-wave applied voltage with the frequency of 500 Hz, a long time constant is (a) 1 ms; (b) 2 ms; (c) 3.7 ms; (d) 5 ms.

10. An *RC* circuit has a 2-μF *C* in series with a 1-MΩ *R*. The time of 6 s equals how many time constants? (a) one; (b) two; (c) three; (d) six.

Essay Questions

1. Give the formula, with units, for calculating the time constant of an *RL* circuit.

2. Give the formula, with units, for calculating the time constant of an *RC* circuit.

3. Redraw the *RL* circuit and graph in Fig. 23-2 for a 2-H *L* and a 100-Ω *R*.

4. Redraw the graphs in Fig. 23-4 to fit the circuit in Fig. 23-5 with a 100-μF *C*. Use a 3000-Ω *R* for charge but a 3-Ω *R* for discharge.

5. List two comparisons of *RC* and *RL* circuits for nonsinusoidal voltage.

6. List two comparisons between *RC* circuits with nonsinusoidal voltage and sine-wave voltage applied.

7. Define the following: (a) a long time constant; (b) a short time constant; (c) an RC differentiating circuit.
8. Redraw the horizontal time axis of the universal curve in Fig. 23-9, calibrated in absolute time units of milliseconds for an RC circuit with a time constant equal to 2.3 ms.
9. Redraw the circuit and graphs in Fig. 23-7 with everything the same except that R is 20 kΩ, making the RC time constant shorter.
10. Redraw the circuit and graphs in Fig. 23-8 with everything the same except that R is 500 kΩ, making the RC time constant longer.
11. Invert the equation $T = RC$, in two forms, to find R or C from the time constant.
12. Show three types of nonsinusoidal waveforms.
13. Give an application in electronic circuits for an RC circuit with a long time constant and with a short time constant.
14. Why can arcing voltage be a problem with coils used in switching circuits?

Problems
Answers to Odd-Numbered Problems at Back of Book

1. Calculate the time constant of the following inductive circuits: (a) L is 20 H and R is 400 Ω; (b) L is 20 μH and R is 400 Ω; (c) L is 50 mH and R is 50 Ω; (d) L is 40 μH and R is 2 Ω.
2. Calculate the time constant of the following capacitive circuits: (a) C is 0.001 μF and R is 1 MΩ; (b) C is 1 μF and R is 1000 Ω; (c) C is 0.05 μF and R is 250 kΩ; (d) C is 100 pF and R is 10 kΩ.
3. A 100-V source is in series with a 2-MΩ R and a 2-μF C. (a) How much time is required for v_C to be 63 V? (b) How much is v_C after 20 s?
4. The C in Prob. 3 is allowed to charge for 4 s and then made to discharge for 8 s. How much is v_C?
5. A 100-V source is applied in series with a 1-MΩ R and a 4-μF C that has already been charged to 63 V. How much is v_C after 4 s?
6. What value of R is needed with a 0.05-μF C for an RC time constant of 0.02 s? For 1 ms?
7. An RC circuit has a time constant of 1 ms. V_T applied is 20 V. How much is v_C on charge after 1.4 ms?
8. A 0.05-μF C charges through a 0.5-MΩ R but discharges through a 2-kΩ R. Calculate the time constants for charge and discharge. Why will the capacitor discharge faster than charge?
9. A 0.05-μF C is charged to 264 V. It discharges through a 40-kΩ R. How much is the time for v_C to discharge down to 132 V?
10. Referring to Fig. 23-6b, calculate the value of v_C on the next charge, starting from 26.5 V.
11. Use the slope line in Fig. 23-9b to calculate dv/dt at the start of the decay in v_C for the circuit in Fig. 23-5b.
12. Use the slope line in Fig. 23-9b to calculate di/dt at the start of the decay in i_L for the circuit in Fig. 23-3b. (Hint: You can ignore the steady 8 V and 100-Ω R_1 because they do not change the di/dt value.)

Answers to Practice Problems

23-1 **a.** T
b. T

23-2 **a.** 0.02 s
b. 0.5 ms

23-3 **a.** Shorter
b. Faster

23-4 **a.** 940 μs
b. 470 ns

23-5 **a.** 63.2 V
b. 36.8 V

23-6 **a.** Shorter
b. Faster

23-7 **a.** $v_R = 37$ V
b. $v_C = 23.3$ V

23-8 **a.** Short
b. Long

23-9 **a.** Short
b. Across *C*

23-10 **a.** Long
b. Across *R*

23-11 **a.** T
b. T
c. T

23-12 **a.** Small X_C
b. Long time constant

Review Chapters 20 to 23

Summary

1. A capacitor, or condenser, consists of two conductors separated by an insulator, which is a dielectric material. With voltage applied to the conductors, charge is stored in the dielectric. One coulomb of charge stored with one volt applied corresponds to one farad of capacitance C. The common units of capacitance are microfarads (μF = 10^{-6} F) or picofarads (pF = 10^{-12} F).

2. Capacitance increases with plate area and larger values of dielectric constant but decreases with the distance between plates.

3. The most common types of capacitors are air, paper, mica, ceramic, and electrolytic. Electrolytics must be connected in the correct polarity. The color coding for mica and ceramic tubular capacitors is illustrated in Figs. 20-9 and 20-10.

4. The total capacitance of parallel capacitors is the sum of the individual values; the combined capacitance of series capacitors is found by the reciprocal formula. These rules are opposite from the formulas for resistors and inductors in series or parallel.

5. In checking with an ohmmeter, a good capacitor shows charging current and then the ohmmeter reads a very high value of ohms equal to the insulation resistance. A short-circuited capacitor reads zero ohms; an open capacitor does not show any charging current.

6. $X_c = 1/(2 \pi fC) \, \Omega$, with f in hertz and C in farads. The higher the frequency and the greater the capacitance, the smaller X_c is.

7. A common application of X_C is in AF or RF coupling capacitors, which have low reactance for one group of frequencies but more reactance for lower frequencies. This is just the opposite of an inductance used as a choke.

8. Reactance X_C is a phasor quantity where the voltage across the capacitor lags 90° behind its charge and discharge current. This phase angle of X_C is exactly opposite from the phase angle for X_L.

9. In series circuits, R and X_C are added by phasors because their voltage drops are 90° out of phase. Therefore, the total impedance Z equals $\sqrt{R^2 + X_C^2}$; the current I equals V_T/Z.

10. For parallel circuits, the resistive and capacitive branch currents are added by phasors: $I_T = \sqrt{I_R^2 + I_C^2}$; the impedance $Z = V_A/I_T$.

11. The time constant of a capacitive circuit equals $R \times C$. With R in ohms and C in farads, the RC product is the time in seconds for the voltage across C to change by 63 percent.

12. Capacitive charge or discharge current i_C is equal to C (dv/dt) for any waveshape of v_C.

13. The time constant of an inductive circuit equals L/R. With L in henrys and R in ohms, L/R is the time in seconds for the current through L to change by 63 percent.

14. Induced voltage v_L is equal to L (di/dt) for any waveshape of i_L.

Review Self-Examination
Answers at Back of Book

Answer True or False.

1. A capacitor can store charge because it has a dielectric between two conductors.

2. With 100 V applied, a 0.01-μF capacitor stores 1 μC of charge.

3. The smaller the capacitance, the higher the potential difference across it for a given amount of charge stored in the capacitor.

4. A 250-pF capacitance equals 250×10^{-12} F.

5. The thinner the dielectric, the more the capacitance and the lower the voltage breakdown rating for a capacitor.

6. Larger plate area increases the capacitance.

7. Capacitors in series provide less capacitance but a higher voltage breakdown rating for the combination.

8. Capacitors in parallel increase the total capacitance with the same voltage rating.

9. Two 0.01-μF capacitors in parallel have a total C of 0.005 μF.

10. A good 0.01-μF paper capacitor will show charging current and read 500 MΩ or more on an ohmmeter.

11. If the capacitance is doubled, the reactance is one-half.

12. If the frequency is doubled, the reactance is one-half.

13. The reactance of a 0.1-μF capacitor at 60 Hz is approximately 60 Ω.

14. In a series circuit, the voltage across X_C lags 90° behind the current.

15. The phase angle of a series circuit can be any angle between 0 and 90°, depending on the ratio of X_C to R.

16. In a parallel circuit, the voltage across X_C lags 90° behind its capacitive branch current.

17. In a parallel circuit of two resistances with 1 A in each branch, the total line current equals 1.414 A.

18. A 1000-Ω X_C in parallel with a 1000-Ω R has a combined Z of 707 Ω.

19. A 1000-Ω X_C in series with a 1000-Ω R has a total Z of 1414 Ω.

20. Neglecting its sign, the phase angle is 45° for both circuits in Probs. 18 and 19.

21. Reactances X_L and X_C are opposite.

22. The total impedance of a 1-MΩ R in series with a 5-Ω X_C is approximately 1 MΩ with a phase angle of 0°.

23. The combined impedance of a 5-Ω R in shunt with a 1-MΩ X_C is approximately 5 Ω with a phase angle of 0°.

24. Reactances X_L and X_C change with frequency, but L and C do not depend on the frequency.

25. A long RC time constant corresponds to a large C and R.

26. When the RC time constant for discharge is calculated, R must be the resistance in the path for discharge current.

27. Resistance and impedance are both measured in ohms.

28. Reactances X_L and X_C are both measured in ohms.

29. Impedance Z can change with frequency because it includes reactance.

30. With 100 V applied, a 1-μF capacitor in series with a 1-MΩ R will have the transient value of $v_C = 63$ V in 1 s, while charging to the steady-state value of 100 V in 5 s.

31. A 1-μF capacitor charged to 2000 V has stored energy equal to 1 J.

32. A 2-H L is in series with a 1-kΩ R and a 100-V source. After 2 ms, the transient value of i_L is 63 mA, while rising to the steady-state value of 100 mA in 10 ms.

33. When the applied voltage increases, charging current can flow as the capacitor takes on additional charge.

34. When the applied voltage decreases, a charged capacitor can discharge because it has a higher potential difference than the source.

35. Capacitors in series have the same charge and discharge current.

36. Capacitors in parallel have the same voltage.

37. The phasor combination of a 30-Ω R in series with a 40-Ω X_C equals 70 Ω impedance.

38. A six-dot mica capacitor color-coded white, green, black, and brown has the capacitance value of 500 pF.

39. Capacitive current can be considered leading current in a series circuit.

40. In a series circuit, the higher the value of X_C, the greater is its voltage drop compared with the IR drop.

References
Additional References at Back of Book

Bell, D. A.: *Fundamentals of Electric Circuits,* Reston Publishing Company, Inc., Reston, Va.

Gilli, A. C.: *Electrical Principles for Electronics,* McGraw-Hill Book Company, New York.

Jackson, H. W.: *Introduction to Electric Circuits,* Prentice-Hall, Inc., Englewood Cliffs, N.J.

Ridsdale, R. E.: *Electric Circuits,* McGraw-Hill Book Company, New York.

Zbar, P. B.: *Basic Electricity: A Text-Lab Manual,* McGraw-Hill Book Company, New York.

Chapter 24
Alternating-Current Circuits

This chapter shows how to analyze sine-wave ac circuits that have R, X_L, and X_C. How do we combine these three types of ohms of opposition, how much current flows, and what is the phase angle? These questions are answered for both series and parallel circuits.

The problems are simplified by the fact that in series circuits X_L is at 90° and X_C at −90°, which are opposite phase angles. Then all of one reactance can be canceled by part of the other reactance, resulting in only a single net reactance. Similarly, in parallel circuits, I_L and I_C have opposite phase angles. These currents oppose each other for one net reactive line current.

Finally, the idea of how ac power and dc power can differ because of ac reactance is explained. Also, types of ac current meters are described including the wattmeter.

Important terms in this chapter are:

apparent power	VAR unit
power factor	voltampere unit
real power	wattmeter

More details are explained in the following sections:

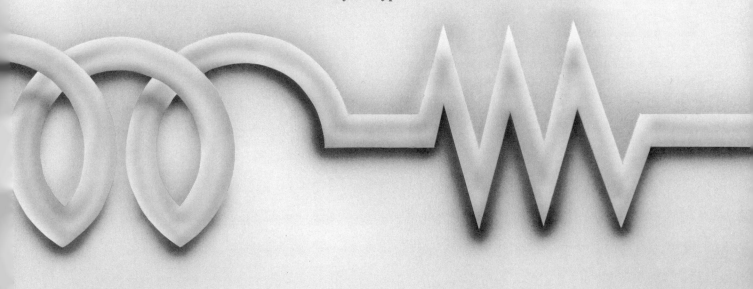

24-1 AC Circuits with Resistance but No Reactance

Combinations of series and parallel resistances are shown in Fig. 24-1. In Fig. 24-1a and b, all voltages and currents throughout the resistive circuit are in the same phase as the applied voltage. There is no reactance to cause a lead or lag in either current or voltage.

Series Resistances For the circuit in Fig. 24-1a, with two 50-Ω resistances in series across the 100-V source, the calculations are as follows:

$$R_T = R_1 + R_2 = 50 + 50 = 100 \ \Omega$$

$$I = \frac{V_T}{R_T} = \frac{100}{100} = 1 \text{ A}$$

$$V_1 = IR_1 = 1 \times 50 = 50 \text{ V}$$

$$V_2 = IR_2 = 1 \times 50 = 50 \text{ V}$$

Note that the series resistances R_1 and R_2 serve as a voltage divider, as in dc circuits. Each R has one-half the applied voltage for one-half the total series resistance.

The voltage drops V_1 and V_2 are both in phase with the series current I, which is the common reference. Also I is in phase with the applied voltage V_T because there is no reactance.

Parallel Resistances For the circuit in Fig. 24-1b, with two 50-Ω resistances in parallel across the 100-V source, the calculations are

$$I_1 = \frac{V_A}{R_1} = \frac{100}{50} = 2 \text{ A}$$

$$I_2 = \frac{V_A}{R_2} = \frac{100}{50} = 2 \text{ A}$$

$$I_T = I_1 + I_2 = 2 + 2 = 4 \text{ A}$$

With a total current of 4 A in the main line from the 100-V source, the combined parallel resistance is 25 Ω. This R_T equals 100 V/4 A for the two 50-Ω branches.

Each branch current has the same phase as the applied voltage. Voltage V_A is the reference because it is common to both branches.

Practice Problems 24-1
Answers at End of Chapter

a. In Fig. 24-1a, what is the phase angle between V_T and I?

b. In Fig. 24-1b, what is the phase angle between I_T and V_A?

24-2 Circuits with X_L Alone

The circuits with X_L in Figs. 24-2 and 24-3 correspond to the series and parallel circuits in Fig. 24-1, with ohms of X_L equal to the R values. Since the applied voltage is the same, the values of current correspond because ohms of X_L are just as effective as ohms of R in limiting the current or producing a voltage drop.

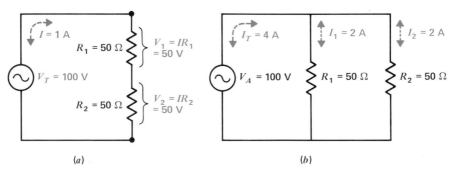

(a)

(b)

Fig. 24-1 AC circuits with resistance but no reactance. (a) Resistances R_1 and R_2 in series. (b) Resistances R_1 and R_2 in parallel.

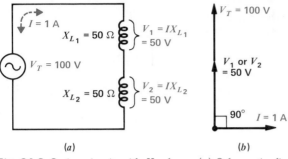

Fig. 24-2 Series circuit with X_L alone. (*a*) Schematic diagram. (*b*) Phasor diagram.

Although X_L is a phasor quantity with a 90° phase angle, all the ohms of opposition are the same kind of reactance in this example. Therefore, without any R or X_C, the series ohms of X_L can be combined directly. Similarly, the parallel I_L currents can be added.

X_L Values in Series   For Fig. 24-2a, the calculations are

$$X_{L_T} = X_{L_1} + X_{L_2} = 50 + 50 = 100 \ \Omega$$
$$I = \frac{V_T}{X_{L_T}} = \frac{100}{100} = 1 \ A$$
$$V_1 = IX_{L_1} = 1 \times 50 = 50 \ V$$
$$V_2 = IX_{L_2} = 1 \times 50 = 50 \ V$$

Note that the two series voltage drops of 50 V each add to equal the total applied voltage of 100 V.

With regard to the phase angle for the inductive reactance, the voltage across any X_L always leads the current through it by 90°. In Fig. 24-2b, I is the reference phasor because it is common to all the series com-

ponents. Therefore, the voltage phasors for V_1 and V_2 across either reactance, or V_T across both reactances, are shown leading I by 90°.

I_L Values in Parallel   For Fig. 24-3a the calculations are

$$I_1 = \frac{V_A}{X_{L_1}} = \frac{100}{50} = 2 \ A$$
$$I_2 = \frac{V_A}{X_{L_2}} = \frac{100}{50} = 2 \ A$$
$$I_T = I_1 + I_2 = 2 + 2 = 4 \ A$$

These two branch currents can be added because they both have the same phase. This angle is 90° lagging the voltage reference phasor as shown in Fig. 24-3b.

Since the voltage V_A is common to the branches, this voltage is across X_{L_1} and X_{L_2}. Therefore V_A is the reference phasor for parallel circuits.

Note that there is no fundamental change between Fig. 24-2b, which shows each X_L voltage leading its current by 90°, and Fig. 24-3b, showing each X_L current lagging its voltage by −90°. The phase angle between the inductive current and voltage is still the same 90°.

Practice Problems 24-2
Answers at End of Chapter

a. In Fig. 24-2, what is the phase angle of V_T with respect to I?

b. In Fig. 24-3, what is the phase angle of I_T with respect to V_A?

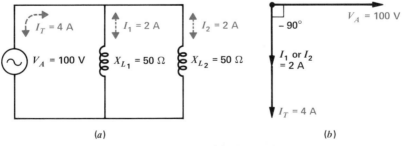

Fig. 24-3 Parallel circuit with X_L alone. (*a*) Schematic diagram. (*b*) Phasor diagram.

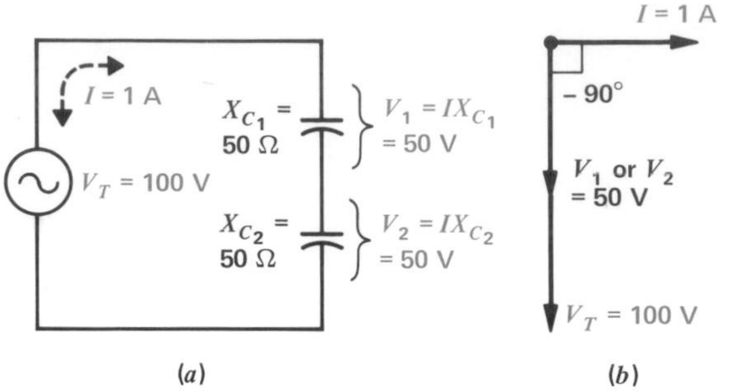

Fig. 24-4 Series circuit with X_C alone. (a) Schematic diagram. (b) Phasor diagram.

24-3   Circuits with X_C Alone

Again, reactances are shown in Figs. 24-4 and 24-5 but with X_C values of 50 Ω. Since there is no R or X_L, the series ohms of X_C can be combined directly. Also the parallel I_C currents can be added.

X_C Values in Series   For Fig. 24-4a, the calculations for V_1 and V_2 are the same as before. These two series voltage drops of 50 V each add to equal the total applied voltage.

With regard to the phase angle for the capacitive reactance, the voltage across any X_C always lags its capacitive charge and discharge current I by 90°. For the series circuit in Fig. 24-4, I is the reference phasor. The capacitive current leads by 90°. Or, we can say that each voltage lags I by −90°.

I_C Values in Parallel   For Fig. 24-5, V_A is the reference phasor. The calculations for I_1 and I_2 are the same as before. However, now each of the capacitive branch currents or the I_T leads V_A by 90°.

Practice Problems 24-3
Answers at End of Chapter

a. In Fig. 24-4, what is the phase angle of V_T with respect to I?

b. In Fig. 24-5, what is the phase angle of I_T with respect to V_A?

24-4   Opposite Reactances Cancel

In a circuit with both X_L and X_C, the opposite phase angles enable one to cancel the effect of the other. For X_L and X_C in series, the net reactance is the difference between the two series reactances, resulting in less reactance than either one. In parallel circuits, the I_L and I_C branch currents cancel. The net line current then is the difference between the two branch currents, resulting in less total line current than either branch current.

X_L and X_C in Series   For the example in Fig. 24-6, the series combination of a 60-Ω X_L and a 40-Ω X_C in Fig. 24-6a and b is equivalent to the net reactance of the 20-Ω X_L shown in Fig. 24-6c. Then, with 20 Ω as the net reactance across the 120-V source, the current is 6 A. This current lags the applied voltage V_T by 90° because the net reactance is inductive.

For the two series reactances in Fig. 24-6a, the current is the same through both X_L and X_C. Therefore, the voltage drops can be calculated as

$$V_L \text{ or } IX_L = 6 \text{ A} \times 60 \text{ Ω} = 360 \text{ V}$$
$$V_C \text{ or } IX_C = 6 \text{ A} \times 40 \text{ Ω} = 240 \text{ V}$$

Note that each individual reactive voltage drop can

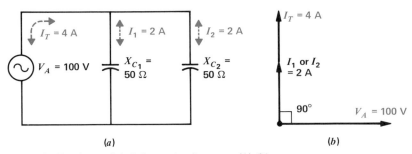

Fig. 24-5 Parallel circuit with X_C alone. (a) Schematic diagram. (b) Phasor diagram.

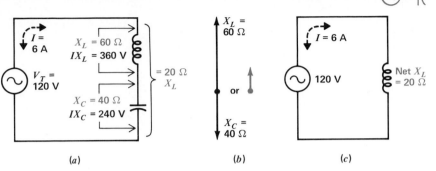

Fig. 24-6 When X_L and X_C are in series, their ohms of reactance cancel. (a) Series circuit. (b) Phasors for X_L and X_C with net resultant. (c) Equivalent circuit with net reactance of 20 Ω of X_L.

be more than the applied voltage. The sum of the series voltage drops still is 120 V, however, equal to the applied voltage. This results because the IX_L and IX_C voltages are opposite. The IX_L voltage leads the series current by 90°; the IX_C voltage lags the same current by 90°. Therefore, IX_L and IX_C are 180° out of phase with each other, which means they are of opposite polarity and cancel. Then the total voltage across the two in series is 360 V minus 240 V, which equals the applied voltage of 120 V.

If the values in Fig. 24-6 were reversed, with an X_C of 60 Ω and an X_L of 40 Ω, the net reactance would be a 20-Ω X_C. The current would be 6 A again, but with a lagging phase angle of −90° for the capacitive voltage. The IX_C voltage would then be larger at 360 V, with an IX_L value of 240 V, but the difference still equals the applied voltage of 120 V.

X_L and X_C in Parallel   In Fig. 24-7, the 60-Ω X_L and 40-Ω X_C are in parallel across the 120-V source. Then the 60-Ω X_L branch current I_L is 2 A, and the 40-Ω X_C branch current I_C is 3 A. The X_C branch has more current because its reactance is less than X_L.

In terms of phase angle, I_L lags the parallel voltage V_A by 90°, while I_C leads the same voltage by 90°. Therefore, the opposite reactive branch currents are 180° out of phase with each other and cancel. The net line current then is the difference between 3 A for I_C and 2 A for I_L, which equals the net value of 1 A. This resultant current leads V_A by 90° because it is capacitive current.

If the values in Fig. 24-7 were reversed, with an X_C of 60 Ω and an X_L of 40 Ω, I_L would be larger. The I_L then equals 3 A, with an I_C of 2 A. The net line current is 1 A again but inductive, with a net X_L.

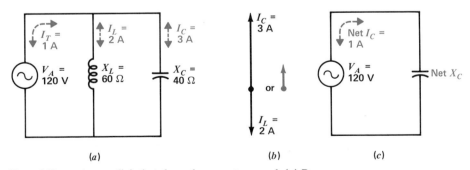

Fig. 24-7 When X_L and X_C are in parallel, their branch currents cancel. (a) Parallel circuit. (b) Phasors for branch currents I_C and I_L with net resultant. (c) Equivalent circuit with net line current of 1 A for I_C.

Practice Problems 24-4
Answers at End of Chapter

a. In Fig. 24-6, how much is the net X_L?
b. In Fig. 24-7, how much is the net I_C?

24-5 Series Reactance and Resistance

In this case, the resistive and reactive effects must be combined by phasors. For series circuits, the ohms of opposition are added to find Z. First add all the series resistances for one total R. Also combine all the series reactances, adding the same kind but subtracting opposites. The result is one net reactance, indicated X. It may be either capacitive or inductive, depending on which kind of reactance is larger. Then the total R and net X can be added by phasors to find the total ohms of opposition for the entire series circuit.

Magnitude of Z After the total R and net reactance X are found, they can be combined by the formula

$$Z = \sqrt{R^2 + X^2} \qquad (24\text{-}1)$$

The circuit's total impedance Z is the phasor sum of the series resistance and reactance. Whether the net X is at $+90°$ for X_L or $-90°$ for X_C does not matter in calculating the magnitude of Z.

An example is illustrated in Fig. 24-8. Here the net series reactance in Fig. 24-8b is a 30-Ω X_C. This value

is equal to a 60-Ω X_L subtracted from a 90-Ω X_C as shown in Fig. 24-8a. The net 30-Ω X_C in Fig. 24-8b is in series with a 40-Ω R. Therefore

$$Z = \sqrt{R^2 + X^2}$$
$$= \sqrt{(40)^2 + (30)^2}$$
$$= \sqrt{1600 + 900} = \sqrt{2500}$$
$$Z = 50 \ \Omega$$

$I = V/Z$ The current is 100 V/50 Ω in this example, or 2 A. This value is the magnitude, without considering the phase angle.

Series Voltage Drops All the series components have the same 2-A current. Therefore, the individual drops in Fig. 24-8a are

$$V_R = IR = 2 \times 40 = 80 \ \text{V}$$
$$V_C = IX_C = 2 \times 90 = 180 \ \text{V}$$
$$V_L = IX_L = 2 \times 60 = 120 \ \text{V}$$

Since IX_C and IX_L are voltages of opposite polarity, the net reactive voltage is 180 minus 120 V, which equals 60 V. The phasor sum of IR at 80 V and the net reactive voltage IX of 60 V equals the applied voltage V_T of 100 V.

Phase Angle of Z The phase angle of the series circuit is the angle whose tangent equals X/R. The angle is negative for X_C but positive for X_L.

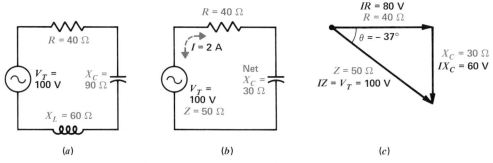

Fig. 24-8 Impedance Z of series circuit. (a) Resistance R, X_L, and X_C in series. (b) Equivalent circuit with one net reactance. (c) Phasor diagram.

In this example, X is the net reactance of 30 Ω for X_C and R is 40 Ω. Then $\tan \theta = -0.75$ and θ is $-37°$, approximately.

The negative angle for Z indicates lagging capacitive reactance for the series circuit. If the values of X_L and X_C were reversed, the phase angle would be $+37°$, instead of $-37°$, because of the net X_L. However, the magnitude of Z would still be the same.

More Series Components How to combine any number of series resistances and reactances is illustrated by Fig. 24-9. Here the total series R of 40 Ω is the sum of 30 Ω for R_1 and 10 Ω for R_2. Note that the order of connection does not matter, since the current is the same in all series components.

The total series X_C is 90 Ω, equal to the sum of 70 Ω for X_{C_1} and 20 Ω for X_{C_2}. Similarly, the total series X_L is 60 Ω. This value is equal to the sum of 30 Ω for X_{L_1} and 30 Ω for X_{L_2}.

The net reactance X equals 30 Ω, which is 90 Ω of X_C minus 60 Ω of X_L. Since X_C is larger than X_L, the net reactance is capacitive. The circuit in Fig. 24-9 is equivalent to Fig. 24-8, therefore, since a 40-Ω R is in series with a net X_C of 30 Ω.

Practice Problems 24-5
Answers at End of Chapter

a. In Fig. 24-8, how much is the net reactance?
b. In Fig. 24-9, how much is the net reactance?

24-6 Parallel Reactance and Resistance

With parallel circuits, the branch currents for resistance and reactance are added by phasors. Then the total line current is found by the formula

$$I_T = \sqrt{I_R^2 + I_X^2} \qquad (24\text{-}2)$$

Calculating I_T As an example, Fig. 24-10a shows a circuit with three branches. Since the voltage across all the parallel branches is the applied 100 V, the individual branch currents are

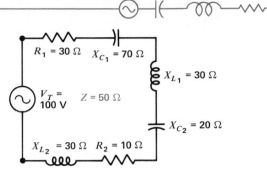

Fig. 24-9 Series ac circuit with more components than Fig. 24-8, but the same values of Z, I, and θ.

$$I_R = \frac{V_A}{R} = \frac{100 \text{ V}}{25 \text{ Ω}} = 4 \text{ A}$$

$$I_L = \frac{V_A}{X_L} = \frac{100 \text{ V}}{25 \text{ Ω}} = 4 \text{ A}$$

$$I_C = \frac{V_A}{X_C} = \frac{100 \text{ V}}{100 \text{ Ω}} = 1 \text{ A}$$

The net reactive branch current I_X is 3 A, then, equal to the difference between the 4-A I_L and the 1-A I_C, as shown in Fig. 24-10b.

The next step is to calculate I_T as the phasor sum of I_R and I_X. Then

$$\begin{aligned}
I_T &= \sqrt{I_R^2 + I_X^2} \\
&= \sqrt{4^2 + 3^2} \\
&= \sqrt{16 + 9} = \sqrt{25} \\
I_T &= 5 \text{ A}
\end{aligned}$$

The phasor diagram for I_T is shown in Fig. 24-10c.

$Z_T = V_A/I_T$ This gives the total impedance of a parallel circuit. In this example, Z_T is 100 V/5 A, which equals 20 Ω. This value is the equivalent impedance of all three branches in parallel across the source.

Phase Angle The phase angle of the parallel circuit is found from the branch currents. Now θ is the angle whose tangent equals I_X/I_R.

For this example, I_X is the net inductive current of the 3-A I_L. Also, I_R is 4 A. These phasors are shown in

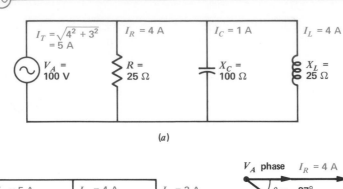

(a)

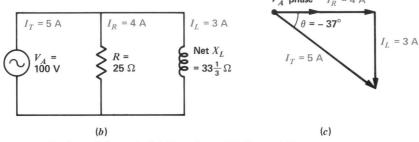

(b)   (c)

Fig. 24-10 Total line current I_T of parallel circuit. (a) Branches of R, X_L, and X_C in parallel. (b) Equivalent circuit with I_R and net reactive branch current. (c) Phasor diagram.

Fig. 24-10c. Then θ is a negative angle with the tangent of $-\frac{3}{4}$ or -0.75. This phase angle is $-37°$, approximately.

The negative angle for I_T indicates lagging inductive current. The value of $-37°$ is the phase angle of I_T with respect to the voltage reference V_A.

When Z_T is calculated as V_A/I_T for a parallel circuit, the phase angle of Z_T is the same value as for I_T but with opposite sign. In this example, Z_T is 20 Ω with a phase angle of $+37°$, for an I_T of 5 A with an angle of $-37°$. We can consider that Z_T has the phase of the voltage source with respect to I_T.

More Parallel Branches Figure 24-11 illustrates how any number of parallel resistances and reactances can be combined. The total resistive branch current I_R

of 4 A is the sum of 2 A each for the R_1 branch and the R_2 branch. Note that the order of connection does not matter, since the parallel branch currents add in the main line. Effectively, two 50-Ω resistances in parallel are equivalent to one 25-Ω resistance.

Similarly, the total inductive branch current I_L is 4 A, equal to 3 A for I_{L_1} and 1 A for I_{L_2}. Also, the total capacitive branch current I_C is 1 A, equal to $\frac{1}{2}$ A each for I_{C_1} and I_{C_2}.

The net reactive branch current I_X is 3 A, then, equal to a 4-A I_L minus a 1-A I_C. Since I_L is larger, the net current is inductive.

The circuit in Fig. 24-11 is equivalent to the circuit in Fig. 24-10, therefore. Both have a 4-A resistive current I_R and a 3-A net inductive current I_L. These values added by phasors make a total of 5 A for I_T in the main line.

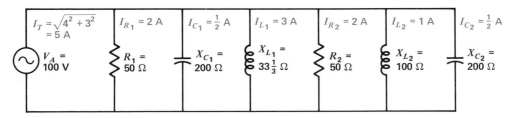

Fig. 24-11 Parallel ac circuit with more components than Fig. 24-10, but with the same values of I_T, Z, and θ.

Practice Problems 24-6
Answers at End of Chapter

a. In Fig. 24-10, what is the net reactive branch current?

b. In Fig. 24-11, what is the net reactive branch current?

24-7 Series-Parallel Reactance and Resistance

Figure 24-12 shows how a series-parallel circuit can be reduced to a series circuit with just one reactance and one resistance. The method is straightforward as long as resistance and reactance are not combined in one parallel bank or series string.

Working backward toward the generator from the outside branch in Fig. 24-12a, we have an X_{L_1} and an X_{L_2} of 100 Ω each in series, which total 200 Ω. This string in Fig. 24-12a is equivalent to X_{L_5} in Fig. 24-12b.

In the other branch, the net reactance of X_{L_3} and X_C is equal to 600 Ω minus 400 Ω. This is equivalent to the 200 Ω of X_{L_4} in Fig. 24-12b. The X_{L_4} and X_{L_5} of 200 Ω each in parallel are combined for an X_L of 100 Ω.

In Fig. 24-12c, the 100-Ω X_L is in series with the 100-Ω $R_{1\text{-}2}$. This value is for R_1 and R_2 in parallel.

The phasor diagram for the equivalent circuit in Fig. 24-12d shows the total impedance Z of 141 Ω for a 100-Ω R in series with a 100-Ω X_L.

With a 141-Ω impedance across the applied V_T of 100 V, the current in the generator is 0.7 A. The phase angle θ is 45° for this circuit.[1]

---

[1]More complicated ac circuits with series-parallel impedances are analyzed with complex numbers, as explained in Chap. 25.

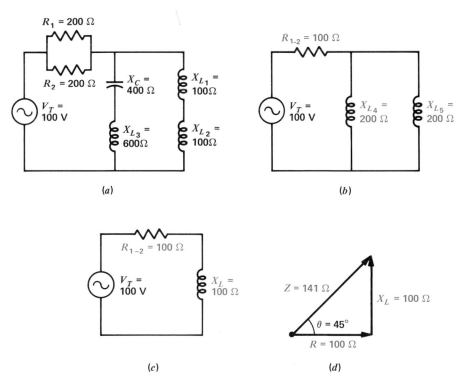

Fig. 24-12 Reducing a series-parallel circuit with R, X_L, and X_C to a series circuit with one R and one X. (a) Actual circuit. (b) Simplified arrangement. (c) Series equivalent circuit. (d) Phasor diagram.

Practice Problems 24-7
Answers at End of Chapter

Refer to Fig. 24-12.
a. How much is $X_{L_1} + X_{L_2}$?
b. How much is $X_{L_3} - X_C$?
c. How much is X_{L_4} in parallel with X_{L_5}?

24-8 Real Power

In an ac circuit with reactance, the current I supplied by the generator either leads or lags the generator voltage V. Then the product VI is not the real power produced by the generator, since the voltage may have a high value while the current is near zero, or vice versa. The real power, however, can always be calculated as I^2R, where R is the total resistive component of the circuit, because current and voltage have the same phase in a resistance. To find the corresponding value of power as VI, this product must be multiplied by the cosine of the phase angle θ. Then

$$\text{Real power} = I^2R \qquad \text{(24-3)}$$

or

$$\text{Real power} = VI \cos \theta \qquad \text{(24-4)}$$

where V and I are in rms values, to calculate the real power, in watts. Multiplying VI by the cosine of the phase angle provides the resistive component for real power equal to I^2R.

For example, the ac circuit in Fig. 24-13 has 2 A through a 100-Ω R in series with the X_L of 173 Ω. Therefore

$$\text{Real power} = I^2R = 4 \times 100$$
$$\text{Real power} = 400 \text{ W}$$

Furthermore, in this circuit the phase angle is 60° with a cosine of 0.5. The applied voltage is 400 V. Therefore

$$\text{Real power} = VI \cos \theta = 400 \times 2 \times 0.5$$
$$\text{Real power} = 400 \text{ W}$$

In both examples, the real power is the same 400 W, because this is the amount of power supplied by the generator and dissipated in the resistance. Either formula can be used for calculating the real power, depending on which is more convenient.

Real power can be considered as resistive power, which is dissipated as heat. A reactance does not dissipate power but stores energy in the electric or magnetic field.

Power Factor Because it indicates the resistive component, $\cos \theta$ is the power factor of the circuit, converting the VI product to real power. For series circuits, use the formula

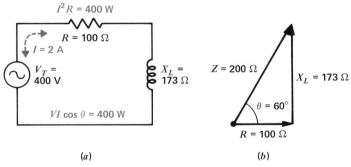

Fig. 24-13 Real power in series circuit. (*a*) Schematic diagram. (*b*) Phasor diagram.

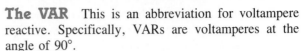

Power factor $= \cos \theta = \dfrac{R}{Z}$ (24-5)

or for parallel circuits

Power factor $= \cos \theta = \dfrac{I_R}{I_T}$ (24-6)

In Fig. 24-13, as an example of a series circuit, we use R and Z for the calculations:

Power factor $= \cos \theta = \dfrac{R}{Z} = \dfrac{100\ \Omega}{200\ \Omega} = 0.5$

For the parallel circuit in Fig. 24-10, we use the resistive current I_R and the I_T:

Power factor $= \cos \theta = \dfrac{I_R}{I_T} = \dfrac{4\ \text{A}}{5\ \text{A}} = 0.8$

The power factor is not an angular measure but a numerical ratio, with a value between 0 and 1, equal to the cosine of the phase angle.

With all resistance and zero reactance, R and Z are the same for a series circuit, or I_R and I_T are the same for a parallel circuit, and the ratio is 1. Therefore, unity power factor means a resistive circuit. At the opposite extreme, all reactance with zero resistance makes the power factor zero, meaning that the circuit is all reactive.

Apparent Power   When V and I are out of phase because of reactance, the product of $V \times I$ is called *apparent power*. The unit is *voltamperes* (VA) instead of watts, since the watt is reserved for real power.

For the example in Fig. 24-13, with 400 V and the 2-A I, 60° out of phase, the apparent power is VI, or $400 \times 2 = 800$ VA. Note that apparent power is the VI product alone, without considering the power factor $\cos \theta$.

The power factor can be calculated as the ratio of real power to apparent power, as this ratio equals $\cos \theta$. As an example, in Fig. 24-13, the real power is 400 W, and the apparent power is 800 VA. The ratio of $^{400}/_{800}$ then is 0.5 for the power factor, the same as $\cos 60°$.

The VAR   This is an abbreviation for voltampere reactive. Specifically, VARs are voltamperes at the angle of 90°.

In general, for any phase angle θ between V and I, multiplying VI by $\sin \theta$ gives the vertical component at 90° for the value of the VARs. In Fig. 24-13, the value of $VI \sin 60°$ is $800 \times 0.866 = 692.8$ VAR.

Note that the factor $\sin \theta$ for the VARs gives the vertical or reactive component of the apparent power VI. However, multiplying VI by $\cos \theta$ as the power factor gives the horizontal or resistive component for the real power.

Correcting the Power Factor   In commercial use, the power factor should be close to unity for efficient distribution. However, the inductive load of motors may result in a power factor of 0.7, as an example, for the phase angle of 45°. To correct for this lagging inductive component of the current in the main line, a capacitor can be connected across the line to draw leading current from the source. To bring the power factor up to 1.0, that is, unity PF, the value of capacitance is calculated to take the same amount of voltamperes as the VARs of the load.

Practice Problems 24-8
Answers at End of Chapter

a. What is the unit for real power?
b. What is the unit for apparent power?

24-9   AC Meters

The D'Arsonval moving-coil type of meter movement will not read if it is used in an ac circuit because the average value of an alternating current is zero. Since the two opposite polarities cancel, an alternating current cannot deflect the meter movement either up-scale or down-scale. An ac meter must produce deflection of the meter pointer up-scale regardless of polarity. This deflection is accomplished by one of the following three methods for ac meters. *careless*

1. *Thermal type.* In this method, the heating effect of the current, which is independent of polarity, is

used to provide meter deflection. Two examples are the thermocouple type and hot-wire meter.

2. *Electromagnetic type*. In this method, the relative magnetic polarity is maintained constant although the current reverses. Examples are the iron-vane meter, dynamometer, and wattmeter.

3. *Rectifier type*. The rectifier changes the ac input to dc output for the meter, which is usually a D'Arsonval movement. This type is the most common for ac voltmeters generally used for audio and radio frequencies.

All ac meters have scales calibrated in rms values, unless noted otherwise on the meter.

A thermocouple consists of two dissimilar metals joined together at one end but open at the opposite side. Heat at the short-circuited junction produces a small dc voltage across the open ends, which are connected to a dc meter movement. In the hot-wire meter, current heats a wire to make it expand, and this motion is converted into meter deflection. Both types are used as ac meters for radio frequencies.

The iron-vane meter and dynamometer have very low sensitivity, compared with a D'Arsonval movement. They are used in power circuits, for either direct current or 60-Hz alternating current.

Practice Problems 24-9
Answers at End of Chapter

Answer True or False.
a. The iron-vane meter can read alternating current.
b. The D'Arsonval meter movement is for direct current only.

24-10 Wattmeters

The wattmeter uses fixed coils to indicate current in the circuit, while the movable coil indicates voltage (Fig. 24-14). The deflection then is proportional to power. Either dc power or real ac power can be read directly by the wattmeter.

In Fig. 24-14*a*, the coils L_{I_1} and L_{I_2} in series are the stationary coils serving as an ammeter to measure current. The two I terminals are connected in one side of the line in series with the load. The movable coil L_V and its multiplier resistance R_M are used as a voltmeter, with the V terminals connected across the line in parallel with the load. Then the current in the fixed coils is proportional to I, while the current in the movable coil is proportional to V. As a result, the deflection is proportional to the VI product, which is power.

Furthermore, it is the VI product for each instant of time that produces deflection. For instance, if the V value is high when the I value is low, for a phase angle close to 90°, there will be little deflection. The meter deflection is proportional to the watts of real power, therefore, regardless of the power factor in ac circuits. The wattmeter is commonly used to measure power from the 60-Hz power line. For radio frequencies, however, power is generally measured in terms of heat transfer.

Practice Problems 24-10
Answers at End of Chapter

a. Does a wattmeter measure real or apparent power?
b. In Fig. 24-14, does the movable coil of the wattmeter measure V or I?

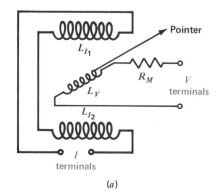

(a)

(b)

Fig. 24-14 Wattmeter. (*a*) Schematic of voltage and current coils. (*b*) Meter for range of 0 to 200 W. (*W. M. Welch Mfg. Co.*)

24-11 Summary of Types of Ohms in AC Circuits

The differences in R, X_L, X_C, and Z are listed in Table 24-1, but the following general features should also be noted. Ohms of opposition limit the amount of current in dc circuits or ac circuits. Resistance R is the same for either case. However, ac circuits can have ohms of reactance because of the variations in alternating current or voltage. Reactance X_L is the reactance of an inductance with sine-wave changes in current. Reactance X_C is the reactance of a capacitor with sine-wave changes in voltage.

Both X_L and X_C are measured in ohms, like R, but reactance has a 90° phase angle, while the phase angle for resistance is 0°. A circuit with steady direct current cannot have any reactance.

Ohms of X_L or X_C are opposite, as X_L has a phase angle of +90°, while X_C has the angle of −90°. Any individual X_L or X_C always has a phase angle that is exactly 90°.

Ohms of impedance Z result from the phasor combination of resistance and reactance. In fact, Z can be considered the general form of any ohms of opposition in ac circuits.

Z can have any phase angle, depending on the relative amounts of R and X. When Z consists mostly of R with little reactance, the phase angle of Z is close to 0°. With R and X equal, the phase angle of Z is 45°. Whether the angle is positive or negative depends on whether the net reactance is inductive or capacitive. When Z consists mainly of X with little R, the phase angle of Z is close to 90°.

The phase angle is θ_Z for Z or V_T with respect to the common I in a series circuit. With parallel branch currents, θ_I is for I_T in the main line with respect to the common voltage.

Practice Problems 24-11
Answers at End of Chapter

a. Which of the following does not change with frequency: Z, X_L, X_C, or R?
b. Which has lagging current: R, X_L, or X_C?
c. Which has leading current: R, X_L, or X_C?

24-12 Summary of Types of Phasors in AC Circuits

The phasors for ohms, volts, and amperes are shown in Fig. 24-15. Note the similarities and differences:

Series Components In series circuits, ohms and voltage drops have similar phasors. The reason is the common I for all the series components. Therefore:

V_R or IR has the same phase as R.

V_L or IX_L has the same phase as X_L.

V_C or IX_C has the same phase as X_C.

Resistance The R, V_R, and I_R always have the same angle because there is no phase shift in a resistance. This applies to R in either a series or a parallel circuit.

Table 24-1. Types of Ohms in AC Circuits

	Resistance R, Ω	Inductive reactance X_L, Ω	Capacitive reactance X_C, Ω	Impedance Z, Ω
Definition	In-phase opposition to alternating or direct current	90° leading opposition to alternating current	90° lagging opposition to alternating current	Phasor combination of resistance and reactance $Z = \sqrt{R^2 + X^2}$
Effect of frequency	Same for all frequencies	Increases with higher frequencies	Decreases with higher frequencies	X_L component increases, but X_C decreases
Phase angle θ	0°	I_L lags V_L by 90°	V_C lags I_C by 90°	Tan $\theta = \pm \dfrac{X}{R}$ in series, or $\pm \dfrac{I_X}{I_R}$ in parallel

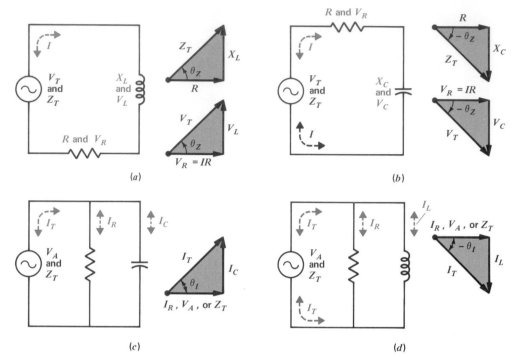

Fig. 24-15 Summary of phasor relations in ac circuits. (*a*) Reactance X_L and R in series. (*b*) Reactance X_C and R in series. (*c*) Parallel branches with I_C and I_R. (*d*) Parallel branches with I_L and I_R.

Reactance   Reactances X_L and X_C are 90° phasors in opposite directions. The X_L or V_L has the angle of +90° with an upward phasor, while the X_C or V_C has the angle of −90° with a downward phasor.

Reactive Branch Currents   The phasor of a parallel branch current is opposite from its reactance. Therefore, I_C is upward at +90°, opposite from X_C downward at −90°. Also, I_L is downward at −90°, opposite from X_L upward at +90°.

In short, I_C and I_L are opposite from each other, and both are opposite from their corresponding reactances.

Phase Angle θ_Z   The phasor resultant for ohms of reactance and resistance is the impedance Z. The phase angle θ for Z can be any angle between 0 and 90°. In a series circuit θ_Z for Z is the same as θ for V_T with respect to the common current I.

Phase Angle θ_I   The phasor resultant of branch currents is the total line current I_T. The phase angle of I_T can be any angle between 0 and 90°. In a parallel circuit, θ_I is the angle of I_T with respect to the applied voltage V_A.

The θ_I is the same value but of opposite sign from θ_Z for Z, which is the impedance of the combined parallel branches.

The reason for the change of sign is that θ_I is for I_T with respect to the common V, but θ_Z is for V_T with respect to the common current I.

Such phasor combinations are necessary in sine-wave ac circuits in order to take into account the effect of reactance. The phasors can be analyzed either graphically, as in Fig. 24-15, or by the shorter technique of complex numbers, with a j operator that corresponds to a 90° phasor.

Practice Problems 24-12
Answers at End of Chapter

a. Of the following three phasors, which two are 180° opposite: V_L, V_C, or V_R?
b. Of the following three phasors, which two are out of phase by 90°: I_R, I_T, or I_L?

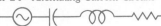
Summary

1. In ac circuits with resistance alone, the circuit is analyzed the same way as for dc circuits, generally with rms ac values. Without any reactance, the phase angle is zero.
2. When capacitive reactances alone are combined, the X_C values are added in series and combined by the reciprocal formula in parallel, just like ohms of resistance. Similarly, ohms of X_L alone can be added in series or combined by the reciprocal formula in parallel, just like ohms of resistance.
3. Since X_C and X_L are opposite reactances, they cancel each other. In series, the ohms of X_C and X_L cancel. In parallel, the capacitive and inductive branch currents I_C and I_L cancel.
4. In ac circuits with R, X_L, and X_C, they can be reduced to one equivalent resistance and one net reactance.
5. In series, the total R and net X at 90° are combined as $Z = \sqrt{R^2 + X^2}$. The phase angle of the series R and X is the angle with tangent $\pm X/R$. First we calculate Z_T and then divide into V_T to find I.
6. For parallel branches, the total I_R and net reactive I_X at 90° are combined as $I_T = \sqrt{I_R^2 + I_X^2}$. The phase angle of the parallel R and X is the angle with tangent $\pm I_X/I_R$. First we calculate I_T and then divide into V_A to find Z_T.
7. The quantities R, X_L, X_C, and Z in ac circuits all are ohms of opposition. The differences with respect to frequency and phase angle are summarized in Table 24-1.
8. The phasor relations for resistance and reactance are summarized in Fig. 24-15.
9. In ac circuits with reactance, the real power in watts equals I^2R, or $VI \cos \theta$, where θ is the phase angle. The real power is the power dissipated as heat in resistance. Cos θ is the power factor of the circuit.
10. The wattmeter measures real ac power or dc power.

Self-Examination
Answers at Back of Book

Choose (a), (b), (c), or (d).

1. In an ac circuit with resistance but no reactance, (a) two 1000-Ω resistances in series total 1414 Ω; (b) two 1000-Ω resistances in series total 2000 Ω; (c) two 1000-Ω resistances in parallel total 707 Ω; (d) a 1000-Ω R in series with a 400-Ω R totals 600 Ω.
2. An ac circuit has a 100-Ω X_{C_1}, a 50-Ω X_{C_2}, a 40-Ω X_{L_1}, and a 30-Ω X_{L_2}, all in series. The net reactance is equal to (a) an 80-Ω X_L; (b) a 200-Ω X_L; (c) an 80-Ω X_C; (d) a 220-Ω X_C.
3. An ac circuit has a 40-Ω R, a 90-Ω X_L, and a 60-Ω X_C, all in series. The impedance Z equals (a) 50 Ω; (b) 70.7 Ω; (c) 110 Ω; (d) 190 Ω.
4. An ac circuit has a 100-Ω R, a 100-Ω X_L, and a 100-Ω X_C, all in series. The impedance Z of the series combination is equal to (a) 33⅓ Ω; (b) 70.7 Ω; (c) 100 Ω; (d) 300 Ω.
5. An ac circuit has a 100-Ω R, a 300-Ω X_L, and a 200-Ω X_C, all in series. The phase angle θ of the circuit equals (a) 0°; (b) 37°; (c) 45°; (d) 90°.

6. The power factor of an ac circuit equals (a) the cosine of the phase angle; (b) the tangent of the phase angle; (c) zero for a resistive circuit; (d) unity for a reactive circuit.

7. Which phasors in the following combinations are *not* in opposite directions? (a) X_L and X_C; (b) X_L and I_C; (c) I_L and I_C; (d) X_C and I_C.

8. In Fig. 24-8a, the voltage drop across X_L equals (a) 60 V; (b) 66⅔ V; (c) 120 V; (d) 200 V.

9. In Fig. 24-10a, the combined impedance of the parallel circuit equals (a) 5 Ω; (b) 12.5 Ω; (c) 20 Ω; (d) 100 Ω.

10. The wattmeter (a) has voltage and current coils to measure real power; (b) has three connections, two of which are used at a time; (c) measures apparent power because the current is the same in the voltage and current coils; (d) can measure dc power but not 60-Hz ac power.

Essay Questions

1. Why can series or parallel resistances be combined in ac circuits the same way as in dc circuits?

2. (a) Why do X_L and X_C reactances in series cancel each other? (b) With X_L and X_C reactances in parallel, why do their branch currents cancel?

3. Give one difference in electrical characteristics comparing R and X_C, R and Z, X_C and C, X_L and L.

4. Name three types of ac meters.

5. Make a diagram showing a resistance R_1 in series with the load resistance R_L, with a wattmeter connected to measure the power in R_L.

6. Make a phasor diagram for the circuit in Fig. 24-8a showing the phase of the voltage drops IR, IX_C, and IX_L with respect to the reference phase of the common current I.

7. Explain briefly why the two opposite phasors at +90° for X_L and −90° for I_L both follow the principle that any self-induced voltage leads the current through the coil by 90°.

8. Why is it that a reactance phasor is always at exactly 90° but an impedance phasor can be less than 90°?

9. Why must the impedance of a series circuit be more than either its X or R?

10. Why must I_T in a parallel circuit be more than either I_R or I_X?

Problems

Answers to Odd-Numbered Problems at Back of Book

1. Refer to Fig. 24-1a. (a) Calculate the total real power supplied by the source. (b) Why is the phase angle zero? (c) What is the power factor of the circuit?

2. In a series ac circuit, 2 A flows through a 20-Ω R, a 40-Ω X_L, and a 60-Ω X_C. (a) Make a schematic diagram of the series circuit. (b) Calculate the voltage drop across each series component. (c) How much is the applied voltage? (d) Calculate the power factor of the circuit. (e) What is the phase angle θ?

3. A parallel circuit has the following five branches: three resistances of 30 Ω each; an X_L of 600 Ω; an X_C of 400 Ω. (a) Make a schematic diagram of the circuit. (b) If 100 V is applied, how much is the total line current? (c) What is the total impedance of the circuit? (d) What is the phase angle θ?

4. Referring to Fig. 24-8, assume that the frequency is doubled from 500 to 1000 Hz. Find X_L, X_C, Z, I, and θ for this higher frequency. Calculate L and C.

5. A series circuit has a 300-Ω R, a 500-Ω X_{C_1}, a 300-Ω X_{C_2}, an 800-Ω X_{L_1}, and 400-Ω X_{L_2}, all in series with an applied voltage V of 400 V. (a) Draw the schematic diagram with all components. (b) Draw the equivalent circuit reduced to one resistance and one reactance. (c) Calculate Z_T, I, and θ.

6. Repeat Prob. 5 for a circuit with the same components in parallel across the voltage source.

7. A series circuit has a 600-Ω R, a 10-μH inductance L, and a 4-μF capacitance C, all in series with the 60-Hz 120-V power line as applied voltage. (a) Find the reactance of L and of C. (b) Calculate Z_T, I, and θ_Z.

8. Repeat Prob. 7 for the same circuit, but the 120-V source has $f = 10$ MHz.

9. (a) Referring to the series circuit Fig. 24-6, what is the phase angle between the IX_L voltage of 360 V and the IX_C voltage of 240 V? (b) Draw the two sine waves for these voltages, showing their relative amplitudes and phase corresponding to the phasor diagram in Fig. 24-6b. Also show the resultant sine wave of voltage across the net X_L.

10. How much resistance dissipates 600 W of ac power, with 4.3-A rms current?

11. How much resistance must be inserted in series with a 0.95-H inductance to limit the current to 0.25 A from the 120-V 60-Hz power line?

12. How much resistance must be inserted in series with a 10-μF capacitance to provide a phase angle of $-45°$? The source is the 120-V 60-Hz power line.

13. With the same R as in Prob. 12, what value of C is necessary for the angle of $-45°$ at the frequency of 2 MHz?

14. A parallel ac circuit has the following branch currents: $I_{R_1} = 4.2$ mA; $I_{R_2} = 2.4$ mA; $I_{L_1} = 7$ mA; $I_{L_2} = 1$ mA; $I_C = 6$ mA. Calculate I_T.

15. With 420 mV applied, an ac circuit has the following parallel branches: $R_1 = 100$ Ω; $R_2 = 175$ Ω; $X_{L_1} = 60$ Ω; $X_{L_2} = 420$ Ω; $X_C = 70$ Ω. Calculate I_T, θ_I, and Z_T.

16. The same components as in Prob. 15 are in series. Calculate Z_T, I, and θ_Z.

17. What R is needed in series with a 0.01-μF capacitor for a phase angle of $-64°$, with f of 800 Hz?

Answers to Practice Problems

24-1 a. 0°
   b. 0°

24-2 a. 90°
   b. −90°

24-3 a. −90°
   b. 90°

24-4 a. 20 Ω
   b. 1 A

24-5 a. $X_C = 30$ Ω
   b. $X_C = 30$ Ω

24-6 a. $I_L = 3$ A
   b. $I_L = 3$ A

24-7 a. 200 Ω
   b. 200 Ω
   c. 100 Ω

24-8 a. Watt
   b. Voltampere

24-9 a. T
   b. T

24-10 a. Real power
   b. V

24-11 a. R
   b. X_L
   c. X_C

24-12 a. V_L and V_C
   b. I_R and I_L

Chapter 25

Complex Numbers for AC Circuits

Complex numbers form a numerical system that includes the phase angle of a quantity, with its magnitude. Therefore, complex numbers are useful in ac circuits when the reactance of X_L or X_C makes it necessary to consider the phase angle.

Any type of ac circuit can be analyzed with complex numbers, but they are especially convenient for solving series-parallel circuits that have both resistance and reactance in one or more branches. Actually, the use of complex numbers is probably the best way to analyze ac circuits with series-parallel impedances.

Important terms in this chapter are:

admittance   real numbers
imaginary numbers   rectangular form
j operator   susceptance
polar form

More details are explained in the following sections:

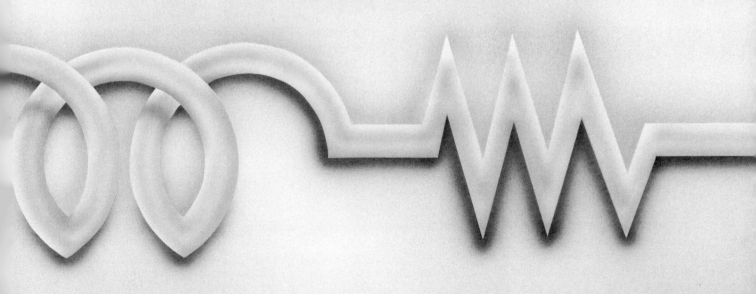

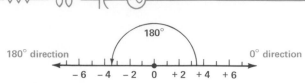

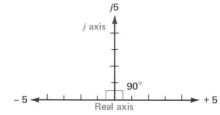

Fig. 25-1 Positive and negative numbers.

Fig. 25-2 The *j* axis at 90° from real axis.

25-1 Positive and Negative Numbers

Our common use of numbers as either positive or negative represents only two special cases. In their more general form, numbers have both quantity and phase angle. In Fig. 25-1, positive and negative numbers are shown as corresponding to the phase angles of 0 and 180°, respectively.

For example, the numbers 2, 4, and 6 represent units along the horizontal or *x* axis, extending toward the right along the line of zero phase angle. Therefore, positive numbers really represent units having the phase angle of 0°. Or this phase angle corresponds to the factor of +1. To indicate 6 units with zero phase angle, then, 6 is multiplied by +1 as a factor for the positive number 6. The + sign is often omitted, as it is assumed unless indicated otherwise.

In the opposite direction, negative numbers correspond to 180°. Or, this phase angle corresponds to the factor of −1. Actually, −6 represents the same quantity as 6 but rotated through the phase angle of 180°. The angle of rotation is the *operator* for the number. The operator for −1 is 180°; the operator for +1 is 0°.

Practice Problems 25-1
Answers at End of Chapter

a. What is the angle for the number +5?
b. What is the angle for the number −5?

25-2 The *j* Operator

The operator for a number can be any angle between 0 and 360°. Since the angle of 90° is important in ac circuits, the factor *j* is used to indicate 90°. See Fig. 25-2. Here, the number 5 means 5 units at 0°, the number −5 is at 180°, while *j*5 indicates the 90° angle.

The *j* is usually written before the number. The reason is that the *j* sign is a 90° operator, just as the + sign is a 0° operator and the − sign is a 180° operator. Any quantity at right angles to the zero axis, therefore, 90° counterclockwise, is on the +*j* axis.

In mathematics, numbers on the horizontal axis are real numbers, including positive and negative values. Numbers on the *j* axis are called *imaginary numbers,* only because they are not on the real axis. Also, in mathematics the abbreviation *i* is used in place of *j*. In electricity, however, *j* is used to avoid confusion with *i* as the symbol for current. Furthermore, there is nothing imaginary about electrical quantities on the *j* axis. An electric shock from *j*500 V is just as dangerous as 500 V positive or negative.

More features of the *j* operator are shown in Fig. 25-3. The angle of 180° corresponds to the *j* operation of 90° repeated twice. This angular rotation is indicated by the factor j^2. Note that the *j* operation multiplies itself, instead of adding.

Since j^2 means 180°, which corresponds to the factor of −1, we can say that j^2 is the same as −1. In short, the operator j^2 for a number means multiply by −1. For instance, $j^2 8$ is −8.

Furthermore, the angle of 270° is the same as −90°, which corresponds to the operator −*j*. These characteristics of the *j* operator are summarized as follows:

$$0° = 1$$
$$90° = j$$
$$180° = j^2 = -1$$
$$270° = j^3 = j^2 \times j = -1 \times j = -j$$
$$360° = \text{same as } 0°$$

As examples, the number 4 or −4 represents 4 units on the real horizontal axis; *j*4 means 4 units with a leading phase angle of 90°; −*j*4 means 4 units with a lagging phase angle of −90°.

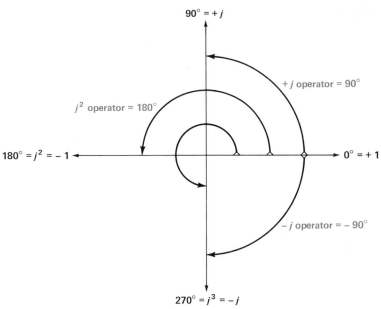

$90° = + j$

$+ j$ operator $= 90°$

j^2 operator $= 180°$

$180° = j^2 = - 1$

$0° = + 1$

$- j$ operator $= - 90°$

$270° = j^3 = - j$

Fig. 25-3 The j operator shows 90° rotation from real axis; $-j$ operator is $-90°$; j^2 operator is 180° rotation back to real axis.

Practice Problems 25-2
Answers at End of Chapter

a. What is the angle for the operator j?
b. What is the angle for the operator $-j$?

25-3   Definition of a Complex Number

The combination of a real and imaginary term is a complex number. Usually, the real number is written first. As an example, $3 + j4$ is a complex number including 3 units on the real axis added to 4 units 90° out of phase on the j axis. The name *complex number* just means that its terms must be added as phasors.

Phasors for complex numbers are shown in Fig. 25-4 on the next page. The $+j$ phasor is up for 90°; the $-j$ phasor is down for $-90°$. The phasors are shown with the end of one joined to the start of the next, to be ready for addition. Graphically, the sum is the hypotenuse of the right triangle formed by the two phasors. Since a number like $3 + j4$ specifies the phasors in rectangular

coordinates, this system is the *rectangular form* of complex numbers.

Be careful to distinguish a number like $j2$, where 2 is a coefficient, from j^2, where 2 is the exponent. The number $j2$ means 2 units up on the j axis of 90°. However, j^2 is the operator of -1, which is on the real axis in the negative direction.

Another comparison to note is between $j3$ and j^3. The number $j3$ is 3 units up on the j axis, while j^3 is the same as the $-j$ operator, which is down on the $-90°$ axis.

Also note that either the real term or j term can be the larger of the two. When the j term is larger, the angle is more than 45°; when the j term is smaller, the angle is less than 45°. If the j term and the real term are equal, the angle is 45°.

Practice Problems 25-3
Answers at End of Chapter

Answer True or False.
a. For $7 + j6$, the 6 is at 90° leading the 7.
b. For $7 - j6$, the 6 is at 90° lagging the 7.

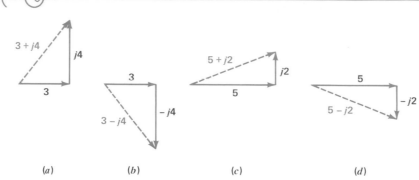

Fig. 25-4 Phasors corresponding to real terms and *j* terms, in rectangular coordinates.

25-4   How Complex Numbers Are Applied to AC Circuits

The applications are just a question of using a real term for 0°, $+j$ for 90°, and $-j$ for $-90°$, to denote the phase angles. Specifically, Fig. 25-5 below illustrates the following rules:

An *angle of 0°* or a real number without any *j* operator is used for resistance *R*. For instance, 3 Ω of *R* is stated just as 3 Ω.

An *angle of 90° or +j* is used for inductive reactance X_L. For instance, a 4-Ω X_L is $j4$ Ω. This rule always applies to X_L, whether it is in series or parallel with *R*. The reason is the fact that X_L represents voltage across an inductance, which always leads the current through the inductance by 90°. The $+j$ is also used for V_L.

An *angle of $-90°$ or $-j$* is used for capacitive reactance X_C. For instance, a 4-Ω X_C is $-j4$ Ω. This rule always applies to X_C, whether it is in series or parallel with *R*. The reason is the fact that X_C represents voltage across a capacitor, which always lags the charge and discharge current of the capacitor by $-90°$. The $-j$ is also used for V_C.

With reactive branch currents, the sign for *j* is reversed, compared with reactive ohms, because of the opposite phase angle. As shown in Fig. 25-6a and b, $-j$ is used for inductive branch current I_L and $+j$ for capacitive branch current I_C.

Practice Problems 25-4
Answers at End of Chapter

a. Write 3 kΩ of X_L with the *j* operator.
b. Write 5 mA of I_L with the *j* operator.

25-5   Impedance in Complex Form

The rectangular form of complex numbers is a convenient way to state the impedance of series resistance and reactance. In Fig. 25-5a, the impedance is $3 + j4$, as Z_a is the phasor sum of a 3-Ω *R* in series with $j4$ Ω for X_L. Similarly, Z_b is $3 - j4$ for a 3-Ω *R* in series with $-j4$ Ω for X_C. The minus sign results from adding the negative term for $-j$. More examples are:

For a 4-kΩ *R* and a 2-kΩ X_L in series,

$$Z_T = 4000 + j2000$$

For a 3-kΩ *R* and a 9-kΩ X_C in series,

$$Z_T = 3000 - j9000$$

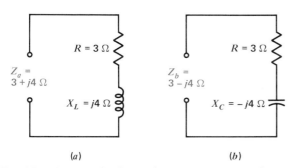

Fig. 25-5 Rectangular form of complex numbers for impedances. (*a*) Reactance X_L is $+j$. (*b*) Reactance X_C is $-j$.

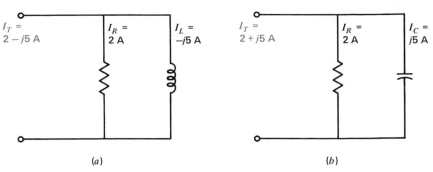

Fig. 25-6 Rectangular form of complex numbers for branch currents. (a) Current I_L is $-j$. (b) Current I_C is $+j$.

For a zero R and a 7-Ω X_L in series,

$Z_T = 0 + j7$

For a 12-Ω R and a zero reactance in series,

$Z_T = 12 + j0$

Note the general form of stating $Z = R \pm jX$. If one term is zero, substitute 0 for this term, in order to keep Z in its general form. This procedure is not required, but there is usually less confusion when the same form is used for all types of Z.

The advantage of this method is that multiple impedances written as complex numbers can then be calculated as follows:

$Z_T = Z_1 + Z_2 + Z_3 + \cdots + \text{etc.}$

for series impedances

$\dfrac{1}{Z_T} = \dfrac{1}{Z_1} + \dfrac{1}{Z_2} + \dfrac{1}{Z_3} + \cdots + \text{etc.}$

for parallel impedances

or

$Z_T = \dfrac{Z_1 \times Z_2}{Z_1 + Z_2}$ for two parallel impedances

Examples are shown in Fig. 25-7. The circuit in Fig. 25-7a is just a series combination of resistances and reactances. Combining the real terms and j terms separately, $Z_T = 12 + j4$. The calculations are $3 + 9 = 12\ \Omega$ for R and $j6$ added to $-j2$ equals $j4$ for the net X_L.

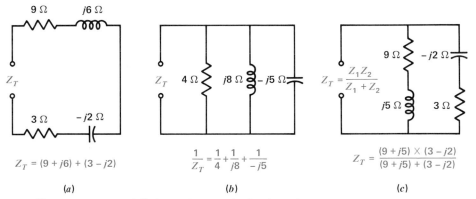

Fig. 25-7 Reactance X_L is a $+j$ term and X_C is a $-j$ term, whether in series or in parallel. (a) Series circuit. (b) Parallel branches. (c) Complex branch impedances Z_1 and Z_2 in parallel.

The parallel circuit in Fig. 25-7b shows that X_L is $+j$ and X_C is $-j$ even though they are in parallel branches, as they are reactances, not currents.

So far, these types of circuits can be analyzed with or without complex numbers. For the series-parallel circuit in Fig. 25-7c, however, the notation of complex numbers is necessary to state the complex impedance Z_T, consisting of branches with reactance and resistance in one or more of the branches. Impedance Z_T is just stated here in its form as a complex impedance. In order to calculate Z_T, some of the rules described in the next section must be used for combining complex numbers.

Practice Problems 25-5
Answers at End of Chapter

Write the following impedances in complex form.
a. X_L of 7 Ω in series with R of 4 Ω.
b. X_C of 7 Ω in series with zero R.

25-6   Operations with Complex Numbers

Real numbers and j terms cannot be combined directly because they are 90° out of phase. The following rules apply:

For Addition or Subtraction   Add or subtract the real and j terms separately:

$$(9 + j5) + (3 + j2) = 9 + 3 + j5 + j2$$
$$= 12 + j7$$
$$(9 + j5) + (3 - j2) = 9 + 3 + j5 - j2$$
$$= 12 + j3$$
$$(9 + j5) + (3 - j8) = 9 + 3 + j5 - j8$$
$$= 12 - j3$$

The answer should be in the form of $R \pm jX$, where R is the algebraic sum of all the real or resistive terms and X is the algebraic sum of all the imaginary or reactive terms.

To Multiply or Divide a j Term by a Real Number   Just multiply or divide the numbers. The answer is still a j term. Note the algebraic signs in the following examples. If both factors have the same sign, either $+$ or $-$, the answer is $+$; if one factor is negative, the answer is negative.

$$4 \times j3 = j12 \qquad j12 \div 4 = j3$$
$$j5 \times 6 = j30 \qquad j30 \div 6 = j5$$
$$j5 \times (-6) = -j30 \qquad -j30 \div (-6) = j5$$
$$-j5 \times 6 = -j30 \qquad -j30 \div 6 = -j5$$
$$-j5 \times (-6) = j30 \qquad j30 \div (-6) = -j5$$

To Multiply or Divide a Real Number by a Real Number   Just multiply or divide the real numbers, as in arithmetic. There is no j operation. The answer is still a real number.

To Multiply a j Term by a j Term   Multiply the numbers and the j coefficients to produce a j^2 term. The answer is a real term because j^2 is -1, which is on the real axis. Multiplying two j terms shifts the number 90° from the j axis to the real axis of 180°. As examples:

$$j4 \times j3 = j^2 12 = (-1)(12)$$
$$= -12$$
$$j4 \times (-j3) = -j^2 12 = -(-1)(12)$$
$$= 12$$

To Divide a j Term by a j Term   Divide the j coefficients to produce a real number; the j factors cancel. For instance:

$$j12 \div j4 = 3 \qquad -j12 \div j4 = -3$$
$$j30 \div j5 = 6 \qquad j30 \div (-j6) = -5$$
$$j15 \div j3 = 5 \qquad -j15 \div (-j3) = 5$$

To Multiply Complex Numbers   Follow the rules of algebra for multiplying two factors, each having two terms:

$$(9 + j5) \times (3 - j2) = 27 + j15 - j18 - j^2 10$$
$$= 27 - j3 - (-1)10$$
$$= 27 - j3 + 10$$
$$= 37 - j3$$

Note that $-j^2 10$ equals $+10$ because the operator j^2 is -1 and $-(-1)10$ becomes $+10$.

To Divide Complex Numbers   This process becomes more involved because division of a real number by an imaginary number is not possible. Therefore, the denominator must first be converted to a real number without any j term.

Converting the denominator to a real number without any j term is called *rationalization* of the fraction. To do this, multiply both numerator and denominator by the *conjugate* of the denominator. Conjugate complex numbers have equal terms but opposite signs for the j term. For instance, $(1 + j2)$ has the conjugate $(1 - j2)$.

Rationalization is permissible because the value of a fraction is not changed when both numerator and denominator are multiplied by the same factor. This procedure is the same as multiplying by 1. In the following example of division with rationalization the denominator $(1 + j2)$ has the conjugate $(1 - j2)$:

$$\frac{4 - j1}{1 + j2} = \frac{4 - j1}{1 + j2} \times \frac{(1 - j2)}{(1 - j2)}$$
$$= \frac{4 - j8 - j1 + j^2 2}{1 - j^2 4}$$
$$= \frac{4 - j9 - 2}{1 + 4}$$
$$= \frac{2 - j9}{5}$$
$$= 0.4 - j1.8$$

As a result of the rationalization, $4 - j1$ has been divided by $1 + j2$ to find the quotient that is equal to $0.4 - j1.8$.

Note that the product of a complex number and its conjugate always equals the sum of the squares of the numbers in each term. As another example, the product of $(2 + j3)$ and its conjugate $(2 - j3)$ must be $4 + 9$, which equals 13. Simple numerical examples of division and multiplication are given here because when the required calculations become too long, it is easier to divide and multiply complex numbers in polar form, as explained in Sec. 25-8.

Practice Problems 25-6
Answers at End of Chapter

a. $(2 + j3) + (3 + j4) = ?$
b. $(2 + j3) \times 2 = ?$

25-7   Magnitude and Angle of a Complex Number

In electrical terms a complex impedance $(4 + j3)$ means 4 Ω of resistance and 3 Ω of inductive reactance with a leading phase angle of 90°. See Fig. 25-8a on the next page. The magnitude of Z is the resultant, equal to $\sqrt{16 + 9} = \sqrt{25} = 5$ Ω. Finding the square root of the sum of the squares is vector or phasor addition of two terms in quadrature, 90° out of phase.

The phase angle[1] of the resultant is the angle whose tangent is ¾ or 0.75. This angle equals 37°. Therefore, $4 + j3 = 5 \underline{/37°}$.

When calculating the tangent ratio, note that the j term is the numerator and the real term is the denominator because the tangent of the phase angle is the ratio of the opposite side to the adjacent side. With a negative j term, the tangent is negative, which means a negative phase angle.

Note the following definitions: $(4 + j3)$ is the complex number in rectangular coordinates. The real term is 4. The imaginary term is $j3$. The resultant 5 is the magnitude, absolute value, or modulus of the complex number. Its phase angle or argument is 37°. The resultant value by itself can be written as $|5|$, with vertical lines to indicate it is the magnitude without the phase angle. The magnitude is the value a meter would read.

---

[1]Appendix E, ''Trigonometric Functions,'' explains the sine, cosine, and tangent of an angle. See also B. Grob, *Mathematics: Outline and Review Problems for Basic Electronics,* McGraw-Hill Book Company, New York.

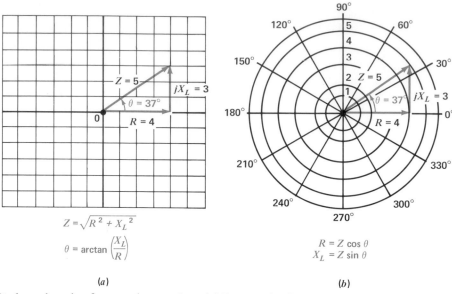

$$Z = \sqrt{R^2 + X_L^2}$$

$$\theta = \arctan\left(\frac{X_L}{R}\right)$$

$$R = Z \cos \theta$$
$$X_L = Z \sin \theta$$

(a)   (b)

Fig. 25-8 Magnitude and angle of a complex number. (a) Rectangular form. (b) Polar form.

For instance, with a current of $5\,\underline{/\,37°}$ A in a circuit, an ammeter reads 5 A. As additional examples:

$$2 + j4 = \sqrt{4 + 16} \ (\arctan 2) = 4.47\,\underline{/\,63°}$$
$$4 + j2 = \sqrt{16 + 4} \ (\arctan 0.5) = 4.47\,\underline{/\,26.5°}$$
$$8 + j6 = \sqrt{64 + 36} \ (\arctan 0.75) = 10\,\underline{/\,37°}$$
$$8 - j6 = \sqrt{64 + 36} \ (\arctan -0.75) = 10\,\underline{/\,-37°}$$
$$4 + j4 = \sqrt{16 + 16} \ (\arctan 1) = 5.66\,\underline{/\,45°}$$
$$4 - j4 = \sqrt{16 + 16} \ (\arctan -1) = 5.66\,\underline{/\,-45°}$$

Note that arctan 2, for example, means the angle with a tangent equal to 2. This can also be indicated as $\tan^{-1} 2$. In either case, the angle is specified as having 2 for its tangent, and the angle is 63.4°.

Practice Problems 25-7
Answers at End of Chapter

For the complex impedance $10 + j10$ Ω,
a. Calculate the magnitude.
b. Calculate the phase angle.

25-8 Polar Form of Complex Numbers

Calculating the magnitude and phase angle of a complex number is actually converting to an angular form in polar coordinates. As shown in Fig. 25-8, the rectangular form $4 + j3$ is equal to $5\,\underline{/\,37°}$ in polar form. In polar coordinates, the distance out from the center is the magnitude of the vector Z. Its phase angle θ is counterclockwise from the 0° axis.

To convert any complex number to polar form:

1. Find the magnitude by phasor addition of the j term and real term.
2. Find the angle whose tangent is the j term divided by the real term. As examples:

$$2 + j4 = 4.47\,\underline{/\,63°}$$
$$4 + j2 = 4.47\,\underline{/\,26.5°}$$
$$8 + j6 = 10\,\underline{/\,37°}$$
$$8 - j6 = 10\,\underline{/\,-37°}$$
$$4 + j4 = 5.66\,\underline{/\,45°}$$
$$4 - j4 = 5.66\,\underline{/\,-45°}$$

These examples are the same as those given before for finding the magnitude and phase angle of a complex number.

The magnitude in polar form must be more than either term in rectangular form, but less than the arithmetic sum of the two terms. For instance, in $8 + j6 = 10 \angle 37°$ the magnitude of 10 is more than 8 or 6 but less than their sum of 14.

Applied to ac circuits with resistance for the real term and reactance for the j term, then, the polar form of a complex number states the resultant impedance and its phase angle. Note the following cases for an impedance where either the resistance or reactance is reduced to zero.

$$0 + j5 = 5 \angle 90°$$
$$0 - j5 = 5 \angle -90°$$
$$5 + j0 = 5 \angle 0°$$

The polar form is much more convenient for multiplying or dividing complex numbers. The reason is that multiplication in polar form is reduced to addition of the angles, and the angles are just subtracted for division in polar form. The following rules apply.

For Multiplication   Multiply the magnitudes but add the angles algebraically:

$$24 \angle 40° \times 2 \angle 30° = 48 \angle +70°$$
$$24 \angle 40° \times (-2 \angle 30°) = -48 \angle +70°$$
$$12 \angle -20° \times 3 \angle -50° = 36 \angle -70°$$
$$12 \angle -20° \; 4 \angle 5° = 48 \angle -15°$$

When you multiply by a real number, just multiply the magnitudes:

$$4 \times 2 \angle 30° = 8 \angle 30°$$
$$4 \times 2 \angle -30° = 8 \angle -30°$$
$$-4 \times 2 \angle 30° = -8 \angle 30°$$
$$-4 \times (-2 \angle 30°) = 8 \angle 30°$$

This rule follows from the fact that a real number has an angle of 0°. When you add 0° to any angle, the sum equals the same angle.

For Division   Divide the magnitudes but subtract the angles algebraically:

$$24 \angle 40° \div 2 \angle 30° = 12 \angle 40° - 30°$$
$$= 12 \angle 10°$$
$$12 \angle 20° \div 3 \angle 50° = 4 \angle 20° - 50°$$
$$= 4 \angle -30°$$
$$12 \angle -20° \div 4 \angle 50° = 3 \angle -20° - 50°$$
$$= 3 \angle -70°$$

To divide by a real number, just divide the magnitudes:

$$12 \angle 30° \div 2 = 6 \angle 30°$$
$$12 \angle -30° \div 2 = 6 \angle -30°$$

This rule is also a special case that follows from the fact that a real number has a phase angle of 0°. When you subtract 0° from any angle, the remainder equals the same angle.

For the opposite case, however, when you divide a real number by a complex number, the angle of the denominator changes its sign in the answer in the numerator. This rule still follows the procedure of subtracting angles for division, since a real number has a phase angle of 0°. As examples,

$$\frac{10}{5 \angle 30°} = \frac{10 \angle 0°}{5 \angle 30°}$$
$$= 2 \angle 0° - 30° = 2 \angle -30°$$
$$\frac{10}{5 \angle -30°} = \frac{10 \angle 0°}{5 \angle -30°}$$
$$= 2 \angle 0° - (-30°) = 2 \angle +30°$$

Stated another way, we can say that the reciprocal of an angle is the same angle but with opposite sign. Note that this operation is similar to working with powers of 10. Angles and powers of 10 follow the general rules of exponents.

Practice Problems 25-8
Answers at End of Chapter

a. $6 \angle 20° \times 2 \angle 30° = ?$
b. $6 \angle 20° \div 2 \angle 30° = ?$

25-9 Converting Polar to Rectangular Form

Complex numbers in polar form are convenient for multiplication and division, but they cannot be added or subtracted. The reason is that changing the angle corresponds to the operation of multiplying or dividing. When complex numbers in polar form are to be added or subtracted, therefore, they must be converted back into rectangular form.

Consider the impedance $Z\underline{/\theta}$ in polar form. Its value is the hypotenuse of a right triangle with sides formed by the real term and j term in rectangular coordinates. See Fig. 25-9. Therefore, the polar form can be converted to rectangular form by finding the horizontal and vertical sides of the right triangle. Specifically:

Real term for $R = Z \cos \theta$

j term for $X = Z \sin \theta$

In Fig. 25-9a, assume that $Z\underline{/\theta}$ in polar form is $5\underline{/37°}$. The sine of 37° is 0.6 and its cosine is 0.8. To convert to rectangular form:

$R = Z \cos \theta = 5 \times 0.8 = 4$

$X = Z \sin \theta = 5 \times 0.6 = 3$

Therefore,

$5\underline{/37°} = 4 + j3$

This example is the same as the illustration in Fig. 25-8. The + sign for the j term means it is X_L, not X_C.

In Fig. 25-9b, the values are the same, but the j term is negative when θ is negative. The negative angle has a negative j term because the opposite side is in the fourth quadrant, where the sine is negative. However, the real term is still positive because the cosine is positive.

Note that R for $\cos \theta$ is the horizontal phasor, which is an adjacent side of the angle. The X for $\sin \theta$ is the vertical phasor, which is opposite the angle. The $+X$ is X_L; the $-X$ is X_C. You can ignore the sign of θ in calculating $\sin \theta$ and $\cos \theta$ because the values are the same up to $+90°$ or down to $-90°$.

These rules apply for angles in the first or fourth quadrant, from 0 to 90° or from 0 to $-90°$. As examples:

$$14.14\underline{/45°} = 10 + j10$$
$$14.14\underline{/-45°} = 10 - j10$$
$$10\underline{/90°} = 0 + j10$$
$$10\underline{/-90°} = 0 - j10$$
$$100\underline{/30°} = 86.6 + j50$$
$$100\underline{/-30°} = 86.6 - j50$$
$$100\underline{/60°} = 50 + j86.6$$
$$100\underline{/-60°} = 50 - j86.6$$

When going from one form to the other, keep in mind whether the angle is smaller or greater than 45° and if the j term is smaller or larger than the real term.

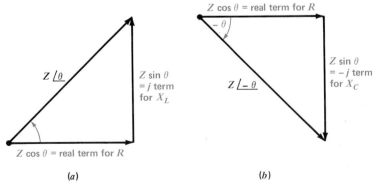

Fig. 25-9 Converting polar form of $Z\underline{/\theta}$ to rectangular form of $R \pm jX$. (a) Positive angle θ in first quadrant has $+j$ term. (b) Negative angle $-\theta$ in fourth quadrant has $-j$ term.

For angles between 0 and 45°, the opposite side, which is the j term, must be smaller than the real term. For angles between 45 and 90°, the j term must be larger than the real term.

To summarize how complex numbers are used in ac circuits in rectangular and polar form:

1. For addition or subtraction, complex numbers must be in rectangular form. This procedure applies to the addition of impedances in a series circuit. If the series impedances are in rectangular form, just combine all the real terms and j terms separately. If the series impedances are in polar form, they must be converted to rectangular form to be added.

2. For multiplication and division, complex numbers are generally used in polar form because the calculations are faster. If the complex number is in rectangular form, convert to polar form. With the complex number available in both forms, then you can quickly add or subtract in rectangular form and multiply or divide in polar form. Sample problems showing how to apply these methods in the analysis of ac circuits are illustrated in the following sections.

Practice Problems 25-9
Answers at End of Chapter

Convert to rectangular form.
a. $14.14 \underline{/45°}$.
b. $14.14 \underline{/-45°}$.

25-10 Complex Numbers in Series AC Circuits

Refer to the diagram in Fig. 25-10 on the next page. Although a circuit like this with only series resistances and reactances can be solved just by phasors, the complex numbers show more details of the phase angles.

Z_T in Rectangular Form The total Z_T in Fig. 25-10a is the sum of the impedances:

$$Z_T = 2 + j4 + 4 - j12$$
$$= 6 - j8$$

The total series impedance then is $6 - j8$. Actually, this amounts to adding all the series resistances for the real term and finding the algebraic sum of all the series reactances for the j term.

Z_T in Polar Form We can convert Z_T from rectangular to polar form as follows:

$$Z_T = 6 - j8$$
$$= \sqrt{36 + 64} \underline{/\arctan -8/6}$$
$$= \sqrt{100} \underline{/\arctan -1.33}$$
$$Z_T = 10 \underline{/-53°} \ \Omega$$

The angle of $-53°$ for Z_T means this is the phase angle of the circuit. Or the applied voltage and the current are $53°$ out of phase.

Calculating I The reason for the polar form is to divide Z_T into the applied voltage V_T to calculate the current I. See Fig. 25-10b. Note that the V_T of 20 V is a real number without any j term. Therefore, the applied voltage is $20 \underline{/0°}$. This angle of $0°$ for V_T makes it the reference phase for the following calculations. We can find the current as

$$I = \frac{V_T}{Z_T} = \frac{20 \underline{/0°}}{10 \underline{/-53°}}$$
$$= 2 \underline{/0° - (-53°)}$$
$$I = 2 \underline{/53°} \ \text{A}$$

Note that Z_T has the negative angle of $-53°$ but the sign changes to $+53°$ for I because of the division into a quantity with the angle of $0°$. In general, the reciprocal of an angle in polar form is the same angle with opposite sign.

Phase Angle of the Circuit The fact that I has the angle of $+53°$ means it leads V_T. The positive angle for I shows the series circuit is capacitive, with leading current. This angle is more than $45°$ because the net reactance is more than the total resistance, resulting in a tangent function greater than 1.

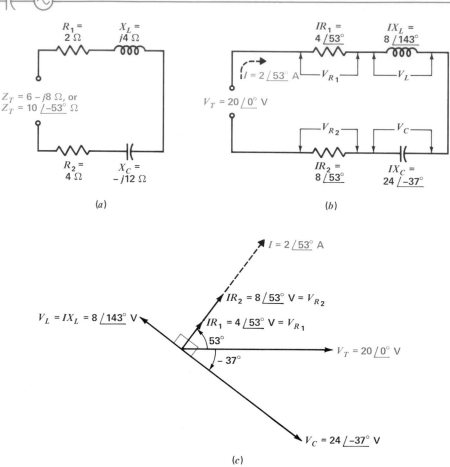

Fig. 25-10 Complex numbers applied to series ac circuits. See text for analysis.
(a) Circuit with series impedances. (b) Current and voltages. (c) Phasor diagram
of current and voltages.

Finding Each *IR* Drop

To calculate the voltage drops around the circuit, each resistance or reactance can be multiplied by *I*:

$$V_{R_1} = IR_1 = 2\,\underline{/53^\circ} \times 2\,\underline{/0^\circ} = 4\,\underline{/53^\circ}\ \text{V}$$

$$V_L = IX_L = 2\,\underline{/53^\circ} \times 4\,\underline{/90^\circ} = 8\,\underline{/143^\circ}\ \text{V}$$

$$V_C = IX_C = 2\,\underline{/53^\circ} \times 12\,\underline{/-90^\circ} = 24\,\underline{/-37^\circ}\ \text{V}$$

$$V_{R_2} = IR_2 = 2\,\underline{/53^\circ} \times 4\,\underline{/0^\circ} = 8\,\underline{/53^\circ}\ \text{V}$$

Phase of Each Voltage

The phasors for these voltages are in Fig. 25-10c. They show the phase angles using the applied voltage V_T as the zero reference phase.

The angle of 53° for V_{R_1} and V_{R_2} shows that the voltage across a resistance has the same phase as *I*. These voltages lead V_T by 53° because of the leading current.

For V_C, its angle of −37° means it lags the generator voltage V_T by this much. However, this voltage across X_C still lags the current by 90°, which is the difference between 53° and −37°.

The angle of 143° for V_L in the second quadrant is still 90° leading the current at 53°, as 143° − 53° = 90°. With respect to the generator voltage V_T, though, the phase angle of V_L is 143°.

V_T Equals the Phasor Sum of the Series Voltage Drops

If we want to add the voltage drops around the circuit to see if they equal the applied voltage, each *V* must be converted to rectangular form. Then these values can be added. In rectangular form then the individual voltages are

$$V_{R_1} = 4 \angle 53° = 2.408 + j3.196 \text{ V}$$
$$V_L = 8 \angle 143° = -6.392 + j4.816 \text{ V}$$
$$V_C = 24 \angle -37° = 19.176 - j14.448 \text{ V}$$
$$\underline{V_{R_2} = 8 \angle 53° = 4.816 + j6.392 \text{ V}}$$
$$\text{Total } V = 20.008 - j0.044 \text{ V}$$

or converting to polar form,

$$V_T = 20 \angle 0° \text{ V} \qquad \text{approximately}$$

Note that for $8 \angle 143°$ in the second quadrant, the cosine is negative for a negative real term but the sine is positive for a positive j term.

Practice Problems 25-10
Answers at End of Chapter

Refer to Fig. 25-10.
a. What is the phase of I to V_T?
b. What is the phase of V_L to V_T?
c. What is the phase of V_L to V_R?

25-11   Complex Numbers in Parallel AC Circuits

A useful application here is converting a parallel circuit to an equivalent series circuit. See Fig. 25-11, with a 10-Ω X_L in parallel with a 10-Ω R. In complex notation, R is $10 + j0$ while X_L is $0 + j10$. Their combined parallel impedance Z_T equals the product over the sum. For Fig. 25-11a, then:

$$Z_T = \frac{(10 + j0) \times (0 + j10)}{(10 + j0) + (0 + j10)}$$
$$= \frac{10 \times j10}{10 + j10}$$
$$Z_T = \frac{j100}{10 + j10}$$

Converting to polar form for division,

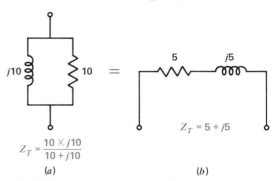

$$Z_T = \frac{10 \times j10}{10 + j10}$$

$$Z_T = 5 + j5$$

Fig. 25-11 Complex numbers used for parallel ac circuit to convert a parallel bank to an equivalent series impedance.

$$Z_T = \frac{j100}{10 + j10} = \frac{100 \angle 90°}{14.14 \angle 45°} = 7.07 \angle 45°$$

Converting the Z_T of $7.07 \angle 45°$ into rectangular form to see its resistive and reactive components,

$$\text{Real term} = 7.07 \cos 45°$$
$$= 7.07 \times 0.707 = 5$$
$$j \text{ term} = 7.07 \sin 45°$$
$$= 7.07 \times 0.707 = 5$$

Therefore,

$$Z_T = 7.07 \angle 45° \qquad \text{in polar form}$$
$$Z_T = 5 + j5 \qquad \text{in rectangular form}$$

The rectangular form of Z_T means that 5-Ω R in series with 5-Ω X_L is the equivalent of 10-Ω R in parallel with 10-Ω X_L, as shown in Fig. 25-11b.

Admittance Y and Susceptance B   In parallel circuits, it is usually easier to add branch currents than to combine reciprocal impedances. For this reason, branch conductance G is often used instead of branch resistance, where $G = 1/R$. Similarly, reciprocal terms can be defined for complex impedances. The two main types are *admittance Y*, which is the reciprocal of impedance, and *susceptance B*, which is the reciprocal of reactance. These reciprocals can be summarized as follows:

$$\text{Conductance} = G = \frac{1}{R} \qquad \text{S}$$

$$\text{Susceptance} = B = \frac{1}{\pm X} \quad \text{S}$$

$$\text{Admittance} = Y = \frac{1}{Z} \quad \text{S}$$

With R, X, and Z in units of ohms, the reciprocals G, B, and Y are in siemens (S) units.

The phase angle for B or Y is the same as current. Therefore, the sign is opposite from the angle of X or Z because of the reciprocal relation. An inductive branch has susceptance $-jB$, while a capacitive branch has susceptance $+jB$, with the same angle as branch current.

With parallel branches of conductance and susceptance the total admittance $Y_T = G \pm jB$. For the two branches in Fig. 25-11a, as an example, G is $\frac{1}{10}$ or 0.1 and B is also 0.1. In rectangular form,

$$Y_T = 0.1 - j0.1 \ \text{S}$$

In polar form,

$$Y_T = 0.14 \underline{/-45°} \ \text{S}$$

This value for Y_T is the same as I_T with 1 V applied across Z_T of $7.07 \underline{/45°}$ Ω.

As another example, suppose that a parallel circuit has 4 Ω for R in one branch and $-j4$ Ω for X_C in the other branch. In rectangular form, then, Y_T is $0.25 + j0.25$ S. Also, the polar form is $Y_T = 0.35 \underline{/45°}$ S.

Practice Problems 25-11
Answers at End of Chapter

a. A Z of $3 + j4$ Ω is in parallel with an R of 2 Ω. State Z_T in rectangular form.
b. Do the same as in Prob. **a** for X_C instead of X_L.

25-12   Combining Two Complex Branch Impedances

A common application is a circuit with two branches Z_1 and Z_2, where each is a complex impedance with both reactance and resistance. See Fig. 25-12. A circuit like

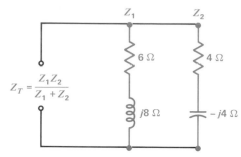

Fig. 25-12 Finding Z_T for any two complex impedances Z_1 and Z_2 in parallel. See text for solution.

this can be solved only graphically or by complex numbers. Actually, using complex numbers is the shortest method.

The procedure here is to find Z_T as the product divided by the sum for Z_1 and Z_2. A good way to start is to state each branch impedance in both rectangular and polar forms. Then Z_1 and Z_2 are ready for addition, multiplication, and division. The solution of this circuit follows:

$$Z_1 = 6 + j8 = 10 \underline{/53°}$$
$$Z_2 = 4 - j4 = 5.66 \underline{/-45°}$$

The combined impedance

$$Z_T = \frac{Z_1 \times Z_2}{Z_1 + Z_2}$$

Use the polar form of Z_1 and Z_2 to multiply, but add in rectangular form:

$$Z_T = \frac{10 \underline{/53°} \times 5.66 \underline{/-45°}}{6 + j8 + 4 - j4}$$
$$= \frac{56.6 \underline{/8°}}{10 + j4}$$

Converting the denominator to polar form for easier division,

$$10 + j4 = 10.8 \underline{/22°}$$

Then

$$Z_T = \frac{56.6 \underline{/8°}}{10.8 \underline{/22°}}$$

Therefore,

$$Z_T = 5.24 \ \Omega \ \underline{/\!-14°}$$

We can convert Z_T into rectangular form. The R component is $5.24 \times \cos(-14°)$ or $5.24 \times 0.97 = 5.08$. Note that $\cos \theta$ is positive in the first and fourth quadrants. The j component equals $5.24 \times \sin(-14°)$ or $5.24 \times (-0.242) = -1.127$. In rectangular form, then,

$$Z_T = 5.08 - j1.27$$

Therefore, this series-parallel circuit combination is equivalent to 5.08 Ω of R in series with 1.27 Ω of X_C. This problem can also be done in rectangular form by rationalizing the fraction for Z_T.

Practice Problems 25-12
Answers at End of Chapter

Refer to Fig. 25-12.
a. Add $(6 + j8) + (4 - j4)$ for the sum of Z_1 and Z_2.
b. Multiply $10 \underline{/53°} \times 5.66 \underline{/\!-45°}$ for the product of Z_1 and Z_2.

25-13   Combining Complex Branch Currents

An example with two branches is shown in Fig. 25-13, to find I_T. The branch currents can just be added in rectangular form for the total I_T of parallel branches. This method corresponds to adding series impedances in rectangular form to find Z_T. The rectangular form is necessary for the addition of phasors.

Adding the branch currents in Fig. 25-13,

$$I_T = I_1 + I_2$$
$$= (6 + j6) + (3 - j4)$$
$$I_T = 9 + j2 \text{ A}$$

Note that I_1 has $+j$ for the $+90°$ of capacitive current, while I_2 has $-j$ for inductive current. These current phasors have the opposite signs from their reactance phasors.

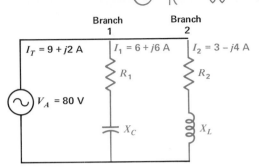

Fig. 25-13 Finding I_T for two complex branch currents in parallel.

In polar form the I_T of $9 + j2$ A is calculated as the phasor sum of the branch currents.

$$I_T = \sqrt{9^2 + 2^2}$$
$$= \sqrt{85} = 9.22 \text{ A}$$
$$\tan \theta = 2/9 = 0.22$$
$$\theta = 12.53°$$

Therefore, I_T is $9 + j2$ A in rectangular form or $9.22 \underline{/12.53°}$ A in polar form. The complex currents for any number of branches can be added in rectangular form.

Practice Problems 25-13
Answers at End of Chapter

a. Find I_T in rectangular form for I_1 of $0 + j2$ A and I_2 of $4 + j3$ A.
b. Find I_T in rectangular form for I_1 of $6 + j7$ A and I_2 of $3 - j9$ A.

25-14   Parallel Circuit with Three Complex Branches

Because the circuit in Fig. 25-14 has more than two complex impedances in parallel, the method of branch currents is used. There will be several conversions between rectangular and polar form, since addition must be in rectangular form, but division is easier in polar form. The sequence of calculations is:

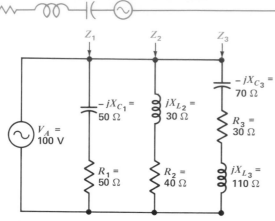

Fig. 25-14 Finding Z_T for any three complex impedances in parallel. See text for solution by means of complex branch currents.

1. Convert each branch impedance to polar form. This is necessary for dividing into the applied voltage V_A to calculate the individual branch currents. If V_A is not given, any convenient value can be assumed. Note that V_A has a phase angle of $0°$ because it is the reference.
2. Convert the individual branch currents from polar to rectangular form so that they can be added for the total line current. This step is necessary because the resistive and reactive components must be added separately.
3. Convert the total line current from rectangular to polar form for dividing into the applied voltage to calculate Z_T.
4. The total impedance can remain in polar form with its magnitude and phase angle, or can be converted to rectangular form for its resistive and reactive components.

These steps are used in the following calculations to solve the circuit in Fig. 25-14. All the values are in A, V, or Ω units.

Branch Impedances Each Z is converted from rectangular form to polar form:

$$Z_1 = 50 - j50 = 70.7 \underline{/-45°}$$
$$Z_2 = 40 + j30 = 50 \underline{/+37°}$$
$$Z_3 = 30 + j40 = 50 \underline{/+53°}$$

Branch Currents Each I is calculated as V_A divided by Z in polar form:

$$I_1 = \frac{V_A}{Z_1} = \frac{100}{70.7 \underline{/-45°}} = 1.414 \underline{/+45°} = 1 + j1$$
$$I_2 = \frac{V_A}{Z_2} = \frac{100}{50 \underline{/37°}} = 2.00 \underline{/-37°} = 1.6 - j1.2$$
$$I_3 = \frac{V_A}{Z_3} = \frac{100}{50 \underline{/53°}} = 2.00 \underline{/-53°} = 1.2 - j1.6$$

The polar form of each I is converted to rectangular form, for addition of the branch currents.

Total Line Current In rectangular form,

$$I_T = I_1 + I_2 + I_3$$
$$= (1 + j1) + (1.6 - j1.2) + (1.2 - j1.6)$$
$$= 1 + 1.6 + 1.2 + j1 - j1.2 - j1.6$$
$$I_T = 3.8 - j1.8$$

Converting $3.8 - j1.8$ into polar form,

$$I_T = 4.2 \underline{/-25.4°}$$

Total Impedance In polar form,

$$Z_T = \frac{V_A}{I_T} = \frac{100}{4.2 \underline{/-25.4°}}$$
$$Z_T = 23.8 \underline{/+25.4°}$$

Converting $23.8 \underline{/+25.4°}$ into rectangular form,

$$Z_T = 21.5 + j10.2$$

Therefore, the complex ac circuit in Fig. 25-14 is equivalent to the combination of 21.5 Ω of R in series with 10.2 Ω of X_L.

This problem can also be done by combining Z_1 and Z_2 in parallel as $Z_1 Z_2/(Z_1 + Z_2)$. Then combine this value with Z_3 in parallel to find the total Z_T of the three branches.

Practice Problems 25-14
Answers at End of Chapter

Refer to Fig. 25-14.
a. State Z_2 in rectangular form for branch 2.
b. State Z_2 in polar form.
c. Find I_2.

Summary

1. In complex numbers, resistance R is a real term and reactance is a j term. Thus, an 8-Ω R is 8; an 8-Ω X_L is $j8$; an 8-Ω X_C is $-j8$. The general form of a complex impedance with series resistance and reactance then is $Z = R \pm jX$, in rectangular form.
2. The same notation can be used for series voltages where $V = V_R \pm jV_X$.
3. For branch currents $I_T = I_R \pm jI_X$, but the reactive branch currents have signs opposite from impedances. Capacitive branch current is jI_C, while inductive branch current is $-jI_L$.
4. The complex branch currents are added in rectangular form for any number of branches to find I_T.
5. To convert from rectangular to polar form: $R \pm jX = Z \underline{/\theta}$. The magnitude of Z is $\sqrt{R^2 + X^2}$. Also, θ is the angle with $\tan = X/R$.
6. To convert from polar to rectangular form, $Z\underline{/\theta} = R \pm jX$, where R is $Z \cos \theta$ and the j term is $Z \sin \theta$. A positive angle has a positive j term; a negative angle has a negative j term. Also, the angle is more than 45° for a j term larger than the real term; the angle is less than 45° for a j term smaller than the real term.
7. The rectangular form must be used for addition or subtraction of complex numbers.
8. The polar form is usually more convenient in multiplying and dividing complex numbers. For multiplication, multiply the magnitudes and add the angles; for division, divide the magnitudes and subtract the angles.
9. To find the total impedance Z_T of a series circuit, add all the resistances for the real term and find the algebraic sum of the reactances for the j term. The result is $Z_T = R \pm jX$. Then convert Z_T to polar form for dividing into the applied voltage to calculate the current.
10. To find the total impedance Z_T of two complex branch impedances Z_1 and Z_2 in parallel, Z_T can be calculated as $Z_1Z_2/(Z_1 + Z_2)$.

Self-Examination
Answers at Back of Book

Match the values in the column at the left with those at the right.

1. $24 + j5 + 16 + j10$
2. $24 - j5 + 16 - j10$
3. $j12 \times 4$
4. $j12 \times j4$
5. $j12 \div j3$
6. $(4 + j2) \times (4 - j2)$
7. 1200 Ω of R + 800 Ω of X_C
8. 5 A of I_R + 7 A of I_C
9. 90 V of V_R + 60 V of V_L
10. $14\underline{/28°} \times \underline{/22°}$
11. $14\underline{/28°} \div 2\underline{/22°}$
12. $15\underline{/42°} \times 3\underline{/0°}$
13. $6\underline{/-75°} \times 4\underline{/30°}$

a. $14\underline{/50°}$
b. $7\underline{/6°}$
c. $1200 - j800 \ \Omega$
d. $40 + j15$
e. $90 + j60$ V
f. $45\underline{/42°}$
g. $24\underline{/-45°}$
h. 4
i. $j48$
j. -48
k. $5 + j7$ A
l. 20
m. $40 - j15$

Essay Questions

1. Give the mathematical operator for the angles of 0°, 90°, 180°, 270°, and 360°.
2. Define the sine, cosine, and tangent functions of an angle.
3. How are mathematical operators similar for logarithms, exponents, and angles?
4. Compare the following combinations: resistance R and conductance G, reactance X and susceptance B, impedance Z and admittance Y.
5. What are the units for admittance Y and susceptance B?
6. Why do Z_T and I_I for a circuit have angles with opposite signs?

Problems

Answers to Odd-Numbered Problems at Back of Book

1. State Z in rectangular form for the following series circuits: (a) 4-Ω R and 3-Ω X_C; (b) 4-Ω R and 3-Ω X_L; (c) 3-Ω R and 6-Ω X_L; (d) 3-Ω R and 3-Ω X_C.
2. Draw the schematic diagrams for the impedances in Prob. 1.
3. Convert the following impedances to polar form: (a) $4 - j3$; (b) $4 + j3$; (c) $3 + j$; (d) $3 - j3$.
4. Convert the following impedances to rectangular form: (a) $5\angle{-27°}$; (b) $5\angle{27°}$; (c) $6.71\angle{63.4°}$; (d) $4.24\angle{-45°}$.
5. Find the total Z_T in rectangular form for the following three series impedances: (a) $12\angle{10°}$; (b) $25\angle{15°}$; (c) $34\angle{26°}$.
6. Multiply the following, in polar form: (a) $45\angle{24°} \times 10\angle{54°}$; (b) $45\angle{-24°} \times 10\angle{54°}$; (c) $18\angle{-64°} \times 4\angle{14°}$; (d) $18\angle{-64°} \times 4\angle{-14°}$.
7. Divide the following, in polar form: (a) $45\angle{24°} \div 10\angle{10°}$; (b) $45\angle{24°} \div 10\angle{-10°}$; (c) $500\angle{-72°} \div 5\angle{12°}$; (d) $500\angle{-72°} \div 5\angle{-12°}$.
8. Match the four phasor diagrams in Fig. 25-4a, b, c, and d with the four circuits in Figs. 25-5 and 25-6.
9. Find Z_T in polar form for the series circuit in Fig. 25-7a.
10. Find Z_T in polar form for the series-parallel circuit in Fig. 25-7c.
11. Solve the circuit in Fig. 25-12 to find Z_T in rectangular form by rationalization.
12. Solve the circuit in Fig. 25-12 to find Z_T in polar form, using the method of branch currents. Assume an applied voltage of 56.6 V.
13. Show the equivalent series circuit of Fig. 25-12.
14. Solve the circuit in Fig. 25-14 to find Z_T in polar form, without using branch currents. (Find the Z of two branches in parallel; then combine this Z with the third branch Z.)
15. Show the equivalent series circuit of Fig. 25-14.
16. Refer to Fig. 25-13. (a) Find Z_1 and Z_2 for the two branch currents given. (b) Calculate the values needed for R_1, R_2, X_C, and X_L for these impedances. (c) What are the L and C values for a frequency of 60 Hz?
17. Solve the series ac circuit in Fig. 24-8 in the previous chapter by the use of complex numbers. Find $Z\angle{\theta}$, $I\angle{\theta}$, and each $V\angle{\theta}$. Prove that the sum of the complex voltage drops around the circuit equals the applied voltage V_T. Make a phasor diagram showing all phase angles with respect to V_T.

18. The following components are in series: $L = 100\ \mu$H, $C = 20$ pF, $R = 2000\ \Omega$. At the frequency of 2 MHz calculate X_L, X_C, Z_T, I, θ, V_R, V_L, and V_C. The applied $V_T = 8$ V.

19. Solve the same circuit as in Prob. 18 for the frequency of 4 MHz. Give three effects of the higher frequency.

20. In Fig. 25-15, show that $Z_T = 4.8\ \Omega$ and $\theta = 36.9°$ by **(a)** the method of branch currents; **(b)** calculating Z_T as $Z_1Z_2/(Z_1 + Z_2)$.

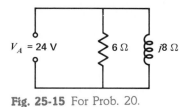

Fig. 25-15 For Prob. 20.

21. In Fig. 25-16, find $Z_T\ \underline{/\theta}$ by calculating Z_{bc} of the parallel bank and combining with the series Z_{ab}.

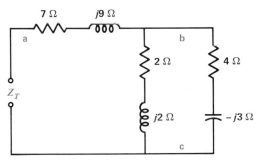

Fig. 25-16 For Prob. 21.

Answers to Practice Problems

25-1 a. 0°
  b. 180°

25-2 a. 90°
  b. −90 or 270°

25-3 a. T
  b. T

25-4 a. $j3$ kΩ
  b. $-j5$ mA

25-5 a. $4 + j7$
  b. $0 - j7$

25-6 a. $5 + j7$
  b. $4 + j6$

25-7 a. 14.14 Ω
  b. 45°

25-8 a. $12\ \underline{/50°}$
  b. $3\ \underline{/-10°}$

25-9 a. $10 + j10$
  b. $10 - j10$

25-10 a. 53°
  b. 143°
  c. 90°

25-11 a. $(6 + j8)/(5 + j4)$
  b. $(6 - j8)/(5 - j4)$

25-12 a. $10 + j4$
  b. $56.6\ \underline{/8°}$

25-13 a. $4 + j5$ A
  b. $9 - j2$ A

25-14 a. $40 + j30$
  b. $50\ \underline{/37°}\ \Omega$
  c. $2\ \underline{/-37°}$ A

Chapter 26

Network Analysis for AC Circuits

This chapter extends the principles of Kirchhoff's laws and network theorems, as explained for dc circuits in Chaps. 9 and 10, to the solution of ac circuits. A dc circuit is simpler because it has only resistance R, with a steady dc voltage source. However, ac circuits can also include

capacitive reactance X_C and inductance reactance X_L, with ac voltage applied. We assume sine-wave sources in solving these ac problems. If more than one source is involved, they all have the same frequency.

In sine-wave ac circuits, the voltage, current, and impedance are complex quantities, with magnitude and phase angle. They can be expressed in either polar or rectangular form, as explained in Chap. 25, "Complex Numbers for AC Circuits." In this chapter, the complex numbers are used for network analysis. Graphical methods are also explained.

Important terms in this chapter are:

double subscripts	Norton current I_N
graphical addition	Norton resistance R_N
Kirchhoff's current law	Thevenin resistance R_{Th}
Kirchhoff's voltage law	Thevenin voltage V_{Th}

More details are explained in the following sections:

26-1 Double-Subscript Notation
26-2 Kirchhoff's Voltage Law
26-3 Kirchhoff's Current Law
26-4 Thevenin's Theorem
26-5 Norton's Theorem
26-6 Superposition Theorem

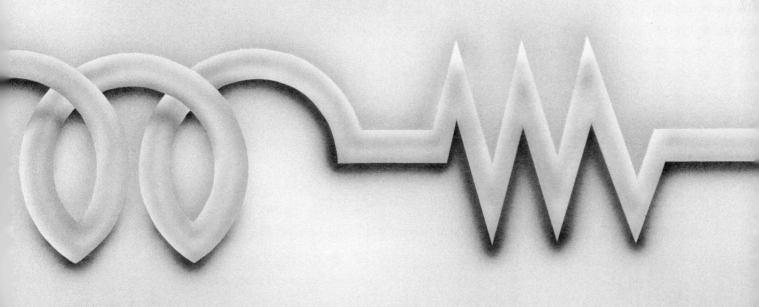

26-1 Double-Subscript Notation

The use of double subscripts to specify a voltage is a powerful tool in analyzing electronic circuits. For an example with transistors, V_{CE} is the dc voltage at the collector with respect to the emitter. The first subscript C is the point of measurement of the voltage, while the second subscript E is the reference point. If V_{CE} measures $+5$ V, it means the collector is more positive (or less negative) than the emitter by 5 V. To measure this voltage, the common lead of the voltmeter is connected to the emitter, while the probe lead is connected to the collector.

If the leads of the voltmeter are interchanged, so that the voltage now being measured is V_{EC}, the voltmeter would now read -5 V. This results because the emitter is less positive (or more negative) than the collector by 5 V.

From these examples we see that V_{EC} is equal to $-V_{CE}$. Another way of saying this is, if the leads of the voltmeter are connected to the same two points of the circuit but interchanged, the magnitude of the voltage remains the same but the polarity will reverse.

As a general rule, for double-subscript notation for dc and ac voltages, V_{XY} equals $-V_{YX}$. Furthermore, in the case of ac voltages, this means that V_{XY} and V_{YX} will be equal in magnitude but 180° out of phase.

Refer to the example in Fig. 26-1, which illustrates two opposite voltages. Here

V_{XY} is 200 $\angle +45°$.
V_{YX} is 200 $\angle +225°$.

Note that 225° is equal to 45° + 180°.
The V_{YX} can also be specified as 200 $\angle -135°$, since 45° − 180° is equal to −135°.

Practice Problems 26-1
Answers at End of Chapter

Answer True or False.
a. Voltages V_{AB} and V_{BA} are opposite.
b. Double-subscript notation is not necessary to specify resistance values.

26-2 Kirchhoff's Voltage Law

The rule for dc circuits states that the algebraic sum of all voltages around any closed path must be zero. For ac circuits, though, Kirchhoff's law states that the *phasor sum* of the voltages is zero. The phase angles must be included. To illustrate this law, the ac circuit in Fig. 26-2 can be used as an example. The points around the closed path are marked A and B across the resistor, with B and C across the capacitor. Voltage V_{AC} specifies the voltage from A to C, which is the source voltage across both the resistor and the capacitor. The values given in Fig. 26-2 are

$V_{AC} = 50 \ \angle 0°$ V
$R_1 = 15 \ \Omega$
$X_{C_1} = 20 \ \Omega$

This circuit can be solved for the following values of the individual voltage drops:

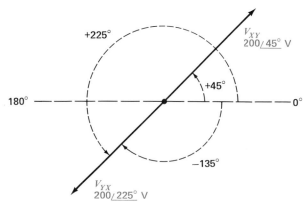

Fig. 26-1 Double-subscript notation for two opposite voltages V_{XY} and V_{YX}.

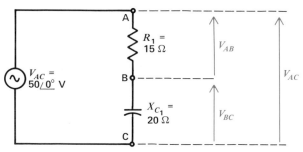

Fig. 26-2 Example to illustrate Kirchhoff's voltage law for ac circuits. Voltage phasors shown in Fig. 26-3.

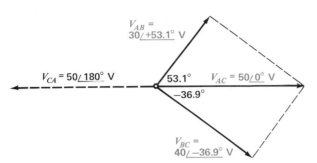

Fig. 26-3 Voltage phasors for circuit in Fig. 26-2.

$$V_{R_1} = V_{AB} = 30 \;/\!\!\underline{+53.1°}\; V$$
$$V_{C_1} = V_{BC} = 40 \;/\!\!\underline{-36.9°}\; V$$

It is important to note the directions shown for the voltage arrows in the schematic diagram of Fig. 26-2. The conventional method is to have the arrowhead at the potential being measured, with respect to the other point for the reference. As examples, the tip of the arrow is at A for V_{AB}, at B for V_{BC}, and at A for V_{AC}. The first letter of the subscript has the arrow tip.

All three voltage phasors are drawn to scale in Fig. 26-3, with their phase angles. Since V_{AC} is the source voltage, it is shown as the horizontal reference at 0°. Then V_{AB} is counterclockwise for the positive angle of 53.1°, and V_{BC} is clockwise at $-36.9°$. Note that $53.1 + 36.9 = 90°$ as the quadrature angle between V_{BC} across C and V_{AB} across R. The dotted phasor for V_{CA} is at 180° to show that it is opposite from V_{AC}.

Adding Voltages Around a Closed Path
To see how the voltages are added, we can go around the loop in a counterclockwise direction, as in Fig. 26-4.

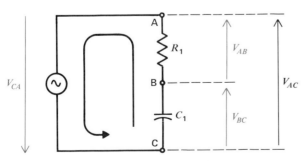

Fig. 26-4 Adding voltages in path CBAC, going in counterclockwise direction.

Start at point C and end at the same point. The counterclockwise direction is the same direction as the voltage arrows for V_{BC} and V_{AB}. However, going through the voltage source, the counterclockwise direction corresponds to a voltage arrow for V_{CA}, opposite to V_{AC}.

For Kirchhoff's law, the voltages around the loop are added. Then

$$V_{BC} + V_{AB} + V_{CA} = 0$$

Another method is to say that the sum of the voltage drops must equal the applied voltage. Then

$$V_{BC} + V_{AB} = V_{AC}$$

Note that V_{AC} is used here for the applied voltage. Both equations are equivalent because V_{AC} is equal to $-V_{CA}$. Substituting for V_{AC},

$$V_{BC} + V_{AB} = -V_{CA}$$
$$V_{BC} + V_{AB} + V_{CA} = 0$$

Either way, we can say that the sum of the voltage drops equals the applied voltage, or the phasor sum of voltages around the loop is equal to zero.

Phasor Addition   Phasor arrows can be added graphically tail to tip, as in Fig. 26-5. In Fig. 26-5a, the correct addition is shown for V_{BC}, V_{AB}, and V_{CA}. The phasors are drawn to scale, with the required angles, in the following steps:

1. Voltage V_{BC} is at $-36.9°$ to the horizontal.
2. Voltage V_{AB} is in quadrature with V_{BC}. Then V_{AB} has the positive angle of 53.1° to the horizontal.
3. Voltage V_{CA} is at 180°.

The added phasors form a closed loop, going back to the start at point C. As a result, we can say that the addition is correct in the equation $V_{BC} + V_{AB} + V_{CA} = 0$.

An incorrect addition is shown in Fig. 26-5b, in order to indicate the need for the correct angles in the phasors. The mistake here is adding V_{AC} at 0° instead of V_{CA} at 180°. In this case, the incorrect result is $2V_{AC}$ for the total voltage around the loop when it should be zero.

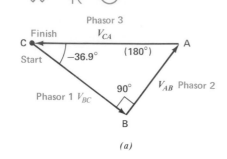

(a)

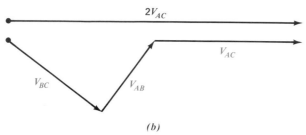

(b)

Fig. 26-5 Phasor addition for voltages in Fig. 26-4. (a) Correct addition with zero for closed path. (b) Incorrect addition.

Mathematical Addition The voltages around the loop in Fig. 26-4 can be added using complex numbers[1] to see that the sum equals zero. Remember that the capacitor voltage V_{BC} is $40 \angle -36.9°$ in polar form. This equals $32 - j24$ in rectangular form. Also, the resistor voltage V_{AB} is $30 \angle 53.1°$ or $18 + j24$.

For the potential difference between A and B, we are adding V_{CA} in the counterclockwise direction around the loop. The V_{CA} is equal to $-V_{AC}$. In order to obtain the negative value of any ac phasor in voltage, current, or impedance, the following rules apply.

1. In polar form, add 180° to the angle.
2. In rectangular form, change the sign of both terms.

For V_{AC} equal to $50 \angle 0°$ or $50 + j0$, then, V_{CA} becomes $50 \angle 180°$ or $-50 \pm j0$. (With a j term of zero, its sign does not matter.)

Now we can add the voltages around the loop. The sum is $V_{BC} + V_{AB} + V_{CA} = 0$. Tabulating the values in rectangular form for addition:

$$V_{BC} = 40 \angle -36.9° = \quad 32 - j24$$
$$V_{AB} = 30 \angle 53.1° = \quad 18 + j24$$

---

[1]Complex numbers for ac circuits are explained in Chap. 25.

$$V_{CA} = 50 \angle 180° \quad = -50 + j0$$
$$\text{Sum} = 0 + j0 = 0 \text{ V}$$

The addition illustrates Kirchhoff's voltage law for ac circuits, which states that the phasor sum of voltages around a closed path must equal zero. Only the rectangular form of the phasors can be added.

Finding an Unknown Voltage If all but one of the voltages in a closed loop are given, the unknown voltage can be obtained either mathematically with complex numbers or graphically. The graphical solution with phasors drawn to scale is shown in Fig. 26-6. The scale is 1 cm = 5 V. Either way, the problem is

Given:
$$V_{BA} = 30 \angle 60° = 15 + j26$$
$$V_{CB} = 20 \angle -90° = 0 - j20$$
$$V_{DC} = 30 \angle -45° = 21.2 - j21.2$$

Find V_{AD}.

For the mathematical solution,

$$V_{BA} + V_{CB} + V_{DC} + V_{AD} = 0$$

or

$$V_{AD} = -(V_{BA} + V_{CB} + V_{DC})$$

Adding the three voltages,

$$V_{AD} = -(36.2 - j15.2) \quad \text{in rectangular form}$$
$$V_{AD} = -(39.25 \angle -22.8°) \quad \text{in polar form}$$

Taking the negative value,

$$V_{AD} = -36.2 + j15.2 \quad \text{in rectangular form}$$
$$V_{AD} = 39.25 \angle 157.2° \quad \text{in polar form}$$

Note that 157.2° is obtained by adding 180° to −22.8°. In Fig. 26-6, the value of $39.25 \angle 157.2°$ is arrived at by drawing the phasors to scale. First, V_{BA} is drawn at the angle of 60°, starting from point A. Then from the arrowhead at B, the phasor for V_{CB} is drawn at −90°. From point C, the phasor for V_{DC} is at −45°.

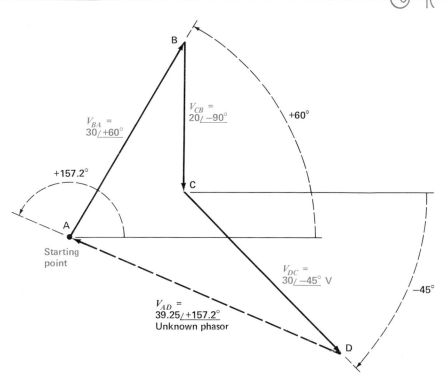

Fig. 26-6 Graphical solution for finding unknown voltage V_{AD} in a loop with V_{AB}, V_{BC}, and V_{CD}. Scale is 1 cm for 5 V. See text for mathematical solution.

The line returning from point D to the starting point at A completes the phasor sum of the voltages. The phasor for V_{AD} as the unknown voltage is equal to 39.2 $\angle\, 157.2°$. This angle is indicated by the dotted line, with the phasor pointing upward. The angle of 157.2° is indicated by the arc near point A, counterclockwise from the horizontal reference for 0°.

It is interesting to note that the phasors can be added in any order. The reason is that the potential difference between any two points in a loop is independent of the path used in arriving at the points. The same principle applies to adding the phasors mathematically.

Practice Problems 26-2
Answers at End of Chapter

a. Give the phasor for $-(8 \angle -10°)$ with a positive angle.

b. In Fig. 26-3, is the phasor 50 $\angle 0°$ for V_{AC} or V_{CA}?

c. In Fig. 26-6, what is the sum of all the voltages around the loop ABCDA?

26-3 Kirchhoff's Current Law

This rule for dc circuits states that the algebraic sum of all the currents at any point in a circuit must be zero. In ac circuits, the *phasor sum* equals zero. There must be a branch point, such as X or Y in Fig. 26-7, for a circuit to have different currents. By convention, current approaching a point is considered a positive phasor, while current leaving is a negative phasor. Kirchhoff's current law can also be stated as, the current out from a point must equal the current in. Otherwise, charge would accumulate, which does not happen in conductors.

For the example in Fig. 26-7, note that I_A and I_B into point X are considered positive. However, I_D and I_E out from point Y are negative with respect to Y. The sign for I_C depends on the reference. With respect to X, the I_C is negative, as it is leaving this point. We can label this current I_{C_x}. However, with respect to point Y, the I_C is positive, as it is into Y. This current is I_{C_y}.

Kirchhoff's current law for point X in Fig. 26-7 can be written with the equation

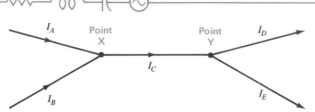

Fig. 26-7 Illustration for Kirchhoff's current law. See text for definitions.

$$I_A + I_B + I_{C_X} = 0$$

All these currents are phasors with the required magnitudes and correct angles. To make this example more specific, let us assume values for I_A and I_B:

$$I_A = 2.83 + j2.83 = 4 \angle 45° \text{ A}$$
$$I_B = 2.95 - j0.52 = 3 \angle -10° \text{ A}$$

Adding I_A and I_B in rectangular form and converting the sum to polar form,

$$I_A + I_B = 5.78 + j2.31$$
$$= 6.22 \angle 21.8° \text{ A}$$

As a result, we can say that the sum of the currents into point X is $6.22 \angle 21.8°$ A. This total current is approaching point Y, so that I_{C_Y} has the same value. However, I_{C_X} is negative because it is out from point X. Therefore,

$$I_{C_X} = -(I_A + I_B) = -(6.22 \angle 21.8° \text{ A})$$

Taking the negative value by adding 180°,

$$I_{C_X} = 6.22 \angle 180° + 21.8°$$
$$= 6.22 \angle 201.8° \text{ A}$$

To summarize the phasor values,

$$I_A = 4 \angle 45° \text{ A}$$
$$I_B = 3 \angle -10° \text{ A}$$
$$I_{C_Y} = 6.22 \angle 21.8° \text{ A}$$
$$I_{C_X} = 6.22 \angle 201.8° \text{ A}$$

All four phasors are shown in Fig. 26-8, with their relative magnitudes and correct angles.

Now we can add the currents in and out of point X to see that the sum is zero.

$I_A = 4 \angle 45°$	=	$2.83 + j2.83$
$I_B = 3 \angle -10°$	=	$2.95 - j0.52$
$I_{C_X} = 6.22 \angle 201.8°$	=	$-5.78 - j2.31$
Sum of $I_A + I_B + I_{C_X} =$		$0 + j0$ = 0 A

Closed Loop of Phasors for a Point Refer to Fig. 26-9. This diagram shows the current phasors added tail to tip. First, I_A is drawn at 45°. Then I_B is added at $-10°$. The loop of phasors is closed with I_{C_X} at 201.8°. The closed loop of phasors for currents at a point, such as X, corresponds to a closed loop of voltages around a circuit path. In either case, the closed loop means that the phasor sum is equal to zero, because we are back to the starting point.

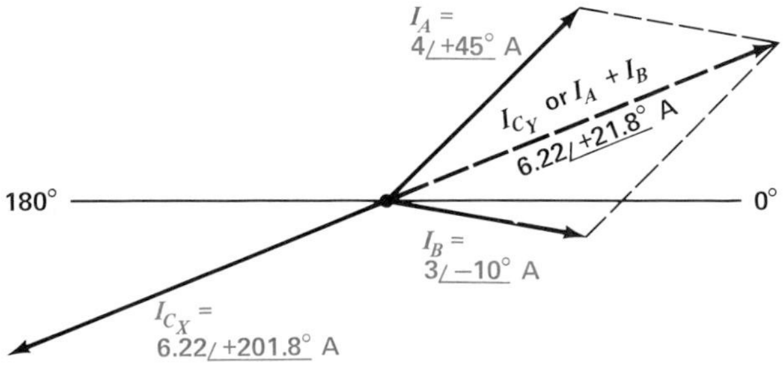

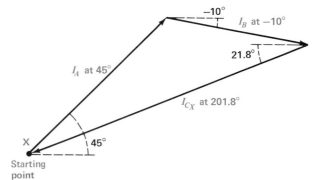

Note: $I_{C_X} = I_C$ with respect to point X
    $I_{C_Y} = I_C$ with respect to point Y

Fig. 26-8 Phasor diagram for currents in Fig. 26-7. Assumed values are $I_A = 4 \angle 45°$ A and $I_B = 3 \angle -10°$ A.

Fig. 26-9 Closed loop of phasors for adding currents into and out of point X in Fig. 26-7. The 201.8° indicates the angle of $180 + 21.8 = 201.8°$.

Finding an Unknown Current Value If all but one of the currents at a point are known, the unknown current can be found either mathematically or graphically with a phasor diagram. An example of a graphical solution is shown in Fig. 26-10. The unknown current is the phasor that is drawn from the head of the last phasor back to the starting point.

In Fig. 26-10, the I_A is 4 A or 4 cm at the angle of 90°. Next, I_B is drawn for 5 A at −20°. Then I_C is 4.5 A at 240°. This angle is made by going around counterclockwise past 180°, almost to 270°. The I_D phasor is 6 A at 30°. These four phasors for I_A, I_B, I_C, and I_D are the known values. Finally, the dashed line for the unknown I_E is drawn from the tip of I_D to the start at P. With the scale of 1 cm for 1 A, therefore, the magnitude of I_E is 7.8 A at the angle of 190°. Note that the 190° angle is just a little past the 180° direction.

The same problem of finding I_E in Fig. 26-10 can be solved mathematically. First, convert the given currents into rectangular form for addition. These values are

$$I_A = 4 \,\underline{/90°}\, \text{A} = 0 + j4$$
$$I_B = 5 \,\underline{/-20°}\, \text{A} = 4.7 - j1.71$$
$$I_C = 4.5 \,\underline{/240°}\, \text{A} = -2.25 - j3.9$$
$$I_D = 6 \,\underline{/30°}\, \text{A} = 5.2 + j3.0$$

According to Kirchhoff's current law, the phasor sum of all the currents must be zero. Then,

$$I_A + I_B + I_C + I_D + I_E = 0$$
$$I_E = -(I_A + I_B + I_C + I_D)$$
$$= -(7.65 + j1.39)$$
$$= -7.65 - j1.39$$
$$I_E = 7.78 \,\underline{/190.3°}\, \text{A}$$

The exact value of 7.78 $\underline{/190.3°}$ A for I_E agrees with the approximate current phasor of 7.8 $\underline{/190°}$ A in the graphical solution of Fig. 26-10.

More practice in phasor problems can be obtained from the next three examples given here. They should be solved mathematically with complex numbers and graphically by the method in Fig. 26-10. The graphical method requires a ruler with a centimeter scale and a protractor for the angles. In Examples 2 and 3, the phasors must be added tail to tip, as in Fig. 26-10.

Example 1 Using a protractor and a ruler, determine the currents for I_A, I_B, I_C, and I_D in Fig. 26-10 in polar form. The scale is 1 cm equals 1 A.

Answer $I_A = 4 \,\underline{/90°}\, \text{A}$
$I_B = 5 \,\underline{/-20°}\, \text{A}$
$I_C = 4.5 \,\underline{/240°}\, \text{A}$
$I_D = 6 \,\underline{/30°}\, \text{A}$

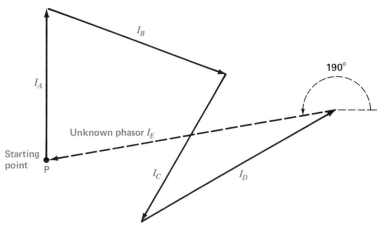

Fig. 26.10 Finding unknown current I_E by graphical addition of phasors. The I_E phasor returns to P. Note that the 190° angle is for I_E. See text for analysis with values for I_A, I_B, I_C, and I_D.

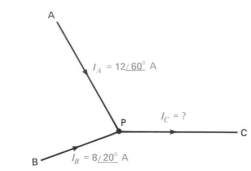

Fig. 26-11 For Example 2 in text.

Example 2   Wires A, B, and C are connected to point P, as shown in Fig. 26-11. Current I_A is 12 $\angle 60°$ A, and I_B is 8 $\angle 20°$ A. **(a)** Determine the phasor sum of I_A and I_B. **(b)** What is I_C with respect to point P?

Answer   a. $I_A + I_B = 18.8 \angle 44.2°$ A
        b. $I_C = 18.8 \angle 224.2°$ A

Example 3   Four wires are connected to point P in a circuit. The currents in three of the wires are known. With respect to P, the values are 4 $\angle 50°$ A, 6 $\angle -10°$ A, and 10 $\angle 120°$ A. Determine the unknown current with respect to point P.

Answer $I = 11.23 \angle 252°$ A

Practice Problems 26-3
Answers at End of Chapter

Answer True or False for Fig. 26-7.
a. $I_A + I_B + I_{C_x} = 0$.
b. If $I_A + I_B = 6 \angle 10°$, then $I_{C_x} = 6 \angle 190°$.
c. If $I_{C_v} = 6 \angle 190°$, then $I_{C_v} = 6 \angle 10°$.

26-4   Thevenin's Theorem

This important principle can be used to reduce a complicated network to a simpler equivalent circuit that has just one voltage source and its series impedance. The

advantage is that the same Thevenin equivalent circuit can be used with different values of load impedance to analyze the effect of changing the load. How to apply Thevenin's theorem in dc circuits is explained in Chap. 10. For ac circuits, an example is illustrated in Fig. 26-12. Thevenin's theorem for ac circuits states that any two-terminal network with fixed impedances and sine-wave energy sources may be replaced with an equivalent series circuit that has:

1. A sine-wave voltage source equal to the open-circuit V at the output terminals of the original network. This derived voltage source for the Thevenin equivalent circuit is V_{Th}.
2. A series impedance equal to the Z looking back from the open-circuit output terminals. All energy sources in the original network are replaced by their internal impedances, without generating any output. This derived series impedance for the Thevenin source is indicated as Z_{Th}.

In Fig. 26-12, the original circuit in a with its specified values of V and Z is reduced to the Thevenin equivalent in d that has derived values of V_{Th} and Z_{Th}. Everything in the circuit to the left of terminals B and C is included in the Thevenin equivalent. No values are shown for Z_3 to the right of terminals B and C because the Thevenin equivalent circuit can be used to feed any value of load impedance connected to these terminals. Impedance Z_3 can be a single impedance or another ac circuit. In effect, the network to the left of terminals B and C is thevenized to determine the equivalent voltage source.

For the first step shown in Fig. 26-12b, remove everything to the right of terminals B and C. The result is an open-circuit load for the circuit to be thevenized.

Now the problem is to find the voltage across Z_2. This voltage is also across terminals B and C as the Thevenin equivalent voltage. In other words, V_{Th} is the open-circuit voltage across terminals B and C. The method requires finding I because the unknown voltage is calculated as $I \times Z_2$.

The calculations are

$$I = \frac{V_{AC}}{Z_1 + Z_2} = \frac{26 \angle 0°}{5 \angle 0° + 12 \angle 90°}$$

$$= \frac{26 \angle 0°}{5 + j0 + 0 + j12} = \frac{26 \angle 0°}{5 + j12} = \frac{26 \angle 0°}{13 \angle 67°}$$

$$I = 2 \angle -67° \text{ A}$$

For the value of IZ_2, V_{BC}, or V_{Th},

$$V_{Th} = IZ_2$$
$$= (2 \ \angle -67°)(12 \ \angle 90°)$$
$$V_{Th} = 24 \ \angle 23° \text{ V}$$

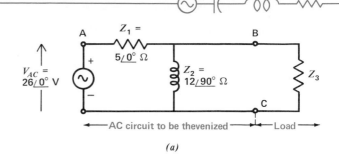

(a)

Note this value of 24 $\angle 23°$ for V_{Th} is shown in Fig. 26-12d.

The next step is to find Z_{Th} for the equivalent series impedance of the Thevenin source. As shown in Fig. 26-12c, remove the original voltage source V_{AC}. A voltage source is "killed" by assuming a short circuit that removes its voltage output, so that only its internal impedance remains. In Fig. 26-12, a perfect voltage source with zero internal impedance is assumed for V_{AC}. If there were any internal impedance, it would be in series with Z_1. Actually, Z_1 could be the internal impedance. For the values in Fig. 26-12c, Z_{Th} is calculated as Z_{BC} with terminals B and C still open but V_{AC} replaced by a short circuit. Then Z_1 and Z_2 are in parallel. Their product divided by their sum is the combined parallel impedance needed for Z_{BC} or Z_{Th}. The calculations are

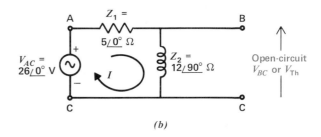

(b)

$$Z_{Th} = \frac{Z_1 Z_2}{Z_1 + Z_2}$$

$$= \frac{5 \ \angle 0° \times 12 \ \angle 90°}{5 \ \angle 0° + 12 \ \angle 90°}$$

$$= \frac{60 \ \angle 90°}{5 \ \angle 0° + 12 \ \angle 90°}$$

$$= \frac{60 \ \angle 90°}{5 + j0 + 0 + j12} = \frac{60 \ \angle 90°}{5 + j12}$$

$$Z_{Th} = \frac{60 \ \angle 90°}{13 \ \angle 67°} = 4.6 \ \angle 23° \ \Omega$$

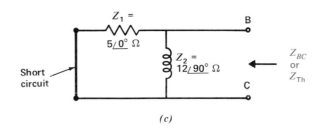

(c)

Convert Z_{Th} to rectangular form to see its resistance and reactive components. Then, $Z_{Th} = 4.23 + j1.8$. In Fig. 26-12d, note that Z_{Th} combines R of 4.23 Ω and inductive reactance X_L of 1.8 Ω.

The final form for the thevenized circuit in Fig. 26-12d shows a voltage source V_{Th} equal to 24 $\angle 23°$ V, with an internal impedance equal to 4.23 + j1.8 Ω, feeding the open terminals B and C. The load Z_3 can be reconnected to the terminals B and C to determine the voltage and current supplied to the

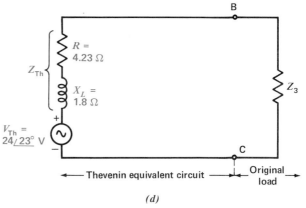

(d)

Fig. 26-12 Example of thevenizing an ac network to make an equivalent series circuit. (a) Original circuit that will be thevenized with respect to terminals B and C. (b) Disconnect load Z_3 to find open-circuit voltage across B and A for V_{Th}. (c) Short-circuit the original source voltage to find impedance between B and C for Z_{Th}. (d) Thevenin equivalent circuit with V_{Th} and Z_{Th} in series connected to original load Z_3.

load by the original circuit in Fig. 26-12a. Furthermore, any other load connected to terminals C and B will still have the same Thevenin equivalent circuit feeding it.

Practice Problems 26-4
Answers at End of Chapter

Answer True or False.

a. Thevenizing a network produces an equivalent series circuit.

b. In Fig. 26-12, the circuit to the right of terminals C and B is kept open to find both V_{Th} and Z_{Th}.

c. The original voltage source is killed to find either V_{Th} or Z_{Th}.

d. The Z_{Th} of $4.23 + j1.8$ Ω in Fig. 26-12d is equal to $4 \angle 48°$ Ω.

26-5 Norton's Theorem

This theorem, like Thevenin's theorem, may also be used to simplify the solution of most ac networks. An example of nortonizing a circuit is shown in Fig. 26-13. The difference between the two methods is that a Thevenin equivalent has a voltage source V_{Th} in series with Z_{Th}, while a Norton equivalent has a current source I_N in parallel with Z_N. However, the method of finding Z is the same for either Z_{Th} or Z_N.

Norton's theorem states that any two-terminal network which has fixed impedances and constant sine-wave energy sources may be replaced with an equivalent parallel circuit that has the following:

1. A current source with I_N equal to the current that would flow between the two output terminals of the network if they were short-circuited

2. An impedance Z_N looking back from the two open output terminals of the network with all energy sources replaced by their internal impedance; Z_N is equal to Z_{Th}

In Fig. 26-13, the original network in a is nortonized with respect to terminals C and D. The final result in Fig. 26-13d shows the Norton equivalent.

The first step in nortonizing the network is shown in Fig. 26-13b. Remove the load impedance Z_4 and replace it by a short circuit between terminals C and D.

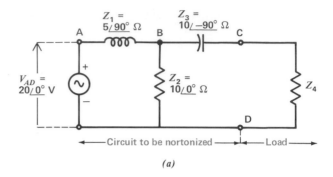

(a)

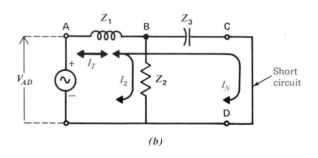

(b)

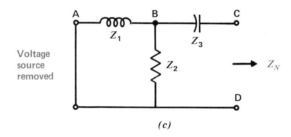

(c)

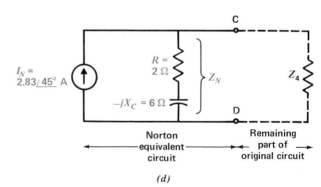

(d)

Fig. 26-13 Example of nortonizing an ac network to make an equivalent parallel circuit. (a) Original circuit that will be nortonized with respect to terminals C and D. (b) Replace the load Z_4 with a short circuit from C to D to find the equivalent Norton current I_N that flows through the short-circuited terminals. (c) Short-circuit the original voltage source to find impedance Z_N looking back from open terminals C and D. (d) Norton equivalent circuit with current source I_N in parallel with Z_N.

Solve for the current flowing through this short circuit, as it equals I_N. The solution is done here in five steps.

1. Solve for Z_2 and Z_3 in parallel.

$$Z_{23} = \frac{Z_2 Z_3}{Z_2 + Z_3} = \frac{(10\ \angle 0°)\ (10\ \angle -90°)}{10 + j0 + 0 - j10}$$

$$= \frac{100\ \angle -90°}{14.1\ \angle -45°}$$

$$Z_{23} = 7.07\ \angle -45° = 5 - j5\ \Omega$$

2. Solve for Z_T.

$$Z_T = Z_1 + Z_{23}$$
$$= 0 + j5 + 5 - j5$$
$$= 5 + j0$$
$$Z_T = 5\ \angle 0°\ \Omega$$

3. Solve for I_T.

$$I_T = \frac{V_{AD}}{Z_T} = \frac{20\ \angle 0°}{5\ \angle 0°} = 4\ \angle 0°\ A$$

4. Solve for V_{BD} as $I_T \times Z_{23}$. Note that I_T flows through the parallel combination of Z_2 and Z_3.

$$V_{BD} = I_T \times Z_{23}$$
$$= 4\ \angle 0° \times 7.07\ \angle -45°$$
$$V_{BD} = 28.28\ \angle -45°\ V$$

5. Solve for I_N.

$$I_N = \frac{V_{BD}}{Z_3}$$

$$= \frac{28.28\ \angle -45°}{10\ \angle -90°}$$

$$I_N = 2.828\ \angle 45°\ A$$

This value of $2.838\ \angle 45°$ A is the Norton current for the equivalent circuit in Fig. 26-13d.

Next, we solve for the Norton impedance Z_N, as shown in Fig. 26-13c. Take away the short circuit across the terminals C and D. Kill the voltage source V_{AD} and replace it with a short circuit with an assumed value of zero for its internal impedance. Find Z looking back from terminals C and D. This method is the same as that for finding Z_{Th}.

In Fig. 26-13c, Z_1, and Z_2 are in parallel. The combined impedance is

$$Z_{12} = 4.46\ \angle 63.4° = 2 + j4\ \Omega$$

Then add Z_3 to Z_{12} for the Norton impedance, equal to

$$Z_N = 6.32\ \angle -71.6° = 2 - j6\ \Omega = Z_N$$

The $2 - j6\ \Omega$ is Z_N in Fig. 26-13d in parallel with the current source I_N. The R is $2\ \Omega$ and the $-jX_C$ is $6\ \Omega$ for the capacitive reactance.

The final result in Fig. 26-13d shows the Norton equivalent of the ac network in Fig. 26-13a. The load impedance Z_4 can be reconnected to terminals C and D. Any value of Z can be connected across the output terminals of the Norton equivalent. Since I_N is a current source with shunt impedance, the current through the load can be determined by the current divider rule. (See Secs. 7-2 and 7-3 in Chap. 7.)

Practice Problems 26-5
Answers at End of Chapter

Answer True or False.
a. Nortonizing a network produces an equivalent parallel circuit.
b. In Fig. 26-13 terminals C and D are short-circuited to find I_N.
c. Terminals C and D are also short-circuited to find Z_N.
d. The $2 - j6\ \Omega$ for Z_N in Fig. 26-13d is equal to an impedance of $6.32\ \angle -71.6°\ \Omega$.

26-6 Superposition Theorem

This theorem simplifies the solution of networks with more than one energy source. Voltage or current sources can be included. For ac networks, sine-wave sources are assumed. They operate at the same frequency but may be out of phase. The superposition theorem for ac circuits can be stated as follows: *The current that flows in any branch of a network containing more than one energy source is equal to the phasor sum of the individual currents produced by each source acting by itself.* The internal impedances of all the sources remain in the circuit.

The ac circuit in Fig. 26-14a can now be solved to illustrate the superposition theorem. Voltages V_1 and V_2

are the ac voltage sources. The polarity marks for the ac sources only indicate the phase reference. The impedance Z_1 for R_1 and X_{L_1} in series is the internal impedance of the source V_1. Also, the impedance Z_2 for R_2 and X_{L_2} is the internal impedance of the source V_2. Finally, Z_3 is the impedance for R_3 and X_{L_3} between branch points X and Y.

The problem here is to find the voltage V_{XY} across the center branch and the current through Z_3. The

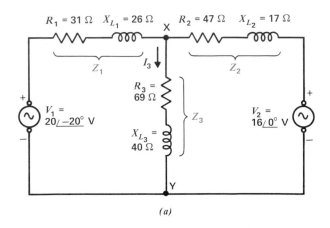

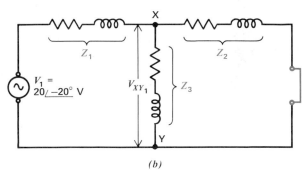

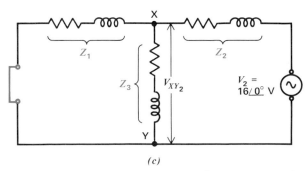

Fig. 26-14 Example of superposition for an ac network. (a) Original circuit. (b) Source V_2 killed to find V_{XY_1}. (c) Source V_1 killed to find V_{XY_2}.

method requires the following steps. First, the source V_2 is killed to find the amount of V_{XY_1} produced by V_1. This equivalent circuit is shown in Fig. 26-14b. Then the source V_1 is killed to find the amount of V_{XY_2} produced by V_2. This equivalent circuit is shown in Fig. 26-14c. After V_{XY_1} and V_{XY_2} are known, they are added to find the net voltage across Z_3 produced by both sources in the actual network in Fig. 26-14a. Then the current can be calculated as the actual potential difference V_{XY} divided by the impedance Z_3.

Remember that a voltage source is killed by replacing it with a short circuit. A current source is killed by replacing it with an open circuit. For both types of sources, the internal impedance remains in the circuit.

In order to find the voltages, the method of voltage division is used here. For the equivalent circuit in Fig. 26-14b, Z_2 and Z_3 can be combined in parallel for Z_{23}. This impedance is also Z_{XY_1}. The V_1 source voltage is divided between Z_1 and Z_{XY_1} to find the component of V_{XY} produced by V_1.

Similarly, in Fig. 26-14c, Z_1 and Z_3 can be combined in parallel for Z_{13}, which is also Z_{XY_2}. The V_2 source voltage is divided between Z_2 and Z_{XY_2} to find the component of V_{XY} produced by V_2.

The individual impedances given in Fig. 26-14a can be stated in rectangular and polar form, to be ready for addition or multiplication and division. The values are

$$Z_1 = 31 + j26 \quad \text{or} \quad 40.4 \; \underline{/40°}$$
$$Z_2 = 47 + j17 \quad \text{or} \quad 50 \; \underline{/20°}$$
$$Z_3 = 69 + j40 \quad \text{or} \quad 79.8 \; \underline{/30.1°}$$

Use the impedances of Z_2 and Z_3 to calculate the impedance Z_{XY_1} for the equivalent circuit in Fig. 26-14b. Then

$$Z_{XY_1} = \frac{Z_2 Z_3}{Z_2 + Z_3}$$

$$Z_2 + Z_3 = 47 + j17 + 69 + j40$$
$$= 116 + j57$$
$$= 129 \; \underline{/26.2°}$$

$$Z_{XY_1} = \frac{(50 \; \underline{/20°})\,(79.8 \; \underline{/30.1°})}{(129 \; \underline{/26.2°})}$$

$$Z_{XY_1} = 30.9 \; \underline{/23.9°} \quad \text{or} \quad 28.2 + j12.5 \; \Omega$$

This Z_{XY_1} is also the value for Z_{23}.

Use the voltage divider rule to find V_{XY_1}. Then

$$V_{XY_1} = V_1\left(\frac{Z_{23}}{Z_1 + Z_{23}}\right)$$

$$Z_1 + Z_{23} = 31 + j26 + 28.2 + j12.5$$

$$= 59.2 + j38.5$$

$$= 70.6 \underline{/33°}$$

$$V_{XY_1} = \frac{(20 \underline{/-20°})(30.9 \underline{/23.9°})}{(70.6 \underline{/33°})}$$

$$V_{XY_1} = 8.75 \underline{/-29.1°} \quad \text{or} \quad 7.65 - j4.26 \text{ V}$$

This value of V_{XY_1} is the voltage produced by V_1 without V_2 for the equivalent circuit in Fig. 26-14b.

Next, we calculate Z_{XY_2} and V_{XY_2} for the equivalent circuit without V_1 in Fig. 26-14c.

$$Z_{XY_2} = \frac{Z_1 Z_3}{Z_1 + Z_3}$$

$$Z_1 + Z_3 = 31 + j26 + 69 + j40$$

$$= 100 + j66$$

$$= 120 \underline{/33.4°}$$

$$Z_{XY_2} = \frac{(40.4 \underline{/40°})(79.8 \underline{/30.1°})}{(120 \underline{/33.4°})}$$

$$Z_{XY_2} = 26.9 \underline{/36.7°} \quad \text{or} \quad 21.6 + j16 \ \Omega$$

This Z_{XY_2} is also the value of Z_{13}.

Again, use the voltage divider rule to find V_{XY_2}. Then

$$V_{XY_2} = V_2\left(\frac{Z_{13}}{Z_2 + Z_{13}}\right)$$

$$Z_2 + Z_{13} = 47 + j17 + 21.6 + j16$$

$$= 68.6 + j33$$

$$= 76.1 \underline{/25.7°}$$

$$V_{XY_2} = \frac{(16 \underline{/0°})(26.9 \underline{/36.7°})}{(76.1 \underline{/25.7°})}$$

$$V_{XY_2} = 5.7 \underline{/11°} \quad \text{or} \quad 5.6 + j1.09 \text{ V}$$

Now we can add the two voltages V_{XY_1} and V_{XY_2} to obtain the actual V_{XY}. The result is

$$V_{XY_1} = \quad 7.65 - j4.26$$
$$V_{XY_2} = \quad 5.6 \ + j1.09$$
$$\overline{V_{XY} = \ 13.25 - j3.17 \text{ V}}$$

In polar form, V_{XY} is $13.6 \underline{/-13.5°}$. This value is used to find the current I_3 through the branch between points X and Y. Then

$$I_3 = \frac{V_{XY}}{Z_3}$$

$$= \frac{13.6 \underline{/-13.5°}}{79.8 \underline{/30.1°}}$$

$$I_3 = 0.17 \underline{/-43.6°} \text{ A}$$

The complete solution for the ac network with two sources in Fig. 26-14a shows that the net voltage between points X and Y is $13.6 \underline{/-13.5°}$ V, to produce I_3 of $0.17 \underline{/-43.6°}$ A through the impedance of Z_3. This result is obtained by superposition, combining the effects of each voltage source without the other.

Practice Problems 26-6
Answers at End of Chapter

Refer to Fig. 26-14.
a. Is the V_1 source killed in Fig. 26-14b or c?
b. Is R_2 the internal resistance of source V_1 or V_2?
c. Is I_3 the net current through L_3 or L_1?

Summary

1. In double-subscript notation, such as V_{CE}, the voltage is measured at C with respect to E. For the opposite voltage, V_{EC} is 180° out of phase.
2. Kirchhoff's voltage law for ac circuits states that the phasor sum of all voltages around any closed path must equal zero.

3. Kirchhoff's current law for ac circuits states that the phasor sum of all currents at any point in a circuit must equal zero. By convention, current directed into point P is considered positive, while current out is negative with respect to P.

4. Thevenin's theorem for ac circuits states that a network can be reduced to a single sine-wave voltage source with V_{Th} equal to the open-circuit voltage at the output terminals. Its internal impedance Z_{Th} is equal to the impedance looking back from the open output terminals with the original voltage source replaced by its internal impedance. In the Thevenin equivalent, Z_{Th} is in series with V_{Th}.

5. Norton's theorem for ac circuits states that a network can be replaced by a current source I_N equal to the current that would flow through a short circuit between the output terminals. The internal impedance Z_N of the current source is the same as Z_{Th}. In the Norton equivalent, Z_N is in parallel with I_N.

6. The superposition theorem for ac circuits states that the current in any branch of a network with more than one energy source is equal to the phasor sum of the individual currents produced by each source acting by itself. One source at a time is considered, with the other sources killed temporarily.

7. A voltage source is killed by replacing the output voltage with its series internal impedance. A current source is killed by replacing it with its parallel internal impedance.

Self-Examination
Answers at Back of Book

Answer True or False.

1. When V_{AB} is 14 $\angle 45°$, then V_{BA} is 14 $\angle 225°$.
2. When V_{AB} is $10 + j10$, then V_{BA} is $10 - j10$.
3. For ac voltages, the sum of the voltage drops must equal the applied voltage around a closed loop.
4. In ac circuits, the current into any branch point must equal the current out.
5. A Thevenin equivalent is a series circuit.
6. A Norton equivalent is a parallel circuit.
7. For any one ac network, the value of Z_N is usually less than Z_{Th}.
8. A closed loop of voltage phasors means that they add to equal zero.
9. A current source is killed by opening the source and using its parallel internal impedance.
10. The superposition theorem cannot be used with current sources.

Essay Questions

1. How does the voltage V_{AB} differ from V_{BA}?
2. Explain Kirchhoff's voltage law for ac circuits. Give two ways of stating this law.
3. What is meant by a closed path?
4. Explain Kirchhoff's current law for ac circuits. Give two ways of stating this law.

5. Describe how phasors are added graphically.
6. Explain Thevenin's theorem. Why is this theorem helpful in the solution of many network problems?
7. Explain Norton's theorem. How is this theorem different from Thevenin's theorem? In what way are the two theorems similar?
8. Give two differences between a voltage source and a current source.

Problems

Answers to Odd-Numbered Problems at Back of Book

1. There are four ac voltages around a closed loop. Three of them are $V_{BA} = 25 \angle 45°$, $V_{CB} = 50 \angle -45°$, and $V_{DC} = 100 \angle 180°$. Determine the fourth unknown voltage in polar form and give V in double-subscript notation.
2. Draw a phasor diagram to approximate scale for the four voltages in Prob. 1.
3. Five ac voltages form a closed loop in a circuit. The four known voltages are $300 \angle 80°$, $100 \angle -65°$, $180 \angle 225°$, and $200 \angle 135°$. Determine the unknown loop voltage in rectangular and polar form.
4. An equation for a closed loop is written as $-V_{AB} - V_{BC} = V_{CD} + V_{DE} - V_{AE}$. Can this equation be correct for a closed loop? (Hint: Place all terms on one side of the equation and rearrange the subscripts for a closed loop of phasors.)
5. Determine which of the following are correct loop equations:

   (a) $V_{CA} + V_{FC} + V_{BF} + V_{EB} + V_{AE} = 0$
   (b) $-V_{CA} + V_{BA} - V_{BC} = 0$
   (c) $V_{AC} = -V_{BA} - V_{CB}$
   (d) $V_{TX} + V_{ZT} + V_{AZ} - V_{AY} = -V_{XY}$
   (e) $V_{CD} + V_{DA} + V_{BC} = -V_{AB}$

6. Three currents are approaching point Y in a circuit. They are $I_A = 25 \angle -40°$ A, $I_B = 10 \angle 75°$ A, and $I_C = 5 \angle 10°$ A. Only one current I_D is leaving point Y. Find I_D with respect to point Y, in polar form.
7. The following three currents are approaching point X in a circuit: $I_1 = 12 \angle -60°$ A, $I_2 = 4 \angle 30°$ A, and $I_3 = 15 \angle 70°$ A. Two currents I_4 and I_5 are leaving point X. If I_4 is $6 \angle 222°$ A, determine the unknown current I_5 in polar form.
8. Convert $(12 + j6)$ V to polar form and give the voltage 180° out of phase in polar and rectangular form.
9. Thevenize the circuit in Fig. 26-15 with respect to terminals A and B.

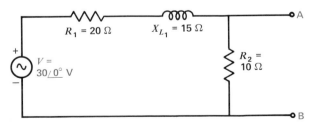

Fig. 26-15 For Probs. 9 and 11.

10. Convert $(7 + j3)$ A to polar form and give the current 180° out of phase in polar and rectangular form.
11. Nortonize the circuit in Fig. 26-15.
12. Given the Thevenin equivalent circuit shown in Fig. 26-16, show the Norton equivalent.
13. Refer to Fig. 26-17. Show the Thevenin and Norton equivalents for this circuit with a current source I as the energy source.

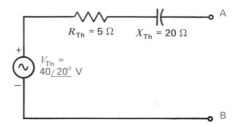

Fig. 26-16 For Prob. 12.

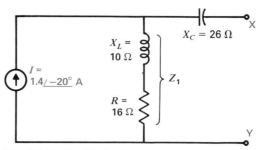

Fig. 26-17 For Prob. 13.

14. (**a**) Convert the voltage source of 24 $\angle 20°$ V with an internal series impedance of 6 $\angle 8°$ Ω to an equivalent current source. (**b**) Convert the current source of 4 $\angle 12°$ A with a shunt parallel impedance of 6 $\angle 8°$ Ω to an equivalent voltage source.
15. Refer to Fig. 26-18. Nortonize or thevenize the circuit to solve for the voltage developed across the load.

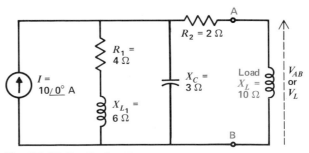

Fig. 26-18 For Prob. 15.

16. Refer to Fig. 26-19. Thevenize the circuit to find the current in Z_L.

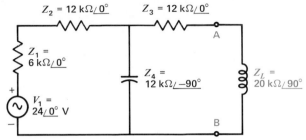

Fig. 26-19 For Prob. 16.

17. Use the superposition theorem for the circuit in Fig. 26-20 to find the current through the load Z_L at the center.

18. Show the Thevenin and Norton equivalents for the circuit in Fig. 26-21.

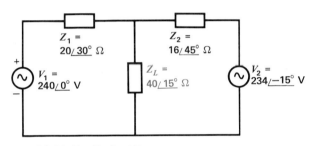

Fig. 26-20 For Prob. 17.

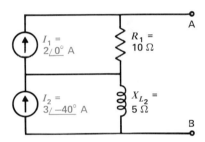

Fig. 26-21 For Prob. 18.

Answers to Practice Problems

26-1 a. T	**26-3 a.** T	**26-5 a.** T	
b. T	**b.** T	**b.** T	
26-2 a. 8 $\angle 170°$	**c.** T	**c.** F	
b. V_{AC}	**26-4 a.** T	**d.** T	
c. 0 V	**b.** T	**26-6 a.** Fig. 26-14c	
	c. F	**b.** V_2	
	d. F	**c.** L_3	

Review
Chapters
24 to 26

Summary

1. Reactances X_C and X_L are opposite. In series, the ohms of X_C and X_L cancel. In parallel, the branch currents I_C and I_L cancel.

2. As a result, circuits with R, X_C, and X_L can be reduced to one net reactance X and one equivalent R.

3. In series circuits, the net X is added with the total R by phasors for the impedance: $Z = \sqrt{R^2 + X^2}$. Then $I = V_T/Z$.

4. For the branch currents in parallel circuits, the net I_X is added with I_R by phasors for the total line current: $I_T = \sqrt{I_R^2 + I_X^2}$. Then $Z = V/I_T$.

5. The characteristics for ohms of R, X_C, X_L, and Z in ac circuits are compared in Table 24-1 on page 479.

6. In ac circuits with reactance, the real power in watts equals I^2R. This value equals $VI \cos \theta$, where θ is the phase angle of the circuit and $\cos \theta$ is the power factor.

7. The wattmeter uses an ac meter movement to read V and I at the same time, measuring watts of real power.

8. In complex numbers, R is a real term at $0°$ and reactance is a $\pm j$ term at $\pm 90°$. In rectangular form, $Z = R \pm jX$. For example, 10 Ω of R in series with 10 Ω of X_L is $10 + j10$ Ω.

9. The polar form of $10 + j10$ Ω is 14 $\underline{/45°}$ Ω. The angle of $45°$ is arctan X/R. The magnitude of 14 is $\sqrt{R^2 + X^2}$.

10. The rectangular form of complex numbers must be used for addition and subtraction. Add or subtract the real terms and the j terms separately.

11. The polar form of complex numbers is easier for multiplication and division. For multiplication, multiply the magnitudes and add the angles. For division, divide the magnitudes and subtract the angle of the divisor.

12. In double-subscript notation for a voltage, such as V_{BE}, the first letter in the subscript is the point of measurement with respect to the second letter. So V_{BE} is the base voltage with respect to the emitter, in a transistor.

13. Reversing the subscripts is equivalent to reversing the phase by $180°$. For

complex numbers in rectangular form, change the sign for both terms to reverse the phase. In polar form, add 180° to the angle.

14. Kirchhoff's laws and the network theorems are applied in ac circuits the same way as in dc circuits, except that V, I, and Z are considered as complex values. See Chaps. 9 and 10 for the methods in dc circuits and Chap. 26 for ac circuits.

Review Self-Examination
Answers at Back of Book

Fill in the numerical answer.

1. An ac circuit with 100 Ω R_1 in series with 200 Ω R_2 has R_T of _____ Ω.
2. With 100 Ω X_{L_1} in series with 200 Ω X_{L_2}, the total X_L is _____ Ω.
3. For 200 Ω X_{C_1} in series with 100 Ω X_{C_2}, the total X_C is _____ Ω.
4. Two X_C branches of 500 Ω each in parallel have combined X_C of _____ Ω.
5. Two X_L branches of 500 Ω each in parallel have combined X_L of _____ Ω.
6. A 500-Ω X_L is in series with a 300-Ω X_C. The net X_L is _____ Ω.
7. For 500 Ω X_C in series with 300 Ω X_{L_1}, the net X_C is _____ Ω.
8. A 10-Ω X_L is in series with a 10-Ω R. The total Z_T is _____ Ω.
9. With a 10-Ω X_C in series with a 10-Ω R, the total Z_T is _____ Ω.
10. With 14 V applied across 14 Ω Z_T, the I is _____ A.
11. For 10 Ω X_L and 10 Ω R in series, the phase angle θ is _____ degrees.
12. For 10 Ω X_C and 10 Ω R in series, the phase angle θ is _____ degrees.
13. A 10-Ω X_L and a 10-Ω R are in parallel across 10 V. The amount of each branch I is _____ A.
14. In question 13, the total line current I_T equals _____ A.
15. In questions 13 and 14, Z_T of the parallel branches equals _____ Ω.
16. With 120 V, an I of 10 A, and θ of 60°, a wattmeter reads _____ W.
17. The Z of $4 + j4$ Ω converted to polar form is _____ Ω.
18. The impedance value of $8 \angle 40°/2 \angle 30°$ is equal to _____ Ω.
19. If V is $42 \angle 0°$ and Z is $21 \angle 17°$, the I is _____ A.
20. For V_{AB} equal to $60 \angle 20°$, the V_{BA} is _____ V.

Answer True or False.

21. In an ac circuit with X_C and R in series, if the frequency is raised, the current will increase.
22. In an ac circuit with X_L and R in series, if the frequency is increased, the current will be reduced.
23. The voltampere is a unit of apparent power.
24. The polar form of complex numbers is best for adding impedance values.
25. Kirchhoff's laws, Thevenin's theorem, and Norton's theorem cannot be applied to ac circuits because of the complex impedances.

References
Additional References at Back of Book

Gilli, A. C.: *Electrical Principles for Electronics*, McGraw-Hill Book Company, New York.
Jackson, H. W.: *Introduction to Electric Circuits*, Prentice-Hall, Inc., Englewood Cliffs, N.J.
Ridsdale, R. E.: *Electric Circuits*, McGraw-Hill Book Company, New York.

Chapter 27
Resonance

This chapter explains how X_L and X_C can be combined to favor one particular frequency, the resonant frequency to which the LC circuit is tuned. The resonance effect occurs when the inductive and capacitive reactances are equal.

The main application of resonance is in RF circuits for tuning to an ac signal of the desired frequency. All examples of tuning in radio and television receivers, transmitters, and electronics equipment in general are applications of resonance.

Important terms in this chapter are:

antiresonance Q of resonant circuit
bandwidth   ringing
damping   series resonance
flywheel effect   tank circuit
half-power frequencies tuning
parallel resonance

More details are explained in the following sections:

27-1   The Resonance Effect

Inductive reactance increases as the frequency is increased, but capacitive reactance decreases with higher frequencies. Because of these opposite characteristics, for any *LC* combination there must be a frequency at which the X_L equals the X_C, as one increases while the other decreases. This case of equal and opposite reactances is called *resonance,* and the ac circuit is then a *resonant circuit.*

Any *LC* circuit can be resonant. It all depends on the frequency. At the resonant frequency, an *LC* combination provides the resonance effect. Off the resonant frequency, either below or above, the *LC* combination is just another ac circuit.

The frequency at which the opposite reactances are equal is the *resonant frequency*. This frequency can be calculated as $f_r = 1/(2\pi\sqrt{LC})$ where *L* is the inductance in henrys, *C* is the capacitance in farads, and f_r is the resonant frequency in hertz that makes $X_L = X_C$.

In general, we can say that large values of *L* and *C* provide a relatively low resonant frequency. Smaller values of *L* and *C* allow higher values for f_r. The resonance effect is most useful for radio frequencies, where the required values of microhenrys for *L* and picofarads for *C* are easily obtained.

The most common application of resonance in RF circuits is called *tuning*. In this use, the *LC* circuit provides maximum voltage output at the resonant frequency, compared with the amount of output at any other frequency either below or above resonance. This idea is illustrated in Fig. 27-1*a*, where the *LC* circuit resonant at 1000 kHz magnifies the effect of this particular frequency. The result is maximum output at

1000 kHz, compared with lower or higher frequencies.

For the wavemeter in Fig. 27-1*b*, note that the capacitance *C* can be varied to provide resonance at different frequencies. The wavemeter can be tuned to any one frequency in a range depending on the *LC* combination.

All examples of tuning in radio and television are applications of resonance. When you tune a radio to one station, the *LC* circuits are tuned to resonance for that particular carrier frequency. Also, when you tune a television receiver to a particular channel, the *LC* circuits are tuned to resonance for that station. There are almost unlimited uses for resonance in ac circuits.

Practice Problems 27-1
Answers at End of Chapter

Refer to Fig. 27-1.
a. Give the resonant frequency.
b. Give the frequency that has maximum output.

27-2   Series Resonance

In the series ac circuit in Fig. 27-2*a*, when the frequency of the applied voltage is 1000 kHz, the reactance of the 239-μH inductance equals 1500 Ω. At the same frequency, the reactance of the 106-pF capacitance also is 1500 Ω. Therefore, this *LC* combination is resonant at 1000 kHz. This is f_r, because the inductive reactance and capacitive reactance are equal at this frequency.

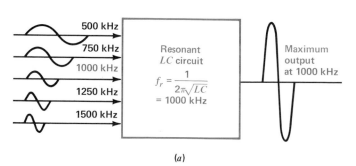

(a)   (b)

Fig. 27-1 (a) Circuit resonant at 1000 kHz to provide maximum output at this f_r. (b) Wavemeter as an example of tuning an *LC* circuit to resonance. (*James Millen Mfg. Co. Inc.*)

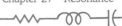

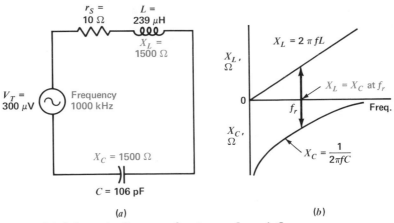

(a)   (b)

Fig. 27-2 Series resonance. (a) Schematic diagram of series r_S, L, and C.
(b) Reactances X_C and X_L are equal and opposite at the resonant frequency f_r.
Inductive reactance X_L is shown up for jX_L, and X_C is down for $-jX_C$.

In a series ac circuit, inductive reactance leads by 90°, compared with the zero reference angle of the resistance, while capacitive reactance lags by 90°. Therefore, X_L and X_C are 180° out of phase. The opposite reactances cancel each other completely when they are equal.

Figure 27-2b shows X_L and X_C equal, resulting in a net reactance of zero ohms. The only opposition to current then is the coil resistance r_S, which is the limit on how low the series resistance in the circuit can be. With zero reactance and just the low value of series resistance, the generator voltage produces the greatest amount of current in the series LC circuit at the resonant frequency. The series resistance should be as small as possible for a sharp increase in current at resonance.

Maximum Current at Series Resonance

The main characteristic of series resonance is the resonant rise of current to its maximum value of V_T/r_S at the resonant frequency. For the circuit in Fig. 27-2a, the maximum current at series resonance is 30 μA, equal to 300 μV/10 Ω. At any other frequency either below or above the resonant frequency, there is less current in the circuit.

This resonant rise of current to 30 μA at 1000 kHz is illustrated in Fig. 27-3. In Fig. 27-3a, the amount of

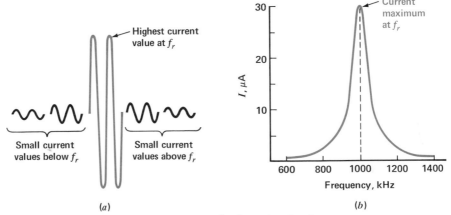

(a)   (b)

Fig. 27-3 Graphs showing maximum current at resonance, for the series circuit in Fig. 27-2. (a) Amplitudes of individual cycles. (b) Response curve showing amount of I below and above resonance. Values of I in Table 27-1.

current is shown as the amplitude of individual cycles of the alternating current produced in the circuit by the ac generator voltage. Whether the amplitude of one ac cycle is considered in terms of peak, rms, or average value, the amount of current is greatest at the resonant frequency. In Fig. 27-3b, the current amplitudes are plotted on a graph for frequencies at and near the resonant frequency, producing a typical *response curve* for a series resonant circuit. The response curve in Fig. 27-3b can be considered as an outline of the increasing and decreasing amplitudes for the individual cycles shown in Fig. 27-3a.

The response curve of the series resonant circuit shows that the current is small below resonance, rises to its maximum value at the resonant frequency, and then drops off to small values above resonance. To prove this fact, Table 27-1 lists the calculated values of impedance and current in the circuit of Fig. 27-2 at the resonant frequency of 1000 kHz and at two frequencies below and two frequencies above resonance.

Below resonance, at 600 kHz, X_C is more than X_L and there is appreciable net reactance, which limits the current to a relatively low value. At the higher frequency of 800 kHz, X_C decreases and X_L increases, making the two reactances closer to the same value. The net reactance is then smaller, allowing more current.

At the resonant frequency, X_L and X_C are equal, the net reactance is zero, and the current has its maximum value equal to V_T/r_S.

Above resonance at 1200 and 1400 kHz, X_L is greater than X_C, providing net reactance that limits the current to much smaller values than at resonance.

In summary:

1. Below the resonant frequency, X_L is small, but X_C has high values that limit the amount of current.
2. Above the resonant frequency, X_C is small, but X_L has high values that limit the amount of current.
3. At the resonant frequency, X_L equals X_C, and they cancel to allow maximum current.

Minimum Impedance at Series Resonance

Since the reactances cancel at the resonant frequency, the impedance of the series circuit is minimum, equal to just the low value of series resistance. This minimum impedance at resonance is resistive, resulting in zero phase angle. At resonance, therefore, the resonant current is in phase with the generator voltage.

Resonant Rise in Voltage Across Series L or C

The maximum current in a series LC circuit at resonance is useful because it produces maximum voltage across either X_L or X_C at the resonant frequency. As a result, the series resonant circuit can select one frequency by providing much more voltage output at the resonant frequency, compared with frequencies above and below resonance. Figure 27-4 illustrates the resonant rise in voltage across the capacitance in a series ac circuit. At the resonant frequency of 1000 kHz, the voltage across C rises to the value of 45,000 μV, while the input voltage is only 300 μV.

In Table 27-1, the voltage across C is calculated as IX_C and across L as IX_L. Below the resonant frequency,

Table 27-1. Series-Resonance Calculations for the Circuit in Fig. 27-2*

Frequency, kHz	$X_L = 2\pi fL$, Ω	$X_C = 1/(2\pi fC)$, Ω	Net reactance, Ω		Z_T, Ω†	$I = V_T/Z_T$, μA†	$V_L = IX_L$, μV	$V_C = IX_C$, μV
			$X_C - X_L$	$X_L - X_C$				
600	900	2500	1600		1600	0.19	171	475
800	1200	1875	675		675	0.44	528	825
$f_r \rightarrow$ 1000	1500	1500	0	0	10	30	45,000	45,000
1200	1800	1250		550	550	0.55	990	688
1400	2100	1070		1030	1030	0.29	609	310

*$L = 239\ \mu$H, $C = 106$ pF, $V_T = 300\ \mu$V, $r_S = 10\ \Omega$.
†Z_T and I calculated without r_S when its resistance is very small compared with the net X_L or X_C. Z_T and I are resistive at f_r.

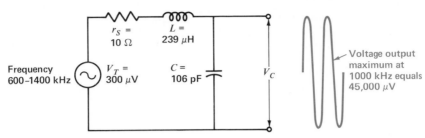

Fig. 27-4 Series circuit selects frequency by producing maximum IX_C voltage output across C at resonance.

X_C has a higher value than at resonance, but the current is small. Similarly, above the resonant frequency, X_L is higher than at resonance, but the current has a low value because of the inductive reactance. At resonance, although X_L and X_C cancel each other to allow maximum current, each reactance by itself has an appreciable value. Since the current is the same in all parts of a series circuit, the maximum current at resonance produces maximum voltage IX_C across C and an equal IX_L voltage across L for the resonant frequency.

Although the voltage across X_C and X_L is reactive, it is an actual voltage that can be measured. In Fig. 27-5, the voltage drops around the series resonant circuit are 45,000 μV across C and 45,000 μV across L, with 300 μV across r_S. The voltage across the resistance is equal to the generator voltage and has the same phase.

Across the series combination of both L and C, the voltage is zero because the two series voltage drops are equal and opposite. In order to use the resonant rise of voltage, therefore, the output must be connected across either L or C alone. We can consider the V_L and V_C voltages as similar to the idea of two batteries con-

nected in series opposition. Together, the net resultant is zero for the equal and opposite voltages, but each battery still has its own potential difference.

In summary, for a series resonant circuit the main characteristics are:

1. The current I is maximum at the resonant frequency f_r.
2. The current I is in phase with the generator voltage, or the phase angle of the circuit is 0°.
3. The voltage is maximum across either L or C alone.
4. The impedance is minimum at f_r, equal only to the low r_S.

Practice Problems 27-2
Answers at End of Chapter

Answer True or False, for series resonance.
a. Impedances X_L and X_C are maximum.
b. Impedances X_L and X_C are equal.
c. Current I is maximum.

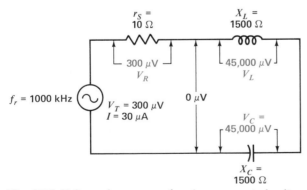

Fig. 27-5 Voltage drops around series resonant circuit.

27-3 Parallel Resonance

With L and C in parallel as shown in Fig. 27-6, when X_L equals X_C, the reactive branch currents are equal and opposite at resonance. Then they cancel each other to produce minimum current in the main line. Since the line current is minimum, the impedance is maximum. These relations are based on r_S being very small compared with X_L at resonance. In this case, the branch currents are practically equal when X_L and X_C are equal.

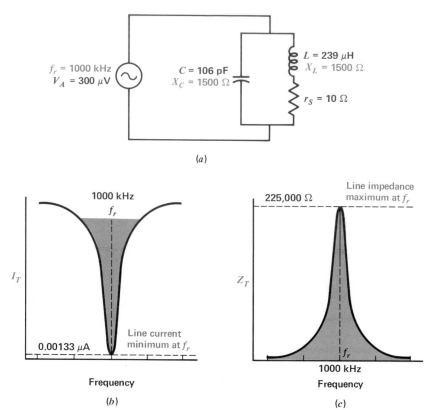

Fig. 27-6 Parallel resonant circuit. (*a*) Schematic diagram of *L* and *C* in parallel branches. (*b*) Response curve of I_T shows that line current dips to minimum at f_r. (*c*) Response curve of Z_T shows that it rises to maximum at f_r.

Minimum Line Current at Parallel Resonance

To show how the current in the main line dips to its minimum value when the parallel *LC* circuit is resonant, Table 27-2 lists the values of branch currents and the total line current for the circuit in Fig. 27-6.

With *L* and *C* the same as in the series circuit of Fig. 27-2, X_L and X_C have the same values at the same frequencies. Since *L*, *C*, and the generator are in parallel, the voltage applied across the branches equals the generator voltage of 300 μV. Therefore, each reactive branch current is calculated as 300 μV divided by the reactance of the branch.

The values in the top row of Table 27-2 are obtained as follows: At 600 kHz the capacitive branch current equals 300 μV/2500 Ω, or 0.12 μA. The inductive branch current at this frequency is 300 μV/900 Ω, or 0.33 μA. Since this is a parallel ac circuit, the capacitive current leads by 90° while the inductive current lags by 90°, compared with the reference angle of the

generator voltage, which is applied across the parallel branches. Therefore, the opposite currents are 180° out of phase, canceling each other in the main line. The net current in the line, then, is the difference between 0.33 and 0.12, which equals 0.21 μA.

Following this procedure, the calculations show that as the frequency is increased toward resonance, the capacitive branch current increases because of the lower value of X_C, while the inductive branch current decreases with higher values of X_L. As a result, there is less net line current as the two branch currents become more nearly equal.

At the resonant frequency of 1000 kHz, both reactances are 1500 Ω, and the reactive branch currents are both 0.20 μA, canceling each other completely.

Above the resonant frequency, there is more current in the capacitive branch than in the inductive branch, and the net line current increases above its minimum value at resonance.

Table 27-2. Parallel-Resonance Calculations for the Circuit in Fig. 27-6*

Frequency, kHz	$X_C = 1/(2\pi fC)$, Ω	$X_L = 2\pi fL$, Ω	$I_C = V/X_C$, μA	$I_L = V/X_L$, μA†	Net reactive line current, μA		I_T, μA†	$Z_T = V_A/I_T$, Ω†
					$I_L - I_C$	$I_C - I_L$		
600	2500	900	0.12	0.33	0.21		0.21	1400
800	1875	1200	0.16	0.25	0.09		0.09	3333
$f_r \rightarrow$ 1000	1500	1500	0.20	0.20	0	0	0.001 33	225,000‡
1200	1250	1800	0.24	0.17		0.08	0.08	3800
1400	1070	2100	0.28	0.14		0.14	0.14	2143

*$L = 239\ \mu H$, $C = 106$ pF, $V_A = 300\ \mu V$.
†I_L, I_T, and Z_T calculated approximately without r_S when its resistive component of the line current is very small compared with I_L.
‡At resonance, Z_T calculated by Formula (27-8). Z_T and I_T are resistive at f_r.

The dip in I_T to its minimum value at f_r is shown by the graph in Fig. 27-6b. At parallel resonance, I_T is minimum and Z_T is maximum.

The in-phase current due to r_S in the inductive branch can be ignored off resonance because it is so small compared with the reactive line current. At the resonant frequency when the reactive currents cancel, however, the resistive component is the entire line current. Its value at resonance equals 0.001 33 μA in this example. This small resistive current is the minimum value of the line current at parallel resonance.

Maximum Line Impedance at Parallel Resonance The minimum line current resulting from parallel resonance is useful because it corresponds to maximum impedance in the line across the generator. Therefore, an impedance that has a high value for just one frequency but a low impedance for other frequencies, either below or above resonance, can be obtained by using a parallel LC circuit resonant at the desired frequency. This is another method of selecting one frequency by resonance. The response curve in Fig. 27-6c shows how the impedance rises to maximum for parallel resonance.

The main application of parallel resonance is the use of an LC tuned circuit as the load impedance Z_L in the output circuit of RF amplifiers. Because of the high impedance, then, the gain of the amplifier is maximum at f_r. The voltage gain of an amplifier is directly proportional to Z_L. The advantage of a resonant LC circuit is that Z is maximum only for an ac signal at the reso-

nant frequency. Also, L has practically no dc resistance, which means practically no dc voltage drop.

Referring to Table 27-2, the total impedance of the parallel ac circuit is calculated as the generator voltage divided by the total line current. At 600 kHz, for example, Z_T equals 300 $\mu V/0.21\ \mu A$, or 1400 Ω. At 800 kHz, the impedance is higher because there is less line current.

At the resonant frequency of 1000 kHz, the line current is at its minimum value of 0.001 33 μA. Then the impedance is maximum and is equal to 300 $\mu V/0.001\ 33\ \mu A$, or 225,000 Ω.

Above 1000 kHz, the line current increases, and the impedance decreases from its maximum value.

The idea of how the line current can have a very low value even though the reactive branch currents are appreciable is illustrated in Fig. 27-7. In Fig. 27-7a, the resistive component of the total line current is shown as though it were a separate branch drawing an amount of resistive current from the generator in the main line equal to the current resulting from the coil resistance. Each reactive branch current has its value equal to the generator voltage divided by the reactance. Since they are equal and of opposite phase, however, in any part of the circuit where both reactive currents are present, the net amount of electron flow in one direction at any instant of time corresponds to zero current. The graph in Fig. 27-7b shows how equal and opposite currents for I_L and I_C cancel.

If a meter is inserted in series with the main line to indicate total line current I_T, it dips sharply to the minimum value of line current at the resonant frequency.

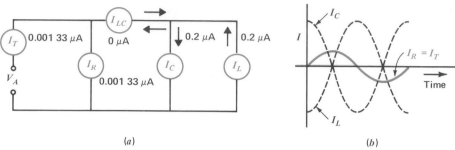

Fig. 27-7 Distribution of currents in parallel circuit at resonance. Resistive current shown as an equivalent branch for I_R. (a) Circuit with branch currents for R, L, and C. (b) Graph of equal and opposite reactive currents I_L and I_C.

With minimum current in the line, the impedance across the line is maximum at the resonant frequency. The maximum impedance at parallel resonance corresponds to a high value of resistance, without reactance, since the line current is then resistive with zero phase angle.

In summary, for a parallel resonant circuit, the main characteristics are:

1. The line current I_T is minimum at the resonant frequency.
2. The current I_T is in phase with the generator voltage V_A, or the phase angle of the circuit is 0°.
3. The impedance Z_T, equal to V_A/I_T, is maximum at f_r because of the minimum I_T.

The LC Tank Circuit It should be noted that the individual branch currents are appreciable at resonance, although I_T is minimum. For the example in Table 27-2, at f_r either the I_L or I_C equals 0.2 μA. This current is greater than the I_C values below f_r or the I_L values above f_r.

The branch currents cancel in the main line because I_C is at 90° with respect to the source V_A while I_L is at −90°, making them opposite with respect to each other.

However, inside the LC circuit, I_L and I_C do not cancel because they are in separate branches. Then I_L and I_C provide a circulating current in the LC circuit, which equals 0.2 μA in this example. For this reason, a parallel resonant LC circuit is often called a *tank circuit*.

Because of the energy stored by L and C, the circulating tank current can provide full sine waves of current and voltage output when the input is only a pulse. The sine-wave output is always at the natural resonant frequency of the LC tank circuit. This ability of the LC circuit to supply complete sine waves is called the *flywheel effect*. Also, the process of producing sine waves after a pulse of energy has been applied is called *ringing* of the LC circuit.

Practice Problems 27-3
Answers at End of Chapter

Answer True or False, for parallel resonance.
a. Currents I_L and I_C are maximum.
b. Currents I_L and I_C are equal.
c. Current I_T is minimum.

27-4 The Resonant Frequency $f_r = 1/(2\pi\sqrt{LC})$

This formula is derived from $X_L = X_C$. Using f_r to indicate the resonant frequency in the formulas for X_L and X_C, we have

$$2\pi f_r L = \frac{1}{2\pi f_r C}$$

Inverting the factor f_r gives

$$2\pi L (f_r)^2 = \frac{1}{2\pi C}$$

Inverting the factor $2\pi L$ gives

$$f_r^2 = \frac{1}{(2\pi)^2 LC}$$

The square root of both sides is then

$$f_r = \frac{1}{2\pi\sqrt{LC}} \qquad (27\text{-}1)$$

With units of henrys for L and farads for C, the resonant frequency f_r is in hertz. Since $1/2\pi$ is a numerical value equal to $1/6.28$, or 0.159, a more convenient form for calculations is

$$f_r = \frac{0.159}{\sqrt{LC}} \qquad \text{Hz} \qquad (27\text{-}2)$$

For example, to find the resonant frequency of the LC combination in Fig. 27-2, the values of 239×10^{-6} H and 106×10^{-12} F are substituted for L and C:

$$f_r = \frac{0.159}{\sqrt{239 \times 10^{-6} \times 106 \times 10^{-12}}}$$

$$= \frac{0.159}{\sqrt{253 \times 10^{-16}}}$$

Taking the square root of the denominator,

$$f_r = \frac{0.159}{15.9 \times 10^{-8}} = \frac{0.159}{0.159 \times 10^{-6}}$$

$$= \frac{0.159 \times 10^6}{0.159}$$

$$f_r = 1 \times 10^6 \text{ Hz} = 1 \text{ MHz} = 1000 \text{ kHz}$$

For any LC circuit, series or parallel, $f_r = 1/(2\pi\sqrt{LC})$ is the resonant frequency that makes the inductive and capacitive reactances equal.

How f_r Varies with L or C   It is important to note that higher values of L or C result in lower values of f_r. Also, an LC circuit can be resonant at any frequency from a few hertz to many megahertz, depending upon the inductance and capacitance values.

As examples, an LC combination with the relatively large values of an 8-H inductance and a 20-μF capaci-

tance is resonant at the low audio frequency of 12.6 Hz; a small inductance of 2 μH will resonate with the small capacitance of 3 pF at the high radio frequency of 64.9 MHz. These examples are solved in the next two problems for more practice with the resonant frequency formula. Such calculations are often used in practical applications of tuned circuits. Probably the most important feature of any LC combination is its resonant frequency.

Example 1   Calculate the resonant frequency for an 8-H L and a 20-μF C.

Answer $f_r = \dfrac{1}{2\pi\sqrt{LC}} = \dfrac{0.159}{\sqrt{8 \times 20 \times 10^{-6}}}$

$$= \frac{0.159 \times 10^3}{\sqrt{160}} = \frac{159}{12.65}$$

$$f_r = 12.6 \text{ Hz}$$

Example 2   Calculate the resonant frequency for a 2-μH L and a 3-pF C.

Answer $f_r = \dfrac{1}{2\pi\sqrt{LC}}$

$$= \frac{0.159}{\sqrt{2 \times 10^{-6} \times 3 \times 10^{-12}}}$$

$$= \frac{0.159}{\sqrt{6 \times 10^{-18}}} = \frac{0.159 \times 10^9}{\sqrt{6}}$$

$$= \frac{159 \times 10^6}{\sqrt{6}} = \frac{159}{2.45} \times 10^6$$

$$f_r = 64.9 \text{ MHz}$$

More specifically, f_r decreases inversely as the square root of L or C. For instance, if L or C is quadrupled, f_r is reduced by one-half. Suppose f_r is 6 MHz with an LC combination. If L or C is made four times larger, then f_r will be reduced to 3 MHz. Or to take the opposite case, to double f_r, the value of L or C must be reduced by one-fourth, or both reduced by one-half.

LC Product Determines f_r   There are any number of LC combinations that can be resonant at one frequency. Table 27-3 lists five possible combinations

Table 27-3. *LC* Combinations Resonant at 1000 kHz

L, μH	C, pF	LC product	X_L, Ω at 1000 kHz	X_C, Ω at 1000 kHz
23.9	1060	25,334	150	150
119.5	212	25,334	750	750
239	106	25,334	1500	1500
478	53	25,334	3000	3000
2390	10.6	25,334	15,000	15,000

of L and C resonant at 1000 kHz. The resonant frequency is the same because when either L or C is decreased by the factor of 10 or 2, the other is increased by the same factor, resulting in a constant value for the LC product.

The reactance at resonance changes with different combinations of L and C, but in all five cases X_L and X_C are equal to each other at 1000 kHz. This is the resonant frequency determined by the value of the LC product in $f_r = 1/(2\pi\sqrt{LC})$.

Measuring *L* or *C* by Resonance   Of the three factors L, C, and f_r in the resonant-frequency formula, any one can be calculated when the other two are known. The resonant frequency of the LC combination can be found experimentally by determining the frequency that produces the resonant response in an LC combination. With a known value of either L or C, and the resonant frequency determined, the third factor can be calculated. This method is commonly used for measuring inductance or capacitance. A test instrument for this purpose is the Q *meter*, which also measures the Q of a coil.

Calculating *C* from f_r   The C can be taken out of the square root sign or radical in the resonance formula, as follows:

$$f_r = \frac{1}{2\pi\sqrt{LC}}$$

Squaring both sides to eliminate the radical gives

$$f_r^2 = \frac{1}{(2\pi)^2 LC}$$

Inverting C and f_r^2 gives

$$C = \frac{1}{(2\pi)^2 f_r^2 L} = \frac{1}{4\pi^2 f_r^2 L} = \frac{0.0254}{f_r^2 L} \qquad (27\text{-}3)$$

With f_r in hertz, the units are farads for C and henrys for L.

The constant factor 0.0254 in the numerator is the reciprocal of 39.44 for $4\pi^2$ in the denominator. These numbers remain the same for any values of f_r, L, and C.

Calculating *L* from f_r   Similarly, the resonance formula can be transposed to find L. Then

$$L = \frac{1}{(2\pi)^2 f_r^2 C} = \frac{1}{4\pi^2 f_r^2 C} = \frac{0.0254}{f_r^2 C} \qquad (27\text{-}4)$$

With Formula (27-4), L is determined by its f_r with a known value of C. Similarly, C is determined from Formula (27-3) by its f_r with a known value of L.

Example 3   What value of C resonates with a 239-μH L at 1000 kHz?

Answer $C = \dfrac{0.0254}{f_r^2 L}$

$$= \frac{0.0254}{(1 \times 10^6)^2 \times 239 \times 10^{-6}}$$

$$= \frac{0.0254}{1 \times 10^{12} \times 239 \times 10^{-6}}$$

$$= \frac{0.0254}{239 \times 10^6} = \frac{0.0254}{239} \times 10^{-6}$$

$$= \frac{25,400}{239} \times 10^{-12}$$

$$= 106 \times 10^{-12} \text{ F}$$

$$C = 106 \text{ pF}$$

Example 4   What value of L resonates with a 106-pF C at 1000 kHz?

Answer $L = \dfrac{0.0254}{f_r^2 C}$

$= \dfrac{0.0254}{1 \times 10^{12} \times 106 \times 10^{-12}}$

$= \dfrac{0.0254}{106}$

$= \dfrac{25,400}{106} \times 10^{-6}$

$= 239 \times 10^{-6}$ H

$L = 239 \; \mu$H

These values are from the LC circuit illustrated in Fig. 27-2 for series resonance and Fig. 27-6 for parallel resonance.

Practice Problems 27-4
Answers at End of Chapter

a. To increase f_r, must the C be more or less?
b. If C is increased from 100 to 400 pF, L must be decreased from 800 μH to what value of microhenrys for the same f_r?

27-5   Q Magnification Factor of Resonant Circuit

The quality, or *figure of merit,* of the resonant circuit, in sharpness of resonance, is indicated by the factor Q. In general, the higher the ratio of the reactance at resonance to the series resistance, the higher is the Q and the sharper the resonance effect.

Q of Series Circuit   In a series resonant circuit we can calculate Q from the following formula:

$$Q = \frac{X_L}{r_S} \tag{27-5}$$

where Q is the figure of merit, X_L is the inductive react-

ance at the resonant frequency, and r_S is the resistance in series with X_L. For the series resonant circuit in Fig. 27-2,

$$Q = \frac{1500 \; \Omega}{10 \; \Omega} = 150$$

The Q is a numerical factor without any units, because it is a ratio of reactance to resistance and the ohms cancel. Since the series resistance limits the amount of current at resonance, the lower the resistance, the sharper the increase to maximum current at the resonant frequency, and the higher the Q. Also, a higher value of reactance at resonance allows the maximum current to produce a higher value of voltage for the output.

The Q has the same value if it is calculated with X_C instead of X_L, since they are equal at resonance. However, the Q of the circuit is generally considered in terms of X_L, because usually the coil has the series resistance of the circuit. In this case, the Q of the coil and the Q of the series resonant circuit are the same. If extra resistance is added, the Q of the circuit will be less than the Q of the coil. The highest possible Q for the circuit is the Q of the coil.

The value of 150 can be considered as a high Q. Typical values are 50 to 250, approximately. Less than 10 is a low Q; more than 300 is a very high Q.

Higher L/C Ratio Can Provide Higher Q   As shown before in Table 27-3, different LC combinations can be resonant at the same frequency. However, the amount of reactance at resonance is different. More X_L can be obtained with a higher L and smaller C for resonance, although X_L and X_C must be equal at the resonant frequency. Therefore, both X_L and X_C are higher with a higher L/C ratio for resonance.

More X_L can allow a higher Q if the ac resistance does not increase as much as the reactance. With typical RF coils, an approximate rule is that maximum Q can be obtained when X_L is about 1000 Ω. In many cases, though, the minimum C is limited by the stray capacitance in the circuit.

Q Rise in Voltage Across Series L or C   The Q of the resonant circuit can be considered a magnification factor that determines how much the voltage across

L or C is increased by the resonant rise of current in a series circuit. Specifically, the voltage output at series resonance is Q times the generator voltage:

$$V_L = V_C = Q \times V_{gen} \qquad (27\text{-}6)$$

In Fig. 27-4, for example, the generator voltage is 300 μV and Q is 150. The resonant rise of voltage across either L or C then equals 300 μV × 150, or 45,000 μV. Note that this is the same value calculated in Table 27-1 for V_C or V_L at resonance.

How to Measure Q in a Series Resonant Circuit

The fundamental nature of Q for a series resonant circuit is seen from the fact that the Q can be determined experimentally by measuring the Q rise in voltage across either L or C and comparing this voltage with the generator voltage. As a formula,

$$Q = \frac{V_{out}}{V_{in}} \qquad (27\text{-}7)$$

where V_{out} is the ac voltage measured across the coil or capacitor and V_{in} is the generator voltage.

Referring to Fig. 27-5, suppose that you measure with an ac voltmeter across L or C and this voltage equals 45,000 μV at the resonant frequency. Also, measure the generator input of 300 μV. Then

$$Q = \frac{V_{out}}{V_{in}} = \frac{45,000 \ \mu V}{300 \ \mu V}$$
$$Q = 150$$

This method is better than the X_L/r_S formula for determining Q because r_S is the ac resistance of the coil, which is not so easily measured. Remember that the coil's ac resistance can be more than double the dc resistance measured with an ohmmeter. In fact, measuring Q with Formula (27-7) makes it possible to calculate the ac resistance. These points are illustrated in the following examples.

Example 5   A series circuit resonant at 0.4 MHz develops 100 mV across a 250-μH L with a 2-mV input. Calculate Q.

Answer $Q = \dfrac{V_{out}}{V_{in}} = \dfrac{100 \ mV}{2 \ mV}$

$Q = 50$

Example 6   How much is the ac resistance of the coil in the preceding example?

Answer   The Q of the coil is 50. We need to know the reactance of this 250-μH coil at the frequency of 0.4 MHz. Then,

$$X_L = 2\pi f L = 2\pi \times 0.4 \times 10^6 \times 250 \times 10^{-6}$$
$$= 2\pi \times 100$$
$$X_L = 628 \ \Omega$$

Also, $Q = \dfrac{X_L}{r_S}$ or

$$r_S = \frac{X_L}{Q}$$
$$= \frac{628 \ \Omega}{50}$$
$$r_S = 12.56 \ \Omega$$

Q of Parallel Circuit

In a parallel resonant circuit, where r_S is very small compared with X_L, the Q also equals X_L/r_S. Note that r_S is still the resistance of the coil in series with X_L (see Fig. 27-8). The Q of the coil determines the Q of the parallel circuit here because it is less than the Q of the capacitive branch. Capacitors used in tuned circuits generally have a very

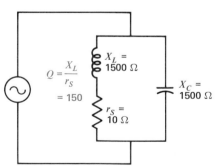

Fig. 27-8 The Q of a parallel resonant circuit in terms of X_L and its series resistance r_S.

high Q because of their low losses. In Fig. 27-8, the Q is 1500 Ω/10 Ω, or 150, the same as the series resonant circuit with the same values.

This example assumes that the generator resistance is very high and that there is no other resistance branch shunting the tuned circuit. Then the Q of the parallel resonant circuit is the same as the Q of the coil. Actually, shunt resistance can lower the Q of a parallel resonant circuit, as analyzed in Sec. 27-10.

Q Rise in Impedance Across Parallel Resonant Circuit

For parallel resonance, the Q magnification factor determines by how much the impedance across the parallel LC circuit is increased because of the minimum line current. Specifically, the impedance across the parallel resonant circuit is Q times the inductive reactance at the resonant frequency:

$$Z_T = Q \times X_L \qquad (27\text{-}8)$$

Referring back to the parallel resonant circuit in Fig. 27-6, as an example, X_L is 1500 Ω and Q is 150. The result is a rise of impedance to the maximum value of 150 × 1500 Ω, or 225,000 Ω, at the resonant frequency.

Since the line current equals V_A/Z_T, the minimum value of line current is 300 μV/225,000 Ω, which equals 0.001 33 μA.

At f_r the minimum line current is $1/Q$ of either branch current. In Fig. 27-7, I_L or I_C is 0.2 μA and Q is 150. Therefore, I_T is 0.2/150, or 0.001 33 μA, which is the same answer as V_A/Z_T. Or, stated another way, the circulating tank current is Q times the minimum I_T.

How to Measure Z_T of a Parallel Resonant Circuit

Formula (27-8) is also useful in its inverted version as $Q = Z_T/X_L$. We can measure Z_T by the method illustrated in Fig. 27-9. Then Q can be calculated.

To measure Z_T, first tune the LC circuit to resonance. Then adjust R_1 to the resistance that makes its ac voltage equal to the ac voltage across the tuned circuit. With equal voltages, Z_T must be the same as R_1.

For the example here, which corresponds to the parallel resonance shown in Figs. 27-6 and 27-8, Z_T is 225,000 Ω. Therefore, Q equals Z_T/X_L with values of 225,000/1500, which equals 150.

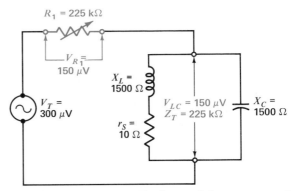

Fig. 27-9 How to measure Z_T of a parallel resonant circuit. Adjust R_1 to make its V_R equal V_{LC}. Then $Z_T = R_1$.

Example 7 In Fig. 27-9, assume that with 4 V ac input signal for V_T, the voltage across R_1 is 2 V when R_1 is 225 kΩ. Determine Z_T and Q.

Answer Because they divide V_T equally, Z_T is 225 kΩ, the same as R_1. The amount of input voltage does not matter, as the voltage division determines the relative proportions between R_1 and Z_T. With 225 kΩ for Z_T and 1.5 kΩ for X_L, the Q is $^{225}/_{1.5}$, or $Q = 150$.

Example 8 A parallel LC circuit tuned to 200 kHz with a 350-μH L has a measured Z_T of 17,600 Ω. Calculate Q.

Answer First, calculate X_L as $2\pi fL$ at f_r:

$$X_L = 2\pi \times 200 \times 10^3 \times 350 \times 10^{-6} = 440\ \Omega$$

Then,

$$Q = \frac{Z_T}{X_L} = \frac{17,600}{440}$$

$$Q = 40$$

Practice Problems 27-5
Answers at End of Chapter

a. In a series resonant circuit, V_L is 300 mV with input of 3 mV. Calculate Q.

b. In a parallel resonant circuit, X_L is 500 Ω. With a Q of 50, calculate Z_T.

27-6 Bandwidth of Resonant Circuit

When we say an *LC* circuit is resonant at one frequency, this is true for the maximum resonance effect. However, other frequencies close to f_r also are effective. For series resonance, frequencies just below and above f_r produce increased current, but a little less than the value at resonance. Similarly, for parallel resonance, frequencies close to f_r can provide a high impedance, although a little less than the maximum Z_T.

Therefore, any resonant frequency has an associated band of frequencies that provide resonance effects. How wide the band is depends on the *Q* of the resonant circuit. Actually, it is practically impossible to have an *LC* circuit with a resonant effect at only one frequency. The width of the resonant band of frequencies centered around f_r is called the *bandwidth* of the tuned circuit.

Measurement of Bandwidth The group of frequencies with a response 70.7 percent of maximum, or more, is generally considered the bandwidth of the tuned circuit, as shown in Fig. 27-10b. The resonant response here is increasing current for the series circuit in Fig. 27-10a. Therefore, the bandwidth is measured between the two frequencies, f_1 and f_2, producing 70.7 percent of the maximum current at f_r.

For a parallel circuit, the resonant response is increasing impedance Z_T. Then the bandwidth is measured between the two frequencies allowing 70.7 percent of the maximum Z_T at f_r.

The bandwidth indicated on the response curve in Fig. 27-10b equals 20 kHz. This value is the difference between f_2 at 60 kHz and f_1 at 40 kHz, both with 70.7 percent response.

Compared with the maximum current of 100 mA for f_r at 50 kHz, f_1 below resonance and f_2 above resonance each allow a rise to 70.7 mA. All frequencies in this band 20 kHz wide allow 70.7 mA, or more, as the resonant response in this example.

Bandwidth Equals f_r/Q Sharp resonance with high *Q* means narrow bandwidth. The lower the *Q*, the broader the resonant response and the greater the bandwidth.

Also, the higher the resonant frequency, the greater is the range of frequency values included in the bandwidth for a given sharpness of resonance. Therefore, the bandwidth of a resonant circuit depends on the factors f_r and *Q*. The formula is

$$f_2 - f_1 = \Delta f = \frac{f_r}{Q} \qquad (27\text{-}9)$$

where Δf is the total bandwidth in the same units as the

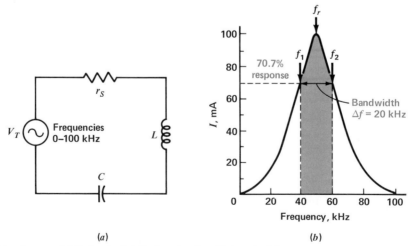

(a)   (b)

Fig. 27-10 Bandwidth of a tuned *LC* circuit. (a) Series circuit with input of 0 to 100 kHz. (b) Response curve with bandwidth Δf equal to 20 kHz between f_1 and f_2.

resonant frequency f_r. The bandwidth Δf can also be abbreviated BW.

For example, a series circuit resonant at 800 kHz with a Q of 100 has a bandwidth of $800/100$, or 8 kHz. Then the I is 70.7 percent of maximum, or more, for all frequencies for a band 8 kHz wide. This frequency band is centered around 800 kHz, from 796 to 804 kHz.

With a parallel resonant circuit having a Q higher than 10, Formula (27-9) also can be used for calculating the bandwidth of frequencies which provide 70.7 percent or more of the maximum Z_T. However, the formula cannot be used for parallel resonant circuits with low Q, as the resonance curve then becomes unsymmetrical.

High Q Means Narrow Bandwidth   The effect for different values of Q is illustrated in Fig. 27-11. Note that a higher Q for the same resonant frequency results in less bandwidth. The slope is sharper for the sides or *skirts* of the response curve, in addition to its greater amplitude.

High Q is generally desirable for more output from the resonant circuit. However, it must have enough bandwidth to include the desired range of signal frequencies.

The Edge Frequencies   Either f_1 or f_2 is separated from f_r by one-half of the total bandwidth. For the top curve in Fig. 27-11, as an example, with a Q of 80, Δf is ± 5 kHz centered around 800 kHz for f_r. To determine the edge frequencies:

$$f_1 = f_r - \frac{\Delta f}{2} = 800 - 5 = 795 \text{ kHz}$$

$$f_2 = f_r + \frac{\Delta f}{2} = 800 + 5 = 805 \text{ kHz}$$

These examples assume the resonance curve is symmetrical. This is true for a high-Q parallel resonant circuit and a series resonant circuit with any Q.

Example 9   An LC circuit resonant at 2000 kHz has a Q of 100. Find the total bandwidth Δf and the edge frequencies f_1 and f_2.

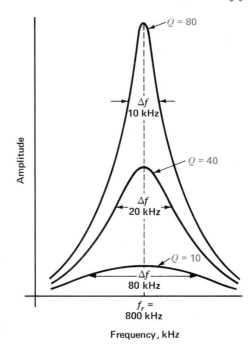

Fig. 27-11   Higher Q provides a sharper resonant response. Amplitude is I for series resonance or Z_T for parallel resonance. Bandwidth at half-power frequencies is Δf.

Answer $\Delta f = \dfrac{f_r}{Q} = \dfrac{2000 \text{ kHz}}{100} = 20 \text{ kHz}$

$$f_1 = f_r - \frac{\Delta f}{2} = 2000 - 10$$

$$= 1990 \text{ kHz}$$

$$f_2 = f_r + \frac{\Delta f}{2} = 2000 + 10$$

$$= 2010 \text{ kHz}$$

Example 10   Do the same as in Example 9 for an f_r equal to 6000 kHz and the same Q of 100.

Answer $f = \dfrac{f_r}{Q} = \dfrac{6000 \text{ kHz}}{100} = 60 \text{ kHz}$

$$f_1 = 6000 - 30 = 5970 \text{ kHz}$$

$$f_2 = 6000 + 30 = 6030 \text{ kHz}$$

Notice that Δf is three times as wide for the same Q because f_r is three times higher.

Half-Power Points It is simply for convenience in calculations that the bandwidth is defined between the two frequencies having 70.7 percent response. At each of these frequencies, the net capacitive or inductive reactance equals the resistance. Then the total impedance of the series reactance and resistance is 1.4 times greater than R. With this much more impedance, the current is reduced to $\frac{1}{1.414}$, or 0.707, of its maximum value.

Furthermore, the relative current or voltage value of 70.7 percent corresponds to 50 percent in power, since power is I^2R or V^2/R and the square of 0.707 equals 0.50. Therefore, the bandwidth between frequencies having 70.7 percent response in current or voltage is also the bandwidth in terms of half-power points. Formula (27-9) is derived for Δf between the points with 70.7 percent response on the resonance curve.

Measuring Bandwidth to Calculate Q The half-power frequencies f_1 and f_2 can be determined experimentally. For series resonance, find the two frequencies at which the current is 70.7 percent of maximum I. Or, for parallel resonance, find the two frequencies that make the impedance 70.7 percent of the maximum Z_T. The following method uses the circuit in Fig. 27-9 for measuring Z_T, but with different values to determine its bandwidth and Q.

1. Tune the circuit to resonance and determine its maximum Z_T at f_r. In this example, assume that Z_T is 10,000 Ω at the resonant frequency of 200 kHz.

2. Keep the same amount of input voltage, but change its frequency slightly below f_r to determine the frequency f_1 which results in a Z_1 equal to 70.7 percent of Z_T. The required value here is $0.707 \times 10,000$, or 7070 Ω, for Z_1 at f_1. Assume this frequency f_1 is determined to be 195 kHz.

3. Similarly, find the frequency f_2 above f_r that results in the impedance Z_2 of 7070 Ω. Assume f_2 is 205 kHz.

4. The total bandwidth between the half-power frequencies equals $f_2 - f_1$ or $205 - 195$. Then the value of $\Delta f = 10$ kHz.

5. Then $Q = f_r/\Delta f$ or 200 kHz/10 kHz = 20 for the calculated value of Q.

Practice Problems 27-6
Answers at End of Chapter

a. An LC circuit with f_r of 10 MHz has a Q of 40. Calculate the half-power bandwidth.

b. For an f_r of 500 kHz and bandwidth Δf of 10 kHz, calculate Q.

27-7 Tuning

This means obtaining resonance at different frequencies by varying either L or C. As illustrated in Fig. 27-12, the variable capacitance C can be adjusted to

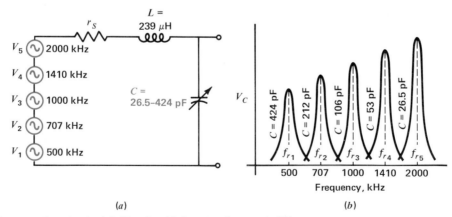

(a)   (b)

Fig. 27-12 Tuning a series circuit. (a) Circuit with input voltages at different frequencies. (b) Resonant responses at different frequencies when C is varied. Relative amplitudes not to scale.

tune the series LC circuit to resonance at any one of the five different frequencies. Each of the voltages V_1 to V_5 indicates an ac input with a specific frequency. Which one is selected for maximum output is determined by the resonant frequency of the LC circuit.

When C is set to 424 pF, for example, the resonant frequency of the LC circuit is 500 kHz for f_{r_1}. The input voltage that has the frequency of 500 kHz then produces a resonant rise of current which results in maximum output voltage across C. At other frequencies, such as 707 kHz, the voltage output is less than the input. With C at 424 pF, therefore, the LC tuned to 500 kHz selects this frequency by providing much more voltage output compared with other frequencies.

Suppose that we want maximum output for the ac input voltage that has the frequency of 707 kHz. Then C is set at 212 pF to make the LC circuit resonant at 707 kHz for f_{r_2}. Similarly, the tuned circuit can resonate at a different frequency for each input voltage. In this way, the LC circuit is tuned to select the desired frequency.

The variable capacitance C can be set at the values listed in Table 27-4 to tune the LC circuit to different frequencies. Only five frequencies are listed here, but any one capacitance value between 26.5 and 424 pF can tune the 239-μH coil to resonance at any frequency in the range of 500 to 2000 kHz. It should be noted that a parallel resonant circuit also can be tuned by varying C or L.

Tuning Ratio When an LC circuit is tuned, the change in resonant frequency is inversely proportional to the square root of the change in L or C. Referring to Table 27-4, notice that when C is decreased to one-fourth, from 424 to 106 pF, the resonant frequency doubles from 500 to 1000 kHz. Or the frequency is increased by the factor $1/\sqrt{\frac{1}{4}}$, which equals 2.

Suppose that we want to tune through the whole frequency range of 500 to 2000 kHz. This is a tuning ratio of 4:1 for the highest frequency to the lowest frequency. Then the capacitance must be varied from 424 to 26.5 pF, which is a 16:1 capacitance ratio.

Radio Tuning Dial Figure 27-13 illustrates a common application of resonant circuits in tuning a receiver to the carrier frequency of a desired station in the band. The tuning is done by the air capacitor C, which can be

Table 27-4. Tuning LC Circuit by Varying C

L, μH	C, pF	f_r, kHz
239	424	500
239	212	707
239	106	1000
239	53	1410
239	26.5	2000

varied from 360 pF with the plates completely in mesh to 40 pF out of mesh. The fixed plates form the *stator*, while the *rotor* has the plates that move in and out.

Note that the lowest frequency F_L at 540 kHz is tuned in with the highest C at 360 pF. Resonance at the highest frequency F_H at 1620 kHz results with the lowest C at 40 pF.

The capacitance range of 40 to 360 pF tunes through the frequency range from 1620 kHz down to 540 kHz. Frequency F_L is one-third F_H because the maximum C is nine times the minimum C. The tuning dial, in kHz, usually omits the last zero to save space.

The same idea applies to tuning through the commercial FM broadcast band of 88 to 108 MHz, with smaller values of L and C. Also, television receivers are tuned to a specific broadcast channel by resonance at the desired frequencies.

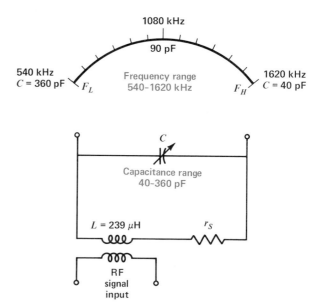

Fig. 27-13 Application of tuning LC circuit through the AM radio band of 540 to 1620 kHz.

For electronic tuning, the C is varied by a *varactor*. This is a semiconductor diode that varies in capacitance when its voltage is changed.

Practice Problems 27-7
Answers at End of Chapter

a. When a tuning capacitor is completely in mesh, is the station tuned in the highest or lowest frequency in the band?

b. A tuning ratio of 2:1 in frequency requires what ratio of variable L or C?

27-8 Mistuning

For example, suppose that a series LC circuit is tuned to 1000 kHz but the frequency of the input voltage is 17 kHz, completely off resonance. The circuit could provide a Q rise in output voltage for current having the frequency of 1000 kHz, but there is no input voltage and therefore no current at this frequency.

The input voltage produces current that has the frequency of 17 kHz. This frequency cannot produce a resonant rise in current, however, because the current is limited by the net reactance. When the frequency of the input voltage and the resonant frequency of the LC circuit are not the same, therefore, the mistuned circuit has very little output compared with the Q rise in voltage at resonance.

Similarly, when a parallel circuit is mistuned, it does not have a high value of impedance. Furthermore, the net reactance off resonance makes the LC circuit either inductive or capacitive.

Series Circuit Off Resonance When the frequency of the input voltage is lower than the resonant frequency of a series LC circuit, the capacitive reactance is greater than the inductive reactance. As a result, there is more voltage across the capacitive reactance than across the inductive reactance. The series LC circuit is capacitive below resonance, therefore, with capacitive current leading the generator voltage.

Above the resonant frequency, the inductive reactance is greater than the capacitive reactance. As a re-

sult, the circuit is inductive above resonance, with inductive current that lags the generator voltage. In both cases, there is much less output voltage than at resonance.

Parallel Circuit Off Resonance With a parallel LC circuit, the smaller amount of inductive reactance below resonance results in more inductive branch current than capacitive branch current. The net line current is inductive, therefore, making the parallel LC circuit inductive below resonance, as the line current lags the generator voltage.

Above the resonant frequency, the net line current is capacitive because of the higher value of capacitive branch current. Then the parallel LC circuit is capacitive, with line current leading the generator voltage. In both cases the total impedance of the parallel circuit is much less than the maximum impedance at resonance. Note that the capacitive and inductive effects off resonance are opposite for series and parallel LC circuits.

Practice Problems 27-8
Answers at End of Chapter

a. Is a series resonant circuit inductive or capacitive below resonance?

b. Is a parallel resonant circuit inductive or capacitive below resonance?

27-9 Analysis of Parallel Resonant Circuits

Parallel resonance is more complex than series resonance because the reactive branch currents are not exactly equal when X_L equals X_C. The reason is that the coil has its series resistance r_S in the X_L branch, while the capacitor has only X_C in its branch.

For high-Q circuits, we consider r_S to be negligible. In low-Q circuits, however, the inductive branch must be analyzed as a complex impedance with X_L and r_S in series. This impedance is in parallel with X_C, as shown in Fig. 27-14. The total impedance Z_T can then be calculated by using complex numbers, as explained in Chap. 25.

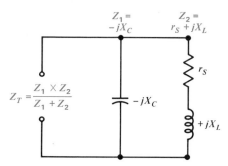

Fig. 27-14 General method of calculating Z_T for a parallel resonant circuit as $Z_1 \times Z_2/(Z_1 + Z_2)$ with complex numbers.

$$Z_T = \frac{Z_1 \times Z_2}{Z_1 + Z_2} = \frac{-j1.5 \times (j1.5 + 0.3)}{-j1.5 + j1.5 + 0.3}$$

$$= \frac{-j^2 2.25 - j0.45}{0.3} = -j^2 7.5 - j1.5$$

$$= 7.5 - j1.5 = 7.65 \underline{/-11.3°} \ k\Omega$$

$$Z_T = 7650 \underline{/-11.3°} \ \Omega$$

The phase angle θ is not zero because the reactive branch currents are unequal, even though X_L and X_C are equal. The appreciable value of r_S in the X_L branch makes this branch current smaller than I_C in the X_C branch.

High-Q Circuit We can apply the general method in Fig. 27-14 to the parallel resonant circuit shown before in Fig. 27-6 to see if Z_T is 225,000 Ω. In this example, X_L and X_C are 1500 Ω and r_S is 10 Ω. The calculations are

$$Z_T = \frac{Z_1 \times Z_2}{Z_1 + Z_2} = \frac{-j1500 \times (j1500 + 10)}{-j1500 + j1500 + 10}$$

$$= \frac{-j^2 2.25 \times 10^6 - j15,000}{10}$$

$$= -j^2 2.25 \times 10^5 - j1500$$

$$= 225,000 - j1500$$

$$Z_T = 225,000 \underline{/0°} \ \Omega$$

Note that $-j^2$ is +1. Also, the reactive $j1500$ Ω is negligible compared with the resistive 225,000 Ω. This answer for Z_T is the same as $Q \times X_L$, or $150 \times 15,000$, because of the high Q with negligibly small r_S.

Low-Q Circuit We can consider a Q less than 10 as low. For the same circuit in Fig. 27-6, if r_S is 300 Ω with an X_L of 1500 Ω, the Q will be 1500/300, which equals 5. For this case of appreciable r_S, the branch currents cannot be equal when X_L and X_C are equal because then the inductive branch will have more impedance and less current.

With a low-Q circuit Z_T must be calculated in terms of the branch impedances. For this example, the calculations are simpler with all impedances stated in kilohms:

Criteria for Parallel Resonance The frequency f_r that makes $X_L = X_C$ is always $1/(2\pi\sqrt{LC})$. However, for low-Q circuits f_r does not necessarily provide the desired resonance effect. The three main criteria for parallel resonance are

1. Zero phase angle and unity power factor.
2. Maximum impedance Z_T and minimum line current.
3. $X_L = X_C$. This is the resonance at $f_r = 1/(2\pi\sqrt{LC})$.

These three effects do not occur at the same frequency in parallel circuits that have a low Q. The condition for unity power factor is often called *antiresonance* in a parallel LC circuit to distinguish it from the case of equal X_L and X_C.

It should be noted that when Q is 10 or higher, though, the parallel branch currents are practically equal when $X_L = X_C$. Then at $f_r = 1/(2\pi\sqrt{LC})$, the line current is minimum with zero phase angle, and the impedance is maximum.

For a series resonant circuit there are no parallel branches to consider. Therefore, the current is maximum at exactly f_r, whether the Q is high or low.

Practice Problems 27-9
Answers at End of Chapter

a. Is the Q of 8 a high or low value?
b. With this Q, will the I_L be more or less than I_C in the parallel branches when $X_L = X_C$?

27-10   Damping of Parallel Resonant Circuits

In Fig. 27-15a, the shunt R_P across L and C is a damping resistance because it lowers the Q of the tuned circuit. The R_P may represent the resistance of the external source driving the parallel resonant circuit, or R_P can be an actual resistor added for lower Q and greater bandwidth. Using the parallel R_P to reduce Q is better than increasing the series resistance r_S because the resonant response is more symmetrical with shunt damping.

The effect of varying the parallel R_P is opposite from the series r_S. A lower value of R_P lowers the Q and reduces the sharpness of resonance. Remember that less resistance in a parallel branch allows more current. This resistive branch current cannot be canceled at resonance by the reactive currents. Therefore, the resonant dip to minimum line current is less sharp with more resistive line current. Specifically, when Q is determined by parallel resistance

$$Q = \frac{R_P}{X_L} \qquad (27\text{-}10)$$

This relation with shunt R_P is the reciprocal of the Q formula with series r_S. Reducing R_P decreases Q, but reducing r_S increases Q. The damping can be done by series r_S, parallel R_P, or both.

Parallel R_P without r_S   In Fig. 27-15a, Q is determined only by the R_P, as no series r_S is shown. We can consider that r_S is zero or very small. Then the Q of the coil is infinite or high enough to be greater than the damped Q of the tuned circuit, by a factor of 10 or more. The Q of the damped resonant circuit here is $R_P/X_L = 50,000/500 = 100$.

Series r_S without R_P   In Fig. 27-15b, Q is determined only by the coil resistance r_S, as no shunt damping resistance is used. Then $Q = X_L/r_S = 500/5 = 100$. This value is the Q of the coil, which is also the Q of the parallel resonant circuit without shunt damping.

Conversion of r_S or R_P   For the circuits in both Fig. 27-15a and b, Q is 100 because the 50,000-Ω R_P is

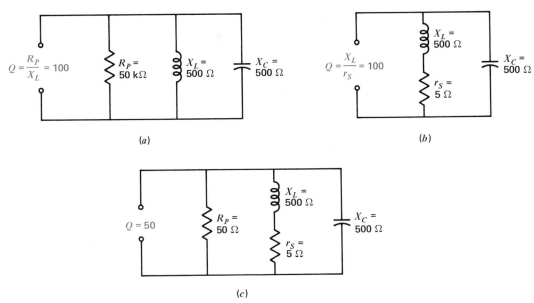

(a)

(b)

(c)

Fig. 27-15 The Q of parallel resonant circuit in terms of coil resistance r_S and parallel damping resistor R_P. (a) Parallel R_P but negligible r_S. (b) Series r_S but no R_P branch. (c) Both R_P and r_S. See Formula (27-11) for calculating Q.

equivalent to the 5-Ω r_S as a damping resistance. One value can be converted to the other. Specifically,

$$r_S = X_L^2/R_P \qquad \text{or} \qquad R_P = X_L^2/r_S$$

In this example, r_S equals $250{,}000/50{,}000 = 5\ \Omega$, or R_P is $250{,}000/5 = 50{,}000\ \Omega$.

Damping with Both r_S and R_P Figure 27-15c shows the general case of damping where both r_S and R_P must be considered. Then the Q of the circuit can be calculated as

$$Q = \frac{X_L}{r_S + X_L^2/R_P} \tag{27-11}$$

For the values in Fig. 27-15c,

$$Q = \frac{500}{5 + 250{,}000/50{,}000} = \frac{500}{5+5} = \frac{500}{10}$$

$$Q = 50$$

The Q is lower here compared with Fig. 27-15a or b because this circuit has both series and shunt damping.

It should be noted that for an r_S of zero, Formula (27-11) can be inverted and simplified to $Q = R_P/X_L$. This is the same as Formula (27-10) for shunt damping alone.

For the opposite case of R_P being infinite for an open parallel path, Formula (27-11) reduces to X_L/r_S. This is the same as Formula (27-5) without shunt damping.

Practice Problems 27-10
Answers at End of Chapter

a. A parallel resonant circuit has an X_L of 1000 Ω and an r_S of 20 Ω, without any shunt damping. Calculate Q.
b. A parallel resonant circuit has an X_L of 1000 Ω, negligible r_S, and shunt R_P of 50 kΩ. Calculate Q.
c. How much is Z_T at f_r for the circuits in (a) and (b)?

27-11   Choosing L and C for a Resonant Circuit

The following example illustrates how resonance is really just an application of X_L and X_C. Suppose that we have the problem of determining the inductance and capacitance for a circuit to be resonant at 159 kHz. First, we need a known value for either L or C, in order to calculate the other. Which one to choose depends on the application. In some cases, particularly at very high frequencies, C must be the minimum possible value, which might be about 10 pF. At medium frequencies, though, we can choose L for the general case where an X_L of 1000 Ω is desirable and can be obtained. Then the inductance of the required L, equal to $X_L/2\pi f$, is 0.001 H or 1 mH, for the inductive reactance of 1000 Ω.

For resonance at 159 kHz with a 1-mH L, the required C is 0.001 μF or 1000 pF. This value of C can be calculated for an X_C of 1000 Ω, equal to X_L at the f_r of 159 kHz, or from Formula (27-3). In either case, if you substitute 1×10^{-9} F for C and 1×10^{-3} H for L in the resonant frequency formula, f_r will be 159 kHz.

This combination is resonant at 159 kHz whether L and C are in series or parallel. In series, the resonant effect is to produce maximum current and maximum voltage across L or C at 159 kHz. The effect is desirable for the input circuit of an RF amplifier tuned to f_r because of the maximum signal. In parallel, the resonant effect at 159 kHz is minimum line current and maximum impedance across the generator. This effect is desirable for the output circuit of an RF amplifier, as the gain is maximum at f_r because of the high Z.

If we assume the 1-mH coil used for L has an internal resistance of 20 Ω, the Q of the coil is 50. This value is also the Q of the series resonant circuit. If there is no shunt damping resistance across the parallel LC circuit, its Q is also 50. With a Q of 50 the bandwidth of the resonant circuit is 159 kHz/50, which equals 3.18 kHz for Δf.

Practice Problems 27-11
Answers at End of Chapter

a. What is f_r for 1000 pF of C and 1 mH of L?
b. What is f_r for 250 pF of C and 1 mH of L?

Summary

Series and parallel resonance are compared in Table 27-5. The main difference is that series resonance produces maximum current and very low impedance at f_r, but with parallel resonance the line current is minimum to provide a very high impedance. Remember that these formulas for parallel resonance are very close approximations that can be used for circuits with a Q higher than 10. For series resonance, the formulas apply whether the Q is high or low.

Table 27-5. Comparison of Series and Parallel Resonance

Series resonance	Parallel resonance (high Q)
$f_r = \dfrac{1}{2\pi\sqrt{LC}}$	$f_r = \dfrac{1}{2\pi\sqrt{LC}}$
I maximum at f_r with θ of $0°$	I_T minimum at f_r with θ of $0°$
Impedance Z minimum at f_r	Impedance Z maximum at f_r
$Q = X_L/r_S$, or	$Q = X_L/r_S$, or
$Q = V_{out}/V_{in}$	$Q = Z_{max}/X_L$
Q rise in voltage $= Q \times V_{gen}$	Q rise in impedance $= Q \times X_L$
Bandwidth $\Delta f = f_r/Q$	Bandwidth $\Delta f = f_r/Q$
Capacitive below f_r, but inductive above f_r	Inductive below f_r, but capacitive above f_r
Needs low-resistance source for low r_S, high Q, and sharp tuning	Needs high-resistance source for high R_P, high Q, and sharp tuning
Source is inside LC circuit	Source is outside LC circuit

Self-Examination
Answers at Back of Book

Choose (a), (b), (c), or (d).

1. For a series or parallel LC circuit, resonance occurs when (a) X_L is ten times X_C or more; (b) X_C is ten times X_L or more; (c) $X_L = X_C$; (d) the phase angle of the circuit is $90°$.
2. When either L or C is increased, the resonant frequency of the LC circuit (a) increases; (b) decreases; (c) remains the same; (d) is determined by the shunt resistance.
3. The resonant frequency of an LC circuit is 1000 kHz. If L is doubled but C is reduced to one-eighth of its original value, the resonant frequency then is (a) 250 kHz; (b) 500 kHz; (c) 1000 kHz; (d) 2000 kHz.
4. A coil has a 1000-Ω X_L and a 5-Ω internal resistance. Its Q equals (a) 0.005; (b) 5; (c) 200; (d) 1000.
5. In a parallel LC circuit, at the resonant frequency, the (a) line current is maxi-

mum; (b) inductive branch current is minimum; (c) total impedance is minimum; (d) total impedance is maximum.

6. At resonance, the phase angle equals (a) 0°; (b) 90°; (c) 180°; (d) 270°.

7. In a series LC circuit, at the resonant frequency, the (a) current is minimum; (b) voltage across C is minimum; (c) impedance is maximum; (d) current is maximum.

8. A series LC circuit has a Q of 100 at resonance. When 5 mV is applied at the resonant frequency, the voltage across C equals (a) 5 mV; (b) 20 mV; (c) 100 mV; (d) 500 mV.

9. An LC circuit resonant at 1000 kHz has a Q of 100. The bandwidth between half-power points equals (a) 10 kHz between 995 and 1005 kHz; (b) 10 kHz between 1000 and 1010 kHz; (c) 5 kHz between 995 and 1000 kHz; (d) 200 kHz between 900 and 1100 kHz.

10. In a low-Q parallel resonant circuit, when $X_L = X_C$ (a) I_L equals I_C; (b) I_L is less than I_C; (c) I_L is more than I_C; (d) the phase angle is 0°.

Essay Questions

1. (a) State two characteristics of series resonance. (b) With a microammeter measuring current in the series LC circuit of Fig. 27-2, describe the meter readings for the different frequencies from 600 to 1400 kHz.

2. (a) State two characteristics of parallel resonance. (b) With a microammeter measuring current in the main line for the parallel LC circuit in Fig. 27-6a, describe the meter readings for the different frequencies from 600 to 1400 kHz.

3. State the Q formula for the following LC circuits: (a) series resonant; (b) parallel resonant, with series resistance r_S in the inductive branch; (c) parallel resonant, with zero series resistance but shunt R_P.

4. Explain briefly why a parallel LC circuit is inductive but a series LC circuit is capacitive below f_r.

5. What is the effect on Q and bandwidth of a parallel resonant circuit if its shunt damping resistance is decreased from 50,000 to 10,000 Ω?

6. Describe briefly how you would use an ac meter to measure the bandwidth of a series resonant circuit for calculating its Q.

7. Why is a low-resistance generator good for a high Q in series resonance, while a high-resistance generator is needed for a high Q in parallel resonance?

8. Referring to Fig. 27-13, why is it that the middle frequency of 1080 kHz does not correspond to the middle capacitance value of 200 pF?

9. (a) Give three criteria for parallel resonance. (b) Why is the antiresonant frequency f_a different from f_r with a low-Q circuit? (c) Why are they the same for a high-Q circuit?

10. Show how Formula (27-11) reduces to R_P/X_L when r_S is zero.

11. (a) Specify the edge frequencies f_1 and f_2 for each of the three response curves in Fig. 27-11. (b) Why does lower Q allow more bandwidth?

12. (a) Why does maximum Z_T for a parallel resonant circuit correspond to minimum line current? (b) Why does zero phase angle for a resonant circuit correspond to unity power factor?

13. Explain how manual tuning of an LC circuit can be done with a capacitor or a coil.

14. What is meant by electronic tuning?

Problems
Answers to Odd-Numbered Problems at Back of Book

1. Find f_r for a series circuit with a 20-μF C, a 8-H L, and a 5-Ω r_S.
2. Find f_r for a parallel circuit with a 32-μF C, a 2-H L, and a 5-Ω r_S.
3. Find f_r for L of 320 μH and C of 30 pF.
4. The f_r is 1.6 MHz with L of 80 μH. Find C.
5. In a series resonant circuit, X_L is 1500 Ω and the internal coil resistance is 15 Ω. At the resonant frequency: **(a)** How much is the Q of the circuit? **(b)** How much is X_C? **(c)** With a generator voltage of 15 mV, how much is I? **(d)** How much is the voltage across X_C?
6. In a parallel resonant circuit, X_L is 1200 Ω, the resistance of the coil is practically zero, but there is a 36,000-Ω resistance across the LC circuit. At the resonant frequency: **(a)** How much is the Q of the circuit? **(b)** How much is X_C? **(c)** With a generator voltage of 12 mV and zero resistance in the main line, how much is the main-line current? **(d)** How much is the voltage across L, C, and R? **(e)** How much is the impedance across the main line?
7. What value of L is necessary with a C of 50 pF for series resonance at 1 MHz? At 4 MHz?
8. Calculate the C needed with a 350-μH L for a 200-kHz f_r.
9. Calculate the lowest and highest values of C needed with 0.1-μH L to tune through the commercial FM broadcast band of 88 to 108 MHz.
10. **(a)** At what frequency will a 200-μH coil with a 20-Ω r_S have a 1000-Ω X_L? **(b)** What size C is needed for a 1000-Ω X_C at this frequency? **(c)** What is f_r for this LC combination? **(d)** How much is the Q of the coil?
11. Draw the schematic diagram of a parallel resonant circuit with the L, C, and r_S of Prob. 10. Let the applied voltage be 5 V. Calculate the values of main-line current I_T, Z_T, and θ at: **(a)** the resonant frequency f_r; **(b)** 0.1 MHz above f_r; **(c)** 0.1 MHz below f_r.
12. For the series resonant circuit in Fig. 27-16: **(a)** How much is X_L? **(b)** Calculate L in millihenrys and C in microfarads. **(c)** Calculate the Q and bandwidth. **(d)** Calculate V_C and V_L. **(e)** If L is doubled and C is one-half, what is f_r? Calculate Q and bandwidth for this case. **(f)** If the original values of both L and C are doubled, calculate f_r.

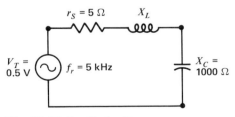

Fig. 27-16 For Prob. 12.

13. Redraw Fig. 27-16 as a parallel resonant circuit with the original values and r_S in series with the coil. **(a)** What are the Q and bandwidth? **(b)** Calculate Z_T at f_r.

14. For the same circuit as Prob. 13, let r_S increase to 500 Ω. **(a)** What is the Q now? **(b)** Calculate Z_T at f_r, one-half f_r, and twice f_r. **(c)** What value of R_P would be used for the same Q if r_S were zero?

15. For the series circuit in Fig. 27-16: **(a)** Tabulate the I and θ values every kilohertz from 2 to 9 kHz. **(b)** Draw the response curve showing I vs. frequency.

16. For the values in Fig. 27-16 connected for parallel resonance: **(a)** Calculate Z_T and θ every kilohertz from 2 to 9 kHz. **(b)** Draw the response curve showing Z_T vs. frequency.

17. A series resonant circuit produces 240 mV across L with a 2-mV input. **(a)** How much is the Q of the coil? **(b)** Calculate r_S, if L is 5 mH and f_r is 0.3 MHz. **(c)** How much C is needed for this f_r?

18. Refer to the relative response curve in Fig. 27-17. **(a)** For $f_r = 10.7$ MHz and $Q = 50$, determine the bandwidth Δf and the edge frequencies f_1 and f_2. **(b)** Do the same for a lower f_r of 456 kHz. Q is still 50.

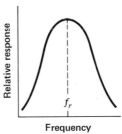

Fig. 27-17 For Prob. 18.

Answers to Practice Problems

27-1	**a.** 1000 kHz		**27-7**	**a.** Lowest
	b. 1000 kHz			**b.** 1:4
27-2	**a.** F		**27-8**	**a.** Capacitive
	b. T			**b.** Inductive
	c. T		**27-9**	**a.** Low
27-3	**a.** F			**b.** Less
	b. T		**27-10**	**a.** $Q = 50$
	c. T			**b.** $Q = 50$
27-4	**a.** Less			**c.** $Z_T = 50$ kΩ
	b. 200 μH		**27-11**	**a.** $f_r = 159$ kHz
27-5	**a.** $Q = 100$			**b.** $f_r = 318$ kHz
	b. $Z_T = 25$ kΩ			
27-6	**a.** $\Delta f = 0.25$ MHz			
	b. $Q = 50$			

Chapter 28
Filters

A filter separates different components that are mixed together. For instance, a mechanical filter can separate particles from liquid, or small particles from large particles. An electrical filter can separate different frequency components.

Generally, inductors and capacitors are used for filtering because of their opposite frequency characteristics. Reactance X_L increases but X_C decreases with higher frequencies. In addition, their filtering action depends on whether L and C are in series or parallel with the load. The

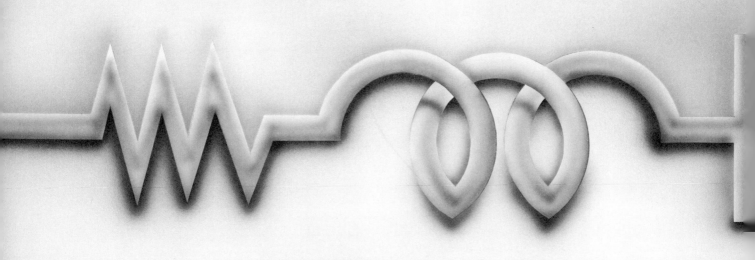

most common filtering applications are separating audio from radio frequencies, or vice versa, and separating ac variations from the average dc level.

Important terms in this chapter are:

ac component	fluctuating dc values
attenuation of filter	high-pass filter
bandpass filter	low-pass filter
bandstop filter	L-type filter
bypass capacitor	π-type filter
capacitive coupling	pulsating dc values
cutoff frequency	T-type filter
dc component	transformer coupling

More details are explained in the following sections:

28-1 Examples of Filtering

Electronic circuits often have currents of different frequencies corresponding to voltages of different frequencies. The reason is that a source produces current with the same frequency as the applied voltage. As examples, the ac signal input to an audio circuit can have high and low audio frequencies; an RF circuit can have a wide range of radio frequencies in its input; the audio detector in a radio has both radio frequencies and audio frequencies in the output. Finally, the rectifier in a power supply produces dc output with an ac ripple superimposed on the average dc level.

In such applications where the current has different frequency components, it is usually necessary either to favor or to reject one frequency or a band of frequencies. Then an electrical filter is used to separate higher or lower frequencies.

The electrical filter can pass the higher-frequency component to the load resistance, which is the case of a high-pass filter, or a low-pass filter can be used to favor the lower frequencies. In Fig. 28-1a, the high-pass filter allows 10 kHz to produce output, while rejecting or attenuating the lower frequency of 100 Hz. In Fig. 28-1b, the filtering action is reversed to pass the lower frequency of 100 Hz, while attenuating 10 kHz. These examples are for high and low audio frequencies.

For the case of audio mixed with radio frequencies, a low-pass filter allows the audio frequencies in the output. Or, a high-pass filter allows the radio frequencies to be passed to the load.

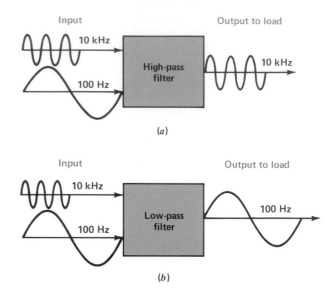

Fig. 28-1 Function of electrical filters. (*a*) High-pass filter couples higher frequencies to the load. (*b*) Low-pass filter couples lower frequencies to the load.

Practice Problems 28-1
Answers at End of Chapter

A high-pass filter will pass which of the following:
a. 10 or 500 kHz.
b. 60 Hz or a steady dc level.

28-2 Direct Current Combined with Alternating Current

Current that varies in amplitude but does not reverse in polarity is considered *pulsating* or *fluctuating* direct current. It is not a steady direct current because its value fluctuates. However, it is not alternating current because the polarity remains the same, either positive or negative. The same idea applies to voltages.

Figure 28-2 illustrates how a circuit can have pulsating direct current or voltage. Here, the steady dc voltage of the battery V_B is in series with the ac voltage V_A. Since the two series generators add, the voltage across R_L is the sum of the two applied voltages, as shown by the waveshape of v_R in Fig. 28-2b.

If values are taken at opposite peaks of the ac variation, when V_A is at $+10$ V, it adds to the $+20$ V of the battery to provide $+30$ V across R_L; when the ac voltage is -10 V, it bucks the battery voltage of $+20$ V to provide $+10$ V across R_L. When the ac voltage is at zero, the voltage across R_L equals the battery voltage of $+20$ V.

The combined voltage v_R then consists of the ac variations fluctuating above and below the battery voltage as the axis, instead of the zero axis for ac voltage. The result is a pulsating dc voltage, since it is fluctuating but always has positive polarity with respect to zero.

The pulsating direct current i through R_L has the same waveform, fluctuating above and below the steady dc level of 20 A. The i and v values are the same because R_L is 1 Ω.

Another example is illustrated in Fig. 28-3. If the 100-Ω R_L is connected across the 120-V 60-Hz ac power line in Fig. 28-3a, the current in R_L will be V/R_L.

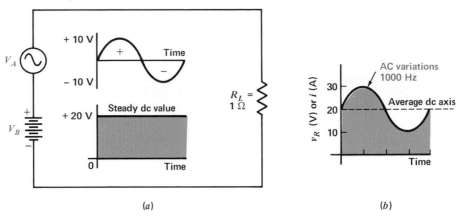

Fig. 28-2 Pulsating direct current and voltage. (a) Circuit. (b) Graph of voltage across R_L equal to V_B plus V_A. Frequency of ac voltage is 1000 Hz.

This alternating current is a sine wave, with an rms value of $^{120}\!/_{100}$ or 1.2 A.

Also, if you connect the same R_L across the 200-V dc source in Fig. 28-3b, instead of using the ac source, the steady direct current in R_L will be $^{200}\!/_{100}$, or 2 A. The battery source voltage and its current are considered steady dc values because there are no variations.

However, suppose that the ac source V_A and dc source V_B are connected in series with R_L, as in Fig. 28-3c. What will happen to the current and voltage for R_L? Will V_A or V_B supply the current? The answer is that both sources will. Each voltage source produces current as though the other were not there, assuming the sources have negligibly small internal impedance. The result then is the fluctuating dc voltage or current shown, with the ac variations of V_A superimposed on the average dc level of V_B.

DC and AC Components   The pulsating dc voltage v_R in Fig. 28-3c is just the original ac voltage V_A with its axis shifted to a dc level by the battery voltage V_B. In effect, a dc component has been inserted into the ac variations. This effect is called *dc insertion*.

Referring back to Fig. 28-2, if you measure across R_L with a dc voltmeter, it will read the dc level of 20 V. An ac voltmeter will read the rms value of the variations, which is 7.07 V.

It is convenient, therefore, to consider the pulsating or fluctuating voltage and current in two parts. One is the steady dc component, which is the axis or average level of the variations; the other is the ac component, consisting of the variations above and below the dc axis. Here the dc level for V_T is +20 V, while the ac component equals 10 V peak or 7.07 V rms value. The ac component is also called ac *ripple*.

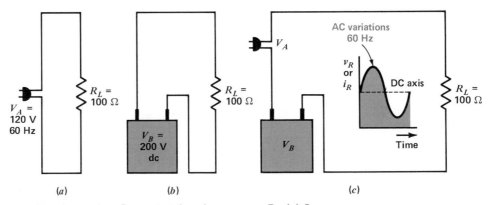

Fig. 28-3 An example of providing fluctuating dc voltage across R_L. (a) An ac source alone. (b) A dc source alone. (c) An ac source in series with dc source.

ᴡᴡᴡ ⌒⌒ ⊣⊢

It should be noted that with respect to the dc level the fluctuations represent alternating voltage or current that actually reverses in polarity. For example, the change of v_R from $+20$ to $+10$ V is just a decrease in positive voltage compared with zero. However, compared with the dc level of $+20$ V, the value of $+10$ V is 10 V more negative than the axis.

Typical Examples of DC Level with AC Component

As a common application, transistors always have fluctuating dc voltage or current when used for amplifying an ac signal. The transistor amplifier needs steady dc voltages to operate. The signal input is an ac variation, usually with a dc axis to establish the desired operating level. The amplified output is also an ac variation superimposed on a dc supply voltage that supplies the required power output. Therefore, the input and output circuits have fluctuating dc voltage.

The examples in Fig. 28-4 illustrate two possibilities, in terms of polarities with respect to chassis ground. In Fig. 28-4a, the waveform is always positive, as in the previous examples. This example could apply to collector voltage on an NPN transistor amplifier. Note the specific values. The average dc axis is the steady dc level. The positive peak equals the dc level plus the peak ac value. The minimum point equals the dc level minus the peak ac value. The peak-to-peak value of the ac component and its rms value are the same as for the ac signal alone. However, it is better to subtract the minimum from the maximum for the peak-to-peak value, in case the waveform is unsymmetrical.

In Fig. 28-4b, all the values are negative. Notice that here the positive peak of the ac component subtracts

from the dc level because of the opposite polarities. Now the negative peak adds to the negative dc level to provide a maximum point of negative voltage.

Separating the AC Component

In many applications, the circuit has pulsating dc voltage, but only the ac component is desired. Then the ac component can be passed to the load, while the steady dc component is blocked, either with transformer coupling or with capacitive coupling. A transformer with a separate secondary winding isolates or blocks steady direct current in the primary. A capacitor isolates or blocks a steady dc voltage.

Practice Problems 28-2
Answers at End of Chapter

For the fluctuating dc waveform in Fig. 28-4a, specify the following voltages:
a. Average dc level.
b. Maximum and minimum values.
c. Peak-to-peak of ac component.
d. Peak and rms of ac component.

28-3 Transformer Coupling

Remember that a transformer produces induced secondary voltage just for variations in primary current. With pulsating direct current in the primary, the secondary has output voltage only for the ac variations, therefore. The steady dc component in the primary has no effect in the secondary.

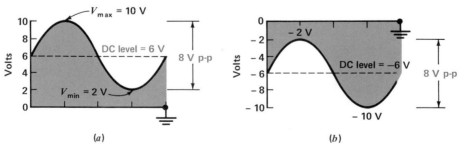

(a)   (b)

Fig. 28-4 Typical examples of dc voltage with ac component. (a) Positive dc values. (b) Negative dc values.

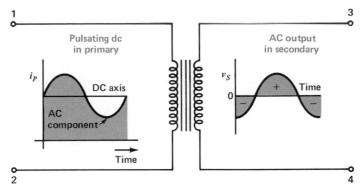

Fig. 28-5 Transformer coupling blocks dc component. With pulsating direct current in primary, the output voltage in secondary has ac component only.

In Fig. 28-5, the pulsating dc voltage in the primary produces pulsating primary current. The dc axis corresponds to a steady value of primary current that has a constant magnetic field, but only when the field changes can secondary voltage be induced. Therefore, only the fluctuations in the primary can produce output in the secondary. Since there is no output for the steady primary current, this dc level corresponds to the zero level for the ac output in the secondary.

When the primary current increases above the steady level, this increase produces one polarity for the secondary voltage as the field expands; when the primary current decreases below the steady level, the secondary voltage has reverse polarity as the field contracts. The result in the secondary is an ac variation having opposite polarities with respect to the zero level.

The phase of the ac secondary voltage may be as shown or 180° opposite, depending on the connections and direction of the windings. Also, the ac secondary output may be more or less than the ac component in the primary, depending on the turns ratio. This ability to isolate the steady dc component in the primary while providing ac output in the secondary applies to all transformers with a separate secondary winding, whether iron-core or air-core.

Practice Problems 28-3
Answers at End of Chapter

a. Is transformer coupling an example of a high-pass or low-pass filter?

b. In Fig. 28-5, what is the level of v_S for the average dc level of i_P?

28-4 Capacitive Coupling

This method is probably the most common type of coupling in amplifier circuits. The coupling means connecting the output of one circuit to the input of the next. The requirements are to include all frequencies in the desired signal, while rejecting undesired components. Usually, the dc component must be blocked from the input to ac amplifiers. The purpose is to maintain a specific dc level for the amplifier operation.

In Fig. 28-6, the pulsating dc voltage across input terminals 1 and 2 is applied to the RC coupling circuit. Capacitance C_C will charge to the steady dc level, which is the average charging voltage. The steady dc component is blocked, therefore, since it cannot produce voltage across R. However, the ac component is developed across R, between the output terminals 3 and 4. The reason is that the ac voltage allows C to produce charge and discharge current through R. Note that the zero axis of the ac voltage output corresponds to the average level of the pulsating dc voltage input.

The DC Component Across C The voltage across C_C is the steady dc component of the input voltage because the variations of the ac component are symmetrical above and below the average level. Furthermore, the series resistance is the same for charge and discharge. As a result, any increase in charging voltage above the average level is counteracted by an equal discharge below the average.

In Fig. 28-6, for example, when v_{in} increases from 20 to 30 V, this effect on charging C_C is nullified by the discharge when v_{in} decreases from 20 to 10 V. At

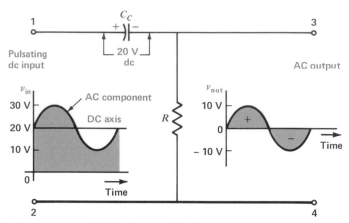

Fig. 28-6 *RC* coupling blocks dc component. With pulsating dc voltage applied, the output voltage across *R* has ac component only.

all times, however, v_{in} has a positive value that charges C_C in the polarity shown.

The net result is that only the average level is effective in charging C_C, since the variations from the axis neutralize each other. After a period of time, depending on the *RC* time constant, C_C will charge to the average value of the pulsating dc voltage applied, which is 20 V here.

The AC Component Across R   Although C_C is charged to the average dc level, when the pulsating input voltage varies above and below this level, the charge and discharge current produces *IR* voltage corresponding to the fluctuations of the input. When v_{in} increases above the average level, C_C takes on charge, producing charging current through R. Even though the charging current may be too small to affect the voltage across C_C appreciably, the *IR* drop across a large value of resistance can be practically equal to the ac component of the input voltage. In summary, a long *RC* time constant is needed for good coupling. (See Fig. 23-8 on p. 454.)

If the polarity is considered, in Fig. 28-6, the charging current produced for an increase of v_{in} produces electron flow from the low side of R to the top, adding electrons to the negative side of C_C. The voltage at the top of R is then positive with respect to the line below.

When v_{in} decreases below the average level, C loses charge. The discharge current then is in the opposite direction through R. The result is negative polarity for the ac voltage output across R.

When the input voltage is at its average level, there is no charge or discharge current, resulting in zero voltage across R. The zero level in the ac voltage across R corresponds to the average level of the pulsating dc voltage applied to the *RC* circuit.

The end result is that with positive pulsating dc voltage applied, the values above the average produce the positive half-cycle of the ac voltage across R; the values below the average produce the negative half-cycle. Only this ac voltage across R is coupled to the next circuit, as terminals 3 and 4 provide the output from the *RC* coupling circuit.

It is important to note that there is practically no phase shift. This rule applies to all *RC* coupling circuits, since R must be ten or more times X_C. Then the reactance is negligible compared with the series resistance, and the phase angle of less than 5.7° is practically zero.

Voltages Around the RC Coupling Circuit
If you measure the pulsating dc input voltage across points 1 and 2 in Fig. 28-6 with a dc voltmeter, it will read the average level of 20 V. A voltmeter that reads only ac values across these same points will read the fluctuating ac component, equal to 7 V rms.

Across points 1 and 3, a dc voltmeter reads the steady dc value of 20 V across C_C. An ac voltmeter across 1 and 3 reads zero.

However, an ac voltmeter across the output R between points 3 and 4 will read the ac voltage of 7 V rms; a dc voltmeter across R reads zero.

Table 28-1. Typical AF and RF Coupling Capacitors*

Frequency	Values of C_C			Frequency band
	$R = 1.6$ kΩ	$R = 16$ kΩ	$R = 160$ kΩ	
100 Hz	10 μF	1 μF	0.1 μF	Low AF
1000 Hz	1 μF	0.1 μf	0.01 μF	Medium AF
10 kHz	0.1 μF	0.01 μF	0.001 μF	High AF
100 kHz	0.01 μF	0.001 μF	100 pF	Low RF
1 MHz	0.001 μF	100 pF	10 pF	Medium RF
10 MHz	100 pF	10 pF	1 pF	High RF
100 MHz	10 pF	1 pF	0.1 pF	VHF

*For coupling circuit in Fig. 28-6; $X_{C_C} = \frac{1}{10} R$.

Typical Coupling Capacitors

Common values of RF and of coupling capacitors for different sizes of series R are listed in Table 28-1. In all cases the coupling capacitor blocks the steady dc component of the input voltage, while the ac component is passed to the resistance.

The size of C_C required depends on the frequency of the ac component. At each frequency listed at the left in Table 28-1, the values of capacitance in the horizontal row have an X_C equal to one-tenth the resistance value for each column. The R increases from 1.6 to 16 to 160 kΩ for the three columns, allowing smaller values of C_C. Typical audio coupling capacitors, then, are about 0.1 to 10 μF, depending on the lowest audio frequency to be coupled and the size of the series resistance. Similarly, typical RF coupling capacitors are about 1 to 100 pF.

Values of C_C more than about 1 μF are usually electrolytic capacitors, which must be connected in the correct polarity. These can be very small, many being ½ in. long, with a low voltage rating of 3 to 25 V for transistor circuits. Also, the leakage current of electrolytic capacitors is not a serious problem in this application because of the low voltage and small series resistance for transistor coupling circuits.

Practice Problems 28-4
Answers at End of Chapter

a. In Fig. 28-6, what is the level of v_{out} across R corresponding to the average dc level of v_{in}?
b. Which of the following is a typical audio coupling capacitor with a 1-kΩ R: 1 pF, 0.001 μF, or 5 μF?

28-5   Bypass Capacitors

A bypass is a path around a component. In circuits, the bypass is a parallel or shunt path. Capacitors are often used in parallel with resistance to bypass the ac component of a pulsating dc voltage. The result, then, is steady dc voltage across the RC parallel combination, if the bypass capacitance is large enough to have little reactance for the lowest frequency of the ac variations.

As illustrated in Fig. 28-7, the capacitance C_1 in parallel with R_1 is an ac bypass capacitor for R_1. For any

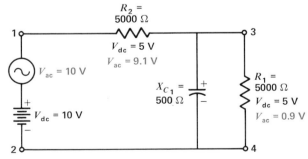

Fig. 28-7 Bypass capacitor C_1 short-circuits R_1 for ac component of pulsating dc input voltage.

frequency at which X_{C_1} is one-tenth of R_1, or less, the ac component is bypassed around R_1 through the low reactance in the shunt path. The result is practically zero ac voltage across the bypass capacitor because of its low reactance.

Since the voltage is the same across R_1 and C_1 because they are in parallel, there is also no ac voltage across R_1 for the frequency at which C_1 is a bypass capacitor. We can say that R is bypassed for the frequency at which X_C is one-tenth of R. The bypassing also applies to higher frequencies where X_C is less than one-tenth of R. Then the ac voltage across the bypass capacitor is even closer to zero because of its lower reactance.

Bypassing the AC Component of a Pulsating DC Voltage

The voltages in Fig. 28-7 are calculated by considering the effect of C_1 separately for V_{dc} and for V_{ac}. For direct current, C_1 is practically an open circuit. Then its reactance is so high compared with the 5000-Ω R_1 that X_{C_1} can be ignored as a parallel branch. Therefore, R_1 can be considered as a voltage divider in series with R_2. Since R_1 and R_2 are equal, each has 5 V, equal to one-half V_{dc}. Although this dc voltage division depends on R_1 and R_2, the dc voltage across C_1 is the same 5 V as across its parallel R_1.

For the ac component of the applied voltage, however, the bypass capacitor has very low reactance. In fact, X_{C_1} must be one-tenth of R_1, or less. Then the 5000-Ω R_1 is so high compared with the low value of

X_{C_1} that R_1 can be ignored as a parallel branch. Therefore, the 500-Ω X_{C_1} can be considered as a voltage divider in series with R_2.

With an X_{C_1} of 500 Ω, this value in series with the 5000-Ω R_2 allows approximately one-eleventh of V_{ac} to be developed across C_1. This ac voltage, equal to 0.9 V here, is the same across R_1 and C_1 in parallel. The remainder of the ac applied voltage, equal to approximately 9.1 V, is across R_2. In summary, then, the bypass capacitor provides an ac short circuit across its shunt resistance, so that little or no ac voltage can be developed, without affecting the dc voltages.

Measuring voltages around the circuit in Fig. 28-7, a dc voltmeter reads 5 V across R_1 and 5 V across R_2. An ac voltmeter across R_2 reads 9.1 V, which is almost all the ac input voltage. Across the bypass capacitor C_1 the ac voltage is only 0.9 V.

In Table 28-2, typical sizes for RF and AF bypass capacitors are listed. The values of C have been calculated at different frequencies for an X_C one-tenth the shunt resistance given in each column. The R decreases for the three columns, from 16 to 1.6 kΩ and 160 Ω. Note that smaller values of R require larger values of C for bypassing. Also, when X_C equals one-tenth of R at one frequency, X_C will be even less for higher frequencies, improving the bypassing action. Therefore, the size of bypass capacitors should be considered on the basis of the lowest frequency to be bypassed.

It should be noted that the applications of coupling and bypassing for C are really the same, except that C_C is in series with R and the bypass C is in parallel with

Table 28-2. Typical AF and RF Bypass Capacitors*

Frequency	Values of C			Frequency band
	$R = 16$ kΩ	$R = 1.6$ kΩ	$R = 160$ Ω	
100 Hz	1 μF	10 μF	100 μF	Low AF
1000 Hz	0.1 μF	1 μF	10 μF	Medium AF
10 kHz	0.01 μF	0.1 μF	1 μF	High AF
100 kHz	0.001 μF	0.01 μF	0.1 μF	Low RF
1 MHz	100 pF	0.001 μF	0.01 μF	Medium RF
10 MHz	10 pF	100 pF	0.001 μF	High RF
100 MHz	1 pF	10 pF	100 pF	VHF

*For RC bypass circuit in Fig. 28-7; $X_{C_1} = \frac{1}{10} R$.

R. In both cases X_C must be one-tenth or less of R. Then C_C couples the ac signal to R. Or the shunt bypass short-circuits R for the ac signal.

Bypassing Radio Frequencies but Not Audio Frequencies

See Fig. 28-8. At the audio frequency of 1000 Hz, C_1 has a reactance of 1.6 MΩ. This reactance is so much higher than R_1 that the impedance of the parallel combination is essentially equal to the 16,000 Ω of R_1. Then R_1 and R_2 serve as a voltage divider for the applied AF voltage of 10 V. Each of the equal resistances has one-half the applied voltage, equal to the 5 V across R_2 and 5 V across R_1. This 5 V at 1000 Hz is also present across C_1, since it is in parallel with R_1.

For the RF voltage at 1 MHz, however, the reactance of the bypass capacitor is only 1600 Ω. This is one-tenth of R_1. Then X_{C_1} and R_1 in parallel have a combined impedance equal to approximately 1600 Ω.

Now, with a 1600-Ω impedance for the R_1C_1 bank in series with the 16,000 Ω of R_2, the voltage across R_1 and C_1 is one-eleventh the applied RF voltage. Then there is 0.9 V across the lower impedance of R_1 and C_1, with 9.1 V across the larger resistance of R_2. As a result, the RF component of the applied voltage can be considered bypassed. C_1 is the RF bypass capacitor across R_1.

Practice Problems 28-5
Answers at End of Chapter

a. In Fig. 28-8, is C_1 an AF or RF bypass?
b. Which of the following is a typical audio bypass capacitor across a 1-kΩ R: 1 pF, 0.001 μF, or 5 μF?

28-6   Filter Circuits

In terms of their function, filters can be classified as either low-pass or high-pass. A low-pass filter allows the lower-frequency components of the applied voltage to develop output voltage across the load resistance, while the higher-frequency components are attenuated or reduced in the output. A high-pass filter does the opposite, allowing the higher-frequency components of the applied voltage to develop voltage across the output load resistance.

The case of an RC coupling circuit is an example of a high-pass filter because the ac component of the input voltage is developed across R while the dc voltage is blocked by the series capacitor. Furthermore, with higher frequencies in the ac component, more ac voltage is coupled. For the opposite case, a bypass capacitor is an example of a low-pass filter. The higher frequencies are bypassed, but the lower the frequency, the less the bypassing action. Then lower frequencies can develop output voltage across the shunt bypass capacitor.

In order to make the filtering more selective in terms of which frequencies are passed to produce output voltage across the load, filter circuits generally combine inductance and capacitance. Since inductive reactance increases with higher frequencies, while capacitive reactance decreases, the two opposite effects improve the filtering action.

With combinations of L and C, filters are named to correspond to the circuit configuration. Most common types of filters are the L, T, and π. Any one of the three can function as either a low-pass filter or a high-pass filter.

For either low-pass or high-pass filters with L and C the reactance X_L must increase with higher frequencies, while X_C decreases. The frequency characteristics of X_L

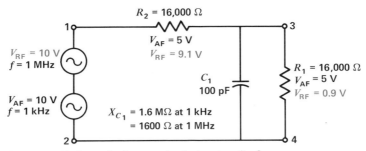

Fig. 28-8 Capacitance C_1 bypasses R_1 for radio frequencies but not audio frequencies.

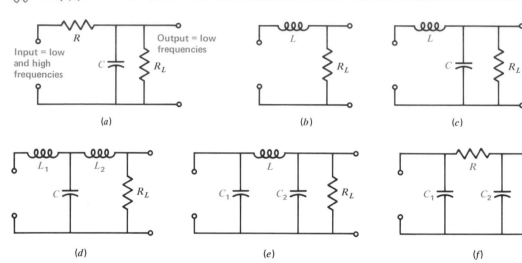

Fig. 28-9 Low-pass filter circuits. (*a*) Bypass capacitor *C* in parallel with R_L. (*b*) Choke *L* in series with R_L. (*c*) The inverted-L type with choke and bypass capacitor. (*d*) The T type with two chokes and one bypass capacitor. (*e*) The π type with one choke and two bypass capacitors. (*f*) The π type with series resistor instead of choke.

and X_C cannot be changed. However, the circuit connections are opposite to reverse the filtering action.

In general, high-pass filters use:

1. Coupling capacitance *C* in series with the load. Then X_C can be low for high frequencies to be passed to R_L, while low frequencies are blocked.
2. Choke inductance *L* in parallel across R_L. Then the shunt X_L can be high for high frequencies to prevent a short circuit across R_L, while low frequencies are bypassed.

The opposite characteristics for low-pass filters are:

1. Inductance *L* in series with the load. The high X_L for high frequencies can serve as a choke, while low frequencies can be passed to R_L.
2. Bypass capacitance *C* in parallel across R_L. Then high frequencies are bypassed by a small X_C, while low frequencies are not affected by the shunt path.

Practice Problems 28-6
Answers at End of Chapter

a. Does high-pass or low-pass filtering require series *C*?

b. Which filtering requires parallel *C*?

28-7   Low-Pass Filters

Figure 28-9 illustrates low-pass circuits from the case of a single filter element with a shunt bypass capacitor in *a* or a series choke in *b*, to the more elaborate combinations of an L-type filter in *c*, a T type in *d*, and a π type in *e* and *f*. With an applied input voltage having different frequency components, the low-pass filter action results in maximum low-frequency voltage across R_L, while most of the high-frequency voltage is developed across the series choke or resistance.

In Fig. 28-9*a*, the shunt capacitor *C* bypasses R_L for high frequencies. In Fig. 28-9*b*, the choke *L* acts as a voltage divider in series with R_L. Since *L* has maximum reactance for the highest frequencies, this component of the input voltage is developed across *L*, with little across R_L. For lower frequencies, *L* has low reactance, and most of the input voltage can be developed across R_L.

In Fig. 28-9*c*, the use of both the series choke and bypass capacitor improves the filtering by providing sharper cutoff between the low frequencies that can develop voltage across R_L and the higher frequencies stopped from the load by producing maximum voltage across *L*. Similarly, the T-type circuit in Fig. 28-9*d* and the π-type circuits in *e* and *f* improve filtering.

Using the series resistance in Fig. 28-9*f* instead of a

choke provides an economical π filter needing less space.

The ability to reduce the amplitude of undesired frequencies is the *attenuation* of the filter. The frequency at which the attenuation reduces the output to 70.7 percent response is the *cutoff frequency*.

Passband and Stop Band As illustrated in Fig. 28-10, a low-pass filter attenuates frequencies above the cutoff frequency of 15 kHz in this example. Any component of the input voltage having a frequency lower than 15 kHz can produce output voltage across the load. These frequencies are in the *passband*. Frequencies of 15 kHz or more are in the *stop band*. The sharpness of filtering between the passband and the stop band depends on the type of circuit. In general, the more L and C components, the sharper the response of the filter can be. Therefore, π and T types are better filters than the L type and the bypass or choke alone.

The response curve in Fig. 28-10 is illustrated for the application of a low-pass filter attenuating RF voltages while passing audio frequencies to the load. This is necessary where the input voltage has RF and AF components but only the audio voltage is desired for the AF circuits that follow the filter.

A good example is filtering the audio output of the detector circuit in a radio receiver, after the RF-modulated carrier signal has been rectified. Another common application of low-pass filtering is where the steady dc component of pulsating dc input must be separated from the higher frequency 60-Hz ac component, as in the pulsating dc output of the rectifier in a power supply.

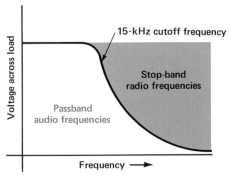

Fig. 28-10 Response of low-pass filter with cutoff at 15 kHz, passing audio signal but attenuating radio frequencies.

Circuit Variations The choice between the T-type filter with a series input choke and the π type with a shunt input capacitor depends upon the internal resistance of the generator supplying input voltage to the filter. A low-resistance generator needs the T filter so that the choke can provide a high series impedance for the bypass capacitor. Otherwise, the bypass must have extremely large values to short-circuit the low-resistance generator for high frequencies.

The π filter is more suitable with a high-resistance generator where the input capacitor can be effective as a bypass. For the same reasons, the L filter can have the shunt bypass either in the input for a high-resistance generator or across the output for a low-resistance generator.

For all the filter circuits, the series choke can be connected either in the high side of the line, as in Fig. 28-9, or in series in the opposite side of the line, without any effect on the filtering action. Also, the series components can be connected in both sides of the line for a *balanced filter* circuit.

Passive and Active Filters All the circuits here are passive filters, as they use only capacitors, inductors, and resistors, which are passive components. An active filter, however, uses the operational amplifier (op amp) on an IC chip, with R and C. The purpose is to eliminate the need for inductance L. This feature is important in filters for low frequencies when large coils would be necessary. More details of active filters are described in most electronics handbooks.

Practice Problems 28-7
Answers at End of Chapter

a. Which diagrams in Fig. 28-9 show a π-type filter?
b. Does the response curve in Fig. 28-10 show low-pass or high-pass filtering?

28-8 High-Pass Filters

As illustrated in Fig. 28-11, the high-pass filter passes to the load all frequencies higher than the cutoff frequency, while lower frequencies cannot develop appreciable voltage across the load. The graph in Fig.

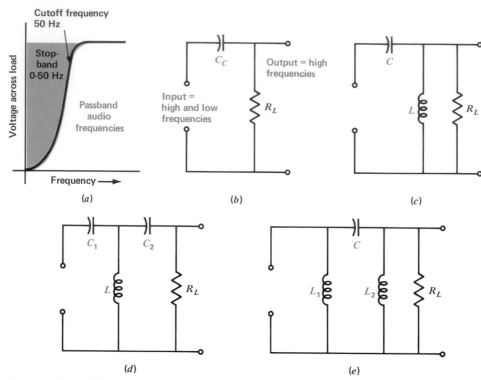

Fig. 28-11 High-pass filters. (*a*) Response curve for AF filter cutting off at 50 Hz. (*b*) *RC* coupling circuit. (*c*) Inverted L type. (*d*) T type. (*e*) π type.

28-11*a* shows the response of a high-pass filter with a stop band of 0 to 50 Hz. Above the cutoff frequency of 50 Hz, the higher audio frequencies in the passband can produce AF voltage across the output load resistance.

The high-pass filtering action results from using C_C as a coupling capacitor in series with the load, as in Fig. 28-11*b*. The L, T, and π types use the inductance for a high-reactance choke across the line. In this way the higher-frequency components of the input voltage can develop very little voltage across the series capacitance, allowing most of this voltage to be produced across R_L. The inductance across the line has higher reactance with increasing frequencies, allowing the shunt impedance to be no lower than the value of R_L.

For low frequencies, however, R_L is effectively short-circuited by the low inductive reactance across the line. Also, C_C has high reactance and develops most of the voltage at low frequencies, stopping these frequencies from developing voltage across the load.

Bandpass Filtering   A high-pass filter can be combined with a low-pass filter. Then the net result is to pass the band of frequencies that are not stopped by either circuit. Such a bandpass response is shown in Fig. 28-12 for audio frequencies. In this example, the

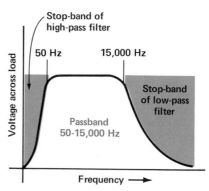

Fig. 28-12 Bandpass response curve for audio frequencies.

only frequencies passed by both filters are 50 to 15,000 Hz. It should be noted, though, that filters for a specific band of frequencies are most often used at radio frequencies, as another application of resonant circuits.

Constant-*k* Filter

If we consider an L type, as a basic example, the values of inductance and capacitance can be designed to make the product of X_L and X_C constant at all frequencies. The purpose is to have the filter present a constant impedance at the input and output terminals. The constant-*k* filter can be high-pass or low-pass.

The *m*-derived Filter

This is a modified form of the constant-*k* filter. The design is based on the ratio of the filter cutoff frequency to the frequency of infinite attenuation. This ratio determines the *m* factor, which is generally between 0.8 and 1.25. The *m*-derived filter also can be high-pass or low-pass. The advantage is very sharp cutoff. Details on the design of filters can be found in most electronics handbooks.

Practice Problems 28-8
Answers at End of Chapter

a. Which diagram in Fig. 28-11 shows a T-type filter?
b. Does the response curve in Fig. 28-11a show high-pass or low-pass filtering?

28-9   Resonant Filters

Tuned circuits provide a convenient method of filtering a band of radio frequencies because relatively small values of *L* and *C* are necessary for resonance. A tuned circuit provides filtering action by means of its maximum response at the resonant frequency.

The width of the band of frequencies affected by resonance depends on the *Q* of the tuned circuit, a higher *Q* providing narrower bandwidth. Because resonance is effective for a band of frequencies below and above f_r, resonant filters are called *band-stop* or *bandpass* filters. Series or parallel *LC* circuits can be used for either function, depending on the connections with respect to R_L.

Series Resonance Filters

A series resonant circuit has maximum current and minimum impedance at the resonant frequency. Connected in series with R_L, as in Fig. 28-13a, the series-tuned *LC* circuit allows frequencies at and near resonance to produce maximum output across R_L. Therefore, this is a case of bandpass filtering. When the series *LC* circuit is connected across R_L as in Fig. 28-13b, however, the resonant circuit provides a low-impedance shunt path that short-circuits R_L. Then there is minimum output. This action corresponds to a shunt bypass capacitor, but the resonant circuit is more selective, short-circuiting R_L just for frequencies at and near resonance. For the bandwidth of the tuned circuit, therefore, the series resonant circuit in shunt with R_L provides band-stop filtering.

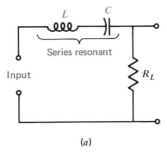
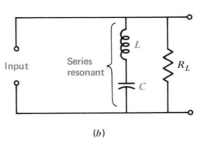

Fig. 28-13 Filtering action of series resonant circuit. (*a*) Bandpass in series with R_L. (*b*) Band stop in shunt with R_L.

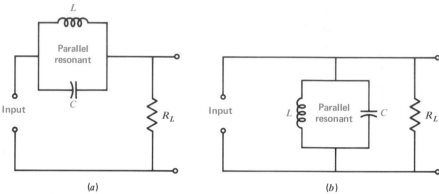

Fig. 28-14 Filtering action of parallel resonant circuit. (a) Band stop in series with R_L. (b) Bandpass in shunt with R_L.

Parallel Resonance Filters A parallel resonant circuit has maximum impedance at the resonant frequency. Connected in series with R_L, as in Fig. 28-14a, the parallel-tuned LC circuit provides maximum impedance in series with R_L, at and near the resonant frequency. Then these frequencies produce maximum voltage across the LC circuit but minimum output voltage across R_L. This is a band-stop filter, therefore, for the bandwidth of the tuned circuit.

The parallel LC circuit connected across R_L, however, as in Fig. 28-14b, provides a bandpass filter. At resonance, the high impedance of the parallel LC circuit allows R_L to develop its output voltage. Below resonance, R_L is short-circuited by the low reactance of L; above resonance, R_L is short-circuited by the low reactance of C. For frequencies at or near resonance, though, R_L is shunted by a high impedance, resulting in maximum output voltage.

L-type Resonant Filter Series and parallel resonant circuits can be combined in L, T, or π sections to improve the filtering. Figure 28-15 illustrates the L-type filter, with band-stop filtering for the circuit arrangement in a but bandpass filtering in b. The circuit in Fig. 28-15a is a band-stop filter because the parallel resonant circuit is in series with the load, while the series resonant circuit is in shunt with the load. In Fig. 28-15b the bandpass filtering results from connecting the series resonant circuit in series with the load, while the parallel resonant circuit is across the load.

Crystal Filters A thin slice of quartz can provide a resonance effect by mechanical vibrations at a particular frequency, like an LC circuit. The reason why is the *piezoelectric effect* in a quartz crystal. The crystal can be excited by voltage input or produce voltage output

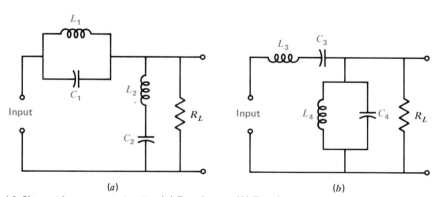

Fig. 28-15 Inverted-L filter with resonant circuits. (a) Band stop. (b) Bandpass.

when it is compressed, expanded, or twisted. As a result, crystals are often used in place of resonant LC circuits. In fact, the Q of a resonant crystal is much higher. Furthermore, special ceramic materials can also be used as crystal resonators. A crystal filter has sharp filter characteristics because of the high Q. More details of the piezoelectric effect and crystal resonators are explained in books on electronic circuits.[1]

Practice Problems 28-9
Answers at End of Chapter

Answer True or False.
a. A parallel resonant LC circuit in series with the load is a band-stop filter.
b. A series resonant LC circuit in series with the load is a bandpass filter.

28-10 Interference Filters

Voltage or current not at the desired frequency represents interference. Usually, such interference can be eliminated by a filter. Some typical applications are (1)

---

[1]See Grob, B.: *Electronic Circuits and Applications*, McGraw-Hill Book Company, New York.

low-pass filter to eliminate RF interference from the 60-Hz power-line input to a receiver, (2) high-pass filter to eliminate RF interference from the signal picked up by a television receiving antenna, and (3) resonant filter to eliminate an interfering radio frequency from the desired RF signal. The resonant band-stop filter is called a *wavetrap*.

Power-Line Filter Although the power line is a source of 60-Hz voltage, it is also a conductor for interfering RF currents produced by motors, fluorescent lighting circuits, and RF equipment. When a receiver is connected to the power line, the RF interference can produce noise and whistles in the receiver output. To minimize this interference, the filter shown in Fig. 28-16 can be used. The filter is plugged into the wall outlet for 60-Hz power, while the receiver is plugged into the filter. An RF bypass capacitor across the line with two series RF chokes forms a low-pass balanced L-type filter. Using a choke in each side of the line makes the circuit balanced to ground.

The chokes provide high impedance for interfering RF current but not for 60 Hz, isolating the receiver input connections from RF interference in the power line. Also, the bypass capacitor short-circuits the receiver input for radio frequencies but not for 60 Hz. The unit then is a low-pass filter for 60-Hz power applied to the receiver while rejecting higher frequencies. The current rating means the filter can be used for equipment that draws 3 A or less from the power line without excessive heat in the chokes.

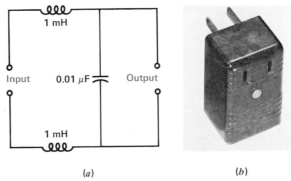

(a) (b)

Fig. 28-16 Power-line filter to pass 60 Hz but filter out radio frequencies. (a) Circuit of balanced L-type low-pass filter. (b) Filter unit, rated 3 A 120 V. (*P. R. Mallory and Co., Inc.*)

Television Antenna Filter When a television receiver has interference in the picture resulting from radio frequencies below the television broadcast band, picked up by the receiving antenna, this RF interference can be reduced by the high-pass filter shown in Fig. 28-17. The filter attenuates frequencies below 54 MHz, which is the lowest frequency for channel 2.

At frequencies lower than 54 MHz the series capacitances provide increasing reactance with a larger voltage drop, while the shunt inductances have less reactance and short-circuit the load. Higher frequencies are passed to the load as the series capacitive reactance decreases and the shunt inductive reactance increases.

Connections to the filter unit are made at the receiver end of the line from the antenna. Either end of the filter is connected to the antenna terminals on the receiver, with the opposite end connected to the antenna line.

(a)

(b)

Fig. 28-17 Television antenna filter to pass channel frequencies above 54 MHz but attenuate lower frequencies. (a) Filter unit. (b) Underside of cover showing series capacitors and shunt chokes. (*R. L. Drake Co.*)

Practice Problems 28-10
Answers at End of Chapter

Answer True or False.
a. A wavetrap is a band-stop filter.
b. The television antenna filter in Fig. 28-17 is a high-pass filter with series capacitors.

Summary

1. A filter separates high and low frequencies. With input of different frequencies, the high-pass filter allows the higher frequencies to produce output voltage across the load; a low-pass filter provides output voltage for the lower frequencies.
2. Pulsating or fluctuating direct current varies in amplitude but does not reverse its direction. Similarly, a pulsating or fluctuating dc voltage varies in amplitude but maintains one polarity, either positive or negative.
3. The pulsating direct current or voltage consists of a steady dc level, equal to the average value, and an ac component that reverses in polarity with respect to the average level. The dc and ac components can be separated by filters.
4. An *RC* coupling circuit is effectively a high-pass filter for pulsating direct current. Capacitance C_C blocks the steady dc voltage but passes the ac component.

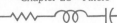

5. A transformer with an isolated secondary winding also is effectively a high-pass filter. With pulsating direct current in the primary, only the ac component produces output voltage in the secondary.

6. A bypass capacitor in parallel with R provides a low-pass filter.

7. Combinations of L, C, and R can be arranged as L, T, or π filters for more selective filtering. All three arrangements can be used for either low-pass or high-pass action. See Figs. 28-9 and 28-11.

8. In high-pass filters, the capacitance must be in series with the load as a coupling capacitor, with shunt R or L across the line.

9. For low-pass filters, the capacitance is across the line as a bypass capacitor, while R or L then must be in series with the load.

10. A bandpass or band-stop filter has in effect two cutoff frequencies. The band-pass filter passes to the load those frequencies in the band between the cutoff frequencies, while attenuating all other frequencies higher and lower than the passband. A band-stop filter does the opposite, attenuating the band between the cutoff frequencies, while passing to the load all other frequencies higher and lower than the stop band.

11. Resonant circuits are generally used for bandpass or band-stop filtering with radio frequencies.

12. For bandpass filtering, the series resonant LC circuit must be in series with the load, for minimum series opposition, while the high impedance of parallel resonance is across the load.

13. For band-stop filtering, the circuit is reversed, with the parallel resonant LC circuit in series with the load, while the series resonant circuit is in shunt across the load.

14. A wavetrap is an application of the resonant band-stop filter.

Self-Examination
Answers at Back of Book

Choose (a), (b), (c), or (d).

1. With input frequencies from direct current up to 15 kHz, a high-pass filter allows the most output voltage to be developed across the load resistance for which of the following frequencies? (a) Direct current; (b) 15 Hz; (c) 150 Hz; (d) 15,000 Hz.

2. With input frequencies from direct current up to 15 kHz a low-pass filter allows the most output voltage to be developed across the load resistance for which of the following frequencies? (a) Direct current; (b) 15 Hz; (c) 150 Hz; (d) 15,000 Hz.

3. An $R_C C_C$ coupling circuit is a high-pass filter for pulsating dc voltage because: (a) C_C has high reactance for high frequencies; (b) C_C blocks dc voltage; (c) C_C has low reactance for low frequencies; (d) R_C has minimum opposition for low frequencies.

4. A transformer with an isolated secondary winding is a high-pass filter for pulsating direct primary current because: (a) the steady primary current has no magnetic field; (b) the ac component of the primary current has the strongest

field; (c) only variations in primary current can induce secondary voltage; (d) the secondary voltage is maximum for steady direct current in the primary.

5. Which of the following is a low-pass filter? (a) L type with series C and shunt L; (b) π type with series C and shunt L; (c) T type with series C and shunt L; (d) L type with series L and shunt C.

6. A bypass capacitor C_b across R_b provides low-pass filtering because: (a) current in the C_b branch is maximum for low frequencies; (b) voltage across C_b is minimum for high frequencies; (c) voltage across C_b is minimum for low frequencies; (d) voltage across R_b is minimum for low frequencies.

7. An ac voltmeter across C_C in Fig. 28-6 reads (a) practically zero; (b) 7.07 V; (c) 10 V; (d) 20 V.

8. Which of the following L-type filters is the best band-stop filter? (a) Series resonant LC circuit in series with the load and parallel resonant LC circuit in shunt; (b) parallel resonant LC circuit in series with the load and series resonant LC circuit in shunt; (c) series resonant LC circuits in series and in parallel with the load; (d) parallel resonant LC circuits in series and in parallel with the load.

9. A 455-kHz wavetrap is a resonant LC circuit tuned to 455 kHz and connected as a (a) band-stop filter for frequencies at and near 455 kHz; (b) bandpass filter for frequencies at and near 455 kHz; (c) band-stop filter for frequencies from direct current up to 455 kHz; (d) bandpass filter for frequencies from 455 kHz up to 300 MHz.

10. A power-line filter for rejecting RF interference has (a) RF coupling capacitors in series with the power line; (b) RF chokes in shunt across the power line; (c) 60-Hz chokes in series with the power line; (d) RF bypass capacitors in shunt across the power line.

Essay Questions

1. What is the function of an electrical filter?
2. Give two examples where the voltage has different frequency components.
3. (a) What is meant by pulsating direct current or voltage? (b) What are the two components of a pulsating dc voltage? (c) How can you measure the value of each of the two components?
4. Define the function of the following filters in terms of output voltage across the load resistance: (a) High-pass filter. Why is an $R_C C_C$ coupling circuit an example? (b) Low-pass filter. Why is an $R_b C_b$ bypass circuit an example? (c) Bandpass filter. How does it differ from a coupling circuit? (d) Band-stop filter. How does it differ from a bandpass filter?
5. Draw circuit diagrams for the following filter types. No values are necessary. (a) T-type high-pass and T-type low-pass; (b) π-type low-pass, balanced with a filter reactance in both sides of the line.
6. Draw the circuit diagrams for L-type bandpass and L-type band-stop filters. How do these two circuits differ from each other?
7. Draw the response curve for each of the following filters: (a) low-pass cutting off at 20,000 Hz; (b) high-pass cutting off at 20 Hz; (c) bandpass for 20 to 20,000 Hz; (d) bandpass for 450 to 460 kHz.
8. Give one similarity and one difference in comparing a coupling capacitor and a bypass capacitor.

9. Give two differences between a low-pass filter and a high-pass filter.
10. Explain briefly why the power-line filter in Fig. 28-16 passes 60-Hz alternating current but not 1-MHz RF current.

Problems
Answers to Odd-Numbered Problems at Back of Book

1. Refer to the RC coupling circuit in Fig. 28-6, with R equal to 16,000 Ω. (**a**) Calculate the required value for C_C at 1000 Hz. (**b**) How much is the steady dc voltage across C_C and across R? (**c**) How much is the ac voltage across C_C and across R?

2. Refer to the R_1C_1 bypass circuit in Fig. 28-8. (**a**) Why is 1 MHz bypassed but not 1 kHz? (**b**) If C_1 were doubled in capacitance, what is the lowest frequency that could be bypassed, maintaining a 10:1 ratio of R to X_C?

3. Calculate the C_C needed to couple frequencies of 50 to 15,000 Hz with a 50-kΩ R.

4. Show the fluctuating collector current i_c of a transistor that has an average dc axis of 24 mA and a square-wave ac component with a 10-mA peak value. Label the dc axis, maximum and minimum positive values, and the peak-to-peak alternating current.

5. (**a**) Draw an inverted L-type band-stop filter used as a wavetrap for 455 kHz. (**b**) Give the inductance necessary with an 80-pF C.

6. (**a**) Referring to Fig. 28-6, calculate the value of C_C necessary for coupling 50 Hz when R is 50 kΩ. (**b**) Referring to Fig. 28-7, calculate the value of C_1 necessary for bypassing R_1 to 50 Hz.

7. Referring to the audio tone-control switch in Fig. 28-18, calculate the required capacitance values for the following: (**a**) C_1 to bypass R_1 at 10,000 Hz; (**b**) C_2 to bypass R_1 at 5000 Hz; (**c**) C_3 to bypass R_1 at 2000 Hz.

8. Referring to the RC low-pass filter in Fig. 28-9a, draw the schematic diagram with values of 75 kΩ for R, 0.001 μF for C, and 10 MΩ for R_L. (**a**) For 10 V input, calculate the values of V_C at 1, 2, 5, 10, and 15 kHz. (**b**) Draw the response curve of the filter, plotting V_C vs. frequency.

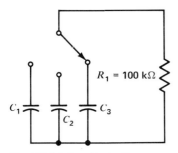

$R_1 = 100$ kΩ

C_1 C_2 C_3

Fig. 28-18 For Prob. 7.

Answers to Practice Problems

28-1 **a.** 500 kHz
  b. 60 Hz

28-2 **a.** 6 V
  b. 10 and 2 V
  c. 8 V
  d. 4 and 2.8 V

28-3 **a.** High-pass
  b. 0 V

28-4 **a.** 0 V
  b. 5 μF

28-5 **a.** RF
  b. 5 μF

28-6 **a.** High-pass
  b. Low-pass

28-7 **a.** e and f
  b. Low-pass

28-8 **a.** d
  b. High-pass

28-9 **a.** T
  b. T

28-10 **a.** T
  b. T

Review

Chapters 27 and 28

Summary

1. Resonance results when the reactances X_L and X_C are equal. In series, the net reactance is zero. In parallel, the net reactive branch current is zero. The specific frequency that makes $X_L = X_C$ is the resonant frequency $f_r = 1/(2\pi\sqrt{LC})$.
2. Larger values of L and C mean lower resonant frequencies, as f_r is inversely proportional to the square root of L or C. If the value of L or C is quadrupled, for instance, f_r will decrease by one-half.
3. For a series resonant LC circuit, the current is maximum. The voltage drop across the reactances is equal and opposite; the phase angle is zero. The reactive voltage at resonance is Q times greater than the applied voltage.
4. For a parallel resonant LC circuit, the impedance is maximum with minimum line current, since the reactive branch currents cancel. The impedance at resonance is Q times the X_L value, but it is resistive with a phase angle of zero.
5. The Q of the resonant circuit equals X_L/r_S for resistance in series with X_L, or R_P/X_L for resistance in parallel with X_L.
6. The bandwidth between half-power points if f_r/Q.
7. A filter uses inductance and capacitance to separate high or low frequencies. A low-pass filter allows low frequencies to develop output voltage across the load; a high-pass filter does the same for high frequencies. Series inductance or shunt capacitance provides low-pass filtering; series capacitance or shunt inductance provides high-pass filtering.
8. A fluctuating or pulsating dc is equivalent to an ac component varying in opposite directions around the average-value axis.
9. An RC coupling circuit is effectively a high-pass filter for pulsating dc voltage, passing the ac component but blocking the dc component.
10. A transformer with an isolated secondary is a high-pass filter for pulsating dc, allowing ac in the secondary but no dc output level.
11. A bypass capacitor in parallel with R is effectively a low-pass filter, since its low reactance reduces the voltage across R for high frequencies.

12. The main types of filter circuits are π type, L type, and T type. These can be high-pass or low-pass, depending on how L and C are connected.

13. Resonant circuits can be used as bandpass or band-stop filters. For bandpass filtering, series resonant circuits are in series with the load or parallel resonant circuits are across the load. For band-stop filtering, parallel resonant circuits are in series with the load or series resonant circuits are across the load.

14. A wavetrap is an application of a resonant band-stop filter.

Review Self-Examination
Answers at Back of Book

Fill in the numerical answer.

1. An L of 10 H and C of 40 μF has f_r of _____ Hz.
2. An L of 100 μH and C of 400 pF has f_r of _____ MHz.
3. In question 2, if $C = 400$ pF and L is increased to 400 μH, the f_r decreases to _____ MHz.
4. In a series resonant circuit with 10 mV applied across a 1-Ω R, a 1000-Ω X_L, and a 1000-Ω X_C, at resonance the current is _____ mA.
5. In a parallel resonant circuit with a 1-Ω r_S in series with a 1000-Ω X_L in one branch and a 1000-Ω X_C in the other branch, with 10 mV applied, the voltage across X_C equals _____ mV.
6. In question 5, the Z of the parallel resonant circuit equals _____ MΩ.
7. An LC circuit resonant at 500 kHz has a Q of 100. Its total bandwidth between half-power points equals _____ kHz.
8. A coupling capacitor for 40 to 15,000 Hz in series with a 0.5-MΩ resistor has the capacitance of _____ μF.
9. A bypass capacitor for 40 to 15,000 Hz in shunt with a 1000-Ω R has the capacitance of _____ μF.
10. A pulsating dc voltage varying in a symmetrical sine wave between 100 and 200 V has the average value of _____ V.

Answer True or False.

11. A series resonant circuit has low I and high Z.
12. A steady direct current in the primary of a transformer cannot produce any ac output voltage in the secondary.
13. A π-type filter with shunt capacitances is a low-pass filter.
14. An L-type filter with a parallel resonant LC circuit in series with the load is a band-stop filter.
15. A resonant circuit can be used for a band-stop filter.

References
Additional References at Back of Book

Grob, B.: *Electronic Circuits and Applications*, McGraw-Hill Book Company, New York.

Jackson, H. W.: *Introduction to Electric Circuits*, Prentice-Hall, Inc., Englewood Cliffs, N.J.

Ridsdale, R. E.: *Electric Circuits*, McGraw-Hill Book Company, New York.

Chapter 29
Electronic Devices

This chapter starts the discussion of active components for electronic circuits. Essentially, the *active devices* are used for amplifiers and rectifiers or for electronic switching and as a visual display of any desired information. Such active components are mainly semiconductor diodes, transistor amplifiers, integrated circuit (IC) chips, vacuum tubes, and display devices. Conversion from electrical signals to light output can be done by the light-emitting diode (LED), the liquid-crystal display (LCD), or the cathode-ray tube (CRT) used for the picture tube in television receivers.

The passive components for resistance R, inductance I, and capacitance C are still necessary to form complete circuits with the active components. When the active and passive components are combined, the result is almost unlimited possibilities for using electronic circuits in many functions. The applications include digital circuits as used in calculators and computers and analog circuits for audio, radio, and television communications.

Important terms in this chapter are:

amplifier	junction FET (JFET)
anode and cathode	junction transistor
barrier voltage	light-emitting diode (LED)
base, emitter, collector	liquid-crystal display (LCD)
bias voltage	NPN and PNP transistors
bias current	oscillator
bipolar transistor	peak inverse voltage
breakover voltage	pentode
cathode-ray tube (CRT)	phase inversion
covalent bond	PNPN device
depletion mode	rectifier
depletion zone	reverse voltage
diac	silicon controlled rectifier (SCR)
diode	source electrode
drain electrode	thyristor
enhancement mode	triac
field-effect transistor (FET)	triode
forward voltage	tunnel diode
gain	unijunction transistor (UJT)
gate electrode	varactor
hole current and charge	varistor
insulated-gate FET (IGFET)	zener diode

More details are explained in the following sections:

29-1   Rectifiers, Amplifiers, and Oscillators

Most electronic circuits can be classified in these three groups. A rectifier changes its ac input to dc output. An amplifier circuit amplifies its input signal. An oscillator circuit is a special case of the amplifier, but the oscillator generates ac output from its dc power supply, without any ac input signal. In addition, the logic switching circuits for digital electronics are explained in Chap. 31.

Amplifiers   The block diagram in Fig. 29-1 illustrates an amplifier for audio signal. In general, the term *signal* is used for the variations in voltage or current corresponding to the desired information. The ac input of 0.1 V here could be the audio signal from a phonograph. By means of a transistor, with its associated passive components, this audio amplifier circuit provides an output of 4 V. Then this signal can be used to drive a power amplifier stage that provides enough current to drive a loudspeaker. Amplifiers connected with the output driving the input of the next stage are *cascaded stages*.

For the amplifier in Fig. 29-1, the ac output signal of 4 V is forty times more than the input signal of 0.1 V. Note that the output signal is shown with reversed polarity, as many amplifier circuits invert the phase of the signal by 180°.

For the passive components, *R* is always useful in providing a desired *IR* voltage, for either direct current or alternating current. Capacitance *C* is commonly used for capacitive coupling to pass the ac signal while blocking any dc component. Or a transformer can be used for coupling. Capacitance *C* can also be used as a bypass capacitor, and *L* can be used as an audio or RF choke, and for transformers. The *LC* resonant circuits are generally used in RF amplifiers.

It should be noted that nonsinusoidal ac voltages can be amplified. In addition, dc amplifiers are used.

Although the amplifier in Fig. 29-1 is amplifying ac signal, the circuit requires a steady dc supply voltage. The reason is that transistors need dc voltage for the electrodes in order to conduct any current at all. In general, any amplifier operates by having the input signal variations control a larger amount of dc power in the output circuit. As a result, amplifier circuits have currents and voltages that include ac variations on an average dc axis.

Oscillators   The block diagram in Fig. 29-2 illustrates an RF feedback oscillator. There is no ac input signal, but the oscillator circuit generates RF output at the natural resonant frequency of the *LC* circuit tuned to 1 MHz.

The only difference between this circuit and an RF amplifier for a 1-MHz signal is that the oscillator has positive feedback from the output to the input. This polarity reinforces the signal variations of the input circuit. The positive feedback can be obtained by two phase reversals of 180°.

Although a tuned RF oscillator circuit is shown here for a 1-MHz sine-wave output, oscillator circuits also can be used to generate pulses, square waves, or sawtooth waves. The frequency can be almost any value, from low audio frequencies to extremely high radio frequencies.

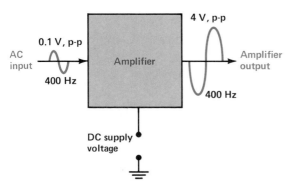

Fig. 29-1 Function of an amplifier circuit. The ac signal input of 0.1 V is amplified to 4 V p-p.

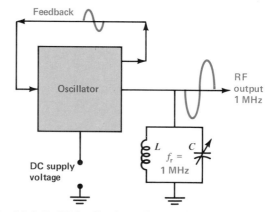

Fig. 29-2 An RF feedback oscillator circuit, tuned to 1 MHz as the resonant frequency.

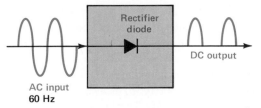

Fig. 29-3 A rectifier circuit converts its ac input to pulsating dc output.

Rectifiers   The block diagram in Fig. 29-3 illustrates a rectifier, changing the 60-Hz ac input voltage to fluctuating dc output voltage. The diode allows current in only one direction, for one polarity of applied voltage. Therefore, current flows in the output circuit only during the half-cycles of the ac input voltage that turn the diode on. This circuit is a half-wave rectifier.

Although not shown here, the output circuit usually has shunt filter capacitors, with a series choke or resistor, to remove the fluctuations in the dc output. The entire circuit then is a dc power supply.

The function of the power supply could be to provide the dc supply voltage needed for the amplifier and oscillator in Figs. 29-1 and 29-2. Diode semiconductors are used as rectifiers. Although 60 Hz is illustrated here for the ac power line, it should be noted that the ac input to the rectifier can have any frequency.

Practice Problems 29-1
Answers at End of Chapter

a. In Fig. 29-1, how much is the voltage gain for the audio signal?

b. In Fig. 29-2, what is the resonant frequency of the *LC* circuit for the tuned RF oscillator?

c. A component called a _____ (diode, transistor) is most commonly used as a rectifier.

29-2   Semiconductors

The main features of pure semiconductors without doping are: (1) *R* more than metal conductors but less than insulators; (2) negative temperature coefficient α, as *R* decreases with heat; and most important, (3) electron valence of ± 4. This valence means the atom has four electrons in the outside ring, halfway to the goal of 8

electrons. The semiconductor elements generally used for transistors and diodes are silicon (Si) and germanium (Ge).

As an example of atomic structure, Si has the atomic number 14. This atom has 14 protons in the nucleus balanced by 14 electrons in the outer shells. The electrons are distributed in rings of 2, 8, and 4.

Covalent Bonds   Such atoms do not usually gain or lose the 4 valence electrons but share them with neighboring atoms. The result is a stable configuration with 8 outer electrons. This union of atoms sharing the valence electrons is called a *covalent bond*.[1]

Crystal Structure   Si and Ge have atoms in a regular network or lattice of cubes to form a crystal solid. A crystal has a definite geometrical form of the internal atoms. A diamond is an example of the crystalline structure for pure carbon. When a crystal is broken into smaller segments, each has the same structure as the original crystal.

If we visualize atoms of Si interconnected by covalent bonds, as in Fig. 29-4, the result is a lattice of atoms to illustrate the crystalline structure. Because of this crystal lattice of covalent bonds, it is possible to add impurity atoms for doping. The purpose is to alter the electrical characteristics of the semiconductor.

Intrinsic Semiconductors   A pure semiconductor without any doping is called an *intrinsic* semiconductor. Silicon itself is an element in most common

---

[1]More details of covalent bonds between atoms are explained on pages 223 and 224.

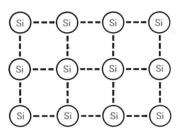

Fig. 29-4 Crystal lattice structure of a semiconductor, with covalent bonds between Si atoms.

(a)   *(b)*

Fig. 29-5 Forms of pure silicon. (*a*) Solid bar. (*b*) Wafer disks. Diameter 1½ in. (*Dow Corning*)

rocks. Sand is silicon dioxide. The element Si was discovered in 1823 and Ge in 1886. Germanium is recovered from the ash of certain coals. The oxides of both Ge and Si are reduced chemically to produce the elements with almost 100 percent purity. Figure 29-5 shows a solid bar of silicon and the slices or discs that are doped for use in semiconductor devices. It should be noted that intrinsic Ge has only ¹⁄₁₀₀₀ the resistance of Si, but silicon is used for most semiconductor devices because it allows less leakage current.

Extrinsic Semiconductors   The doping results in an *extrinsic* semiconductor, meaning that it is not in the natural form. The difference is that an intrinsic semiconductor has only covalent bonds for all the atoms, but an extrinsic semiconductor also has free charges as a result of the doping.

N-Type and P-Type Doping   The elements used for doping generally have an electron valence of either 5 or 3. As a result, the doped semiconductor has either an excess or a shortage of electrons in the covalent bond structure. A semiconductor with excess electrons is an N type, while a P type has a shortage of electrons. The doping elements are listed in Table 29-1. Those in the top group, including arsenic, have a valence of 5. For the bottom group, the valence is 3.

Free Electron Charges in N-Type Semiconductor   The doping elements arsenic, antimony, and phosphorus have a valence of 5. For each of these atoms there are 5 electrons in the outermost ring. In a covalent bond with Ge or Si atoms having 4 valence electrons, each impurity atom provides an extra electron.

This idea is illustrated in Fig. 29-6, where the crystal lattice of Si atoms includes one phosphorus atom. Four of the five valence electrons of the impurity element become part of the covalent bond structure. However, the 1 extra electron can be considered a free charge because it is not needed for a covalent bond.

Free Hole Charges in P-Type Semiconductors   The doping elements aluminum, boron, gallium, and indium have a valence of 3. For each of these atoms there are 3 electrons in the outermost ring. In a covalent bond with Ge or Si atoms, there are 7 electrons instead of 8 for each bond with an impurity

Table 29-1. Doping Elements and Semiconductors

Element	Symbol	Atomic number	Valence electrons	Applications
Antimony	Sb	51	5	*Donor* impurity elements. Give electrons to form
Arsenic	As	33	5	N-type semiconductor. As and Sb used for Ge;
Phosphorus	P	15	5	P for Si
Germanium	Ge	32	4	*Intrinsic* semiconductors used in pure crystal form to
Silicon	Si	14	4	be doped with impurity elements. Doped semiconductors are *extrinsic*
Aluminum	Al	13	3	*Acceptor* impurity elements. Take electrons to form
Boron	B	5	3	P-type semiconductor. Ga and In used for Ge; Al
Gallium	Ga	31	3	and B for Si
Indium	In	49	3	

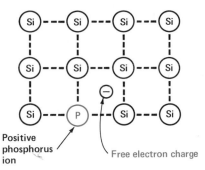

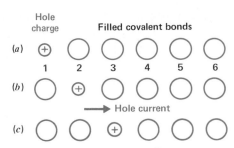

Fig. 29-8 Hole charge moving in a P-type semiconductor to provide hole current. Movement of charge from location 1 to 2 and 3 illustrated in (a), (b), and (c).

Fig. 29-6 Crystal lattice structure of Si doped with phosphorus. The covalent bonds have one free electron for each phosphorus atom.

element. The 1 missing electron in such a covalent bond can be considered as a free positive charge called a *hole charge*. Figure 29-7 illustrates a hole charge in the crystal lattice of doped silicon.

A hole has the same amount of positive charge as a proton, equal to that of an electron but with opposite polarity. However, a hole charge is not a proton. A proton is a stable charge in the nucleus that is not free to move. A hole is a positive charge outside the nucleus present only in semiconductors because of unfilled covalent bonds. Hole charges can be moved by an applied potential difference.

Hole Current The idea of hole charges moving to provide hole current is illustrated in Fig. 29-8. In Fig. 29-8a, along the top row, a hole charge is shown at point 1, along with some filled covalent bonds. Suppose that a valence electron from the filled bond at point 2 moves to point 1. As shown in Fig. 29-8b, the

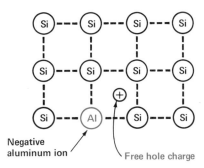

Fig. 29-7 Crystal lattice structure of Si doped with aluminum. The covalent bonds have one free positive hole charge for each aluminum atom.

bond at point 1 becomes filled and there is a hole charge at point 2. Similarly, an electron can move from point 3 to point 2 to fill this bond. With this sequence, the hole charge is moving from point 1 to point 6 to provide hole current, from left to right here.

To produce this hole current, voltage could be applied across the semiconductor with the positive terminal at point 1. The direction of hole current is the same as that of conventional current, opposite from electron flow. *All arrow symbols for current in semiconductors are shown for the direction of hole current.*

It should be noted, though, that hole current flows only in P-type semiconductors. The current is electron flow in N-type semiconductors and all wire conductors.

Fixed Ion Charges The free charges in a doped semiconductor are balanced by ions of the impurity element. For example, in Fig. 29-6, the phosphorus atom with a valence of 5 supplies 1 free electron that can easily move through the crystal. As a result, the nucleus of the phosphorus atom has 1 extra proton.

Any atom with unbalanced charges is an ion. The phosphorus atom in Fig. 29-6 is a positive ion, therefore.

In Fig. 29-7, the aluminum atom effectively has an extra electron as part of a covalent bond structure, with 8 electrons whenever the free hole charge moves away from the impurity atom. Therefore, the aluminum atom becomes a negative ion.

The ions are fixed charges that are not easily moved in the solid crystal. Since the ions are impurity atoms, they are present only in the doped semiconductor. The fixed charges of the ions are important because they provide an internal contact potential across a PN junction.

It should be noted that the doping really does not add or subtract charges. The semiconductor is still neutral, with equal positive and negative charges. However, the doping redistributes the valence electrons so that more free charges are available.

Current in Semiconductors With doping, an N-type semiconductor has a large supply of free electrons, a result of the added impurity atoms. Then the electrons become the dominant or *majority charges*. Still there are *minority charges* of holes. Similarly, a P-type semiconductor has majority hole charges, but there are minority electron charges.

When the majority charges are made to move in the semiconductor by an applied voltage, the result is a relatively large amount of *forward current* or *easy current*. This easy current is a flow of electrons as the majority carriers in N-type semiconductors, or hole current in P-type semiconductors. The amount of I is in the range of milliamperes and amperes.

When minority charges move, this current is in the reverse direction, compared with the forward current of the majority charges. The reason is simply that the polarity of minority charges is always opposite from that of the majority charges.

This very small current of minority charges is called *reverse current* or *leakage current*. It is in the order of microamperes.

Furthermore, the reverse current increases with higher temperatures, as more minority charges are produced by an increase of thermal energy. This increase in the reverse current of minority charges is the reason why temperature is very important in the operation of NPN and PNP transistors.

Practice Problems 29-2
Answers at End of Chapter

a. What is the valence of silicon and germanium?
b. Is the doped silicon illustrated in Fig. 29-6 N-type or P-type semiconductor?
c. Is the doped Si shown in Fig. 29-7 N-type or P-type semiconductor?
d. Is hole flow in the opposite or same direction as electron flow?
e. Free electrons are the majority charge carriers in N-type semiconductors. What are the carriers in P-type semiconductors?

29-3 The PN Junction

Figure 29-9 illustrates a magnified view of the junction between successive layers of P and N semiconductors in a single crystal. Some electrons in the N side are attracted to the P side, while at the same time an equal number of hole charges move in the opposite direction. These charges produce neutral electron-hole pairs. However, this effect at the junction uncovers the charged ions of the impurity atoms. These ions provide an internal barrier potential V_b across the two sides of the junction. This V_b has the polarity to prevent any more free electrons or hole charges from crossing the junction. In effect, V_b maintains the free electrons in the N semiconductor and the hole charges in the P semiconductor, to prevent the opposite sides from becoming neutralized.

It should be noted that the junction is only an electrical boundary between the P and N types. There is no actual physical separation.

The Internal Barrier Potential (V_b) Although it is an internal contact potential that cannot be measured directly, the effect can be overcome by 0.3 V for a Ge junction or 0.7 V for Si. The barrier voltage is more for Si because its lower atomic number allows more stability in the covalent bonds.

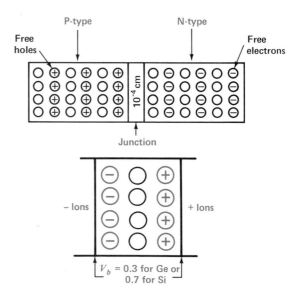

Fig. 29-9 PN junction with depletion zone magnified to show how ion charges produce internal contact potential V_b at the barrier.

The V_b is a characteristic of the element. Therefore these values apply to all PN junctions for semiconductor diodes and transistors of any size or power rating. The V_b is 0.7 V for any junction in all semiconductors made of silicon; the V_b is 0.3 V for any junction made of germanium.

The barrier voltage is what makes the junction useful because the effect of V_b can be controlled by an external voltage. With forward voltage applied in the polarity to cancel V_b, forward current flows. The V_b can be neutralized either completely or partially. Reverse voltage is applied in the polarity that does not cancel V_b.

Depletion Zone   Because of its neutral electron-hole pairs, the junction area is the depletion zone, meaning that it has no free charges. However, it still has the ion charges anchored in position to produce V_b.

Effect of Temperature on V_b   The values of 0.3 V for Ge and 0.7 V for Si are at normal room temperature of 25°C. However, V_b decreases at higher temperatures. The reason is more minority charge carriers. Specifically, V_b decreases 2.5 mV/°C rise in temperature, for both Ge and Si.

Forward Voltage   See Fig. 29-10a. The forward voltage is applied by wire conductors to the P and N electrodes for the bulk materials. Such a connection without any barrier potential is called an *ohmic contact*. Then the external voltage is applied through the bulk

materials to the PN junction. Forward current flows as the forward voltage neutralizes V_b. The required polarity is

$+V_F$ to the P electrode

$-V_F$ to the N electrode   forward voltage

In other words, the polarity of V_F corresponds to the semiconductor type.

This polarity of the external V_F can neutralize the internal V_b. The reason is that $+V_F$ at the P electrode repels hole charges to the side of the junction that has negative ion charges for V_b. Then the negative ions are neutralized. Also, $-V_F$ repels electrons to the junction at the positive side of V_b to neutralize the positive ions.

Reverse Voltage   In Fig. 29-10b, the reverse voltage V_R attracts minority charges away from the junction. Then the ion charges at the junction remain intact to maintain V_b. The polarity of V_R is reversed from V_F. Specifically, the required polarity is

$-V_R$ to the P electrode

$+V_R$ to the N electrode   reverse voltage

Volt-Ampere Characteristic   Figure 29-11 summarizes the effects of forward and reverse voltages for Ge and Si. Separate graphs are shown because the forward characteristics in Fig. 29-11a are for V_F in tenths of a volt and high currents in the ampere range, while the reverse characteristics in Fig. 29-11b are for higher voltages with very small reverse currents.

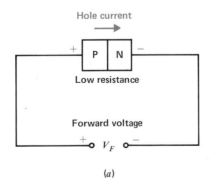

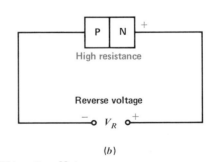

(a)   (b)

Fig. 29-10 (a) Polarity for applying forward voltage V_F across PN junction. Note direction of hole current in the forward direction from P to N. (b) Polarity for reverse voltage V_R. There is no forward current.

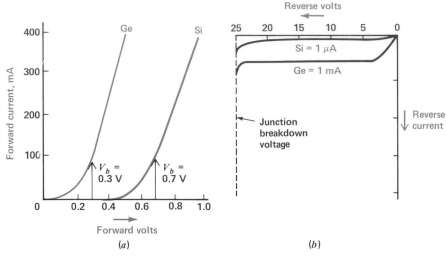

Fig. 29-11 Volt-ampere characteristic curves for Ge and Si junctions. (*a*) Small forward voltage and high current. (*b*) High reverse voltage and small reverse leakage current.

Forward Current   As shown in Fig. 29-11*a*, for Si with V_b of 0.7 V, forward current flows when V_F approaches 0.5 V. With 0.7 V or more for V_b, forward current increases sharply to the maximum value at saturation. Then the external V_F completely neutralizes the internal V_b.

A middle value of V_F for forward current is 0.6 V for a Si junction. Similarly, for a Ge junction, the range of V_F for forward current is 0.1 to 0.3 V. Typical values of forward current are 0.2 to 20 A.

Reverse Current   With reverse voltage, only a small reverse current of minority charges can flow, as shown in Fig. 29-11*b*. The separate curves indicate typical values of 1 mA for Ge and 1 μA for Si. Note the advantage of Si with practically zero reverse current.

This current has the symbol I_{CO}, indicating a small cutoff current. The junction is practically an open circuit with reverse voltage, compared with a short circuit with forward voltage.

Current I_{CO} is in the opposite direction from the forward current. Since I_{CO} consists of minority charges, the reverse current increases with temperature. For every 10°C rise, the I_{CO} doubles approximately.

Junction Breakdown   The reverse current is constant at a very small value until V_R reaches the junction breakdown voltage, shown as 25 V in Fig. 29-11*b*. Then a relatively large reverse current can flow. Most important, the reverse voltage across the junction is constant at the breakdown value. This effect is used for voltage regulator diodes.

For transistor amplifiers and rectifier diodes, though, the forward characteristics in Fig. 29-11*a* are used. Then the forward current can be controlled to provide amplification and rectification.

Practice Problems 29-3
Answers at End of Chapter

a. Assume a V_F of 1 V. The $+V_F$ connects the P electrode, and $-V_F$ connects the N electrode; then does a PN junction have high or low resistance?

b. The barrier potential in a silicon PN junction is about _____ V.

c. Refer to Fig. 29-11. A reverse voltage of about 10 V applied to an Si junction will cause a leakage current of about _____.

d. Does the depletion region of a PN junction contain free electrons, hole charges, or ions?

29-4 Semiconductor Diodes

The standard symbol for a semiconductor diode is an arrow showing the direction of hole flow and a bar, as shown in Fig. 29-12c. The anode or arrow side of the symbol contains the P-type material, whereas the cathode or bar side has the N-type semiconductor. In terms of free electrons, we can think of the diode as a one-way valve that will permit *electron flow against the arrow* in the symbol.

Two small semiconductor diodes are pictured in Fig. 29-12. The cathode end of the diode may be marked with a colored band around the diode. Sometimes the diode schematic symbol is printed on the unit, as in Fig. 29-12a. Occasionally the cathode end is marked with a + to show that this is the terminal for positive dc output voltage in a rectifier circuit.

The only numbering system is the letter N for semiconductors and the prefix 1 for diodes with one junction. The 1 indicates only one junction, compared with two or more junctions for transistors and thyristors. As an example, 1N4003 is a popular silicon diode. The 1N prefix specifies that it is a semiconductor diode, while the 4003 part specifies the exact characteristics of the diode. In schematic diagrams, semiconductor diodes are commonly labeled D, CR, or even X and Y. The CR stands for crystal rectifier.

Practical Diode Specifications Semiconductor diodes are sometimes specified as to their use. For in-

stance, the silicon diode in Fig. 29-12a might be referred to as a *rectifier* because it is typically used to convert ac power input to dc output. Other tiny diodes are called *signal diodes* because they are used to process signals, such as in the detector stage of an AM radio. The glass-enclosed germanium diode shown in Fig. 29-12b looks like many of the signal diodes.

The two most important electrical specifications for a rectifier or power diode are its PIV and I_F. The PIV specification stands for peak inverse voltage. This value is the maximum voltage the diode can tolerate in the reverse direction. The I_F specification stands for forward current. The I_F is the maximum forward current through the diode when it is conducting.

Rectifier Packaging The plastic package shown in Fig. 29-12a is very common for low-current rectifiers, up to 1000 mA. Two other common power diode or rectifier packages are shown in Fig. 29-13. The metal can "top hat" style is pictured in Fig. 29-13a. The heavy-duty stud-mounted style is shown in Fig. 29-13b. These types are used for current of 1 to 1500 A. The schematic symbol is typically printed on the side of these two rectifier packages.

Using an arrangement of four diodes called a full-wave bridge rectifier in a power supply is very common. A schematic for the full-wave bridge is drawn in Fig. 29-14a. Because of its wide use, manufacturers produce a variety of bridge-rectifier modules, such as those pictured in Fig. 29-14. These modules contain the circuitry shown in Fig. 29-14a. Like other rectifi-

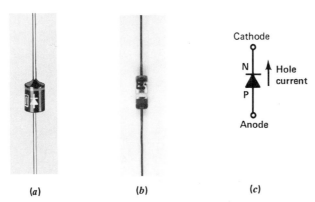

Fig. 29-12 Semiconductor diode rectifiers. Length about ½ in. without leads. (a) Silicon power diode rated at 1 A. Arrow shows hole current. (b) Germanium detector diode. Band at cathode end. (c) Schematic symbol showing direction of hole current.

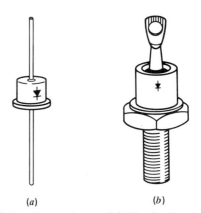

Fig. 29-13 Rectifier packages. (a) "Top hat" style package. Height without leads is ⅜ in. (b) Heavy-duty stud rectifier package. Height without stud is ½ to 1 in.

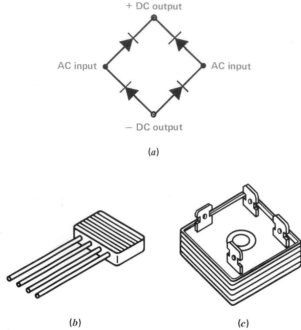

(a)

(b)   *(c)*

Fig. 29-14 Bridge rectifier. (*a*) Schematic for diodes wired as a full-wave bridge rectifier. (*b*) Medium-duty bridge rectifier package. Length is ¾ in. (*c*) Heavy-duty bridge rectifier package for mounting on heat sink. Size is 1 in. square.

ers, they are specified as to peak inverse voltage PIV and forward current I_F. The rectifier module in Fig. 29-14*b* might have a current rating I_F of about 10 A or less, whereas the unit in Fig. 29-14*c* would have a rating of about 25 A or more.

Testing Diodes with an Ohmmeter   An approximate test of a diode can be done with an ohmmeter. The procedure for checking the forward and reverse resistance is shown in Fig. 29-15. The forward resistance is being measured in Fig. 29-15*a*. It should be low, perhaps from 20 Ω to 1 kΩ. Next, the reverse resistance of the diode is measured in Fig. 29-15*b*. It should be very high, near infinity on all silicon diodes. This test can also be used to identify which lead of the diode is the cathode end if it cannot be determined from the bands or symbols. The cathode takes negative voltage in the forward direction for low resistance.

When R is very high in both directions, the diode is open. When R is low in both directions, the diode is shorted. If parallel paths confuse the reading, disconnect one side of the diode from the circuit. Do not use the low-power ohms position for these tests. The ohmmeter must supply enough voltage to turn on the diode in the forward polarity.

It should be noted that when a digital multimeter (DMM) has a diode position, it can be used for a dynamic test of the diode. The meter reads normal forward voltage of approximately 0.7 V for a silicon junction. This test is best.

Practice Problems 29-4
Answers at End of Chapter

a. A power diode might be used to process signals in an AM radio detector or as a _____ in a dc power supply.

b. Diodes used as rectifiers in a power supply are probably made of _____ (Ge, Si).

c. List the two most important electrical specifications for a power diode.

d. Refer to Fig. 29-15. If both resistance readings are very high, the diode is said to be _____ (open, short-circuited).

e. Refer to Fig. 29-14*c*. This package contains how many power diodes wired as a full-wave bridge rectifier?

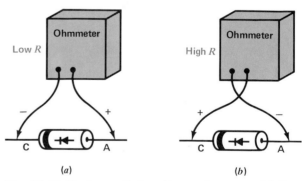

(a)   *(b)*

Fig. 29-15 Testing a diode with an ohmmeter. (*a*) Forward-biasing the diode causes a low resistance reading. (*b*) Reverse-biasing the diode causes a high (infinite) resistance reading.

29-5 The Bipolar Transistor

This name means NPN and PNP transistors, as they have two opposite polarities of doped semiconductors. As shown in Fig. 29-16, the bipolar transistor consists of a PN junction and an NP junction, by making either a P or N semiconductor between opposite types. The purpose is to have the first section supply charges, either holes or electrons, to be collected by the third section, through the middle section. The electrode that supplies charges is the *emitter;* the electrode at the opposite end to collect the charges is the *collector*. The *base* in the middle forms two junctions between emitter and collector, to control the collector current. This type is also called a *junction transistor.*

Emitter The emitter-base junction is biased with forward voltage. Typical values are 0.2 V for Ge or 0.6 V for Si.

As shown for the PNP transistor in Fig. 29-16a, the P emitter supplies hole charges to its junction with the base. This direction is indicated by the emitter arrow for forward hole current in the schematic symbol. The arrow pointed into the base shows a PN junction between emitter and base, corresponding to the symbol for a PN diode.

For the NPN transistor in Fig. 29-16b, the emitter supplies electrons to the base. Therefore, the symbol for the N emitter shows the arrow out from the base, opposite to the direction of electron flow.

In the schematic symbols, only the emitter has an arrow. The arrow pointing into the base means a PNP transistor; the arrow out from the base means an NPN transistor. Practically all small transistors for audio and RF amplifiers are NPN, made of silicon, with a typical forward bias of 0.6 V between base and emitter.

Collector Its function is to remove charges from the junction with the base. In Fig. 29-16a, the PNP transistor has a P collector receiving hole charges. For the NPN transistor in Fig. 29-16b, the N collector receives electrons.

The collector-base junction always has reverse voltage. Typical values are 4 to 100 V. This polarity means that no majority charges can flow from collector to base. However, in the opposite direction, from base to collector, the collector voltage attracts the charges in the base supplied by the emitter.

Base The base in the middle separates the emitter and collector. The base-emitter junction is forward-biased. As a result, the resistance for the emitter circuit is very low. The base-collector junction is reverse-biased, providing a much higher resistance in the collector circuit.

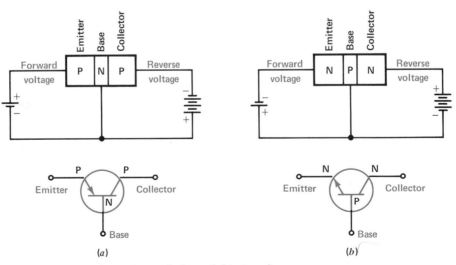

Fig. 29-16 Junction transistors schematic symbols and biasing diagrams. (a) PNP. (b) NPN.

Collector Current   The final requirement for transistor action is to have the collector current controlled by the emitter-base circuit. The emitter has heavy doping to supply majority charges. However, the base has only light doping and is very thin, so that its charges can move to the collector junction. The collector voltage is relatively high. Because of these factors, practically all the charges supplied by the emitter to the base are made to flow in the collector circuit. Typically, 98 to 99 percent or more of the emitter charges provide collector current I_C. The remaining 1 to 2 percent or less becomes base current I_B.

Consider the currents for an NPN transistor. The N emitter supplies electrons to the P base. Here the electrons are minority charges. Because of light doping in the base, though, very few of the electrons can recombine with hole charges. Any recombination of charges in the base provides the very small I_B returning from base to emitter.

There is a heavy concentration of free electron charges in the base at the emitter junction, because of the forward bias. As a result, almost all the electrons diffuse through the thin base to the collector junction.

The N collector has reverse voltage of positive polarity. For electrons moving from the base, however, the positive collector voltage attracts these free charges. As a result, the electrons diffused from the emitter side of the base move into the collector to provide a drift current of electrons for I_C in the collector circuit. For a PNP transistor, all voltage polarities are reversed and I_C at the collector electrode consists of hole charges.

Electrode Currents   As an example, in Fig. 29-17, the emitter supplies 10 mA of forward current. This is I_E. Of this, 9.8 mA is injected into the collector circuit. This is I_C. Only 0.2 mA or 200 μA of I_B flows through the base terminal to return to the emitter. As a formula,

$$I_E = I_C + I_B \qquad (29\text{-}1)$$

Note that I_E in Fig. 29-17 is marked negative, only to indicate that its direction is opposite from that of I_C and I_B. *It is standard practice to consider hole current into a semiconductor as the positive direction of I.* Since I_E is electron flow into the transistor, this I is in the negative direction.

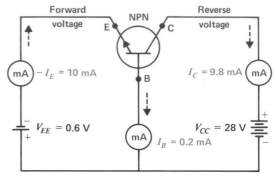

Fig. 29-17 Electrode currents. I_E, I_B, and I_C shown for NPN transistor. Dotted arrows show direction of electron flow for I in external circuit.

Algebraically, the values are

$$-10 \text{ mA} + 9.8 \text{ mA} + 0.2 \text{ mA} = 0$$

In practical terms, the formula states that the collector and base currents must add to equal the emitter current, which is the source.

Example 1   How much is I_E with 100 mA for I_C and 800 μA or 0.8 mA for I_B?

Answer
$$I_E = I_C + I_B$$
$$= 100 + 0.8$$
$$I_E = 100.8 \text{ mA}$$

Example 2   With 4 A for I_E and 3.8 A for I_C, how much is I_B?

Answer   Transposing Formula (29-1),
$$I_B = I_E - I_C$$
$$= 4 - 3.8$$
$$I_B = 0.2 \text{ A or } 200 \text{ mA}$$

For power transistors, I_E and I_C are generally in amperes. With smaller transistors, these currents are in milliamperes, with base current I_B in microamperes. In all cases, I_C and I_B return through the external circuit to the emitter, which is the source of the charges.

The Base Current Controls the Collector

Current When I_B is increased by more forward voltage, this means more majority charges are in the base to be injected into the collector. Therefore, increasing I_B means more I_C. For the opposite effect with less forward voltage and less I_B, the collector current is reduced.

Practice Problems 29-5
Answers at End of Chapter

a. Does the base-emitter junction have forward or reverse bias?

b. Does the collector-base junction have forward or reverse bias?

c. If I_C is 1 mA and I_B is 5 μA, how much is I_E?

d. An Si transistor has 0.1-V forward bias. Is it conducting or cut off?

29-6 Field-Effect Transistor (FET)

This device is a semiconductor amplifier with the same function as a junction transistor, but the FET construction provides a very high input resistance, in megohms.

The FET can take several volts for the input circuit, compared with tenths of a volt for junction transistors.

As illustrated in Fig. 29-18a, the FET operation depends on controlling current through a semiconductor channel of one polarity. An N channel is shown here, but a P channel can be used instead. The bulk or substrate material is neutral or lightly doped silicon. This only serves as a platform on which the other electrodes are diffused.

When voltage is applied between the drain and the source at opposite ends of the channel, the current through the channel is controlled by the gate. The terms anode and cathode are not used here because the channel can be either N or P, but current always flows from the source to the drain.

The FET is a unipolar device, as the charge carriers in the channel have only one polarity. A conventional PNP or NPN transistor is bipolar, with both majority and minority charge carriers.

The FET has very high input resistance from gate to source. Also, the FET is less sensitive to temperature, x-rays, and cosmic radiation, which can produce minority carriers in bipolar transistors. The disadvantages are less gain for a given bandwidth and smaller power ratings. Also, the switching speed is slower, compared with bipolar transistors.

In summary, the electrodes of the FET correspond to

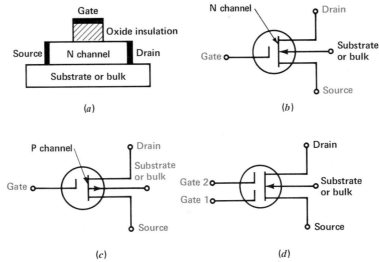

Fig. 29-18 Insulated-gate field-effect transistor. (a) Construction of IGFET. (b) Schematic symbol for N-channel IGFET. (c) Symbol for P-channel IGFET. (d) Symbol for N-channel IGFET with two gate inputs.

the emitter, base, and collector in a junction transistor, as follows:

1. *Source.* This is the terminal where the charge carriers enter the channel bar to provide current through the channel. Source current is I_S. The source corresponds to the emitter.
2. *Drain.* This is the terminal where current leaves the channel. Drain current is I_D. The drain corresponds to the collector.
3. *Gate.* This electrode controls the conductance of the channel between the source and the drain. Input signal voltage is generally applied to the gate. The gate voltage is V_G. The gate corresponds to the base, but the gate voltage controls the electric field in the channel, while the base controls the current in a bipolar transistor.
4. *Substrate or bulk.* This electrode is usually connected to the source, with both grounded in a common-source circuit.

The arrow for hole charges into the channel indicates an N channel. The source and drain have no polarity, as they are just ohmic contacts.

Function of the Gate   Input signal is generally applied to the gate, with amplified output from the drain. In the input circuit, the gate and channel act like two plates of a capacitor. A charge of one polarity on the gate induces an equal and opposite charge in the channel. As a result, the conductivity of the channel can be increased or decreased by the gate voltage. With an N channel, positive voltage at the gate induces negative charges in the channel to allow more electron flow from source to drain.

IGFET   The insulated-gate FET consists of a metal electrode separated from the channel by a thin layer of glassy silicon dioxide, as shown in Fig. 29-18a. Voltage applied to the gate can induce charges in the channel to control the drain current, but there is no PN junction. The insulated gate has very high input resistance independent of the polarity of the gate voltage. Also, the characteristics are not sensitive to temperature. This type is a metal-oxide semiconductor field-effect transistor (MOSFET) or an IGFET.

The insulated gate is essentially an electrostatic device with very high resistance. A typical r_{in} is 15 MΩ.

An IGFET not connected in a circuit may require a shorting ring on the leads to protect against a buildup of static charge. The IGFET has many uses in low-power amplifiers and in counting circuits. This method of construction is also applied in IC units.

The schematic diagram in Fig. 29-18b is for an IGFET with an N channel. However, the schematic in Fig. 29-18c is for a P-channel IGFET. Also, the diagram in Fig. 29-18d is for an N-channel IGFET that has two gates.

Depletion or Enhancement Mode   IGFETs come as either depletion or enhancement types. An *N-channel depletion-mode IGFET* is shown connected in series with a battery and ammeter in Fig. 29-19a. The gate has been left disconnected or open. Under these conditions, the depletion-mode IGFET acts like a *normally on* device and conducts through the N channel from source to drain. The ammeter shows a current flow.

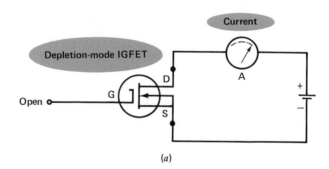

(a)

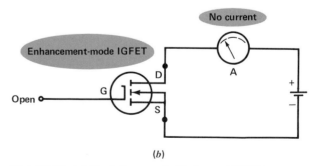

(b)

Fig. 29-19 (a) Depletion-mode N-channel IGFET operating as a normally on switch with the gate open. (b) Enhancement-mode N-channel IGFET operating as a normally off switch with the gate open.

Applying a negative voltage at the gate terminal of the depletion-mode IGFET in Fig. 29-19a would cause the current carriers in the N channel to be *depleted*. The current from source to drain would decrease as the gate became more negative.

An *N-channel enhancement-mode IGFET* is connected in series with a battery and an ammeter in Fig. 29-19b. Carefully note the broken line in the N-channel line of the symbol, compared with the depletion-mode IGFET in Fig. 29-19a. The broken N-channel line on the schematic symbol suggests that the N channel is *normally open* for the enhancement-mode IGFET. With the gate disconnected as in Fig. 29-19b, no current flows from source to drain of the FET. When a positive voltage is applied to the gate of the FET in Fig. 29-19b, current carriers are *enhanced* in the N channel. More current carriers in the N channel causes reduced resistance, which produces more current flow. An increasing positive voltage at the gate in Fig. 29-19b causes a proportional increase in current from source to drain of the N-channel enhancement-mode IGFET.

In summary, the depletion-mode FET acts like a normally on device, whereas the enhancement-mode FET is considered a normally off unit.

JFET A silicon dioxide insulating layer is used in Fig. 29-18a to isolate the gate from the N channel in the IGFET. However, the *junction field-effect transistor (JFET)* uses a different method of creating this isolation. The JFET uses a reverse-biased junction between the gate and the channel.

A cross section of an N-channel JFET is shown in Fig. 29-20a. The gate and N channel form a PN junction. The JFET is a depletion-mode device in that the source-to-drain resistance is low until a negative voltage (reverse bias) is placed on the gate terminal. The negative voltage will *deplete* the current carriers in the N channel and "pinch off" the current.

A schematic symbol for a JFET is shown in Fig. 29-20b. The channel here is N-type because the arrow is pointing toward the channel. Therefore, the schematic in Fig. 29-20b is for an N-channel JFET.

Both junction and insulated-gate FETs have very high input impedance, whereas bipolar transistors have a low input resistance. Field-effect transistors are considered voltage-driven devices, whereas bipolar transistors are current-driven. FETs are generally low-power devices, while bipolar transistors can be used in either low- or high-power applications.

FET Amplifier Circuit The common-source (CS) circuit in Fig. 29-21 is similar to a common-emitter (CE) amplifier. This FET is an N-channel depletion type. In the input circuit the gate has zero bias. Then the ac input swings the gate voltage ± 0.5 V, or 1 V p-p. As a result, the channel conductance changes to vary the drain current I_D.

In the output circuit, the varying I_D produces corresponding changes in the voltage drop across the 2-kΩ R_L. The result is voltage amplification. In this example, v_{out} is 10 V p-p. Then A_V is $^{10}/_1 = 10$.

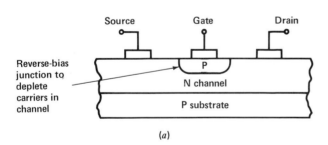

(a)

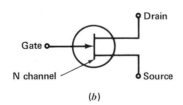

(b)

Fig. 29-20 Junction field-effect transistor. (a) Construction of the JFET. Note that the source-gate junction must be reverse-biased. (b) Schematic symbol for the N-channel JFET.

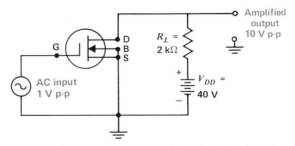

Fig. 29-21 Common-source amplifier circuit, with N-channel, depletion type of IGFET.

Practice Problems 29-6
Answers at End of Chapter

Answer True or False.
a. Compared with bipolar transistors, FETs have a high input resistance.
b. The gate lead on an FET compares with the emitter lead on a bipolar transistor.
c. The schematic symbol in Fig. 29-21 is for an N-channel depletion-mode IGFET.

29-7 Thyristors

A *thyristor* is a general term used to include such devices as SCRs and triacs. Thyristors can be thought of as solid-state switches with three or more PN junctions. They are often called *PNPN devices*.

Silicon Controlled Rectifier (SCR)
The SCR is a four-layer, three-junction device. A diagram showing SCR construction is drawn in Fig. 29-22a. The three leads are labeled *cathode, anode,* and *gate*. The schematic symbol illustrated in Fig. 29-22b shows the diode symbol with an extra control lead for the gate. The function of the gate is to control the start of conduction in the SCR. One common case style for medium-power SCRs is the TO-220 package in Fig. 29-22c.

An SCR is connected as a simple switch in Fig. 29-23 to help explain a few of its characteristics. Assume that the light bulb is not lit, which means that the SCR is not conducting from cathode to anode. The SCR is acting like an open switch in the power circuit at this time, even though S_2 is closed. Next, press S_1 to permit a *small current* to flow in the control circuit. This "turns on" the SCR, and the lamp lights. The SCR is now acting like a closed switch. Next, release the normally open push-button switch S_1. Current stops flowing in the control circuit, but current continues to flow in the power circuit. The SCR is acting like a latching relay. Briefly press and release the push-button switch S_2 in the power circuit. This causes the lamp to go out and stay out. The SCR again is acting like an open switch.

In Fig. 29-23, a small current in the control circuit is turning on a greater current in the power circuit. The SCR remains turned on until the current in the power circuit drops below a minimum level called the *holding current*. When current is interrupted for even an instant, the SCR turns off. With the SCR off, the gate signal (current in the control circuit) can turn it on again.

Observe in Fig. 29-23 that the diode part of the SCR is forward-biased. That is, the negative of B_2 is connected to the cathode and the positive to the anode. The

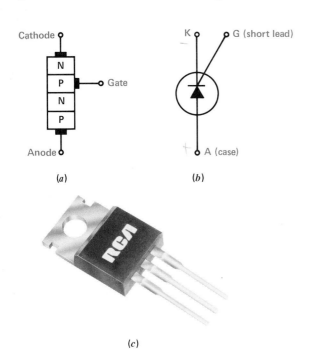

Fig. 29-22 Silicon controlled rectifier. (a) Four-layer PNPN construction of SCR. (b) Schematic symbol for SCR. (c) Common TO-220 package used for medium-power SCRs.

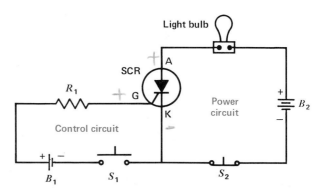

Fig. 29-23 Circuit for observing action of SCR used as a switch.

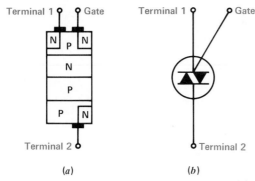

Fig. 29-24 The triac, a gate-controlled rectifier. (a) Construction of triac. (b) Symbol for triac.

Diac   This is also bidirectional between main terminals 1 and 2, but it does not have a gate. There are just three layers.

Unijunction Transistor (UJT)   As shown in Fig. 29-25, the UJT has an emitter and two connections to the base, without a collector terminal. The UJT is not used as a transistor amplifier, but is used in switching and timing applications. In effect, the emitter serves as a gate to control the resistive voltage divider between B_1 and B_2.

SCR can conduct only when it is connected in this manner *and* when the *forward-breakover voltage* is exceeded. This forward-breakover voltage is extremely low (about 0.7 V) for regular silicon diodes but is variable on an SCR. The forward-breakover voltage for the SCR is very high (100 V or more) when no current is flowing in the gate circuit. However, this forward-breakover voltage drops to a very low value when a gate signal is present. The gate current determines the forward-breakover voltage.

The PUT   This abbreviation is for a programmable unijunction transistor. However, the construction is a four-layer device with a gate, like the SCR. A resistive voltage divider in the gate circuit determines the firing voltage, though, similar to the idea of the divided base resistances in the UJT.

Practice Problems 29-7
Answers at End of Chapter

Answer True or False.
a. The SCR is used as a gate-controlled switch.
b. Thyristor is a general category which includes devices like SCRs and triacs.
c. Unlike the triac, the SCR is bidirectional.
d. The emitter terminal on a UJT functions as a control electrode like the gate terminal on an SCR.

Triac   As shown in Fig. 29-24, the construction enables the triac to conduct for either polarity of load voltage in the main circuit. Also, the gate can have either positive or negative forward voltage. In effect, the triac is a bidirectional SCR. There is no anode or cathode in the triac, as current can flow in either direction between main terminals 1 and 2.

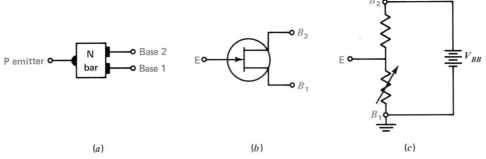

(a)   (b)   (c)

Fig. 29-25 The unijunction transistor (UJT). (a) Construction of the UJT. (b) Symbol for the UJT. (c) Equivalent circuit for double-base connections.

29-8 Special-Purpose Diodes

A semiconductor diode is just a PN junction. Since current flows only one way, the main application is rectification. However, the junction has additional properties that are useful. With reverse bias, the junction has capacitance which can be controlled by an external voltage. Also, the PN junction is photosensitive. Furthermore, the reverse breakdown voltage can be used for voltage-reference diodes. The schematic symbols for these diodes are shown in Fig. 29-26.

Tunnel Diodes These are also called *Esaki diodes,* named after the man who discovered that heavy doping can cause a tunneling effect of charge carriers through the depletion zone at the junction. One important feature of the tunnel diode is its *negative resistance* for a specific range of forward voltage. With negative resistance, the current decreases for an increase of applied voltage. Because of this characteristic, the tunnel diode can operate as an amplifier or oscillator. Essentially, this diode is a low-power device for microwave frequencies with relative freedom from radiation effects.

Varistor As shown in Fig. 29-26, two junctions of opposite polarities are used. The varistor is often used as a shunt for the collector to protect against excessive voltage spikes, either positive or negative.

Zener Diodes These are also called *voltage-reference diodes*. They are named after C. A. Zener, who analyzed the voltage breakdown of insulators. They are silicon diodes designed for a specific reverse breakdown voltage. When the *avalanche current*

flows, the voltage across the diode remains constant. This characteristic makes the diode useful as a voltage regulator. Multiple diodes in series increase the voltage rating.

Varactor Diodes The barrier voltage at a reverse-biased junction enables the junction to serve as a capacitance because of the separated charges in the depletion zone. Most important, the amount of junction capacitance can be controlled by the reverse voltage. The C changes as the depletion zone is made wider or narrower by changes in V. Diodes made for this use as a *voltage-sensitive capacitor* are *varactors* or *varicaps*. A common application is electronic tuning for television receivers.

Photoconductive Diodes These are made of a photosensitive material, such as cadmium sulfide, where the resistance decreases with more light. The applications include many light-control devices.

Light-Emitting Diode (LED) A PN junction radiates light when current passes through the unit. For best efficiency, special compounds of gallium (Ga) are used.

The schematic symbol for the LED is drawn in Fig. 29-26. Typical construction of a very common style of LED is shown in Fig. 29-27. The plastic lens is very important in directing and magnifying the small amount of light emitted by the LED chip. The cathode of the LED is identified on the package in Fig. 29-27 by the flat side on the plastic.

Light is emitted from an LED when the diode is forward-biased. A typical LED may require a forward

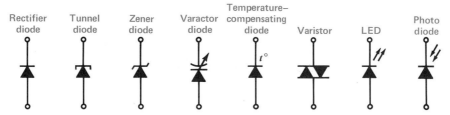

Fig. 29-26 Schematic symbols for types of semiconductor diodes. Note that the light-emitting diode (LED) generates light output while the photodiode requires light input.

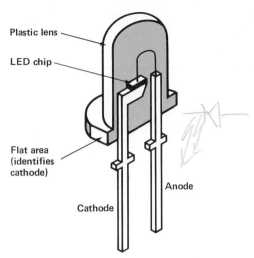

Fig. 29-27 Construction of a common light-emitting diode. Note that the flat area on the plastic shows the cathode lead. Width is about ¼ in.

voltage of about 1.6 V to produce a forward current of about 20 mA. LEDs come in several colors, including red, green, yellow, and white. The color of the LED is associated mostly with the material used in the LED chip. For instance, gallium arsenide phosphide (GaAsP) is commonly used in producing red LEDs.

Practice Problems 29-8
Answers at End of Chapter

Answer True or False.
a. A tunnel diode is also called an Esaki diode.
b. A zener diode is widely used for voltage regulation and as a voltage reference.
c. An LED is also called a photodiode and changes resistance as light shines on the PN junction.
d. Varactor diodes are used as voltage-sensitive capacitors in electronic TV tuning circuits.

29-9 Simple Visual Displays

A display produces light output to show information in visual form. Displays can be divided into two categories. *Character displays* give visual indications of numbers and letters. *Graphic displays* are more complex and can give pictorial as well as alphanumeric information.

Two types of character display formats are common—the 7-segment and dot matrix displays. Typical fonts for both a 7-segment and a dot matrix display are shown in Fig. 29-28. (A *font* is the term for a complete assortment of characters of one size and style.) The 7-segment display in Fig. 29-28*a* is used where numbers and a limited amount of other symbols are required. The 5 × 7 dot matrix display shown in Fig. 29-28*b* can represent most alphanumeric characters. Note the five columns of seven dots in the matrix for each character.

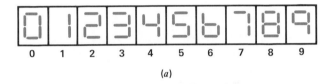

(a)

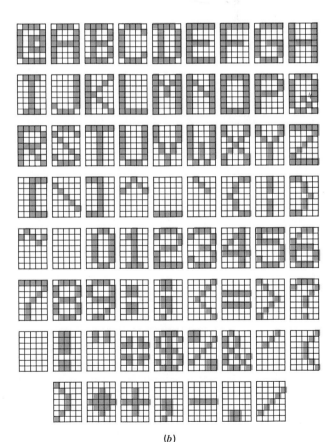

(b)

Fig. 29-28 (a) Numbers formed by a 7-segment display. (b) Alphanumeric font for a typical 5 × 7 dot matrix display. (*Texas Instruments, Inc.*)

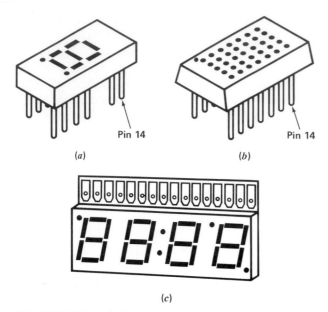

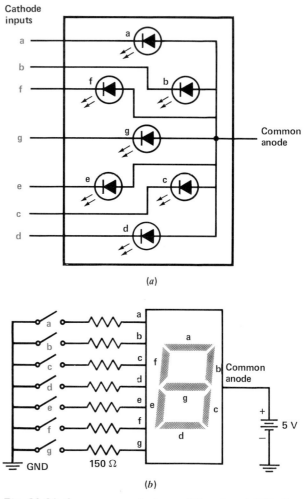

Fig. 29-29 Typical character display packages. (*a*) The 7-segment display fits in standard 14-pin DIP socket. (*b*) The 5 × 7 dot matrix display fits in 14-pin DIP socket. (*c*) Multidigit 7-segment display.

Typical packages for character displays are sketched in Fig. 29-29. The 7-segment display in Fig. 29-29*a* fits a standard 14-pin DIP socket. The 5 × 7 dot matrix display in Fig. 29-29*b* also can be inserted in a 14-pin DIP socket. The multidigit display shown in Fig. 29-29*c* is common in digital clocks and other equipment.

Displays may be constructed in several ways. *Incandescent* displays use thin filaments for each segment, similar to regular lamps. Another type of display uses the gas-discharge tube. This older type of unit operates at high voltages and emits an orange glow. A *fluorescent* display gives off a greenish glow and operates at low voltages. In the past, the fluorescent display was widely used on calculators. The common LED display typically gives off a reddish glow. The newer *liquid-crystal* display (LCD) creates black characters on a silver background.

Light-Emitting Diode (LED) See Fig. 29-30*a*. Each segment is shown as a light-emitting diode. The anodes of all the LEDs are connected; therefore, this unit is referred to as a *common-anode* 7-segment display. The standard segment lettering (a to g) is shown

in Fig. 29-30*a*. The cathode of each segment is connected to an external input.

Testing of the common-anode 7-segment LED display is illustrated in Fig. 29-30*b*. A standard +5-V power supply is connected to the display anode. In this example, switches a through g can activate (ground) any of the display inputs. Note the use of 150-Ω limiting resistors at each input. These limit the current to a safe level of about 20 mA. Activating the display is called driving the display. ICs called *decoder/drivers* are typically used to activate numbers on the display.

Which switches would have to be closed in Fig. 29-30*b* for the decimal number 7 to appear on the LED display? Decimal 7 is formed when segments a, b, and

Fig. 29-30 Common anode type of 7-segment LED display. (*a*) Construction. (*b*) Driving the display with switches.

c are lit. Switches a, b, and c would have to be closed, therefore, to light these three segments.

LED displays come in a variety of packages, including single-digit and multidigit units, as in Fig. 29-29. Both common-anode and common-cathode 7-segment LED displays are available. LEDs are used to produce both the dot matrix and 7-segment types of displays.

Liquid-Crystal Display (LCD)

While LEDs give off light, *liquid-crystal displays* (LCDs) are not light sources but control light. Liquid-crystal displays reflect part of the surrounding light, while other parts of the display absorb light. For example, the display on the hand-held digital multimeter in Fig. 29-31 shows the silvery background reflecting light, while the activated symbols and numbers are absorbing light. Since LCDs radiate no light, they must be used in lighted areas.

Liquid-crystal displays are either the *dynamic-scattering* or the *field-effect* type. The older dynamic-scattering type of LCD produces frosty white letters on a dark background. The field-effect LCD produces black letters on a silvery background. As an example, the display used on the DMM in Fig. 29-31 is the field-effect type of LCD.

Figure 29-32 shows how a typical field-effect LCD is assembled. The thickness is greatly magnified. The display consists of two glass plates with a special liquid crystal or *nematic fluid* filling the space between. The under surface of the top plate has nearly invisible metallized shapes where the segments and symbols are

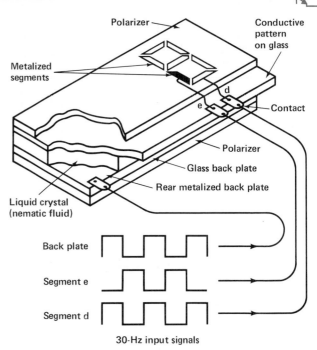

Fig. 29-32 Construction of field-effect LCD with ac signals driving segment e.

to appear. The glass back plate is also metallized. A polarizer forms the top and bottom of the sandwich. Contacts are attached to the back plate and to each segment of the display.

Segments on the LCD are driven by low-frequency ac signals. Figure 29-32 shows that the back plate receives a 30-Hz square-wave input. In this example, segment e receives a 30-Hz signal that is *180° out of phase* with the back-plate signal. This causes segment e to be activated and to appear dark on the silvery background. In Fig. 29-32, segment d receives a 30-Hz signal that is *in phase* with the signal on the back plate. This does not activate segment d, and it remains nearly invisible.

Direct current must not be used to drive LCDs, as it will damage them. LCDs are widely used in battery power applications such as calculators because of their extremely low power consumption. They are easy to read in sunlight and other areas of high light intensity. For this reason, they are widely used on service station pumps. The field-effect LCD is the most widely used because it consumes the least power and is easy to read. A steady backup light can be used for the LCD in applications where the light level is too low.

Fig. 29-31 Digital multimeter using a field-effect LCD for the display. (*Simpson Electric Company*)

Practice Problems 29-9
Answers at End of Chapter

a. The _____ (LCD, LED) display does not emit light but controls light.

b. Two types of character displays are the dot matrix and the very popular _____ type.

c. Refer to Fig. 29-30*b*. The number _____ will appear on the display when switches b, c, f, and g are closed.

d. The _____ (LCD, LED) display uses the least power.

29-10 The Cathode-Ray Tube (CRT)

As illustrated in Fig. 29-33, this tube consists of an electron gun and a fluorescent screen inside the evacuated glass envelope. The electron beam is focused for a small, sharp spot. Also, the beam can be deflected horizontally and vertically to any position on the screen.

The cathode, control grid, focusing grid, and accelerating grid form the electron gun. These electrodes are constructed as metal cylinders, with a small aperture or hole for the electron beam to pass through to the screen. The cathode is heated to emit electrons. The voltage of the control grid determines how many electrons can be attracted by the positive potential of the accelerating grid and the final anode. High voltage is used for the anode. All the cylinders are in one assembly as the electron gun to form a beam of electrons aimed at the screen. The CRT is a vacuum tube, but its function is that of a visual display device, not an amplifier. The front glass faceplate is coated on the inside surface with a fluorescent phosphor that emits light when bombarded by electrons. Green and white are two common colors for the phosphors, depending on the chemical composition. A green screen is generally used for the CRT in oscilloscopes because green phosphors are most efficient.

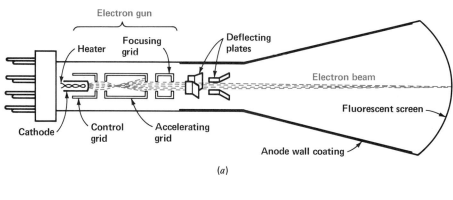

(a)

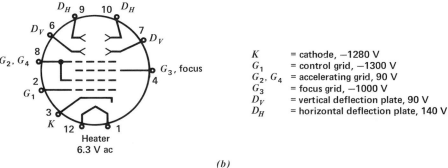

(b)

Fig. 29-33 (*a*) Construction of a cathode-ray tube (CRT) using electrostatic deflection and focusing. (*b*) Basing diagram for 12-pin socket of 3AQP1 or 5AQP1, with functions of electrodes and typical voltages.

When the electron beam hits the screen, it produces a spot of light visible through the glass faceplate. To deflect the beam and change the position of the light spot on the screen, deflection voltage can be applied to the deflection plates, as in Fig. 29-33. A pair of deflection plates left and right of the beam provides the potential difference needed to move the beam horizontally. Also, there are vertical deflection plates above and below the beam. The method with deflection plates is electrostatic deflection. This type of CRT is used in the oscilloscope (Fig. 29-34). The oscilloscope is a test instrument that shows the waveform of voltage applied to its input terminals.

Picture Tubes The image reproduced on the screen of the picture tube in a television receiver is another application of the CRT. However, picture tubes use magnetic deflection. In this case, a yoke with deflection coils is slipped over the neck of the tube, against the wide glass bell. The external deflection yoke is used instead of internal deflection plates. Magnetic deflection is necessary because picture tubes with a large screen need higher anode voltage, which would require too much deflection voltage unless the tube was extremely long.

In a monochrome or black-and-white receiver, the picture tube has one electron gun with a white screen. Color picture tubes have three electron beams for a tricolor screen with red, green, and blue phosphors. Magnetic deflection is used, but the focusing is electrostatic.

Anode Voltage For monochrome picture tubes, positive anode voltage of 10 to 18 kV is used for screen sizes of 12 to 19 in. With color picture tubes, the anode voltage is generally 25 to 30 kV for screen sizes of 15 to 25 in. Higher anode voltage is necessary for sufficient brightness with a bigger screen. Also, color picture tubes are less efficient than monochrome tubes because some of the beam current is blocked by an internal shadow mask necessary to keep the colors separate.

The anode is also called the *ultor,* as it has the highest positive potential. High voltage is generally connected to the picture tube at a ¼-in. recessed cavity at the center of the wide glass bell.

For the CRT in an oscilloscope, the required high voltage is generally 1 to 2 kV. This voltage is usually inverted as negative voltage at the cathode of the electron gun. The purpose is to have the deflection plates close to ground potential. Typical electrode voltages for an oscilloscope CRT with electrostatic deflection are listed with the basing diagram in Fig. 29-33*b*. Note that the pin numbers are counted clockwise looking at the tube base, or counterclockwise for a bottom view of the socket.

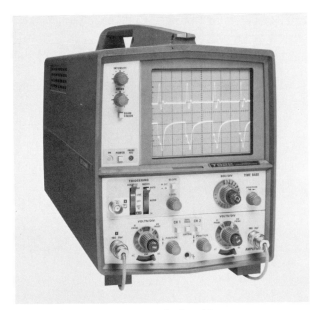

Fig. 29-34 Oscilloscope with 8 × 10 cm green screen. This unit is a dual-trace "scope," with two input channels. (*Tektronix*)

CRT Type Numbers The first digits give the screen size for the diameter of a round screen or the diagonal of a rectangular screen. The phosphor number at the end of the designation is generally 1 for a green screen, 4 for white, and 22 for the red, blue, and green screen in color picture tubes. Letters P and B are used for the phosphor type. For instance, a white screen can be P4 or B4. Also, the screen size may be given as inches, centimeters, or millimeters. As an example, the 330WB22 is a color picture tube with a screen diagonal of 330 mm, 33 cm, or approximately 13 in. Heater voltage is generally 6.3 V.

CRT Applications   In its use for an oscilloscope, the CRT is turned on to provide a spot of light on the screen. Internal horizontal deflection is usually provided to produce a horizontal line. This horizontal reference is the *time base*. Then the oscilloscope input signal voltage is amplified and used for vertical deflection. The screen display then shows the waveform of the input signal as vertical changes in amplitude with respect to time on the horizontal axis.

For the television picture tube, the receiver provides constant horizontal and vertical deflection to produce a pattern of horizontal lines on the screen. This rectangular area of lines is the scanning raster. Then the video signal input for the picture tube is applied to the cathode-grid circuit of the electron gun to control the beam intensity. The resulting variation in light from the screen corresponds to the picture information in the video signal.

Fig. 29-35 Typical amplifier tube.

Practice Problems 29-10
Answers at End of Chapter

a. Which electrode in the CRT has the highest positive potential?
b. What is the P number for a green phosphor?
c. Give the screen size in inches for a 25GP22 picture tube.

29-11   Vacuum Tubes

Vacuum tubes dominated electronics until the early 1960s, when solid-state devices took over. Tubes are still used in special applications, such as CRTs, high-power RF amplifiers, and some microwave devices. Generally tubes consume more power, are larger, produce more heat, use high voltages, and are not as reliable as semiconductor devices. Therefore, semiconductor devices have all but taken over the jobs vacuum tubes used to perform as amplifiers and rectifiers.

As illustrated in Fig. 29-35, a vacuum tube has a glass envelope enclosing metal electrodes in a vacuum. The electrodes make it possible to have electrons flow through the evacuated space inside the tube.

Diode Tube   The construction is shown in Fig. 29-36*a*. The schematic symbol is drawn in Fig. 29-36*b*. The *heater* is typically connected to an ac power source. The filament heats the cathode element inside the tube to a high temperature. Then the *cathode* becomes surrounded by a space charge of electrons due to the "boiling off" of electrons by thermionic emission. If the *plate* or *anode* becomes positive with respect to the cathode, the emitted electrons flow to the plate. However, if the plate becomes negative compared with the cathode, the emitted electrons are pushed back to the cathode and no current flows. The vacuum tube diode can perform rectification just as a semiconductor diode can. The word *diode* implies two terminals. In this case, only the cathode and plate are counted as terminals, even though filament power must be supplied.

Heater Voltage   When the heaters of multiple tubes are connected in parallel, the heater voltage is generally 6.3 V or 12.6 V. For heaters in series, typical values are 3 V, 5 V, 11 V, 25 V, 35 V or 50 V, to make all the series voltages add to equal the 120 V of the ac power line.

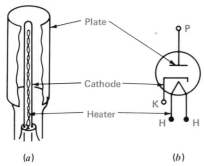

Fig. 29-36 Diode vacuum tube. (a) Construction. (b) Schematic symbol.

Triodes As illustrated in Fig. 29-37, a cathode and a plate are needed to provide plate current, as in a diode, but in addition the triode has the control-grid electrode. The grid is a fine metal wire, usually nickel, molybdenum, or iron, wrapped around two supports placed in the space between the cathode and the plate. All electrons attracted to the plate from the cathode go through the openings in the grid. The grid is connected to a base pin, however, so that it can have voltage applied to determine the amount of electrons that travel from cathode to plate. Now the plate current depends on two factors: the plate voltage and the control-grid voltage.

C^- Voltage The potential applied to the control grid is usually a small negative voltage with respect to the cathode, as shown in Fig. 29-38. This voltage completes the alphabet, with A voltage for the heater, B^+ voltage for the plate, and C^- for the control grid. The grid voltage is usually made negative so that it cannot

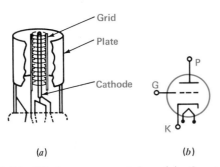

Fig. 29-37 Triode vacuum tube. (a) Construction. (b) Schematic symbol.

attract electrons. Then there will be no grid current. The function of the control grid is not to provide current but to have its voltage control the plate current.

The effect of the grid voltage on plate current can be summarized as follows:

1. Less negative grid voltage increases the plate current.
2. More negative grid voltage decreases the plate current.
3. If the grid voltage is made negative enough, there will be no plate current even with the plate positive.

Grid-Cutoff Voltage This is the amount of negative grid voltage required to cut off the plate current with a specified amount of positive voltage on the plate. For some tubes, the grid-cutoff voltage may be as little as -1 V; for others, the grid-cutoff voltage may be -60 V, as typical examples. The grid-cutoff voltage is a characteristic of the tube construction, but also depends on the amount of plate voltage.

Triode Circuit Plate current flows in Fig. 29-38 because the plate has its positive voltage and the control-grid potential is less than the grid-cutoff voltage. The electron flow inside the tube is from the cathode, through the spaces between the grid wire, to the anode.

In the external plate circuit, the electrons flow through the 20-kΩ plate-load resistor R_L and the B supply, returning to the cathode through the chassis. Assuming a 5-mA plate current value, the $I_b R_L$ drop is $0.005 \times 20,000 = 100$ V.

The plate-to-cathode circuit of the tube and R_L are in series as a voltage divider across the 300-V supply V_{bb}. Therefore, the plate-to-cathode voltage is 300 V minus the 100-V drop across R_L, or V_b is $300 - 100 = 200$ V. As a formula,

$$V_b = V_{bb} - I_b R_L \tag{29-2}$$

This formula is used to determine the actual plate-to-cathode voltage.

> **Example 3** Calculate V_b for a 20-mA I_b with a 2000-Ω R_L and a plate supply of 250 V.

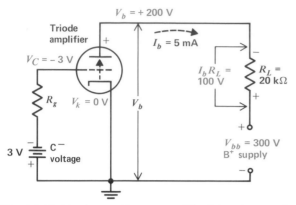

Fig. 29-38 Triode amplifier circuit. The B⁺ voltage is for the plate and C⁻ voltage for the grid.

Answer $V_b = V_{bb} - I_b R_L$
$$= 250 - (20 \times 10^{-3} \times 2 \times 10^3)$$
$$= 250 - 40$$
$$V_b = 210 \text{ V}$$

Control-Grid Bias A bias voltage is a constant voltage that is used to obtain the desired method of operation. In Fig. 29-38 the V_C of -3 V is a negative dc grid bias. Its function is to maintain the average grid voltage negative, even with an ac signal input. The ac signal swings the grid voltage in a positive direction, but the negative dc bias is more than the positive peak of the ac signal.

How a Triode Amplifies the Control-Grid Voltage The main advantage in having the control grid is the fact that it enables the vacuum tube to amplify a small voltage at the grid to provide much more voltage at the plate. The amplification results from the ability of the control-grid voltage to vary the plate current.

Consider the numerical examples listed in Table 29-2. For the circuit in Fig. 29-38, let the average plate current I_b be 5 mA, with -3 V for the control-grid bias voltage V_C. These values are listed in the middle row of Table 29-2. With a 5-mA I_b, the plate voltage equals 200 V. The 300-V B⁺ voltage drops to 200 V for the plate because of the 100-V IR drop across the 20-kΩ R_L.

Now suppose the grid voltage v_c is reduced from -3 to -2 V. The grid, being less negative, allows more plate current. For the case of i_b increasing to 6 mA, then, the voltage drop across R_L increases to 120 V. The voltage for v_b is less, however, falling to 180 V, that is, the 300-V B⁺ minus the 120 V $i_b R_L$ drop.

For the third case, in the bottom row, when v_c is made more negative, from -3 to -4 V, less plate current flows, Then i_b is reduced to 4 mA. This change results in a smaller $i_b R_L$ drop of 80 V. The resulting plate voltage, equal to 300 V minus 80 V, rises to 220 V.

Note the effect of the change as v_c varies ±1 V, above and below the middle value of -3 V. Similarly, i_b varies ±1 mA around the 5-mA middle value.

So far, all that the control-grid voltage has done is vary the plate current. In terms of plate voltage, however, v_b varies ±20 V above and below the middle value of 200 V. Here is where the voltage amplification is accomplished.

The changes in plate current through the plate-load resistance allow the plate voltage to vary by a much larger amount than the grid-voltage variation. In this example, the grid voltage of ±1 V has been multiplied by a factor of 20 to produce a corresponding plate-voltage variation of ±20 V.

For voltage amplification, the requirements are that (1) the grid voltage must vary the plate current and (2) the plate circuit must have a plate-load resistance that can vary the plate voltage as the plate current varies.

Table 29-2. Amplification in the Plate Circuit*

	v_c, V	i_b, mA	$i_b R_L$, V	$v_b = 300 - i_b R_L$, V
	-2	6	120	180
Average level →	-3	5	100	200
	-4	4	80	220

*$R_L = 20$ kΩ and B⁺ or $V_{bb} = 300$ V.

Phase Inversion If we consider the variations in Table 29-2 as peak values of a sine-wave ac signal, the resultant waveforms are shown in Fig. 29-39. In Fig. 29-39a the grid voltage v_c is a fluctuating dc voltage with variations above and below the bias axis of -3 V. These variations in v_c produce the variations of i_b in waveform Fig. 29-39b. The plate current varies above and below the average dc axis of 5 mA. The voltage across R_L has similar variations, as $v_L = i_b R_L$.

However, the plate voltage v_b shown in Fig. 29-39c has polarity opposite from i_b and v_L. The reason is that v_b is the difference between V_{bb} and v_L. As v_L increases, the value of positive dc plate voltage v_b decreases. Therefore, the variations in v_b have opposite polarity from the variations in v_c. Or the input signal is inverted by 180° in the amplified output.

Voltage Gain The symbol is A_V. As a formula,

$$A_V = \frac{V_{\text{out}}}{V_{\text{in}}} \tag{29-3}$$

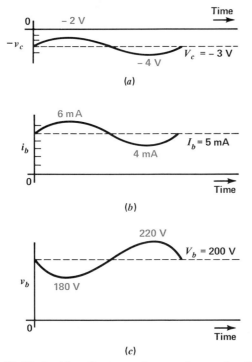

(a)

(b)

(c)

Fig. 29-39 Ladder diagram of waveshapes for Table 29-2. (a) Sine-wave grid voltage v_c. (b) Plate current i_b. (c) Plate voltage v_b equal to $V_{bb} - i_b R_L$.

These are ac signal voltages, in rms, peak, or peak-to-peak values, as long as the same measure is used for both. The A_V has no units, since it is a ratio of two voltages. For this example, A_V is 20 V/1 V = 20.

Tube Types Additional types of amplifier tubes are the pentode, which has five electrodes, and the tetrode. See Table 29-3 for a summary of types of vacuum tubes. In the type numbers for tubes, the first digits give the heater voltage. As examples, for the 6BQ7, the heater voltage is 6.3 V; the 12BQ7 has a heater voltage of 12.6 V; and the 38HK7 heater voltage is 38 V.

Comparing Tubes and Semiconductor Devices The vacuum tube triode is a voltage-controlled device which has high input impedance. This is also true of the FET. However, the bipolar transistor is a current-controlled device with a lower input impedance.

Table 29-4 compares the function of the terminals of the vacuum tube, junction transistor, and FET. For instance, the control element on the tube is the control grid, while the base lead of a junction transistor or the gate on an FET serves this purpose.

It should be noted that the analysis of a triode amplifier circuit really applies to all types of amplifier devices. The amplification results from making a small input signal control a larger amount of power in the output circuit. The amplified output is not the original signal but a close duplicate, with minimum distortion. The complete circuit is necessary to take out the internal effect of the amplifying device itself. A dc bias for the amplifier is necessary if it is to use the best part of its operating characteristic. With transistors and tubes, the amplifier needs dc voltage at the output electrode in order to conduct any current. As a result, the amplified output for ac signal input consists of ac variations with a dc component. This is the reason why dc values are important in amplifiers for ac signals.

Practice Problems 29-11
Answers at End of Chapter

a. List three main internal parts of a vacuum tube triode.

b. While diodes are used as rectifiers, triodes are used as _____.

c. The part of a triode comparable to the emitter of a junction transistor is the _____.

Table 29-3. Types of Vacuum Tubes

Tube	Symbol*	Electrodes	Characteristics	Applications
Diode		Plate Cathode	When plate is positive, conducts current only from cathode to plate	One diode for half-wave rectifier; two diodes for full-wave rectifier
Triode		Plate Control grid Cathode	Negative grid controls plate current; has large C_{gp}	Triode amplifier with external R_L, for AF amplifiers; low tube noise
Tetrode		Plate Screen grid Control grid Cathode	Screen grid reduces C_{gp}; screen must have +dc voltage and bypass C to cathode for ac signal	Tetrode little used because of secondary emission from plate
Pentode		Plate Suppressor Screen grid Control grid Cathode	Suppressor grid eliminates secondary emission; screen grid provides high μ and high r_p	For AF or RF amplifiers; sharp cutoff except variable-μ tubes

*Tubes shown with indirectly heated cathode.

Table 29-4. Electrodes in Transistors and Vacuum Tubes

Function	Vacuum Tube	Junction Transistor	Field-Effect Transistor
Source of charges	Cathode	Emitter	Source
Control of charges	Control grid	Base	Gate
Collector of charges	Plate	Collector	Drain

Summary

1. The main types of electronic circuits are rectifiers, amplifiers, oscillators, and digital logic circuits. These circuits can be implemented with tubes or solid-state devices, mainly transistors, diodes, and integrated circuits.
2. The rectifier is a one-way conductor. It changes ac input to dc output.
3. An amplifier increases the amplitude of its signal input.
4. An oscillator generates ac signal output without any ac input.
5. Doping forms a PN junction in semiconductor materials. The PN junction is the basis for diodes, bipolar transistors, SCRs, and other similar devices.
6. The main types of semiconductor devices and their schematic symbols are listed in Table 29-5.
7. Diodes, SCRs, and junction transistors are formed using bipolar technology, whereas IGFETs are made using MOS technology.
8. NPN and PNP transistors are junction transistors or bipolar transistors.
9. Thyristors are solid-state switches, including such devices as SCRs and triacs.
10. Visual displays are divided into character and graphics types. The 7-segment and dot matrix displays are common character displays. The CRT is a common graphics display.
11. Light-emitting diodes and liquid-crystal displays are used as character displays. The LED operates on low-voltage dc, while the LCD requires ac power.
12. Vacuum tubes are still used for a few applications, such as the CRT. Generally, they have been displaced by solid-state devices. See Table 29-4 for a comparison of the electrodes in transistors and tubes.

Self-Examination
Answers at Back of Book

1. Transistors would be classified as _____ (active, passive) electronic devices.
2. An electronic circuit that generates an ac output with no input signal is called an _____ (amplifier, oscillator).
3. Germanium and _____ are the two semiconductors used in producing transistors.
4. A doped semiconductor with a shortage of electrons for the covalent bond structure is called a(n) _____ (N, P)-type material.
5. The majority current carriers in an N-type material are _____ (free electrons, holes).
6. The barrier potential for a silicon PN junction is about _____ (0.3, 0.7) V.
7. If a PN junction is reverse-biased, _____ (a large, no) current will flow through the unit.
8. Hole flow occurs in _____ (N, P)-type semiconductor materials.
9. The _____ (anode, cathode) of a semiconductor diode is made of an N-type material.
10. The arrow on diode and transistor schematic symbols points _____ (away from, toward) the N-type material.
11. The very thin, lightly doped control terminal on a bipolar transistor is the _____ (base, emitter).

Table 29-5. Semiconductor Devices

Symbol	Type	Symbol	Type
B—C, E (PNP transistor)	PNP transistor; needs negative V_C	G, MT$_1$, MT$_2$ (triac)	Triac thyristor; MT is main terminal; uses either polarity of MT voltage
B—C, E (NPN transistor)	NPN transistor; needs positive V_C	(diac)	Diac trigger diode; uses either polarity of voltage
G—D, S (JFET)	Junction field-effect transistor (JFET); N-channel	A—K (diode)	Diode rectifier; arrow shows hole current
G—D, Sub, S (IGFET)	Field-effect transistor; N-channel, depletion IGFET	(tunnel diode)	Tunnel diode or Esaki diode
G—D, Sub, S (IGFET)	Field-effect transistor; P-channel, enhancement IGFET	(zener diode)	Zener diode or voltage-reference diode
G$_1$, G$_2$—D, Sub, S	IGFET with two gates	(varactor)	Capacitive diode or varactor
A—G, K (SCR)	Silicon controlled rectifier (SCR); gate G is shorter lead	$t°$ (temp diode)	Temperature-dependent diode
E—B$_1$, B$_2$ (UJT)	Unijunction transistor (UJT); thyristor with B$_1$ and B$_2$ as gates for N bar	(photosensitive diode)	Photosensitive diode; light in
		(LED)	Photoemissive diode; light out (LED)

12. The emitter-base junction of a bipolar transistor is always _____ (forward, reverse) biased.

13. The FET is a _____ (current, voltage) driven device with high input impedance.

14. The base on a junction transistor serves the same purpose as the _____ (gate, source) on the FET.

15. An FET that is normally on is said to be a(n) _____ (depletion, enhancement)-mode device.

16. The SCR is a diode that _____ (will, will not) conduct from cathode to anode when a positive voltage is applied to the gate, with positive anode voltage.

17. A _____ (tunnel, zener) diode is used for voltage regulation.
18. A diode that radiates light is an _____ (LCD, LED).
19. The _____ (dot matrix, 7-segment) display is an example of an alphanumeric display.
20. The _____ (cathode, plate) of a tube has a function similar to the collector on a transistor.
21. A vacuum tube triode contains a heater, cathode, anode, and _____ inside the glass envelope.
22. While an LCD controls light, the LED _____ (emits, stops) light.
23. The CRT in an oscilloscope uses electrostatic deflection of the _____ (electron, proton) beam.
24. The 19VP4 picture tube has a heater voltage of _____ (6.3 V, 19 V).

Essay Questions

1. Give the functions for rectifier, amplifier, and oscillator circuits.
2. Define the following: (**a**) doping; (**b**) N-type silicon; (**c**) P-type silicon; (**d**) PN junction; (**e**) internal barrier potential; (**f**) depletion zone.
3. Compare electron current with hole current.
4. For a PN junction, show a battery applying forward voltage and reverse voltage.
5. List the two most important specifications for a rectifier.
6. Draw the schematic symbol for PNP and NPN transistors. Label the emitter, base, and collector leads.
7. Explain the biasing on the emitter-base and base-collector junctions of a transistor.
8. Refer to Fig. 29-17. What will happen to the resistance from E to C if the base current decreases?
9. Draw the schematic symbol for an N-channel FET, depletion type and enhancement type.
10. Draw the schematic symbol for an N-channel JFET.
11. What do we mean when we say that an FET has high input impedance or that it is voltage-driven?
12. Refer to Fig. 29-19a. What polarity voltage must be applied to the gate of the FET to decrease the current through the ammeter?
13. Why is it said that an SCR connected as in Fig. 29-23 acts like a latching relay?
14. Draw the schematic symbols for the following: (**a**) zener diode; (**b**) LED; (**c**) rectifier diode; (**d**) varistor; (**e**) varactor diode.
15. How do you identify the cathode lead on a common plastic lens LED?
16. Refer to Fig. 29-30b. What switches must be closed to cause decimal 9 to light on the display?
17. What are the advantages of liquid-crystal displays?
18. What parts are contained in an oscilloscope-type CRT?
19. Draw the schematic symbols for the following: (**a**) Pentode tube; (**b**) CRT.
20. Why have vacuum tubes been displaced by semiconductor devices for amplifiers, rectifiers, and electronic switching?

Problems

Answers to Odd-Numbered Problems at Back of Book

1. Refer to Fig. 29-1. If the input voltage = 0.01 V p-p and the output voltage = 1 V p-p, then the gain of the amplifier is _____.

2. Calculate the resistance of a silicon diode that has 0.5 A flowing through it with a 0.85-V drop across it.

3. Refer to Fig. 29-17. If I_B increases to 0.3 mA and I_E to 15 mA, the I_C will increase to _____ mA.

4. Refer to Fig. 29-21. What is the voltage gain for this FET amplifier circuit?

5. Refer to Fig. 29-23. If the current flow from G to A is zero, then the resistance of the SCR from K to A equals _____.

6. Refer to Fig. 29-30. If each LED segment draws 15 mA, how much current is drawn when the number 8 is displayed?

7. Refer to Fig. 29-38. Calculate the voltage drop across the load resistor R_L if the plate current I_b increases to 6 mA.

8. Refer to Fig. 29-38. With 6 mA for I_b, how much is V_b?

9. For each of the circuits in Fig. 29-40, indicate whether the diode and transistors are cut off or conducting.

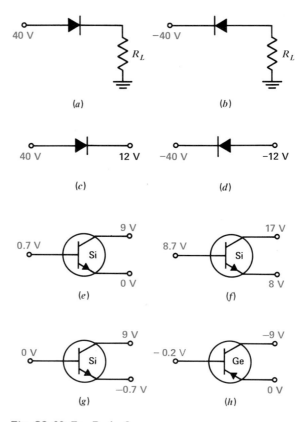

Fig. 29-40 For Prob. 9.

Answers to Practice Problems

29-1 **a.** 40
  b. 1 MHz
  c. Diode

29-2 **a.** 4
  b. N
  c. P
  d. Opposite
  e. Holes

29-3 **a.** Low
  b. 0.7 V
  c. 1 μA
  d. Ions

29-4 **a.** Rectifier
  b. Si (silicon)
  c. I_F (forward current)
    PIV (peak inverse voltage)
  d. Open
  c. 4

29-5 **a.** Forward bias
  b. Reverse bias
  c. 1.005 mA
  d. Cut off

29-6 **a.** T
  b. F
  c. T

29-7 **a.** T
  b. T
  c. F
  d. T

29-8 **a.** T
  b. T
  c. F
  d. T

29-9 **a.** LCD
  b. 7-segment
  c. 4
  d. LCD

29-10 **a.** Anode or ultor
  b. P1
  c. 25 in.

29-11 **a.** Cathode
    Control grid
    Plate
  b. Amplifier
  c. Cathode

Chapter 30
Transistor Amplifiers

The transistor is probably the most important active device for electronic circuits. Because the tiny solid-state semiconductor can perform amplification, rectification, and switching, it has revolutionized the design of all electronic equipment. In addition to their use as separate or discrete components, transistors are the main components in integrated circuits. The transistor was invented in 1948 at the Bell Telephone Laboratories. The Nobel prize in physics was awarded to William Shockley, Walter H. Brattain, and John Bardeen for their work on the transistor.

The transistor is a solid-state device using the element silicon (Si) or germanium (Ge). Instead of depending on thermal emission of electrons, the solid semiconductor is doped to provide free charges, which can be either negative electrons or positive holes. The charges are controlled by the input circuit to control the output current. Since there is no heated filament, the operation of transistors and all solid-state semiconductor devices is instantaneous, without any warmup time.

The name *transistor* is derived from "trans resistor," meaning that it changes resistance. As an active device, the transistor can be used as an electronic switch. It changes between very low R as a short circuit for a closed switch and very high R as an open circuit for an open switch. For amplification, the transistor resistance and current are made to vary with changes of the input signal. As discussed in Chap. 29, there are two main types of transistors, the bipolar junction transistor and the field-effect transistor (FET). This chapter analyzes the circuits for junction transistors, which include NPN and PNP types. The FET circuits are similar.

Important terms in this chapter are:

alpha characteristic	cutoff
beta characteristic	Darlington circuit
bias current	emitter-follower
bias stabilization	leakage current
cascaded stages	power gain
CB circuit	saturation
CC circuit	TO packages
CE circuit	voltage gain
current gain	

More details are explained in the following sections:

30-1 The Transistor as a Switch
30-2 The Transistor as an Amplifier
30-3 Transistor Amplifier Circuits
30-4 Analysis of the Common-Emitter (CE) Amplifier
30-5 Transistor Specifications
30-6 Practical Testing and Troubleshooting
30-7 Troubleshooting Examples

30-1 The Transistor as a Switch

The switching action for $Q1$ in Fig. 30-1 illustrates how the output current can be controlled at the input. Note the following important operating characteristics:

1. The transistor is normally off, without any output current unless forward voltage is applied in the base-emitter circuit.
2. The forward voltage controlling the base current determines the amount of output current.

In Fig. 30-1, the control circuit for the input determines base current. For the power circuit, the output is collector current. An NPN transistor is used for $Q1$. This type requires positive V_{BE} for forward voltage. The emitter is common to both the control circuit at the input and the power output circuit. The common-emitter (CE) circuit is the most popular transistor circuit arrangement.

The base-emitter junction $Q1$ in Fig. 30-1 can be forward-biased by the battery B_1. Switch S_1 must be closed to apply the forward voltage. Reverse voltage for the collector of $Q1$ is supplied by B_2. The reverse polarity means that the N collector is more positive than the base.

With switch S_1 open, no current flows in the base-emitter or control circuit. The reason is that the forward voltage is not applied. Therefore, the resistance from the emitter to the collector of the transistor in Fig. 30-1 is very high. No current flows in the power circuit, and the lamp does not light.

Next, assume that switch S_1 is closed. This causes a *small current* to flow in the control circuit. R_1 is a current-limiting resistor for the base circuit. Therefore, the resistance from the emitter to the collector of the

transistor drops. Consequently, a *large current* flows in the power circuit, causing the lamp to light.

Finally, opening switch S_1 in the control circuit causes the lamp in the power circuit to go out. This is because the resistance from the emitter (E) to the collector (C) of $Q1$ has again increased to near infinity.

In summary, a small current in the control circuit causes a large current to flow in the power circuit. With no current in the control circuit, the transistor acts like an open switch. With some current in the control circuit, the transistor acts like a closed switch.

The circuit using a transistor as a switch has been redrawn in Fig. 30-2. To aid in calculations, a 100-Ω resistor R_2 has been substituted for the lamp in Fig. 30-1. The lamp had a resistance of about 100 Ω. The push-button switch has been omitted and the B_1 power supply is a variable unit in Fig. 30-2.

The schematic in Fig. 30-2a shows the measured voltages and collector current I_C in this circuit. Note that only a tiny leakage current of 1 μA flows from the

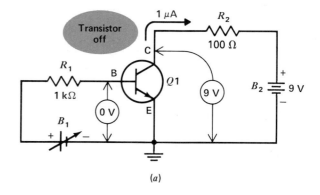

(a)

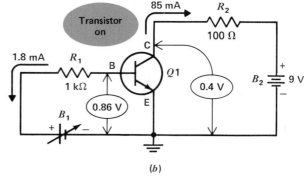

(b)

Fig. 30-2 Analysis of transistor switching circuit. (*a*) Off condition. (*b*) On condition.

Fig. 30-1 NPN transistor operating as a switch.

emitter to the collector. The resistance from E to C is calculated as

$$R = \frac{V}{I} = \frac{9 \text{ V}}{0.000001 \text{ A}} = 9 \text{ M}\Omega$$

The transistor has a resistance of 9 MΩ, which is like the open or off condition of a switch.

The schematic in Fig. 30-2b shows the measured voltages and currents in the "transistor on" circuit. First, the voltage from emitter to base has been increased by adjusting B_1. The value of 0.86 V forward-biases the emitter-base junction of the transistor, causing 1.8 mA to flow in the control circuit. This current in turn causes the resistance of the transistor from E to C to drop. The effect is that a large current of 85 mA flows from the collector of the transistor. The resistance from E to C in Fig. 30-2b is calculated as

$$R = \frac{V}{I} = \frac{0.4 \text{ V}}{0.085 \text{ A}} = 4.7 \text{ } \Omega$$

The resistance of the transistor from E to C has dropped from its previous high of 9 MΩ to a low of 4.7 Ω. As a result, the transistor is acting like a closed switch in Fig. 30-2b.

The transistor in Fig. 30-2a is said to be at *cutoff*. It has reached its maximum resistance from E to C and has cut off the current. The very tiny current still flowing is due to minority current carriers in the transistor, which is leakage current.

The transistor in Fig. 30-2b is said to be at *saturation*. It has reached its minimum resistance from E to C, which produces the maximum collector current. When used as a switch, the transistor is driven to cutoff or to saturation by the emitter-base voltage.

Practice Problems 30-1
Answers at End of Chapter

Answer True or False.
a. In Fig. 30-1, with S_1 open, the lamp would light.
b. In Fig. 30-2a, the collector voltage is high at 9 V because Q1 is at cutoff.
c. In Fig. 30-2b, the collector voltage is low at 0.4 V because Q1 is at saturation.

30-2 The Transistor as an Amplifier

Consider the circuit shown in Fig. 30-3. The dc operating voltages are shown in Fig. 30-3a, without any input signal. The dc biasing levels are set by the feedback resistor R_2. The NPN transistor is biased so that the collector-to-emitter voltage V_C is half of the supply voltage. For the supply voltage of 10 V, therefore, the collector voltage is set at one-half the total, or +5 V. The 0.7 V at the base is *partially turning on* the transistor. The transistor acts as an amplifier when in this partially turned on condition. It is the amount of dc forward bias that determines the operating level of the transistor.

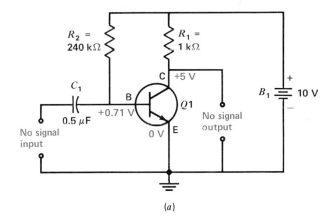

(a)

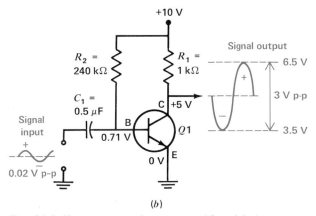

(b)

Fig. 30-3 Transistor used as an amplifier. (a) Common-emitter amplifier circuit with bias level halfway between cutoff and saturation. (b) Same CE amplifier circuit with input and output signals.

An input signal has been added to the amplifier in Fig. 30-3b. Input is coupled to the base by C_1. Amplified output is taken from the collector. The input signal is 0.02 V p-p as measured on an oscilloscope. The measured output signal voltage is 3 V p-p. The ac gain of the amplifier, therefore, is calculated as

$$A_V = \frac{V_{out}}{V_{in}} = \frac{3 \text{ V p-p}}{0.02 \text{ V p-p}} = 150$$

The output signal (3 V p-p) is 150 times greater than the input signal of 0.02 V p-p. This amplifier stage is said to have a voltage gain of 150.

Consider the positive (+) alternation of the input signal entering the base of the transistor in Fig. 30-3b. The forward bias on the emitter-base junction of $Q1$ becomes greater, causing more base current to flow. This causes the resistance from E to C of the transistor to decrease. Consequently, V_C will drop to the 3.5-V level, as shown by the waveform of output signal at the right in Fig. 30-3b. The output voltage went in a negative direction while the input was going in a positive direction.

Next consider the negative (−) alternation of the input signal entering the base of the transistor in Fig. 30-3b. The forward bias on the emitter-base junction of $Q1$ is decreased, causing less base current to flow. This causes the resistance from E to C of the transistor to increase. Consequently, V_C will increase to the 6.5-V level, as shown on the right in Fig. 30-3b. The output voltage went in a positive direction while the input signal went in a negative direction.

In summary, an amplifier is biased between cutoff and saturation. The input and output signals on this CE amplifier are 180° out of phase. That is, when the input signal goes positive, the output will go negative. The voltage gain is 150 for this amplifier circuit.

Practice Problems 30-2
Answers at End of Chapter

Answer True or False, for Fig. 30-3.
a. Feedback resistor R_2 is used to provide the proper bias for the base of the transistor.
b. This is called a common-base circuit because the signal enters the base of the transistor.
c. The average dc voltage across R_1 is 5 V.

30-3   Transistor Amplifier Circuits

Since a PNP or NPN transistor has only three electrodes, one must be a common connection for two pairs of terminals for input and output signals. The general case of three electrodes, with one common, is illustrated in Fig. 30-4a. Specifically, the three possibilities for amplifier circuits are: common base (CB) in Fig. 30-4b, common emitter (CE) in Fig. 30-4c, and common collector (CC) in Fig. 30-4d. The circuits are shown for NPN transistors, but they are the same with all polarities reversed for PNP transistors. Although the common electrode is shown grounded here, it need not be connected to chassis ground. The main features of these circuits are compared in Table 30-1.

For all these amplifier circuits, the collector has reverse voltage and the emitter-base junction requires forward bias. It is labeled either V_{BE} or V_{EB}. This average forward voltage is called *bias voltage,* and the corresponding current is *bias current.* The bias sets the average dc values in the amplifier. Then ac signal varies the values around the bias level.

The common-emitter circuit is the one generally used for transistor amplifiers because it has the most gain with cascaded stages. However, each type of circuit has special features.

Common-Base (CB) Circuit   In Fig. 30-4b, the input voltage is applied to the emitter, with respect to the grounded base. The amplified output is taken from the collector. Resistance R_L is in series with the collector supply V_{CC}. The positive side of V_{CC} supplies reverse voltage for the N collector.

In the CB circuit the emitter input has low resistance r_i because I_E is high. The output resistance r_o for the collector is high. Typical values for a small-signal transistor with I_E of 1.5 mA are 20 Ω for r_i and 1 MΩ for r_o. These values are for the internal resistance of each electrode to the common base.

The CB circuit is seldom used. It has no current gain from input to output because I_C must be less than I_E. The voltage gain can be high, but the output is shunted by the low input resistance of the next stage. The only advantage of the CB circuit is that it has the best stability with an increase in temperature. The reason is that reverse leakage current from collector to base is not amplified in the CB circuit.

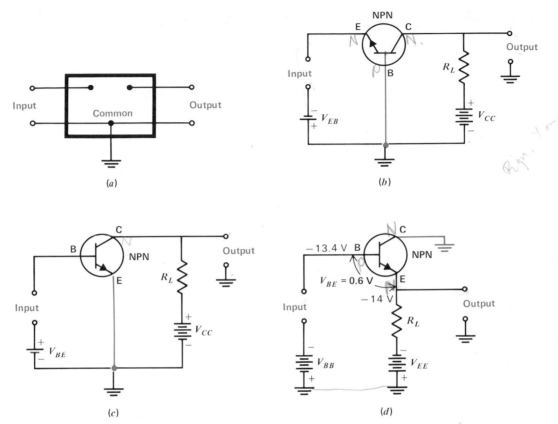

Fig. 30-4 Types of amplifier circuits, shown with NPN transistors. All polarities are reversed for PNP transistors. (*a*) General case of a common terminal for two pairs of connections. (*b*) Common base (CB). (*c*) Common emitter (CE). (*d*) Common collector (CC) or emitter-follower.

Common-Emitter (CE) Circuit

See Fig. 30-4*c*. Input voltage is applied to the base instead of to the emitter, which is now the grounded electrode. Note that the emitter is shown at the bottom of the schematic symbol. The input circuit here involves I_B instead of I_E. As a result, the r_i for the CE circuit is much higher than for the CB circuit. A typical value is 1000 Ω for r_i.

The output voltage is still taken from the collector with its R_L. The V_{CC} is positive for reverse bias on the N collector. A typical value for r_o in the collector output circuit is 50 kΩ.

In the input circuit, the forward bias V_{BE} is applied to the base instead of the emitter. Note the polarity. Positive V_{BE} to the P base corresponds to negative V_{EB} at the negative emitter. Both are forward-bias voltages, with the polarity the same as the N or P electrode.

Furthermore, the positive V_{BE} at the base for forward bias uses the same voltage polarity as positive V_{CC} for reverse voltage at the N collector. This feature allows the practical convenience of using one voltage supply for both forward bias in the input and reverse bias in the output. As an example, the CE circuit in Fig. 30-5 uses R_1 to drop the collector supply of 12 V to 0.6 V for the base.

The CE circuit has current gain because I_C is much larger than I_B. The voltage gain is the same as for the CB circuit. With a higher r_i, however, the CE circuit can be used in cascaded amplifiers where the collector output of one stage drives the base input of the next. The CE circuit is the amplifier generally used for transistors because it has the best combination of voltage gain and current gain. The disadvantage is that reverse leakage current is amplified in the CE circuit, but bias stabilization methods can be used.

Table 30-1. Comparison of Transistor Circuits

Characteristic	Common base (CB)	Common emitter (CE)	Common collector (CC)
Signal into	Emitter	Base	Base
Signal out of	Collector	Collector	Emitter
Advantage	Stability	High gain	High r_i
Phase inversion	No	Yes	No
Similar vacuum-tube circuit	Grounded grid	Grounded cathode	Grounded plate (cathode-follower)
Input resistance* r_i	20 Ω	1000 Ω	150 kΩ
Output resistance* r_o	1 MΩ	50 kΩ	80 Ω

*Typical values of r for small-signal transistor with I_E of 1.5 mA.

Only the CE amplifier inverts the polarity of signal voltage. This phase inversion of 180° is neither good nor bad, but just a result of the circuit connections. When the base input signal increases the forward voltage, the collector voltage of the same polarity decreases because of the voltage drop across R_L.

Common-Collector (CC) Circuit
See Fig. 30-4d. Signal input is applied to the base, as in the CE circuit. However, the collector is grounded, instead of the emitter. Therefore, the emitter has R_L for output signal.

Now there are two questions for the electrode voltages: how to apply reverse voltage for the grounded collector and forward bias for the base. Note that the emitter is at -14 V with respect to chassis ground. The collector is at chassis ground. In effect, the collector is connected to the positive side of the emitter supply voltage. This makes the N collector positive with respect to the emitter, as reverse voltage for V_{CE}.

For forward voltage, the base must be more positive than the emitter. Voltage V_E in this example is -14 V. Therefore V_B is made -13.4 V. The actual bias V_{BE} then is 0.6 V.

In the CC circuit, the input circuit has high r_i. A typical value is 150 kΩ. The output in the emitter circuit has low r_o of about 80 Ω. Note that for the CC circuit, the input resistance is high and output resistance is low, compared with low r_i and high r_o for the other circuits.

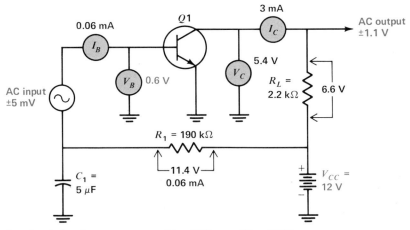

Fig. 30-5 Example of a circuit for the common-emitter (CE) amplifier. NPN takes $+V_{CC}$ for reverse voltage and $+V_B$ for forward bias. The V_B is 0.6 V for silicon transistor. Meters shown for V and I indicate average dc values.

There is no voltage gain in the CC circuit because the output signal voltage across R_L in the emitter circuit provides negative feedback to the base input circuit. However, there is appreciable current gain.

Emitter-Follower This is the name generally used for the CC circuit, because it corresponds to the cathode-follower circuit with vacuum tubes. The output signal follows the polarity of the input signal. These circuits are often used for impedance matching, from a high-impedance source to a low-impedance load.

Darlington Pair This circuit consists of two emitter-followers in cascade. The two stages are usually packaged in one unit, with dc coupling internally and just three external leads. The package provides higher input resistance and more current gain than just one stage.

Practice Problems 30-3
Answers at End of Chapter

a. Which circuit has input signal to the base and output from the collector?
b. Which circuit has the most gain?
c. Which circuit has the highest input resistance?

30-4 Analysis of the Common-Emitter (CE) Amplifier

Typical values of V_{BE} are in tenths of a volt for a junction transistor. The required bias at the base for a class A amplifier is 0.6 to 0.7 V for Si or 0.2 to 0.3 V for Ge. Class A operation means that the amplifier conducts current for 360° of the signal cycle, for minimum distortion. Furthermore, the maximum ac input signal without overload distortion is ±0.1 V. These values are summarized in Table 30-2. Note that 0.1 V is 100 mV.

Without any forward bias a junction transistor is cut off by its own internal barrier potential. The cut-in voltage in the first column of Table 30-2 is the lowest V_{BE} that allows appreciable I_C.

The saturation voltage in the second column of Table 30-2 is the highest V_{BE} that allows it to produce proportional changes in I_C. At saturation, the maximum I_C does not increase with an increase of forward voltage.

The input voltages listed in Table 30-2 apply to all junction transistors, regardless of size and power rating. However, the difference is in the amount of base current I_B and collector current I_C. As an example, a small-signal transistor may have I_B of 60 μA with V_{BE} of 0.6 V at the input, for I_C output of 3 mA. In a medium-power transistor I_B could be 20 mA for the same V_{BE} of 0.6 V, to produce I_C of 1 A.

Circuit Components The transistor amplifier itself is usually labeled Q, as for $Q1$ in Fig. 30-5. The 2.2-kΩ R_L is the collector load. It is in series with the positive V_{CC} of 12 V for reverse collector voltage on the NPN transistor.

One supply voltage is used for both collector and base in the CE circuit, as positive base bias and positive collector voltage are needed. However, the collector voltage is too high for base bias. Therefore, the 190-kΩ R_1 is connected in series as a voltage-dropping resistor for the base.

The required forward-bias voltage for Si is 0.6 V. The corresponding bias current for this transistor is taken as 60 μA or 0.06 mA, as an example. Therefore, the voltage drop across R_1 is 0.06 mA × 190 kΩ = 11.4 V. Then $12 - 11.4 = 0.6$ V remains for V_{BE} as forward voltage for the base.

The bypass capacitor C_1 allows the ac input signal voltage to vary the base current without the series re-

Table 30-2. Input Voltages V_{BE} at 25°C for Junction Transistors

	Cut-in voltage	Saturation voltage	Active region	Average bias voltage
Ge	0.1	0.4	0.1–0.4	0.2–0.3
Si	0.5	0.8	0.5–0.8	0.6–0.7

sistance of R_1. Then very small changes of input voltage can produce appreciable changes in base current. We are assuming an ac input of ±5 mV for the base input signal, or 10 mV p-p.

Signal Variations The input signal of base current i_B is shown in Fig. 30-6a. This waveform shows i_B varies by 10 μA above the 60-μA bias axis, up to the peak of 70 μA. On the down side, i_B decreases by 10 μA, from 60 to 50 μA. Positive signal voltage in the forward direction increases i_B, while negative signal voltage reduces i_B. The peak-to-peak signal in i_B then is 70 − 50 = 20 μA, or 0.02 mA.

The variations in i_B cause corresponding variations in i_C, as shown in the waveform in Fig. 30-6b. Let the current transfer ratio be 50, meaning that this is the ratio of collector output current to base input current. Then the i_B variations of ±10 μA swing i_C by 50 × 10 = 500 μA or 0.5 mA.

The average dc level for I_C is taken as 3 mA for a small-signal low-power transistor. Then the ac signal swing of i_C is ±0.5 mA above and below the axis of

3 mA. In i_C, the peak-to-peak signal for the output current is 3.5 − 2.5 = 1 mA.

The signal changes in i_C produce variations in the voltage drop $i_C R_L$ across the collector load for the output circuit. As a result, V_C varies because it is the difference between V_{CC} of the supply and the voltage drop across R_L. As a formula,

$$V_C = V_{CC} - i_C R_L \qquad (30\text{-}1)$$

For example, with an average level of 3-mA I_C through the 2.2-kΩ R_L, this voltage drop is 0.003 × 2200 = 6.6 V. Subtracting 6.6 V from 12 V, the difference is 5.4 V for the average V_C.

The variations of v_C are in the waveform in Fig. 30-6c. This shows the amplified signal output voltage. The average dc level or axis is 5.4 V. When i_B increases, the v_C decreases to 4.3 V because of a larger voltage drop across R_L. On the next half-cycle i_C decreases. Then less voltage across R_L allows v_C to rise to 6.5 V. Then peak-to-peak signal voltage is 6.5 − 4.3 = 2.2 V for v_C at the collector. This amplified output voltage is 180° out of phase with the signal input voltage at the base.

The basis for Formula (30-1) is just the fact that R_L and the collector-emitter circuit of the transistor are in series with each other as a voltage divider from the high side of V_{CC} to chassis ground. The equivalent collector circuit is shown in Fig. 30-7 with R_L as the external load resistor. The R_Q is the internal resistance of the transistor conducting current from emitter to collector. In this example, $Q1$ is conducting 3 mA, resulting in 5.4 V for V_{CE} in the divider with V_L. Note that V_Q is the same as V_{CE}. These values are 12 − 6.6 = 5.4 V for V_Q. Voltages V_C and V_{CE} are the same here because the emitter is grounded.

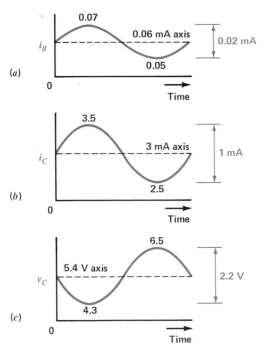

(a)

(b)

(c)

Fig. 30-6 Ladder diagram of signal waveforms for circuit in Fig. 30-5. (a) Base current i_B. (b) Collector current i_C. (c) Output voltage $V_C = V_{CC} - i_C R_L$.

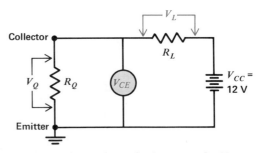

Fig. 30-7 Equivalent voltage divider circuit for V_C in series with V_L.

Typical Base Bias Voltage In a class A amplifier with collector output current for the full signal cycle, the general requirements are to have the value of I_C that allows the voltage drop $I_C R_L$ to equal about one-half the dc supply voltage. Then the average dc collector voltage V_C is also approximately $V_{CC}/2$. *The base bias V_{BE} is made the value, around 0.6 to 0.7 V, that produces the required I_C with the specified R_L.*

Gain To calculate the voltage amplification, or gain, with peak-to-peak values for Fig. 30-5,

$$A_V = \frac{V_{out}}{V_{in}} \tag{30-2}$$

$$A_V = \frac{2.2 \text{ V}}{0.01 \text{ V}} = 220$$

The current amplification is

$$A_I = \frac{i_{out}}{i_{in}} \tag{30-3}$$

$$A_I = \frac{i_C}{i_B} = \frac{1 \text{ mA p-p}}{0.02 \text{ mA p-p}} = 50$$

The power amplification is

$$A_P = A_V \times A_I \tag{30-4}$$
$$A_P = 220 \times 50 = 11,000$$

Since the CE circuit has both voltage and current amplification, the power gain is high. Power gain is desirable because it means the voltage output can drive a low-impedance circuit without too much loss of voltage amplification.

Cascaded Stages As shown in Fig. 30-8, the amplifiers $Q1$, $Q2$, $Q3$, and $Q4$ are in cascade. With CE amplifiers, the collector output of one stage drives the base input of the next. It is important to realize that the voltage amplification need not build up the signal to a level greater than ± 0.1 V. This is the maximum signal swing without distortion for junction transistors. However, the cascaded stages provide enough signal current to drive the base of the output stage $Q4$.

As an example, suppose that $Q4$ is a power output stage to drive a loudspeaker that needs 5 A as the load. With an average level of 5 A for I_C in $Q4$, its I_B would be of the order of 250 mA or 5 A/20, with a current transfer ratio of 20. This I_B of 250 mA can be supplied by the collector output of the driver stage $Q3$. Similarly, $Q3$ with an I_C of 250 mA would have I_B of 12.5 mA with a current ratio of 20. Also, with a current ratio of 50 for $Q2$, its I_B would be 12.5 mA/50 = 0.25 mA, or 250 μA. This drive for base current in $Q3$ can be provided by the collector current of the input stage $Q1$. In summary, each of the cascaded amplifiers increases the signal current enough to drive the next stage.

Practice Problems 30-4
Answers at End of Chapter

a. In Fig. 30-5, how much is the base-bias V_{BE}?
b. In Fig. 30-7, how much is V_Q with V_L of 5.5 V?
c. In Fig. 30-6, how much is the peak-to-peak signal output in i_C?

30-5 Transistor Specifications

For type numbers registered with the Electronic Industries Association (EIA), the letter N indicates a semiconductor with a numerical prefix for the number of junctions. Diodes are numbered 1N, junction transistors are 2N, and an FET with one gate is 3N. The digits that follow indicate specific types. An example is the 2N3904, which is a silicon NPN transistor. Most general-purpose transistors are NPN silicon.

Fig. 30-8 Amplifier stages in cascade. Note increase in drive current for each successive transistor stage.

Japanese transistors have different type numbers. The following designations are generally used:

2SA   PNP, high frequency

2SB   PNP, low frequency (power)

2SC   NPN, high frequency

2SD   NPN, low frequency (power)

2SE   P-gate thyristor

2SG   N-gate thyristor

2SH   N-gate UJT

2SJ   P-channel FET

2SK   N-channel FET

3SK   Dual-gate FET

2SM   Triac

Note that S is used to indicate a semiconductor. The first digit is the number of junctions or the number of external connections minus one.

Electrical Characteristics

Manufacturers publish data sheets with specifications for their transistors. The information is available in their application notes and in transistor manuals.

Characteristics for some general-purpose transistor amplifiers are listed in Fig. 30-9. Consider the popular 2N3904 in the middle row. This transistor is the type used for the circuits shown in Figs. 30-2 and 30-3. The electrical specifications are defined as follows:

Collector-emitter voltage V_{CEO}. The maximum voltage which may be applied to the collector-emitter terminals with the base terminal open.

Collector-base voltage V_{CBO}. The maximum voltage which may be applied to the collector-base terminals with the emitter terminal open.

Current gain h_{FE}. The ratio of collector current I_C to base current I_B at a specified collector-emitter voltage.

Collector-emitter saturation voltage $V_{CE(sat)}$. The dc voltage measured between the collector and emitter terminals at specified current conditions.

Base-emitter saturation voltage $V_{BE(sat)}$. The base-emitter voltage as measured with conditions as specified for $V_{CE(sat)}$.

Current-gain-bandwidth product f_T. The frequency at which the small-signal forward-current transfer ratio becomes 1, meaning no gain. The f_T specifies the upper frequency limit for the transistor as an amplifier.

Other electrical characteristics commonly listed in summaries or data sheets are:

Material—Ge or Si

Application—general-purpose, RF, switching, power amplifier, or high-voltage

Power dissipation—maximum power in watts or milliwatts

Maximum collector current I_C

Leakage current I_{CBO} from base to collector at maximum V_{CB}

Electrical characteristics are typically quoted for an ambient temperature of 25°C, which corresponds to normal room temperature of 77°F.

Beta (β) Characteristic

This specifies how much I_C changes for a change in I_B. Typical values of β

NPN General-purpose Amplifiers

Device	Package	V_{CEO} V, min	V_{CBO} V, min	h_{FE}	@	I_C/V_{CE} mA/V	$V_{CE(sat)}$ V*, max	$V_{BE(sat)}$ V*, max	f_T MHz, min
2N3860	TO-92	40	40	150–300		2.0/4.5	0.125		90
2N3903	TO-92	40	60	50–150		10/1.0	0.2	0.85	250
2N3904	TO-92	40	60	100–300		10/1.0	0.2	0.85	300
2N3923	TO-39	150	150	30–120		25/10	1.0	0.9	40
2N3946	TO-18	40	60	59–150		10/1.0	0.2	0.9	250

*Specified for $(I_C/I_B) = 10$

Fig. 30-9 Partial page from selection guide in manufacturer's transistor data book. The 2N3904 in middle row is typical.

are 10 to 300. The lower values are for power transistors.

Beta is the forward-current transfer ratio for the CE circuit, the same as h_{FE}. The dynamic or ac values of both β and h_{FE} specify the current gain for a CE amplifier. The h indicates *hybrid* parameters, which are combinations of voltage and current ratios.

Alpha (α) Characteristic   This specifies how much I_C changes for a change in I_E, in the common-base circuit. Alpha is less than 1, as I_C must be less than I_E. Typical values of α are 0.98 or 0.99.

Heat Sink   Although the transistor does not have a heater for thermionic emission, the collector becomes hot because of I_C, especially in power transistors where the collector current is in amperes. Therefore, it is necessary to use a heat sink, which is a metal structure to radiate heat out of the collector.

Physical Characteristics   According to the chart in Fig. 30-9, the 2N3904 transistor is packaged in the TO-92 case. The TO stands for "transistor outline." These packages are standardized by the *Joint Electron Device Engineering Council*. The JEDEC TO-92 case style is shown in detail in Fig. 30-10*a*, with a basing diagram in Fig. 30-10*b*. Both the transistor package dimensions and the basing diagram are given on specific data sheets and in transistor data manuals.

Case Styles   Several common forms are illustrated in Fig. 30-11. The JEDEC TO-92 plastic case is suited for small-signal transistors. The TO-5 case has a metal can on the outside, for medium-power transistors. A typical basing diagram for a TO-5 transistor is shown in Fig. 30-11*b*.

The JEDEC TO-220 case, shown in Fig. 30-11*c*, is used for power transistors. Its metal tab is meant to be attached to a heat sink. The larger TO-3 case, shown in Fig. 30-11*d*, is used for higher-power applications. The TO-3 case is mounted on a heat sink. Since the case is the collector terminal of the transistor, the unit is commonly *isolated* from the heat sink by a mica washer and appropriate mounting hardware. Both sides of the mica washer are coated with silicone grease when mounting

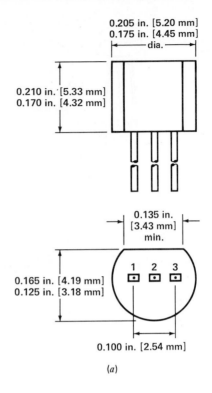

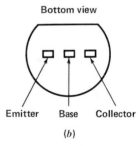

Fig. 30-10 JEDEC TO-92 case style. (*a*) Package dimensions and details. (*b*) Typical basing diagram. (*Fairchild Camera and Instrument Corp.*)

the TO-3 case to help conduct heat away from the transistor and into the heat sink. A basing diagram for a TO-3 transistor is also shown in Fig. 30-11*d*.

Power Ratings   Small-signal transistors are generally rated at 200 to 800 mW with an I_C of 1 to 50 mA. A typical medium-power transistor in the TO-5 case may be rated at 5 W with I_C of 2.5 A. For the TO-220 style with a metal tab, a typical power rating is 10 W with I_C of 4 A. With the larger TO-3 case, power rat-

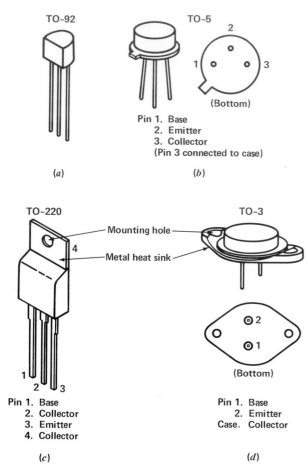

Fig. 30-11 Typical transistor case styles. (a) TO-92 plastic small-signal package. (b) TO-5 metal can small-signal package and typical basing diagram. (c) TO-220 metal tab power package and typical basing diagram. (d) TO-3 metal high-power package and typical basing diagram.

ings are 25 W to more than 100 W, with I_C generally more than 5 A.

Practice Problems 30-5
Answers at End of Chapter

Refer to Fig. 30-9.
a. The 2N3946 transistor comes packaged in what standard case?
b. It takes _____ V applied to the emitter-base junction of the 2N3904 transistor for it to reach saturation.
c. The 2N3903 transistor has a current gain of about _____ with a collector current of 10 mA.

30-6 Practical Testing and Troubleshooting

With all the advances in modern electronic equipment, troubles are still caused most of the time by a short circuit, open circuit, or low dc supply voltage. The open circuit can be in the transistor itself, resistors, capacitors, coils, or the wiring, especially at plugs, sockets, and connectors. A short circuit can occur in transistors and capacitors. It is a good idea to check the dc supply voltage first, especially with batteries.

Transistor Troubles Failures generally result from an open weld at the wire leads to the semiconductor, a short circuit at a junction caused by momentary overloads, and circuit failures that cause transistor overheating. In most cases, a defective transistor is internally short-circuited or open, and simple tests will reveal the trouble.

The cause of some problems, like an increase in leakage, a drop in breakdown voltage, or excessive noise, is more difficult to detect, and direct substitution may be the easiest way to localize the fault. Transistor testers are available to check the transistor in or out of the circuit for an open circuit, short circuit, leakage, and β. However, open and short circuits in a transistor can also be checked by using a multimeter to test the junction as a diode. If a DMM has a diode test position, it can be used for a dynamic test of each junction.

In-Circuit Tests These tests are helpful because transistors are usually soldered into the circuit board. Dc voltage measurements can determine if the junctions are intact and if the transistor is conducting normal current.

Refer to Figs. 30-12 and 30-13 for examples of a CE amplifier. This circuit is typical because of the method of obtaining base-emitter bias V_{BE}. In the emitter circuit, self-bias of 1.9 V is obtained as the voltage drop of I_E through the 470-Ω R_E. This voltage is *self-bias* because V_E depends on the transistor's own emitter current. However, V_E biases $Q1$ in the reverse direction, as this voltage is positive at the N emitter. The purpose of R_E is to provide *bias stabilization,* by preventing I_C from increasing with more leakage current. In order to provide forward bias on $Q1$, the voltage divider with R_1 and R_2 is used. The divider provides 2.6 V for V_B. It should be noted that I_B returns to the emitter through R_1. Both V_B and V_E are dc bias voltages, positive with

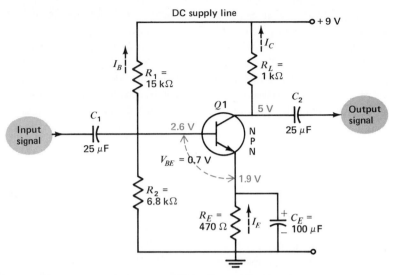

Fig. 30-12 Typical audio amplifier circuit with emitter self-bias for stabilization (R_E and C_E) and R_1R_2 voltage divider for base bias.

respect to chassis ground. The net voltage for V_{BE}, then, is $2.6 - 1.9 = 0.7$ V, which provides forward bias for the transistor amplifier.

In the collector circuit of $Q1$, the 1-kΩ R_L is the collector load resistor connected to the V^+ supply of 9 V. R_L has a dc voltage drop of 4 V, which allows 5 V for the collector voltage to chassis ground. The actual V_{CE}, though, is $5 - 1.9 = 3.1$ V. The collector has a positive dc supply for reverse voltage on the NPN transistor. As a CE amplifier, $Q1$ has input signal from the coupling capacitor C_1, with amplified output signal coupled by C_2.

Check Forward Bias Measure the base-to-emitter voltage by putting the dc voltmeter leads directly across these terminals. In Fig. 30-12, as an example, V_{BE} should read 0.7 V.

If the reading for V_{BE} is zero, the base-emitter junction is short-circuited. For the opposite case, if V_{BE} is 0.8 V or higher, the junction is probably open. One word of caution, though. These voltage readings apply to class A amplifiers, which require a middle value of forward bias. However, in pulse circuits it may be normal to have reverse bias of several volts for V_{BE} in order to cut off I_C until the input pulse drives the transistor into conduction.

Checking $I_C R_L$ Voltage To check for correct collector current, put the voltmeter between the collector

and the supply voltage to read the voltage drop across the load. In Fig. 30-12, a VOM across the 1-kΩ R_L should read 4 V. Then divide this reading by the dc resistance in the collector circuit to calculate the current I_C. For this example, the normal I_C value is 4 V/1 kΩ = 4 mA.

If there is no voltage drop across R_L, then I_C must be zero. Then V_C at the collector will have the same value as the supply voltage V_{CC}. For the opposite trouble, excessive I_C can cause excessive voltage across R_L, resulting in zero or very low V_C.

If I_C is zero or very low, the transistor may be open. However, check for opens in the emitter circuit before making a replacement.

If excessive I_C is flowing, short-circuit the base-to-emitter voltage and repeat the test. Under these conditions only the small leakage current should flow. If I_C is still high, the transistor collector is probably short-circuited.

Measuring I_C Calculation of collector current may be difficult if the collector circuit contains little resistance or an unknown value, such as a transformer primary. In this case current measurements may be better. You can open the collector circuit easily by cutting the foil of the printed-circuit board with a razor blade. Then put the leads of your VOM, set to read milliamperes or amperes, across the cut. You can bridge the cut with solder when the test is finished.

Checking the Emitter Circuit   You can also check current by measuring the voltage drop across the emitter resistance and dividing by the value of the resistor. However, be careful in your analysis. This voltage will read almost normal even if the emitter resistor is open. The reason is that the voltmeter resistance then completes the emitter circuit. So check the value of the emitter resistor first.

Resistance Tests   Resistance measurements in transistor circuits are often misleading, as the internal battery of the ohmmeter may forward-bias a junction and provide a parallel path for current. Know the voltage polarity at your ohmmeter leads and make sure all junctions are reverse-biased for ohmmeter readings. Or, best of all, use an ohmmeter with low-power ohms, which does not have enough voltage for forward bias on transistors.

Out-of-Circuit Tests   Ohmmeter checks on a transistor are useful for finding opens or short circuits. Avoid the $R \times 1$ scale when checking low-power transistors. On this scale the current in the leads is usually more than 100 mA when the leads are short-circuited, even though the current through the meter movement may be less. Check the resistance between base and collector and reverse the leads. Do the same for base and emitter. You should get a very high reading when the junction is reverse-biased and a very low reading when the junction is forward-biased. With forward bias, the P side of the junction takes the positive voltage. These tests really check each transistor junction as a diode. Do not use the low-power ohms position, as it does not supply enough voltage to turn on the junction.

The actual reading in ohms depends upon your ohmmeter, the range selected, and the type of transistor. For conclusive tests, make a comparison with another transistor of the same type that is known to be good. It is helpful to realize that the reverse resistance of a silicon junction is usually infinite.

The red lead on a VOM is usually the positive side of the ohmmeter battery, but check the polarity with another voltmeter.

The diode function on a digital multimeter (DMM) can also be used to check for normal forward voltage at each junction of a transistor.

Replacing Transistors   Small-signal and medium-power transistors are usually soldered on the printed-circuit board. However, the transistor can be removed easily, without damage to the PC board, by using desoldering techniques to suck out the old solder. The trick is to remove all the solder from the terminals so that the leads can be straightened to lift out the transistor. Desoldering methods for PC boards are described in App. H.

What Type of Transistor?   With junction transistors that are not marked, some practical points can help identify the type. In a circuit where the dc supply voltage is positive for the collector or negative for the emitter, the transistor is NPN. This polarity is needed as reverse voltage for V_{CE}. When the forward voltage for V_{BE} is 0.6 to 0.7 V, the transistor is silicon. Germanium transistors take 0.2 to 0.3 V.

Consider a transistor by itself, out of the circuit. Some ohmmeter tests can determine which leads are for the emitter, base, and collector. On the $R \times 100$ or $R \times 10,000$ scale for small transistors, measure the resistance for the three combinations between two pins. Then reverse the ohmmeter leads and repeat the measurements. The lowest reading is the resistance between base and emitter in the forward direction. With an NPN type, the base has the positive side of V_{BE}. The next-to-lowest reading is between collector and emitter, also in the forward direction. Incidentally, the reverse readings for silicon are practically infinite ohms. As a definite final check, you can monitor I_C with a 10-kΩ R_L in series with a reverse V_{CC} and see how a change in V_{BE} changes the collector current.

Practice Problems 30-6
Answers at End of Chapter

a. Refer to Fig. 30-12. If V_{BE} is very low, perhaps from 0 to 0.1 V, the transistor's emitter-base junction may be _____ (open, short-circuited).

b. Refer to Fig. 30-12. If the collector lead became open inside the transistor case, V_C would measure about _____ V and I_C would be about _____ mA.

c. Refer to Fig. 30-12. If V_C is much too low, then I_C would be _____ (higher, lower) than usual.

d. Out-of-circuit tests for open and short circuits can be performed on a transistor using an _____ (ohmmeter, oscilloscope).

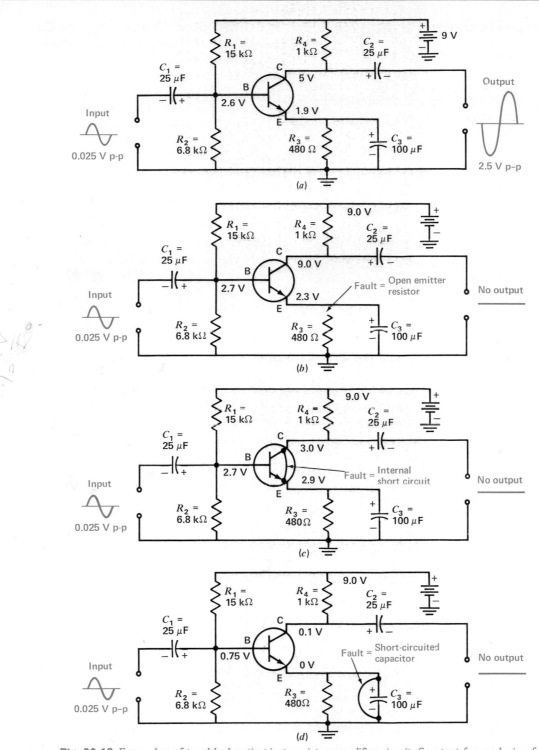

Fig. 30-13 Examples of troubleshooting in transistor amplifier circuit. See text for analysis of abnormal dc voltages. (*a*) Normal CE amplifier with typical bias and signal levels. (*b*) No signal output because of open emitter resistor. (*c*) No signal output because of internal short circuit in transistor. (*d*) No signal output because of short circuit in emitter bypass capacitor.

30-7   Troubleshooting Examples

Consider the normal CE amplifier circuit in Fig. 30-13*a*, before we analyze the troubles in Fig. 30-13*b*, *c*, and *d*. The circuit is the same as in Fig. 30-12, but with signal input of 0.025 V p-p and amplified output of 2.5 V p-p. The voltage gain is 100.

It should be noted that ac signal voltages are usually measured with an oscilloscope. The dc voltages are measured with a multimeter, either a VOM or DMM.

Open Emitter Resistor

As a troubleshooting exercise, consider the *faulty* amplifier circuit in Fig. 30-13*b*. The measured input signal is 0.025 V p-p with no output signal. The voltage at the collector has increased from its normal 5.1 V to 9 V. The base voltage has increased slightly, to 2.7 V. The voltage is 9 V at the top and bottom of load resistor R_4. This means that no current is flowing through R_4 or that there is no collector current I_C. Next, the technician makes some guesses as to what may cause this problem. The fault might be an open circuit in the transistor itself or in R_3. Another possible cause might be a short-circuited emitter-base junction; however, if this were the case, the emitter and base voltages would be the same. In this case, careful observation of the circuit or perhaps an ohmmeter check reveals an *open circuit* in the emitter resistor lead R_3. This fault is identified on the schematic in Fig. 30-13*b*. With the emitter return circuit open, there cannot be any I_C.

Short-Circuited Transistor

A second troubleshooting example is illustrated in Fig. 30-13*c*. The input signal is 0.025 V p-p with no output signal. The collector voltage has dropped from 5 V to 3 V, which means that I_C must have increased. This could be caused by an increase in forward bias on the emitter-base junction; however, this is not the case, since this junction is reverse-biased according to the voltage readings. Note that V_B is less positive than V_E. The voltage readings at the collector and emitter are almost the same; therefore, it is assumed that the fault is in the transistor itself. In this case, there is an internal short circuit between the emitter and collector of the transistor. Suspect transistors are often checked in the circuit with a transistor checker or by substitution of a new transistor.

Short-Circuited Emitter Bypass Capacitor

A third troubleshooting example is shown in Fig. 30-13*d*. The input signal is 0.025 V p-p with no output signal. The collector voltage is 0.1 V, indicating a very high I_C. The emitter is at 0 V, which would indicate that it is grounded or that no current is flowing through R_3. However, I_C is high. Therefore, the emitter current must be high. The emitter must be grounded, which means C_3 is short-circuited.

In summary, typical troubleshooting tests on a CE amplifier can include the following steps:

1. A signal is injected into the base circuit to see if normal amplified output is produced at the collector. This step localizes the trouble to one stage.
2. A transistor checker can be used to find a defective transistor.
3. Direct-current voltage readings are taken at the emitter, base, and collector leads. These tests are important because most troubles change the dc values.
4. From the dc tests, some educated guesses can be made about the problem.
5. Finally, the suspected section or component can be tested with an ohmmeter.

It is interesting to note that the final result of all the checking usually is an ohmmeter test to find an open circuit or short circuit.

Practice Problems 30-7
Answers at End of Chapter

Refer to Fig. 30-13*a*.

a. If the emitter-base junction of the transistor is short-circuited, then V_C would measure __9V__ (less than 1 V, about 9 V).

b. If R_2 becomes open, then V_C would __de'__ (decrease, increase) as a result of the increase in forward bias on the emitter-base junction of the transistor.

c. If R_1 becomes open, then V_C would increase to about 9 V as a result of the __lack of__ (increase in, lack of) forward bias on the emitter-base junction of the transistor.

Summary

1. A transistor used as a switch is either cut off to be an open circuit or conducting saturation current for a short circuit.
2. A transistor used as an amplifier is biased between the extremes of cutoff and saturation. Typical forward bias for silicon transistors is V_{BE} of 0.6 to 0.7 V.
3. Transistor amplifiers can be connected in common-base (CB), common-emitter (CE), or common-collector (CC) circuits. The CC circuit is an emitter-follower. See Fig. 30-4.
4. The CE amplifier is most common because it has the most gain. In this amplifier, a small input current in the base-to-emitter circuit controls a larger output current in the collector-to-emitter circuit.
5. The emitter-base junction of a transistor amplifier is forward-biased, while the base-collector junction has reverse voltage.
6. Manufacturers' data sheets contain information on the electrical and thermal characteristics of transistors, with specifications for case size, transistor outline (TO) style, and pin connections. See Figs. 30-10 and 30-11.
7. Typical amplifier checks in troubleshooting include signal injection, checking transistors, and dc voltage tests.
8. Each junction in a PNP or NPN transistor can be checked as a semiconductor diode.

Self-Examination
Answers at Back of Book

Answer True or False.

1. Transistors are classified as either the junction or field-effect type.
2. Refer to Fig. 30-1. Closing switch S_1 causes the emitter-to-collector resistance of the transistor to increase.
3. When the switch is closed in Fig. 30-1, the control and power circuits both have about the same amount of current flowing.
4. Refer to Fig. 30-1. Opening switch S_1 causes the light to go off.
5. Refer to Fig. 30-2a. The transistor is said to be at saturation in this example.
6. Refer to Fig. 30-2b. The forward bias on the emitter-base junction causes the transistor to conduct current in the power circuit.
7. Refer to Fig. 30-3a. This CE circuit is biased to perform as an amplifier.
8. A bias voltage is usually ac, whereas a signal voltage is dc.
9. A typical forward bias on the emitter-base junction of a Ge transistor is 0.6 V.
10. The CB circuit is the most popular transistor amplifier configuration.
11. Typically, in amplifier circuits, the emitter-base junction is reverse-biased, while the base-collector junction of the transistor is forward-biased.
12. A CE transistor circuit has both current gain and voltage gain.
13. A CC transistor circuit has good current gain but no voltage gain.
14. A CB transistor circuit has good current gain but no voltage gain.
15. The collector of a CC transistor circuit is grounded for ac signal.
16. Voltage gain may be calculated by dividing the output signal by the input signal.

17. Refer to Fig. 30-5. If I_B increases, then I_C will increase and the voltage across R_L will increase.
18. Refer to Fig. 30-9. The 2N3903 NPN transistor has a current gain of from 50 to 150 at an I_C of 10 mA.
19. Besides the specifications listed in Fig. 30-9, it is common for a data sheet to also give the material, applications, power dissipation, maximum collector current, and leakage current for a transistor.
20. The TO-3 case style is best suited for low-power transistors.
21. The value of β or h_{FE} gives the current gain for a CE amplifier.
22. Bias voltage measurements in transistor circuits can be made with a VOM.
23. A transistor packaged in a TO-92 case would be mounted on a large heat sink to dissipate heat.
24. Refer to Fig. 30-13a. If load resistor R_4 is open, V_C will be near +9 V.
25. Refer to Fig. 30-13a. If the base lead of the transistor is disconnected or open, V_C will be near +9 V.

Essay Questions

1. Refer to Fig. 30-2a. Why is there still 1 μA of collector current flowing even though the transistor is in the off condition?
2. Refer to Fig. 30-3b. Explain why a positive-going input signal causes the output signal to go in the negative direction.
3. List several advantages of the CE transistor circuit.
4. Describe at least one advantage of the CC transistor circuit.
5. Explain the transistor terms *cutoff* and *saturation*.
6. Which transistor listed in Fig. 30-9 seems to have both high voltage and high current ratings?
7. Which transistor package shown in Fig. 30-11 would be most adaptable to high-power applications?
8. Refer to Fig. 30-13d. Is this transistor at cutoff or at saturation?
9. In Fig. 30-13d, why has the voltage at the base dropped to about 0.75 V instead of the normal 2.6 V?
10. Compare the operation of a transistor as a class A amplifier and as a switch.

Problems

Answers to Odd-Numbered Problems at Back of Book

1. Refer to Fig. 30-2b. If $V_C = 0.2$ V and $I_C = 88$ mA, calculate the resistance of the transistor from E to C.
2. Refer to Fig. 30-2b. Calculate the current gain of the transistor in this circuit.
3. Refer to Fig. 30-3b. If the signal input is 0.01 V p-p and the output signal is 1 V p-p, calculate the voltage gain of the transistor.
4. Refer to Fig. 30-3b. If the voltage gain = 150 and the current gain = 40, calculate the power gain of the transistor.
5. Refer to Fig. 30-7. If $V_C = 5.6$ V, calculate V_L.

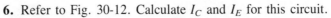

6. Refer to Fig. 30-12. Calculate I_C and I_E for this circuit.
7. Refer to Fig. 30-13a. If C_1 became short-circuited and $+9$ V were applied to the base of the transistor, what would V_C measure?

Answers to Practice Problems

30-1 a. F
     b. T
     c. T

30-2 a. T
     b. F
     c. T

30-3 a. Common emitter (CE)
     b. Common emitter (CE)
     c. Common collector (CC)

30-4 a. 0.6 V
     b. 6.5 V
     c. 1 mA

30-5 a. TO-18
     b. 0.85 V
     c. 50 to 150

30-6 a. Short-circuited
     b. 9, 0
     c. Higher
     d. Ohmmeter

30-7 a. About 9 V
     b. Decrease
     c. Lack of

Chapter 31
Digital Electronics

The entire world of electronics can be divided into two broad groups of circuits, analog and digital. In analog circuits, the voltage and current waveforms have variations similar to the signal information. The sine wave of *V* or *I* is a perfect example. A digital signal, however, is a group of pulses that have the same level and are either on or off. These conditions are also labeled HIGH or LOW and 1 or 0. This is a *binary* system, as it has just two bits of information. Variations are indicated by a HIGH or LOW pulse in the digital signal.

The logical use of digital circuits is for numerical counting, as in electronic calculators and test equipment. Digital electronics also includes alphanumeric information, though, with letters and numbers. The pulse information can be encoded to represent letters in the alphabet, punctuation marks, and instructions.

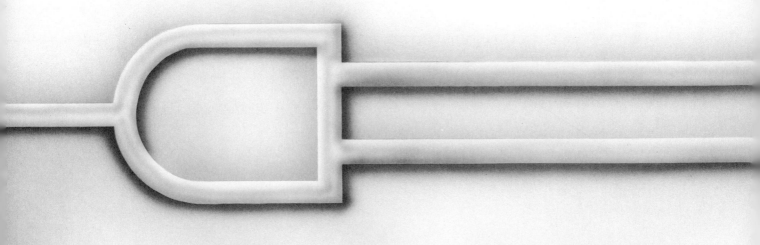

Digital signals have an advantage over analog signals in that they reduce the effect of noise interference. It is relatively easy for a digital circuit to recognize a specific level of the pulses.

Because of these features, practically all modern electronic equipment contains a mixture of analog and digital circuits. Changing from one form of signal to the other is convenient with digital-to-analog (D/A) converters.

Important terms in this chapter are:

AND gate	decade counter	multiplexer	read
bit	demultiplexer	NAND gate	ROM
Boolean algebra	full-adder	NOR gate	shift register
byte	half-adder	NOT gate	toggle
carry in	hexadecimal	OR gate	truth table
carry out	inverter	radix	write
clock	microprocessor	RAM	XOR gate
counter	modulus		

More details are explained in the following sections:

31-1   Binary Numbers
31-2   Binary Arithmetic
31-3   Basic Logic Gates
31-4   Other Logic Gates
31-5   Combining Logic Gates
31-6   Practical Logic Gates
31-7   Practical TTL Gating Circuit
31-8   Practical CMOS Gating Circuit
31-9   Flip-Flops
31-10 Using Flip-Flops
31-11 Other Digital Devices
31-12 Troubleshooting Digital Circuits

31-1   Binary Numbers

An analog meter measures quantities over an infinite number of steps. For example, the pointer in Fig. 31-1*a* can come to rest at any one of an infinite number of points along the scale. The digital meter (Fig. 31-1*b*), on the other hand, indicates readings in discrete steps of digits. In digital electronics the discrete steps are characterized by high or low voltage, open or closed circuits, and the digits 0 to 1. For this reason, digital electronics lends itself to the binary number system.

The binary number system uses only the digits 0 and 1, which represent the two levels in a digital signal. Examples of binary numbers are 0001, 0101, and 1001. Any number of places is possible, each with a 0 or 1. Figure 31-2 illustrates how the two levels of a digital waveform correspond to the binary digits 0 and 1.

The *base* or *radix* of a number system specifies how many different digits can be used for counting in each place. For binary numbers, the base is two. The base or radix is often indicated in a subscript. For example, 111_2 represents a base-two number. Read this as one, one, one to base two.

Our conventional decimal system for counting uses all ten digits from 0 to 9. Therefore, the base or radix for decimal numbers is ten.

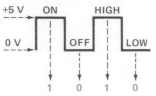

Fig. 31-2 Specifying binary digits 1 and 0 for HIGH and LOW conditions of voltage pulses.

The highest digit for any place is one less than the base. In decimal numbers, 9 is the highest digit, while 1 is the highest digit in binary numbers, for counting in successive places.

How Each Place Is Used for Different Counts

To count numbers more than the base itself, the complete number must have an order of places for larger counts. The order is indicated by the *decimal point* in decimal numbers or the *binary point* in binary numbers.

Figure 31-3 illustrates the decimal orders. For numbers greater than 1, the first place to the left of the decimal point is for the count of the digits alone, from 0 to 9. The second place is for counts of ten. Next is the place for counts of 10^2, or hundreds, and then for 10^3, or thousands. The counts for successive places increase in multiples of 10.

For decimal fractions less than 1, the places are to the right of the decimal point. These values decrease in multiples of $\frac{1}{10}$ or 0.1.

The same method is used with binary numbers. However, the first place to the left of the point can be only 0 or 1. As usual, higher counts are carried over to the next place. The second place is for counts of two. Successive places to the left are for counts of 2^2, or fours; 2^3, or eights; etc.

Places to the right of the binary point can be used for fractions less than 1, as in decimal numbers. However, the values decrease in multiples of $\frac{1}{2}$ instead of $\frac{1}{10}$.

For the binary example in Fig. 31-4, the $(111)_2$ is $4 + 2 + 1 = (7)_{10}$. The binary number 111 is actually 111.0, but the point is usually omitted, as in decimal numbers.

As another example, $(1100)_2$ is $8 + 4 + 0 + 0 = (12)_{10}$. Binary numbers need more places for counting because the base is smaller.

Fig. 31-1 (*a*) Volt-ohm-milliammeter (VOM) is an example of an analog meter. (*b*) Digital multimeter (DMM).

$10^3 =$ thousands	$10^2 =$ hundreds	$10^1 =$ tens	Digits 0 to 9	Quantity
0	1	1	1	$(111)_{10}$

Fig. 31-3 Order of places in a decimal number. This number is one hundred and eleven.

$2^3 =$ eights	$2^2 =$ fours	$2^1 =$ twos	$2^0 =$ ones	Quantity
0	1	1	1	$(7)_{10}$

Fig. 31-4 Order of places in a binary number. The number is one, one, one to base 2.

$(Base)^3$	$(Base)^2$	Base	Digits	Base or radix	Name of number system
1000	100	Ten	0 to 9	10	Decimal
8	4	Two	0 or 1	2	Binary
512	64	Eight	0 to 7	8	Octal
4096	256	Sixteen	0 to 9 and A, B, C, D, E, F	16	Hexadecimal

Fig. 31-5 Order of places for numbers with bases 10, 2, 8, and 16.

The general method of counting in number systems with a different base or radix is illustrated in Fig. 31-5. Included with 10 and 2 as the base are base 8, or *octal*, and base 16, or *hexadecimal* numbers. Binary, octal, and hexadecimal numbers are all widely used in digital electronics.

Converting Binary to Decimal   Just make the count for each place and add the values. As an example, the seven places in $(1010101)_2$ show the count

$$64 + 0 + 16 + 0 + 4 + 0 + 1 = (85)_{10}$$

Converting Decimal to Binary   There are several methods, but a common system is as follows:

1. Divide the decimal number successively by 2 until the quotient is zero.
2. Note whether each remainder is 0 or 1.

3. The remainders of 0 and 1 are the digits in the binary number.

The places of the digits 0 or 1 in the binary number are to the left of the binary point. Continue the divisions until the last quotient is zero.

As an example, to convert $(13)_{10}$ to binary form:

$$13 \div 2 = 6 \quad \text{with remainder of } 1$$
$$6 \div 2 = 3 \quad \text{with remainder of } 0$$
$$3 \div 2 = 1 \quad \text{with remainder of } 1$$
$$1 \div 2 = 0 \quad \text{with remainder of } 1$$

$$(13)_{10} = (1 \quad 1 \quad 0 \quad 1)_2$$

The remainders are assembled into the binary number as shown above. Therefore, $(13)_{10} = (1101)_2$. Note that the last division, with a quotient of zero, is most important because it may give a 1 for the highest place in the binary number. This system of dividing by 2 for converting to decimal form is called *double dabbling*.

Shifting the Decimal or Binary Point In decimal numbers moving the point one place to the right multiplies the number by 10. For instance, 10×53.7 and 537.0 are equal. Moving the point one place to the left divides by 10.

In binary numbers, shifting the point one place multiplies or divides by 2. For instance, $(111.0)_2$ is $(7)_{10}$, but $(1110.0)_2$ is $(14)_{10}$. The 7 is doubled to 14 by moving the binary point one place to the right.

For division by 2, the point is moved one place to the left. For instance, $(111.0)_2$ is $(7)_{10}$, but $(11.10)_2$ is $(3.5)_{10}$. The binary places in 11.1 have the values of $2 + 1 + \frac{1}{2}$, which equals $3\frac{1}{2}$ or 3.5. The 3.5 is one-half of 7.

Binary Bits, Bytes, Nibbles, and Words A 0 or 1 is a *bit* because it has one item of information. A string of bits with 0s and 1s forms a word. An 8-bit word is a *byte*. One-half of a byte (four bits) is a *nibble*. Some examples follow:

0 or 1 = bit
0101 = nibble
0101 0011 = byte
0101 0011 0010 1010 = two bytes for a 16-bit word

Note that the bits are in groups of four, for convenience in reading. The binary information is generally processed with 8-bit or 16-bit words.

Binary-Coded Decimal (BCD) In the BCD system, each digit in a decimal number is encoded as a binary number. Four bits take care of digits up to 9. For example, 3 is coded into 0011. Also, 9 becomes 1001. Then decimal 39 is 0011 1001 in BCD form. As an application, when you punch in 39 on a calculator keyboard, this number is converted to 0011 1001.

Hexadecimal Values This system uses 16 characters with digits 0 to 9 and letters A to F for 10 to 15. Hexadecimal values are commonly used in microcomputers because they are much shorter than binary numbers. As an example, hexadecimal 9F corresponds to binary 1001 1111.

Practice Problems 31-1
Answers at End of Chapter

a. Convert $(1111)_2$ to decimal form.
b. Convert 10 in decimal form to a binary number.
c. How much is $(000)_2$ as a decimal number?
d. How many bits are in one byte?
e. Give the binary equivalent of B in hexadecimal notation.

31-2 Binary Arithmetic

Only the digits 0 and 1 can be used in each place. For binary addition, all possible sums of 0 and 1 are

$$0 + 0 = 0 \qquad 1 + 0 = 1$$
$$0 + 1 = 1 \qquad 1 + 1 = (10)_2$$

The last sum is not decimal number ten but is the binary equivalent of decimal number two.

We cannot have the digit 2 as a bit. However, 1 in the second binary place has a weight of two. The general rule is that for binary $1 + 1$ the sum is 0 with a carry of 1 to the next higher order.

To add numbers with more than one digit, start next to the binary point. As an example,

$$\begin{array}{r} (110)_2 \\ +(111)_2 \\ \hline 1101 \end{array}$$

In the third column from the point, the binary addition of $1 + 1 +$ carry of 1 results in the sum of 1 in this column and carry of 1 to the fourth column.

As a check on the answer, this problem in decimal numbers is $6 + 7 = 13$.

The only two possible combinations with a carry are

a. $1 + 1 =$ sum of 0 with carry of 1. This is binary 10.
b. $1 + 1 +$ carry of $1 =$ sum of 1 with carry of 1. This is binary 11.

It should be noted that a column with $1 + 1$ has a *carry out* of 1. The next column has a *carry in* of 1.

For binary multiplication, all possible products of 0 and 1 are

$$0 \times 0 = 0 \qquad 1 \times 0 = 0$$
$$0 \times 1 = 0 \qquad 1 \times 1 = 1$$

These values are the same as in decimal multiplication. For long hand multiplication with more digits, the partial products are added.

More details on binary numbers and computer arithmetic can be found in books listed at the end of "Review: Chapters 29 to 32" and also in the bibliography near the end of the book. It is common for the arithmetic logic units (ALUs) of microprocessors to have only addition circuitry. Subtraction, multiplication, and division are performed by the digital circuitry by following special sequences of operations called *algorithms*. Sometimes the algorithms are programmed into the microprocessor and sometimes the programmer must furnish the algorithms. For one example, subtraction can be done by adding complementary numbers.

Practice Problems 31-2
Answers at End of Chapter

a. Add $(101)_2 + (010)_2$.
b. Add $(111)_2 + (111)_2$.

31-3   Basic Logic Gates

A circuit that determines whether an input pulse can pass through to the output is called a *logic gate*. The three main types are the AND, OR, and NOT gates, illustrated in Figs. 31-6 to 31-8. These circuits are also the building blocks for other types of logic gates.

The AND Gate   Consider the series circuit shown in Fig. 31-6a. The bulb will be lit only if both switches S_1 AND S_2 are closed. This circuit action is called the *logical AND function*. The *logic symbol* for an AND gate is shown in Fig. 31-6b.

A common method used to describe the action of a gate is the *truth table*. It tabulates all the possible combinations of input and output levels. A truth table for the AND function (AND gate) is shown in Fig. 31-6c. The inputs to A and B of the AND gate are shown as either LOW or HIGH voltages in Fig. 31-6c. Note that the *unique output* for the AND gate is shown in the bottom line of the truth table—output Y is HIGH only when both inputs are HIGH.

A shorthand method used to describe the AND function is called a *Boolean expression*. The Boolean ex-

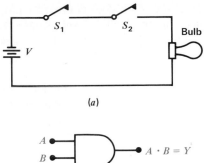

(a)

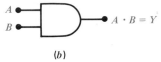

$A \cdot B = Y$

(b)

Inputs		Output
A	*B*	*Y*
LOW	LOW	LOW
LOW	HIGH	LOW
HIGH	LOW	LOW
HIGH	HIGH	HIGH

(c)

Inputs		Output
A	*B*	*Y*
0	0	0
0	1	0
1	0	0
1	1	1

(d)

Fig. 31-6 The AND gate. (a) Equivalent circuit. (b) Logic symbol and Boolean expression. (c) Truth table. (d) Truth table using 0 and 1 for LOW and HIGH.

pression gets its name from George Boole, an English mathematician (1815–1864). He started the system we call Boolean algebra. Claude Shannon of Bell Telephone Laboratories adapted Boolean algebra for logic circuitry.

The Boolean expression for the AND function is shown on the right side of the AND gate logic symbol in Fig. 31-6b. The Boolean expression for the 2-input AND gate is $A \cdot B = Y$. The dot (·) means to logically

AND inputs A and B. It does not mean multiply, as in regular algebra. In Boolean expressions, the dot ($\cdot$) may be omitted, as in regular algebra. Therefore, the Boolean expression $AB = Y$ is the same as $A \cdot B = Y$.

A common form of the AND truth table is shown in Fig. 31-6d. Instead of describing the input voltage levels as HIGH and LOW, they are just referred to as 1 or 0. A logical 1 means HIGH, and a logical 0 means LOW. Some manufacturers use truth tables with 0s and 1s, while others describe the inputs and outputs as HIGHS or LOWS.

In summary, the AND logic function can be described using a logic symbol, truth table, or Boolean expression. An AND gate produces a HIGH at the output only when *all* inputs are HIGH.

We can think of the AND gate as a coincidence detector for the HIGH condition. The HIGH output is produced only when all inputs coincide at the HIGH level.

The OR Gate   Consider the circuit shown in Fig. 31-7a. The bulb will be lit when either or both of the

parallel switches are closed. This circuit action is called the *logical* OR *function*. The logic symbol for an OR gate is drawn in Fig. 31-7b. Also, the exact circuit action for a 2-input OR gate is described in detail in the OR *truth table* in Fig. 31-7c. Note that the *unique output* from the OR gate is a 0 (LOW) when both inputs are 0 or LOW.

The OR gate can be considered as a coincidence detector for the LOW condition. A LOW output results only when all inputs coincide at the LOW level.

The Boolean expression used to represent the OR function is $A + B = Y$. This Boolean expression is shown on the output side of the OR logic symbol in Fig. 31-7b. Notice that the plus sign ($+$) is used to represent the OR function in Boolean algebra. This plus sign *does not* mean to add as in regular algebra.

The OR logic function described in Fig. 31-7 is sometimes called the *inclusive* OR *function*. In the next section you will learn about an exclusive OR gate.

The NOT Gate   The logic symbol for a NOT gate is shown in Fig. 31-8. This gate has only one input and one output. Its job is to invert the input so that the output is always opposite. If the input is 1, the output will be 0; if the input is 0, the output will be 1. The NOT gate is commonly called an *inverter*.

The Boolean expression for the invert function is shown on the right side of the NOT gate logic symbol in Fig. 31-8. If the input is A, then the output from the inverter is $\overline{A}$ (read as not A). The overbar ($^{-}$) represents the invert function in the Boolean expression.

Notice the small bubble on the output side of the inverter logic symbol in Fig. 31-8. This is called an *invert bubble*. Inversion is also called *negating* or *complementing*.

In actual logic circuits, the phase inversion of the common-emitter (CE) amplifier corresponds to the NOT function. Also, the common-source circuit with field-effect transistors is an inverter.

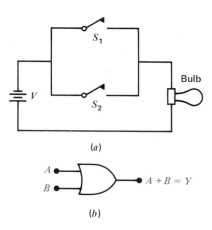

(a)

(b)

Inputs		Output
A	**B**	**Y**
0	0	0
0	1	1
1	0	1
1	1	1

(c)

Fig. 31-7 The OR gate. (a) Equivalent circuit. (b) Logic symbol and Boolean expression. (c) Truth table.

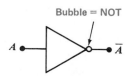

Fig. 31-8 Inverter logic symbol and Boolean expression.

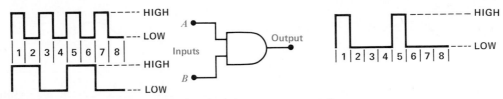

Fig. 31-9 Output from AND gate is HIGH only when both inputs are HIGH at the time periods marked 1 and 5. This 1 is not a logic level.

Using the AND Gate

For a practical application, Fig. 31-9 shows two groups of input pulses for the AND gate. The waveform at the right is for the resulting output. Remember that the AND gate produces a HIGH output only when both inputs are HIGH. Therefore, the output Y in Fig. 31-9 is HIGH at the time intervals marked 1 and 5. At the time interval 3, for instance, A is HIGH but B is LOW.

Diode Circuit for AND Gate

Figure 31-10 illustrates how two diodes can be used for an AND gate to implement the logic function $Y = AB$. For both diodes D_A and D_B, the anode is connected to +5 V through R_1. The diodes are conducting, therefore, assuming a low level for the inputs A and B. Each diode cathode returns to chassis ground through the input source. Then point Y in the output is pulled down to a low level, close to ground potential. A conducting diode is practically a short circuit.

When a HIGH level in the input makes a cathode positive, the diode is cut off. Then the diode is practically an open circuit. However, both diodes must be off to isolate point Y from ground. With only one diode off,

the other diode pulls down Y to the LOW level. When both diodes are off, though, Y is pulled up to the +5 V of the supply for the HIGH level. Since both inputs A and B must be HIGH to make Y HIGH, the diode circuit performs the logic function $Y = AB$.

Only two inputs are shown in Fig. 31-10, but more inputs can be used with additional diodes. Still, all the diodes would have to be cut off for a HIGH in the Y output.

Practice Problems 31-3
Answers at End of Chapter

Answer True or False.
a. The AND function corresponds to two switches in series.
b. The OR function corresponds to two switches in parallel.
c. The NOT gate operates as an inverter.
d. The plus sign (+) is used in Boolean algebra to represent the AND function.
e. A CE amplifier can perform the NOT function.
f. In Fig. 31-10, both diodes must be off for Y to be HIGH.

31-4 Other Logic Gates

These include the NAND and NOR gates. The AND circuit negated with an inverter makes a NOT AND or NAND gate. Also, the OR circuit negated with an inverter results in the NOR gate.

NAND Gate

The logic symbol for the NAND gate is illustrated in Fig. 31-11a. It consists of the AND symbol with an invert bubble added to the output.

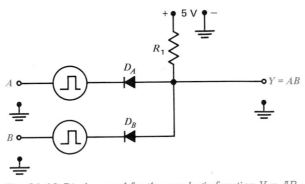

Fig. 31-10 Diodes used for the AND logic function $Y = AB$.

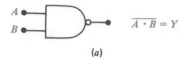

$$\overline{A \cdot B} = Y$$

(a)

Inputs		Output
A	B	Y
0	0	1
0	1	1
1	0	1
1	1	0

(b)

Fig. 31-11 The NAND gate. (a) Logic symbol and Boolean expression. (b) Truth table.

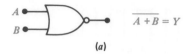

$$\overline{A + B} = Y$$

(a)

Inputs		Output
A	B	Y
0	0	1
0	1	0
1	0	0
1	1	0

(b)

Fig. 31-12 The NOR gate. (a) Logic symbol and Boolean expression. (b) Truth table.

The NAND logic function is detailed in the NAND *truth table* in Fig. 31-11b. The truth table shows that the output column is the opposite or complement of that of the AND truth table. While the AND gate generates a HIGH output only with both inputs HIGH, the NAND gate generates a LOW when both inputs are HIGH. The unique output for the NAND gate is shown in the bottom line of the truth table.

The Boolean expression used to represent the NAND function is shown to the right of the logic symbol in Fig. 31-11a. It consists of the AND expression with a bar drawn over the entire input part of the expression. The Boolean expression for the NAND function is shown as $\overline{A \cdot B} = Y$. As with the AND Boolean expression, the AND dot (·) may be omitted from the expression. Therefore, the Boolean expression $\overline{AB} = Y$ may be used to represent a 2-input NAND gate.

NOR Gate   The logic symbol for the NOR gate is drawn in Fig. 31-12a. It consists of the OR symbol with an invert bubble added to the output.

The NOR logic function is detailed in the NOR *truth table* in Fig. 31-12b. The truth table shows that the output column is the opposite or complement of that of the OR truth table. The unique output for the NOR gate is a HIGH when both inputs are LOW. This is shown in the first line of the NOR truth table.

The Boolean expression for the NOR function is shown to the right of the logic symbol in Fig. 31-12a. It

consists of the OR expression with a bar drawn over the entire input part of the expression. The Boolean expression for the NOR function becomes $\overline{A + B} = Y$.

The Exclusive OR (XOR) Gate   The exclusive feature means that the XOR circuit selects only certain combinations of input pulses. Remember that an *inclusive* or conventional OR gate produces a 1 in the output for any 1, or all 1s, at the input. This operation is like parallel switches. However, for the XOR gate:

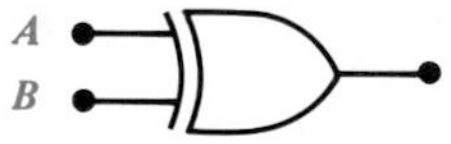

$$A \oplus B = Y$$

(a)

	Inputs		Output
	A	B	Y
	0	0	0
Odd	0	1	1
Odd	1	0	1
Even	1	1	0

(b)

Fig. 31-13 The exclusive OR (XOR) gate. (a) Logic symbol and Boolean expression. (b) Truth table.

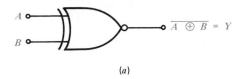

(a)

| Inputs | | Output |
A	B	Y
0	0	1
0	1	0
1	0	0
1	1	1

Odd (row with A=0, B=1)
Odd (row with A=1, B=0)
Even (row with A=1, B=1)

(b)

Fig. 31-14 The exclusive NOR (XNOR) gate. (a) Logic symbol and Boolean expression. (b) Truth table.

1. An even number of 1s at the input produces 0 or LOW output.
2. An odd number of 1s at the input produces 1 or HIGH output.

The XOR logic function is illustrated in Fig. 31-13. The symbol in Fig. 31-13a has the OR symbol with an extra curved line at the input side to indicate the exclusive feature. In the Boolean expression shown at the output side, the circled + sign is used for the XOR function.

The truth table in Fig. 31-13b shows the exclusive principle of the XOR gate. Only an even number of 1s, such as two or four, can produce the LOW output of 0. An odd number of 1s, such as one or three, produces the HIGH output of 1.

The Exclusive NOR (XNOR) Gate This logic function is illustrated in Fig. 31-14. In Fig. 31-14a, the logic symbol consists of the XOR symbol with an invert

bubble added to the output. The Boolean expression has an overbar for the inversion.

The truth table in Fig. 31-14b shows that the output is the opposite or complement of that of the XOR truth table. For the XNOR logic function:

1. An odd number of 1s at the input produces 0 at the output.
2. An even number of 1s at the input produces 1 at the output.

A comparison between the XNOR and XOR functions can be seen in Table 31-1, which lists the unique features for the main types of logic gates. Note that 1 is the HIGH level. This system, with 1 for the higher positive voltage, is called *positive logic*. With bipolar transistors that use a dc supply of $+5$ V, the HIGH level can be approximately 2 to 5 V, while the LOW level is 0 to 0.8 V.

Using the NAND Gate For a practical application, Fig. 31-15 shows two groups of pulses as inputs. Remember that the NAND gate will produce a LOW output when all inputs are HIGH. For all other combinations of input, the output is HIGH. These include any LOW input or all inputs LOW. In Fig. 31-15, note that only during the time periods 1 and 5 are both inputs HIGH. Then the output is LOW. For the other time periods, either the A or the B input is LOW, producing HIGH output.

Practice Problems 31-4
Answers at End of Chapter

a. Inverting the output of an OR gate will produce the NAND function.
b. Inverting the output of an AND gate will produce the NAND function.
c. With an even number of 1s for input to an XOR gate, the output is LOW.

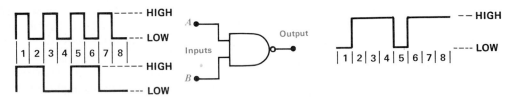

Fig. 31-15 Outputs from the NAND gate are LOW only when both inputs are HIGH.

Table 31-1. Unique Features of Logic Gates

Gate	Symbol	Input	Output
AND	In ───⊐D─── Out	All 1 Any 0	1 0
NOT	In ──▷o─── Out	1 0	0 1
NAND	In ───⊐Do─── Out	All 1 Any 0	0 1
OR	In ───⊃D─── Out	All 0 Any 1	0 1
NOR	In ───⊃Do─── Out	All 0 Any 1	1 0
XOR	In ──⊐⊃D─── Out	Even 1s Odd 1s	0 1
XNOR	In ──⊐⊃Do─── Out	Odd 1s Even 1s	0 1

31-5 Combining Logic Gates

For most practical circuits in digital electronics, the logic gates must be combined in a specific pattern. Examples of such logic diagrams are shown in Figs. 31-16 to 31-19.

Converting a Boolean Expression to a Logic Diagram Boolean expressions are used in building logic circuits. Consider the more complicated Boolean expression shown in Fig. 31-16a. This is called a *minterm* or *sum-of-products* type. The Boolean expression suggests that we OR ($\overline{A} \cdot B$) with ($A \cdot \overline{B}$), because of the + sign. Step 1 in Fig. 31-16b shows these terms written at the inputs to an OR gate. Step 2 in Fig. 31-16c shows gates 2 and 3 being added to the OR gate to generate the $A \cdot \overline{B}$ term. Step 3 in Fig. 31-16d shows gates 4 and 5 being added to produce the $\overline{A} \cdot B$ part of the expression. Finally, the logic symbol diagram in

Fig. 31-16d is equal to the desired Boolean expression $(\overline{A} \cdot B) + (A \cdot \overline{B}) = Y$.

You have drawn an AND-OR logic symbol diagram from a minterm Boolean expression. Note that you must work from the output towards the inputs to solve these problems.

Converting a Truth Table to a Boolean Expression Refer to the truth table in Fig. 31-17a. Look only at the lines in the truth table that have outputs of 1. These are the terms you will use in your minterm Boolean expression. The first term from line 2 of the truth table is shown as $\overline{A} \cdot B$, whereas the second term from line 3 is $A \cdot \overline{B}$. Next, the terms are ORed together. The purpose is to make a minterm expression, which is the OR type. In the example in Fig. 31-17, this yields the minterm expression $(\overline{A} \cdot B) + (A \cdot \overline{B}) = Y$.

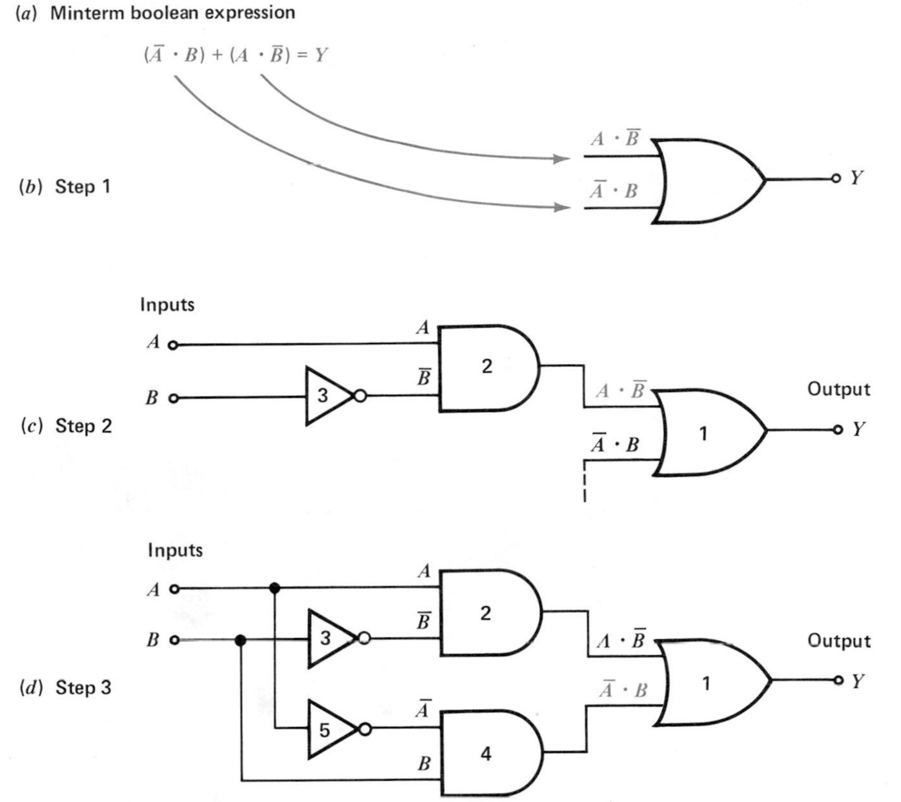

(a) Minterm boolean expression

$$(\overline{A} \cdot B) + (A \cdot \overline{B}) = Y$$

(b) Step 1

(c) Step 2

(d) Step 3

Fig. 31-16 Converting a Boolean expression to a logic diagram.

In summary, remember to look for combinations of variables that generate an output of 1 in the truth table. Then OR these terms together to form a minterm Boolean expression.

You may have noticed that the example we used in Fig. 31-17a was the truth table for the XOR gate. The Boolean expression for the XOR truth table is $(\overline{A} \cdot B) + (A \cdot \overline{B}) = Y$ or the simpler $A \oplus B = Y$ that we learned earlier. Both mean the same logic function, as the minterm expression indicates an odd number of 1s in the input for 1 in the output.

Most truth tables have more than one Boolean expression that will describe them accurately. To save electronic hardware, we usually use the simplest Boolean expression when drawing a logic symbol diagram.

Converting a Logic Diagram to a Truth Table Consider the AND-OR logic diagram in Fig. 31-18a on the next page. Notice that this logic circuit has three inputs for two 3-input AND gates. These are

gates 2 and 3. Their outputs are ORed by gate 1, yielding the minterm Boolean expression shown in Fig. 31-18b. The resulting truth table is in Fig. 31-18c.

Inputs		Output
A	**B**	**Y**
0	0	0
0	1	1
1	0	1
1	1	0

(a)

$\longrightarrow \overline{A} \cdot B$ (for row 0 1)

$\longrightarrow A \cdot \overline{B}$ (for row 1 0)

$$(\overline{A} \cdot B) + (A \cdot \overline{B}) = Y$$

(b)

Fig. 31-17 Converting (a) a truth table to (b) a minterm Boolean expression.

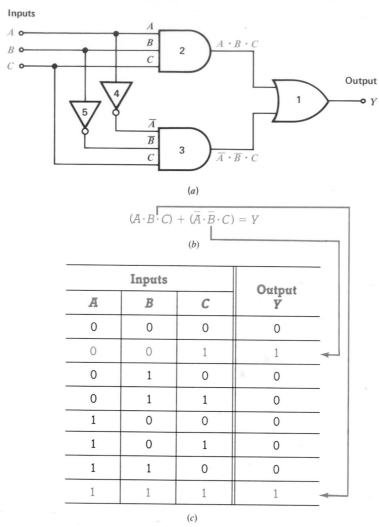

Fig. 31-18 Converting (a) a logic symbol diagram to (b) an equivalent minterm Boolean expression to (c) an equivalent truth table.

The term $(A \cdot B \cdot C)$ generates a 1 in the last line of the truth table in Fig. 31-18c. The other term from the Boolean expression, $(\overline{A} \cdot \overline{B} \cdot C)$, generates a 1 in the second line of the truth table. These two terms specify the only conditions for 1 in the output. The remaining outputs of the truth table (lines 1, 3, 4, 5, 6, and 7) are filled in with 0s. For the input columns, the eight possible combinations are listed for the three inputs.

Minterm and Maxterm Boolean Expressions   As we have learned, $(\overline{A} \cdot B) + (A \cdot \overline{B}) = Y$ is an example of a *minterm* or *sum-of-products* type of Boolean expression. When implemented with gates, a minterm Boolean expression forms an AND-OR pattern of logic symbols. Figure 31-16d is an example of the AND-OR pattern of gates.

For a different combination, the expression $(A + B) \cdot (C + D) = Y$ is referred to as a *maxterm* or *product-of-sums* type Boolean expression. When implemented with gates, it will form an OR-AND pattern of logic gates. Unlike the AND-OR circuit, the OR-AND logic circuit will have an AND gate nearest the output. An example of the OR-AND pattern of gates is shown in Fig.

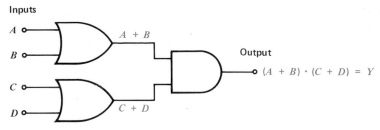

Inputs

$A + B$

$C + D$

Output

$(A + B) \cdot (C + D) = Y$

Fig. 31-19 Example of OR-AND pattern of gates used to implement maxterm Boolean expressions.

Table 31-2. Minterm and Maxterm Expressions

	Definition	Example
Minterm	Sum of products	$(AB) + (CD)$
Maxterm	Product of sums	$(A + B) \cdot (C + D)$

31-19. This logic circuit implements the maxterm expression $(A + B) \cdot (C + D) = Y$. See Table 31-2 for a comparison of minterm and maxterm expressions.

Practice Problems 31-5
Answers at End of Chapter

a. The expression $(A + B) \cdot (A + \overline{C}) = Y$ is a _____ (maxterm, minterm) type Boolean expression.
b. When implemented, $(A \cdot \overline{B}) + (A \cdot C) = Y$ will produce a circuit with an _____ (AND-OR, OR-AND) pattern of logic gates.
c. In forming a minterm Boolean expression from a truth table, each _____ (0, 1) in the truth table output column generates a term for the expression.

31-6   Practical Logic Gates

Modern digital logic gates are usually packaged in integrated circuit (IC) form. Several compatible groups called *families* have been developed by manufacturers. The most common families of digital ICs are the *TTL* and *CMOS* types.

TTL stands for *transistor-transistor logic*. TTL gates are based on a bipolar technology similar to that used for producing NPN and PNP transistors. TTL ICs are

probably the most popular and are noted for their very fast switching speed.

Several TTL *subfamilies* are manufactured. These include the *low-power TTL* family, which greatly reduces the power consumed by the IC. Another is the *high-power TTL* family, which uses more power but is faster. The *low-power Schottky TTL* family reduces power consumption and increases speed. Because of their improved characteristics, low-power Schottky TTL ICs are currently very popular.

CMOS stands for *complementary-symmetry metal-oxide semiconductor*. CMOS digital ICs are based on the MOS technology. This technology is similar to that used in producing insulated-gate field-effect transistors (IGFETs). CMOS digital ICs are rapidly gaining in popularity because of their *extremely low power consumption*. CMOS ICs also are able to operate on a wide range of dc supply voltages, usually 3 to 18 V.

Other logic families include the obsolete RTL (resistor-transistor logic) and DTL (diode-transistor logic) types. Also, ECL (emitter-coupled logic) ICs are used when extremely high speed is required.

Besides CMOS, the metal-oxide semiconductor technology is used to produce both PMOS and NMOS devices. PMOS (P type) and NMOS (N type) devices are often found in large-scale and very large-scale integration (*LSI* and *VLSI*). Microprocessors and large memories are typical examples.

Logic Levels   How do we define a HIGH or LOW logic level? Figure 31-20 shows HIGH and LOW logic levels defined for both TTL and CMOS devices.

TTL ICs use a +5-V power supply. Using the 5-V supply, a logical LOW would appear to be any voltage from 0 to about 0.8 V. A HIGH would be any voltage from about 2.0 to 5 V. Any TTL input or output voltage between about 0.8 V and 2.0 V would be in the undefined region, as shown in Fig. 31-20. Voltages in the undefined region cause unpredictable results and usually point to problems in a logic circuit.

CMOS ICs can operate on a wide range of supply voltages (from about 3 to 18 V). If the CMOS supply voltage were 10 V, a HIGH would mean any voltage from about 7 to 10 V, whereas a LOW would be any voltage from 0 to about 3 V. Voltages from 3 to 7 V would be in the undefined region.

Fan-in and Fan-out   The number of inputs that can be connected to the logic circuit is its fan-in factor. The maximum number of circuits that can be driven by the output is the fan-out factor. This effectively specifies how many parallel loads can be connected without too much current in the driver circuit.

Speed of Gates   The speed of a logic gate has to do with how much time it takes the IC's internal circuitry to change the output once the input has changed logic states. The longer the *propagation delay*, the slower the speed of the IC.

Generally, bipolar devices such as TTL ICs are faster (superior speed) than devices such as CMOS ICs. Whereas a low-power Schottky TTL gate may have a propagation delay of only 5 to 10 ns, a CMOS gate may have a delay of 30 to 40 ns.

Practice Problems 31-6
Answers at End of Chapter

a. CMOS digital ICs are produced using _____ (bipolar, FET) technology.
b. The _____ (CMOS, TTL) family of ICs has the shortest propagation delay and is therefore considered the _____ (fastest, slowest).
c. A dc supply of 12 V would be for _____ (TTL, CMOS) logic circuits.

31-7   Practical TTL Gating Circuit

The TTL family of ICs is typically numbered 74XXX, with the numbers in the X position specifying the exact function. A military grade of the TTL family is numbered 54XXX. Thus the TTL 7400 IC is a commercial-grade quad 2-input NAND gate, while the 5400 IC is the military grade. The quad description means that there are four of the same gates in the one IC chip. Many combinations are available.

The pin diagram for a 7400 quad 2-input NAND gate TTL IC is drawn in Fig. 31-21. Each NAND gate is independent. However, they do share common power

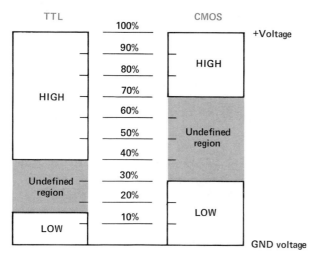

Fig. 31-20 Logic levels defined for TTL and CMOS digital logic families.

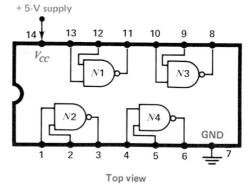

Fig. 31-21 A TTL IC with four NAND gates. Each gate has two inputs. Top view of 14 pins, numbered counterclockwise from notch at left. This IC is a 14-pin dual inline package (DIP).

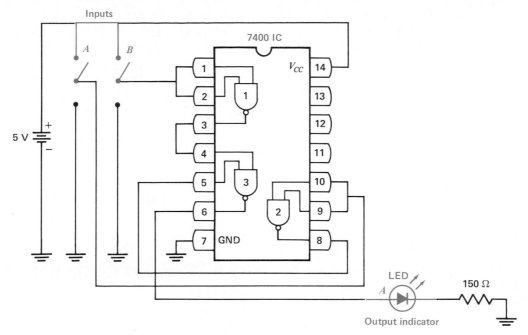

Fig. 31-22 Practical wiring diagram of a logic circuit using NAND gates from the 7400 TTL IC. Output LED indicator will light only if pin 6 is HIGH. Anode of LED is marked A.

connections at pins 7 and 14 of the DIP IC. The +5 V is connected to the V_{CC} input to the IC and the negative of the power supply is connected to the GND (ground connection). Note the location of pin 1, *immediately counterclockwise* from the notch on the DIP IC (when viewed from the top). The diagram in Fig. 31-21 is also called a *pin-out*. DIP means dual-inline package for the pin connections.

Using the 7400 TTL IC

Consider the gating circuit wired in Fig. 31-22. A +5-V power supply is shown at the left. Two switches A and B provide inputs to the gating circuit. When the switches are up, they are HIGH, whereas if they are down, they generate a LOW. The output indicator is a light-emitting diode (LED). A HIGH at the anode of the LED will cause the indicator to light. However, a LOW at the anode of the LED indicator will cause it to not light.

Circuit Action

Assume that switches A and B in Fig. 31-22 are both up or HIGH. Will the output indicator light? Follow input A to NAND gate 2 and note that

the HIGH or 1 is applied to both inputs at pins 9 and 10. According to the NAND truth table, two HIGH inputs cause the output of gate 2 to be LOW at pin 8. This places a LOW at pin 5 of the IC for input to gate 3. Gate 2 produced the invert function.

Next trace the HIGH from input switch B to pins 1 and 2 as inputs of NAND gate 1. With both inputs of NAND gate 1 HIGH, the output will be LOW. This LOW output at pin 3 is the LOW input to gate 3 at pin 4. Gate 1 produced the invert function.

Now both inputs to NAND gate 3 are LOW, causing the output to go HIGH. The HIGH at pin 6 becomes the source of current to light the LED output indicator.

According to the analysis, when both inputs are HIGH, the output from the gating circuit is HIGH to turn on the LED. You may analyze the other three possible conditions of switches A and B to determine the outputs of this gating circuit.

Circuit for TTL NAND Gate

Refer to Fig. 31-23. The transistors $Q1$ and $Q2$ are in series for the dc supply of 5 V. Both must be on for there to be current through the collector load resistor R_3. With current in

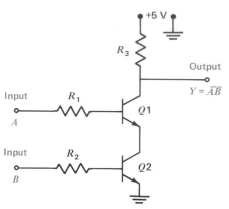

Fig. 31-23 Circuit for bipolar NAND gate.

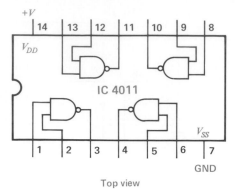

Fig. 31-24 Pin diagram of 4011 CMOS quad 2-input NAND gate IC.

R_3, the output at Y is LOW because of the IR_3 voltage drop. Without the output current, Y is HIGH at 5 V.

The transistors are NPN. A positive pulse at the HIGH level applied to the base makes the transistor conduct. However, both inputs must be HIGH for there to be current in R_3 because the transistors are in series for collector current. Therefore this circuit implements the NAND function $Y = \overline{AB}$. The inverter function in the output results because the collector output voltage goes less positive, or in the negative direction, when the base input voltages go positive.

Practice Problems 31-7
Answers at End of Chapter

a. Refer to Fig. 31-22. When both switches are down (both LOW), the output indicator will _____ (light, not light) because the output of the gating circuit is _____ (HIGH, LOW).

b. Refer to Fig. 31-22. When switch A is LOW and switch B is HIGH, the output from the gating circuit will be _____ (HIGH, LOW).

c. Refer to Fig. 31-22. When switch A is HIGH and switch B is LOW, the output from the gating circuit will be _____ (HIGH, LOW).

d. Pin 1 on a DIP IC is located immediately _____ (clockwise, counterclockwise) from the notch when viewed from the top.

e. In Fig. 31-23, how much is the Y voltage when input A is HIGH but B is LOW?

31-8   Practical CMOS Gating Circuit

One family of CMOS ICs is numbered 40XXX, with the numbers in the X positions specifying the exact function. Thus the CMOS 4011 IC is a quad 2-input NAND gate, while the 4071 is a quad 2-input OR gate.

The pin diagram for the CMOS *4011 quad 2-input* NAND *gate* is drawn in Fig. 31-24. Each NAND gate is independent, but they share a common power supply, connected at pins 7 and 14 of the DIP IC. The positive ($+$) of the power supply is connected to V_{DD} (pin 14), whereas the ground or negative ($-$) is connected to V_{SS} (pin 7) of the IC. DIP pins are numbered counterclockwise from the notch in the IC when viewed from the top.

Using the 4011 CMOS IC   Consider the gating circuit wired in Fig. 31-25. A +12-V power supply is shown at the left. Two switches A and B provide inputs to the gating circuit. When the switches are up, they produce a HIGH, and when they are down, they generate a LOW logic level. The output indicator is an LED. The LED lights when a HIGH is present at its anode and does not light when a LOW is connected to its anode.

Circuit Action   Assume that both switches A and B in Fig. 31-25 are up or HIGH. Will the output indicator

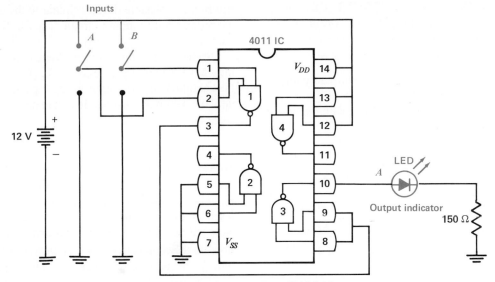

Fig. 31-25 Practical wiring circuit using NAND gates from a 4011 CMOS IC. Output indicator will light only when pin 10 is HIGH.

light? Follow inputs A and B to NAND gate 1. Both inputs to NAND gate 1 are HIGH, causing the output to go LOW. The LOW output from gate 1 is applied to both inputs of NAND gate 3. These LOW inputs cause the output of NAND gate 3 to go HIGH; therefore the LED output indicator will light.

According to this analysis, when both inputs are HIGH, the output from the gating circuit is HIGH. You may analyze the other three possible conditions of switches A and B to determine the outputs of this gating circuit.

Circuit for CMOS NAND Gate   Refer to Fig. 31-26. Four transistors are used in two pairs. Transistors $Q1$ and $Q2$ are in parallel. These have a P-type channel, which requires a LOW input to conduct. Transistors $Q3$ and $Q4$ are in series. The N channel for these transistors needs a HIGH input for conduction.

Conduction in either $Q1$ or $Q2$ connects the Y output terminal to 12 V. This action is pulling Y up to the HIGH level. Either $Q3$ or $Q4$ must be off to isolate Y from chassis ground.

For the opposite case, conduction in both $Q3$ and $Q4$ connects the Y output terminal to chassis ground. This

action is pulling Y down to the LOW level. Both $Q1$ and $Q2$ must be off to isolate Y from the 12-V supply.

When any input or every input is at a LOW, the output is HIGH. Only when both inputs A and B are HIGH can the output Y be LOW. Then $Q3$ and $Q4$ are conducting while $Q1$ and $Q2$ are off. Therefore, this circuit implements the NAND function $Y = \overline{AB}$.

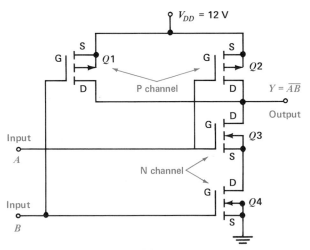

Fig. 31-26 Circuit for CMOS NAND gate.

Precautions When Using CMOS ICs CMOS
ICs are sensitive to static electricity as a result of their very high input impedance. Manufacturers typically suggest special handling techniques such as:

1. Store CMOS ICs in conductive foam pad until installation in circuit.
2. Do not insert or remove CMOS ICs with the power on.
3. Ground soldering iron tips and other metal tools. (Some suggest using a battery-powered soldering iron when working on CMOS circuits.)
4. Unused *input* leads should be tied to the (+) on the power supply. Note in Fig. 31-25 how the unused inputs to gates 2 and 4 are tied to GND and +12 V, respectively.

Practice Problems 31-8
Answers at End of Chapter

a. Refer to Fig. 31-25. When both switches are down (both LOW), the output indicator will _____ (light, not light) because the output of the gating circuit is _____ (HIGH, LOW).
b. Refer to Fig. 31-25. When switch *A* is LOW and switch *B* is HIGH, the output from the gating circuit will be _____ (HIGH, LOW).
c. Refer to Fig. 31-25. When switch *A* is HIGH and switch *B* is LOW, the output from the gating circuit will be _____ (HIGH, LOW).
d. Several precautions must be observed when handling CMOS ICs because of their sensitivity to _____ electricity.
e. In Fig. 31-26, how much is the *Y* output voltage when input *A* is HIGH but *B* is LOW?

31-9 Flip-Flops

The name *flip-flop* refers to the ability of the circuit to remain in one of two stable states. The action is like that of a latching relay. An input pulse sets one state for the flip-flop. It can remain that way indefinitely. Then the next input pulse resets the flip-flop to the opposite state, which is also stable. The two opposite states are considered stable conditions because an input pulse is needed to change the level in the output. Therefore, the

flip-flop is a bistable device, similar to the operation of the bistable multivibrator circuit. The flip-flop abbreviation is FF.

Flip-flops are important in logic circuits because they have a memory characteristic. The circuit holds the conditions in the output for each pulse in the input, until the next pulse arrives. It should be noted that logic gates do not have this memory.

Reset-Set Flip-Flop The most basic form of flip-flop may be formed by interconnecting two NAND gates as shown in Fig. 31-27a. This forms what is called a reset-set flip-flop, usually shortened to *R-S flip-flop*. Note the two inputs are labeled *S* and *R*. The *R-S* flip-flop has the characteristic two outputs, labeled *Q* and $\overline{Q}$ (not *Q*). Under normal operation, the outputs will always be opposite or complementary.

The truth table in Fig. 31-27b shows the four possible combinations of the *S* and *R* inputs to the *R-S* flip-flop. Three of these input combinations produce a useful *mode of operation*. These are:

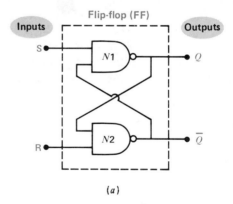

(a)

Mode of operation	Inputs		Outputs	
	S	R	Q	$\overline{Q}$
Prohibited	0	0	1	1
Set	0	1	1	0
Reset	1	0	0	1
Hold	1	1	No change	

(b)

Fig. 31-27 *R-S* flip-flop. (a) Wired using two NAND gates. (b) Truth table.

1. *Set mode.* To *set* means to cause output Q to be set to 1. This results from having input $S = 0$ while input $R = 1$.
2. *Reset mode.* To *reset* means to cause output Q to be reset or cleared to 0. This is caused by having input $R = 0$ while input $S = 1$.
3. *Hold mode.* This mode is the "do nothing" or "remembering" state. The outputs stay as they were (Q at either 1 or 0). According to the truth table in Fig. 31-27b, the hold mode results when both inputs S and $R = 1$.

Synchronous or Asynchronous Types

The previous *R-S* flip-flop was an *asynchronous* device in that it did not have a clock input. All other flip-flops we will study are *synchronous* devices with a *clock input* for timing. Synchronous means to operate in step with the system clock. The clock is an accurate pulse generator.

Clocked *R-S* Flip-Flop

A typical simplified logic diagram for a *clocked R-S flip-flop* is drawn in Fig. 31-28a. Data enter through inputs R and S while the clock input is attached to the system clock. The customary Q and complementary $\bar{Q}$ (not Q) outputs are shown on the right of the clocked *R-S* flip-flop logic symbol.

The truth table in Fig. 31-28b gives three useful modes of operation for the clocked *R-S* flip-flop. These are:

1. *Reset mode.* Output Q is reset to 0 when data input $R = 1$ while $S = 0$ *and* a clock pulse appears at the clock input. Note that the reset action does not occur until the clock pulse reaches the *CK* input. This flip-flop is operating in step with the clock, or synchronously.
2. *Set mode.* Output Q of the FF will be set to 1 when data input $S = 1$ while $R = 0$ *and* when a clock pulse reaches the *CK* input.
3. *Hold mode.* The outputs do not change. They hold their data when both data inputs S and $R = 0$. In the hold mode, repeated clock pulses to the *CK* input of the clocked *R-S* flip-flop will have *no effect* on the outputs. The FF is serving as a memory cell while in the hold mode.

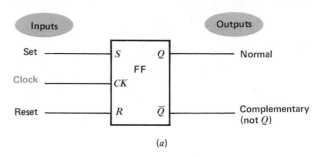

(a)

Mode of operation	Inputs			Outputs	
	*CK**	*S*	*R*	*Q*	$\bar{Q}$
Hold	⎍	0	0	No change	
Reset	⎍	0	1	0	1
Set	⎍	1	0	1	0
Disabled	⎍	1	1	1	1

*⎍ is positive clock pulse.

(b)

Fig. 31-28 Clocked *R-S* flip-flop. *CK* is abbreviation for clock. (*a*) Logic symbol. (*b*) Truth table.

D Flip-Flop

The logic symbol for a *D* type flip-flop is shown in Fig. 31-29. It has only one data input D and a clock input *CK*. However, it has the Q and complementary $\bar{Q}$ outputs that are normal for all flip-flops.

The name *D* flip-flop means a *data flip-flop*. It is also sometimes called a *delay flip-flop* because it is often used to delay data from reaching output Q by one clock pulse. This time delay of a few nanoseconds can be very important in applications with a long chain of circuits.

The arrowhead ⟩ symbol next to the clock input in Fig. 31-29 shows that this is an *edge-triggered* flip-

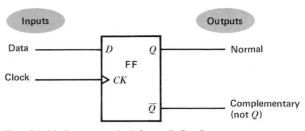

Fig. 31-29 Logic symbol for a *D* flip-flop.

flop. In this case, the *D* flip-flop is a positive edge-triggered flip-flop, which means that data are transferred from the *D* input to the *Q* output on the LOW-to-HIGH transition of the clock pulse. A bubble is used with the arrow to indicate triggering on the opposite transition, from HIGH to LOW.

J-K Flip-Flop The *J-K* flip-flop is considered the universal FF. The logic symbol for one *J-K* flip-flop is shown in Fig. 31-30a. This FF has two data inputs labeled *J* and *K* as well as a clock input (*CK*). It has the normal *Q* and complementary $\bar{Q}$ outputs. The arrowhead ⟩ at the *CK* input shows that it is edge-triggered, and the bubble shows that it is negative edge-triggered. Negative edge triggering means that data are transferred from the data inputs to output *Q* on the HIGH-to-LOW transition of the clock pulse.

The truth table in Fig. 31-30b shows four useful modes of operation for the *J-K* flip-flop. These are:

1. *Hold mode*. This is the memory state. Clock pulses at the *CK* input have no effect on the outputs.

2. *Reset mode*. Output *Q* is reset to 0 on the HIGH-to-LOW transition of the clock pulse when *J* = 0 and *K* = 1.
3. *Set mode*. Output *Q* is set to 1 on the HIGH-to-LOW transition of the clock pulse when *J* = 1 and *K* = 0.
4. *Toggle mode*. The toggle action means that output *Q* is alternately switched from 1 to 0 to 1 to 0, etc., on repeated clock pulses. The *J-K* flip-flop is in the toggle mode when both data inputs *J* and *K* are held at 1. The toggle mode is a very useful mode of operation.

Like logic gates, flip-flops are typically packaged in convenient DIP ICs. While gates are used to form *combinational logic* circuits, flip-flops are the fundamental building block for *sequential logic* circuits. Sequential means that the output of each FF feeds the input to the next FF to determine its action. The main types of flip-flop circuits are summarized in Table 31-3.

Practice Problems 31-9
Answers at End of Chapter

a. The *R-S* flip-flop is classified as a(n) _____ (asynchronous, synchronous) device.
b. The _____ (*CK*, data) input to a flip-flop ensures that the unit operates in step with the system clock.
c. To *set* a flip-flop means to cause the normal *Q* output to be at a _____ (HIGH, LOW) logic level.
d. If both *J* and *K* data inputs to the *J-K* flip-flop are 1, the unit is in its _____ mode of operation.

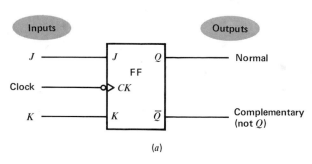

(a)

Mode of operation	Inputs			Outputs	
	CK*	J	K	Q	$\bar{Q}$
Hold	⎍	0	0	No change	
Reset	⎍	0	1	0	1
Set	⎍	1	0	1	0
Toggle	⎍	1	1	To opposite state	

*⎍ is positive clock pulse.

(b)

Fig. 31-30 *J-K* flip-flop. (a) Logic symbol. (b) Truth table.

31-10 Using Flip-Flops

The memory characteristic of the FF is probably its most important feature. This ability to maintain a stable state is used for counter circuits and shift registers.

Counters The basic binary counter is a chain of flip-flops. Each divides by 2. Successive flip-flops divide in multiples of 2. The division is in the frequency of the clock-pulse input. For this reason, it is common to refer to binary counters as divide-by-2, divide-by-4, divide-by-8, or divide-by-16 counters.

Table 31-3. Types of Flip-Flop Circuits

Name	Symbol	Notes
Reset-set (R-S)	S — FF — Q R — $\bar{Q}$	Q and $\bar{Q}$ opposite; set is $Q = 1$
Clocked R-S	S — FF — Q CK — R — $\bar{Q}$	CK is clock pulse; no bubble means trigger on rising edge
D-type	D — FF — Q CK —▷ $\bar{Q}$	Also called delay FF
J-K	J — FF — Q CK — K — $\bar{Q}$	J and K high for toggle on CK pulses; bubble indicates CK trigger on falling edge

Three *J-K* flip-flops are wired in Fig. 31-31*a*, forming a divide-by-8 counter. This counter may also be called a *3-bit counter* or a *mod-8 counter*. The modulus (mod) of a counter is the number of different states the counter can sequence through. The counter in Fig. 31-31 counts from binary 000 to 111 (0 to 7), or eight counts. This type is also called a *ripple counter* because each flip-flop is cascaded with the next and the clock pulses ''ripple'' through the counter.

In Fig. 31-31*a* the *J* and *K* inputs to the flip-flops are held HIGH at 5 V, which puts each FF in its toggle mode. An *asynchronous reset* input has been added to the 3-bit counter for clearing all outputs to 0. The bubble at the top of each *J-K* flip-flop symbol means that it takes a logical 0 to activate the reset inputs.

Waveform diagrams, such as the one shown in Fig. 31-31*b* for the mod-8 counter, are commonly used to show the detailed operation of counters. The counter is first reset to 000. Then each HIGH-to-LOW transition of the clock pulses toggles FF-*A*. The output of FF-*A* then causes FF-*B* to toggle on the HIGH-to-LOW edge of clock pulse b. During clock pulse h the binary count is 111. The clock pulses are marked along the top of Fig. 31-31*b*.

Ripple counters, such as the one in Fig. 31-31, are relatively slow. More complicated *synchronous* *counters* are available when higher counter frequencies are required.

Four-bit (mod-16) and mod-10 counters are both very popular. Mod-10 counters are also called *decade counters*. Decade counters contain four flip-flops plus a logic gate to reset the count to 0000 after the maximum count of 1001 for nine is reached.

Shift Registers   A register is a memory device. The counter circuit in Fig. 31-31 also functions as a register in that it holds the count between clock pulses.

A *4-bit shift register* is shown in Fig. 31-32. Note the use of data flip-flops, with all clock inputs tied together. An asynchronous reset input clears all flip-flops to 0 when activated with a LOW. The *serial data* input is for entering a bit of data at a time.

Suppose you wanted to load 0100 into the shift register in Fig. 31-32. You would use the following procedure:

1. Reset all flip-flops to 0. The result would be 0000 in the register.
2. Place a 1 at the serial data input.
3. Pulse the clock input *once*. The result would be

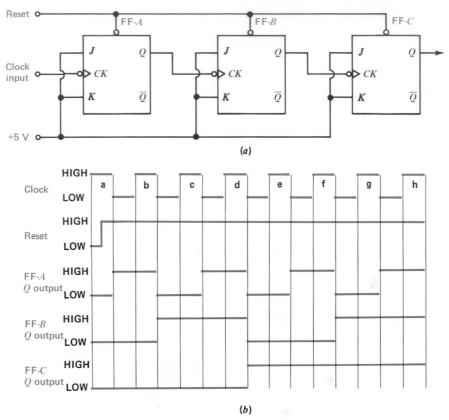

Fig. 31-31 Divide-by-8 ripple counter. (a) Logic diagram using three *J-K* flip-flops. (b) Waveforms for the counter.

1000 in the register at the outputs *A, B, C,* and *D.*
4. Place a 0 at the serial data input.
5. Pulse the clock input once. The result would be 0100 in the register.

In this example, data were entered one bit at a time, or *serially*. Many shift registers allow all bits to be entered at one time. This is called *parallel* or *broadside* loading. Shift registers may be 4, 8, or more bits wide.

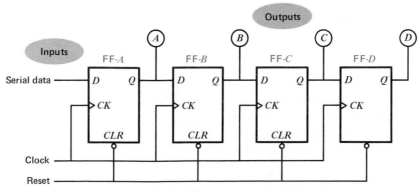

Fig. 31-32 Logic symbol diagram of 4-bit serial-load shift register.

Practice Problems 31-10
Answers at End of Chapter

a. The counter shown in Fig. 31-31 is called a mod-_____ ripple counter.

b. Refer to Fig. 31-31*b*. What is the binary count *immediately after* clock pulse c?

c. Refer to Fig. 31-32. How many clock pulses would it take to load binary 0001 into this shift register?

d. The shift register shown in Fig. 31-32 is a _____ (parallel, serial)-load 4-bit shift register.

31-11   Other Digital Devices

Functional digital devices such as counters and shift registers are typically packaged in easy-to-use DIP ICs. Other common functional digital devices are adders, display decoders, and multiplexers.

Adders Arithmetic operations are usually performed by adders. A half-adder circuit is used to add the 1s column in binary addition. Full-adder circuits are used to add the more significant columns (2s, 4s, 8s, etc.).

The block symbol for a *half-adder* circuit is shown in Fig. 31-33*a*. The inputs *A* and *B* are the bits to be

added. The results are displayed at the *sum* and *carry* outputs of the half-adder.

A block symbol for a *full-adder* circuit is shown in Fig. 31-33*b*. Three inputs (carry in, *A*, and *B*) are required. Inputs *A* and *B* are the bits to be added, while the *carry in* input is the carry bit from the less significant column. The results are displayed at the *sum* and *carry out* outputs from the full-adder circuit.

Half- and full-adders can be combined to add longer binary numbers. Figure 31-34 shows a logic diagram for a *4-bit parallel adder* circuit. The bits from the problem are entered at the left, and the sum is immediately displayed at the lower right. Note the use of a

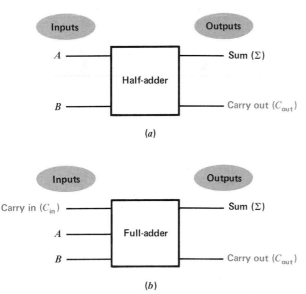

Fig. 31-33 (*a*) Logic symbol for a half-adder. (*b*) Logic symbol for a full-adder.

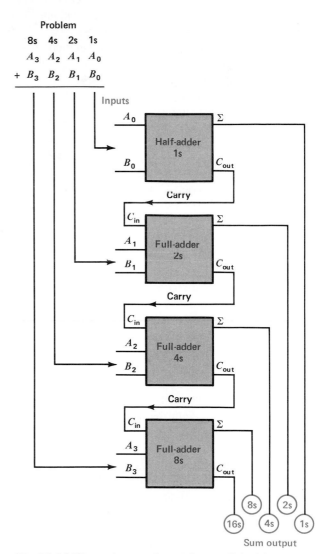

Fig. 31-34 Wiring diagram for a 4-bit parallel adder circuit.

649

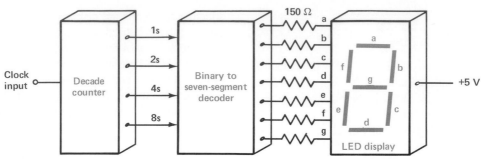

Fig. 31-35 Counter-decoder-display circuit. Decoder translates BCD input to a decimal output shown on the seven-segment LED display.

half-adder for the 1s column and full-adders for all other columns. The carries from column to column are handled by the *carry lines* between adders.

Adders are combinational logic circuits composed of gates. As such, they do not have a memory characteristic. In larger systems, registers called *latches* would be added to hold the binary numbers at the inputs and outputs of the adder circuits.

Using special mathematical techniques, adders can be used to do binary subtraction. Adders are typically found in the *arithmetic logic unit* (ALU) of microprocessors and computers. These adders can be used to perform addition, subtraction, multiplication, and division.

Display Decoder   Digital circuits are widely used where numbers (or letters) must be displayed. Figure 31-35 shows a typical counter-decoder-display circuit. A *binary-to-seven-segment-display decoder* is shown converting the counter's output to a readable decimal output on the *seven-segment LED display*. The decoder outputs a to g determine which segments on the display light up.

The three blocks shown in Fig. 31-35 come in easy-to-use DIP form. Internally, decoders are combinational logic circuits, with the required gates.

Multiplexers and Demultiplexers   The basic idea of how *multiplexers* and *demultiplexers* operate is shown in Fig. 31-36. With rotation of the control, the 4-input multiplexer shown in Fig. 31-36a is converting *parallel data* to *serial data*. The position of the "control" is selected electronically. A multiplexer is also called a *data selector*.

The *1-of-4 demultiplexer* shown in Fig. 31-36b is converting serial data to parallel data. A demultiplexer is also called a *distributor* or *decoder*. The position of the "control" in the demultiplexer is selected electronically.

Multiplexers and demultiplexers can be used in pairs for data transmission. The multiplexer converts parallel data to serial data for transmission. The demultiplexer reassembles the serial data into parallel form at the receiving end of the transmission line.

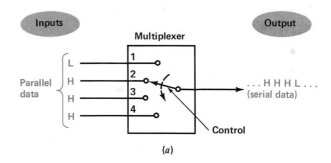

(a)

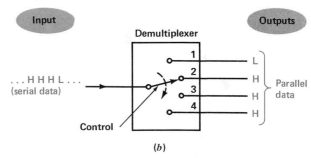

(b)

Fig. 31-36 (a) Mechanical rotary-switching-type action of a multiplexer. Note many inputs and only one output. H and L are high and low pulses. (b) Mechanical rotary-switching-type action of demultiplexer. Note one input and many outputs.

Semiconductor Memories Most digital equipment contains a variety of simple memory devices, such as flip-flops, registers, and latches. When flip-flop units are organized into larger groups, they are usually called RAMs, ROMs, or PROMs.

A *RAM* (*random access memory*) is a digital storage device which can store many bits of data. A RAM is also referred to as a *read-write memory* because you can read data from the memory or write data into it. Figure 31-37*a* shows that data can be placed into a RAM using the *write mode*. The data can also be retrieved from a RAM using the *read mode*. The *address input* to the RAM specifies the exact location or address of the data to be read or location to be written into. RAMs have the disadvantage of losing their data when the power is turned off for even an instant. Therefore, RAMs are classed as *volatile* memories.

The ROM is a nonvolatile *read-only memory*. ROMs are permanently programmed by the manufacturer. Figure 31-37*b* shows that ROMs can be used only in the *read mode*. As with the RAM, the ROM has an address input to specify the exact location of the data to be read. A special ROM that can be programmed by the user is called a *programmable read-only memory* (*PROM*). Several types of *erasable programmable read-only memories* (*EPROMs*) are also widely used.

Your personal computer uses both ROM and RAM types of semiconductor memories. If your computer has 64K of memory, it has 64,000 (actually 2^{16} or 65,536) bytes of memory. A *byte* is an 8-bit group of 1s and 0s. This 64K of memory in a microcomputer will be composed of perhaps 4K of ROM and 60K of RAM. While RAMs and ROMs are used for microcomputer internal memory, other devices, such as magnetic tape or discs, are used for external bulk storage. A typical 5¼-in. floppy disc used with many personal computers will hold over 100,000 bytes of data.

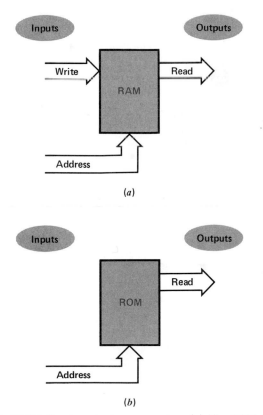

(a)

(b)

Fig. 31-37 Semiconductor memories. (a) The RAM has both read and write capabilities. (b) The ROM has only read capabilities.

Practice Problems 31-11
Answers at End of Chapter

a. Refer to Fig. 31-34. If binary 1111 is being added to 0001, the sum would be _____ and the carry line between the 1s and 2s adders would be _____ (HIGH, LOW).

b. Refer to Fig. 31-35. If the binary count at the output of the decade counter is 0111, the decimal number _____ will appear on the display and segments _____ will be lit.

c. A digital device called a _____ is used to convert parallel data to serial data.

d. The _____ is a semiconductor memory with both read and write capabilities.

e. A _____ (RAM, ROM) is a permanent type of semiconductor memory.

31-12   Troubleshooting Digital Circuits

The basic tools used are the logic probe and pulser probe (Fig. 31-38) and the oscilloscope (Fig. 31-39).

Digital Logic Probe The probe has LED indicators that light to indicate 1 or 0. Most logic probes have a switch for selecting either TTL or CMOS logic. In

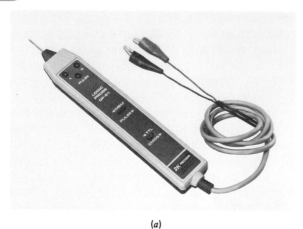

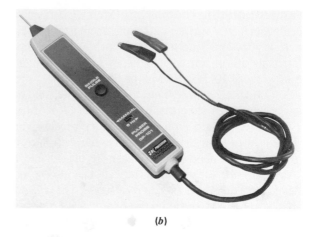

(a)

(b)

Fig. 31-38 Digital test probes. (a) Logic probe. (b) Pulser probe. Length is 6.5 in. or 16.25 cm. (*Dynascan Corporation*)

TTL circuits, the logic thresholds are 2.4 V for logic 1 and 0.8 V for logic 0. For CMOS, logic 1 is 70 percent of supply voltage; logic 0 is 30 percent. Typical specifications are 2-MΩ input impedance and maximum input frequency of 40 MHz.

Digital Pulser Probe   The probe in Fig. 31-39b can generate output of one pulse or a continuous train of pulses at the frequency of 5 Hz. When the output is applied to a digital circuit, the probe senses the logic state and immediately pulls it to the opposite level. The probe is compatible with TTL and CMOS circuits. Operating power is obtained from the circuit under test so that a battery is not required.

Oscilloscope   A *dual-trace oscilloscope* is pictured in Fig. 31-39. The dual-trace oscilloscope is particularly useful in comparing waveforms in sequential logic circuits, where, along with logic levels, timing becomes a critical problem.

Troubleshooting Problem   Consider the combinational logic circuit in Fig. 31-40. The circuit is a simple TTL 2-input AND gating circuit with LED output indicator and series current-limiting resistor.

The steps in troubleshooting this circuit might be as follows:

1. Use your senses. Touch the top of the IC to see if it is too hot. Look carefully for breaks in the PC board traces or bent pins.
2. Use a TTL logic probe. Connect power to the logic probe from the circuit under test. Check pin 14 (V_{CC}) and the +5-V trace line to make sure they are HIGH (see point 1 in Fig. 31-40).
3. With the logic probe, check IC pin 7 and the ground trace on the printed circuit board. Both should be LOW (see point 2 in Fig. 31-40).
4. Manipulate the inputs to generate the *unique output* from the circuit. In this case, both inputs to the AND gate must be HIGH for the output to be HIGH.

Fig. 31-39 Dual-trace oscilloscope. (*Simpson Electric Company*)

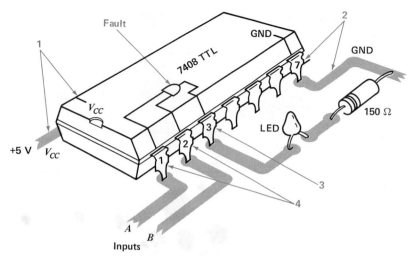

Fig. 31-40 Using the logic probe to locate a fault in a simple gating circuit.

5. Using a logic probe, check the output. With two HIGH inputs, it should indicate a HIGH. See point 3.
6. Suppose the indication at IC pin 3 is LOW instead of HIGH. This is caused by the fault inside the IC, as shown in Fig. 31-40. In this case the fault is an open in the output lead from the AND gate.
7. To be sure, you may want to double-check IC pins 1 and 2 as shown at point 4 in Fig. 31-40 to make sure they are both HIGH.
8. The fault is in the 7408 TTL IC. This IC must be replaced. After replacement, the circuit is tested for proper operation.

Practice Problems 31-12
Answers at End of Chapter

a. Most logic probes have a switch for selecting either TTL or _____.
b. The logic probe is an important tool in trouble-shooting combinational logic circuits, whereas the _____ is important in checking sequential logic circuits where timing is very critical.
c. (True or False) Logic probes are usually powered by their own internal batteries.

Summary

Table 31-4 shows the basic building blocks for digital circuits. In addition, the terminology used in digital electronics is summarized in the following alphabetical list.

AND gate Logic function comparable to series switches. All inputs must be HIGH for HIGH output.
BCD Binary-coded decimal. Each decimal digit is equated to a 4-bit binary group.
Binary Number system using only the symbols 0 and 1.
Bit Abbreviation for binary digit, 0 or 1.
Byte An 8-bit grouping of binary digits used as a group to symbolize a number, letter, or computer operation.
Clear Same as reset on a flip-flop.
Clock Usually a multivibrator to generate pulses at an exact frequency to synchronize the logic circuits.

Table 31-4. Symbols for Digital Circuits

Function	Symbol
AND gate	A, B → AB
OR gate	A, B → $A + B$
XOR gate	A, B → $A \oplus B$
Inverter	A → $\overline{A}$
NAND gate	A, B → $\overline{AB}$
NOR gate	A, B → $\overline{A + B}$
XNOR gate	A, B → $\overline{A \oplus B}$
Reset-set (R-S) flip-flop	S, R → FF → Q, $\overline{Q}$
J-K flip-flop	J, CK, K → FF → Q, $\overline{Q}$

CMOS   Complementary-symmetry metal-oxide semiconductor. A semiconductor technology and a common digital logic family that uses very little power.

Complement   A binary opposite, changing between 0 and 1.

Counter circuit   Digital circuit based on flip-flops used to accumulate the binary count of incoming pulses.

Decimal   Customary number system using the digits 0 through 9.

Decoder   Electronic device for converting from one code to another, such as from BCD to decimal.

Demultiplexer   Electronic device used to convert serial data to parallel data. Also known as a distributor or decoder.

D flip-flop   Data or delay flip-flop.

Digit   One symbol in a number. Binary numbers have two digits, 0 and 1. Decimal numbers have ten digits, from 0 through 9.

Exclusive NOR gate   Logic function in which the output is HIGH only when an even number of inputs are HIGH.

Exclusive OR gate   Logic function in which the output is HIGH only when an odd number of inputs are HIGH.

Fan-out   A logic gate's maximum number of output circuits that can be used as a load.

Flip-flop (FF)   Bistable circuit that can store one bit of information. The outputs at the Q and $\overline{Q}$ terminals must be opposite.

Gate   Logic circuit with two or more inputs designed to give one output only for specific combinations of input pulses.

Hexadecimal   Base 16 number system using the digits 0 to 9, A, B, C, D, E, and F. Widely used in microcomputer work.

Inverter   Logic function that negates the input to provide its complement at the output. Also called the NOT function.

J-K flip-flop   Includes two data inputs J and K and a clock input, with Q and $\bar{Q}$ outputs. Has set, reset, hold, and toggle modes of operation.

Logic   A form of mathematics based on two-state truth tables. Electronic logic circuits use gates and flip-flops to perform logic functions.

Memory   Also called storage. The bistable FF is a basic memory cell. RAMs, ROMs, PROMs, and EPROMs are larger semiconductor memory devices. Mass bulk storage uses magnetic tapes and discs or paper tape or cards.

Microprocessor   Complex programmable digital system on an IC. Used as the central processing unit (CPU) in modern microcomputers.

Multiplexer   Electronic device for converting parallel data to serial data. Also called a data selector.

NAND gate   Logic function in which all inputs must be HIGH for output to be LOW; negation of AND gate.

NOR gate   Logic function in which output is HIGH only when all inputs are LOW; negation of OR gate.

Octal   Base 8 number system, using the digits 0 through 7.

OR gate   Logic function comparable to parallel switches. All inputs must be LOW for LOW output.

Radix   The base for a number system. Radix is 2 for binary numbers and 10 for decimal numbers.

RAM   Random access memory. A volatile read/write type of semiconductor memory.

Read   Acquire information for use or from memory.

Readout   Visual display of information.

Reset   To clear normal Q output of FF to 0. A mode of operation on most flip-flops.

ROM   A nonvolatile read-only semiconductor memory.

R-S flip-flop   Simple FF having set and reset inputs and Q and $\bar{Q}$ outputs.

Set   To cause normal Q output of FF to go HIGH or to 1. A mode of operation on most flip-flops.

Shift register   A memory or register that shifts data right and/or left.

Storage   Also called memory. Holds data to be retrieved at a later time.

Truth table   Lists all possible states of input and output for a logic function.

TTL   Transistor-transistor logic. Currently the most commonly used digital logic family.

Write   Introduce information into memory.

Self-Examination
Answers at Back of Book

1. What is the radix for binary numbers?
2. The digit 0 with carry of 1 is the sum for what binary addition?
3. Which logic gate is similar to the function of two series switches?

4. Which logic gate is similar to the function of two parallel switches?
5. Which logic function has the output LOW only when both inputs are HIGH?
6. In any flip-flop, when the $\bar{Q}$ output is at 1, what is the state of the Q terminal?
7. Which circuit is used for a clock generator, a free-running MV or a *J-K* flip-flop?
8. What type of circuit is used for counters?
9. How many flip-flop circuits are needed to divide by 16?
10. Causing the Q output of a flip-flop to go HIGH is called what?
11. A flip-flop that operates in step with the system clock is said to operate _____.
12. What is another name for a data selector?
13. What is the abbreviation for a programmable read-only memory?
14. Which digital circuits require less power, TTL or CMOS?
15. The *J-K* flip-flop is one type of XOR gate. (True or False)

Essay Questions

1. Describe the difference between digital and analog signals.
2. Give the base or radix for the following number systems: binary, octal, decimal, hexadecimal.
3. Define fan-in and fan-out for a logic gate.
4. Give at least two comparisons between the CMOS and TTL families of digital circuits.
5. Compare the following pairs of logic gates: (a) AND with OR; (b) AND with NAND; (c) OR with NOR; (d) XOR with XNOR.
6. Write the Boolean expression for the Y output of the following gates: (a) AND; (b) NAND; (c) OR; (d) XOR.
7. Give the difference between a half-adder and a full-adder.
8. What precautions must be taken when working with CMOS digital IC units?
9. What are the useful modes of operation for a *J-K* flip-flop?
10. What is the difference between a multiplexer and a demultiplexer?
11. What do the following abbreviations stand for: RAM, ROM, PROM, EPROM?
12. Describe the operation of a seven-segment LED digital display.
13. What is the purpose of a truth table?
14. How does a flip-flop circuit differ from a logic gate?
15. Name two types of flip-flops.
16. What is the function of the clock input to a flip-flop circuit?
17. What is meant by the toggle mode of operation for a flip-flop?
18. Give at least two basic tools used in troubleshooting digital circuits.
19. Define minterm and maxterm Boolean expressions.
20. Draw the logic-symbol diagram for the following minterm Boolean expression: $A \cdot \bar{B} + \bar{A} \cdot B + A \cdot B = Y$.

Problems
Answers to Odd-Numbered Problems at Back of Book

1. Convert the following decimal numbers to binary equivalents: (a) 39; (b) 62.
2. Convert $(187)_{10}$ to binary form.

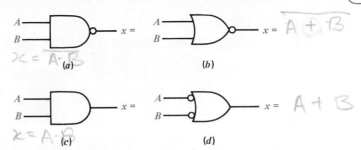

Fig. 31-41 For Prob. 5.

3. Convert $(1110)_2$ to a decimal number.
4. Add $(16)_{10}$ and $(9)_{10}$ in binary form and check that the sum equals $(25)_{10}$.
5. Give the output equation for each of the logic symbols in Fig. 31-41.

Answers to Practice Problems

31-1 a. 15
    b. 1010
    c. Zero
    d. Eight
    e. 1011

31-2 a. 111
    b. 1110

31-3 a. T
    b. T
    c. T
    d. F
    e. T
    f. T

31-4 a. F
    b. T
    c. T

31-5 a. Maxterm
    b. AND-OR
    c. 1

31-6 a. FET
    b. TTL, fastest
    c. CMOS

31-7 a. Not light, LOW
    b. HIGH
    c. HIGH
    d. Counterclockwise
    e. 5 V

31-8 a. Not light, LOW
    b. LOW
    c. LOW
    d. Static
    e. 12 V

31-9 a. Asynchronous
    b. *CK*
    c. HIGH
    d. Toggle

31-10 a. 8
    b. 011 (decimal 3)
    c. Four
    d. Serial

31-11 a. 10000, HIGH
    b. 7 (a, b, and c)
    c. Multiplexer (or data selector)
    d. RAM
    e. ROM

31-12 a. CMOS
    b. Dual-trace oscilloscope
    c. F

Chapter 32
Integrated Circuits

Microelectronic technology has greatly reduced the size of most electronic circuits. Microscopic transistors, diodes, resistors, and even capacitors are formed on a single piece of silicon. Hence the name *monolithic integrated circuit*. Both bipolar and MOS technologies can be used to form the tiny transistors in integrated circuits (ICs). An IC unit in a package 1 in. long by ⅜ in. wide can contain from a dozen to thousands of transistors.

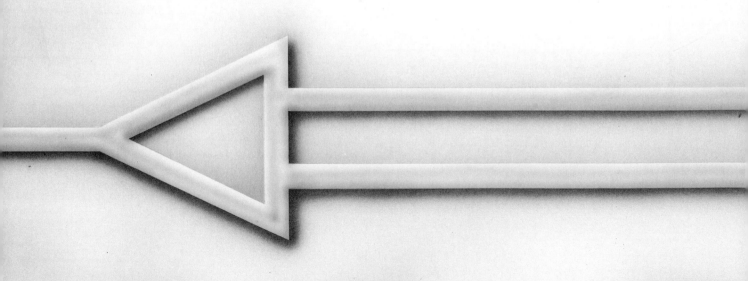

All new electronic equipment is full of integrated circuits. ICs are used almost exclusively in many digital devices, such as clocks and watches, microcomputers, games, and calculators. Integrated circuits are also widely used in modern analog devices, such as audio amplifiers and television receivers. The power rating for IC units varies widely, from a few microwatts for CMOS devices to 5 W and more for bipolar power amplifiers.

Important terms in this chapter are:

A/D converter
analog switch
astable MV
bistable MV
D/A converter
hybrid IC
integrated circuit (IC)
inverting amplifier
linear IC

monolithic IC
monostable MV
multivibrator (MV)
noninverting amplifier
operational amplifier (op amp)
sample-and-hold circuit
voltage comparator
voltage regulator

More details of integrated circuits are explained in the following sections:

32-1 Types of Integrated Circuits
32-2 Production of Integrated Circuits
32-3 Integrated Components
32-4 Linear IC Applications
32-5 Operational Amplifier Circuits
32-6 Multivibrators

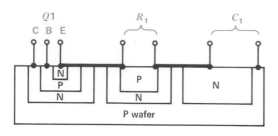

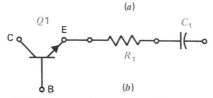

Fig. 32-1 Integrated circuit on silicon wafer. (*a*) Construction with NPN transistor, resistors, and capacitors. (*b*) Schematic diagram.

32-1   Types of Integrated Circuits

The example in Fig. 32-1 illustrates a transistor, resistor, and capacitor integrated on a single silicon wafer. Inductors are generally not integrated because they take too much space. Transistors and diodes are the easiest components to integrate. Resistors are not precise, but the circuits are designed to depend on the ratios of resistances rather than absolute values. The range of R values is 100 to 25,000 Ω, approximately. The range of C values is approximately 3 to 30 pF. External components are connected to the IC unit for larger values of R and C, with any L required for tuning.

For the construction in Fig. 32-1, the diffused layers at the left form an NPN transistor. Note that the collector has a reverse NP junction with the wafer to isolate the transistor from the silicon substrate or platform. In the middle section, R between the two ohmic contacts is determined by the length and width of the strip. The C at the right is the capacitance across the reverse-biased junction of the N material to the P wafer.

Monolithic IC   In this type, all the components are formed as part of a single P-type or N-type wafer. This is the construction shown in Fig. 32-1.

Thin-Film IC   In this type, the substrate is ceramic or glass, which is an insulator. All the components are evaporated on this insulating platform. The advantage is good isolation between the integrated components.

Thick-Film IC   In this type, R and C are formed on the substrate, but the transistors are added as discrete chips.

Hybrid IC   This type combines monolithic and thin-film units on a single ceramic platform. Discrete transistors may also be included, where high power rating is needed.

Digital and Linear IC Units   In terms of their function, IC chips are classified in either of these two broad groups. *Digital ICs* process digital signals and are pulse circuits. A digital signal is one that uses two discrete voltage levels, sometimes referred to as logical 0 and 1. Digital ICs were described in Chap. 31, "Digital Electronics."

Linear ICs process analog signals. An analog signal *varies continuously* in proportion to the input. Linear IC chips generally contain several amplifier circuits for either audio or RF signals. A linear amplifier has an output that varies in direct proportion to the input.

VLSI, LSI, MSI, and SSI Chips   Digital IC units are classified according to the complexity of their circuitry. Texas Instruments defines a VLSI (very-large-scale integration) chip as one that contains circuitry equal to 1000 or more logic gates. An LSI (large-scale integration) chip contains circuitry equal to 100 to 1000 logic gates. MSI (medium-scale integration) chips contain circuitry equal to from 12 to 100 gates, while SSI (small-scale integration) chips contain less than 12 gates. Manufacturers' definitions for VLSI and LSI chips do vary.

Practice Problems 32-1
Answers at End of Chapter

Answer True or False.
a. Digital ICs process signals that have two discrete voltage levels.
b. A linear IC would be used if an audio amplifier were needed.
c. Figure 32-1*a* illustrates hybrid IC construction.

32-2   Production of Integrated Circuits

For the monolithic type, IC chips are made by the batch (of 25 to hundreds of chips) from one silicon wafer. Figure 32-2 illustrates the steps in construction, from the wafer slices of a silicon ingot in *a*, to the scribed chips on the wafer in *b*, and the diffused components in *c* and *d*.

The individual silicon wafers are about 10 to 25 mils thick, with a diameter of 1.5 to 4 in. First, each wafer is polished to a mirror finish by acid etching. Then a layer of silicon dioxide (SiO_2) is formed on the wafer. This thin glass insulator protects the silicon surface and serves as a barrier to the doping for semiconductor junctions. The oxide coating is opened in a window pattern by photochemical techniques, to allow doping where desired. A sequence of oxidation and diffusion operations results in layers of the doped semiconductors. The oxidation forms silicon dioxide as a barrier and insulator. Then the diffusion forms the junctions.

Metallic paths are needed to interconnect the components in each IC chip. These are essentially aluminum printed wiring. They terminate at the edges of the chip, where tiny wires are bonded for connection to the external leads.

Automatic equipment with test probes is used to check the aluminum electrode pads. Each complete chip is tested sequentially by a large number of needle-tipped probes. This wafer probing on many electrodes for all the chips is controlled and monitored by computers.

Fig. 32-3 IC packages. Round IC unit is TO-5, with diameter of ½ in. The 16-pin dual inline package (DIP) is ¾ in. long. Flat pack at bottom is 1 in.

Since the complete silicon wafer has the physical properties of a thin sheet of glass, the chips are separated by methods similar to those used for cutting glass. A fine diamond point is used to scribe the wafer into chips (Fig. 32-2*b*). All the chips on one wafer are the same. The slice is held firmly and then stressed to break along the scribe lines. This division of the slice into chips is called "dicing."

Chips that are defective in probe testing are sorted out. Also, microscopic examination shows chips with imperfections resulting from dicing.

Then each chip is mounted in its package, which is generally a plastic case. Typical IC packaged units are shown in Fig. 32-3. The chip is essentially glued with

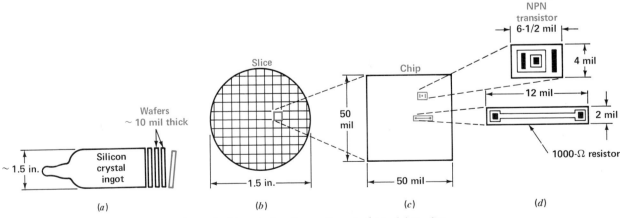

Fig. 32-2 Construction of integrated circuit chips starting from silicon ingot in (*a*) to slice in (*b*) and IC chip in (*c*), with magnified view of diffused components in (*d*).

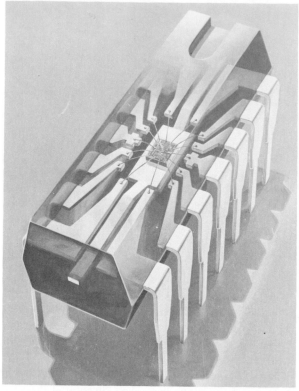

Fig. 32-4 IC chip in DIP unit. Length of package is 1 in.

epoxy. The terminal leads on the package are connected to the internal electrodes on the chip with 1.5-mil wire, aluminum or gold. Connections are made by either ultrasonic bonding or thermocompression bonding. The ultrasonic energy vibrates the bonding tip to produce heat. The final IC package is shown in Fig. 32-4. Note that most of the space in the package is needed for the connecting leads and pins.

The IC package is generally the round TO-5 case, ceramic flat pack, or dual inline form (DIP). The TO-5 style may be a glass-metal package, hermetically sealed in dry nitrogen, or a plastic case. The flat pack and DIP packages may be ceramic or plastic. Ceramic packages have a better hermetic seal than plastic.

Typical pin connections for IC units are shown in Fig. 32-5. The round TO-5 package generally has 8, 10, or 12 leads. The dual inline and flat pack may have 8, 14, 16, 24, 28, or 40 pins. Corresponding sockets are available for all the IC packages, but they are often soldered on the printed-circuit board.

The pins are numbered clockwise as you view the IC from the top. It should be noted that temporary connections can usually be made to the top of the pins on the IC package.

Practice Problems 32-2
Answers at End of Chapter

Answer True or False.
a. Dicing is the process of breaking the silicon wafer into individual chips.
b. Silicon dioxide (SiO_2) is an insulator, like glass.
c. Pin 1 is usually the center pin on an IC package.

32-3 Integrated Components

These include transistors, diodes, resistors, and capacitors. Isolation between adjacent components is provided by a reverse-biased PN junction. The isolation R can be 100 MΩ or more. The advantages of integrated components are miniaturization and matched character-

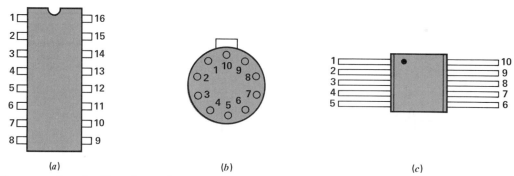

(a)   *(b)*   *(c)*

Fig. 32-5 Pin connections for IC packages. Dot or notch at pin 1. (*a*) A 16-pin DIP in TO-116 case. Top view. (*b*) A 10-pin TO-5 case. Bottom view. (*c*) A 10-pin ceramic flat pack in TO-81 case. Top view.

istics for the components with respect to operating values and temperature gradient.

Integrated Bipolar Transistors
Most common is the NPN construction shown before in Fig. 32-1. Note that the P substrate forms a reverse diode with the N collector. Also, the P substrate could be the collector for a PNP transistor with the adjacent N and P layers. Then the combination would form a complementary pair of NPN and PNP transistors.

Integrated Diodes
An N-type diffusion into the P substrate can produce diode junctions where they are needed on the chip. However, this is essentially the same as the emitter-base junction on a transistor. Therefore, the diode function can be provided by joining the transistor collector to base, serving as the anode (Fig. 32-6). The collector should not be left open, to prevent parasitic circuits with other components on the chip. Furthermore, the diode characteristics are matched to the transistors in the chip, as they are all made the same way. These diodes can be used for transistor bias in the base circuit, temperature compensation, and isolating the incorrect polarity of voltage for the collector on a separate transistor.

The collector can also be connected to the emitter for diode operation. This way both junctions are used in parallel for the diode.

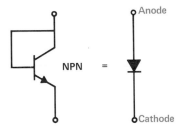

Fig. 32-6 Bipolar transistor connected as a diode.

Integrated Zener Diodes
By using reverse voltage on the emitter-base junction, it can serve as a zener diode. The zener voltage is typically 7 V. Remember that zener diodes are used for voltage regulation, to stabilize the supply voltage against changes with different load currents.

Integrated Resistors
A common method uses a strip of lightly doped P material, as shown in Fig. 32-7.

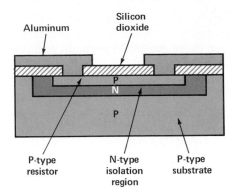

Fig. 32-7 Cross section of P-type diffused resistor in IC chip. (*RCA Manual IC-42*)

The N layer below the P layer will be reverse-biased for isolation from the substrate. The silicon dioxide provides insulation from the next layer. External connections to the resistor are made with the aluminum pads.

This type of monolithic R has a resistivity of about 200 Ω for the entire chip. However, the ratio of length to width of the R strip can provide values of 100 to 25,000 Ω. Higher values of 10 to 50 kΩ can be obtained by special doping techniques that reduce the cross-sectional area of the R strip. This type of resistor is a *pinch resistor*.

The tolerance on monolithic resistors is ± 30 to ± 50 percent, which is poor for precise values. Also, the temperature coefficient of resistivity is high. However, the resistance ratios on the same chip can be controlled within 2 to 3 percent. Therefore, the circuits are designed to depend on a resistance ratio, rather than on the absolute value of R.

Integrated Capacitors
The two types are the junction type illustrated for C in Fig. 32-1 and the MOS type in Fig. 32-8. In the junction type, the capacitance is across the depletion zone of a PN junction with re-

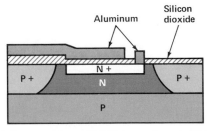

Fig. 32-8 Cross section of MOS type of monolithic capacitor. (*RCA Manual IC-42*)

verse bias. Typical values are 0.1 to 0.4 pF per square mil, with reverse bias of 5 V. If an area of 100 square mil is used out of the 2500 square mil on a chip, the C then will be 10 to 40 pF.

For the MOS capacitor in Fig. 32-8, the N^+ layer forms the bottom plate, the SiO_2 insulating layer is the dielectric, and the aluminum metallization serves as the opposite plate. The C value of 3 to 30 pF depends on just the dielectric and plate area. The + sign for N or P indicates heavy doping.

CMOS Transistors This abbreviation stands for complementary symmetry with P-channel and N-channel MOSFET or IGFET units, generally of the enhancement type. Other abbreviations used are COS/MOS and C/MOS. Figure 32-9 illustrates a CMOS inverter circuit in digital logic. The function of the inverter is to change the conditions in a train of pulse signals from high to low, or vice versa.

In Fig. 32-9a, the FET at the right uses the N substrate for the N channel. The gate, drain, and source electrodes are shown with heavy lines. The P^+ areas are used as guard bands to separate the different electrodes from each other. For the FET at the left, the P well forms the P channel. In this FET, the gate, drain, and source electrodes are separated from each other by N^+ guard bands.

In Fig. 32-9b, the inverter circuit makes use of com-

plementary symmetry. The substrate of the P channel is at $+V_{DD}$, and the N channel is at ground or 0 V. When the input voltage is zero, the P-channel FET is conducting but the N-channel FET is off. Then the output voltage is close to V_{DD}. When the input has a positive voltage pulse, however, conditions are reversed. Then the P-channel FET is off and the N-channel FET is on. These conditions make the output voltage close to 0 V. The net result is that V_{out} is high when V_{in} is low and V_{out} is low when V_{in} is high, which is the inverting function.

Supply Voltage Typical values are 15 to 40 V for linear IC units and 3 to 15 V for digital IC units, with monolithic construction. Usually, positive polarity is required, as for collector voltage on NPN transistors. For bipolar digital IC chips, +5 V is a common supply voltage, with 12 V for CMOS types.

Practice Problems 32-3
Answers at End of Chapter

Answer True or False.
a. A typical value for an integrated monolithic C is 80 μF.
b. To use an NPN transistor as a diode, join the collector and base.

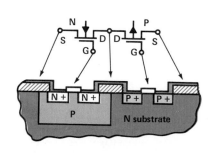

(a)

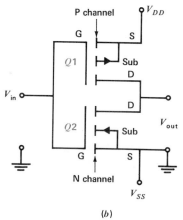

(b)

Fig. 32-9 (a) Cross section of CMOS diffused transistors on IC chip. (b) Inverter circuit for this construction. (RCA)

32-4 Linear IC Applications

As examples, National Semiconductor Corporation, a manufacturer of ICs, lists the following categories of linear integrated circuits in its databook:

1. Voltage regulators
2. Voltage references
3. Operational amplifiers
4. Instrumentation amplifiers
5. Voltage comparators
6. Analog switches
7. Sample-and-hold circuits
8. Analog-to-digital and digital-to-analog converters
9. Industrial, automotive, functional blocks and tele-communications
10. Audio, radio, and TV circuits
11. Transistor/diode arrays

These types of linear circuits are enclosed in flat pack, dual inline, or metal can packages like those shown in Fig. 32-3. Many linear ICs are monolithic integrations, while a few are hybrid circuits. Although most are bipolar, some use MOS technology. Some linear ICs combine bipolar and field-effect devices on the same substrate.

Voltage Regulators A voltage regulator is a circuit that holds an output voltage at a predetermined level regardless of input voltage changes or output load changes. The LM309K IC is an example of a popular +5-V voltage regulator. The TO-3 case style is shown in Fig. 32-10a. The pin connections for the LM309K IC are shown in Fig. 32-10b. A typical +5-V power supply used with TTL digital ICs and using the LM309K IC is drawn in Fig. 32-10c.

The half-cycles of sine waves shown for the input in Fig. 32-10c represent the output of a full-wave rectifier, using both half-cycles of the ac input for the fluctuating dc output. It has only one polarity, but the amplitudes vary. Filtering is needed to smooth out the amplitude values. The regulator circuit also has filtering, while maintaining a steady dc value of 5 V in the regulator output. A regulated supply voltage is important for digital circuits to provide specific values for the HIGH and LOW voltages. Note that the input and output filter capacitors are external components.

Voltage References A voltage reference is a highly regulated voltage source. This application is for circuits where the accuracy of the voltage source is very important.

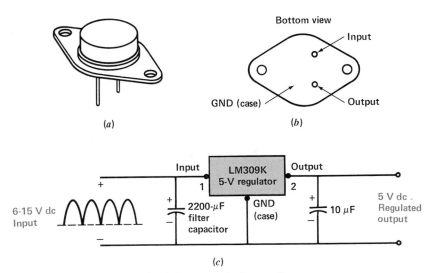

Fig. 32-10 LM309K +5-V voltage regulator. (a) TO-3 case style for regulator. (b) Base diagram for LM309K voltage regulator. (c) Use of LM309K regulator in a 5-V regulated power supply.

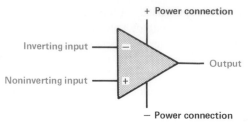

Fig. 32-11 Schematic symbol for an operational amplifier (op amp).

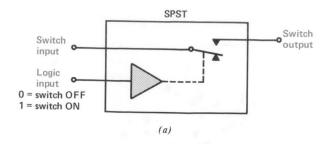

(a)

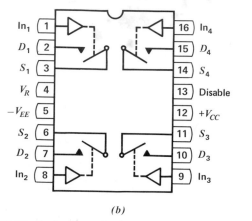

(b)

Operational Amplifiers An operational amplifier (op amp) is a high-gain direct-coupled amplifier that depends on external feedback from output to input to determine the operating characteristics. The gain of the op amp is set with external resistors. The word *operational* comes from the fact that early op amps were used in analog computers to perform mathematical operations, such as addition or multiplication.

The schematic symbol for an op amp is shown in Fig. 32-11. The op amp has two inputs, labeled inverting ($-$) and noninverting ($+$). The output is shown on the right side of the symbol in Fig. 32-11. Power-supply connections are shown at the top and bottom of the symbol.

The operational amplifier is probably the most adaptable linear IC. Because of its popularity, methods of using the op amp will be explained in Sec. 32-5.

Voltage Comparators A voltage comparator IC compares two analog input voltages and produces a logical 0 or 1 output when the voltages being compared are equal or are greater or less than the reference level. A voltage comparator uses the op amp schematic symbol because op amps are often used as voltage comparators.

Analog Switches An analog switch is an IC that replaces a mechanical switch. A logic input controls when the bilateral switch is on or off. Analog switches come in SPST, SPDT, normally open, normally closed, 1-of-8 multiplexer, and other customary switching arrangements.

A schematic diagram for a single analog switch is shown in Fig. 32-12a. When a logical 1 is placed on

Fig. 32-12 Analog switch. (*a*) Schematic symbol for a SPST analog switch. (*b*) Pin diagram for a typical IC containing four SPST analog switches. Diagram for the LF11331 quad SPST JFET analog switch. Top view. (*National Semiconductor Corporation*)

the logic input, the switch is ON. When the logic input drops to a logical 0, the switch opens or is OFF. A pin diagram for a typical DIP quad SPST analog switch is shown in Fig. 32-12b.

Sample-and-Hold Circuits A sample-and-hold IC is used to ''sample'' an analog input signal. This sample input voltage is frozen in time by being stored in an external capacitor. While the input is disconnected, this frozen sample voltage is output to another device such as an A/D converter for processing.

A typical schematic of an LM198 sample-and-hold IC with external hold capacitor is drawn in Fig. 32-13. The LM198 IC takes a sample of the analog input voltage when the logic input is at logical 1. The LM198 IC holds the voltage, stored in the capacitor at the output, when the logic input is 0.

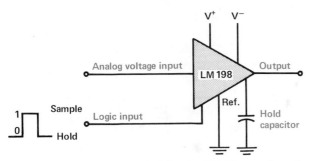

Fig. 32-13 Schematic symbol for the LF198 sample-and-hold IC. (*National Semiconductor Corporation*)

A/D and D/A Converters

A digital-to-analog IC (D/A converter) converts a digital (binary) input into a voltage whose magnitude is proportional to the value of the digital signal. For instance, if the digital input of 0001 causes a 1-V output, then a digital input of binary 0011 would cause a 3-V analog output.

An analog-to-digital IC (A/D converter) converts a continuously varying analog input voltage into a proportional digital output. For instance, if a 1-V analog input equals a binary 0001, then a 5-V input would equal a digital output of binary 0101.

Industrial, Automotive, Functional Blocks, and Telecommunications

Linear ICs in this miscellaneous category include some of the following:

1. Timers
2. Dot/bar display driver
3. Fluid detector
4. Temperature controller
5. Voltage/temperature to frequency converter
6. Ground-fault interrupter
7. Tone decoder
8. Phase-locked loop
9. Adjustable current sources
10. Voltage-controlled oscillator
11. Smoke detector
12. Light-to-current converter

The phase-locked loop is a basic element in many types of communications systems, including radio and television.

Audio, Radio, and TV Circuits

Some of the linear ICs in the audio circuits category include:

1. Preamplifiers
2. Audio amplifiers

3. Power amplifiers
4. FM detector, limiter, and audio amplifiers
5. Dolby B-type noise reduction processors
6. AGC/squelch amplifiers

Some of the linear ICs in the radio circuits category include:

1. RF/IF amplifiers
2. AM radio system
3. AM/CB receiver system
4. FM receiver IF system
5. IF amp/detector
6. Phase-locked loop FM stereo demodulator
7. Balanced modulator-demodulator

Some of the linear ICs in the TV circuits category include:

1. Monolithic TV sound system
2. TV video modulator
3. TV automatic fine tuning
4. IF video amp/detector
5. Color TV chroma demodulator
6. Digital tuning station detector
7. TV chroma IF amplifier
8. Monolithic video IF amplifier
9. RGB (red, green, blue) matrix preamplifier
10. TV signal processing circuit
11. Color demodulator

Transistor and Diode Arrays

Groups of transistors and diodes are integrated on a common monolithic substrate. They are packaged in dual inline and metal can packages. The arrays are useful where matched pairs of devices are necessary.

Practice Problems 32-4
Answers at End of Chapter

Answer True or False.

a. The op amp is probably the linear IC most adaptable to many uses.

b. The schematic symbol shown in Fig. 32-12*a* is for a sample-and-hold gate.

c. The sample-and-hold IC has both an analog voltage and a logic input.

d. On a D/A converter, if binary input 0011 equals an analog output of 3 V, then an input of binary 0111 would equal an output of 10 V.

32-5   Operational Amplifier Circuits

The practical features to note about the op amp circuit are that it has two opposite input terminals and one output terminal. A signal at the inverting input produces an output of opposite polarity. Signal input to the noninverting terminal is not inverted. Feedback from the output terminal to one of the input terminals is generally used for the purpose of stabilizing the gain. The gain is reduced, but the feedback increases the bandwidth. The more the gain is reduced, the wider the frequency response is.

Inverting Amplifier   The op amp is shown wired as an inverting amplifier in Fig. 32-14a. Note the use of a feedback resistor R_F and an input resistor R_{in}. The signal is applied to the inverting (−) input of the op amp. Look at the waveforms in Fig. 32-14a and notice that the signal has been both *inverted* and *amplified*.

The *voltage gain* of the op amp is set by the value of the resistors R_{in} (input resistor) and R_F (feedback resis-

tor). A simple method of calculating the voltage gain A_V of the op amp is shown in Fig. 32-14b. In this example, $R_{in} = 1$ kΩ, while $R_F = 10$ kΩ. Using the voltage gain formula for op amp inverting amplifiers,

$$A_V = \frac{R_F}{R_{in}} = \frac{10,000}{1000} = 10$$

The voltage gain of the inverting amplifier circuit in Fig. 32-14b equals 10.

Suppose the input voltage V_{in} in Fig. 32-14b is + 0.5 V. The output voltage V_{out} would be calculated as

$$V_{out} = A_V \times V_{in}$$
$$= 10 \times 0.5$$
$$V_{out} = 5 \text{ V}$$

The output voltage for the inverting amplifier shown in Fig. 32-14b would be −5 V. The output voltage is negative because the inverting input to the op amp was used, causing a polarity reversal.

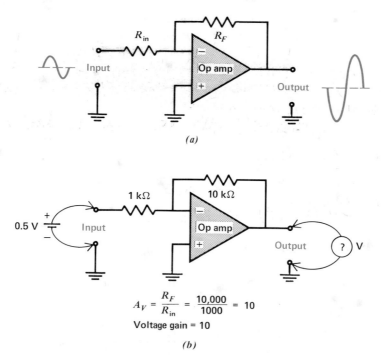

Fig. 32-14 Inverting amplifier circuit. (a) Op amp wired as an inverting amplifier. Note the phase reversal from input to output. (b) Calculating the voltage gain for the op amp circuit using the values of external resistors R_F and R_{in}.

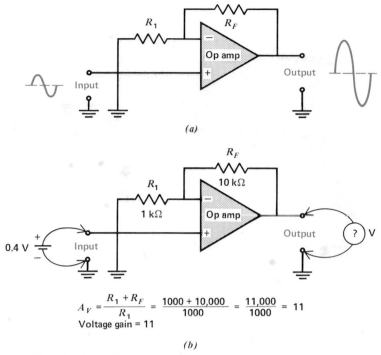

$$A_V = \frac{R_1 + R_F}{R_1} = \frac{1000 + 10,000}{1000} = \frac{11,000}{1000} = 11$$

Voltage gain = 11

(b)

Fig. 32-15 Noninverting amplifier circuit. (a) Op amp wired as a noninverting amplifier. Note that the input and output waveforms are in phase. (b) Calculating the voltage gain of the noninverting op amp circuit using the values of the external resistors.

Noninverting Amplifier   A schematic diagram for a noninverting amplifier circuit using an op amp is shown in Fig. 32-15a. Note that the signal goes into the noninverting input (+) of the op amp. However, R_F and R_1 are still connected in the same way as in Fig. 32-14. The waveforms in Fig. 32-15a show some voltage gain but no phase reversal. The input and output signals are said to be in phase.

The voltage gain of a noninverting amplifier using an op amp is calculated in Fig. 32-15b. Using the voltage gain formula for noninverting amplifiers,

$$A_V = \frac{R_1 + R_F}{R_1} = \frac{1000 + 10,000}{1000} = \frac{11,000}{1000} = 11$$

The voltage gain of the noninverting amplifier circuit in Fig. 32-15b equals 11. Note that the gain formula is not the same as before because R_1 is not in the path of input signal, as in Fig. 32-14.

Suppose the input voltage V_{in} in Fig. 32-15b is +0.4 V. The output voltage V_{out} would be calculated as

$$V_{out} = A_V \times V_{in}$$
$$= 11 \times 0.4$$
$$V_{out} = 4.4 \text{ V}$$

The output voltage from the noninverting amplifier shown in Fig. 32-15b would be +4.4 V. The output voltage is positive because the noninverting input to the op amp was used.

Note that both amplifiers in Figs. 32-14 and 32-15 can be either dc or ac amplifiers. The voltage gain in both circuits is set by the ratio of external resistors. These amplifier circuits are but two of many applications of operational amplifiers. For clarity, the dc power supply connections were not shown in Figs. 32-14 and 32-15, but are understood to be present for proper operation of the op amps.

Practice Problems 32-5
Answers at End of Chapter

a. The input to the op amp schematic symbol labeled with a _____ (−, +) is the inverting input.

b. Refer to Fig. 32-14a. If $R_F = 10$ kΩ and $R_{in} = 5$ kΩ, the voltage gain A_V for this amplifier will be _____.

c. Refer to Fig. 32-14a. If $R_F = 10$ kΩ, $R_{in} = 5$ kΩ, and $V_{in} = -1$ V, then $V_{out} = $ _____ V (sign and number).

d. Refer to Fig. 32-15a. If $R_F = 4500$ Ω, $R_{in} = 500$ Ω, and $V_{in} = -0.2$ V, then $V_{out} = $ _____ V (sign and number).

32-6   Multivibrators

The multivibrator (MV) is a pulse generator, producing a square-wave or rectangular-wave output. Multivibrators are often classified as to their stability. A stable stage will remain in the OFF condition until it is triggered into operation by external pulses. The three classes are:

Astable MV   This is also called a *free-running multivibrator*. It is an oscillator circuit that generates a continuous flow of digital pulses, as shown in Fig. 32-16a. An astable MV is an MV without a stable state.

Bistable MV   This is also called a *flip-flop*. An input pulse reverses the state of the output. As an example, Fig. 32-16b shows the single input pulse causing the output to drop from HIGH to LOW and remain LOW. The bistable MV is always in one of two stable states.

Monostable MV   This is also called a *one-shot multivibrator*. With any length input pulse, the monostable MV produces a single output pulse of a given time duration. Figure 32-16c shows a long input pulse producing a short output pulse from the monostable MV.

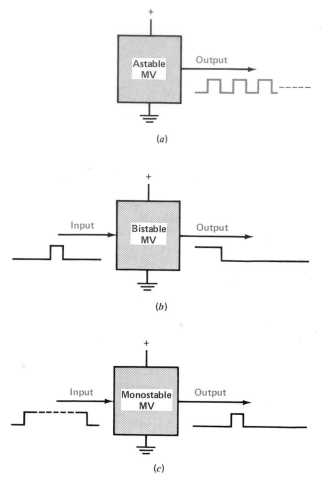

Fig. 32-16 Types of multivibrators (MV). (*a*) Astable MV produces a continuous string of pulses. (*b*) Bistable MV or flip-flop. Output changes state with input trigger pulse. (*c*) Monostable or one-shot MV. Low to high input causes short output pulse.

555 Timer IC   This IC can be connected with external resistors and capacitors to form astable, bistable, or monostable multivibrators. The 555 IC is wired as a free-running MV in Fig. 32-17a. This astable MV produces a continuous square-wave output. The frequency and duty cycle of the output can be changed by adjusting the values of the resistors and capacitor.

The 555 timer is a very popular and extremely adaptable IC. It is packaged in an 8-pin mini DIP, pictured in Fig. 32-17b, or in a metal can package. The 8-pin DIP pin connection diagram is shown in Fig. 32-17c.

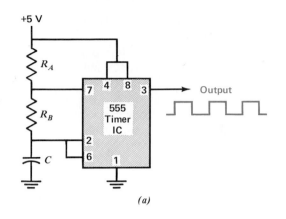

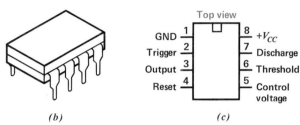

Fig. 32-17 555 timer IC. (*a*) Connected as an astable pulse generator. Values of *R* and *C* determine the frequency and duty cycle of the multivibrator. (*b*) Typical IC package. This is 8-pin DIP or mini DIP. (*c*) Pinout diagram.

Practice Problems 32-6
Answers at End of Chapter

Answer True or False.

a. A multivibrator is a pulse generator that produces a sine-wave output.

b. A one-shot multivibrator is called a bistable MV.

c. A monostable MV is also called a flip-flop.

d. The 555 IC in Fig. 32-17*a* is wired as a free-running multivibrator.

Summary

The following terms and definitions summarize the main features of integrated circuits.

A/D converter Analog-to-digital converter; translates analog inputs to proportional binary outputs.

Analog switch IC form of a mechanical switch activated by either a HIGH or LOW logic level.

Bipolar transistors NPN and PNP transistors. Also junction transistors.

Chip Tiny semiconductor containing integrated circuits.

CMOS (*also* COS/MOS *and* C/MOS) Complementary symmetry with MOSFET devices.

D/A converter Digital-to-analog converter translates binary inputs to proportional analog outputs.

Dicing Breaking or cutting the semiconductor slice into chips.

Digital IC Circuit which contains gates or flip-flops and processes digital signals.

DIP Dual inline package, commonly used for ICs.

Hybrid IC Chip which combines monolithic and thin-film construction techniques.

IC Integrated circuit with semiconductors, resistors, and capacitors on one chip.

Linear IC IC unit which processes analog signals.

LSI Large-scale integration chip; contains circuitry for 100 to 1000 gates.

Microelectronics Microscopic components used to make ICs and thus miniaturize electronic equipment.

Monolithic IC Chip with circuit components integrated on a single silicon substrate.

MOS Metal-oxide semiconductor type of IC construction. Components are like IGFETs.

MSI Medium-scale integration chip; contains circuitry for 12 to 100 gates.

Multivibrator (MV) A pulse generator circuit producing a square wave. May be of the astable, bistable, or monostable type.

Operational amplifier High-gain amplifier used with feedback network to determine its operating characteristics.

Sample-and-hold circuit Freezes a "sample" analog input voltage and presents the sample voltage at the output for a time.

Substrate Platform on which IC components are formed.

Thick-film IC Circuit in which R and C are formed on glass or ceramic substrate with transistors added as discrete chips.

Thin-film IC Circuit in which R, C, and transistors are deposited on glass or ceramic substrate.

VLSI Very-large-scale integration chips; contain circuitry for more than 1000 gates.

Voltage regulator IC that maintains output voltage constant regardless of input voltage or output load conditions.

Self-Examination
Answers at Back of Book

Answer True or False.

1. Figure 32-1 illustrates the construction of a monolithic linear IC.
2. Transistors and diodes are the easiest components to integrate, while inductors are rarely placed on ICs.
3. Most of the chips in a microcomputer would be digital ICs.
4. A silicon wafer is used as the foundation or substrate for most monolithic ICs.
5. The DIP is used to package digital ICs, whereas the metal can package is preferable for linear ICs.
6. MOS technology integrates active devices similar to the IGFET.
7. A typical dc supply voltage for IC units is 3 to 40 V.
8. A voltage comparator is an example of a digital IC.
9. The input to a D/A converter may be binary while its output will be a proportional analog voltage.
10. A modern TV receiver uses all digital ICs.
11. An astable MV is commonly called a free-running multivibrator.
12. The 555 timer IC can be wired as a D/A converter or a voltage comparator.
13. A mini DIP refers to a 16-pin dual inline package IC.
14. The op amp can produce either inverted or noninverted output.
15. Feedback is generally not necessary for the op amp.

Essay Questions

1. What is the difference between linear and digital ICs?
2. Briefly compare the construction of monolithic, thin-film, thick-film, and hybrid IC units.
3. List five types of components that are typically integrated in monolithic construction.
4. What is meant by complementary symmetry?
5. (a) List eight types of linear ICs. (b) List five types of digital ICs.
6. Draw the schematic symbol for an op amp, an analog switch, and a sample-and-hold circuit.
7. What two linear ICs "bridge the gap" or interface digital and analog circuits?
8. Define astable, bistable, and monostable multivibrators.
9. What adaptable IC unit can be wired as an astable, bistable, or monostable type of MV?
10. Describe two features of the op amp.
11. How do you identify the pin numbers on an IC package?
12. What is the purpose of using a voltage regulator IC?

Problems

Answers to Odd-Numbered Problems at Back of Book

1. Calculate the voltage gain A_V of the op amp in Fig. 32-14a when R_{in} is 5 kΩ and R_F is 10 kΩ.
2. In Fig. 32-14a, what is the output voltage when V_{in} is +0.5 V, for $R_{in} = 5$ kΩ and $R_F = 10$ kΩ, as in Prob. 1?
3. What is the output voltage and polarity in Fig. 32-14a with V_{in} of −1.3 V? The R_{in} is 5 kΩ and R_F is 10 kΩ, as in Prob. 1.
4. Calculate the voltage gain A_V of an amplifier with output of 2.7 V and input of 220 mV.
5. Refer to the op amp in Fig. 32-15a. Calculate the voltage gain A_V for $R_1 = 1$ kΩ and $R_F = 100$ kΩ.
6. Increase R_1 in Prob. 5 to 2 kΩ. How much is A_V now?
7. What are the output voltage and polarity for the example in Prob. 5 with V_{in} equal to +0.02 V? The R_1 is 1 kΩ.

Answers to Practice Problems

32-1 a. T	32-3 a. F	32-5 a. −	32-6 a. F
b. T	b. T	b. 2	b. F
c. F	32-4 a. T	c. +2	c. F
32-2 a. T	b. F	d. −2	d. T
b. T	c. T		
c. F	d. F		

Review Chapters 29 to 32

Summary

1. In vacuum tubes the cathode is heated to emit electrons. The anode or plate collects these electrons to provide plate current. The main tube types are diodes, triodes, tetrodes, and pentodes, as listed in Table 29-3. Semiconductor devices are listed in Table 29-5.

2. Diodes are used as rectifiers to change alternating current to direct current. Current can flow through the diode only when the anode is positive.

3. Triodes, tetrodes, and pentodes are used for amplifiers, since the control-grid voltage controls the plate current. Making the control-grid voltage more negative decreases the plate current; making it less negative increases the plate current.

4. Transistors are amplifiers that make use of controlled current flow in solid semiconductors such as germanium and silicon. No heater is necessary.

5. N-type germanium or silicon is doped with an impurity that makes it negative with a surplus of free electrons; P-type has a surplus of positive hole charges.

6. The transition between P and N semiconductors is a PN junction. The junction has an internal barrier voltage of 0.7 V for silicon or 0.3 V for germanium that blocks forward current. However, forward voltage of the same polarity as the electrodes allows current in the forward direction. Reverse voltage of opposite polarity prevents forward current.

7. A current of hole charges is in the direction of conventional current, opposite from electron flow. Hole current into the semiconductor is considered the positive direction of I.

8. A PNP transistor has a wafer of an N-type semiconductor forming two junctions with two P-type semiconductors. An NPN transistor is the reverse. Both types are used as amplifiers. The semiconductor in the center is the base. The

emitter at one end has forward bias to supply charges to its junction with the base; the collector at the other end has reverse bias and receives charges from its junction with the base.

9. In the schematic symbol the arrowhead on the emitter indicates the direction of hole current into the base for a P emitter or out from the base for an N emitter.

10. The main types of transistor circuits are common base (CB), common emitter (CE), and common collector (CC). The CE circuit is generally used for amplifiers because it has the most gain.

11. The two main characteristics of a transistor are its α (alpha), which is the ratio of collector current to emitter current, and β (beta), which is the ratio of collector current to base current.

12. Silicon and germanium diodes are generally used in rectifier circuits.

13. A half-wave rectifier uses one diode. With 60-Hz ac input, the ripple frequency in the dc output is 60 Hz. A full-wave rectifier uses two diodes to rectify both halves of the ac input cycle; the ripple frequency is 120 Hz.

14. The SCR is a silicon-controlled rectifier with a gate electrode to control the start of current between anode and cathode. The triac is a similar power control device, but is bidirectional for current flow. The unijunction transistor (UJT) is a double-base diode used for switching and timing applications. The SCR, triac, and UJT are thyristors.

15. The varactor is a semiconductor diode used as a voltage-controlled capacitor.

16. The zener diode takes reverse voltage for use as a voltage regulator.

17. The field-effect transistor (FET) has a gate electrode to control the electric field in the channel between source and drain. The gate corresponds to the base; the source supplies charges like an emitter; the drain corresponds to a collector receiving the charges. Input signal voltage is applied to the gate, and amplified output is taken from the drain, in the common-source (CS) circuit.

18. Visual display devices include the light-emitting diode (LED) and liquid-crystal display (LCD).

19. The cathode-ray tube (CRT) has an electron gun and fluorescent screen in an evacuated glass envelope. See Fig. 29-33.

20. The operational amplifier (op amp) has one output terminal and two inputs, inverting and noninverting. Feedback is necessary to stabilize the gain and increase bandwidth.

21. Integrated circuits combine semiconductor devices R and C on one silicon chip. The advantages are small space and reliability.

22. Linear IC units are commonly used for AF and RF amplifiers.

23. Digital IC units are commonly used for gates, flip-flops, and additional types of logic circuits.

24. Digital electronics uses binary numbers with only the digits 0 and 1 to represent the two states of a pulse signal.

25. The basic gates for digital logic are AND, NAND, OR, and NOR. See Table 31-1.

26. The flip-flop is a bistable circuit that can be changed from one state to the opposite by input pulses.

27. The astable multivibrator (MV) is a pulse generator used as the clock reference for timing in digital circuits.

28. The terminology for digital electronics is described in the Summary for Chap. 31. The terminology of IC units is described in the Summary for Chap. 32.

Review Self-Examination

Answers at Back of Book

Answer True or False.

1. A thermionic cathode is heated to emit electrons.
2. The anode has a positive potential with respect to the cathode to attract electrons.
3. Plate current can flow in only one direction.
4. An open heater in a CRT results in no beam current.
5. An N-type semiconductor has free electrons, while a P-type has hole charges.
6. In a transistor amplifier, the collector has reverse voltage.
7. The schematic symbol with an arrowhead into the base indicates an NPN transistor.
8. Forward bias is applied to an NPN transistor by the + voltage to the emitter and the − voltage to the base.
9. The β characteristic of transistors compares collector current to base current.
10. Typical values of β for transistors are 0.98 to 0.99.
11. In the NPN transistor, collector current is hole charges supplied by the emitter.
12. The + mark or a band on silicon diode power rectifiers indicates where positive dc output voltage is obtained.
13. Positive voltage applied to the collector of an NPN transistor is the polarity for reverse voltage.
14. Minority charge carriers have opposite polarity from majority carriers.
15. The I_{CBO} in a transistor consists of majority carriers.
16. A typical value of input signal for a PNP transistor is about 8 V.
17. Bias stabilization is used to prevent thermal runaway.
18. The FET has a very high input resistance.
19. The gate electrode in the FET corresponds to the collector in a bipolar transistor.
20. The SCR is a silicon rectifier with a gate electrode to control when current flows from cathode to anode.
21. Typical forward bias on a silicon NPN transistor is 0.6 to 0.7 V.
22. The zener diode is used as a constant-voltage source.
23. A typical capacitance on an IC chip is 8 μF.
24. An N channel is used in NPN transistors.
25. A DIP unit is generally in a TO-3 case.
26. The binary equivalent of decimal 7 is 0111.
27. The sum of $1 + 1$ in the binary system is 0 with carry of 1.
28. The flip-flop circuit is a latch.
29. The AND gate has output of AB with inputs of A and B.
30. In a flip-flop, when the Q output is 1, the $\overline{Q}$ output is 0.
31. The J-K flip-flop uses clock pulses.
32. A clock generator can use the astable MV.
33. The op amp circuit is used in A/D converters.
34. A counter circuit uses flip-flops.
35. The XOR gate is the same as a NAND gate.

References
Additional References at Back of Book

Hall, D.: *Microprocessors and Digital Systems*, McGraw-Hill Book Company, New York.

Hibberd, R. B.: *Integrated Circuits*, McGraw-Hill Book Company, New York.

Ilardi, F.: *Computer Circuit Analysis—Theory and Applications*, Prentice-Hall, Inc., Englewood Cliffs, N.J.

Jung, W. G.: *IC Op-amp Cookbook*, Howard W. Sams and Co., Inc., Indianapolis, Ind.

Malvino, A. P.: *Digital Computer Electronics*, McGraw-Hill Book Company, New York.

Malvino, A. P.: *Transistor Circuit Approximations*, McGraw-Hill Book Company, New York.

Millman, J., and C. Halkias: *Electronic Devices and Circuits*, McGraw-Hill Book Company, New York.

Sams, H. W.: *Semiconductor General-Purpose Replacements*, Howard W. Sams and Co., Inc., Indianapolis, Ind.

Schuler, Charles A. *Electronics—Principles and Applications*, McGraw-Hill Book Company, New York.

Tokheim, R. L.: *Digital Electronics*, McGraw-Hill Book Company, New York.

Pamphlets, Catalogues, and Manuals

Application Notes, General Instrument Corp., Hicksville, N.Y.

Application Notes, Motorola Semiconductor Products Inc., Phoenix, Ariz.

Application Notes, Texas Instruments Inc., Dallas, Tex.

GE Transistor Manual, General Electric Co., Semiconductor Products, Syracuse, N.Y.

Linear Integrated Circuits, National Semiconductor Corporation, Santa Clara, Calif.

RCA Receiving Tube Manual, RCA Electronic Components and Devices, Harrison, N.J.

RCA Solid-State Devices Manual, RCA Solid-State Division, Somerville, N.J.

SCR Designers Handbook, Westinghouse Electric Corporation, Youngwood, Pa.

Semiconductor Cross-reference Guide, Motorola Semiconductor Products Inc., Phoenix, Ariz.

Semiconductor Data Handbook, General Electric Co., Syracuse, N.Y.

Semiconductor Master Replacement Guide, Philips ECG Inc., Williamsport, Pa.

TTL Data Book, Fairchild Camera and Instrument Corporation, Mountain View, Calif.

Appendixes

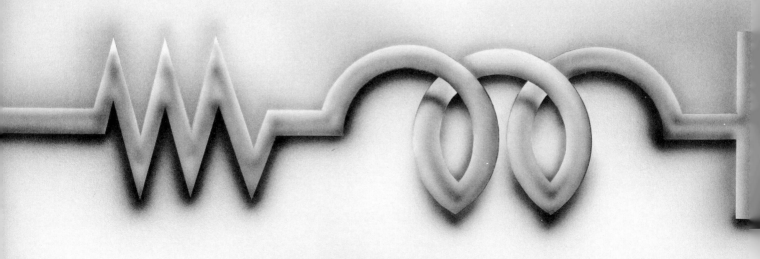

Contents

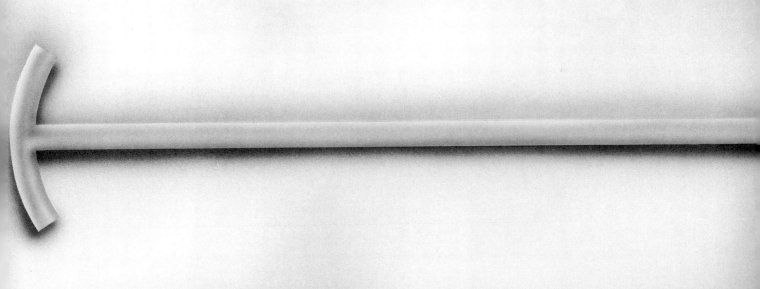

Appendix A

Electronic Frequency Spectrum

The four main categories of electromagnetic radiation and their frequencies are described as follows:

1. Radio frequency waves from 30 kHz to 300,000 MHz.
2. Heat waves or infrared rays from 1×10^{13} to 2.5×10^{14} Hz. Infrared means below the frequency of visible red light.
3. Visible light frequencies from about 2.5×10^{14} Hz for red up to 8×10^{14} Hz for blue and violet.

4. Ionizing radiation such as ultraviolet rays, x-rays, gamma rays, and cosmic rays from about 8×10^{14} Hz for ultraviolet light to above 5×10^{20} Hz for cosmic rays. Ultraviolet means above the frequencies of blue and violet visible light.

Note that radio frequency (RF) waves have the lowest frequencies and longest wavelengths in these categories of electromagnetic radiation.

Frequency or Wavelength*	Name	Applications
0 Hz	Steady direct current or voltage	DC motors; solenoids; relays; dc electrode voltages for tubes, transistors, and integrated circuits
16–16,000 Hz	Audio frequencies	60-Hz power, ac motors, audio amplifiers, microphones, loudspeakers, phonographs, tape recorders, high-fidelity equipment, public address systems, and intercoms
16–30 kHz	Ultrasonic frequencies or very low radio frequencies	Sound waves for ultrasonic cleaning, vibration testing, thickness gaging, flow detection, and sonar; electromagnetic waves for induction heating
30 kHz–30,000 MHz	Radio frequencies (see App. B)	Radio communications and broadcasting, including television, radio navigation, radio astronomy, industrial, medical, scientific, and military radio
30,000–300,000 MHz or 1–0.1 cm	Extra-high frequencies	Experimental, weather radar, amateur, government
30–0.76 μm	Infrared light rays	Heating, infrared photography
0.76–0.39 μm	Visible light rays	Color, illumination, photography
0.39–0.032 μm	Ultraviolet rays	Sterilizing, deodorizing, medical
0.032–1.0 $\times$ 10^{-5} μm	X-rays	Thickness gages, inspection, medical
1.0 $\times$ 10^{-5}–6.0 $\times$ 10^{-7} μm	Gamma rays	Radiation detection; more penetrating than hardest x-rays
Shortest of all electromagnetic waves	Cosmic rays	Exist in outer space; can penetrate 70 m of water or 1 m of lead

*Frequency and wavelength are inversely proportional to each other. The higher the frequency, the shorter the wavelength, and vice versa. AF and RF waves are generally considered in terms of frequency because the wavelength is so long. The exception is microwaves, which are often designated by wavelength because their frequencies are so high. Light waves, x-rays, and gamma rays are also generally considered in terms of wavelength because their frequencies are so high. The units of wavelength are the micrometer, equal to 10^{-6} m; the nanometer, equal to 10^{-9} m; and the angstrom (Å), equal to 10^{-10} m.

Appendix B

FCC Frequency Allocations from 30 kHz to 300,000 MHz

The main categories of radio frequencies can be summarized as follows:

1. Very low frequencies (VLF), below 30 kHz
2. Low frequencies (LF), 30 to 300 kHz
3. Medium frequencies (MF), 0.3 to 3 MHz
4. High frequencies (HF), 3 to 30 MHz
5. Very high frequencies (VHF), 30 to 300 MHz
6. Ultra-high frequencies (UHF), 300 to 3000 MHz or 0.3 to 3 GHz
7. Super-high frequencies (SHF), 3000 to 30,000 MHz or 3 to 30 GHz
8. Extra-high frequencies (EHF), 30,000 to 300,000 MHz or 30 to 300 GHz

The gigahertz (GHz) unit is 10^9 Hz or 10^3 MHz. Note that each band covers a range of 10:1 in frequencies. Details of the frequencies assigned by the Federal Communications Commission for specific services are listed below.

Band	Allocation	Remarks
30–535 kHz	Includes maritime communications and navigation, aeronautical radio navigation	Low and medium radio frequencies
535–1605 kHz	Standard radio broadcast band	AM broadcasting
1605 kHz–30 MHz	Includes amateur radio, loran, government radio, international shortwave broadcast, fixed and mobile communications, radio navigation, industrial, scientific and medical, and the CB radio band	Amateur bands 3.5–4.0 MHz and 28–29.7 MHz; industrial, scientific, and medical band 26.95–27.54 MHz; citizen's band class D for voice is 26.965–27.405 MHz in forty 10-kHz channels
30–50 MHz	Government and nongovernment, fixed and mobile	Includes police, fire, forestry, highway, and railroad services; VHF band starts at 30 MHz
50–54 MHz	Amateur	6-m band
54–72 MHz	Television broadcast channels 2 to 4	Also fixed and mobile services
72–76 MHz	Government and nongovernment services	Aeronautical marker beacon on 75 MHz
76–88 MHz	Television broadcast channels 5 and 6	Also fixed and mobile services
88–108 MHz	FM broadcast	Also available for facsimile broadcast; 88–92 MHz educational FM broadcast
108–122 MHz	Aeronautical navigation	Localizers, radio range, and airport control
122–174 MHz	Government and nongovernment, fixed and mobile, amateur broadcast	144–148 MHz amateur band
174–216 MHz	Television broadcast channels 7 to 13	Also fixed and mobile services
216–470 MHz	Amateur, government and nongovernment, fixed and mobile, aeronautical navigation, citizen's radio band	Radio altimeter, glide path, and meteorological equipment; citizen's radio band 462.5–465 MHz; civil aviation 225–400 MHz; UHF band starts at 300 MHz
470–890 MHz	Television broadcasting	UHF television broadcast channels 14 to 83
890–3000 MHz	Aeronautical radio navigation, amateur broadcast, studio-transmitter relay, government and nongovernment, fixed and mobile	Radar bands 1300–1600 MHz
3000–30,000 MHz	Government and nongovernment, fixed and mobile, amateur broadcast, radio navigation	Super-high frequencies (SHF); communications satellites at 3.7–4.2 GHz, 5.9–6.4 GHz, 12.2–12.7 GHz, and 17.3–17.8 GHz
30,000–300,000 MHz	Experimental, government, amateur	Extra-high frequencies (EHF)

Appendix C

Alphabetical Listing of the Chemical Elements

Element	Symbol	Atomic Number	Year of Discovery	Remarks*
Actinium	Ac	89	1899	
Aluminum	Al	13	1825	Metal conductor
Antimony	Sb	51	Ancient	
Argon	Ar	18	1894	Inert gas
Arsenic	As	33	1649	
Barium	Ba	56	1808	Ends 8-electron O shell
Beryllium	Be	4	1798	
Bismuth	Bi	83	1753	
Boron	B	5	1808	
Bromine	Br	35	1826	
Cadmium	Cd	48	1817	
Calcium	Ca	20	1808	Ends 8-electron M shell
Carbon	C	6	Ancient	Semiconductor
Cerium	Ce	58	1803	Starts rare-earth series 58 to 71
Cesium	Cs	55	1860	Photosensitive
Chlorine	Cl	17	1774	Active gas
Chromium	Cr	24	1798	
Cobalt	Co	27	1735	
Copper	Cu	29	Ancient	Metal conductor
Fluorine	F	9	1771	Active gas
Gallium	Ga	31	1875	
Germanium	Ge	32	1886	Semiconductor
Gold	Au	79	Ancient	Metal conductor
Hafnium	Hf	72	1923	Starts 32-electron N shell
Helium	He	2	1895	Inert gas
Hydrogen	H	1	1766	Active gas
Indium	In	49	1863	
Iodine	I	53	1811	
Iridium	Ir	77	1804	
Iron	Fe	26	Ancient	Magnetic
Krypton	Kr	36	1898	Inert gas
Lanthanum	La	57	1839	
Lead	Pb	82	Ancient	
Lithium	Li	3	1817	
Magnesium	Mg	12	1755	
Manganese	Mn	25	1774	
Mercury	Hg	80	Ancient	
Molybdenum	Mo	42	1781	
Neon	Ne	10	1898	Inert gas
Nickel	Ni	28	1751	
Niobium	Nb	41	1801	

*Most rare-earth elements with atomic numbers 58 to 71 and 90 to 103 are omitted.

Element	Symbol	Atomic Number	Year of Discovery	Remarks*
Nitrogen	N	7	1772	
Osmium	Os	76	1804	
Oxygen	O	8	1774	
Palladium	Pd	46	1803	N shell has 18 electrons
Phosphorus	P	15	1669	
Platinum	Pt	78	1735	
Polonium	Po	84	1898	
Potassium	K	19	1807	
Radium	Ra	88	1898	
Radon	Rn	86	1900	Inert gas
Rhenium	Re	75	1925	
Rhodium	Rh	45	1803	
Rubidium	Rb	37	1861	
Ruthenium	Ru	44	1844	
Scandium	Sc	21	1879	
Selenium	Se	34	1818	Photosensitive
Silicon	Si	14	1823	Semiconductor
Silver	Ag	47	Ancient	Metal conductor
Sodium	Na	11	1807	
Strontium	Sr	38	1790	Ends 8-electron N shell
Sulfur	S	16	Ancient	
Tantalum	Ta	73	1802	
Technetium	Tc	43	1937	
Tellurium	Te	52	1783	
Thallium	Tl	81	1861	
Thorium	Th	90	1829	Starts rare-earth series 90 to 103
Tin	Sn	50	Ancient	
Titanium	Ti	22	1791	
Tungsten	W	74	1783	
Uranium	U	92	1789	
Vanadium	V	23	1831	
Xenon	Xe	54	1898	Inert gas
Yttrium	Y	39	1843	
Zinc	Zn	30	1746	
Zirconium	Zr	40	1789	

*Most rare-earth elements with atomic numbers 58 to 71 and 90 to 103 are omitted.

Appendix D
Physics Units

All the units are based on the fundamental dimensions of length, mass, and time. These are considered basic quantities, compared with derived quantities such as area, force, velocity, and acceleration, which are only different combinations of length, mass, and time. Each of the basic dimensions has units in the U.S. Customary System (USCS) and in the decimal or metric system, as listed in Table D-1.

Systems of Units

The cgs system is an abbreviation for its basic units of centimeters, grams, and seconds. The mks system, based on meters, kilograms, and seconds, provides larger units which are closer to practical values, since the kilogram is 1000 g and the meter is 100 cm. The SI system is based on and very similar to the mks system.

In many cases, it is necessary to convert between USCS and metric units. Then the following conversions can be used:

LENGTH

1 meter = 39.37 inches

1 inch = 2.54 centimeters

MASS

1 kilogram = 2.2 pounds

1 gram = 0.03527 ounce

The basic unit of time is the second in all systems.

Mass

The dimension of mass is often considered to be similar to the weight of an object. However, weight is actually the force due to the acceleration of gravity. To define mass more specifically, it is necessary to use Newton's second law of motion: $F = Ma$. This can be transposed to $M = F/a$, which states that the mass is defined by how much force is necessary for a given amount of acceleration.

Derived Quantities

A simple example is velocity, a combination of the basic dimensions of length and time. The units can be feet per second, meters per second, or centimeters per second. Still, basically velocity is just L/T, meaning it is the time rate of change of length. Sometimes speed and velocity are used interchangeably. However, velocity is a vector quantity that has direction, while speed is a scalar quantity without direction. Another derived quantity is acceleration, the time rate of change of velocity. In the mks and SI systems the unit is meters per second per second, or m/s^2. The basic dimensions are L/T^2, which results from L/T for velocity, divided by T. Additional derived quantities are force, work or energy, and power.

Table D-1. Units for Basic Dimensions

Dimension	USCS unit	Metric unit	
		CGS	MKS and SI
Length (L)	foot	centimeter	meter
Mass (M)	slug*	gram	kilogram
Time (T)	second	second	second

*1 slug is the mass of a 1-lb weight.

Force

Newton's law of acceleration is used to derive the units of force. With the formula $F = Ma$ in the cgs system, one *dyne* is the force needed for an acceleration of one centimeter per second per second with a mass of one gram. In the mks and SI systems, one *newton* is the force needed for an acceleration of one meter per second per second with a mass of one kilogram. One newton equals 10^5 dyne. To convert to USCS units, 1 N = 0.225 lb.

Work and Energy

Work W is the product of force F times the distance s through which the force acts. As a formula, $W = Fs$. For example, if you lift a 20-lb weight through a distance of 2 ft, the work equals 40 ft·lb.

In the cgs system, F is in dyne-centimeters. One dyne-centimeter is an *erg*. A larger cgs unit, also used in SI, is the *joule*, equal to 10^7 ergs.

In the mks system, F is in newton-meters. This unit is the same as 10^7 ergs. Or 1 N·m equals 1 J. The joule unit of work is named after James P. Joule (1818–1889), an important English physicist.

Energy is the ability to do work. Kinetic energy is due to the motion of a mass, as when you throw a ball. Potential energy is stored energy, as in a coiled spring. The units for both kinetic and potential energy are the same as for work.

Power

This is the time rate of doing work, or $P = W/T$. The practical unit in the metric system is the joule per second, equal to 1 *watt*. In the USCS system the unit is foot-pounds per second. For a larger unit, 550 ft·lb/s equal 1 hp.

Temperature Scales

The Celsius scale, formerly known as the centigrade scale, invented by A. Celsius, has 100 divisions between 0° for the freezing point of water and 100° for the boiling point. The Fahrenheit scale, invented by G. D. Fahrenheit, is still used for weather observations and general purposes. On this scale, the freezing point is 32° and the boiling point is 212°, with 180 divisions between. To convert from one scale to the other,

$$T_C = \frac{5}{9}(T_F - 32°) \qquad \text{and} \qquad T_F = \frac{9}{5}T_C + 32°$$

The kelvin or absolute temperature scale was devised by Lord Kelvin. On this scale, the zero point is absolute zero, 273° below 0°C. At 0 K any material loses all its thermal energy. The divisions of the °C scale and the K scale are the same. (Note: The SI unit for temperature is K for kelvin, without the degree symbol.)

To convert from °C to K, just add 273°. Thus, 0°C equals 273 K. To convert from K to °C, subtract 273°. Then 0 K equals −273°C.

Average room temperature is generally considered to be about 20 to 25°C. This equals 68 to 77°F. On the absolute scale, the corresponding temperature range is 293 to 298 K.

Units of Heat Energy

In the cgs system, one *calorie* is the amount of heat needed to raise the temperature of one gram of water by one degree Celsius. In mks units, the amount of heat needed to raise the temperature of one kilogram of water by one degree Celsius is equal to one kilocalorie. The SI system uses the joule as a unit of heat as well as a unit of work. In USCS units, the *British thermal unit (Btu)* is the amount of heat needed to raise the temperature of one pound of water by one degree Fahrenheit. The Btu is the larger unit, as 1 Btu equals 252 calories. The fact that heat is a form of energy can be seen from the use of the joule; 1 calorie is equivalent to 4.19 J. One joule equals 0.24 calorie. Table D-2 summarizes all the units described here.

Metric Conversions

Some of the more common equivalents between metric and USCS units are listed in Table D-3. Note that 1 cm is a little less than ½ in. A meter is a little longer than a yard (approximately 4 in. longer). Also, one kilometer (km) is 1.6 mi.

For volume units, one liter (L) is a little less than one quart, as 1 qt = 0.95 L. Also, a gallon (gal) or four quarts is equal to 3.8 L.

In the units of mass or weight, 28 grams (g) are equal to one ounce. Also, 2.2 kg = 1 lb or 16 oz.

When using the conversion factors in the last column of Table D-3, you can decide whether or not to use the reciprocal form by checking the units in the conversion. The unit you are changing from should cancel, leaving just the unit you need. As an example, suppose that we want to convert 7 in. to centimeter units. Then

$$7 \text{ in.} \times 2.54 \frac{\text{cm}}{\text{in.}} = 17.78 \text{ cm}$$

Note that the units of inches cancel in the numerator and denominator.

As another example, suppose that we want to convert 17.78 cm to inches. If we use the same conversion factor, the units will be wrong because there is no cancellation, as shown here:

$$17.78 \text{ cm} \times 2.54 \frac{\text{cm}}{\text{in.}} = ?$$

Instead, this problem is done with the reciprocal of the conversion factor. Then

$$17.78 \text{ cm} \times \frac{1}{2.54 \text{ cm/in.}}$$

or

$$17.78 \text{ cm} \times \frac{1 \text{ in.}}{2.54 \text{ cm}} = 7 \text{ in.}$$

When we use the reciprocal conversion factor, the centimeter units we are changing from can be cancelled, leaving the inch units for the answer.

References

For more details on physics, chemistry, the periodic table, and atomic structure, the following references can be helpful:

Chemical Education Material Study: *Chemistry, and Experimental Science,* W. H. Freeman and Company, San Francisco, Calif.

Table D-2. Summary of Physics Units

Quantity	USCS (fps) units	CGS units	MKS and SI units	Dimensions
Length	foot	centimeter	meter	L
Mass	slug	gram	kilogram	M
Time	second	second	second	T
Velocity	feet per second	centimeters per second	meters per second	L/T
Acceleration	feet per second per second	centimeters per second per second	meters per second per second	L/T^2
Force	pound	dyne	newton	$Ma = ML/T^2$
Work and energy	foot-pound*	erg = dyne-centimeter; 10^7 ergs = 1 joule	joule = newton-meter	force $\times L$
Power	foot-pounds per second; 1 hp = 550 ft · lb/s	ergs per second; 10^7 ergs/s = 1 W	watt = joule per second	work/T
Heat	Btu	calorie	kilocalorie (mks); joule (SI)	1 calorie = 4.19 joules

*Foot-pound is a unit of work, while pound-foot is used for angular torque.

Table D-3. Metric Conversions

Quantity	Equivalents	Conversion factor
Length	1 in. = 2.54 cm 1 m = 39.39 in. 1 mi = 1.61 km	2.54 cm/in. or reciprocal* 39.37 in./m or reciprocal 1.61 km/mi or reciprocal
Volume	1 L = 0.95 qt 1 gal = 3.8 L	0.95 qt/L or reciprocal 3.8 L/gal or reciprocal
Mass or weight	1 oz = 28 g 1 lb = 2.2 kg	28 g/oz or reciprocal 2.2 kg/lb or reciprocal

*See text for explanation of when to use the reciprocal.

Dull, Metcalfe, and Williams: *Modern Physics,* Holt, Rinehart and Winston, Inc., New York.

Ewing-Meyer: *Chemistry: A Survey of Principles,* John Wiley & Sons, Inc., New York.

Holton and Roller: *Foundations of Modern Physical Science,* Addison-Wesley Publishing Company, Inc., Reading, Mass.

Key to Welch Periodic Chart of the Atoms, Sargent Welch Scientific Co., Chicago, Ill.

Semat, H.: *Fundamentals of Physics,* Holt, Rinehart and Winston, Inc., New York.

White, Manning, and Weber: *Practical Physics,* McGraw-Hill Book Company, New York.

Appendix E
Trigonometric Functions

The six functions of an angle θ are sine, cosine, tangent, cotangent, secant, and cosecant. All are numerical ratios, comparing the sides formed by the angle θ in a right triangle, as in Fig. E-1. Any one function specifies the angle.

The values for sin θ, cos θ, and tan θ are listed in Table E-1. These functions are defined in terms of the right triangle, as follows:

$$\sin \theta = \frac{\text{opposite side}}{\text{hypotenuse}} = \frac{a}{c}$$

$$\cos \theta = \frac{\text{adjacent side}}{\text{hypotenuse}} = \frac{b}{c}$$

$$\tan \theta = \frac{\text{opposite side}}{\text{adjacent side}} = \frac{a}{b}$$

These trigonometric functions are commonly used in ac circuits. Tan θ is especially useful for finding the phase angle θ because the tangent function uses only the sides, without the hypotenuse.

It may be of interest to note that the other three trigonometric functions are reciprocals. These are: cot θ = 1/tan θ; sec θ = 1/cos θ; csc θ = 1/sin θ.

Sine Values

In Table E-1, sin θ increases from 0 for 0° to 1 for 90°. The sine increases with θ as the opposite side becomes longer. However, the maximum sine ratio is 1 because no side of the triangle can be larger than the hypotenuse.

Cosine Values

The values for cos θ start from 1 for 0° as its maximum value. Then cos θ decreases to 0 as θ increases to 90°.

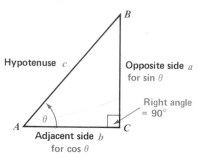

Fig. E-1 The right triangle.

The cosine function becomes smaller because the adjacent side is shorter for larger angles.

Tangent Values

The values for tan θ increase with θ as the opposite side becomes larger. However, tan θ values should be considered below and above 45°, as follows:

Tan θ from 0 to 45° increases from 0 to 1.

At 45°, tan θ = 1 because the opposite and adjacent sides are equal.

Above 45°, tan θ is more than 1 because the opposite side is larger than the adjacent side.

Important Values

Some values are worth memorizing, since they summarize the trend of the entire table of trigonometric functions. These are:

Angle	sin θ	cos θ	tan θ
0°	0	1	0
30°	0.5	0.866	0.577
45°	0.707	0.707	1
60°	0.866	0.5	1.73
90°	1.0	0	∞

Table E-1. Trigonometric Functions

Angle	Sin	Cos	Tan	Angle	Sin	Cos	Tan
0°	0.0000	1.000	0.0000	45°	0.7071	0.7071	1.0000
1	.0175	.9998	.0175	46	.7193	.6947	1.0355
2	.0349	.9994	.0349	47	.7314	.6820	1.0724
3	.0523	.9986	.0524	48	.7431	.6691	1.1106
4	.0698	.9976	.0699	49	.7547	.6561	1.1504
5	.0872	.9962	.0875	50	.7660	.6428	1.1918
6	.1045	.9945	.1051	51	.7771	.6293	1.2349
7	.1219	.9925	.1228	52	.7880	.6157	1.2799
8	.1392	.9903	.1405	53	.7986	.6018	1.3270
9	.1564	.9877	.1584	54	.8090	.5878	1.3764
10	.1736	.9848	.1763	55	.8192	.5736	1.4281
11	.1908	.9816	.1944	56	.8290	.5592	1.4826
12	.2079	.9781	.2126	57	.8387	.5446	1.5399
13	.2250	.9744	.2309	58	.8480	.5299	1.6003
14	.2419	.9703	.2493	59	.8572	.5150	1.6643
15	.2588	.9659	.2679	60	.8660	.5000	1.7321
16	.2756	.9613	.2867	61	.8746	.4848	1.8040
17	.2924	.9563	.3057	62	.8829	.4695	1.8807
18	.3090	.9511	.3249	63	.8910	.4540	1.9626
19	.3256	.9455	.3443	64	.8988	.4384	2.0503
20	.3420	.9397	.3640	65	.9063	.4226	2.1445
21	.3584	.9336	.3839	66	.9135	.4067	2.2460
22	.3746	.9272	.4040	67	.9205	.3907	2.3559
23	.3907	.9205	.4245	68	.9272	.3746	2.4751
24	.4067	.9135	.4452	69	.9336	.3584	2.6051
25	.4226	.9063	.4663	70	.9397	.3420	2.7475
26	.4384	.8988	.4877	71	.9455	.3256	2.9042
27	.4540	.8910	.5095	72	.9511	.3090	3.0777
28	.4695	.8829	.5317	73	.9563	.2924	3.2709
29	.4848	.8746	.5543	74	.9613	.2756	3.4874
30	.5000	.8660	.5774	75	.9659	.2588	3.7321
31	.5150	.8572	.6009	76	.9703	.2419	4.0108
32	.5299	.8480	.6249	77	.9744	.2250	4.3315
33	.5446	.8387	.6494	78	.9781	.2079	4.7046
34	.5592	.8290	.6745	79	.9816	.1908	5.1446
35	.5736	.8192	.7002	80	.9848	.1736	5.6713
36	.5878	.8090	.7265	81	.9877	.1564	6.3138
37	.6018	.7986	.7536	82	.9903	.1392	7.1154
38	.6157	.7880	.7813	83	.9925	.1219	8.1443
39	.6293	.7771	.8098	84	.9945	.1045	9.5144
40	.6428	.7660	.8391	85	.9962	.0872	11.43
41	.6561	.7547	.8693	86	.9976	.0698	14.30
42	.6691	.7431	.9004	87	.9986	.0523	19.08
43	.6820	.7314	.9325	88	.9994	.0349	28.64
44	.6947	.7193	.9657	89	.9998	.0175	57.29
				90	1.0000	.0000	∞

Angles More than 90°

The angles considered so far are *acute* angles, less than 90°. The 90° angle is a right angle. Angles larger than 90° are *obtuse* angles. The complete circle is 360°. After 360° or any multiple of 360°, the angles just repeat the values from 0°.

For angles from 90 to 360°, the full circle is divided into four quadrants, as shown in Fig. E-2. To use the table of trigonometric functions for obtuse angles in quadrants II, III, and IV, convert to equivalent acute angles in quadrant I by the following rules:

In quadrant II, use $180° - \theta$.

In quadrant III, use $\theta - 180°$.

In quadrant IV, use $360° - \theta$.

Also, use the appropriate sign or polarity for the trigonometric functions in different quadrants, as shown in Fig. E-2.

Note that the conversions are only with respect to the horizontal axis, using 180 or 360° as the reference. This way the obtuse angle is always subtracted from a larger angle.

All the functions are positive in quadrant I. Notice that the tangent alternates in polarity through quadrants I, II, III, and IV.

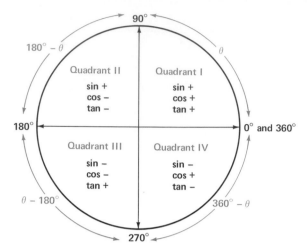

Fig. E-2 The four quadrants of a circle.

In all the quadrants, the sine is + when the vertical ordinate is up or − when the ordinate is down. Similarly, the cosine is + when the horizontal abscissa is to the right or − when it is to the left. The hypotenuse has no polarity.

Finally, the tangent is + when the sine and cosine have the same polarity, either both + in quadrant I or both − in quadrant III.

Appendix F

Electrical Symbols and Abbreviations

Table F-1 summarizes the letter symbols used as abbreviations for electrical characteristics and their basic units. All the metric prefixes for multiple and fractional values are listed in Table F-2. In addition, Table F-3 on the next page shows electronic symbols from the Greek alphabet.

Table F-1. Electrical Characteristics

Quantity	Symbol	Basic unit
Current	I or i	ampere (A)
Charge	Q or q	coulomb (C)
Power	P	watt (W)
Voltage	V or v	volt (V)
Resistance	R	ohm (Ω)
Reactance	X	ohm (Ω)
Impedance	Z	ohm (Ω)
Conductance	G	siemens (S)
Admittance	Y	siemens (S)
Susceptance	B	siemens (S)
Capacitance	C	farad (F)
Inductance	L	henry (H)
Frequency	f	hertz (Hz)
Period	T	second (s)

Capital letters for I, Q, and V are generally used for peak, rms, or dc value; small letters are used for instantaneous values. Small r and g are used for internal values, such as r_p and g_m of a tube.

Table F-2. Multiples and Submultiples of Units*

Value	Prefix	Symbol	Example
$1\ 000\ 000\ 000\ 000 = 10^{12}$	tera	T	$THz = 10^{12}$ Hz
$1\ 000\ 000\ 000 = 10^{9}$	giga	G	$GHz = 10^{9}$ Hz
$1\ 000\ 000 = 10^{6}$	mega	M	$MHz = 10^{6}$ Hz
$1\ 000 = 10^{3}$	kilo	k	$kV = 10^{3}$ V
$100 = 10^{2}$	hecto	h	$hm = 10^{2}$ m
$10 = 10$	deka	da	$dam = 10$ m
$0.1 = 10^{-1}$	deci	d	$dm = 10^{-1}$ m
$0.01 = 10^{-2}$	centi	c	$cm = 10^{-2}$ m
$0.001 = 10^{-3}$	milli	m	$mA = 10^{-3}$ A
$0.000\ 001 = 10^{-6}$	micro	μ	$\mu V = 10^{-6}$ V
$0.000\ 000\ 001 = 10^{-9}$	nano	n	$ns = 10^{-9}$ s
$0.000\ 000\ 000\ 001 = 10^{-12}$	pico	p	$pF = 10^{-12}$ F

*Additional prefixes are exa $= 10^{18}$, peta $= 10^{15}$, femto $= 10^{-15}$, and atto $= 10^{-18}$.

Table F-3. Greek Letter Symbols

Name	Letter Capital	Letter Small	Uses
Alpha	A	α	α for angles, transistor characteristic
Beta	B	β	β for angles, transistor characteristic
Gamma	Γ	γ	
Delta	Δ	δ	Small change in value
Epsilon	E	ϵ	ϵ for permittivity; also base of natural logarithms
Zeta	Z	ζ	
Eta	H	η	
Theta	Θ	θ	Phase angle
Iota	I	ι	
Kappa	K	κ	
Lambda	Λ	λ	λ for wavelength
Mu	M	μ	μ for prefix micro, permeability, amplification factor
Nu	N	ν	
Xi	Ξ	ξ	
Omicron	O	o	
Pi	Π	π	π is 3.1416 for ratio of circumference to diameter of a circle
Rho	P	ρ	ρ for resistivity
Sigma	Σ	σ	Summation
Tau	T	τ	Time constant
Upsilon	Υ	υ	
Phi	Φ	ϕ	Magnetic flux, angles
Chi	X	χ	
Psi	Ψ	ψ	Electric flux
Omega	Ω	ω	Ω for ohms; ω for angular velocity

This table includes the complete Greek alphabet, although some letters are not used for electronic symbols.

Appendix G
Color Codes

Color codes are standardized by the Electronic Industries Association (EIA). Included are codes for chassis wiring connections and the values for carbon resistors and mica and ceramic capacitors.

Chassis Wiring

The color of the wire indicates the function. Either a solid color or helical striping on white insulation can be used.

Red V^+ from the dc power supply

Blue   Plate (amplifier tube), collector (transistor), or drain (FET)

Green   Control grid (tube), base (transistor), or gate (FET)

Yellow   Cathode (amplifier tube), emitter (transistor), or source (FET)

Orange   Screen grid (vacuum tube)

Brown   Heater (vacuum tube)

Black   Chassis ground

White   Return to AVC or AGC bias line

In addition, blue wire is used for the high side of antenna-input connections.

IF Transformers

Used for interstage coupling in IF amplifiers. The terminals may be color-coded as follows:

Blue   High side of the primary to the amplifier plate, collector, or drain electrode

Red   Low side of the primary returning to V^+

Green   High side of the secondary for the signal output

White   Low side of the secondary

Violet   Can be used for an additional secondary output

AF Transformers

The AF transformers used for interstage coupling and power output in audio amplifiers often have color-coded leads as follows:

Blue   Primary to the amplifier plate, collector, or drain electrode. It indicates the end of primary winding.

Red   Connection to V^+. Center tap of a push-pull winding.

Brown   Lead opposite to the blue lead for push-pull. Start of a primary winding.

Green   High side of the secondary for the signal output. The end of the secondary winding.

Black   Ground return.

Yellow   Center tap on the secondary.

Stereo Audio Channels

The wiring of stereo audio channels is color-coded as follows:

Left channel	Right channel
White: High side	Red: High side
Blue: Low side	Green: Low side

Power Transformers

The leads of power transformers are color-coded as follows:

Start and end leads of the primary without a tap are black for both leads.

Tapped primary
    Common: black
    Tap: black and yellow
    End: black and red

High-voltage secondary is red for both start and end of the winding.

Center tap on secondary is red and yellow.

Low-voltage secondary is green and yellow.

Carbon Resistors

For ratings of 2 W or less, carbon resistors are color-coded with either bands or the body-end-dot system; see Fig. G-1 and Table G-1. The colors are listed in Table G-2. They apply to resistor values in ohms and capacitors in picofarads.

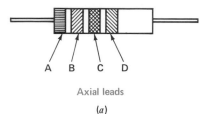

Axial leads

(a)

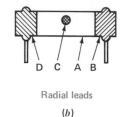

Radial leads

(b)

Fig. G-1 Resistor color codes. (a) Color stripes on R with axial leads. (b) Body-end-dot colors on R with radial leads.

Table G-1. Color Codes for Carbon Resistors

Axial leads	Color	Radial leads
Band A	First significant figure	Body A
Band B	Second significant figure	End B
Band C	Decimal multiplier	Dot C
Band D	Tolerance	End D

Notes: Band A is double width for wire-wound resistors with axial leads. The body-end-dot system is a discontinued standard, but it may still be found on some old resistors. When resistors have color stripes and axial leads, body color is not used for color-coded value. Film resistors have five stripes; the fourth stripe is the multiplier and the fifth is tolerance.

Table G-2. Color Values for Resistor and Capacitor Codes

Color	Significant figure	Decimal multiplier	Tolerance,* %
Black	0	1	20
Brown	1	10	1
Red	2	10^2	2
Orange	3	10^3	3
Yellow	4	10^4	4
Green	5	10^5	5
Blue	6	10^6	6
Violet	7	10^7	7
Gray	8	10^8	8
White	9	10^9	9
Gold		0.1	5
Silver		0.01	10
No color			20

*Tolerance colors other than gold and silver for capacitors only.

The preferred values in Table G-3 also apply to resistors and capacitors. Only the basic value is listed, but multiples are available. For instance, resistors are 47, 470, 4700, and 47,000 Ω. There are more intermediate resistance values for lower tolerances. As an example, the next higher value above 47 can be 68, 56, or 51, depending on the tolerance.

Mica and Ceramic Capacitors

See Fig. G-2 for color coding of mica capacitors. The first dot on a mica capacitor is white to indicate the EIA six-dot code. It may also be black for a military code. In either case, read the capacitance in picofarads from the next three color dots. Table G-2 lists tolerance colors for the fifth dot. The sixth dot indicates classes A to E of leakage and temperature coefficients. The maximum rating for dc working voltage is generally 500 V.

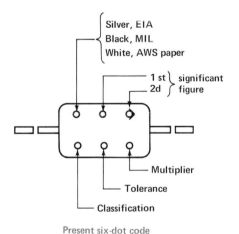

Present six-dot code

Fig. G-2 How to read EIA six-dot code for mica capacitors.

If a color is used in the first dot, the capacitor is marked with the old EIA code. Use that dot and the next two dots for significant figures and the fourth dot as the decimal multiplier.

Ceramic capacitors have stripes or dots with three or five colors. The construction may be tubular with axial leads or a disk with radial leads (Fig. G-3). When there

Table G-3. Preferred Values* for Resistors and Capacitors

20% tolerance	10% tolerance	5% tolerance
10	10	10
		11
	12	12
		13
15	15	15
		16
	18	18
		20
22	22	22
		24
	27	27
		30
33	33	33
		36
	39	39
		43
47	47	47
		51
	56	56
		62
68	68	68
		75
	82	82
		91
100	100	100

*Numbers and decimal multiples for ohms or picofarads.

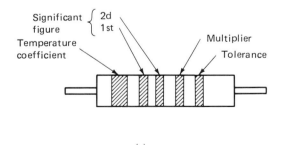

(a)

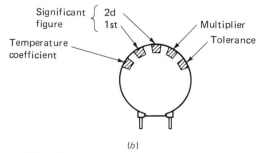

(b)

Fig. G-3 Color code for ceramic capacitors. (a) Tubular type with radial leads. (b) Disk type with radial leads.

are five colors, the first and last indicate the temperature coefficient and tolerance (Table G-4). The middle three colors give the picofarad value.

Ceramic disk capacitors often have the values printed on them. A value more than 1 is in picofarads, such as 47 pF. A value less than 1, say, 0.002, is in microfarads. The letter R may be used to indicate a decimal point; for example, $2R2 = 2.2$.

Table G-4. Color Code for Ceramic Capacitors

Color	Decimal multiplier	Tolerance		Temperature coefficient, ppm per °C
		Above 10 pF, %	Below 10 pF, in pF	
Black	1	20	2.0	0
Brown	10	1		−30
Red	100	2		−80
Orange	1000			−150
Yellow				−220
Green		5	0.5	−330
Blue				−470
Violet				−750
Gray	0.01		0.25	30
White	0.1	10	1.0	500

Appendix H
Soldering and Tools

In addition to the usual workshop tools, working on electronic equipment often requires long-nose or needle-nose pliers to bend the end of a wire, diagonal cutting pliers to cut wire, a ¼-in. socket wrench for machine screws with ¼-in. hexagonal heads, and a Phillips screwdriver. These tools are shown in Fig. H-1.

The cutting pliers can be used when you want to strip about an inch or less of the insulation from the end of stranded wire that is not the push-back type. First crush the insulation by squeezing below the cutting edges of the pliers; then notch the crushed insulation with the cutting edge down to the wire, being careful not to cut the wire; finally pull off the crushed insulation with the cutting edges of the pliers in the notch. A wire stripper is a special tool for this.

With stiff insulation that cannot be crushed, such as the cover on a shielded coaxial cable, slit the insulation with a razor blade along the required length, pull back, and cut it off with the diagonal cutters. The braided shield can be opened by picking it apart with a pointed tool.

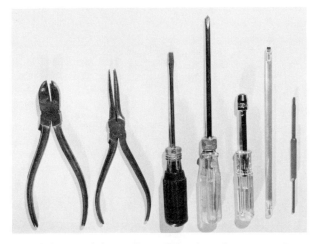

Fig. H-1 From left to right: cutting pliers, long-nose pliers, slotted-head screwdriver, Phillips-head screwdriver, ¼-in socket wrench, plastic alignment tool with slotted head, and hexagonal plastic alignment tool.

Soldering

Solder is an alloy of tin and lead used for fusing metals at relatively low temperatures, about 500 to 600°F. The joint where two metal conductors are to be fused is heated, and then solder is applied so that it can melt and cover the connection. The reason for soldering connections is that it makes a good bond between the joined metals, covering the joint completely to prevent oxidation. The coating of solder provides protection for practically an indefinite period of time.

The trick in soldering is to heat the joint, not the solder. When the joint is hot enough to melt the solder, the solder flows smoothly to fill all the cracks, forming a shiny cover without any air spaces. Do not move the joint until the solder has set, which takes only a few seconds.

Either a soldering iron or a soldering gun can be used, rated at 25 to 100 W. See Fig. H-2. The gun is convenient for intermittent operation, since it heats almost instantaneously when you press the trigger. The small pencil iron of 25 to 40 W is helpful for soldering small connections where excessive heat can cause damage. This precaution is particularly important when working on PC boards, where too much heat can soften the plastic form and loosen the printed wiring. A soldering iron for FET devices should have the tip grounded to eliminate static charge.

The three grades of solder generally used for electronics work are 40-60, 50-50, and 60-40 solder. The first figure is the percentage of tin, while the other is the percentage of lead. The 60-40 solder costs more, but it melts at the lowest temperature, flows more freely, takes less time to harden, and generally makes it easier to do a good soldering job.

In addition to the solder, there must be flux to remove any oxide film on the metals being joined. Otherwise they cannot fuse. The flux enables the molten solder to wet the metals so that the solder can stick. The two types are acid flux and rosin flux. Acid flux is more active in cleaning metals but is corrosive. Rosin flux is always used for the light soldering work in making wire connections.

Generally, the rosin is in the hollow core of solder intended for electronics work, so that a separate flux is unnecessary. Such rosin-core solder is the type generally used. It should be noted, though, that the flux is not a substitute for cleaning the metals to be fused. They must be shiny clean for the solder to stick.

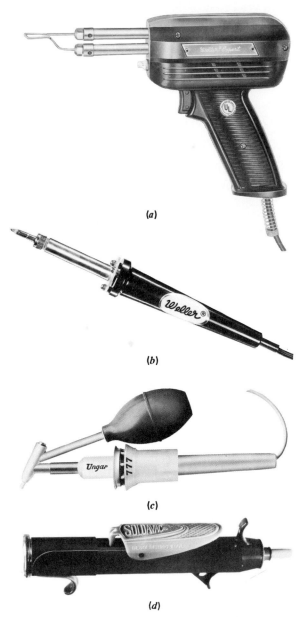

(a)

(b)

(c)

(d)

Fig. H-2 (*a*) Soldering gun. (*b*) Soldering pencil. (*c*) Desoldering iron with vacuum bulb. (*d*) Plunger type of desoldering tool.

Desoldering

On printed-wiring boards, desoldering to remove a defective component can be more important than the soldering. From the wiring side of the PC board, desolder at the holes where the component leads go through the board to join the printed wiring. Three methods are:

1. Soldering gun with vacuum attachment.
2. Soldavac tool with a separate iron. The vacuum plunger pulls the molten solder out of the connection.
3. Soldawick or similar metal braid with a separate iron. Heat the braid on the connection, and the molten solder runs up into the braid. See Fig. H-3. This method is most convenient where there is no room for a vacuum tool.

Remove enough solder so that you can actually see the hole where the leads come through. Then the component will practically fall off the board. If you try to force the component out, the board can be damaged.

For more details on soldering and tools, the following references can be helpful:

Anderson and Tatro: *Shop Theory,* McGraw-Hill Book Company, New York.
Radio Amateur's Handbook (chapter on construction practices), American Radio Relay League, Newington, Conn.
Solder, Kester Solder Co., Chicago, Ill.

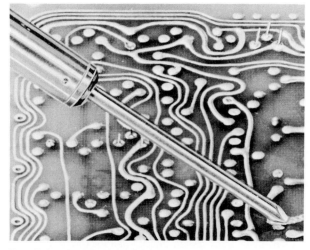

Fig. H-3 Use of metal braid and iron to remove solder from printed-circuit board.

Appendix I

Schematic Symbols

Schematic symbols are organized in separate tables for different types of electronic devices. Logic symbols are in Table I-8.

Table I-1 General Symbols

     I-2 Vacuum Tubes

     I-3 Bipolar Junction Transistors

I-4 Field-Effect Transistors(FETs)

I-5 Semiconductor Diodes

I-6 Thyristors

I-7 Protective Devices for Semiconductors

I-8 Digital Logic Symbols

Table I-1. General Symbols

Device	Symbol	Device	Symbol
AC voltage		Coil or inductance Air-core	
Amplifier		Iron-core	
Antenna General		Variable	
Dipole		Powdered iron or ferrite slug	
Loop		Conductor General	
		Connection	
Battery, cell, or dc voltage Longer line positive		No connection	
		Current source	
Capacitor General, fixed Curved electrode is outside foil, negative, or low-potential side		Crystal, piezoelectric	
		Diodes	See Table I-5.
		Fuse	
Variable		Ground, earth or metal frame Chassis or common return connected to one side of voltage source	
Ganged		Chassis or common return not connected to voltage source Common return	

Table I-1. General Symbols (*Continued*)

Device	Symbol	Device	Symbol

Jack
 	
Plug for jack
	TIP
	SLEEVE
		Resistor, fixed
		  Tapped
		
		  Variable	

Key telegraph		Switch, SPST
		
		  SPDT
		
		  2-pole (DPDT)
		
		  3-pole,
		    3-circuit wafer	

Loudspeaker, general Phones or headset			

Magnet
  Permanent
 	
  Electromagnet
	PM	Shielding
		
		  Shielded conductor	

Microphone		Transformer
		  Air-core
		
		  Iron-core
		
		  Autotransformer
		
		  Link coupling	

Meters; letter or symbol to indicate range or function	A mA V		
Motor	M		
Neon bulb			
Relay, coil Contacts		Transistors	See Tables I-3 and I-4.
		Vacuum tubes	See Table I-2.

Table I-2. Vacuum Tubes

Tube	Symbol	Electrodes
Diode		P = plate K = cathode
Tetrode		P = plate G_2 = screen grid G_1 = control grid K = cathode
Triode		P = plate G_1 = control grid K = cathode
Pentode		P = plate G_3 = suppressor grid G_2 = screen grid G_1 = control grid K = cathode
CRT,* with deflection plates		$G_2 + G_4 + CL$ = anode G_3 = focusing electrode
CRT,* magnetic deflection		$G_3 + G_5 + CL$ = anode G_4 = focusing electrode

*CRT is cathode-ray tube; DJ is deflection plate; IC indicates internal connection; NC means no connection; CL is internal wall coating for collector; C is 500-pF capacitance of external wall coating.

Table I-3. Bipolar Junction Transistors*

Type	Symbol	Electrodes	Notes
NPN		C = collector B = base E = emitter; hole current out from base	Needs $+V_C$ reverse voltage for collector; forward bias V_{BE} of 0.6 V for silicon or 0.2 V for germanium
PNP		C = collector B = base E = emitter; hole current into base	Needs $-V_C$ reverse voltage for collector; forward bias V_{BE} of 0.2 V for germanium or 0.6 V for silicon

*Type numbers for bipolar junction transistors start with 2 N, as in 2N1482, for two semiconductor junctions in the EIA system. Japanese transistors are marked 2SA or 2SB for PNP types, 2SC for NPN high-frequency types, and 2SD for NPN low-frequency power types.

Table I-4. Field-Effect Transistors (FET)*

Type†	Symbol	Electrodes	Notes
JFET		D = drain G = gate S = source	Junction-type FET; arrow pointing to channel indicates P gate and N channel; reverse bias at junction
IGFET or MOSFET N-channel		D = drain G = gate S = source	Insulated-gate; depletion or depletion-enhancement type; B is bulk or substrate connected internally to source
IGFET or MOSFET P-channel		D = drain G = gate S = source	Arrow pointing away from channel indicates P channel
IGFET or MOSFET, N-channel enhancement		D = drain G = gate S = source	Broken lines for channel show enhancement type
Dual-gate IGFET or MOSFET, N-channel		D = drain G_2 = gate 2 G_1 = gate 1 S = source	Either or both gates control amount of drain current I_D

*Classified by EIA as type A, B, or C. For an N-channel: depletion type A takes negative gate bias for a middle value of I_D; depletion-enhancement type B can operate with zero gate bias; enhancement type C needs positive gate bias.
†EIA type number for FETs start with 3N, as in 3N143. Japanese type numbers are 2SJ or 2SK for single-gate and 3SK for dual-gate types.

Table I-5. Semiconductor Diodes*

Type	Symbol	Applications
Rectifier	A = anode K = cathode	Conducts forward current with positive anode voltage
Zener		Voltage regulator; takes reverse voltage
Varactor		Capacitive diode; *C* varies with amount of reverse voltage; used for electronic tuning
Tunnel		Microwave diode; Esaki diode; same symbol for Schottky diode and Gunn diode
Photoconductive		Takes reverse voltage; high dark resistance decreases with light input
Photovoltaic		No voltage applied; light input generates voltage output; solar cell
Photoemissive		Light-emitting diode (LED) produces red, green, or yellow light output; takes forward voltage

*Type numbers for diodes start with 1N, for one semiconductor junction. An example is 1N2864.

Table I-6. Types of Thyristors

Name	Symbol	Electrodes	Notes
Silicon controlled rectifier (SCR)		K = cathode G = gate A = anode	Unidirectional with positive anode voltage; G can turn SCR on but not off
Triac		MT1 = main terminal 1 G = gate MT2 = main terminal 2	Bidirectional with gate; G can turn triac on for either polarity, but not off
Diac		MT1 = main terminal 1 MT2 = main terminal 2	Bidirectional trigger without gate
Unijunction transistor (UJT)		B_2 = base 2 E = emitter B_1 = base 1	Base has standoff voltage from dc supply; V_E applied to start conduction

Table I-7. Protective Devices for Semiconductors

Type	Symbol	Notes
Thermistor		Resistance decreases with temperature; used as a series component
Voltage-dependent resistor (VDR)		Resistance decreases with V continuously; used as a series component
Varistor		Resistance decreases abruptly at specific voltage; used as a shunt component to short-circuit voltage transients

Table I-8 Digital Logic Symbols

Name	Symbol	Notes
AND gate	A, B → Y	$AB = Y$
OR gate	A, B → Y	$A + B = Y$
Inverter*	A → $\bar{A}$	$A = \bar{A}$
NAND gate	A, B → Y	$\overline{AB} = Y$
NOR gate	A, B → Y	$\overline{A + B} = Y$
J-K flip-flop	J, CP, K — FF — Q, $\bar{Q}$	CP is clock pulse; open bubble for CP indicates trigger on falling edge
Clock pulse trigger		Falling edge Rising edge
Operational amplifier, op amp	V_o	For analog circuits and D/A converters

*Circle at output indicates NOT function, complementary or opposite state.

Glossary

The following are technical terms with their definitions, listed in alphabetical order.

ac Abbreviation for alternating current.

active device One that can control voltage or current. Examples are transistor amplifier and diode rectifier.

acute angle Less than 90°.

A/D converter A device that converts analog input signals to digital output.

admittance (Y) Reciprocal of impedance Z in ac circuits. $Y = 1/Z$.

air gap Air space between poles of a magnet.

alkaline cell or battery One that uses alkaline electrolyte.

alpha (α) Characteristic of junction transistors. Ratio of collector current to emitter current. Value is 0.98 to 0.99.

alligator clip Spring clip for temporary wire connection.

alternating current (ac) Current that reverses direction at a regular rate. Alternating voltage reverses in polarity. The rate of reversals is the frequency.

alternator Ac generator.

ampere (A) Basic unit of electric current. Value of one ampere flows when one volt of potential difference is applied across one ohm of resistance.

ampere-turn Unit of magnetizing force equal to $1\,A \times 1$ turn.

amplifier A device that increases the amplitude of a signal.

amplitude modulation (AM) Changing the amplitude of an RF carrier wave in step with a lower-frequency signal that has the desired information.

analog circuits Circuits that use continuous variations in voltage or current, compared with digital pulse circuits.

AND gate Digital logic circuit. Produces HIGH output of 1 only when all inputs are at 1.

antiresonance Term sometimes used for parallel resonance.

apparent power Product of $V \times I$ when they are out of phase. Measured in voltampere units, instead of watts.

armature The part of a generator in which the voltage is produced. In a motor it is commonly the rotating member. Also, the movable part of a relay.

astable MV Multivibrator that has no stable state. Used as an oscillator to generate clock timing pulses.

audio frequency (AF) Within the range of hearing, approximately 16 to 16,000 Hz.

autotransformer A single, tapped, winding used to step up or step down voltage.

average value In sine-wave ac voltage or current, is 0.637 of peak value.

Ayrton shunt Circuit for multiple current ranges in a volt-ohm-milliammeter.

back-off scale Ohmmeter readings from right to left.

bandpass Filter that allows a band of frequencies to be coupled to the load.

bandstop Filter that prevents a band of frequencies from being coupled to the load.

bandwidth A range of frequencies that have a resonant effect in *LC* circuits.

bank Components connected in parallel.

battery Group of cells connected in series or parallel.

BCD Binary coded decimal. Converts a decimal number to a binary equivalent.

beta (β) Current-gain characteristic of junction transistors. Ratio of collector current to base current.

bias Average dc level of amplifier current to set operating characteristic.

binary number system A system with only two digits, 0 and 1.

bipolar transistor NPN or PNP type.

bistable MV Multivibrator that has two stable states. Used as flip-flop circuit.

bit One unit of information, either 0 or 1, in the binary number system.

bleeder current Steady current from source, used to stabilize output voltage with changes in load current.

boolean algebra Logical system of using binary information in digital circuits.

branch Part of a parallel circuit.

bridge Circuit in which voltages or currents can be balanced for a net effect of zero.

brushes In a motor or generator, devices that provide stationary connections to the rotor.

bypass capacitor One that has very low reactance in a parallel path.

byte Digital word with a string of eight bits of 0 and 1.

C Symbol for capacitance.

C Abbreviation for coulomb, the unit of electric charge.

calorie Amount of heat energy needed to raise the temperature of one gram of water by 1°C.

capacitor Device used to store electric charge.

capacitance The ability to store electric charge.

carry-in Number added from the previous lower-order column in addition.

carry-out Number added to the next higher-order column in addition. Carry is necessary when the sum in a column is more than the radix of the number system.

cascaded amplifiers Collector output of CE stage drives the base input of next stage.

CB circuit Common-base amplifier for junction transistors. Signal into base and output from collector.

CB radio Citizen's band radio, 26.965 to 27.405 MHz.

CC circuit Common-collector amplifier for junction transistors. Signal into base and output from emitter. Is emitter-follower stage.

CD circuit Common-drain circuit for field-effect transistors. Is source-follower stage.

CE circuit Common-emitter amplifier for junction transistors. Signal into base and output from collector.

Celsius scale (°C) Temperature scale that uses 0° for the freezing point of water and 100° for the boiling point. Formerly called centigrade.

ceramic Insulator with a high dielectric constant, often used for capacitors.

CG circuit Common-gate amplifier for field-effect transistors.

cgs Centimeter-gram-second system of units.

chip Miniature semiconductor containing the elements of the device.

choke Inductance with high X_L compared with the R of the circuit.

circuit breaker A protective device that opens when excessive current flows in circuit. Can be reset.

circular mil Cross-sectional area of round wire with diameter of 1 mil or 0.001 in.

clamp probe A device used to measure current without opening the circuit.

clear Same as reset on a flip-flop. Puts Q output at logic 0.

clock A device that provides timing pulses for digital circuits; it is usually a multivibrator (MV) oscillator.

closed circuit A continuous path for current.

CMOS Complementary metal-oxide semiconductor. Uses both P-type and N-type channels.

coaxial cable An inner conductor surrounded by an outer conductor that serves as a shield.

coil Turns of wire conductor to concentrate the magnetic field.

color code System in which colors are used to indicate values in resistors and capacitors.

commutator A device that converts reversing polarities to one polarity.

complex number One that has real and j terms. Uses the form $A + jB$.

conductance (G) Ability to conduct current. It is the reciprocal of resistance, $G = 1/R$. The unit of conductance is the siemens (S).

constant-current source One that has high r_i to supply constant I with variations in R_L.

constant-voltage source One that has low r_i to supply constant V with variations in R_L.

continuity Reading of zero ohms with an ohmmeter.

conventional current Direction of flow of positive charges, opposite from electron flow.

corona Effect of ionization of air around a point at high potential.

cosine A trigonometric function of an angle, equal to the ratio of the adjacent side to the hypotenuse in a right triangle.

cosine wave One whose amplitudes vary as the cosine function of an angle. It is 90° out of phase with the sine wave.

coulomb (C) Unit of electric charge. $1 \text{ C} = 6.25 \times 10^{18}$ electrons.

counter Digital circuit using a flip-flop to accumulate the count of binary pulses.

coupling capacitor Has very low X_C in series path.

covalent bond Pairing of atoms with electrical valence of ± 4.

cps Cycles per second. Formerly used as unit of frequency.

CRT Cathode-ray tube. A device that converts electric signals to a visual display on a fluorescent screen.

CS circuit Common-source amplifier for field-effect transistors.

current divider A parallel circuit to provide branch I less than the main-line current.

current source Supplies $I = V/r_i$ to load, with r_i in parallel.

cutoff No current output in an active device such as a transistor.

cycle One complete set of values for a repetitive waveform.

D/A converter A device that converts digital input to analog output.

damping Reducing the Q of a resonant circuit to increase the bandwidth.

Darlington pair Combination of two emitter-follower stages.

D'Arsonval meter A dc analog meter movement commonly used in ammeters and voltmeters.

data selector Digital circuit to convert parallel data to serial data. Also called multiplexer.

dB Abbreviation for decibel, the logarithm of the ratio of two power levels.

dc Abbreviation for direct current.

degaussing Demagnetizing by applying an ac field and gradually reducing it to zero.

delta (Δ) network Three components connected in series in a closed loop. Same as pi (π) network.

demultiplexer Digital circuit to convert serial data to parallel data. Also called distributor or decoder.

detector diode A device that rectifies a modulated signal to recover information in the modulation.

diac Two-terminal semiconductor device used for trigger pulses.

diamagnetic Material that can be weakly magnetized in the opposite direction from the magnetizing field.

dicing Cutting a slice of semiconductor material into tiny chips.

dielectric Insulating material. It cannot conduct current but does store charge.

dielectric constant (k) Ability to concentrate the electric field in a dielectric.

differentiating circuit An *RC* circuit with a short time constant for pulses across *R*.

digital circuit One that uses only two amplitudes for a pulse of voltage or current, either HIGH at 1 or LOW at 0.

digital IC Abbreviation for digital integrated circuit.

DIP Dual inline pins on IC package.

direct current (dc) Current that flows in only one direction. Dc voltage has a steady polarity that does not reverse.

discrete component A single individually packaged component usually with two or three leads.

distributor Digital circuit to convert serial data to parallel data. Also called demultiplexer.

DMM Abbreviation for digital multimeter.

doping Adding impurities to pure semiconductor material to provide positive and negative charges.

double subscripts An example is V_{BA} to indicate voltage at point B with respect to point A.

DPDT Double-pole double-throw switch or relay contacts.

DPST Double-pole single-throw switch or relay contacts.

Dynamometer Type of ac meter, generally for 60 Hz.

ECL Emitter-coupled logic circuits.

eddy current Circulating current induced in the iron core of an inductor by ac variations.

effective value For sine-wave ac waveform, 0.707 of peak value. Corresponds to heating effect of same dc value. Also called rms value.

efficiency Ratio of power output to power input.

electricity Dynamic electricity is the effect of voltage in producing current in conductors. Static electricity is accumulation of charge.

electrolyte Solution that forms ion charges.

electrolytic capacitor Type with very high *C* because electrolyte is used for the dielectric. Must be connected with correct polarity in a circuit.

electromagnet Magnet whose magnetic field is associated with electric current in a coil.

electron Basic particle of negative charge, in orbital rings around the nucleus in an atom.

electron flow Current of negative charges in motion. Direction is from the negative terminal of the voltage source, through the external circuit, and returning to the positive side of the source. Opposite to the direction of conventional current.

electron volt Unit of energy equal to the work done in moving a charge of 1 electron through a potential difference of 1 V.

electronics Based on electrical effects of the electron. Includes applications for amplifiers, oscillators, rectifiers, control circuits, and digital pulse circuits.

emf Electromotive force, voltage to produce current in a circuit.

emitter follower Circuit in which signal input is to base and output is from emitter. Same as common-collector circuit.

EPROM Erasable programmable ROM device.

F connector Plug for coaxial cable that does not need soldering.

Fahrenheit scale (°F) Temperature scale that uses 32° for the freezing point of water and 212° for the boiling point.

fan-in Maximum number of input circuits for a logic gate.

fan-out For a logic gate, the maximum number of output circuits that can be used as a load.

farad (F) Unit of capacitance. Value of one Farad stores one coulomb of charge with one volt applied.

Faraday's law For magnetic induction, the generated voltage increases with more flux and faster rate of change.

FCC Federal Communications Commission.

ferrite Magnetic material that is not a metal conductor.

ferromagnetic Magnetic properties of iron and other metals that can be strongly magnetized in the same direction as the magnetizing field.

FET Field-effect transistor.

field Group of lines of force. Can be a magnetic or an electric field.

field-effect transistor (FET) A device that depends on an electric field to control the current in a silicon channel.

field winding The part of a motor or generator that supplies the magnetic field cut by the armature.

fluctuating dc Varying voltage and current but no change in polarity.

flux (ϕ) Magnetic lines of force.

flux density Amount of flux per unit area.

flywheel effect Ability of an *LC* circuit to continue oscillating after the energy source has been removed.

forward voltage Polarity that allows current of majority carriers through a semiconductor junction.

frequency (*f*) Number of cycles per second for a waveform with periodic variations. The unit of frequency is hertz (Hz).

frequency modulation (FM) Changing the frequency of an RF carrier wave in step with a lower-frequency signal that has the desired information.

full-adder Digital circuit that has provisions for carry-in of 1 and carry-out of 1.

fuse Metal link that melts from excessive current and opens circuit.

gain (*A*) Also amplification. Ratio of amplified output to input.

galvanic cell Electrochemical type of voltage source.

galvanometer Sensitive instrument for measuring electric charge.

gate Logic circuit with two or more inputs but one HIGH or LOW output for specific combinations of input pulses.

gauss (G) Unit of flux density in cgs system, equal to one magnetic line of force per square centimeter.

generator A device that produces voltage output.

germanium (Ge) Semiconductor element used for transistors and diodes.

giga (G) Metric prefix for 10^9.

gilbert (Gb) Unit of magnetomotive force in cgs system. One gilbert equals 0.794 ampere-turn.

ground Common return to earth for ac power lines. Chassis ground in electronic equipment is the common return to one side of the internal power supply.

half-adder Logic circuit that has provision for carry-out of 1 but not carry-in.

half-power frequencies Define bandwidth with 70.7 percent response for resonant LC circuit.

hall effect Small voltage generated by a conductor with current in an external magnetic field.

harmonic Exact multiple of fundamental frequency.

henry (H) Unit of inductance. Current change of one ampere per second induces one volt across an inductance of one henry.

hertz (Hz) Unit of frequency. One hertz equals one cycle per second.

hexadecimal Number system with radix of 16.

h_{FE} Hybrid parameter for junction transistors that specifies current gain for common-emitter circuit.

hole Positive charge that exists only in doped semiconductors because of covalent bonds between atoms. Amount of hole charge is the same as a proton and an electron.

hole current Motion of hole charges. Direction is the same as that of conventional current, opposite from electron flow.

hot resistance The R of a component with its normal load current. Determined by V/I.

hot-wire meter Type of ac meter.

hybrid IC A device that has discrete components with an integrated circuit.

hypotenuse Side of a right triangle opposite the 90° angle.

hysteresis In electromagnets, the effect of magnetic induction lagging in time behind the applied magnetizing force.

Hz Hertz unit of frequency, equal to one cycle per second.

IC Abbreviation for integrated circuit.

I_{CB_o} Leakage current from collector to base with emitter open.

IGFET Insulated-gate field-effect transistor.

imaginary number Value at 90°, indicated by j operator, as in the form jA.

impedance match Making load impedance equal to the internal impedance of the source for maximum power at load.

inductance (L) Ability to produce induced voltage when cut by magnetic flux. Unit of inductance is the henry (H).

induction Ability to generate V or I without physical contact. Electromagnetic induction by magnetic field; electrostatic induction by electric field.

insulator A device that does not allow current to flow when voltage is applied, because of its high resistance.

integrated circuit Contains transistors, diodes, resistors, and capacitors in one miniaturized package. Can use bipolar transistor or FET technology.

integration circuit An RC circuit with a long time constant. Voltage output across C.

internal resistance r_i Limits the current supplied by the voltage source to $I = V/r_i$.

inverse relation Same as reciprocal function. As one variable increases, the other decreases.

inverter Stage that changes signal input to opposite polarity 180° out of phase. In digital circuits, an inverter changes logic 1 or 0 to the opposite state.

ion Atom with net charge. Can be produced in liquids, gases, and doped semiconductors.

IR drop Voltage across a resistor.

iron-vane meter Type of ac meter, generally for 60 Hz.

j operator Indicates 90° phase angle, as in $j8\ \Omega$ for X_L. Also, $-j8\ \Omega$ is at $-90°$ for X_C.

JFET Junction field-effect transistor.

JK flip-flop Popular type that has a clock input to toggle outputs between logic 1 and 0 when the J and K terminals are held HIGH.

joule (J) Practical unit of work or energy. One joule equals one watt-second of work.

k Coefficient of coupling between coils.

keeper Magnetic material placed across the poles of a magnet to form a complete magnetic circuit. Used to maintain strength of magnetic field.

Kelvin (K) scale Absolute temperature scale, 273° below values on Celsius scale.

kilo (k) Metric prefix for 10^3.

Kirchhoff's current law (KCL) The phasor sum of all currents into and out of any branch point in a circuit must equal zero.

Kirchhoff's voltage law (KVL) The phasor sum of all voltages around any closed path must equal zero.

laminations Thin sheets of steel insulated from one another to reduce eddy-current losses in inductors, motors, and transformers.

latch A device that remains in one stable state until activated to the opposite state. Can store binary information as logic 1 or 0. Is flip-flop in digital circuits.

LCD Liquid-crystal display.

leakage current Small reverse current of minority carriers across a PN junction.

Leclanché cell Carbon-zinc primary cell.

LED Light-emitting diode.

linear IC A device that contains analog circuits, such as amplifiers, oscillators, rectifiers, and control circuits, rather than digital circuits.

linear relation Straight-line graph between two variables. As one increases, the other increases in direct proportion.

load Takes current from the voltage source, resulting in load current.

loading effect Source voltage is decreased as amount of load current increases.

loop In a circuit, any closed path.

LSI Large-scale integration for IC chips.

magnetic pole Concentrated point of magnetic flux.

magnetism Effects of attraction and repulsion by iron and similar materials without the need for an external force. Electromagnetism includes the effects of a magnetic field associated with an electric current.

magnetomotive force (MMF) Ability to produce magnetic lines of force. Measured in units of ampere-turns.

magnitude Value of a quantity without any phase angle or at an angle of 0°.

maxwell (Mx) Unit of magnetic flux, equal to one line of force in the magnetic field.

mega (M) Metric prefix for 10^6.

memory device Digital circuit that can store information as bits of logic 1 or 0. Often is flip-flop circuit.

mesh current Assumed current in a closed path, without any current division, for application of Kirchhoff's current law.

micro (μ) Metric prefix for 10^{-6}.

microelectronics Microscopic components used for IC chips to miniaturize size of equipment.

microprocessor (μP) Digital IC package that can be programmed for many functions in a computer.

milli (m) Metric prefix for 10^{-3}.

minidip Miniature IC package with eight pins in DIP form.

mks Meter-kilogram-second system of units, larger than cgs units.

monostable Having one stable state, as in one-shot multivibrator.

motor A device that produces mechanical motion from electric energy.

MSI Medium-scale integration for IC chips.

multiplexer Digital circuit to convert parallel data to serial data. Also called data selector.

multiplier Resistor in series with a meter movement for voltage ranges.

multivibrator (MV) Oscillator circuit to produce pulses as a clock generator for timing in digital circuits.

mutual induction (L_M) Ability of one coil to induce voltage in another coil.

NAND gate Logic circuit that produces a LOW output of 0 only when all inputs are HIGH at 1.

nano (n) Metric prefix for 10^{-9}.

NC Normally closed for relay contacts, or no connection for pinout diagrams.

neutron Particle without electric charge in the nucleus of an atom.

nibble Binary word with four bits, equal to one-half byte.

node A common connection for two or more branch currents.

NOR gate Logic circuit that produces a LOW output of 0 when any of the inputs is HIGH at 1.

Norton's theorem Method of reducing a complicated network to one current source with shunt resistance.

NOT gate Circuit to change binary 1 to 0 or 0 to 1. Same as inverter.

obtuse angle More than 90°.

octal base Eight pins for vacuum tubes. Eight digits for octal number system.

octave Frequency ratio of 2:1.

oersted (Oe) Unit of magnetic field intensity; 1 Oe = 1 Gb/cm.

ohm (Ω) Unit of resistance. Value of one ohm allows current of one ampere with potential difference of one volt.

Ohm's law In electric circuits, $I = V/R$.

ohms per volt Sensitivity rating for a voltmeter. High rating means less meter loading.

open circuit One that has infinitely high resistance, resulting in zero current.

operational amplifier (op amp) High-gain amplifier with negative feedback, commonly used in linear IC chips and A/D converters.

OR gate Digital logic circuit that produces a HIGH output of 1 when any of the inputs is HIGH at 1.

oscillator Circuit that generates ac output from dc power input, without any ac signal input.

parallel circuit One that has two or more branches for separate currents from one voltage source.

paramagnetic Material that can be weakly magnetized in the opposite direction from the magnetizing force.

passive device Components such as resistors, capacitors, and inductors. They do not generate voltage or control current.

PC board A device that has printed circuits.

peak-to-peak value (p-p) Amplitude between opposite peaks.

peak value Maximum amplitude, in either polarity; 1.414 times rms value for sine-wave V or I.

pentode Vacuum-tube amplifier with five electrodes.

permanent magnet (PM) One that has magnetic poles produced by internal atomic structure. No external current necessary.

permeability Ability to concentrate magnetic lines of force.

permeance Reciprocal of magnetic reluctance.

phase angle The angle between a phasor and a reference line denoting a shift in time.

phasor A line representing magnitude and direction of a quantity, such as voltage or current, with respect to time.

pico (p) Metric prefix for 10^{-12}.

pinout Pin numbers for IC package.

polar form Form of complex numbers that gives magnitude and phase angle in the form $A \angle \theta°$.

polarity Property of electric charge and voltage. Negative polarity is excess of electrons. Positive polarity means deficiency of electrons.

potential Ability of electric charge to do work in moving another charge. Measured in volt units.

potentiometer Variable resistor connected as a series voltage divider.

power (P) Rate of doing work. The unit of electric power is the watt.

power factor Cosine of the phase angle for a sine-wave ac circuit. Value is between 1 and 0.

preferred values Common values of resistors and capacitors generally available for replacement purposes.

primary cell or battery Type that cannot be recharged.

primary winding Transformer coil connected to the source voltage.

printed wiring Conducting paths printed on plastic board.

proton Particle with positive charge in the nucleus of an atom.

pulsating dc value Includes ac component on average dc axis.

pulse A sharp rise and decay of voltage or current of a specific peak value for a brief period of time.

Q Figure of quality or merit, in terms of reactance compared with resistance. The Q of a coil is X_L/r_i. For an LC circuit, Q indicates sharpness of resonance. Also used as the symbol for charge: $Q = C/V$.

quadrature A 90° phase angle.

R Symbol for resistance.

radian (rad) Angle of 57.3°. Complete circle includes 2π rad.

radio Wireless communication by electromagnetic waves.

radio frequencies (RF) Those high enough to be radiated efficiently as electromagnetic waves, generally above 30 kHz.

radix Base for a number system; 10 for decimal numbers and 2 for binary numbers.

RAM Read-and-write or random-access memory. Can store input instructions and supply output information.

ramp Sawtooth waveform with linear change in V or I.

read Take out digital information from a memory device.

real number Any positive or negative number not containing j. $(A + jB)$ is a complex number but A and B by themselves are real numbers.

real power The net power consumed by resistance. Measured watts.

reciprocal relation As one variable increases, the other decreases.

rectangular form Representation of a complex number in the form $A + jB$.

rectifier A device that allows current in only one direction.

relay Automatic switch operated by current in a coil.

reluctance ($\mathcal{R}$) Opposition to magnetic flux. Corresponds to resistance for current.

reset Put Q output of flip-flop to logic 0.

resistance (R) Opposition to current. Unit is the ohm (Ω).

resistance wire A conductor having a high resistance value.

resonance Condition of $X_L = X_C$ in an LC circuit to favor the resonant frequency for a maximum in V, I, or Z.

reverse voltage Polarity that prevents forward current through a PN junction.

rheostat Variable resistor.

ringing Ability of an LC circuit to oscillate after a pulse of V or I has been supplied.

rms value For sine-wave ac waveform, 0.707 of peak value. Also called effective value.

ROM Read-only memory. Can supply output information but cannot store input instructions.

rotor Rotating part of generator or motor.

saturation Maximum limit at which changes of input have no control in changing the output.

sawtooth wave One in which amplitude values have a slow linear rise or fall and a sharp change back to the starting value. Same as a linear ramp.

Schmitt trigger Circuit used to produce sharp edges for pulses.

SCR Silicon controlled rectifier. Has anode, cathode, and gate electrodes.

secondary winding Transformer coil that is connected to the load.

secondary cell or battery Type that can be recharged.

self-inductance (L) Inductance produced in a coil by current in the coil itself.

series circuit One that has only one path for current.

set Put Q output of flip-flop at logic 1.

shield Metal enclosure to prevent interference of radio waves.

shift register Memory unit that shifts data left or right in place value.

short-circuit Has zero resistance, resulting in excessive current.

shunt A parallel connection. Also a device used to increase the range of an ammeter.

SI Abbreviation for *Système International,* a system of practical units based on the meter, kilogram, second, ampere, kelvin, mol, and candela.

siemens (S) Unit of conductance. Reciprocal of ohms unit.

silicon (Si) Semiconductor chemical element used for transistors, diodes, and integrated circuits.

sine Trigonometric function of an angle, equal to the ratio of the opposite side to the hypotenuse in a right triangle.

sine wave One in which amplitudes vary in proportion to the sine function of an angle.

slip rings In an ac generator, devices that provide connections to the rotor.

solder Alloy of tin and lead used for fusing wire connections.

solenoid Coil often used for electromagnetic relays.

source follower FET amplifier circuit in which input is to the gate and output from the source electrode. Same as common-drain circuit. Corresponds to emitter follower.

spade lug A type of wire connector.

SPDT Single-pole double-throw switch or relay contacts.

specific gravity Ratio of weight of a substance with that of an equal volume of water.

specific resistance The R for a unit length, area, or volume.

SPST Single-pole single-throw switch or relay contacts.

square wave An almost instantaneous rise and decay of voltage or current in a periodic pattern with time and with a constant peak value. A V or I on at constant value and off for equal times.

static electricity Electric charges not in motion.

stator Stationary part of a generator or motor.

steady-state value The V or I produced by a source, without any sudden changes. Can be dc or ac value.

storage cell or battery Type that can be recharged.

string Components connected in series.

superconductivity Very low R at extremely low temperatures.

superposition theorem Method of analyzing a network with multiple sources by using one at a time and combining their effects.

supersonic Frequency above the range of hearing, generally above 16,000 Hz.

susceptance (B) Reciprocal of reactance in sine-wave ac circuits; $B = 1/X$.

switch Device used to open or close connections of a voltage source to a load circuit.

tangent Trigonometric function of an angle, equal to the ratio of the opposite side to the adjacent side in a right triangle.

tank circuit An LC tuned circuit. Has ability to store energy in L and C.

tantalum Chemical element used for electrolytic capacitors.

taper How R of a variable resistor changes with the angle of shaft rotation.

taut-band meter Type of construction for meter movement often used in VOM.

temperature coefficient For resistance, how R varies with a change in temperature.

terrestrial magnetism Magnetic effects of the earth as a huge magnet with north and south poles.

tesla (T) Unit of flux density, equal to 10^8 lines of force per square meter.

Thevenin's theorem Method of reducing a complicated network to one voltage source with series resistance.

three-phase power Ac voltage generated with three components differing in phase by 120°.

thyristor NPNP four-layer semiconductor used for power-control circuits. Common types are SCR and triac.

time constant Time required to change by 63 percent after a sudden rise or fall in V and I. Results from the ability of L and C to store energy. Equals RC or L/R.

toggle For digital circuits, changing between HIGH at logic 1 and LOW at logic 0.

TO outline Type of package for transistors and integrated circuits. Gives size and pin connections.

toroid Electromagnet with its core in the form of a closed magnetic ring.

transconductance Ratio of current output to voltage input.

transformer A device that has two or more coil windings used to step up or step down ac voltage.

transient Temporary value of V or I in capacitive or inductive circuits caused by abrupt change.

transistor Semiconductor device used for amplifiers. Includes NPN and PNP junction types and field-effect transistors (FETs).

transmission line Method of sending power or signals from one point to another using conductors.

triac Thyristor that can control current in two opposite directions.

trigonometry Analysis of angles and triangles.

triode Electron tube device with three electrodes.

truth table Listing of all possible combinations of inputs and outputs for a digital logic circuit.

TTL Transistor-transistor logic circuits.

tuning Varying the resonant frequency of an LC circuit.

tunnel diode Semiconductor diode used in microwave applications.

turns ratio Comparison of turns in primary and secondary for a transformer.

twin lead Transmission line with two conductors in plastic insulator.

UHF Ultra high frequencies in band of 30 to 300 MHz.

UJT Unijunction transistor.

VAR Unit for voltamperes of reactive power, 90° out of phase with real power.

varactor Semiconductor diode which varies in C with amount of reverse voltage.

variac Transformer with variable turns ratio to provide different amounts of secondary voltage.

vector A line representing magnitude and direction in space.

VHF Very high frequencies, in band of 30 to 300 MHz.

VLSI Very-large-scale integration for IC chips.

volatile memory Memory that loses its stored information when the power is turned off.

volt (V) Practical unit of potential difference. One volt produces one ampere of current in a resistance of one ohm.

voltage divider A series circuit to provide V less than the source voltage.

voltage drop Voltage across each passive component, such as R, L, or C, in a series circuit.

voltage regulator A device that maintains a constant output voltage with changes of input voltage or output load current.

voltage source Supplies potential difference across two terminals. Has series r_i.

voltampere (VA) Unit of apparent power, equal to $V \times I$.

volt-ampere characteristic Graph to show how I varies with V.

voltmeter loading The amount of current taken by the voltmeter acting as a load. As a result the measured voltage is less than the actual value.

VOM Volt-ohm-milliammeter.

watt (W) Unit of real power. Product of I^2R.

wattmeter Measures real power as instantaneous value of $V \times I$.

wavelength (λ) Distance in space between two points with the same magnitude and direction in a propagated wave.

wavetrap An LC circuit tuned to reject the resonant frequency.

weber (Wb) Unit of magnetic flux, equal to 10^8 lines of force.

Wheatstone bridge Balanced circuit used for precise measurements of resistance.

wire gage A system of wire sizes based on the diameter of the wire. Also, the tool used to measure wire size.

word In digital circuits, a group of bits of 0 and 1. Usually written in groups of four, eight, or sixteen bits.

work Corresponds to energy. Equal to power $\times$ time, as in kilowatthour unit. Basic unit is one joule, equal to one volt-coulomb, or one watt-second.

write In digital circuits, to put information into a memory unit.

wye network Three components connected with one end in a common connection and the other ends connected to the line or load. Same as T network.

X_C Capacitive reactance, equal to $1/(2\pi fC)$.

X_L Inductive reactance, equal to $2\pi fL$.

XNOR gate Digital logic circuit for exclusive NOR gate.

XOR gate Digital logic circuit for exclusive OR gate.

Y Symbol for admittance in an ac circuit. Reciprocal of impedance Z; the $Y = 1/Z$.

Y network Another way of denoting a wye network.

Z Symbol for ac impedance. Includes resistance with capacitive and inductive reactance.

zener diode Semiconductor diode used for voltage regulation. Takes reverse voltage.

zero-ohms adjustment Used with ohmmeter of VOM to set the correct reading at zero ohms.

Bibliography

Electronics and Electricity

Adams, J. E., and G. Rockmaker: *Industrial Electricity—Principles and Practices,* McGraw-Hill Book Company, New York.

Boylestad, R., and L. Nashelsky: *Electricity, Electronics and Electromagnetics,* Prentice-Hall, Inc., Englewood Cliffs, N.J.

Chute, G. M., and R. D. Chute: *Electronics in Industry,* McGraw-Hill Book Company, New York.

DeFrance, J. J.: *Electrical Fundamentals,* Prentice-Hall, Inc., Englewood Cliffs, N.J.

Fowler, R. J.: *Electricity, Principles and Applications,* McGraw-Hill Book Company, New York.

Gilmore, C. M.: *Instruments and Measurements,* McGraw-Hill Book Company, New York.

Grob, B.: *Basic Television and Video Systems,* McGraw-Hill Book Company, New York.

Grob, B.: *Electronic Circuits and Applications,* McGraw-Hill Book Company, New York.

Jackson, H. W.: *Introduction to Electric Circuits,* Prentice-Hall, Inc., Englewood Cliffs, N.J.

Kaufman, M.: *Radio Operator's License Q and A Manual,* Hayden Book Company, Rochelle Park, N.J.

Lurch, E. N.: *Fundamentals of Electronics,* John Wiley & Sons, Inc., New York.

Malvino, A. P.: *Electronic Principles,* McGraw-Hill Book Company, New York.

Marcus, A., and C. M. Thomson: *Electricity for Technicians,* Prentice-Hall, Inc., Englewood Cliffs, N.J.

Oppenheimer, S. L., F. R. Hess, and J. P. Borchers: *Direct and Alternating Currents,* McGraw-Hill Book Company, New York.

Radio Amateurs' Handbook, American Radio Relay League, Newington, Conn.

Ridsdale, R. E.: *Electric Circuits,* McGraw-Hill Book Company, New York.

Schuler, C. A.: *Electronics: Principles and Applications,* McGraw-Hill Book Company, New York.

Shrader, R. L.: *Electronic Communication,* McGraw-Hill Book Company, New York.

Zbar, P. B.: *Basic Electricity,* McGraw-Hill Book Company, New York.

Zbar, P. B., and A. P. Malvino: *Basic Electronics,* McGraw-Hill Book Company, New York.

Transistors and Integrated Circuits

Bell, D. A.: *Fundamentals of Electronic Devices,* Reston Publishing Company, Reston, Va.

Coughlin, R. F., and F. F. Driscoll: *Operational Amplifiers and Linear Integrated Circuits,* Prentice-Hall, Inc., Englewood Cliffs, N.J.

G.E. Transistor Manual, General Electric Company, Semiconductor Products, Syracuse, N.Y.

Jung, W. G.: *IC Op-Amp Cookbook,* Howard W. Sams & Co., Inc., Indianapolis, Ind.

Linear Applications Handbook, National Semiconductor Corporation, Santa Clara, Calif.

RCA Solid-State Devices Manual, RCA Solid-State Division, Somerville, N.J.

Seidman, A. H.: *Integrated Circuits Application Handbook,* John Wiley & Sons, Inc., New York.

Tocci, R. J.: *Fundamentals of Electronic Devices,* Charles E. Merrill Publishing Company, Columbus, Ohio.

Digital Electronics and Computers

Gilmore, C. M.: *Introduction to Microprocessors,* McGraw-Hill Book Company, New York.

Ilardi, F.: *Computer Circuit Analysis—Theory and Applications,* Prentice-Hall, Inc., Englewood Cliffs, N.J.

Malvino, A. P.: *Digital Computer Electronics,* McGraw-Hill Book Company, New York.

Nashelsky, L.: *Introduction to Digital Technology,* John Wiley & Sons, Inc., New York.

Oleksy, J. E., and G. B. Rutkowski: *Microprocessor and Digital Computer Technology,* Prentice-Hall, Inc., Englewood Cliffs, N.J.

Pasahow, E.: *Microprocessors and Microcomputers for Electronics Technicians,* McGraw-Hill Book Company, New York.

Tokheim, R. L.: *Digital Electronics,* McGraw-Hill Book Company, New York.

TTL Data Book, Fairchild Camera and Instrument Corporation, Mountain View, Calif.

Ward, D. M.: *Applied Digital Electronics,* Charles E. Merrill Publishing Company, Columbus, Ohio.

Mathematics

Cooke, N. M., H. Adams, and P. B. Dell: *Basic Mathematics for Electronics,* McGraw-Hill Book Company, New York.

Deem, B.: *Electronics Math,* Prentice-Hall, Inc., Englewood Cliffs, N.J.

Grob, B.: *Mathematics Outline and Review Problems for Basic Electronics,* McGraw-Hill Book Company, New York.

Kramer, A. D.: *Fundamentals of Technical Mathematics,* McGraw-Hill Book Company, New York.

Rice, H. S., and R. M. Knight. *Technical Mathematics,* McGraw-Hill Book Company, New York.

Roberts and Stockton: *Elements of Mathematics,* Addison-Wesley Publishing Company, Inc., Reading, Mass.

Manufacturers' Periodical Publications

Aerovox Research Worker, Aerovox Corp., New Bedford, Mass.
GE Service News, General Electric Co., Syracuse, N.Y.
RCA Service News, RCA Service Co., Camden, N.J.
Sylvania News, GTE Sylvania Electric Products, Inc., Batavia, N.Y.
Tekscope, Tektronix, Inc., Beaverton, Ore.

Magazines

Byte, McGraw-Hill Publications, New York.
Computers and Electronics, Ziff-Davis Publishing Co., New York.
Electronics, McGraw-Hill Publications, New York.
Electronic Servicing and Technology, Intertec Publishing Corp., Overland Park, Kan.
QST, American Radio Relay League, West Hartford, Conn.
Radio-Electronics, Gernsback Publications, Inc., New York.
Spectrum, Institute of Electrical and Electronics Engineers, New York.

Sources for Films and Filmstrips

National Audiovisual Center, General Services Administration, Washington, D.C. 20409.
Projected Learning Programs, Inc., Chico, Calif.

Industry Standards

American National Standards Institute, New York.
Electronic Industries Association (EIA), Washington, D.C.
Electronic Industries Association of Japan (EIAJ).
Institute of Electrical and Electronics Engineers (IEEE), New York.
International Society of Certified Electronic Technicians (ISCET), Indianapolis, Ind.
National Fire Protection Association, Quincy, Mass.

Answers to Self-Examinations

1. T	**8.** T	**15.** T
2. T	**9.** T	**16.** T
3. T	**10.** T	**17.** T
4. T	**11.** T	**18.** T
5. T	**12.** T	**19.** F
6. T	**13.** T	**20.** F
7. T	**14.** T	

CHAPTER 2

1. 2	**8.** 10	**15.** 144
2. 4	**9.** 0.4	**16.** 2
3. 16	**10.** 72	**17.** 1.2
4. 0.5	**11.** 8	**18.** 3
5. 2	**12.** 2	**19.** 0.2
6. 25	**13.** 2	**20.** 0.12
7. 25	**14.** 0.83	

CHAPTER 3

1. (d)	**5.** (c)	**9.** (b)
2. (c)	**6.** (d)	**10.** (d)
3. (d)	**7.** (c)	
4. (b)	**8.** (b)	

CHAPTER 4

1. (b)	**5.** (a)	**9.** (c)
2. (a)	**6.** (c)	**10.** (b)
3. (a)	**7.** (c)	
4. (d)	**8.** (b)	

CHAPTER 5

1. (c)	**5.** (d)	**9.** (d)
2. (c)	**6.** (b)	**10.** (d)
3. (c)	**7.** (d)	
4. (c)	**8.** (a)	

CHAPTER 6

1. (d)	**5.** (a)	**9.** (b)
2. (a)	**6.** (b)	**10.** (c)
3. (d)	**7.** (d)	
4. (c)	**8.** (a)	

REVIEW: CHAPTERS 1 TO 6

1. (c)	**7.** (d)	**13.** (a)
2. (c)	**8.** (b)	**14.** (a)
3. (c)	**9.** (a)	**15.** (a)
4. (b)	**10.** (c)	**16.** (a)
5. (c)	**11.** (b)	**17.** (c)
6. (b)	**12.** (a)	**18.** (b)

CHAPTER 7

1. T	**5.** T	**9.** F
2. T	**6.** T	**10.** T
3. T	**7.** T	
4. F	**8.** T	

CHAPTER 8

1. (a)	**5.** (c)	**9.** (d)
2. (c)	**6.** (c)	**10.** (c)
3. (a)	**7.** (a)	
4. (a)	**8.** (c)	

REVIEW: CHAPTERS 7 AND 8

1. T	**5.** F	**9.** T
2. T	**6.** F	**10.** F
3. T	**7.** T	
4. T	**8.** T	

CHAPTER 9

1. T	**5.** T	**9.** T
2. F	**6.** T	**10.** T
3. T	**7.** F	
4. T	**8.** T	

CHAPTER 10

1. T	**5.** T	**9.** T
2. T	**6.** T	**10.** T
3. T	**7.** T	
4. T	**8.** F	

REVIEW: CHAPTERS 9 AND 10

1. T	**6.** F	**11.** T
2. T	**7.** F	**12.** T
3. T	**8.** T	**13.** T
4. T	**9.** T	**14.** T
5. T	**10.** T	**15.** T

CHAPTER 11

1. (a)	**5.** (b)	**9.** (c)
2. (d)	**6.** (a)	**10.** (c)
3. (d)	**7.** (b)	
4. (b)	**8.** (c)	

CHAPTER 12

1. (d)	**5.** (d)	**9.** (a)
2. (c)	**6.** (d)	**10.** (d)
3. (b)	**7.** (a)	
4. (a)	**8.** (c)	

REVIEW: CHAPTERS 11 AND 12

1. (d)	**4.** (c)	**7.** (c)
2. (b)	**5.** (d)	**8.** (b)
3. (b)	**6.** (c)	

CHAPTER 13

1. T	**6.** T	**11.** F
2. T	**7.** T	**12.** T
3. T	**8.** T	**13.** T
4. F	**9.** T	**14.** T
5. T	**10.** T	**15.** F

CHAPTER 14

1. F	**5.** T	**9.** T
2. T	**6.** T	**10.** T
3. T	**7.** T	**11.** T
4. F	**8.** T	**12.** T

CHAPTER 15

1. T	**6.** T	**11.** T
2. T	**7.** T	**12.** T
3. T	**8.** T	**13.** T
4. T	**9.** T	**14.** T
5. T	**10.** T	**15.** T

CHAPTER 16

1. T	**5.** T	**9.** T
2. T	**6.** T	**10.** T
3. T	**7.** F	**11.** 28.28 V
4. T	**8.** F	**12.** 1.2 A

CHAPTER 16 (*Continued*)

13. 70.7 V	**19.** 7.07 V	**25.** 240 Hz
14. 3×10^4 cm	**20.** 40 V	**26.** 240 V
15. 0.001 ms	**21.** 1000 Hz	**27.** 120 V
16. 60 Hz	**22.** 180 Hz	**28.** 208 V
17. 0.01 μs	**23.** 11.1 V	
18. 0.25 MHz	**24.** 120 V	

REVIEW: CHAPTERS 13 TO 16

1. (b)	**5.** (b)	**9.** (c)
2. (a)	**6.** (d)	**10.** (a)
3. (c)	**7.** (a)	**11.** (d)
4. (d)	**8.** (d)	**12.** (a)

CHAPTER 17

1. (b)	**5.** (c)	**9.** (b)
2. (c)	**6.** (d)	**10.** (a)
3. (c)	**7.** (b)	
4. (d)	**8.** (d)	

CHAPTER 18

1. (a)	**5.** (a)	**9.** (d)
2. (c)	**6.** (c)	**10.** (c)
3. (c)	**7.** (c)	
4. (d)	**8.** (b)	

CHAPTER 19

1. (c)	**5.** (b)	**9.** (c)
2. (c)	**6.** (b)	**10.** (d)
3. (c)	**7.** (c)	
4. (c)	**8.** (b)	

REVIEW: CHAPTERS 17 TO 19

1. (c)	**5.** (d)	**9.** (b)
2. (b)	**6.** (d)	**10.** (c)
3. (d)	**7.** (c)	
4. (d)	**8.** (a)	

CHAPTER 20

1. (a)	**5.** (c)	**9.** (c)
2. (b)	**6.** (c)	**10.** (b)
3. (b)	**7.** (c)	
4. (c)	**8.** (b)	

CHAPTER 21

1. (b)	**5.** (d)	**9.** (b)
2. (c)	**6.** (a)	**10.** (a)
3. (c)	**7.** (a)	
4. (b)	**8.** (d)	

CHAPTER 22

1. (d)	**5.** (b)	**9.** (b)
2. (b)	**6.** (b)	**10.** (b)
3. (a)	**7.** (c)	
4. (b)	**8.** (c)	

CHAPTER 23

1. (a)	**5.** (d)	**9.** (d)
2. (c)	**6.** (c)	**10.** (c)
3. (d)	**7.** (d)	
4. (b)	**8.** (c)	

REVIEW: CHAPTERS 20 TO 23

1. T	**15.** T	**29.** T
2. T	**16.** T	**30.** T
3. T	**17.** F	**31.** F
4. T	**18.** T	**32.** T
5. T	**19.** T	**33.** T
6. T	**20.** T	**34.** T
7. T	**21.** T	**35.** T
8. T	**22.** T	**36.** T
9. F	**23.** T	**37.** F
10. T	**24.** T	**38.** T
11. T	**25.** T	**39.** T
12. T	**26.** T	**40.** T
13. F	**27.** T	
14. T	**28.** T	

CHAPTER 24

1. (b)	**5.** (c)	**9.** (c)
2. (c)	**6.** (a)	**10.** (a)
3. (a)	**7.** (b)	
4. (c)	**8.** (c)	

CHAPTER 25

1. (d)	**3.** (i)	**5.** (h)
2. (m)	**4.** (j)	**6.** (l)

CHAPTER 25 (*Continued*)

7. (c)	**10.** (a)	**12.** (f)
8. (k)	**11.** (b)	**13.** (g)
9. (e)		

CHAPTER 26

1. T	**5.** T	**9.** T
2. F	**6.** T	**10.** F
3. T	**7.** F	
4. T	**8.** T	

REVIEW: CHAPTERS 24 TO 26

1. 300	**10.** 1	**19.** $2 \angle -17°$
2. 300	**11.** 45	**20.** $60 \angle 200°$
3. 300	**12.** -45	**21.** T
4. 250	**13.** 1	**22.** T
5. 250	**14.** 1.41	**23.** T
6. 200	**15.** 7.07	**24.** F
7. 200	**16.** 600	**25.** F
8. 14.1	**17.** $5.66 \angle 45°$	
9. 14.1	**18.** $4 \angle 10°$	

CHAPTER 27

1. (c)	**5.** (d)	**9.** (a)
2. (b)	**6.** (a)	**10.** (b)
3. (d)	**7.** (d)	
4. (c)	**8.** (d)	

CHAPTER 28

1. (d)	**5.** (d)	**9.** (a)
2. (a)	**6.** (b)	**10.** (d)
3. (b)	**7.** (a)	
4. (c)	**8.** (b)	

REVIEW: CHAPTERS 27 AND 28

1. 8	**7.** 5	**13.** T
2. 0.8	**8.** 0.08	**14.** T
3. 0.4	**9.** 40	**15.** T
4. 10	**10.** 150	
5. 10	**11.** F	
6. 1	**12.** T	

CHAPTER 29

1. Active
2. Oscillator
3. Silicon
4. P-type
5. Free electrons
6. 0.7 V
7. No
8. P-type
9. Cathode
10. Towards
11. Base
12. Forward
13. Voltage
14. Gate
15. Depletion
16. Will
17. Zener
18. LED
19. Dot matrix
20. Plate
21. Control grid
22. Emits
23. Electron
24. 6.3 V

CHAPTER 30

1. T
2. F
3. F
4. T
5. F
6. T
7. T
8. F
9. F
10. F
11. F
12. T
13. T
14. F
15. T
16. T
17. T
18. T
19. T
20. F
21. T
22. T
23. F
24. F
25. T

CHAPTER 31

1. 2
2. 1 + 1
3. AND
4. OR
5. NAND
6. 0
7. MV
8. FF
9. Four
10. Set
11. Synchronously
12. Multiplexer
13. PROM
14. CMOS
15. F

CHAPTER 32

1. T
2. T
3. T
4. T
5. F
6. T
7. T
8. F
9. T
10. F
11. T
12. F
13. F
14. T
15. F

REVIEW: CHAPTERS 29 TO 32

1. T
2. T
3. T
4. T
5. T
6. T
7. F
8. F
9. T
10. F
11. F
12. T
13. T
14. T
15. F
16. F
17. T
18. T
19. F
20. T
21. T
22. T
23. F
24. F
25. F
26. T
27. T
28. T
29. T
30. T
31. T
32. T
33. T
34. T
35. F

Answers to Odd-Numbered Problems

1. $I = 4$ A
3. See Prob. 4 values

5. 2.2 V
7. 0.2 Ω

1. (a) See Fig. 2-2
   (b) $I = 2$ mA
   (c) $I = 2$ mA
   (d) $I = 1$ mA
3. (b) $R = 7.9$ Ω
5. (a) $V = 12$ V
   (b) $P = 24$ W
   (c) $P = 24$ W

7. 10,000 V
9. (a) 1496 V
   (b) 108.1 V
   (c) 2.84 V
11. 9.84 V

1. $I = 1$ A, $R_2 = 10$ Ω
3. $V_2 = 0.6$ V
5. $V_T = 30$ V, $I = 0.5$ A
7. $R_T = 2,552,470$ Ω
9. Each $R = 5$ kΩ

11. $I = 1$ mA
13. $I = 2.5$ mA
15. $R_2 = 25$ Ω
17. $V_2 = 13$ V

1. (b) 45 V
   (c) $I_1 = 3$ A, $I_2 = 1$ A
   (d) $I_T = 4$ A
   (e) $R_T = 11.25$ Ω
3. (b) 20 V
   (c) $I_2 = 2$ A
       $I_3 = 4$ A
5. (a) $I_2 = 0$
   (b) $I_1 = 1$ A
   (c) $I_T = 1$ A
   (d) $R_T = 10$ Ω
   (e) $P_T = 10$ W

7. (a) 7.14 Ω
   (b) 2 kΩ
   (c) 250 Ω
   (d) 54.6 Ω
   (e) 714 Ω
   (f) 5 kΩ
9. $G_T = 0.038$ S
11. (a) 100 kΩ
    (b) 33.3 kΩ
    (c) 11.1 kΩ

1. (a) $R_T = 25$ Ω
   (b) $I_T = 4$ A
3. (b) $R_T = 15$ Ω
5. (a) $R = 6$ Ω
   (b) $R = 24$ Ω
7. (a) $V_1 = 2.23$ V
       $V_2 = 0.74$ V
       $V_3 = 6.7$ V
       $V_4 = 22.3$ V

   (b) $P_1 = 204$ mW
       $P_2 = 69$ mW
       $P_3 = 620$ mW
       $P_4 = 2.08$ W
   (c) R_4
   (d) R_4
9. $R_1 = 3$ kΩ
   $R_2 = 1$ kΩ

CHAPTER 5 (*Continued*)

11. $V_1 = V_X = 1$ V
    $V_2 = V_S = 10$ V
    $R_X = 4.2$ Ω
13. $R_T = 10.45$ Ω
15. (a) $V_2 = 20$ V
    (b) $V_1 = V_2 = 22.5$ V

17. (a) $V_1 = V_2 = V_3 = 40$ V
    $I_1 = I_2 = 2$ mA
    $I_3 = I_T = 4$ mA
    (b) $V_{AG} = V_{BG} = +40$ V
    I is the same as in (a)

CHAPTER 6

1. (a) 1 W
   (b) 2 W
3. $R_1 = 4700$ Ω $\pm$ 10%
   $R_2 = 2.2$ MΩ $\pm$ 10%
   $R_3 = 33$ Ω $\pm$ 5%

$R_4 = 910$ Ω $\pm$ 5%
$R_5 = 2.2$ Ω $\pm$ 5%
$R_6 = 10,000$ Ω $\pm$ 20%
5. 12.177 to 12.423 Ω
7. 200 Ω, ¼ W

CHAPTER 7

1. $V_1 = 4$ V
   $V_2 = 8$ V
   $V_3 = 40$ V
3. $I_B = 0.6$ mA
5. $I_1 = 3$ mA
   $I_2 = 6$ mA
   $I_3 = 30$ mA

7. $R_D = 555.6$ Ω
   $R_E = 740.7$ Ω
   $R_F = 500$ Ω
9. $R_1 = 75$ Ω
   $R_2 = 86.4$ Ω
   $R_3 = 38.5$ Ω
11. $I_2 = 3.26$ mA

CHAPTER 8

1. (a) $R_S = 50$ Ω
   (b) $R_S = 5.55$ Ω
   (c) $R_S = 0.505$ Ω
   (d) 1 mA, 5 mA, and 50 mA
3. (a) 300 mA
   (b) 60 V
5. (a) 199 kΩ
   599 kΩ
   9999 kΩ

(b) 20,000 Ω/V
(c) 10 MΩ
7. $R_1 = 78.3$ Ω
   $R_2 = 7.83$ Ω
   $R_3 = 0.87$ Ω
9. (a) $R_1 = 145$ Ω
   (b) 150 Ω for 5 mA
11. $V_1 = 40$ V
    $V_2 = 80$ V

CHAPTER 9

1. $I_1 = 1.42$ A
3. $I_1 = 1.42$ A, $V_{R_1} = 11.36$ V
   $I_2 = 0.32$ A, $V_{R_2} = 0.64$ V

$I_3 = 1.1$ A, $V_{R_3} = 4.4$ V
$I_4 = 1.1$ A, $V_{R_4} = 2.2$ V
5. $I_A = I_1 = 1.1$ A

CHAPTER 10

1. $V_{Th} = 15$ V
   $R_{Th} = 3$ Ω
   $V_L = 6$ V
3. $I_S = 5$ A
   $R_S = 4$ Ω
   $I_L = 3$ A

5. R_L not open
7. $V_P = 4.2$ V
9. $V_{R_2} = 19.2$ V
11. $V_{R_2} = 19.2$ V
13. $V_{R_3} = 10.6$ V
15. See Fig. 10-28

CHAPTER 11

1. (a) 1024 cmil
   (b) Gage No. 20
   (c) $R = 1.035\ \Omega$
3. (a) 1-A fuse
   (b) 0 V
   (c) 120 V

5. $R = 96\ \Omega$
7. 5000 ft
9. (a) $4.8\ \Omega$
   (b) 4000 ft
11. 3 V
13. $I = 30$ A

CHAPTER 12

1. 1.5 mA
3. 600 A
5. (a) 2.88×10^5 C
   (b) 40 h

7. 20 kΩ
9. 6 Ω

CHAPTER 13

1. 5×10^3 Mx
   5×10^{-5} Wb
3. 0.4 T

5. 24×10^3 Mx
7. 300
9. $1\ \mu\text{Wb} = 10^{-6} \times 10^8$ Mx

CHAPTER 14

1. (a) 200
   (b) 500
3. (a) 300 G/Oe
   (b) 378×10^{-6} T/(A/m)
   (c) 300
5. (a) 126×10^{-6}
   (b) 88.2×10^{-6}

7. (b) 40 V
   (c) 1000 A/m
   (d) 0.378 T
   (e) 3.02×10^{-4} Wb
   (f) 66×10^4 A/Wb
9. 14.4

CHAPTER 15

1. 8 kV
3. (a) 2 Wb/s
   (b) -2 Wb/s
5. (a) 0.2 A
   (b) 80 ampere-turns

(c) 400 ampere-turns/m
(d) 0.252 T
(e) 1.512×10^{-4} Wb

CHAPTER 16

1. (a) $I = 6$ A
   (b) $f = 60$ Hz
   (c) $0°$
   (d) 120 V
3. (a) $t = 0.25$ ms
   (b) $t = 0.0625\ \mu\text{s}$
5. (a) $f = 20$ Hz
   (b) $f = 200$ Hz
   (c) $f = 0.2$ MHz
   (d) $f = 0.2$ GHz

7. (a) $+10$ and -10 V
   (b) $+10$ and -10 V
   (c) $+10$ and -10 V
   (d) $+15$ and -5 V
9. $I_1 = 40\ \mu\text{A}$
   $I_2 = 20\ \mu\text{A}$
   $V_1 = V_2 = 200$ V
   $P_1 = 8$ mW
   $P_2 = 4$ mW

CHAPTER 16 (*Continued*)

11. $I = 2.5$ A

13. (a) 27.15 V

    (b) 20.8 V

15. $I_1 = 2.553$ A

    $I_2 = 1.765$ A

    $I_3 = 5.455$ A

    $I_T = 9.773$ A

CHAPTER 17

1. (a) 1.5 A/s

    (b) 10,000 A/s

    (c) 10,000 A/s

    (d) $-10,000$ A/s

3. (a) 7.5 mV

    (b) 50 V

    (c) 50 V

    (d) -50 V

5. (a) 60 Hz

    (b) 960 V

    (c) 96 mA

    (d) 0.768 A

7. (a) 300 μH

    (b) 66.7 μH

    (c) 320 and 280 μH

    (d) 0.0707

9. $R = 10.52$ Ω

11. 0.243×10^{-3} J

13. (a) 80 percent

    (b) 500 W

15. 1.26 mH

CHAPTER 18

1. At 100 Hz, $X_L = 314$ Ω

    At 200 Hz, $X_L = 628$ Ω

    At 1000 Hz, $X_L = 3140$ Ω

3. (b) $I = 21.4$ mA

    (c) $V_L = 16$ V

5. $X_L = 1.2$ kΩ

    $L = 3.18$ H

7. $L = 0.159$ H

    $X_L = 10$ kΩ

9. (a) $X_{L_T} = 5$ kΩ

    (b) $I = 2$ mA

    (c) $V_{L_1} = 2$ V

    $V_{L_2} = 8$ V

    (d) $L_1 = 2.65$ H

    $L_2 = 10.6$ H

11. (a) $f = 0.16$ kHz

    (b) $f = 1.27$ kHz

    (c) $f = 0.4$ MHz

    (d) $f = 1.6$ MHz

    (e) $f = 16$ MHz

13. $X_L = 1628.6$ Ω

15. $X_L = 754$ Ω

17. (d) At 500 Hz,

    $X_L = 785$ Ω

    $I = 12.7$ mA

CHAPTER 19

1. (a) $Z = 100$ Ω

    $I = 1$ A

    $\theta = 0°$

    (b) $Z = 100$ Ω

    $I = 1$ A

    $\theta = 90°$

    (c) $Z = 70.7$ Ω

    $I = 1.41$ A

    $\theta = 45°$

3. (b) $X_L = 377$ Ω

    (c) $Z = 390$ Ω

    (d) $I = 0.3$ A

    (e) $I = 47.8$ mA

5. $Z = 400$ Ω

    $X_L = 400$ Ω

7. $R_e = 94$ Ω

9. $Z = 566$ Ω

    $I = 0.177$ A

    $V_L = 70.7$ V

    $V_R = 70.7$ V

    $\theta_z = 45°$

11. At 800 Hz,

    $I_R = 0.25$ A

    $I_L = 0.125$ A

    $\theta_I = -26.6°$

13. $X_L = 500$ Ω

    $L = 15.9$ mH

CHAPTER 19 (*Continued*)

15. $Z_T = 583\ \Omega$
$I = 0.2\ \text{A}$
$\theta_Z = 59°$

17. v_L is a square wave,
$\pm 160\ \text{V p-p}$

CHAPTER 20

1. $Q = 400\ \mu\text{C}$
3. **(a)** $Q = 18\ \mu\text{C}$
   (b) 9 V
5. $C = 1062\ \text{pF}$
7. **(a)** 200 V
   (b) $Q = 200\ \mu\text{C}$
   (c) $C = 1\ \mu\text{F}$

9. **(a)** $2.5 \times 10^{-2}\ \text{J}$
   (b) 12.5 J
   (c) 3.2 J
11. **(a)** $0.06\ \mu\text{F}$
   (b) 74.2 pF

CHAPTER 21

1. 80 pF at 1 MHz
3. **(b)** $I = 4.5\ \text{mA}$
   (c) $f = 1\ \text{kHz}$
5. **(b)** $I = 2\ \text{mA}$
   (c) $V_{C_1} = 2\ \text{V}$
   (d) $V_{C_2} = 8\ \text{V}$
7. $f = 3183\ \text{Hz}$
9. **(a)** $X_{C_T} = 200\ \Omega$

   (b) $C = 333.3\ \text{pF}$
      $C_T = 1000\ \text{pF}$
11. **(a)** $X_C = 300\ \Omega$
     $C = 8.85\ \mu\text{F}$
   (b) $C = 17.7\ \mu\text{F}$
13. $X_C = 206.5\ \Omega$
15. $I = 0.78\ \text{mA}$
17. $C = 422\ \text{pF}$

CHAPTER 22

1. **(b)** $Z = 50\ \Omega$
   (c) $I = 2\ \text{A}$
   (d) $V_R = 80\ \text{V}$
     $V_C = 60\ \text{V}$
   (e) $\theta_Z = -37°$
3. $C = 0.08\ \mu\text{F}$ at 100 Hz
   $C = 80\ \text{pF}$ at 100 kHz
5. At 60 Hz, $C = 26.59\ \mu\text{F}$
   At 1 kHz, $C = 1.59\ \mu\text{F}$
   At 1 MHz, $C = 1590\ \text{pF}$
7. $I_C = 15\ \text{mA}$
   $I_R = 20\ \text{mA}$
   $I_T = 25\ \text{mA}$

   $Z_T = 1.2\ \text{k}\Omega$
   $\theta_I = 37°$
   $V_R = V_C = 30\ \text{V}$
9. For dc or ac,
   $V_1 = 400\ \text{V}$
   $V_2 = 200\ \text{V}$
   $V_3 = 100\ \text{V}$
11. $C = 66\ \mu\text{F}$
   $L = 106\ \text{mH}$
13. $Z_T = 583\ \Omega$
   $I = 0.2\ \text{A}$
   $\theta_Z = -59°$
15. i_C is a square wave, $\pm 4\ \text{mA p-p}$

CHAPTER 23

1. **(a)** 0.05 s
   (b) $0.05\ \mu\text{s}$
   (c) 1 ms
   (d) $20\ \mu\text{s}$
3. **(a)** 4 s
   (b) 100 V

5. $v_C = 86\ \text{V}$
7. $v_C = 15\ \text{V}$
9. 1.4 ms
11. $0.05 \times 10^6\ \text{V/s}$

CHAPTER 24

1. (a) 100 W
   (b) No reactance
   (c) 1
3. (b) $I = 10$ A, approx.
   (c) $Z = 10$ Ω
   (d) $\theta = 0°$
5. (c) $Z_T = 500$ Ω
   $I = 0.8$ A
   $\theta_Z = 53°$
7. (a) $X_L = 0$, approx.
   $X_C = 665$ Ω

(b) $Z_T = 890$ Ω
   $I = 135$ mA
   $\theta_Z = -47.9°$
9. (a) 180°
11. $R = 320$ Ω
13. $C = 300$ pF
15. $I_T = 6.9$ mA, $\theta_I = -16.9°$
   $Z_T = 60.9$ Ω, $\theta_Z = 16.9°$
17. $R = 9704$ Ω

CHAPTER 25

1. (a) $4 - j3$
   (b) $4 + j3$
   (c) $3 + j6$
   (d) $3 - j3$
3. (a) $5 \angle -37°$
   (b) $5 \angle 37°$
   (c) $3.18 \angle 18.5°$
   (d) $4.24 \angle -45°$
5. $Z_T = 65.36 + j23.48$
7. (a) $4.5 \angle 14°$
   (b) $4.5 \angle 34°$
   (c) $100 \angle -84°$
   (d) $100 \angle -60°$
9. $Z_T = 12.65 \angle 18.5°$

11. $Z_T = 5.25 \angle -13.7°$
13. $R = 5.08$ Ω
   $X_C = 1.27$ Ω
15. $R = 21.4$ Ω
   $X_L = 10.2$ Ω
17. $Z_T = 50 \angle -37° = 40 - j30$ Ω
   $I = 2 \angle 37° = 1.6 + j1.2$ A
   $V_R = 80 \angle 37° = 64 + j48$ V
   $V_L = 120 \angle 127° = -72 + j96$ V
   $V_C = 180 \angle -53° = 108 - j144$ V
19. $Z_T = 2.07$ kΩ $\angle 14.6°$ kΩ
   $I = 3.88$ mA $\angle -14.6°$ mA
21. $Z_T = 13.4 \angle 46.5°$

CHAPTER 26

1. $V_{DA} = 50.1 \angle 20.7°$
3. $V = 174 - j219 = 280 \angle -51.5°$
5. (a) Correct
   (b) Correct
   (c) Correct
   (d) Correct
   (e) Incorrect
7. $I_5 = 10.2 \angle 190°$ A
9. $V_{Th} = 8.94 \angle -26.6°$ V
   $Z_{Th} = 7.5 \angle 10.3°$ Ω or
   $7.33 + j1.33$ Ω

11. $I_N = 1.2 \angle -36.9°$ A
   $Z_N = 7.45 \angle 10.3°$ Ω or
   $7.33 + j1.33$ Ω
13. $V_{Th} = V_{Z_1} = 26.5 \angle 12°$ V
   $Z_{Th} = 22.6 \angle -45°$ or $16 - j16$ Ω
   $I_N = 1.17 \angle 57°$ A
   $Z_N = Z_{Th}$
15. $V_{AB} = V_L = 63.3 \angle -40.4°$ V
17. $I_L = 4.75 \angle -27.5°$ A

CHAPTER 27

1. $f_r = 12.6$ Hz
3. $f_r = 1.624$ MHz
5. (a) $Q = 100$
   (b) $X_C = X_L = 1500$ Ω

(c) $I = 1$ mA
   (d) $V_C = 1.5$ V
7. $L = 507$ μH at 1 MHz
   $L = 31.7$ μH at 4 MHz

CHAPTER 27 (*Continued*)

9. $C_{max} = 32.7$ pF
   $C_{min} = 21.7$ pF
11. (a) At f_r of 795 kHz,
       $I_T = 0.1$ mA
       $Z_T = 50$ kΩ
       $\theta = 0°$
    (b) At 895 kHz,
       $I_T = 1.25 \underline{/90°}$ mA
       $Z_T = 4 \underline{/-90°}$ kΩ
       $\theta = 90°$
    (c) At 695 kHz,
       $I_T = 1.3 \underline{/-90°}$ mA
       $Z_T = 3.85 \underline{/90°}$ kΩ
       $\theta = -90°$

13. (a) $Q = 200$
       $BW = 25$ Hz
    (b) $Z_T = 200$ kΩ
15. (a) At 5 kHz,
       $Z_T = 5$ Ω
       $I = 100$ mA
    (b) See Fig. 27-3b.
17. (a) $Q = 120$
    (b) $r_S = 78.4$ Ω
    (c) $C = 56.5$ pF

CHAPTER 28

1. (a) $C = 0.1$ μF
   (b) $V_R \cong 0$ V
       $V_C = 20$ V
   (c) $V_R = 7.07$ V rms value
       $V_C \cong 0$ V

3. $C = 0.64$ μF
5. (b) $L = 1.53$ mH
7. (a) $C_1 = 0.001$ 59 μF
   (b) $C_2 = 0.003$ 18 μF
   (c) $C_3 = 0.007$ 95 μF

CHAPTER 29

1. 100
3. 14.7 mA
5. Infinite

7. 120 V
9. All conducting

CHAPTER 30

1. 2.27 Ω
3. 100

5. 6.4 V
7. 0 V (approx.)

CHAPTER 31

1. (a) 100111
   (b) 111110
3. 14
5. (a) $\overline{AB} = x$

   (b) $\overline{A + B} = x$
   (c) $AB = x$
   (d) $\overline{A} + \overline{B} = x$ or $\overline{AB} = x$

CHAPTER 32

1. 2
3. +2.6 V

5. 101
7. +2.02 V

Index